# Structural and Functional Relationships in Prokaryotes

Larry L. Barton

# Structural and Functional Relationships in Prokaryotes

With 337 Illustrations

Larry L. Barton
Department of Biology
University of New Mexico
Albuquerque, NM 87131
USA
lbarton@unm.edu

*Cover illustration*: The cover illustration is a combination of figures 1.5, 3.30a and b, and 5.10 appearing in the text. The sheathed bacterium is from "Studies on the Filamentous Sheathed Iron Bacterium *Sphaerotilus Natans*." J.L. Stokes (1954), *Journal of Bacteriology* 67:278–291 (used with permission). The star-shaped bacterium is from "*Stella*, a New Genus of Soil Prosthecobacterium, with Proposals for *Stella humosa* sp. nov. and *Stella vacuolata* sp. nov." L.V. Vasilyeva (1985), *International Journal of Systematic Bacteriology* 35:518–521 (used with permission). The square-shaped cell is from "Walsby's Square Bacterium: Fine Structure of an Orthogonal Prokaryote." W. Stoeckenius (1981), *Journal of Bacteriology* 148: 352–360 (used with permission). The electron micrograph of *Escherichia coli* in the foreground is provided by Sandra L. Barton (used with permission).

Library of Congress Cataloging-in-Publication Data
Barton, Larry.
    Structural and functional relationships in prokaryotes / L.L. Barton.
      p.  cm.
    Includes bibliographical references and index.
    ISBN 0-387-20708-2 (alk. paper)
    1. Prokaryotes.  I. Title.
  QR41.2.B37 2004
  579.3—dc22                       2003067331

ISBN 0-387-20708-2      Printed on acid-free paper.

Printed in the United States of America.    (EB)

9  8  7  6  5  4  3  2  1     SPIN 10960493

springeronline.com

This book is dedicated to Sandra L. Barton,
who not only shares my interest in microbiology
but also has assisted me in countless areas
that have enabled me to complete this book.

# Preface

For several decades, bacteria have served as model systems to describe the life processes of growth and metabolism. In addition, it is well recognized that prokaryotes have contributed greatly to the many advances in the areas of ecology, evolution, and biotechnology. This understanding of microorganisms is based on studies of members from both the *Bacteria* and *Archaea* domains. With each issue of the various scientific publications, new characteristics of prokaryotic cells are being reported and it is important to place these insights in the context of the appropriate physiological processes.

*Structural and Functional Relationships in Prokaryotes* describes the fundamental physiological processes for members of the *Archaea* and *Bacteria* domains. The organization of the book reflects the emphasis that I have used in my 30 years of teaching a course of bacterial physiology. The philosophy used in the preparation of this book is to focus on the fundamental features of prokaryotic physiology and to use these features as the basis for comparative physiology. Even though diverse phenotypes have evolved from myriad genetic possibilities, these prokaryotes display considerable functional similarity and support the premise that there is a unity of physiology in the prokaryotes. The variations observed in the chemical structures and biochemical processes are important in contributing to the persistence of microbial strains in a specific environment.

*Structural and Functional Relationships in Prokaryotes* is intended to be used as a textbook for courses in bacterial physiology and as a resource for graduate students and scientists interested in specialized topics. The goals of this book are to review: (1) the structure–function relationships present in prokaryotes; (2) the growth of bacteria, including the mechanisms of cellular control; (3) the function of membranes, with an emphasis on bioenergetics and transport activities; and (4) the unique physiological types of bacteria, with an emphasis on growth in extreme environments. Information presented here is intended to build on a background that students have received from introductory courses of microbiology, cell biology, genetics, and biochemistry. An integrative approach enables the reader to appreciate the aspects of comparative physiology and encourages research in the areas in which critical assessments have yet to be conducted.

## Organization of This Book

The biochemical and physiological activities of prokaryotes are organized into sections that emphasize cellular structures, growth, and physiological processes. It is my intent to provide a basis for understanding the physiology of prokaryotes through appropriate illustrations but not to provide a compendium of all prokaryotic cells. When appropriate, reference is made to early research that served as the basis for defining the underlying concepts. Each chapter is highlighted with a quotation from a recognized microbial physiologist, and the purpose is to provide insight into the personality of individuals attracted to this field of study.

## Section I: An Introduction to the Prokaryotic Cell

The first section provides introductory information about prokaryotic cells and provides the basis for discussion of physiological processes found in subsequent chapters. Chapter 1 reviews cellular organization and describes the isolation of macromolecular structures of prokaryotes. Through characterization of bacterial and archaeal cells in this section, a basis for comparative physiology is established.

## Section II: Cellular Structures and Their Function

Chapter 2 describes the chemical composition of the plasma membrane and emphasizes lipid–protein interactions that are important for transmembrane activities. Chapter 3 describes the cell walls of prokaryotes with respect to organization of the peptidoglycan structure found in bacteria and bacterial endospores. Chemical architecture of the cell wall is presented with respect to providing structural stability against osmotic rupture and describing activities of agents that destabilize the cell wall structure. In addition, comparative features of the cell wall of *Bacteria* and *Archaea* are discussed. Chapter 4 examines the chemical compounds associated with the cell wall that do not contribute directly to structural stability. Protein and polysaccharide structures are presented and a major segment of this chapter is devoted to description of the outer membrane of Gram-negative bacteria. Chapter 5 centers on the physical structures found on the cell surface as well as structures found inside prokaryotic cells. The biochemistry of bacterial capsules is addressed using examples of both heteropolymers and homopolymers. Characteristics and production of pili, including structures previously referred to as fimbrae, are discussed. Included in this chapter is a review of the granules, crystals, globules, and vesicles that are found as internal structures of heterotrophic or chemolithotrophic prokaryotes.

## Section III: Profiles of Bacterial Growth

Chapter 6 is concerned with motility, cell sensing, and signaling contributing to cellular communication. The two-component signal transduction system is used as the model for sensing activities and examples using this sensing include porin regulation, bacterial virulence, sugar phosphate transport, and production of degradative enzymes. This chapter also discusses chemotaxis and highlights several examples of quorum sensing including chemiluminescence, sporulation, and pheromone activities leading to exchange of DNA. Chapter 7 reviews growth of bacterial cells including the changes that occur in individual cells as cultures traverse through the various phases of growth. The production of endospores is addressed in this chapter as well as the activities contributing to formation of secondary products of metabolism. The regulatory activities in stationary phase cells are illustrated through the discussion of responses of cells to starvation and the production of toxins, endospores, and antibiotics. Chapter 8 is an overview of the various physiological types of prokaryotes and includes current information concerning the mechanism for these organisms to grow in harsh environments. This would include molecular changes that enable cells to grow in environments with extremes of temperature, pH, and salinity. Growth in response to oxygen stress, osmotic stress, or radiation stress provides an insight into the capability of prokaryotes to adapt to changing environments.

## Section IV: Bioenergetics

Chapter 9 covers the basic principles of electron transport in chemolithotrophic, heterotrophic, and photosynthetic bacteria. In characterization of these systems for electron movement, emphasis is given to proteins with transition metals, cytochromes, dehydrogenases, oxidases, and reductases. Further emphasis is placed on bioenergetics, including the action of proton pumps and the coupling of ATP production to proton-motive force established through cellular respiration. In the discussion of aerobic and anaerobic bacterial photosynthesis there is a focus on the unifying bioenergetic concepts as well as appropriate illustration of the variations in these light-driven activities. In this chapter, thermodynamic considerations responsible for the energizing of the plasma membrane and supporting oxidative phosphorylation systems are discussed. Chapter 10 addresses movement of various solute molecules across the plasma membrane by passive and active processes. The description of solute transport in bacteria includes proton-motive force-driven activities, group translocation systems, and ATPase transporters. Several examples are presented for the transport of metal cations, mineral anions, sugars, and amino acids. The current status of vitamin $B_{12}$ and DNA uptake in bacteria is presented in this chapter.

## Section V: Metabolic Processes

Chapter 11 characterizes the principal catabolic pathways employed by heterotrophic prokaryotes. In addition to the various organic compounds serving as energy sources, the fundamental types of fermentation found in prokaryotes are described. Reactions associated with acetogenic bacteria and methane-producing archaea are outlined in this chapter. The utilization of C-1 units by prokaryotes is examined through various $CO_2$ fixation reactions and the pathways for methane oxidation are outlined. Chapter 12 serves as the central chapter for protein production starting with biosynthesis of amino acids, transfer of information into nucleic acid macromolecules, and ribosomal-related activities. The production of antibiotics is described and sites for inhibition of protein synthesis by antibiotics are summarized. Four types of protein secretion are discussed and examples of posttranslational processing of proteins in prokaryotes are provided. Chapter 13 addresses the metabolism of inorganic nitrogen and sulfur compounds. The enzymology of dinitrogen fixation is outlined with the formation of ammonium by appropriate prokaryotes. Assimilatory nitrate and ammonia reactions provide the basis for amino acid formation. Dissimilatory nitrate reduction with intermediates of nitrite, nitric oxide, and nitrous oxide is discussed. The oxidation of nitrite, ammonia, and hydroxylamine by prokaryotes completes the discussion of the interaction of microorganisms with the nitrogen cycle. The sulfur cycle activities pursued in this chapter include the reduction and the oxidation of inorganic sulfur compounds. Membrane-associated reactions attributed to dissimilatory reduction of sulfate and sulfur by prokaryotes are described as well as the sulfide oxidation reactions by photosynthetic sulfur bacteria. Chapter 14 focuses on the interaction of metals and prokaryotes. The metal ions required for growth are discussed and the activities of prokaryotes in the geochemical cycles of iron and manganese are highlighted. The role of metal mimicry in metabolic toxicity of metals and metalloids is discussed. Metabolic activities involving bacterial utilization of iron and selenium are presented and microbial reduction of mercury, arsenic, chromium, selenate, and uranium is outlined. In addition, the detoxification processes used by bacteria for resistance to cadmium and silver are presented.

## Appendix

The appendix contains the abbreviations used for amino acids and provides questions related to the chapters.

## Content and Expressions

Readers of this book will find that heavy reference is made to a relatively small number of bacterial species and few archaeal species to support the basic processes discussed. This coverage of the subject is intended to provide a focus for those who are using this book as an introductory text to prokaryotic physiology. The book is constructed primarily from review articles and appropriate use of designated references enables the reader to pursue additional information on each topic. A list of selected reference citations is provided at the end of each chapter to direct the reader to supplemental information on selected topics. The selection and use of information in this book was my responsibility and readers can contact me at lbarton@unm.edu for concerns that they may have with this book.

Each chapter opens with a quotation from a scientist who contributed to the field of microbial physiology. These quotations are selected so as to illustrate the scientific understanding or approach of the author. In some instances these quotations are from textbooks authored by biochemists. Unfortunately, many worthy quotes by outstanding scientists are not included because early literature was frequently written in a style that did not lend itself to brief statements.

Six scientists are presented in this book to highlight their contributions and provide insight into their selection of microbiology as their field of interest. I am most appreciative of their providing a short narrative describing their activities. The purpose of this inclusion of scientists is to highlight the range of specialization within the area of prokaryotic physiology and to encourage students to pursue careers in physiology.

## Acknowledgments

I express my appreciation to those individuals who introduced me to microbial physiology while I was a student at the University of Nebraska–Lincoln and these include Thomas L. Thompson, Jessup M. Shively, and Carl E. Georgi. Over the years, I have had the good fortune of meaningful discussions with colleagues in seeking answers to challenging physiological questions and I appreciate these experiences. The figures in this book were skillfully drafted by Robbie Narang and the illustrations were prepared with the assistance of Richard Plunkett. Special recognition and thanks is given to Sandra Barton for the critiques she provided as the book progressed. The professional support provided by Springer-Verlag has enabled this book to become a reality and I greatly applaud their assistance.

LARRY L. BARTON, Albuquerque, New Mexico

# Contents

## *Perspectives provided for this text*                                    *Page*

# I
# An Introduction to the Prokaryotic Cell

# 1
# The Cellular System

The third form [of life] has . . . revolutionized our notion of the prokaryote.
C.R. Woese and R.S. Wolfe, *The Bacteria*, Vol. VIII, 1985

## 1.1. Introduction

Cells from prokaryotic and eukaryotic organisms can be distinguished on the basis of several structural and biochemical features. Of paramount importance in distinguishing these two life forms is the nature of DNA organization and the extent to which membrane structures compartmentalize activities in cells. Eukaryotic cells have a discrete nucleus with DNA arranged in structurally distinct chromosomes while prokaryotic cells have a circular molecule of double-stranded DNA that is looped throughout the cytoplasm. In eukaryotic cells mitochondria, Golgi, vacuoles, and other internal structures are enclosed by membranes but prokaryotic cells lack internal organelles and have no structures delineated by a typical lipid–protein membrane. Microorganisms that are characterized as prokaryotes include members of the *Bacteria* and *Archaea* domains. Although prokaryotes lack internal structures, they are not to be regarded as a "bag of enzymes" because there considerable organization exists at the molecular level that is responsible for the various cellular processes. Although this type of molecular organization is available to the eukaryotes because of similar specificity of the numerous molecules, in eukaryotes other specialized activities are associated with the internal organelles. The lack of internal organelles in prokaryotes is not detrimental for the prokaryotes and may be beneficial; it may, in part, account for the prokaryotes' domination of numerous environments. For prokaryotes, the rapid growth rate and the rapid response of the cell to changes in the environment can be accomplished because cells are not limited by transport processes across the membrane barriers. In the case of eukaryotes, physiology is the study of structure and function whereas in the case of prokaryotes, physiology involves not only the study of physical structures but also the examination of chemical activities associated with molecular organization.

Historically, courses concerned with microbial physiology were based on the study of bacterial strains that were associated with economic or medical importance. As a result of interest in geomicrobiology, biotechnology, and molecular evolution, many studies have been conducted on prokaryotes that have been isolated only within the last 25 years. In some instances, bacteria obtained from specialized environments have been found to possess distinctive metabolic and physiological processes. Research on bacteria associated with the human environment has resulted in the identification of complex chemical structures, elucidation of metabolic intricacies, and insight into multicomponent sensory processes.

Even though a substantial body of knowledge concerning the physiology and biochemistry of prokaryotes exists, these observations are based on a small number of species. Whitman, Coleman, and Wiebe have reported that there are vast numbers of prokaryotes on planet Earth (Table 1.1), and it is highly probable that many unusual prokaryotes remain to be isolated. Almost all of microbial physiology is based on the study of prokaryotes isolated from terrestrial or aquatic habitats and the organisms examined constitute only a small fraction of prokaryotes from the Earth's surface. Current information underscores the cellular adaptations that organisms have pursued to inhabit the diverse environments, and comparative physiology reveals that there are many different biological possibilities for these cellular adjustments. As the examination of the structure and function relationships in prokaryotic organisms progresses, there will be a greater insight into the fundamental mechanisms of life.

This chapter serves to review the general characteristics of prokaryotes as they relate to macromolecular structure and to fundamental metabolic processes. The general procedures used for isolation of the major structures of prokaryotes are presented and subsequent chapters focus on the function of these structures. Prior to determining the chemical characteristics of the various cell structures and subcellular fractions, purity of isolated structures needs to be established and this chapter reviews some of the characteristics of purified prokaryotic structures. Because electron microscopy is an important tool to examine cell structure, attention is given in this chapter to the different applications of electron microscopy to microbial physiology.

## 1.2. Phylogeny of the Prokaryotes[1]

Biologists have long been interested in the organization of life, and various phylogenic systems have been produced to reflect a progression in evolutionary development. The

Table 1.1. Distribution of prokaryotes on planet Earth.[a]

| Environment | Number of prokaryotic cells ($\times 10^{28}$) | Percent of total number of prokaryotes on earth |
| --- | --- | --- |
| Soil | 26 | 5.0% |
| Aquatic habitats | 12 | 2.3% |
| Oceanic subsurface | 355 | 68.6% |
| Terrestrial subsurface | 125 | 24.1% |

[a]Data from Whitman et al., 1998.[21]

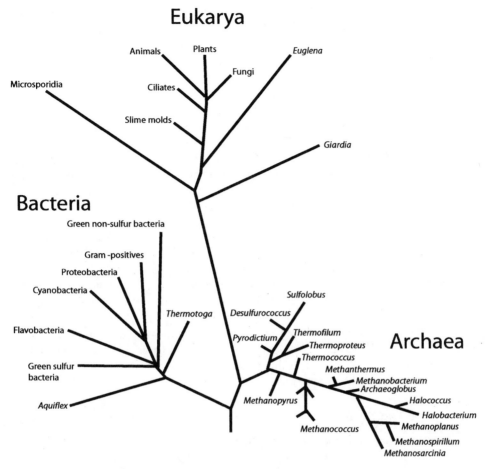

Figure 1.1. The tree of life according to Carl R. Woese.

approach proposed by Carl Woese examines the primary structure of genes for 16S rRNA and 18S rRNA for ribosomal subunits of prokaryotes and eukaryotes, respectively. In many instances, genes for 16S rRNA from mitochondria are used in analysis of eukaryotes instead of genes for 18S rRNA. Woese's investigations resulted in the arrangement of life into three domains: *Archaea*, *Bacteria*, and *Eukarya* (see Figure 1.1). In this classification, members of the *Archaea* and the *Bacteria* are prokaryotic while eukaryotic plants and animals are placed in the *Eukarya*. This phylogenic designation of prokaryotic cells reflects differences involving biochemical and molecular biology characteristics. The concept of bacteria is changing in that it formerly referred to all cells with prokaryotic characteristics as being equal. Now it is recognized that members of the *Archaea* and the *Bacteria* are distinguished by significant variations in biochemical and molecular biology. Members of *Archaea* domain include those organisms that were formerly identified as archaebacteria [Greek *archaios*, ancient] while members of *Bacteria* domain consists of those organisms formerly called eubacteria [Greek *eu*, true]. The term "bacteria" is used only for those organisms that are members of the *Bacteria* while prokaryotes refers to members of both *Archaea* and *Bacteria*.

## 1.2.1. Variation Among the Prokaryotes

There is a tendency in biology to study the differences between species more than the similarities, and this focus also extends to the prokaryotes. Christian Gram in 1884 established that there were distinctive stain reactions that enabled bacterial pathogens to be separated into two major groups: Gram-positive and Gram-negative bacteria. As morphological characteristics and biochemical activities of bacteria became known, organisms were grouped according to production of endospores, presence of prosthecae, fixation of nitrogen, reduction of sulfate, production of methane, and other obvious phenotypic characteristics. This designation of the various groups of prokaryotes results from the focus on a single enzyme or activities of several related genes and is practical in derivative-based research. A limited number of fossilized forms of early cellular life have been discovered. These microfossils have a morphology reminiscent of cyanobacteria that were found in the Warrawoona Group of western Australia. Although only a limited number of prokaryotic fossils are available, metabolic activities and physiological processes have been used to establish evolutionary relationships (see Table 1.2). Numerous assumptions must be made to generate a chronology of events in early life history and at this time one cannot resolve whether the first life forms were heterotrophs or autotrophs.

### 1.2.1.1. The Two Domains

A comprehensive taxonomic coverage of the *Archaea* and the *Bacteria* is found in the multivolume series of *Bergey's Manual of Systematic Bacteriology* and in *The Prokaryotes*. Distinctions between members of the *Archaea* and the *Bacteria* domains are generally centered on cell energetics, processing of genetic information, and composition of cell wall and plasma membrane. Physiological and biochemical features of *Archaea* and *Bacteria* are summarized in Table 1.3. With respect to comparative physiology, there is an obvious difference in the type of macromolecule that provides rigidity of the cell wall. Although both Gram-positive and Gram-negative types of bacteria have a cell wall that contains peptidoglycan with D-forms of amino acids, the archaea examined do not contain typical peptidoglycan but have different types of cross-linked carbohydrate polymers. There is relatively little difference between archaea and bacteria with respect to capsules or S-layers. A major structural difference is observed in membrane composition, with archaea having ether linkages involving long-chain alcohols while bacteria have ester linkages between glycerol and fatty acids. There are several differences between the archaea and bacteria with respect to protein synthesis, and as a consequence antibiotics used to inhibit bacteria are not effective against archaea. Initiation of protein synthesis in bacteria begins with *N*-formylmethionine whereas in archaea methionine is the first amino acid in protein

Table 1.2. Chronology of precambrian evolution of life.[a]

| Time scale (years before present) | Organic solution scenario[b] | Surface metabolism scenario[c] |
|---|---|---|
| $4.6 \times 10^9$ | Origin of the Earth | Origin of the Earth |
| $\sim 4.3 \times 10^9$ | First self-reproducing molecules | First metabolism using mineral surfaces |
| $\sim 4.1 \times 10^9$ | — | First chemoautrophic cells |
| $\sim 4.0 \times 10^9$ | — | First sulfate and sulfur respiring cells |
| $\sim 4.0–3.8 \times 10^9$ | First heterotrophic cells | — |
| $\sim 3.8 \times 10^9$ | First methanogens and/or anoxygenic photosynthetic cells | First photoautotrophs |
| $\sim 3.5 \times 10^9$ | Warrawoona (cellular ?) fossils | Warrawoona (cellular ?) fossils and first heterotrophs |
| $\sim 3.5–3.0 \times 10^9$ | First oxygenic photosynthesis | First oxygenic photosynthesis |
| $\sim 2.7 \times 10^9$ | First anaerobic respiration (Sulfate reducers and others) | — |
| $2.1 \times 10^9$ | Fully oxidized atmosphere | Fully oxidized atmosphere |
| $2.0 \times 10^9$ | First cells with aerobic respiration | First cells with aerobic respiration |
| $1.4–0.85 \times 10^9$ | First eukaryotic cell | First eukaryotic cell |

[a]Modified from Ehrlich, 1996.[22]

[b]Precursor molecules for life are produced in the primordial soup and these include amino acids, sugars, purines, and pyrimidines. The monomers are condensed into polymeric protein, nucleic acids, and polysaccharides by heat, electrical, or radiant energy processes. Forms of prelife would include self-replicating forms of RNA, DNA, or protein.

[c]Precursor biomolecules originated from reactions between inorganic compounds and charged mineral surfaces. With the genesis of a lipid membrane, the mineral surface reactions were incorporated within the membrane. Production of RNA and DNA led to self-replicating life.

synthesis. In addition, ribosomal-mediated protein synthesis is sufficiently distinct in that streptomycin and tetracycline serve as inhibitors in bacteria but not in archaea. A single DNA-dependent RNA polymerase occurs in bacteria while several of these enzymes exist in archaea. With respect to cell energetics, photosynthesis occurs only in bacteria while the photosensitive rhodopsin system is unique for halophilic archaea. A fundamental difference between these two domains is found in ATP synthase with the mitochondrial-type (F-type) found in bacteria but not in archaea. The respiratory ATP synthase found in archaea has been considered by many to be similar to the plant-associated vacuolar (V-type); however, based on new observations there is growing support to designate it as an archaeal-type ATP synthase. Collectively, these differences between bacteria and archaea support the phylogenetic separation of these forms of life. In addition, numerous variations exist within each group of prokaryotes (e.g., bacteria and archaea).

Although there is a distinction between prokaryotes and eukaryotes with respect to cell structure, several features are shared by archaea and eukaryotes. For example, both archaea and eukaryotes use methionine to initiate protein synthesis and both groups have multiple DNA-dependent RNA polymerases with 8 to 12 subunits. In addition, histones and introns are found in both archaea and eukaryotes. Members of

Table 1.3. Distinguishing characteristics for the two domains containing prokaryotes.

| Characteristic | Bacteria | Archaea |
|---|---|---|
| **Cell wall** | | |
| Peptidoglycan is found in most species. | Yes | No |
| Some species lack a cell wall | Yes | Yes |
| Some species are Gram-positive | Yes | Yes |
| Some species are Gram-negative | Yes | Yes |
| S-layers (glycoproteins) in some species | Yes | Yes |
| **Structural characteristics** | | |
| Ether linkage in membrane lipids | No | Yes |
| Some species have histone-like proteins. | No | Yes |
| Some species produce endospores | Yes | No |
| Gas vacuoles in some species | Yes | Yes |
| D-forms of amino acids are present | Yes | No |
| Attachment of flagella | Ring system | Pilin-like system |
| **Molecular biology** | | |
| *E. coli* type of 16S ribosomal RNA | Yes | No |
| Initiator tRNA in protein synthesis | *N*-formylmethionine | Methionine |
| Ribosome is sensitive to diphtheria toxin (elongation factor 2) | No | Yes |
| DNA-dependent RNA polymerase | One | Several |
| Subunits structure of DNA-dependent RNA polymerase | 4 Subunits | 8–12 Subunits |
| **General characteristics** | | |
| Pathogens for animals or plants | Yes | No |
| Capsule production | Yes | Yes |
| Sensitive to penicillin, chloramphenicol, streptomycin, and kanamycin | Yes | No |
| Chlorophyll-based photosynthesis | Yes | No |
| Rhodopsin-like photosystem | No | Yes |
| Methanogenic | No | Yes |
| Nitrogen fixation | Yes | Yes |
| ATP synthase | F-type | V-type[a] |

[a]Considered by some to be sufficiently distinct from V-type that a new category should be employed; however, throughout this text V-type will be used.

the *Bacteria* contain several unique molecular and macromolecular arrangements while the *Archaea* share some molecular characteristics with the *Eukarya*.

An emerging tool for studying the evolution of prokaryotic organisms is the analysis of ribosomal RNA to identify specific conserved sequences. Signature sequences of a few oligonucleotide sequences are present in 16S and 18S rRNA unique for *Bacteria*, *Archaea*, or *Eukarya*. These nucleotide sequences are given in Table 1.4. In effect, these RNA sequences are phylogenic signatures. From an evolutionary perspective,

Table 1.4. Sequences found in 16S or 18S rRNA that are conserved.[a]

| Distribution | Signature oligonucleotide[b] | Approximate position |
|---|---|---|
| Found only in *Bacteria* | CACYYG | 315 |
| | AAACUCAAA | 910 |
| | CAACCYYCR | 1,110 |
| Found only in *Archaea* | CACACACCG | 1,400 |
| Found both in *Archaea* and *Eukarya* but not in *Bacteria* | AAACUUAAAG | 910 |
| | YUYAAUUG | 960 |
| | UCCCUG | 1,380 |
| Found in *Bacteria* and *Eukarya* but not in *Archaea* | UACACACCG | 1,400 |

[a]Modified from Madigan et al., 2003.[23]
Y = any pyrimidine and R = purine.

the lines for development of *Bacteria* and *Archaea* are not isolated but reflect considerable interaction.

### 1.2.1.2. The *Archaea*[2]

Based on 16S rRNA analysis, the cultured species of *Archaea* are clustered into either of two subgroups: Euryarchaeota or Crenarchaeota. The Euryarchaeota subgroup includes extremely halophilic and methanogenic organisms while the Crenarchaeota consist of extreme thermophiles and organisms that grow in cold environments. Specific examples of the Crenarchaeota and Euryarchaeota are given in Table 1.5. A third subgroup, "Korarchaeota," has been proposed for a unique group of prokaryotes that exist in extremely high-temperature environments but thus far have not been cultured in the laboratory.

Members of the *Archaea* domain grow in extreme environments and, as indicated in Table 1.5, comprise only a few unique physiological types. Their primary function is cycling of elements or nutrients under environmental conditions that either preclude the presence of bacteria or promote a rich diversity of microbial species. At this time, there are 69 different genera and 217 species in the *Archaea*. This is a relatively few number of organisms in the archaea in comparison to the larger number of bacteria known. Because only a small fraction of all microorganisms in the biosphere have been cultivated, the number of archaeal species is sure to increase as research continues.

### 1.2.1.3. The *Bacteria*[3,4]

The establishment of the *Archaea* and *Bacteria* domains is based on the use of 16S rRNA information. In an effort to evaluate other phylogenic markers, databases have been constructed using 23S rRNA, RNA polymerase, the F-type ($F_1F_0$) ATPase β-subunit, the elongation factor Tu, the HSP60 heat shock protein, and the RecA protein.

Table 1.5. Selected physiological groups of *Archaea*.

| Group | Representative genera | Gram stain[a] | Sub-group[b] | Autotrophic growth[c] | pH range for growth | Temperature (°C) for growth |
|---|---|---|---|---|---|---|
| Cell wall–less form | *Thermoplasma* | − | E | No | 1.0–2.0 | 33–67 |
| Extreme halophiles | *Halobacterium* | − | E | No | 6.0–7.0 | 40–50 |
| | *Haloferax* | − | E | No | 6.0–7.0 | 40–50 |
| | *Natronobacterium* | − | E | ? | 9.0–10.0 | 40–50 |
| Extreme thermophiles | *Acidianus* | − | C | Yes | 1.0–3.0 | 65–95 |
| capable of oxidizing | *Caldococcus* | ? | U | No | 6.2–6.4 | 80–90 |
| elemental sulfur (S°) | *Desulfurococcus* | − | C | No | 5.5–6.8 | 80–90 |
| | *Pyrodictium* | ? | C | Yes | 6.0–7.0 | |
| | *Pyrococcus* | − | C | No | 6.7–7.5 | 70–103 |
| | *Sulfolobus* | ? | C | Yes | 2.0–3.0 | 50–90 |
| | *Thermococcus* | ? | E | No | 5.0–6.0 | 75–97 |
| | *Thermoproteus* | − | C | Yes | 5.0–6.8 | 75–96 |
| Methanogens | *Methanobacterium* | + | E | Yes | 6.0–8.0 | 25–37 |
| | *Methanococcus* | − | E | Yes | 6.0–8.0 | 75–90 |
| | *Methanothermus* | ? | E | Yes | 6.0–7.0 | 80–95 |
| | *Methanosarcina* | ? | E | Yes | 4.5–7.5 | 64–92 |
| Sulfate reducer | *Archaeoglobus* | − | E | Yes | 5.0–6.0 | 60–95 |
| Iron oxidizer | *Ferroglobus* | − | E | Yes | 6.5–7.5 | 70–95 |

[a]Gram stains: −, negative; +, positive.
[b]Organization of *Archaea* domain based on 16S rRNA sequence: C = *Crenarchaeota*; E = *Euryarchaeota*; U = unclassified.
[c]If one or more species of that genera display autotrophic growth, "yes" is designated.

In the phylogenic tree of *Bacteria* using 16S rRNA analysis, there are at least 18 major lines along which organisms may have evolved (Figure 1.2). From data available, the analysis of a highly conserved region of the $\sigma^{70}$ sigma factor, there are three prominent groups of bacteria and these include the proteobacteria, the cyanobacteria, and the Gram-positive bacteria (see Figure 1.3). Although the data are limited, construction of the individual gene trees using 23S rRNA, the elongation factor Tu, ATPase, and $\sigma^{70}$ sigma factor reflect some of the distinctions between bacterial species as demonstrated by analysis of genes for 16S rRNA. Variation may be expected to be greater in the individual gene trees of enzymes owing to the lateral transfer of genes in prokaryotes. Although the grouping of bacteria by gene analysis of the 16S rRNA is an evaluation of only a segment of the single macromolecule, clearly this information is highly conserved. Perhaps the presence of multiple genes for ribosomal synthesis in each bacterial genome contributes to its stability. As more sequence data are collected from additional essential proteins, the evolutionary relationships of the *Bacteria* will be extended.

In the *Bacteria* domain, there is a diversity of physiological types that inhabit the environments supporting eukaryotic life and certain bacteria have developed associations with plants or animals that range from symbiosis to pathology. At this time, there are 871 different genera and 5,007 species making up the *Bacteria*. The taxonomic subunits of the *Bacteria* frequently follow along the lines of physiological types

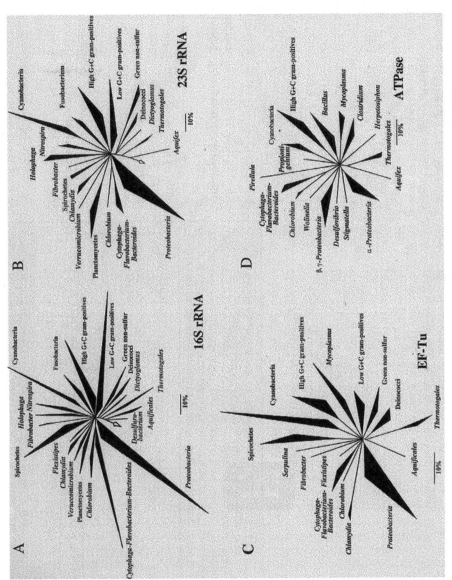

Figure 1.2. Phylogenic relationships of bacteria based on primary structure of (**A**) 16S rRNA, (**B**) 23S rRNA, (**C**) the Tu protein or elongation factor of ribosomal-mediated protein synthesis, and (**D**) the β-subunit of F-type (F₁F₀) ATPase. Phylogenic groups are indicated by the *triangles*. The designation of *Bacillus* includes *Enterococcus* and *Lactobacillus* while *Clostridium* includes *Acetobacterium* and *Ruminococcus*. (From Ludwig, W. and K.-H. Schleifer. 1999. Phylogeny of *Bacteria* beyond the 16S rRNA standard, *ASM News* 65:754. Reprinted with permission from the American Society for Microbiology.)

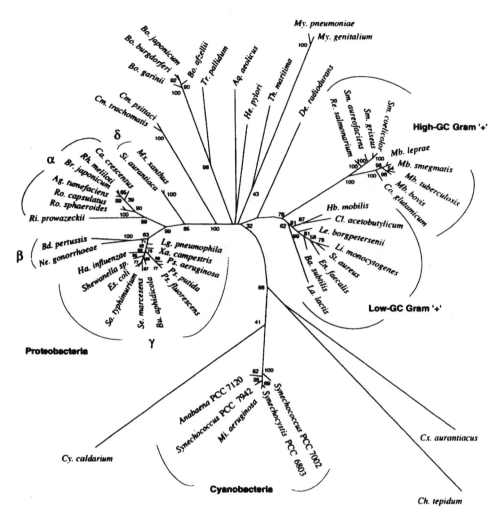

Figure 1.3. Phylogenic analysis of bacteria using information from $\sigma^{70}$ sigma factor in the RNA polymerase. *Ag., Agrobacterium*; *Aq, Aquaflex*; *Ba, Bacillus*; *Bd, Bordetella*; *Bo, Borrelia*; *Br, Bradyrhizobium*; *Bu, Buchnera*; *Cx, Caulobacter*; *Ch, Chlorobium*; *Cl, Clostridium*; *Cm, Chlamydia*; *Co, Corynebacterium*; *Cx, Chloroflexus*; *Cy, Cyanidium*; *De, Deinococcus*; *En, Enterococcus*; *Es, Escherichia*; *Ha, Haemophilus*; *Hb, Heliobacillus*; *He, Helicobacter*; *La, Lactococcus*; *Le, Leptospira*; *Li, Listeria*; *Lg, Legionella*; *Mb, Mycobacterium*; *Mi, Microcystis*; *Mx, Myxococcus*; *My, Mycoplasma*; *Ne, Neisseria*; *Ps, Pseudomonas*; *Rh, Rhizobium*; *Ri, Rickettsia*; *Ro, Rhodobacter*; *Sa, Salmonella*; *Se, Serratia*; *Si, Stigmatella*; *St, Staphylococcus*; *Sm, Streptomuces*; *Th, Thermotoga*; *Tr, Treponema*; *Xa, Xanthomonas*. (From Gruger and Bryant, 1999.[3] Used with permission.)

Table 1.6. Selected examples of bacteria contained in proteobacteria group.

| Subgroup | Representative genera with selected characteristics |
|---|---|
| **α-Proteobacteria** | *Brucella*, a pathogen for animals including humans<br>*Caulobacter*, prosthecate bacterium with holdfast that grows in aquatic environments<br>*Hypomicrobium*, prosthecate bacterium that may divide by budding<br>*Nitrobacter*, aerobic oxidizers of nitrite to nitrate (nitrifying bacteria)<br>*Rhizobium*, fixes nitrogen when grown in leguminous root nodules<br>*Rhodospirillum*, photosynthetic bacteria that have purple photopigments and do not use sulfur as an electron donor (purple nonsulfur)<br>*Rickettsia*, obligate intracellular parasites that may have invertebrate vectors |
| **β-Proteobacteria** | *Bordatella*, a pathogen for animals, especially mammals including humans<br>*Leptothrix*, bacteria with sheaths containing iron oxide and manganese oxide<br>*Neisseria*, saprophytic or pathogenic bacteria that grow in mucous membranes<br>*Nitrosomonas*, an aerobic autotroph that oxidizes ammonia to nitrite<br>*Sphaerotilus*, a sheath-forming bacterium that generally is devoid of metal oxides<br>*Thiobacillus*, autotrophs that oxidize sulfur and thiosulfate to sulfuric acid |
| **γ-Proteobacteria** | *Azotobacter*, aerobic soil bacterium that fixes nitrogen under nonsymbiotic conditions<br>*Chromatium*, photosynthetic bacterium that have purple photopigments and can use sulfur as an electron donor (purple sulfur bacteria)<br>*Escherichia, Enterobacter, Klebsiella, Proteus, Yersinia, Shigella, Salmonella*, and *Serratia*, enteric bacteria of humans and some strains pathogenic<br>*Haemophilus*, many strains pathogenic for mammals and require blood factors for growth<br>*Methylococcus*, aerobic oxidizers of methane, methanol, formaldehyde (methylotrophs)<br>*Photobacterium*, chemiluminescent bacteria often associated with marine fish<br>*Pseudomonas*, broad carbon utilization and some strains produce opportunistic infections in animals |
| **δ-Proteobacteria** | *Bdellovibrio*, may grow in the periplasmic space of Gram-negative bacteria<br>*Desulfovibrio*, uses sulfate and sulfur as electron acceptors with generation of $H_2S$<br>*Myxococcus*, aerobic bacterium that may form fruiting bodies containing resistive cells |
| **ε-Proteobacteria** | *Campylobacter*, many species pathogenic for humans and mammals<br>*Helicobacteria*, found in gastric region of humans and mammals |

and there are several distinct groups, currently the proteobacteria and Gram-positive subgroups contain the largest number of species. Proteobacteria is the largest subgroup, comprising more than 1,300 species from several hundred genera of Gram-negative organisms. There are five subgroups of proteobacteria (α, β, γ, δ, and ε) and the diversity of their physiological types is indicated in Table 1.6. Contained within the proteobacteria group are some of the important human pathogens as well as bacteria of the nitrogen and sulfur cycles and bacteria found in aquatic environments. Although all proteobacteria are Gram-negative, other taxonomic lineages of *Bacteria* such as cyanobacteria, green-sulfur bacteria, and green nonsulfur bacteria also contain Gram-negative bacteria. The Gram-positive group is generally divided into two major segments: the low-GC group and the high-GC group. Other subgroups under the

Table 1.7. Unique 16S rRNA sequences.

| Organisms | Sequence | Target position |
| --- | --- | --- |
| Prokaryotes and eukaryotes | CCGTCAATTCCTTTGAGTTT | 907–928 |
| *Bacteria* but not *Archaea* | CCTACGGGAGGCAGCAG | 341–357 |
| Most δ-proteobacteria and a few Gram-positive bacteria | CGGCGTCGCTGCGTCAGG | 385–402 |

Modified from Barton and Plunkett, 2000.[24]

*Bacteria* include: Cyanobacteria, *Plantomyces*, green sulfur bacteria, spirochetes, *Deinococcus*, green nonsulfur bacteria, *Thermotoga*, environmental noncultured organisms, and *Aquiflex*. Some of the members in each of these subgroups are given in Figure 1.3. Phylogeneticists prefer to deemphasize the use of the Gram-positive and Gram-negative designations because several bacteria including *Deinococcus* and *Mycobacterium* have a cell wall that is distinct from that of either of these two groups.

As a means of characterizing organisms present in the environment, numerous DNA probes have been developed for specific genera of bacteria. As presented in Table 1.7, DNA probes have also been established for broad biological groups. Probes have been developed to bind to DNA from both eukaryotic and prokaryotic forms of life. In addition, specific DNA sequences have been identified for all bacteria and even major subgroups such as the proteobacteria. These DNA sequences are for ribosomal RNA and are reminiscent of those reported in Table 1.5. It is indeed impressive that the 16S rRNA with about 1,500 oligonucleotides has unique regions that have evolved at rates distinct from the rest of the molecule.

## 1.2.2. Information Content

As determined by calculating the guanine and cytosine content in DNA, each prokaryote has a specific nucleic acid content. The amount of guanine and cytosine as compared to the sum of all DNA bases is commonly expressed as a percentage, calculated according to the following formula:

$$\frac{moles\ (G + C)}{moles\ (A + T + G + C)} \times 100 = mol\%\ G + C$$

Adenine (A) and guanine (G) are purines and cytosine (C) and thymine (T) are pyrimidines. Like all biological species, members of both bacteria and archaea have a specific content of mol% G + C which is also known as the GC ratio. The range of mol% G + C for prokaryotes is approximately 20 to 80 which exceeds the 30% to 50% range found in plants and animals. If a large protein is configured to have the lowest mol% G + C and also configured to have the highest possible mol% G + C, the range of mol% G + C would be comparable to the range seen in prokaryotic cells. Thus, prokaryotes represent the biological limits that G + C are found in nature.

Table 1.8. Selected prokaryotes and their G + C content expressed as a percentage of total number of bases in DNA.

| DNA (mol% G + C) | Bacteria | DNA (mol% G + C) | Archaea |
|---|---|---|---|
| 30–36 | Clostridium, Staphylococcus, Bacillus, Streptococcus | 23–31 | Methanobrevibacter, Methanosphaera |
| 38–44 | Proteus, Neisseria, Hemophilus, Lactobacillus, Coxiella, Bacillus Streptococcus | 31–35 | Acidianus, Methanotorris, Methanococcus |
| 46–52 | Escherichia, Enterobacter, Klebsiella, Salmonella, Vibrio, Pasteurella | 36–43 | Methanosacrina, Sulfolobus, Pyrococcus, Methanoplanus |
| 54–60 | Alcaligenes, Corynebacterium Pseudomonas, Lactobacillus, Serratia, Clostridium, Klebsiella | 45–51 | Desulfurococcus, Methanospirillum, Metallosphaera |
| 62–68 | Bacillus, Mycobacterium, Vibrio, Rhizobium, Micrococcus | 56–60 | Methanopyrus, Thermoproteus |
| 70–80 | Mycobacterium, Streptomyces, Nocardia | 62–65 | Stetteria, Pyrodictium |

The mol% G + C represents the genetic constraints of a given species but does not indicate evolutionary development. For example, both humans and *Streptococcus pneumoniae* have the same mol% G + C (38 to 40) and *E. coli* has mol% G + C of 50 to 52. The mol% G + C for several bacteria and archaea is given in Table 1.8. The molecular basis for assigning a prokaryote to a given species requires that organisms have a similar G + C ratio and that DNA/DNA hybridization must exceed 70%.

The mol% G + C in DNA can be measured by high-pressure liquid chromatography, buoyant density centrifugation, or thermal denaturation. The basis for DNA melting is the ease with which hydrogen bonding between the nucleotide bases of AT or GC can be disrupted in the double-stranded DNA. The GC base pairing is stronger than the AT pairing because there are three hydrogen bonds between G and C but only two hydrogen bonds between A and T. On heating, the bond between the individual bases of the double-stranded DNA separates according to the strength of the base pairing. DNA of high G + C ratio shows more thermal stability than DNA of low G + C ratio. From this it may be assumed that thermophilic prokaryotes would have the highest mol% G + C; however, hyperthermophiles have a mol% G + C of 31 to 60, which is comparable to that of many mesophilic bacteria. In this case, the stability of DNA in hyperthermophiles is not attributed to the G + C ratio but rather to the presence of unique DNA binding proteins and the positive supercoiling of DNA attributed to special reverse gyrases.

If mol% G + C ratios are high, the gene information varies with the first and last base in the codon. To illustrate this point, the production of 3-isopropylmalate dehydrogenase in *Thermus thermophilus* is presented in Table 1.9. *Thermus thermophilus* has a mol% G + C of 70 and the second position in the codon has relatively equal use of the four bases. However, with respect to the first position in the codons,

Table 1.9. Codon usage by gene for 3-isopropylmalate in *Thermus thermophilus*.[a]

| Amino acid | Number of residues | Codon | Number of times used in gene | Amino acid | Number of residues | Codon | Number of times used in gene |
|---|---|---|---|---|---|---|---|
| Alanine | 42 | GCT | 1 | Phenylalanine | 12 | TTT[b] | 3 |
| | | GCC[b] | 28 | Proline | 27 | TTC | 9 |
| | | GCA | 1 | | | CCT[b] | 3 |
| | | GCG | 12 | | | CCC[b] | 18 |
| Arginine | 28 | CGT | 1 | Serine | 15 | TCT | 1 |
| | | CGC | 9 | | | TCC[b] | 8 |
| | | CGA[b] | 2 | | | TCG[b] | 1 |
| | | CGG[b] | 10 | | | AGC | 5 |
| | | AGG[b] | 6 | Threonine | 13 | ACC | 6 |
| Asparagine | 6 | AAC | 6 | | | ACG[b] | 7 |
| Glutamine | 3 | CAG | 3 | Tryptophan | 2 | TGG | 1 |
| Glycine | 36 | GGC | 11 | Tyrosine | 6 | TAT[b] | 5 |
| | | GGA[b] | 6 | | | TAC | |
| | | GGG[b] | 19 | Valine | 32 | GTC[b] | 12 |
| Histidine | 5 | CAC | 5 | | | GTG[b] | 20 |
| Isoleucine | 9 | ATC | 8 | | | | |
| | | ATA[b] | 1 | | | | |
| Leucine | 36 | CTT[b] | 6 | | | | |
| | | CTC[b] | 15 | | | | |
| | | CTA[b] | 1 | | | | |
| | | CTG | 10 | | | | |
| | | TTA[b] | 1 | | | | |
| | | TTG[b] | 3 | | | | |
| Lysine | 16 | AAA | 1 | | | | |
| | | AAG | 15 | | | | |
| Methionine | 6 | ATG | 6 | | | | |

[a]From Oshima, T. 1986. The genes and genetic apparatus of extreme thermophiles. In *Thermophiles: General, Molecular and Applied Microbiology* (T.D. Brock, ed.). Wiley-Interscience, John Wiley, New York, pp. 137–157. Copyright 1986 by John Wiley & Sons, Inc. This material is used by permission of Wiley-Liss, Inc., a subsidiary of John Wiley & Sons, Inc.
[b]Nonoptimal codons in *Escherichia coli*.

there is a high G content and this accounts for greater levels of valine, alanine, glutamate, and glycine as compared to codons with lower levels of G in the first position. Of considerable importance is the third position in the codon where the mol% G + C content is about 90. This example demonstrates how codon usage influences polypeptide composition and also that a single gene utilizes the full redundancy of the DNA code.

## 1.2.3. The Size of Bacteria[5]

Since the first observations of bacteria, it was always assumed that these organisms were universally small. However, size cannot be used as a limiting characteristic of classification because *Nanochlorum eukaryotum*, a eukaryote with a nucleus, chloroplast, and mitochondrion, is only 1 to 2 µm. Prokaryotes that are pathogenic for humans, animals, and plants are relatively similar in size with a cell diameter of 1 to 2 µm. The small size of prokaryotes is evident in the cell wall–less bacteria known as mycoplasma ("plastic molds") that have a spherical cell that is 0.2 to 0.8 µm in diameter. Without internal membranes to compartmentalize activities, diffusion through a 1-µm spherical cell is about 10 msec, which is adequate to saturate intracellular enzymes including those that have a turnover rate of a few hundred per second. It was reasoned that large bacterial cells could not achieve mixing of cytoplasmic contents to maintain enzymes loaded with substrates. However, recent observations reveal that some bacterial cells may be extremely large and, in fact, even visible to the naked eye. The size of some prokaryotic cells is given in Table 1.10.

The largest prokaryote thus far reported is *Thiomargarita namibiensis*, a chemolithotrophic, nonmotile, marine bacterium, which has a spherical cell with a diameter of 750 µm. Another bacterium with large cells is *Epulopiscium fishelsoni*, a heterotrophic bacterium found in the intestine of sturgeon fish, with a rod shape that is 80 µm × 600 µm. *Beggiatoa*, a chemolithotrophic bacterium found in marine and freshwater environments, can have multicellular filaments that may be several millimeters in length with cells 40 to 200 µm in diameter. *Thiomargarita namibiensis, Epulopis-*

Table 1.10. Examples of prokaryotes that have a unique size.[a]

| Organism | Characteristic[b] | Size (µm) | Volume of cell or filament (µm³) |
|---|---|---|---|
| *Thiomargarita namibiensis* | Sulfur respiring coccus | 750 | $2 \times 10^8$ |
| *Epulopiscium fishelsoni* | Heterotrophic, oblong cell | 80 × 600 | $3 \times 10^6$ |
| *Beggiatoa* spp. | Filamentous, sulfur respiration | 160 × 50 | $1 \times 10^6$ |
| *Lyngbya majuscula* | Filamentous, cyanobacteria | 80 × 8 | $4 \times 10^4$ |
| *Thiovulum majus* | Sulfur respiring, coccus | 18 | $3 \times 10^3$ |
| *Staphylothermus marinus* | Coccus, archaea | 15 | $1.8 \times 10^3$ |
| *Magnetobacterium bavaricum* | Magnetotactic | 2 × 10 | 30 |
| *Escherichia coli* | Heterotrophic, typical bacterium | 1 × 2 | 2 |
| *Mycoplasma pneumoniae* | Pathogenic | 0.2 | $5 \times 10^{-3}$ |
| *Thermodiscus* | Archaea | 0.2 × 0.08 | $3 \times 10^{-3}$ |

[a]From Schulz, H.N. and B.B. Jørgensen. 2001. Big bacteria. With permission, from the *Annual Review of Microbiology*, Volume 55, ©2001 by Annual Reviews, *www.annualreviews.org*.
[b]Organisms are bacteria unless designated.

*cium fishelsoni*, and other large bacteria listed in Table 1.10 have huge vacuoles in the cytoplasm that fill the center of the cell, with the cytoplasmic constituents occupying a layer that is only 1 to 2 μm thick. Thus, diffusion may be less of a problem in these large bacteria than previously considered. Of physiological importance is that these large cells may extend across transition zones of oxygen and nutrients, thereby enabling the organism to optimize growth depending on the immediate chemical environment. Clearly these large bacteria have evolved some interesting avenues for conducting life-associated processes. At the other end of the size spectrum, there are some bacteria and archaea that have cells as small as 0.1 to 0.2 μm. These minute prokaryotes approach the size of some of the large viruses and are referred to as nanobacteria. While some organisms such as *Mycoplasma* and *Thermodiscus* are characteristically small, nanobacteria from diverse species are found in marine or deep subsurface environments. Because most laboratory studies are conducted with bacteria about the size of *Escherichia coli*, the small as well as the large prokaryotes should be more than biological curiosities and their physiological uniqueness remains to be addressed.

## 1.2.4. The Unity of Physiology

The scope of physiology is broad and includes all the activities associated with living systems. An approach used to examine complex issues such as life processes of cells is to divide the topic into several areas and characterize the components. This analytical approach is then followed by assembly of the information from the various facets being analyzed to provide a unified or comprehensive view of a living cell. With microorganisms displaying different ranges of activities and complexities, it is practical to divide the life processes into different categories (e.g., cell surface, membranes, respiration, or catabolism) and apply analytical approaches to characterize each component. With the assembly of information from the diverse areas that have been examined, an insight is gained with respect to the physiological mechanisms of cells. This synthesis of information provides not only a theme for the process of life but also a basis for variations.

In the early part of the 20th century, it became apparent that bacteria and higher forms of life shared many characteristics. A unity in metabolism was proposed by Albert J. Kluyver in 1924 and this concept was extended by C.B. van Niel in the 1940s. Ultimately, it became known as the "unity of biochemistry" in which all living organisms were found to share numerous features characteristic of living systems. This concept was based on the observation that organisms employed similar enzymes in their metabolic pathways and that cells employed similar systems to achieve cell growth. When it was determined that prokaryotes and eukaryotes had a common basis of heredity, experiments in early molecular biology demonstrated a similarity in mechanisms for DNA replication and protein synthesis. Over the last few decades, microbial cells have been determined to share similar themes in metabolism and growth, response to the environment, cell energetics, and processing of genetic information. Even though there are specific differences between the *Bacteria* and the *Archaea* with respect to molecular biology and structural detail, there is an overriding "unity of physiology." Physiology unifies all of the individual biochemical reactions that result in cellular growth. Microbial physiologists study the biochemistry of cell processes and the mechanisms for cellular adjustments to an ever changing environment. A study of the variations or differences among prokaryotes enables us to understand how fundamental principles essential for life are modified to enable microbes to occupy the various regions of the biosphere. Before one studies the differences between the various species of bacteria and archaea, it is important to identify shared characteristics.

## 1.3. Structural Characteristics of Prokaryotes[6]

For both the eukaryotic and prokaryotic organisms, the structural unit is the cell. However, there are several organizational differences between these organisms. In prokaryotic cells, the genetic material is dispersed throughout the cell and not confined to a membrane-enclosed nucleus. In addition, there are important distinctions concerning the distribution of information in DNA for these two groups; specifically, prokaryotes do not have DNA banded into morphologically discrete chromosomes. Internal organelles are found in eukaryotes but prokaryotic cells can be characterized as devoid of internal membranes. The prokaryotes are distinguished from eukaryotes by certain details of protein synthesis including the size and composition of ribosomes. The characteristics of ribosomes in prokaryotic and eukaryotic organisms are given in Table 1.11.

From the anatomy of the prokaryotic cell, several distinctive physical structures are present. An architectural component found in all prokaryotic cells is the membrane, which provides the selective permeable boundary for the cytoplasm. Although this structure is designated by some as the cytoplasmic membrane of bacteria, throughout this book it will be referred to as the plasma membrane.

Characteristically, bacteria have cell walls that confer osmotic stability and shape. In a strict use of the term, cell wall would refer only to the peptidoglycan or a similar chemical component that encompasses the plasma membrane. However, chemicals may be attached to this structure or may extend through it and would be isolated along

Table 1.11. Nucleic acid and protein composition of prokaryotic and eukaryotic ribosomes.

| Characteristic | Prokaryotic cells[a] | | Eukaryotic cells | |
|---|---|---|---|---|
| Size | 70S | | 80S | |
| Subunits | 50S | 30S | 60S | 40S |
| rRNA present | 23S and 5S | 16S | 23S, 5.8S, and 5S | 18S |
| Number of different proteins present | 31 | 21 | 50 | 33 |

[a]Based on ribosomes from *Escherichia coli*.

with the peptidoglycan. The dilemma is how to designate components found associated with the rigid structure of Gram-negative and Gram-positive cells. Some microbiologists suggest the use of the term *envelope* to describe cellular structures exterior to the bacterial plasma membrane; however, biologists may also use *envelope* to describe the surface of eukaryotic cells and viruses. Thus, the term *cell wall* for prokaryotic organisms gives general description but does not provide the specificity needed when the structure exterior to the plasma membrane is chemically dissected. Throughout this book references to the cell wall will provide a general location and not a discrete structure. A working description of cell wall for Gram-positive bacteria is that it is the structure exterior to the plasma membrane and for Gram-negative bacteria it encompasses the structural layer plus the outer membrane.

## 1.3.1. The Periplasm[7-9]

From observation of bacteria through the electron microscope, a region between the plasma membrane and outer membrane has been identified in Gram-negative bacteria as the periplasmic space. The area between the plasma and outer membranes is not an empty area as would be implied by calling it a space but has considerable activity. It has been theorized that the region between the outer and inner membranes contains a gel-like substance that maintains turgor pressure. The diameter of this region is relatively constant in different electron micrographs and the stains for electron microscopy reveal faint images in this region. In 1961, Peter Mitchell suggested it would be more appropriate to designate this region of the cell as the periplasm to complement the term cytoplasm which has been used to describe cell contents. The size of the periplasm is not trivial but for enteric bacteria has been estimated to account for 10% to 30% of the total cell volume.

It has been suggested that Gram-positive bacteria also have a periplasmic region. Based on electron microscopy, some Gram-positive cells have an electron-transparent region between the plasma membrane and the peptidoglycan cell wall but this should not be considered as a periplasm. From the historical designation, the periplasm is the area between the outer and the inner membranes of bacteria. Clearly, Gram-positive bacteria are devoid of an outer membrane structure and proteins are not freely dispersed in an environment exterior to the plasma membrane, as is the case for Gram-negative bacteria. Proteins and enzymes on the outer surface of plasma membrane of Gram-positive bacteria are tethered into the plasma membrane. Future studies may clarify the activity in the region just exterior to the plasma membrane in Gram-positive prokaryotes and identify it with a name that is distinct from periplasm.

### 1.3.1.1. Function of the Periplasm

The periplasmic region of Gram-negative bacteria is attributed to special metabolic activities or processes associated with organisms having outer and inner (plasma) membranes. As assessed from examining a small number of Gram-negative bacteria, several different activities are associated with the periplasm and they are summarized as follows. (1) Specific electron transport proteins maybe localized in the periplasm.

For example, in *Paracoccus denitrificans*, NO reductase and N₂O reductase function in the denitrification process and dehydrogenases for the oxidation of methanol or methylamine have been shown to involve *c*-type cytochromes in the periplasm. In another example, several strains of sulfate-reducing bacteria have hydrogenase and *c*-type cytochrome in the periplasm where these proteins have been suggested to participate in cell energetics by contributing to hydrogen cycling. (2) Periplasmic binding proteins for solute transport are present in the periplasm. Many different examples of proteins associated with the ABC transport system have been isolated from the periplasm of *Escherichia coli* and *Salmonella typhimurium*. (3) Enzymes for catabolism and cellular development can be found in the periplasm. Alkaline phosphatase, 3' or 5' nucleotidases, proteases, and dextrinases have been found in the periplasm of various bacteria. The periplasm contains the biosynthetic enzymes involved in the formation of the peptidoglycan as well as β-lactamase (penicillinase) and penicillin binding proteins. Proteases are present in the periplasm to degrade leader sequences released from proteins after they have been exported from the cell and to hydrolyze damaged proteins involved with insertion of proteins into the outer membrane. (4) The periplasm may serve as a staging area for the assembly of some of the proteins found on the surface of the cell. This would include chaperone proteins for the synthesis of pili and flagella. (5) Carbohydrate complexes with mixed functions and activities are present in the periplasm. Enteric bacteria have anionic polysaccharides with a backbone structure of 6 to 12 glucose units with β-1,6 or β-1,2 linkages between the glucose moieties. Attached to glucose are phosphoethanolamine, *sn*-1-phosphoglycerol, and *O*-succinyl side chains that produce a molecular mass of about 2.3 kDa. In addition, the periplasm contains small segments of loosely cross-bridged peptidoglycan that is assembled at the outer face of the plasma membrane and not yet incorporated into peptidoglycan. (6) Detoxification reactions occur in the region of the periplasm. $Hg^{2+}$ and other toxic heavy metals are bound to periplasmic proteins and prevent toxic reactions between the metal cation and the outer surface of the plasma membrane. As a consequence of the presence of a large number of different types of compounds, the periplasm consists of a gel-like matrix and some have suggested that this gel has a role in osmotic regulation of the cell. Clearly, the periplasm of the Gram-negative cell is important for a variety of cellular processes.

## 1.3.1.2. Isolation of Periplasmic Proteins

Purification of periplasmic proteins can be greatly facilitated if cytoplasmic proteins are excluded in the isolation process. One means of liberating proteins from the periplasm is to degrade the cell wall of the Gram-negative bacteria under conditions in which protoplast formation is favored. While hydrolysis of the peptidoglycan readily occurs with lysozyme, ethylenediaminetetraacetic acid (EDTA) is required as a pretreatment to chelate divalent cations that stabilize the outer membrane structure. With loss of structural rigidity provided by an intact peptidoglycan, the cell ruptures because osmotic pressures inside the cell are greater than in the aqueous solution outside of the cell. However, if 0.6 *M* sucrose is present in the buffers, cells and protoplasts remain intact with the release of periplasmic proteins.

Some of the nutrient transport proteins associated with the ABC (ATP binding cassette) transport system found in the periplasm are loosely associated with the plasma membrane. H.C. Neu and L.A. Heppel developed an osmotic shock treatment that releases binding proteins from Gram-negative bacteria. For the shock treatment, cells are suspended in a sucrose solution and sucrose diffuses through the water channels in the outer membrane to establish a high concentration in the periplasm. The cells are rapidly diluted into distilled water and the osmotic pressure in the periplasm cause the outer membrane to rupture. Because the peptidoglycan is intact, the protoplasmic membrane is not ruptured. Only about 15% of the total number of cell proteins in Gram-negative bacteria reside in the periplasm and are released by osmotic shock.

## 1.3.2. Genome Structure[10,11]

Physiology is the study of cellular and molecular structures. With the presence of relatively few cellular structures in bacteria to regulate activities or processes, the configuration of molecules becomes extremely important. A good example of how a molecule can have a unique role in terms of function is the special structure of circular DNA of the bacterial chromosome. If one examines the DNA molecule from *Escherichia coli* important functional features can be assessed. The cell of *Escherichia coli* is only 0.5 μm by 1.0 μm while the DNA molecule is about 1.5 mm long. The organization of the DNA molecule is shown in Figure 1.4. Interactions within the DNA macromolecule produce loop structures every 100 kb and the DNA molecule is supercoiled to produce a structure of about 1 μm in length. An obvious function of this supercoiling is the condensation of the DNA molecule so that it can fit inside the *Escherichia coli* cell. However, this molecular architecture promotes stability of information by favoring intraloop DNA–DNA interactions but not interloop exchanges. In this context, bacterial DNA with the supercoiled loop domain structure has numerous functions. Further stability is attributed to proteins associated with DNA in the prokaryotes and histone-like proteins have been reported for members of the archaea.

While models are available to explain the structural configuration of a single haploid genome copy in a cell, the organization of multiple chromosomes in a bacterial cell must also be addressed. Actively growing cultures of *Escherichia coli* will have at least four copies of genome per cell but under certain conditions *Azotobacter vinelandii* contains >20 genome copies per cell. In nitrogen-limited chemostat growth, *Desulfovibrio gigas* has an average of 9 genomes per cell while in log phase in batch culture there are 17 genomes per cell. For a prokaryotic cell to hold all the DNA, there must be an organized state for the DNA. The formation of loops in the DNA certainly satisfies the configuration requirements of DNA in bacteria by means of circular DNA molecules. While most prokaryotes have a single circular DNA molecule, one strain of *Borrelia burgdorferi*, the agent of Lyme disease, has one large DNA structure of 1 Mbp (million base pairs) with 12 linear and 9 circular extrachromosomal elements. Another bacterium with a linear chromosome is *Streptomyces*

**A**  RANDOM COIL STRUCTURE

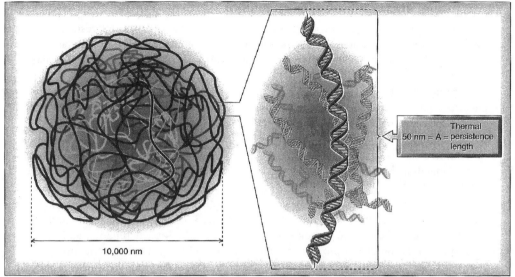

**B**  LOOP DOMAIN STRUCTURE

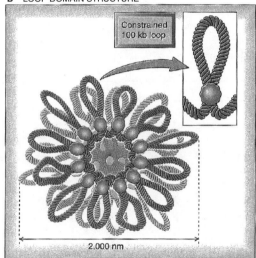

**C**  SUPERCOILED LOOP DOMAIN STRUCTURE

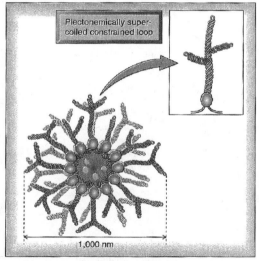

Figure 1.4. Structures to economize the distribution of DNA in prokaryotes. (**A**) The random coil structure which results from a random bend in the DNA every 50 nm or 150 basepairs. (**B**) The loop domain structure in which the DNA attaches to itself every 50 kb to produce a loop domain; see the inset. (**C**) The supercoiled loop domain structure of DNA provides the greatest degree of structure reduction. (From Trun, N.J. and J.F. Marko. 1998. Architecture of a bacterial chromosome. *ASM News* 64:279. Reprinted with permission from the American Society for Microbiology.)

*coelicolor*, which has a very large genome of 8.66 Mbp. It remains to be established how linear DNA molecules are replicated and folded inside the cell.

Genome analysis of numerous prokaryotes has produced some interesting information concerning genotypic information and size of the microbial genome. The amount of DNA in a prokaryote varies considerably and a distinction can be made between prokaryotes as specialists with a genome of ~2 Mbp to ~0.5 Mbp while prokaryotes as generalists have a genome of ~3 Mbp to ~9 Mbp. Examples of bacteria and archaea with different genome sizes is presented in Table 1.12. The specialists have a limited growth range, with some being obligate intracellular pathogens, some are found with a limited number of hosts, and a few archaea are found in a limited number of geological sites. The generalists includes organisms with a diverse metabolic capability as well as those organisms that have special physiological capabilities such as sporulation or photosynthetic system. Even though genome analysis has been conducted on a relatively small number of prokaryotes, some generalizations are resulting from the analysis (see Table 1.13). In both bacteria and archaea, nearly one half of the coding regions have no known biological function and nearly one fourth of the predicted proteins produced are found only in that specific organism. Future research will define the genes common to the prokaryotes as well as expand our knowledge concerning the biological function of all genes in the prokaryotes.

Table 1.12. Examples of prokaryotes reflecting different genome size.[a]

| Group and organism | Domain | Size (Mb) |
|---|---|---|
| **Specialists** | | |
| *Coxiella burnetii* | B | 2.1 |
| *Haemophilus ducreyi* | B | 1.76 |
| *Lactobacillus acidophilus* | B | 1.9 |
| *Mycoplasma hyopneumoniae* | B | 0.89 |
| *Neisseria meningitidis* | B | 2.2 |
| *Streptococcus pyogenes* | B | 1.98 |
| **Generalists** | | |
| *Bacteroides fragilis* | B | 5.3 |
| *Burkholder mallei* | B | 6.0 |
| *Clostridium tetani* | B | 4.4 |
| *Halobacterium salinarium* | A | 4.0 |
| *Magnetospirillum magnetotacticum* | B | 4.5 |
| *Mycobacterium smegmatis* | B | 7.0 |
| *Myxococcus xanthus* | B | 9.45 |
| *Nostoc punctiforme* | B | 10 |
| *Rhodopseudomonas palustris* | B | 5.46 |
| *Salmonella typhi* | B | 5.08 |
| *Xanthomonas campestris* | B | 5.1 |

[a]From Schulz and Jørgensen, 2001.[5]
A = archaea, B = bacteria.

Table 1.13. Genome characteristics of a selected number of microorganisms.[a]

| Organism | Genome size (Mbp) | Total ORFs | Unknown function (%) | Unique ORFs (%) |
|---|---|---|---|---|
| **Archaea** | | | | |
| *Archaeoglobus fulgidus* | 2.18 | 2,409 | 31.8 | 26 |
| *Methanobacterium thermoautotrophicum* | 1.75 | 1,869 | 32.8 | 21.9 |
| *Pyrococcus horikoshii* | 1.74 | 2,064 | 33.3 | 31.7 |
| *Methanococcus jannaschii* | 1.66 | 1,735 | 40.6 | 21.3 |
| **Bacteria** | | | | |
| *Escherichia coli* | 4.60 | 4,270 | 10.3 | 26 |
| *Mycobacterium tuberculosis* | 4.41 | 3,924 | 39 | 15 |
| *Bacillus subtilis* | 4.20 | 4,099 | 38 | 26 |
| *Synechocystis* sp. | 3.57 | 3,217 | 28.9 | 26.9 |
| *Deinococcus radiodurans* | 3.28 | 3,193 | 47 | 31 |
| *Thermotoga maritima* | 1.86 | 1,877 | 46 | 20 |
| *Haemophilus influenza* | 1.83 | 1,692 | 35 | 14 |
| *Helicobacter pylori* | 1.66 | 1,657 | 45 | 33 |
| *Aquifex aeolicus* | 1.50 | 1,521 | 44 | 27 |
| *Borrelia burgdorferi* | 1.44 | 1,751 | 65 | 39 |
| *Chlamydia pneumoniae* | 1.23 | 1,073 | 40 | 17 |
| *Treponema pallidum* | 1.14 | 1,040 | 44 | 27 |
| *Rickettsia prowazekii* | 1.11 | 834 | 38 | 25 |
| *Chlamydia trachomatis* | 1.04 | 894 | 32 | 28 |
| *Mycoplasma pneumoniae* | 0.81 | 677 | 37 | 10 |
| *Mycoplasma genitalium* | 0.58 | 470 | 37 | 2 |

[a]From Cordwell, 1999.[11]

# 1.4. Application of Electron Microscopy[12,13]

Electron microscopy has become an important tool to aid in biochemical and physiological studies. Visualization of a specimen is dependent on the deflection of the electron beam by heavy metals that are added as stains. Chemicals used for staining cellular material would include osmic acid, uranyl acetate, lead citrate, and phosphotungstic acid. For a time there was an intent to find heavy metals that would selectively stain lipids, proteins, and nucleic acids; however, it is now apparent that metal ions do not bind to biomolecules with that level of selectivity. Metal ions associate with the structures to varying degrees, as influenced by the microclimate of the molecules and this variation can be detected because differences in electron scattering produces discrete images. Some of the special staining techniques used to evaluate cell topography, surface appendages, and internal composition are outlined in Table 1.14. Electron microscopy is important in the characterization of cell anatomy and provides structural detail at high resolution. In Figure 1.5, an electron micrograph of a cell of *Escherichia coli* reveals the presence of surface appendages (fimbrae/pili)

## Box 1.1
## Microscopy's Role in Determining Structure and Function in Bacteria

A Perspective by Terry J. Beveridge

In this day and age, when the elucidation of functional relationships draws heavily on molecular biological techniques, the "tried-and-true" techniques of microscopy are often neglected. Indeed, it is not unusual to enter a microbiology research laboratory today and not see a light microscope. Imagine—ongoing research on microbial life forms which are so small (approximately 1.5–2.5 $\mu m^3$) that they can only be seen as individual cells by microscopy, and (frequently) no microscopes are readily available! I wonder how many present-day microbiologists derive enjoyment by viewing their lively cells with a simple phase microscope? Not many, I fear.

In microbiology, we often forget that that our recognition of bacteria, fungi, algae, and protozoa as minute cells was entirely due to the remarkable technical skills of early light microscopists. Shape, form, patterns of division, nucleoid structure, Gram-staining attributes, endospores, cysts, β-polyhydroxybutyrate granules (and other deposits), gas vesicles, capsules, sheaths, and flagella were all first identified by microscopy combined with specialized staining regimens. The advent of electron microscopy in the 1940s and its dedicated application to microbiology in the 1950s and 1960s clearly demonstrated that prokaryotes were entirely different than their eukaryotic counterparts. How exciting it must have been for Chapman and Hillier in 1953 to first observe bacteria (*Bacillus cereus*) in thin section and to see that these small life forms contained no true nucleus, no mitochondria or other organelles, and no cytoskeletal system (Chapman and Hillier, 1953). Only a rather thick cell wall seemed responsible for the rod shape of this bacterium. The very first images of viruses, which are often nanometers in diameter, required the use of electron microscopy and the remarkable use of contrasting agents (i.e., negative stains such as sodium phosphotungstate or uranyl acetate). More recently, electron microscopy has demonstrated dramatic structural differences between bacteria and archaea. The fundamental structural format of all microbial cells was discovered by microscopy in one form or another.

Today we are poised at a renaissance in microscopy which can clearly benefit microbiology. Sophisticated devices such as atomic force microscopes (AFMs), which are unlike any other microscopes since they have no lenses, are providing exquisite molecular details of such important macromolecules as DNA. Since AFMs can probe specimens under water, real-time dynamic molecular movement can be seen (e.g., the opening and closing of porin channels). AFMs are even being used to probe the mechanical properties of bacterial surfaces. Lipopolysaccharides of Gram-negative bacteria have been shown to be in continual rapid motion and their O-side chains are so soft they are difficult to detect even at 0.1 nN probe forces, and the peptidoglycan sacculus is so elastic with high tensile strength that it is much stronger than any known man-made plastic (Yao et al., 1999). New-generation electron microscopes are able to visualize vitreously frozen microorganisms held at—196°C or lower so that all molecular motion has essentially stopped and structure is exquisitely preserved; now we are really getting an accurate look at these important life forms! In-column (or below column) electromagnetic filters and X-ray multichannel analyzers (both are in even more specialized electron microscopes) are providing compositional analyses via electron energy loss spectroscopy (EELS) or energy dispersive X-ray spectroscopy (EDS), and electron spectroscopic imagining (ESI) is providing almost atomic detail of biostructures (e.g., the coiling of rRNA can be visualized in ribosomes by following the phosphorus signal). Even the tremendous energy of huge synchrotron facilities is being harnessed to determine the bonding structure of microbial macromolecules (e.g., microbeam extended X-ray absorption fine structure (EXAFS) analysis).

Microscopy is in a particularly fluid and dynamic stage so that, when combined with the awesome potential of molecular biology, truly exhilarating times are in store for the elucidation of microbial structure and function.

# References

Chapman, C.B. and J. Hillier. 1953. Electron microscopy of ultrathin sections of bacteria: I. Cellular division in *Bacillus cereus. Journal of Bacteriology* 66:362–373.

Yao, X., M. Jericho, D. Pink, and T. Beveridge. 1999. Thickness and elastic properties of Gram-negative murein sacculi measured by atomic force microscopy. *Journal of Bacteriology* 181:6865–6875.

extending from the cell. Considerable information about the number of fimbrae, the distribution of the surface of the cell, and the physical sizes of these structures can be obtained from a single observation.

While metal staining of bacterial structures lacks specificity, the observation of a specific bacterial structure can be accomplished through the use of antibodies for that structure. The antibody molecule is labeled with colloidal gold or ferritin (a 460-kDa protein containing up to 23% iron atoms) that functions to deflect the electron beam and identify the specific localization of a structure or molecule. An example of a structure localized the use of antibody-staining technique is given in Figure 1.6 with ATP synthase (also referred to as ATPase). In Figure 1.6(a), there is no appearance of structures attached to the isolated bacterial membranes; however, when treated with anti-ATPase conjugated to ferritin Figure 1.6(b), the ATPase structures are localized exclusively on one side of the plasma membrane. The individual ATPase units are

Table 1.14. Special procedures applied to microbial specimens.

| Techniques | Applications |
|---|---|
| **Negative staining** | Specimens are dried on a grid and a thin film of phosphotungstic acid is applied. This is used to examine topography of cells, to see flagella, pili, or structural features of cells. |
| **Shadowing** | The specimen dried on a grid is placed in a vacuum chamber and platinum is vaporized by electronic means. The grid is at a fixed angle with respect to platinum. Any specimen with depth will display a shadow when viewed under a TEM. It is ideal for visualizing flagella, pili, DNA, peptidoglycan, and whole cells. |
| **Thin sections** | Cells or specimen is fixed with 2.5% glutaraldehyde, stained with a lead acetate, dehydrated, and embedded in plastic. Sections of the plastic are cut with a diamond knife at a thickness of 30 to 100 nm. Sections are stained with osmic acid, placed on grids, and examined with a TEM. This procedure is used to examine internal structural arrangements at high magnification. |
| **Freeze–fracture** | Specimens are frozen in liquid nitrogen at $-196°C$ and placed in a vacuum chamber where a precooled knife fractures the cell. The exposed surfaces are shadowed with platinum and carbon. Fractures occur along the weakest faces or layers and membranes will be torn from the cells as a discrete smooth layer. If the membrane is torn along the central lipophilic region of the membrane, it will expose a rough internal region attributable to transmembrane proteins. Cytoplasm is rough due to the presence of protein and nucleic acid aggregates. |
| **Energy-dispersive X-ray analysis (EDX)** | Cells or specimen are not stained with a heavy metal but are examined directly through the electron microscope with an appropriate probe. Elemental analysis can detect almost all the elements on the atomic chart and indicate the elements present. If the inorganic aggregates are crystalline, the spacing of the rings can be used to assist in identifying the crystal. These instruments have large databases that assist in analysis. Specimens may be either dried whole mounts or sectioned. |

not seen in the electron micrograph without anti-ATPase treatment because their visualization requires significantly greater magnification and resolution.

Recently, these electron microscopes have been interfaced with powerful detectors to become highly precise analytical instruments. Some of the common electron microscopes and their applications are given in Table 1.15. Both transmission and scanning electron microscopes now have the capability of providing an analysis of elements distributed in the cells as well as to provide high resolution of structures. Initially these instruments were used to identify heavy metal atoms; however, electron microscopes can now establish the charge on metal ions and provide information about the nature of metal ion binding to cellular components. The sensitivity of these element detectors continues to improve so that electron microscopes are now able to detect a few hundred atoms of a heavy metal and can quantify lighter elements such

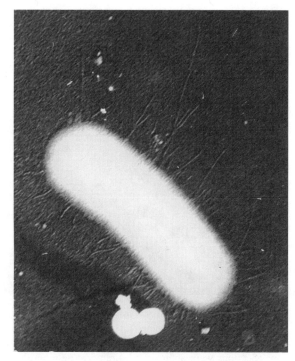

Figure 1.5. Surface structures associated with a cell of *Escherichia coli*. The specimen was dried and shadowed with gold (Au) before it was observed by TEM. Styrofoam beads of 0.3 μm in diameter were added to the grid to assist in cell measurements as well as to indicate the direction of the shadowing. (Courtesy of Sandra L. Barton.)

as oxygen, carbon, and nitrogen. It has been proposed that in the future, electron thermography will be used to produce a three-dimensional image of proteasomes, internal structures, and even individual proteins inside prokaryotic cells.

In addition, many of the research electron microscopes have the capability of X-ray analysis to identify inorganic crystals present inside or on the surface of the cells. Electron microscopes are being used as tools to provide detailed information on cell–cell interactions and metabolic activity. One of the applications of electron microscopy in the transformation of metals is to map cells to determine regions of activity. An example is given in Figure 1.7, in which a bacterial cell has a precipitate collected on the surface that is demonstrated to be crystalline $UO_2$. It is apparent that electron microscopy is a powerful tool to evaluate microbial metabolism as well as cellular structures.

New procedures are being developed for electron microscopes that can examine live cells under low vacuum conditions. One of the microscopes that can function under conditions in which the specimen is wet and not exposed to high vacuum conditions is the environmental electron microscope (ESEM). The specimen is placed under the electron beam in the natural state, which is referred to as "real time" as fixation, dehydration, and staining is not required. Examination of environmental samples of bacteria by ESEM provides a three-dimensional image while specimens prepared for scanning electron microscopy (SEM) lacks the structural detail. Considerable use is being made of ESEM in the fields of biocorrosion and biofilms. With the development

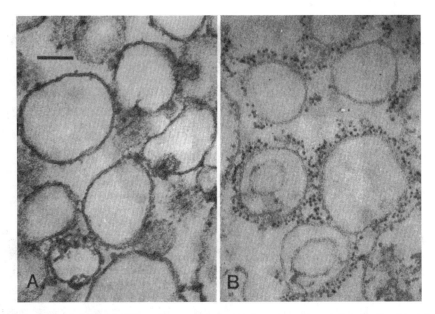

Figure 1.6. Localization of ATPase in membranes of *Micrococcus lysodeikticus*. Ferritin-labeled antibodies specific for F-type ($F_1F_0$) ATPase were reacted with thin sections of bacterial membranes. (**A**) Unlabeled and (**B**) ferritin-labeled antibodies against $F_1$. $\times$ 100,000. Bar = 0.1 μm. Sections were stained with uranyl acetate and lead citrate. (Reprinted from Oppenheim, J.D. and M.R.J. Salton. 1973. Localization and distribution of *Micrococcus lysodeikticus* membrane ATPase determination by ferretin labeling. *Biochimica et Biophysica Acta* 289:297–322. Copyright 1973, with permission from Elsevier Science.)

of high-resolution electron microscopes that can image wet specimens, scientists will some day be able to follow some enzyme reactions in real time.

Atomic force microscopy (AFM) not only provides the capability of examining cell structures at an extremely high resolution but also the specimen may be hydrated. In contrast, conventional high-resolution transmission electron microscopy (TEM) requires the specimen to be anhydrous. As can be observed in Figure 1.8, the cell wall specimen from *E. coli* that was dried appears smooth and without surface features, while the hydrated specimen reveals more of a "real-life image" in which molecules on the surface of the wall structure assume a three-dimensional configuration. While dried specimens have provided considerable insight into microbial structures over the years, the advent of high-resolution microscopy that enables one to examine cell structures under hydrated conditions provides better information about the function of these structures in the cell. The high resolution that can be achieved using AFM will enhance developments in molecular biology by providing visual activity displayed by restriction enzymes and regulator proteins on DNA.

Table 1.15. Types of electron microscopes used in bacterial physiology.

| Instruments | Capabilities |
| --- | --- |
| Transmission electron microscope (TEM) | Resolution of 1 nm, magnification several million. Sample must be dry and on a 3-mm diameter grid. Greatest use is to see detail of bacterial structures. |
| Scanning electron microscope (SEM) | Resolution of 7 nm, magnification of 100,000. Sample must be dry and is placed on a 12-mm diameter stub. Greatest advantage is to provide a three-dimensional image of cell surface that is more realistic than with TEM. |
| Environmental scanning electron microscope (ESEM) | Resolution and magnification similar to the SEM. Sample can be several inches in diameter. Greatest advantage is that the specimen is not dried prior to observation and "real time" measurements are taken. |
| Scanning tunneling microscope | A small voltage is applied to a needle-like probe kept at a given distance from the specimen and voltage flows to the electron cloud of atoms making up the solid material. Magnification of 100 million. Used to create three-dimensional image of various biomolecules including the arrangement of atoms in DNA. |
| Atomic force microscope | The microscope probe is maintained at a constant distance over the atoms and a laser beam is used to measure the deflection of the probe. Magnification of 100 million. Used to localize restriction endonuclease enzymes on DNA, characterize S-layers of bacteria, and examine protein–protein interactions. |

## 1.5. Fractionation of Cells

Histological staining of plant and animal cells has provided a wealth of information about cell activity; however, the small size of bacteria precludes the use of conventional staining methods. Similar application of cell staining has not been useful with prokaryotes owing to the small size of the cell. Therefore, physiologists have pursued their study of microbial cells by isolation of enzymes or macromolecules and subjecting these isolated components to extensive analysis. With eukaryotic cells, considerable advancements were made in understanding cellular regulation with the study of isolated intracellular structures. Because the prokaryotic cells have a relatively simple structure, importance is placed on activities compartmentalized in the three major regions of a cell: (1) in the periplasm of Gram-negative bacteria, (2) in the cytosol of the cell, or (3) in the plasma membrane. Various techniques have been developed for examination of activities in specific regions of the cells and although most of the procedures were established to study bacterial cells, these methods are appropriate also for the archaea.

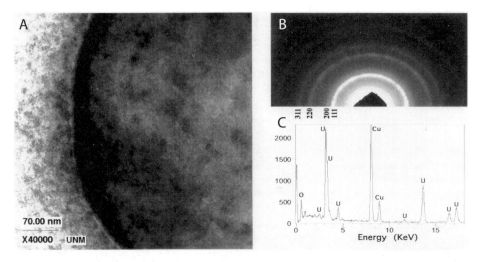

Figure 1.7. Uranium present at the surface of *Desulfovibrio desulfuricans*. Uranyl, U(VI), ions were added to a culture of *Desulfovibrio desulfuricans* and after several minutes a sample was withdrawn and placed directly on a grid for observation by electron microscopy. The bacteria had reduced the uranyl ion to uraninite, U(IV), which was deposited at the periphery of the whole cell; see (**A**). Elemental analysis of material outside the bacterial cell (**B**) reveals the presence of U, which has several energy levels of detection, and of oxygen, O, which is characteristic of $UO_2$, uraninite. Examination of the uranium material for structure by X-ray analysis (**C**) reveals the presence of rings that indicate the presence of crystalline structure. Evaluation of the ring size results in verification of the structure as $UO_2$. (Courtesy of R. Plunkett and H. Xu.)

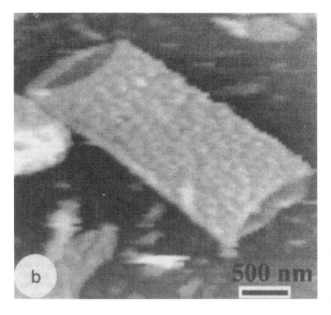

Figure 1.8. Cell wall structure of *Escherichia coli* examined by atomic force microscopy. The wall structure was obtained by ultrasonic treatment and the image is of the hydrated specimen. (From Yao et al., 1999.[28] Used with permission.)

**Box 1.2**
**Isolation of Periplasmic Proteins from Gram-Negative Bacteria**

The application of osmotic shock procedure for the release of periplasmic proteins varies with age of the culture. After bacterial cells are collected from the stationary phase of growth, they are washed several times using a buffer solution containing 0.01 $M$ Tris-HCl, pH 7.4, and 0.03 $M$ NaCl. The cells are resuspended in a solution containing 20% sucrose, 0.03 $M$ Tris-HCl buffer at pH 7.4, and 1 m$M$ EDTA. The ratio of cell wet weight to volume of buffer is 1 g:80 ml of buffer. After 30 min at room temperature, the cells are collected by centrifugation and quickly resuspended in cold distilled water. The cold water shocks the cells and the periplasmic proteins are released. The cells can be removed from this shock fluid by centrifugation and the proteins can be purified by traditional biochemical procedures. If cells are used from the log phase of growth, the concentration of EDTA used in the buffer is 0.1 m$M$ and the cold shock solution contains 0.5 m$M$ and not distilled water. (Modified from Rosen and Heppel, 1973.[29])

## 1.5.1. Disruption of Cells[14]

Except for a few organisms such as the cell wall–less forms, bacteria have extremely rigid cell walls as a result of intermolecular interactions that confer rigidity and great tensile strength. Therefore, to disrupt bacterial cells, considerable effort is required. Chemical methods involving enzymes or antibiotics to destabilize the bacterial cell wall are used in limited or specialized conditions. The group polysaccharide antigen in *Streptococcus* and glycoprotein S-layers from bacteria have been isolated by extraction because of the unique chemical features of these molecules. In general, it is difficult to employ acid extractions of whole cells with the purpose of obtaining a given structure because extraction subjects the whole cell to the treatment.

Various methods of cell disruption have been developed and they are summarized in Table 1.16. The two most commonly employed methods for cell disruption are ultrasonic forces and the French pressure cell. Ultrasonic disruption is performed using commercial instruments producing 20-kHz ultrasonic waves, which is significantly greater energy than that used in sonic cleaning baths. In ultrasonic treatment, small bubbles are generated in the sonicated fluid and the cavitation or collapse of these bubbles results in a shear force that disrupts the cells. The ultrasonic process is most often used to break the cells; however, cooling of bacterial suspension must be rigorously maintained to prevent the formation of hydroxyl and hydroperoxyl free radicals.

Many different types of pressure chambers have been developed over the years and

Table 1.16. Mechanisms that have been used for disruption of bacterial cells.

| Procedure | Application |
|---|---|
| 1. Autolysis | The self-digestion of a bacterium by its own enzymes. This procedure is generally not recommended owing to the changes in structures as a result of enzyme modification. If equipment is not available for disruption of cells this system may be used; however, this procedure is not often used in research. |
| 2. Osmotic lysis | Lysozyme is added to destabilize the peptidoglycan and the resulting protoplast or spheroplast is slowly added to water and osmotic disruption of cells results. This works for small quantities of Gram-positive or Gram-negative bacteria. Dilution of halophilic cells into water also results in osmotic lysis. |
| 3. Ultrasonic treatment | An ultrasonic oscillator is commonly used to disrupt cells for the purpose of isolating membranes, peptidoglycan, or soluble proteins. Cell disruption generally takes only a few minutes. Perhaps the greatest disadvantage is the difficulty of disrupting some of the spherical cells. |
| 4. Grinding or agitation | Bacterial cells can be disrupted by vigorous grinding with abrasives mortar and pestle using sand or glass beads (0.1 to 0.2 mm in diameter). Alternately, cells are disrupted by shaking cells in special agitators in which cells are mixed with glass beads. High-velocity shakers using glass beads have been used to disrupt spores with some success. |
| 5. Pressure cell disruption | The most common unit available is the French press. A bacterial suspension is placed inside a steel cylinder and subjected to 20,000 psi. Cells flow past a needle valve into environment of no pressure and disrupt as cells experience the great change in pressure. The French press is commonly used in the production of plasma membranes for electron transport studies and for the release of soluble proteins. |

the design by C.S. French is one of the devices commonly employed for disruption of bacterial cells. Organisms in the French press are subjected to 20,000 psi, and as the pressurized bacterial cell suspension is slowly released to atmospheric pressures, the bacterial cell ruptures. If several pounds of bacterial cells are to be disrupted, the French pressure chamber is not practical but a Manton–Gaulin homogenizer may be used. This homogenizer works on the same principle as the French pressure system, and was developed by the paint industry to emulsify paint pigment.

## 1.5.2. Isolation of Prokaryotic Cell Structures[15]

It is difficult if not impossible to use one sample to isolate each of the cell structures; thus, it would be appropriate to start with a separate cell suspension to isolate the specific structures. Most of the isolation procedures rely on the use of differential centrifugation which is based on the premise that each structure has a discrete density. In addition to standard centrifugation in which the gravitational force is varied, structures may also be isolated through the use of density gradients in which the centrifugation tubes contain increasing concentrations of sucrose. The velocity of sedimentation in a centrifugal field is the sedimentation coefficient and for macromolecules this sedimentation coefficient is given in Svedberg units (S) with huge macromolecules having a large S value. The distribution of cellular structures in fractions following centrifugation techniques is given in Figure 1.9. In work involving cell suspensions, chemicals are used to stabilize the pH and these are referred to as buffers. Chemicals useful as buffers include Tris and related chemicals but not citrate

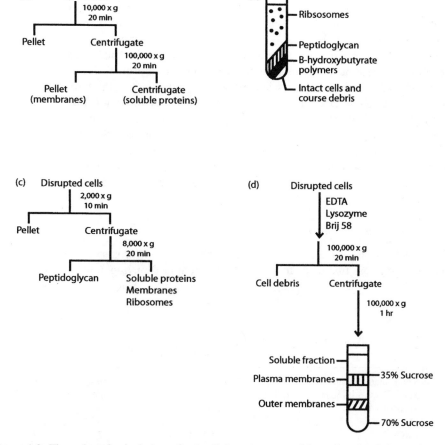

Figure 1.9. Flow chart for isolation of subcellular structures of bacteria. (**a**) Scheme for separation of soluble proteins from membrane-bound proteins. (**b**) Example of distribution of structures of disrupted Gram-positive bacterial cells following centrifugation at 10,000$g$ for 20 min. (**c**) System for centrifugation to obtain peptidoglycan from Gram-positive bacteria. (**d**) Centrifugal system applied to disrupted Gram-negative bacteria to separate outer membranes from plasma membranes.

or phosphate because the latter are readily metabolized. Prokaryotic cells contain proteases to assist in their normal processing of proteinaceous materials; however, in vivo there is considerable regulation of these hydrolytic enzymes. When the cells are disrupted, proteolytic enzymes are released and they can attack virtually any cell protein. To prevent this catalytic digestion of proteins, it is important to add protease inhibitors to all buffers used in the isolation and purification procedures. For best results a mixture of protease inhibitors is used to prevent the hydrolytic action of various proteases in microbial cells including serine proteases, cysteine proteases, aspartic proteases, metallo-proteases, and aminopeptidases.

### 1.5.2.1. Isolation of Ribosomes[16]

The macromolecular units containing ribonucleic acid and protein required for protein synthesis were designated as ribosomes by R.B. Roberts in 1958. Ribosomes are isolated from prokaryotic cells following disruption by osmotic lysis or pressure cell treatment. Ultrasonic rupture of cells is not desirable for ribosome evaluation because the ultrasonic treatment may remove proteins from the ribosome. To maintain integrity of the 70S ribosome, it is necessary to have 10 mm $Mg^{2+}$ in the buffer. When cells are disrupted, a few ribosomes are found in the pellet following low–speed centrifugation along with membranes and polysomes. However, the greatest quantity of ribosomes can be found in the pellet after centrifugation at 100,000$g$ for 2 hr. Purification of ribosomes can be accomplished by the use of sucrose gradients to separate the ribosomes from other cellular materials. Ribosomes from *Escherichia coli* have been extensively analyzed and some of the characteristics of these ribosomes are listed in Table 1.17.

### 1.5.2.2. Isolation of Plasma Membranes[17]

The methods for isolation of plasma membranes from prokaryotic cells generally employ a physical treatment followed by differential centrifugation. Cells disrupted

Table 1.17. Characteristic of *Escherichia coli* ribosomes.[a]

| Characteristic | Small subunit | Large subunit |
|---|---|---|
| Sedimentation coefficient | 30S | 50S |
| Mass (RNA + protein) | 930 kDa | 1,590 kDa |
| Mass (RNA) | 560 kDa | 1,104 kDa |
| Mass (protein) | 370 kDa | 487 kDa |
| rRNA | 16S = 1,542 bases | 23S = 2,904 bases |
|  |  | 5S = 120 bases |
| Number of different proteins present | 21 proteins | 31 proteins |
| RNA (fraction of ribosome) | 60% | 70% |
| Protein (fraction of ribosome) | 40% | 30% |

[a]Data from Noller and Nomura, 1986.

**Box 1.3**
**Experimental Evidence for Protein Synthesis**

The cell-free system for protein synthesis was demonstrated in the early 1960s and was instrumental in our understanding of the translation system in bacteria. The universal system used to study protein synthesis was to use polyuridylic acid as the synthetic message, which resulted in the formation of a peptide consisting entirely of phenylalanine. For the protein synthesizing system, a 0.25-ml reaction consists of the following: a buffer of 2 μmol of Tris-HCl at pH 7.5, 0.2 μmol of ATP, 0.01 μmol of GTP, 0.1 mg of tRNA, 50 μg of polyuridylic acid, 0.05 μCi of [$^{14}$C] phenylalanine, 0.3 mg of the S-100 fraction, and 0.2 mg of ribosomes. The S-100 fraction is a soluble protein fraction that was obtained from washing the ribosomes and is so designated because it was the soluble fraction after centrifugation at 100,000$g$. The reaction is incubated for 30 min and the newly formed protein (polyphenylanaline) is precipitated with trichloroacetic acid. The quantity of radioactivity in the precipitated protein is used to determine the amount of protein synthesized. Careful dissection of the components in this protein synthesizing reaction led to a detailed characterization of bacterial protein synthesis including the role of initiating factors and ribosome specificity.

by the French pressure cell method usually have significant amounts of native DNA, and treatment with a commercially available endonuclease is required to reduce the viscosity of the extract. Highly viscous samples cannot be subjected to centrifugation because most of the material will be pelleted as a result of nonspecific adsorption of proteins to DNA. Because DNA is readily sheared in the ultrasonic method of cell disruption these extracts do not require DNase treatment. A low-speed centrifugation of 10,000$g$ for 10 min is used to remove unbroken cells, polysomes, and dense pieces of cell debris. The centrifugate is subjected to ultracentrifugation at 100,000$g$ for 1 hr to pellet membranes and leave soluble proteins in suspension. A standard criterion in establishing if an enzyme is membrane bound or soluble is to determine its location following centrifugation at 100,000$g$ for 1 hr. In Gram-positive bacteria, the pellet will contain plasma membrane whereas with Gram-negative bacteria, the pellet will contain plasma membrane plus the outer membrane.

   To assess activities or characteristics of membranes, it is necessary to isolate membranes from a disrupted cell. In the native state, membranes have proteins or other molecules adhering to the membrane and these molecules are stripped from membranes with the harsh disruption of the cell. Treatment of cells with 0.2% to 0.5% glutaraldehyde promotes rapid cross linkage of proteins contiguous to the membrane, and thereby increases the protein content of the isolated membrane by about 20%.

Table 1.18. Distribution of proteins in *Micrococcus luteus* (*lysodeikticus*).[a]

| Enzyme/cytochrome/quinone | Localization (expressed as percent of total activity)[b] | |
| --- | --- | --- |
| | Plasma membrane | Cytoplasm |
| Adenosine deaminase | 0 | 100 |
| Glutamic acid-oxalacetic acid transaminase | 0 | 100 |
| Hexokinase | 30 | 70 |
| Adenylate kinase | 30 | 70 |
| Isocitrate dehydrogenase | 35 | 65 |
| Catalase | 50 | 50 |
| Polynucleotide phosphorylase | 65 | 35 |
| NADH dehydrogenase | 85 | 15 |
| Succinic dehydrogenase | 98 | 2 |
| Diphosphatidyl glycerol synthetase | 98 | 2 |
| Cytrochrome *c* | 99 | 1 |
| Cytochromes *a* and *b* | 100 | 0 |
| Quinone | 100 | 0 |

[a]From Salton, 1971.[26]
[b]Membranes were not washed with buffer.

Enzymes loosely adhering to the membrane can be removed by emulsification with mild detergents or ion-exchange reaction using NaCl solutions. The membranes can be stabilized with the presence of $Ca^{2+}$ or $Mg^{2+}$, which neutralizes charges on hydrophilic regions of the membrane. As listed in Table 1.18, certain enzymes or proteins are strongly attached to the plasma membrane and can be used as indicators for presence of plasma membrane.

### 1.5.2.3. Isolation of Peptidoglycan[18]

To isolate peptidoglycan from Gram-positive bacterial cells, the suspension of disrupted cells is subjected to centrifugation at 1,500*g* for 5 min to pellet intact cells and cell debris. The peptidoglycan plus soluble proteins are present in the centrifugate while peptidoglycan is pelleted by centrifugation at 8,000*g* for 20 min. At this stage, the peptidoglycan is not pure but contains membranes, proteins, and nucleic acids that can be digested with proteolytic enzymes, lipases, DNase, and RNase. The treated peptidoglycan fraction would be subjected to centrifugation at 8,000*g* for 20 min and the pellet would contain the peptidoglycan. To remove charged molecules adhering to the peptidoglycan, several washes with 1 *M* NaCl would be employed. Washing of the peptidoglycan is achieved by resuspending the pellets in the salt solution and collecting the peptidoglycan by centrifugation. Finally, the treated peptidoglycan is suspended in dilute buffer and subjected to centrifugation at 8,000*g* for 20 min. The purified peptidoglycan is in the centrifugate.

Owing to the tight association of the peptidoglycan to the outer membrane in Gram-negative bacteria, it is difficult to use differential centrifugation to separate the cell wall structure following cell disruption. The isolation of peptidoglycan can be achieved by boiling the cells in 4% sodium dodecyl sulfate (SDS). The outer membrane and plasma membrane are solublized as a result of the melting of lipids at elevated temperatures and SDS solubilizes the proteins. Other proteins including outer membrane lipoproteins and ribosomal proteins are dissolved by the SDS treatment. The crude peptidoglycan can be collected by centriguZation at 1,500$g$ for 5 min and the SDS is removed by several washes in dilute buffer. Treatment of the peptidoglycan with DNase, RNase, and a protease is used to remove contaminating cellular materials.

## 1.5.2.4. Isolation of the Outer Membrane[19]

On disruption, outer membrane fragments may vary in size and in sedimentation velocity; however, outer membrane fragments all have the same buoyant density. Thus, buoyant density can be used to separate cell structures and if the buoyant density is less than 1.3 g/ml of sucrose gradients in centrifuge tubes are used but for structures of high density cesium chloride is used. The buoyant density of cellular materials is as follows: the plasma membrane is 1.14 to 1.17 g/ml, the outer membrane is 1.222 g/ml, fimbrae/pili from *Escherichia coli* is 1.11 to 1.13 g/ml, flagella from *Escherichia coli* is 1.30 g/ml, and endospores from *Bacillus subtilis* is 1.21 to 1.27 g/ml.

Disrupted Gram-negative bacterial cells are suspended in buffer containing EDTA, lysozyme, and Brij 58. Cells and large aggregates of cellular debris are removed by centrifugation at 2,000$g$ for 10 min. The resulting centrifugate is layered over a buffered gradient of 35% sucrose to 70% sucrose and subjected to centrifugation at 100,000$g$ for several hours. Several bands are present in the centrifugation tube with the one with greatest buoyant density containing the outer membrane fraction and located in the region of highest sucrose density. The fraction that is the least dense would have migrated the shortest distance in the tube and this is the plasma membrane. A band in the sucrose gradient that is intermediate between the outer and plasma membrane fractions contains a mixture of both membranes. Plasma membrane contaminating the outer membrane fraction can be removed by treatment with 0.5% Triton X-100 or sarcosyl. To distinguish the outer membrane from the plasma membrane, several characteristics can be examined and some of these markers are given in Table 1.19.

Table 1.19. Markers for outer membranes and plasma membranes in Gram-negative bacteria.

| Components in outer membrane | Components in plasma membrane |
| --- | --- |
| O-antigen | ATP synthase/ATPase |
| Lipid A | NAD-coupled dehydrogenases |
| Porin proteins | Succinic dehydrogenase |
|  | Quinones (menaquinone or ubiquinone) |
|  | Cytochromes: $a$-type and $b$-type |

### 1.5.2.5. Isolation of Polysaccharide Capsule[20]

To remove capsule from prokaryotic cells containing a polysaccharide capsule, a dilute aqueous suspension of cells is placed in a standard blender and subjected to mixing at a high speed. The blender propels bacterial cells through the aqueous suspension and the capsule is stripped away from the cells. This is a process of low efficiency in that the capsule is only partially removed from the cells. After about a minute in the blender, the cells are collected from the blended suspension by low-speed centrifugation. Capsular material remains in the supernatant and is precipitated by the addition of several volumes of cold ($-20°C$) ethanol. The crude fraction of polysaccharide capsule is collected by centrifugation and suspended in buffer before it is treated with protease, DNase, or RNase to remove cellular materials.

### 1.5.2.6. Criteria for Purity of Isolated Structure[20]

With each structure isolated, it is important to verify that it is indeed the intended structure and to demonstrate that it is free from contamination by other cellular components. The presence of cytoplasmic granules, cell walls, flagella, or pili can be assessed by electron microscopy. Adsorption at 260 nm and 280 nm can be employed to determine the presence of nucleic acid and protein, respectively. Purified structures of plasma membrane, outer membrane, peptidoglycan, and polysaccharide capsule would not contain nucleic acid bases and would have no spectral absorption at 260 nm. Adsorption at 280 nm is attributed to the presence of tryptophan and tyrosine in protein. The estimated occurrence of tryptophan and tyrosine in proteins is 1.1% and 3.5%, respectively. Polysaccharide capsule, peptide capsules, and peptidoglycan do not have tryptophan or tyrosine, and therefore would not have an absorption peak at 280 nm. The quantity of protein can be determined by applying the following formula: Protein concentration (mg/mL) = $(1.5 \times A_{280}) - (0.75 \times A_{260})$. It should be noted that peptidoglycan scatters visible light in a spectrophotometer and this optical density is not wavelength dependent. A way to verify that the structure isolated is indeed peptidoglycan is to subject it to digestion with lysozyme and watch for a loss of light scattering. The small molecular weight sugar products produced by lysozyme are soluble. The use of several criteria for purity of a structure along with direct chemical analysis are needed to provide confidence that the desired structure has indeed been isolated.

# References

1. Kerr, R.A. 1997. Microbiology's sacred revolutionary. *Science* 276:699–704.
2. Sandman, K. and J.N. Reeve. 2001. Chromosome packing by archaeal histones. *Advances in Applied Microbiology* 50:75–99.
3. Gruber, T.M. and D.A. Bryant. 1999. An overview of RNA polymerase sigma factors in phototrophs. In *The Phototrophic Prokaryotes* (G.A. Peschek, W. Löffelhardt, and G. Schmetterer, eds.). Kluwer Academic/Plenum, New York, pp. 791–798.
4. Ludwig, W. and K.-H. Schleifer. 1999. Phylogeny of *Bacteria* beyond the 16S rRNA standard. *ASM News* 65:752–757.

5.  Schulz, H.N. and B.B. Jørgensen. 2001. Big bacteria. *Annual Review of Microbiology* 55: 105–137.

6.  Rogers, H.J. 1983. *Bacterial Cell Structure*. American Society for Microbiology Press, Washington, D.C.

7.  Bayer, M.E. and M.H. Bayer. 1994. Periplasm. In *Bacterial Cell Wall* (J.M. Ghuysen and R. Hakenbeck, eds.). Elsevier-Science, New York, pp. 447–464.

8.  Ferguson, S.J. 1992. The periplasm. In *Prokaryotic Structure and Function* (S. Mohan, C. Dow, and J.A. Coles, eds.). Society for General Microbiology, Symposium 47, Cambridge University Press, New York.

9.  Oliver, D.B. 1996. Periplasm. In Escherichia coli *and* Salmonella *Cellular and Molecular Biology*, 2nd edit., Vol. 1 (F.C. Neidhardt, R. Curtiss III, J.L. Ingraham, E.C.C. Lin, K. Brooks Low, B. Magasanik, W.S. Rezenikoff, M. Riley, M. Schaechter, and H.E. Umbarger, eds.). American Society for Microbiology Press, Washington, D.C., pp. 88–103.

10. Trun, N.J. and J.F. Marko. 1998. Architecture of a bacterial chromosome. *ASM News* 64: 276–283.

11. Cordwell, S.J. 1999. Microbial genomes and "missing" enzymes: Redefining biochemical pathways. *Archives for Microbiology* 172:269–279.

12. Mayer, F. 1988. Electron microscopy in microbiology. *Methods in Microbiology*, Vol. 20. Academic Press, New York.

13. Little, B., P. Wagner, R. Ray, R. Pope, and R. Scheetz. 1991. Biofilms: An ESEM evaluation of artifacts introduced during SEM preparation. *Journal of Industrial Microbiology* 8:213–222.

14. Coakley, W.T., A.J. Bater, and D. Lloyd. 1977. Disruption of micro-organisms. *Advances in Microbial Physiology* 16:279–341.

15. Schnaitman, C.A. 1981. Cell fractionation. In *Manual of Methods for General Microbiology* (P. Gerhardt and R.C.F. Murry, eds.). American Society for Microbiology Press, Washington, D.C., pp. 52–61.

16. Nomura, M., A. Tissières, and P. Lengyel (eds.). 1974. Ribo*somes*. Cold Spring Harbor Laboratory Press, Cold Spring Harbor, NY.

17. Kaback, H.R. 1971. Bacterial membranes. *Methods in Enzymology*, Vol. XXII:99–120. Academic Press, New York.

18. Salton, M.R.J. 1964. *The Bacterial Cell Wall*. Elsevier, Amsterdam.

19. Nikado, H. 1994. Isolation of outer membranes. *Methods in Enzymology*, Vol. 235. Academic Press, New York, pp. 225–234.

20. Hancock, I. and I. Poxton. 1988. *Bacterial Cell Surface Techniques*. John Wiley, New York.

21. Whitman, W.B., D.C. Coleman, and W.J. Wiebe. 1998. Prokaryotes: The unseen majority. *Proceedings of National Academy of Science* USA 95:6578–6583.

22. Ehrlich, H.L. 1996. *Geomicrobiology*. Marcel Dekker, New York.

23. Madigan, M.T., J.M. Martinko, and J. Parker. 2003. *Brock Biology of Microorganisms*, 10th edit. Prentice-Hall, Upper Saddle River, NJ.

24. Barton, L.L. and R. Plunkett. 2002. Sulfate-reducing bacteria: Environmental and technological aspects. In *Encyclopedia for Environmental Microbiology* (G. Bitton, ed.). John Wiley, New York, pp. 3087–3096.

25. Oshima, T. 1986. The genes and genetic apparatus of extreme thermophiles. In *Thermophiles: General, Molecular and Applied Microbiology* (T.D. Brock, ed.). Wiley-Interscience, John Wiley, New York, pp. 137–157.

26. Salton, M.R.J. 1971. Bacterial membranes. *CRC Reviews in Microbiology* 1:161–190.

27. Oppenheim, J.D. and M.R.J. Salton. 1973. Localization and distribution of *Micrococcus*

*lysodeikticus* membrane ATPase determination by ferretin labeling. *Biochimica et Biophysica Acta* 289:297–322.

28. Yao, X., M. Jericho, D. Pink, and T. Beveridge. 1999. Thickness and elasticity of Gram-negative murein sacculi measured by atomic force microscopy. *Journal of Bacteriology* 181:6865–6875.

29. Rosen, B.P. and L.A. Heppel. 1973. Present status of binding proteins that are released from Gram-negative bacteria by osmotic shock. In *Bacterial Membranes and Walls* (L. Leive, ed.). Marcel Dekker, New York, pp. 210–239.

# Additional Reading

## *General References*

Balows, A., H.G. Trüper, M. Dworkin, W. Harder, and K.-H. Schleifer (eds.). 1992. *The Prokaryotes*, 2nd edit. Springer-Verlag, New York.

Herbert, R.A. and R.J. Sharp. 1992. *Molecular Biology and Biotechnology of Extremophiles.* Chapman and Hall, New York.

Hill, W.E., P.B. Moore, A. Dahlberg, D. Schlessinger, R.A. Garrett, and J.R. Warner (eds.). 1990. *The Ribosome.* American Society for Microbiology Press, Washington, D.C.

Garrity, G. (editor-in-chief). 2001. *Bergey's Manual of Systematic Bacteriology.* Springer-Verlag, New York.

Koch, A.L. 1998. The biophysics of the Gram-negative periplasmic space. *Critical Reviews in Microbiology* 24:23–60.

Madigan, M.T., J.M. Martinko, and J. Parker. 2002. *Brock Biology of Microorganisms.* Prentice-Hall, Upper Saddle River, NJ.

Neidhardt, F.C., R. Curtiss III, J.L. Ingraham, E.C.C. Lin, K. Brooks Low, B. Magasanik, W.S. Rezenikoff, M. Riley, M. Schaechter, and H.E. Umbarger (eds.). 1996. Escherichia coli *and* Salmonella *Cellular and Molecular Biology*, 2nd edit., Vols. I and II. American Society for Microbiology Press, Washington, D.C.

Perry, J.J. and J.T. Staley. 1997. *Microbiology: Dynamics and Diversity.* Saunders College Publishing, New York.

Prescott, L.M., J.P. Harley, and D.A. Klein. 2001. *Microbiology.* McGraw-Hill, New York.

Schmidt, M.B. 1988. Structure and functioning of the bacterial chromosome. *Trends in Biochemical Science* 13:131–135.

Woese, C.A. and R.S. Wolfe. 1985. *The Bacteria, Vol. VIII: Archaebacteria.* Academic Press, New York.

## *Evolution and Phylogeny*

Data for phylogenic tree of bacteria can be obtained from the *The Ribosomal Database Project* (*www.cme.msu.edu/RDP/*).

Genome-sequencing projects completed and in progress (*www.tigr.org* and *www.vimss.org*).

Gupta, R.S. 1998. Protein phylogenies and signature sequences: A reappraisal of evolutionary relationships among Archaebacteria, Eubacteria and Eukaryotes. *Microbiology and Molecular Biology Reviews* 62:1435–1491.

Ludwig, W. and K-H. Schleifer. 1999. Phylogeny of Bacteria Beyond the 16S rRNA Standard. *ASM News* 65:752–757.

Margulis, L., M.F. Dolan, and R. Guerro. 2000. The chimeric eukaryote: Origin of the nucleus from the karyomastigont in amitochondriate protists. *Proceedings of the National Academy of Sciences (USA)* 97:6954–6959.

Rosseló-Mora, R. and R. Amann. 2001. The species concept. *FEMS Microbiology Reviews* 25:39–69.

Suzuki, K., M. Goodfellow, and A.G. O'Donnell. 1993. Cell Envelopes and classification. In *Handbook of New Bacterial Systematics* (M. Goodfellow and A.G. O'Donnell, eds.). Academic Press, New York, pp. 195–250.

Woese, C.A. 1987. Bacterial evolution. *Microbiological Reviews* 51:221–271.

## *Techniques and Methods*

Hughes, D.E., J.W.T. Wimpenny, and D. Lloyd. 1971. The disintegration of micro-organisms. *Methods in Microbiology* 5B:1–54.

Sykes, J. 1971. Centrifugal techniques for the isolation and characterization of sub-cellular components of bacteria. *Methods in Microbiology* 5B:55–207.

# II
# Cellular Structures and Their Function

# 2
# The Plasma Membrane

The organized features of living beings reach all the way from the microscopic to the molecular level, and one of the major tasks of biology is to understand and describe this organization, its interdependent levels, its stability, and its variability. S.E. Luria, *The Bacteria*, Vol. I, 1960

## 2.1. Introduction

This chapter examines the structural and molecular features that contribute to the functional processes of prokaryotic plasma membranes. It is important here to describe the molecular composition of the membrane as a prerequisite to understanding how specific molecules contribute to the structural integrity of the membrane unit. It is well recognized that cells are highly organized with organelles having specific functions; examination of the plasma membrane demonstrates this structural precision also at the molecular level. The biologist readily focuses on the roles of lipid–protein interactions; however, it is equally important to recognize the physical and chemical characteristics of lipids that are essential to the formation of the membrane structure. Membrane organization and function are attributed to specific molecular forms of lipids as well as to different types of interactions between proteins and lipids. The content of lipids that are not critical for structure or biological activity may vary and the resulting patterns may reflect the magnitude of lateral gene transfer in nature. Information presented here is intended to provide an introduction to specific functional issues of signal transduction, solute transport, and cellular energetics that are discussed in subsequent chapters. Finally, this chapter provides information about membrane-bound proteins and their isolation.

## 2.2. Membrane Function[1–3]

All prokaryotes are enclosed by a plasma membrane (also called a cytoplasmic membrane) that has an indispensable role in cell viability. The selective permeability of the plasma membrane is a hallmark of this structure. It is important for organisms to exclude certain substances from the cell proper and retain many others within the cytoplasm, thereby providing an appropriate environment for cellular activity. Our current understanding of plasma membranes is that they are actively involved in numerous physiological activities, including the following: Plasma membranes (1) import solutes by active transport processes; (2) generate energy through respiratory directed electron transport with conservation of energy resulting in either charging of the membrane or in ATP synthesis; (3) export proteins for construction of molecular structures outside of the plasma membrane and export polysaccharides for capsular formation; (4) process signals across the plasma membrane to enable quorum sensing attributed to chemicals in the environment, directing cellular activities in sporulation, producing chemilumenescence, and synthesizing virulence factors; (5) respond to temperature changes to promote changes in lipids for maintaining proper gel–liquid crystal character; (6) provide an appropriate motor for flagellar activity in cell motility and for sensory processing such as occurs in chemotaxis; (7) transport precursor molecules for the construction of peptidoglycan and lipopolysaccharide in the cell wall; (8) participate in cell division by providing sites to initiate cross wall or septum formation; and (9) interact with DNA to influence DNA replication and separation of chromosomes during cell division. Some of these activities reflect "gate-keeping processes" that are a result of the barrier imposed by the plasma membrane. Other membrane events reflect "circuit board" activities in which the plasma membrane expedites communication between the extracellular environment and the cytoplasm. Many of these topics are too broad for coverage in this chapter but they are addressed in subsequent sections of this book.

## 2.2.1. Lipids as Anchors for Linear Molecules

Attached into the outer side of the plasma membrane of both Gram-negative and Gram-positive bacteria are linear molecules that extend from the plasma membrane into the cell wall region. These molecules that are secured into the membrane include lipoproteins, lipoteichoic acids, and lipoglycans. The lipoproteins are commonly found in Gram-positive bacteria, and these molecules are binding proteins associated with the ATP binding cassette (ABC) transport for solute uptake. Lipoteichoic acids are polymers of sugar phosphates and they function in establishing a negative charge in the periplasm at the outer face of the membrane. The lipoglycans are sometimes referred to as membrane-derived oligosaccharides (MDOs), and they have been associated with stabilization of the plasma membrane as a stress response process. All molecules have the common feature in that they are tethered into the membrane by a glycerol moiety with carbon atom number one of glycerol covalently bonded to the linear molecule and the other carbon atoms attached to fatty acids through ester bonds. The chemical activity of the fatty acids secures the fatty acids into the lipophilic regions of the plasma membrane and accounts for localization of these linear molecules on the external side of the plasma membrane.

## 2.2.2. Proteins and Transporters

The mechanism whereby a protein is attached into the membrane is important because this determines whether it is an integral or a peripheral protein. Excellent examples

**Box 2.1**
**Membrane–DNA Interactions: More Questions than Answers**

With the absence of internal membranes in prokaryotes, scientists have often speculated that the plasma membrane may function in directing some of the cellular processes. In 1963, F. Jacob and colleagues proposed the Replicon Model in which the plasma membrane had a proposed role in the initiation of DNA synthesis. Other questions arose and they include: (1) What is the involvement of the plasma membrane in the separation of chromosomes as daughter cells are produced? (2) What is the role of the plasma membrane in the sequestering of the chromosome into the prespore? (3) Does the plasma membrane have a role in the regulation of plasmid replication? (4) Does the plasma membrane receive signals on the termination of DNA replication? and (5) What is the role of the plasma membrane in the entry of DNA into the cell during transformation? The mechanism of membrane involvement in each of these activities remains unknown except for the transport of single-stranded DNA across the plasma membrane as covered in Chapter 10. It is known that the initiation of bacterial chromosome replication involves the *oriC* gene and this DNA segment binds strongly to the phospholipids of the plasma membranes. In fact, membranes of bacteria can be isolated with the *oriC* segment attached. However, greater binding occurs between *oriC* and outer membrane than between the *oriC* and the plasma membrane. This produces a geographic paradox because in growing cells, the outer membrane is spatially separated from DNA. Clearly, the role of plasma membranes in regulating some of the cellular activities remains to be established. For additional information, see Funnell, 1996.[35]

of transmembrane proteins are the permeases or transporters for sugars. In general, these uptake transporters have 12 transmembrane domains to secure the protein molecule into the membrane; in addition there are the necessary recognition sites to provide specificity for solutes. Although there are numerous transmembrane proteins in each cell, an example of one such protein is the bacteriorhodopsin molecule (see Figure 2.1). An important feature for proteins to be stabilized in the membrane is the absence of any charge on the $\alpha$-helical transmembrane spanning domains of the polypeptide. The charge on a single amino acid is not relevant to the formation of a transmembrane helix, but usually the charge of 17 amino acids (eight amino acid residues on either side of a central amino acid) is used to determine charge on the domain. Transmembrane helices result when adjacent amino acid residues are neutralized.

Values have been assigned for the various amino acids (see Table 2.1), and the hydrophobicity index can be calculated for membrane proteins. The number of transmembrane helices varies considerably and can range from 1 to 14 depending on the protein. Loosely attached peripheral proteins may be held by charges between the

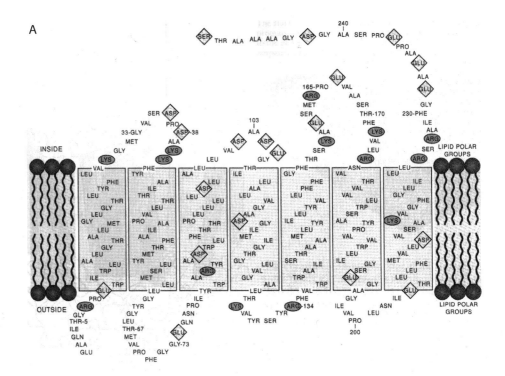

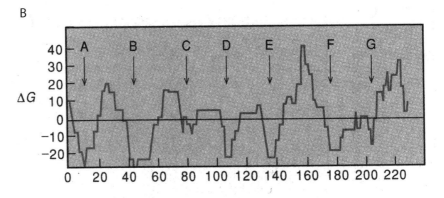

First of 19 amino acids in helix

Figure 2.1. (**A**) Amino acid sequence of bacteriorhodopsin in the membrane of the purple sulfur bacterium *Rhodopseudomonas viridis*. As proposed by D.M. Engelman, A. Goldman, and T.A. Steitz, there are seven transmembrane helices. (**B**) Hydropathy plot for bacteriorhodopsin. For this calculation, the hydrophobic properties of each amino acid in the protein over a sequence of 19 residues around each amino acid were calculated. The water-to-oil transfer free energy ($\Delta G$) is "0" at the lipid–water interface while a positive value indicates water solubility and a negative value reflects lipid solubility. The number of times the line crosses the "0" position corresponds to the number of transmembrane helices for the membrane-bound bacteriorhodopsin. As with most experimental data, there is considerable lack of symmetry in the peaks and there is also a lack of smoothness in the lines. (Transmembrane model from Engelman et al., 1982.[3] Used with permission.)

Table 2.1. Abbreviations and hydrophobicity values for amino acids.

| | Amino acid | Three-letter | One-letter | Hydrophobicity scales | |
| | | | | Method 1[a] | Method 2[b] |
|---|---|---|---|---|---|
| Nonpolar side chains: | Alanine | Ala | A | 1.8 | 1.6 |
| | Cysteine | Cys | C | 2.5 | 2.0 |
| | Glycine | Gly | G | −0.4 | 1.0 |
| | Isoleucine | Ile | I | 4.5 | 3.1 |
| | Leucine | Leu | L | 3.8 | 2.8 |
| | Methionine | Met | M | 1.9 | 3.4 |
| | Phenylalanine | Phe | F | 2.8 | 3.7 |
| | Proline | Pro | P | −1.6 | −0.2 |
| | Tryptophan | Trp | W | −1.3 | −0.7 |
| | Valine | Val | V | 4.2 | 2.6 |
| Uncharged polar side chains: | Asparagine | Asn | N | −3.5 | −4.8 |
| | Glutamine | Gln | Q | −3.5 | −4.1 |
| | Serine | Ser | S | −0.8 | 0.6 |
| | Threonine | Thr | T | −0.7 | 1.2 |
| | Tyrosine | Tyr | Y | −1.3 | −0.7 |
| Acidic side chains: | Aspartic acid | Asp | D | −3.5 | −9.2 |
| | Glutamic acid | Glu | E | −3.5 | −8.2 |
| Basic side chains: | Arginine | Arg | R | −4.5 | −12.3 |
| | Histidine | His | H | −3.2 | −3.0 |
| | Lysine | Lys | K | −3.8 | −8.8 |

[a]Kyte and Doolittle, 1982.[28]
[b]Engelman et al., 1986.[29]

positively charged amino acids and the negatively charged phospholipids. The orientation of the protein so that the correct part of the protein is exposed on the outer side of the membrane is influenced by the mechanism for insertion of the protein into the membrane.

The import and export of proteins across the plasma membrane present specific problems for prokaryotes because the plasma membrane does not contain holes or channels to facilitate the free traffic of proteins across the membrane. Extracellular proteins in the culture media or damaged proteins in the periplasm are not imported across the plasma membrane into the cell but must be degraded to small peptides before uptake transport can occur. On the other hand, some proteins in the cytoplasm are exported across the plasma membrane into the periplasm or into the cell wall region. Examples of structural proteins exported across the plasma membrane include the polypeptides for synthesis of flagella, pili or fimbrae, and cell walls. In addition, proteins found exterior to the plasma membrane include exocellular enzymes to hydrolyze polymers in the environment for nutrients, enzymes in the cell wall region for peptidoglycan synthesis, chaperone-like proteins to assist the assembly of flagella and pili, periplasmic enzymes, and binding proteins for nutrient transport. Therefore, one feature of tactical importance is how the proteins are identified for export as well as the mechanism for this transmembrane movement of the protein. Some proteins targeted for export contain a unique peptide on the N-terminus of the newly synthesized protein which is referred to as the leader peptide. Special peripheral proteins on the inner side of the plasma membrane recognize the leader peptide and direct it through appropriate conduits outward across the plasma membrane. The transmembrane transport of proteins is a fascinating area that involves chaperone proteins. A discussion on protein secretion with leader peptides is found in Chapter 12.

## 2.3. Membrane Composition

The plasma membrane accounts for about 10% to 25% of cell dry weight. In some instances lipoproteins extend outward from the plasma membrane and consist of carbohydrate polymers that are covalently linked to protein in the plasma membrane. The membrane is not rigid with the lipids covalently linked to other lipids or to proteins but forms a closed structure as a result of complex cohesive forces attributed to ionic interactions, hydrogen bonding, van der Waals forces, and hydrophobic (lipophilic) interactions. The outer boundary of the plasma membrane is a phospholipid with the phosphate moiety covalently linked to the lipid. Charges on the phosphate moiety make the boundary of the membrane hydrophilic.

A model of the plasma membrane of prokaryotic cells is given in Figure 2.2. Phospholipids form the layer at both faces of the membrane while globular proteins are either integral (transmembrane) or peripheral (available at either side). Thin sections of osmium-stained bacterial membranes reveal three layers: an electron-dense layer at each surface of the plasma membrane and a central core that is electron transparent. These plasma membranes are 70 to 80 Å in diameter, similar to that of membranes typically found in plant and animal cells.

The plasma membrane is continuous and serves as a barrier to sugars, amino acids, or salts whereas water and gasses freely pass through it. As a result of the membrane composition, $H^+$ and $OH^-$ cannot freely traverse the membrane, which results in a charge across the membrane. Maintaining a charge on the plasma membrane is a premier requirement for living cells, and constant adjustments must be made by the cell to ensure that directional flow of ions is maintained. From the aspect of cell energetics, rapidly growing cells expend a small percentage of their overall energy on maintaining a membrane charge; however, for cells that are growing very slowly, maintaining a charge on the membrane is a major energy commitment. Disruption of membrane integrity by either chemical or physical perturbations leads to cell death. Most quaternary ammonium compounds used as disinfectants produce localized lesions in the bacterial membrane that results in loss of charge on the plasma membrane, and cell death ensues.

### 2.3.1. Membrane Lipids[4-8]

Lipids account for 2% to 20% of dry weight of bacterial cells and 20% to 40% of the dry weight of the plasma membrane. There are many different ways to categorize the lipids in the plasma membranes; one of the more practical systems is to separate them into three classes: neutral lipids, lipoconjugates, and polar lipids. Neutral lipids

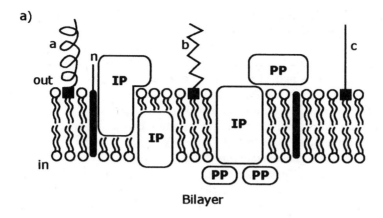

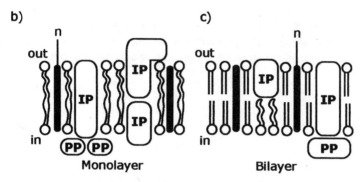

Figure 2.2. Models of plasma membranes in prokaryotes. (**a**) An idealized bacterial membrane containing a lipid bilayer with fatty acids projected toward the center of the membrane. (**b**) Lipid monolayer membrane as seen in certain archaea where the $C_{40}$ tetraethers extend across the membrane. (**c**) Lipid bilayer constructed of $C_{20}$ diethers as seen in many archaea. IP, Integral protein; PP, peripheral proteins; n, hopanoids; a, lipoglycan; b, lipoteichoic acid; c, lipoprotein.

include free fatty acids, carotenoids, triacylglycerols, polyprenols, sterol esters, and other noncharged lipids. Lipoconjugates contain lipids plus polysaccharides or proteins and are amphiphilic molecules in which one end is lipophilic and the other end is hydrophilic. Polar lipids include phospholipids, glycolipids, sulfolipids, chlorophylls, and oxy-carotenoids. For most prokaryotes, acidic lipids consist of phospholipids but a few organisms contain sulfolipids in which sulfate is attached to a diacylglycerol sugar or taurine is attached to the polar head of phospholipid.

## 2.3.1.1. Glycerol Backbone

In the plasma membrane, glycerol commonly serves as a backbone structure to support the attachment of sugars, fatty acids, alcohols, or phosphate moieties. Glycerol lipids commonly have hydrocarbons bound to the C-1 or C-2 position of glycerol by one of the following bond types: (1) acyl esters containing fatty acids, (2) vinyl ethers with fatty aldehydes, and (3) alkyl ethers containing alcohols. The chemical bonds in these glycerol lipids are given in Figure 2.3. Owing to the bonding of glycerol to both lipophilic groups and hydrophilic groups, glycerol lipids are found at the periphery of membranes in all life forms. The interior of the prokaryotic membrane may contain carotenoids, hopenoids, squalenes, or cholesterols. Hopanoids appear broadly distributed throughout the bacteria; however, the presence of squalene, cholesterol, and carotenoids is restricted to a few genera.

A frequently made generalization is that ester bonds are found in the glycerol lipids of bacteria while ether bonds are found in archaea. A notable exception occurs in many anaerobic bacteria that have a vinyl ether bond at C-1 of glycerol and a fatty acid linked to the C-2 of glycerol of phosphatidyl ethanolamine. This glycerol monoether is commonly called a plasmalogen (see Figure 2.3). If the fatty acid is saturated, the bond formed with glycerol is an alkyl ether; however, because the lipid is unsaturated in the plasmalogen, it is an alkenyl ether. Plasmalogens are found not only in bacteria but make up more than 50% of phospholipid in human heart tissue. It has been suggested that the plasmalogen in bacteria provides protection against singlet oxygen in oxidative stress as it contains an unsaturated fatty acid.

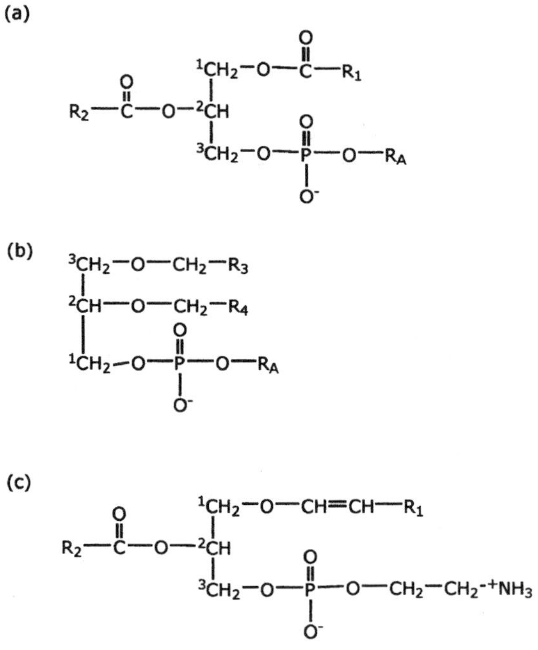

**(a)**

**(b)**

**(c)**

Figure 2.3. Bonding of fatty acids and alcohols to glycerol lipids in prokaryotes. (**a**) Ester bonds in bacteria at positions C-1 and C-2 of glycerol and fatty acids ($R_1$ and $R_2$). Additions at $R_A$ provide specificity for the various phospholipids. (**b**) Ether linkages in archaea between position C-1 and C-2 of glycerol with phytanyl molecules ($R_3$ and $R_4$). (**c**) Mixed ether and ester bonds in certain bacteria. Ether linkage at position C-1 and unsaturated lipid ($R_1$). Plasmalogen with ester bond at position C-1 of glycerol with a fatty acid ($R_2$). At position C-3 is phosphate and ethanolamine.

2.3.1.2. Fatty Acids and Isopranyl Alcohols[4–9]

Fatty acids bound to the glycerol by acyl esters are usually $C_{14}$ to $C_{20}$, with $C_{15}$ to $C_{18}$ being the most common in the bacteria. As illustrated in Figure 2.4, fatty acids are present in various structural forms and include saturated, straight-chain, monounsaturated, branched-chain, and 2- or 3- hydro- or cyclopropane types. Monomethyl branched fatty acids may be in *iso* configuration with the methyl group at the second carbon from the end or *anteiso* configuration with the methyl group on the third carbon from the end. It is unclear why prokaryotes have this great diversity of fatty acids and why certain fatty acids are used by a given bacterial species.

Ether linkages are characteristically present in archaea but less prevalent in bacteria. The plasmalogens are an example of ether-linked lipids in bacteria. There is a difference in chemical stability of these glycerol bonds in that ethers are more stable than esters, which may be important for archaea to grow in extremely harsh environments. In archaea, glycerol diethers are present in which the alcohols are bonded to both the C-1 and the C-2 of glycerol. The types of hydrocarbons may vary in the glycerol ether linkage as seen in the archeon *Thermodesulfobacterium commune* in which five different glycerol diether combinations are found: $C_{16}$:$C_{16}$, $C_{16}$:$C_{17}$, $C_{17}$:$C_{17}$, $C_{17}$:$C_{18}$, and $C_{18}$:$C_{18}$.

The distribution of unsaturated alcohols is limited to the archaea, where they occur as saturated $C_{20}$ or $C_{40}$ alcohols known as phytanyls. The saturated $C_{20}$ alcohols are attached to C-1 and to C-2 of glycerol. Structures with long-chain alcohols attached to glycerol by ether linkages is given in Figure 2.5. Alternately, a $C_{40}$ saturated phytanyl is linked in *trans* configuration to two glycerol molecules and this unit can span the membrane. As is the case with *Sulfolobus*, variation in lipid structure may lead to hybrids of $C_{20}$ and $C_{40}$ hydrocarbons. Another variation involving ether-linked molecules is seen in *Methanococcus jannaschi*, in which a single $C_{40}$ molecule is cyclic with the biphytanyl linked at the terminal ends to the same end of the glycerol molecule. In *Thermoplasma* and *Sulfolobus*, 90% to 95% of glycerol lipids are tetraethers while the remainder are diethers. An additional variable is that some of the $C_{40}$ lipids in *Sulfolobus* may contain one to four cyclopentyl rings.

The synthesis of ether-linked alcohols to a glycerol backbone is currently under study and at this time a tentative scheme has been proposed (see Figure 2.6). The precursor for glycerol phosphate is dihydroxyacetone phosphate in *Halobacterium halobium*. However, in *Methanobacterium* and *Sulfolobus*, the three-carbon unit is from glycerol-1-phosphate and not from dihydroxyacetone phosphate or glycerol-3-phosphate. Geranylgeranyl pyrophosphate is the source of the hydrocarbon bonded into the three-carbon unit and it is produced as a product of isoprenoid biosynthesis. Reactions leading to the establishment of tetraether bonds for archaea remains to be resolved.

In terms of membrane construction, the archaea may have either a bilayer or a monolayer membrane depending on the length and orientation of the phytanyl molecules (see Figure 2.2). With the presence of a $C_{40}$ phytanyl linked to two glycerol molecules, the phytanyl molecule extends across the membrane and produces a monolayer membrane. With the cyclic $C_{40}$ phytanyl and with the $C_{20}$ hydrocarbons, there

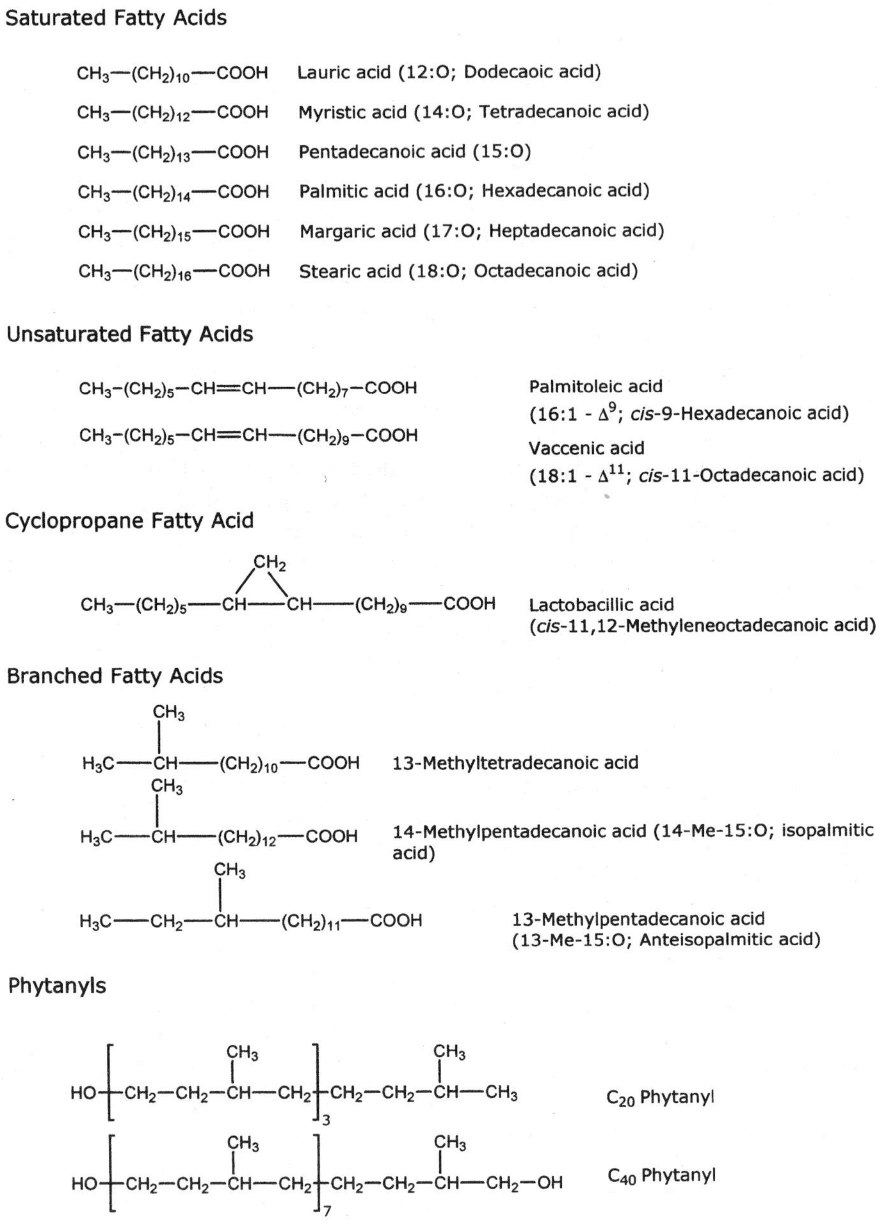

## Saturated Fatty Acids

$CH_3-(CH_2)_{10}-COOH$     Lauric acid (12:0; Dodecaoic acid)

$CH_3-(CH_2)_{12}-COOH$     Myristic acid (14:0; Tetradecanoic acid)

$CH_3-(CH_2)_{13}-COOH$     Pentadecanoic acid (15:0)

$CH_3-(CH_2)_{14}-COOH$     Palmitic acid (16:0; Hexadecanoic acid)

$CH_3-(CH_2)_{15}-COOH$     Margaric acid (17:0; Heptadecanoic acid)

$CH_3-(CH_2)_{16}-COOH$     Stearic acid (18:0; Octadecanoic acid)

## Unsaturated Fatty Acids

$CH_3-(CH_2)_5-CH{=}CH-(CH_2)_7-COOH$     Palmitoleic acid
(16:1 - $\Delta^9$; *cis*-9-Hexadecanoic acid)

$CH_3-(CH_2)_5-CH{=}CH-(CH_2)_9-COOH$     Vaccenic acid
(18:1 - $\Delta^{11}$; *cis*-11-Octadecanoic acid)

## Cyclopropane Fatty Acid

$CH_3-(CH_2)_5-CH-CH-(CH_2)_9-COOH$     Lactobacillic acid
(*cis*-11,12-Methyleneoctadecanoic acid)

## Branched Fatty Acids

$H_3C-CH-(CH_2)_{10}-COOH$ (with $CH_3$ branch)     13-Methyltetradecanoic acid

$H_3C-CH-(CH_2)_{12}-COOH$ (with $CH_3$ branch)     14-Methylpentadecanoic acid (14-Me-15:0; isopalmitic acid)

$H_3C-CH_2-CH-(CH_2)_{11}-COOH$ (with $CH_3$ branch)     13-Methylpentadecanoic acid
(13-Me-15:0; Anteisopalmitic acid)

## Phytanyls

$HO{-}[CH_2-CH_2-CH(CH_3)-CH_2]_3{-}CH_2-CH_2-CH(CH_3){-}CH_3$     $C_{20}$ Phytanyl

$HO{-}[CH_2-CH_2-CH(CH_3)-CH_2]_7{-}CH_2-CH_2-CH(CH_3){-}CH_2{-}OH$     $C_{40}$ Phytanyl

Figure 2.4. Examples of fatty acids and phytanyls found in prokaryotic membranes.

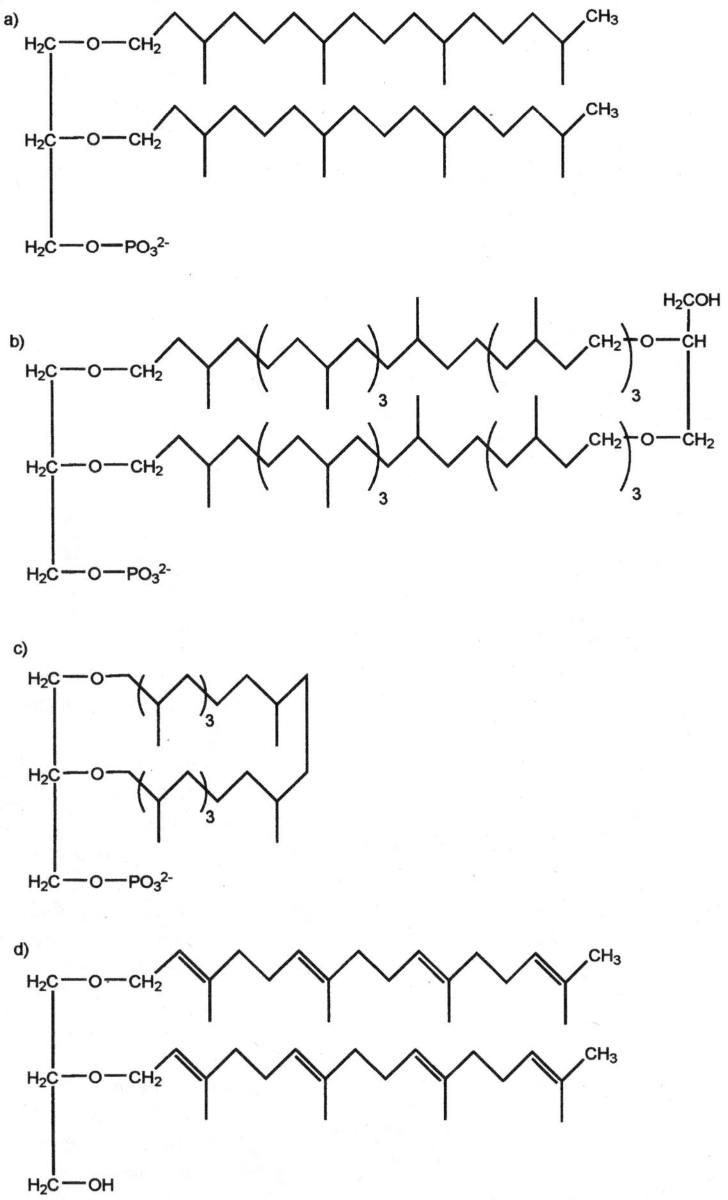

Figure 2.5. Structures of ether-type lipids attached to the glycerol backbone in archaea. (**a**) Phytanyl ($C_{20}$) moiety linked to glycerol phosphate by an ether bond. (**b**) Biphytanyl ($C_{40}$) moiety attached by an ether bond to produce diglycerol tetraether. In *Thermoplasma acidophilum*, the phosphate group is replaced by an oligosaccharide of glucose-[mannose ($\alpha$-1 $\rightarrow$ 2) -mannose ($\alpha$-1 $\rightarrow$ 4) -mannose ($\alpha$-1 $\rightarrow$ 3)$_8$. (**c**) A $C_{40}$ phytanyl moiety coupled to one 3-C unit. (**d**) Unsaturated hydrocarbon ether linked to glycerol as found in *Methanopyrus*.

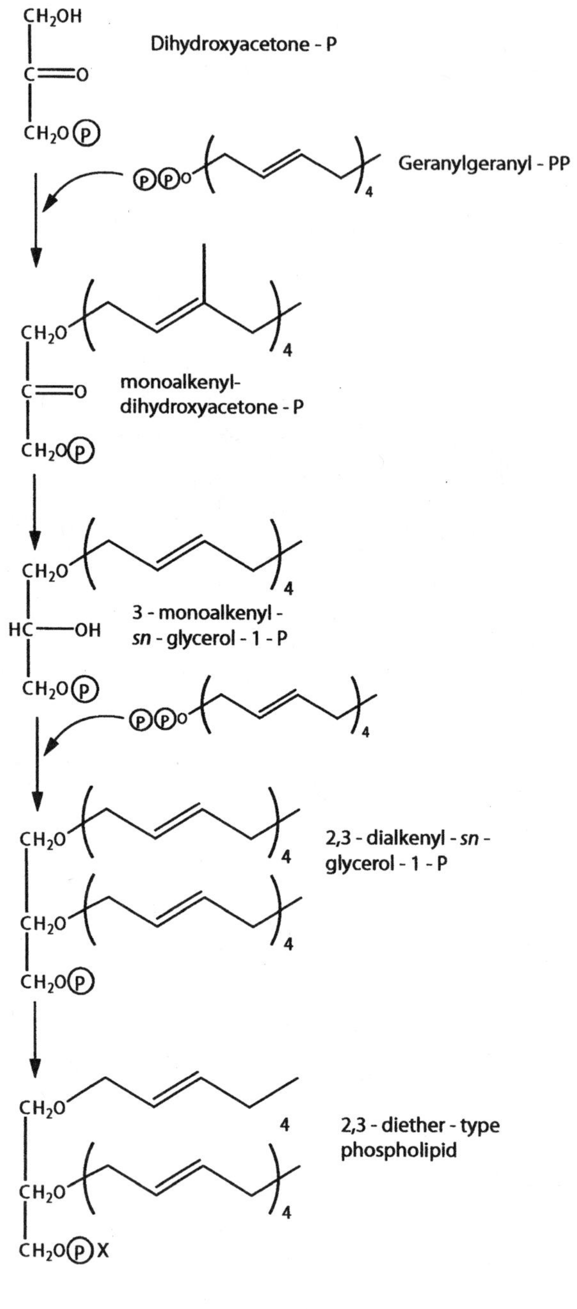

Figure 2.6. Proposed pathway for synthesis of ether-type phospholipids in archaea. (1) Reaction leading to displacement of pyrophosphate and establishing of ether bond. (2) Reduction of the keto group of dihydroxyacetone to give *sn*-glycerol backbone. (3) Addition of geranyl geranyl-PP to produce a diether glycerol phosphate with dialkenyl groups. (4) Formation of 2,3-diether-type phospholipid occurs with reduction of unsaturated hydrocarbon and additions to phosphate on glycerol phosphate to produce phospholipids. The sequence of activities associated with reaction 4 is not resolved at this time. (From Koga et al., 1993.[5])

Table 2.2. Presence of diethers and tetraethers in archaea.[a]

| Organism | Diether (%) | Tetraether (%) |
|---|---|---|
| *Sulfolobus* | 5 | 95 |
| *Thermoplasma* | 10 | 90 |
| *Methanobacterium formicicum* | 11 | 89 |
| *Methanospirillum hungatei* | 41 | 59 |
| *Methanobacterium thermoautotrophicum* | 45 | 55 |
| *Methanosarcina barkeri* | 100 | 0 |
| *Halobacterium halobium* | 100 | 0 |

[a]From Smith, 1988.[33]

is a membrane bilayer similar to that seen with the fatty acids in the bacterial membranes. There is a distinction in the distribution of diether and tetraether lipids in the membranes of the archaea. As presented in Table 2.2, all archaea produce diether lipids but a few in addition produce tetraethers. Although not listed in Table 2.2, most archaea produce only diether lipids in the plasma membrane.

### 2.3.1.3. Phospholipids[4-9]

In Gram-negative bacteria such as *Escherichia coli*, 65% to 75% of total phospholipids are found in the plasma membrane with the remaining phospholipids in the outer membrane. It has been estimated that there are about $2 \times 10^7$ molecules of phospholipid in the plasma membrane of an average bacterial cell. The major lipid fraction found in the plasma membrane belongs to the phospholipid class, and the structures of some of these molecules are given in Figure 2.7. The most common zwitterionic phospholipid found in both Gram-positive and Gram-negative bacteria is phosphatidyl ethanolamine. While many bacteria contain anionic phospholipids of phosphatidyl glycerol and diphosphatidyl glycerol (cardiolipin), some contain small amounts of phosphatidyl serine, sulfolipids, phosphatidyl choline, and sphingolipids. Examples of phospholipids found in bacteria are given in Table 2.3. The distribution of lipids in bacteria is not identical for all species, and these variations can be used to demonstrate the evolutionary changes in microorganisms. A summary of the distribution of lipids in bacteria is given in Table 2.4 and it is evident that the type of lipid present is characteristic of a given biochemical group.

Whereas phospholipids are common in bacteria, phosphoglycolipids are most abundant in Archaea. It should be noted that cells of the archaea frequently produce several phospholipids but the phospholipids with attached sugars are the most prevalent. Each genus of *Archaea* has a distinct ether lipid, making it useful to identify organisms using lipids. For example, the plasma membrane of *Methanospirillum hungatei* contains about 50% protein, 36% lipid, and 12% carbohydrate. About 90% of lipids in this organism are polar lipids with ether linkages and the remaining 10% is primarily squalene. In *Methanospirillum hungatei*, 65% of the lipid is phosphodiglycosyldibi-

**Phosphatidic acid (PA)**

**Phosphatidyl choline (PC)**

**Phosphatidyl ethanolamine (PE)**

**Phosphatidyl serine (PS)**

**Phosphatidyl glycerol (PG)**

**Diphosphatidyl glycerol (DPG)**

**Phosphatidyl inositol (PI)**

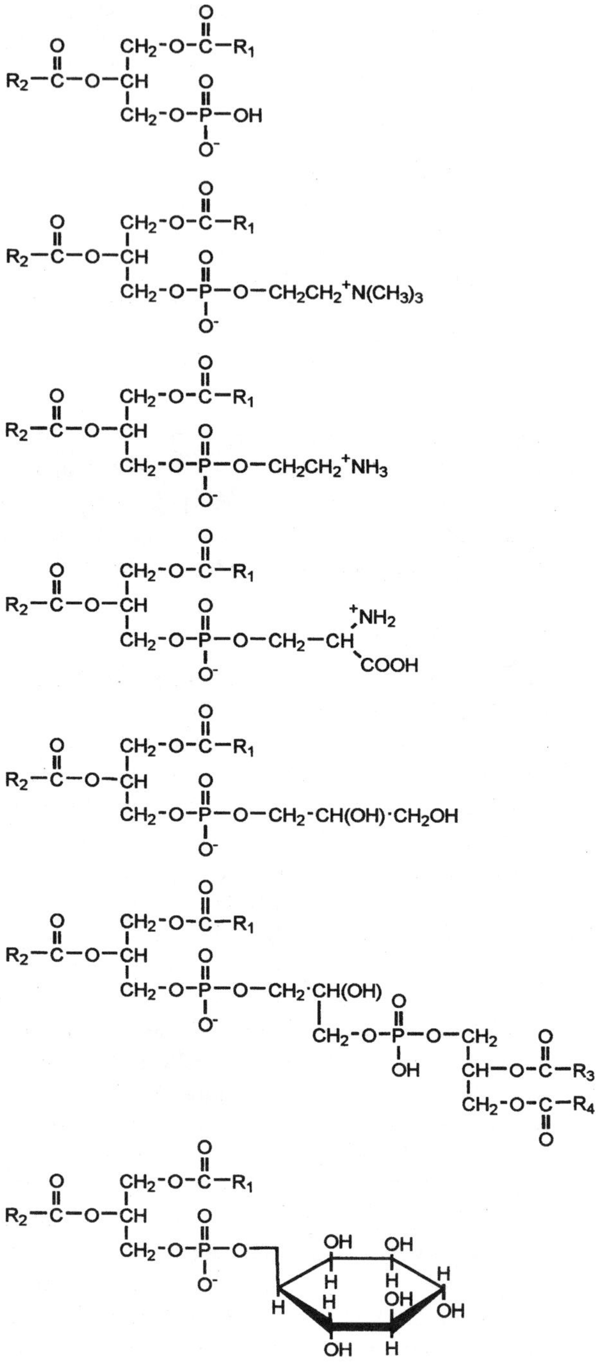

Figure 2.7. Examples of phospholipids in bacteria. Fatty acids are substituted at $R_1$, $R_2$, $R_3$, and $R_4$.

Table 2.3. Examples of phospholipids present in plasma membranes of bacteria.[a]

| Organism | PE | PG | DPG | PS | PC | MGD | PI | Others |
|---|---|---|---|---|---|---|---|---|
| *Bacillus megaterium* | 34% | 49% | 10% | 0 | 0 | 0 | 0 | 7% |
| *Bacillus stearothermophilus* | 23% | 24% | 49% | 0 | 0 | 0 | 0 | 4% |
| *Clostridium butyricum* | 14% | 26% | 0 | 0 | 0 | 0 | 0 | 38% |
| *Escherichia coli* | 76% | 20% | 1% | 0 | 0 | 0 | 0 | 3% |
| *Lactobacillus casei* | 1% | 0 | 0 | 0 | 45% | 0 | 0 | 46% |
| *Myxococcus xanthus* | 76% | 9% | 1% | 0 | 0 | 0 | 0 | 14% |
| *Staphylococcus aureus* | 0 | 40% | 16% | 0 | 0 | 2% | 0 | 42% |
| *Stigmatella aurantiaca* | 48% | 12% | 0 | 0 | 0 | 0 | 23% | 17% |
| *Treponema pallidum* | 0 | 0 | 11% | 0 | 41% | 0 | 0 | 48% |

PE, Phosphatidyl ethanolamine; PI, phosphatidyl inositol; PC, phosphatidyl choline; PG, phosphatidyl glycerol; DPG, diphosphatidyl glycerol; PS, phosphatidyl serine; MGD, monoglucosyldiglyceride.
[a]From Ratledge and Wilkerson, 1988.[8]

phytanyldiglycerol tetraether, which is a polar lipid with the following general composition:

sugar–glycerol–diphytanyl (C-40)–glycerol–phosphate–glycerol

At one of the end of the molecule a sugar moiety is attached to the C-1 of glycerol with the C-2 and C-3 of glycerol attached to phytanyl structures by ether linkages. At the other end of this lipid molecule, phosphate is attached to a C-1 of another glycerol that is attached to the phytanyl group by ether linkages. Finally, there is another glycerol molecule attached to the phosphate. In *Methanobacterium thermoautotrophicum*, the polar group is phosphoethanolamine while in *Methanobrevibacterium arboriphilus* the polar group is phosphoserine. In *Methanothrix concilii*, the phospholipids contain inositol or a disaccharide of mannose and galactose. Thus, the lipids in methanogens differ in core lipid structures as well as in polar head groups.

There is a large array of phospholipids in plasma membranes of the various species of bacteria that have several important roles, including the following: (1) Phospholipids influence charge density on the membrane, which has an important impact on nutrient transport and membrane structure. Basic amino acids such as lysine and ornithine can be attached to phosphatidyl glycerol and alter the charge established on the membrane in acidic conditions. Phosphatidyl ethanolamine influences the transport systems for proline, leucine, tryptophan, lysine, and lactose by direct interaction with the high-affinity transport proteins. (2) In addition to establishing a permeability barrier for bacteria, phospholipids appear to be important in the signal transduction process associated with chemotaxis and with protein kinases. (3) Anionic phospholipids appear to have various roles in translocation of protein across the membrane by systems that involve molecular chaperones. (4) There is strong evidence to suggest that in *E. coli*, anionic phospholipids interact with DNA to influence the initiation, elongation, and termination sites on the bacterial genome. (5) Phosphatidyl ethanolamine, along with *sn*-1-phosphoglycolipid, binds glucose polymers and the tethered carbohydrate polymer (MDO) extends from the plasma membrane into the cell wall.

Table 2.4. Some characteristics of lipids in plasma membranes of prokaryotes.[a]

| Lipid | Organisms | Features of lipids present |
|---|---|---|
| Fatty acid esters | Gram-positive bacteria | Aminoacylphosphatidyl glycerol with amino acids of lysine, alanine, and ornithine are common. Branched fatty acids are common. |
| | Gram-negative bacteria | Saturated fatty acids are commonly $C_{16}$. Few have fatty acids that are branched, contain cyclopropane, or have an odd number of carbons. Phosphomonomethyl ethanolamine and phosphodimethyl ethanolamine are common. |
| | Cyanobacteria | Commonly have phosphatidyl glycerol but not phosphatidyl inositol, phosphatidyl choline, or phosphatidyl ethanolamine. |
| Phytanyl ethers | Archaea | Hydrocarbons of $C_{20}$ or $C_{40}$ linked to phosphoglycerol as mono-, di-, or tetra-ethers. |
| Complex lipids | Bacteria | Most organisms are considered to contain hopanoids. Phototrophs and yellow to red colonies contain carotenoids and xanthophylls. |
| Nonpolar lipids | Archaea | Some have squalenes and modified squalenes. |
| Plasmalogens | *Clostridium, Desulfovibrio, Bacteroides, Treponema, Propionobacterium, Selenomonas,* and other anaerobes | As phosphatidyl ethanolamine or phosphatidyl choline. |
| Cholesterol | *Mycoplasma* and other bacteria without cell wall | Supplement phospholipids in membrane with cholesterols from the environment. |
| Sulfolipids | *Gluconobacter, Cytophaga, Flexibacter,* and *Flavobacterium* | Contain sulfolipids in addition to phospholipids. |
| Ornithine lipids | *Shewanella putrefaciens, Acidithiobacillus thiooxidans, Brucella,* and others | Ornithine is attached as an amide to phospholipid. |

[a]From Ratledge and Wilkerson, 1988.[8]

### 2.3.1.4. Squalene

In the archaea, nonpolar lipids are distributed in the hydrophobic regions of the plasma membrane. Squalene and tetrahydrosqualene are the most prevalent nonpolar lipids, and their structures are given in Figure 2.8. Nonpolar or neutral lipids constitute up to 20% of total lipids as squalenes ($C_{30}$ hexaisoprenoids) in halophiles, thermoacidophiles, and methanogens. Squalene is a precursor of cholesterol and in the membrane may be viewed as a more flexible structure than the rigid cholesterol molecule. The

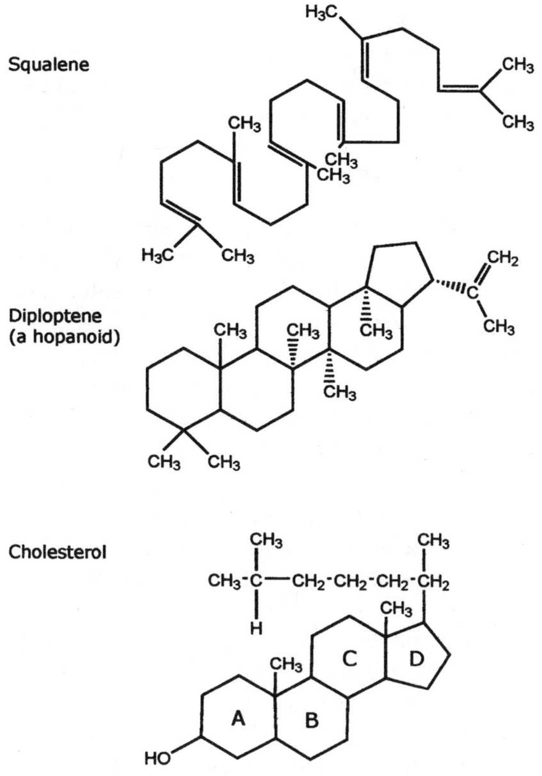

Squalene

Diploptene
(a hopanoid)

Figure 2.8. Structures of squalene and hopanoid in prokaryotic membranes. Many different molecular forms of hopanoids are found in prokaryotes and diploptene is given as an example. The presence of cholesterol appears limited to cell wall–less forms of bacteria.

Cholesterol

function of the squalene molecules in the membrane is not resolved at this time but they appear to have some role in stabilizing the membrane structure.

### 2.3.1.5. Hopanoids[10,11]

In addition to phospholipids and glycolipids, hopanoids are found in plasma membranes of bacterial but not archaeal cells. Although hopanoids constitute an important class of lipids, their detection and analysis is often overlooked owing to the large number of structural variants and molecular destruction during extraction. Each gram of cellular material contains 0.1 to 5.0 mg of hopanoid. As illustrated in Figure 2.8, hopanoids have a pentacyclic molecule with a short side chain while the cholesterol molecule has four rings with an extended side chain. Functionally, hopanoids contribute to the stability of bacterial membranes by a mechanism similar to that of cholesterol in eukaryotic membranes. Hopanoids provide a valuable function by limiting the diffusion of $H^+$ and $OH^-$ through the plasma membrane. The cyclic structure of hopanoids is embedded in the lipophilic region of the membrane while the side chain interacts with the hydrophilic regions of the membrane. In some bacteria such as *Bacillus acidocaldarius*, hopanoids are used to secure sugar moieties to the membrane. It remains to be demonstrated if hopanoids are used by other bacteria to build a diverse molecular structure in the membrane. Cholesterol biosynthesis requires oxygen, but hopanoid synthesis can occur under both aerobic and anaerobic conditions. This may account for a broad distribution of hopanoids in bacteria but a limited presence of cholesterol in prokaryotes.

### 2.3.1.6. Carotenoids[10,11]

Carotenoids are pigments that range from yellow to purple and account for 1% to 2% of lipid in the plasma membranes of different bacterial types. The quantity of carotenoid can be readily determined by solubilizing in hot ethanol and measuring the absorption of the extract in a spectrophotometer. The carotenoids (Figure 2.9) are commonly called tetraterpenoids and have a carbon chain of 40 atoms, which is long enough to span the plasma membrane. Xanthophylls are yellow to orange lipid-soluble pigments that have a molecular structure similar to that of carotenoids in that they are oxygen-containing tetraterpenoids. Both xanthrophylls and carotenoids are considered to be important in protecting against photolytic damage to DNA. Although carotenoids would promote viscosity in membranes, the quantity of carotenoids in plasma membranes may be too small to have a significant impact on membrane fluidity. Carotenoids are not known to be present in the outer membrane of Gram-negative bacteria. When *Mycobacterium kansasii* or other photochromogenic mycobacteria are grown in the presence of light, the cells become yellow-orange as a result of carotenoid production. Although some consider that the quantity of carotenoids is too low to have a significant impact on membrane fluidity, others suggest that the site of carotenoids' interaction and the nature of carotenoids' intercalation with membrane lipids could be important. A great deal is known about the different molecular types of carotenoids present in bacteria but little is known regarding their function.

### 2.3.1.7. Quinones

The plasma membrane contains the components of the electron transport system used in bacterial aerobic and anaerobic respiration. The most obvious but not the only components include flavin mononucleotide (FMN), quinones, dehydrogenases, and cytochromes. The levels of these electron transport components found in some bacteria are given in Table 2.5. The quantity of proteins and lipids in the plasma membrane varies with the strain of the organism as well as the chemical environment under which the organism grows.

The quinones are lipid-soluble molecules that both carry electrons and function as a proton shuttle in the plasma membrane. Their structures are given in Figure 2.9. Quinones produced by bacteria are in the form of ubiquinones or menaquinones, and these molecules reside in the central hydrophobic region of the plasma membrane. Ubiquinones ($U_{-n}$) were formerly called coenzyme Q and are examples of benzoquinones. Generally, $U_{-n}$ are found in Gram-negative bacteria but not Gram-positive bacteria or archaea. Other examples of benzoquinones are plastiquinones ($PQ_{-n}$) and rhodoquinones ($RQ_{-n}$), which are found in cyanobacteria and *Rhodospirillaceae*, respectively. Both $PQ_{-n}$ and $RQ_{-n}$ are associated with photorespiration. Menaquinones ($MQ_{-n}$) have the fundamental structure of naphthoquinones and are widely distributed

Table 2.5. Abundance of respiratory components in plasma membrane of bacteria.

| Bacteria | FAD +FMN | Menaquinones | Ubiquinones | Cytochrome types | | |
| --- | --- | --- | --- | --- | --- | --- |
| | | | | *a* | *b* | *c* |
| *Mycobacterium phlei*[a] | 0.68 | 12.0 | 0 | 0.27 | 0.18 | 0.62 |
| *Escherichia coli* (aerobic)[a] | 0.12 | 0.86 | 1.26 | 0.1 | 0.62 | 0 |
| *Azotobacter vinelandii*[a] | 2.0 | 0 | 8.9 | 0.67 | 1.59 | 1.62 |
| *Desulfovibrio gigas*[b] | 1.1 | 5.9 | 0 | 0 | 0.15 | 1.20 |

Values are in µmol/g of protein.
[a]Gel'man et al., 1967.[30]
[b]Odom and Peck, 1981.[31]

in prokaryotes. Frequently, earlier reports designated menaquinones as naphthoquinones, and throughout this text these redox-active lipids are referred to as menaquinones. Depending on the species of organism, the production of menaquinones may be influenced by the environment. In both menaquinones and ubiquinones, the "$-n$" indicates the number of isoprenoid units attached to the quinone structure. For example, $MK_{-6}$ (also written as MK-6) has six isoprenoid units while $U_{-10}$ (also written as U-10) has 10 isoprenoid units. The number of isoprenoid units in the side chain with ubiquinones or menaquinones ranges from 1 to 16, with 6 to 10 being the most common.

## 2.3.2. Proteins in the Plasma Membrane[1-3]

Proteins in the plasma membrane constitute about 7% to 15% of all cellular proteins. In Gram-negative bacteria such as *E. coli*, 6% to 9% of the total protein is found in the plasma membrane, with at least 100 different protein species. Approximately, 20% to 30% of the membrane protein is integral protein and the remainder are peripheral proteins associated with the membrane surfaces. The integral proteins are held in the plasma membrane as a result of transmembrane integration of nonpolar regions of the protein. Amino acids carrying either a negative or a positive charge are found outside the membrane and not in the hydrophobic region of the plasma membrane. The charge distribution on amino acids adjacent to the polar transmembrane helices determines the orientation of the protein in the membrane. The presence of proteins in membranes is an example of self-assembly, with molecules in the membrane having a configuration that is at the lowest energy level. If the molecules in the membrane were highly charged, their insertion and even perhaps retention of the charged molecules in the lipid region would require a considerable expenditure of energy.

### 2.3.2.1. Solubilization of Membrane Proteins

The ultimate goal in understanding membrane structure is to describe the many functions performed by the membrane at the molecular level. Enzymes loosely adhering

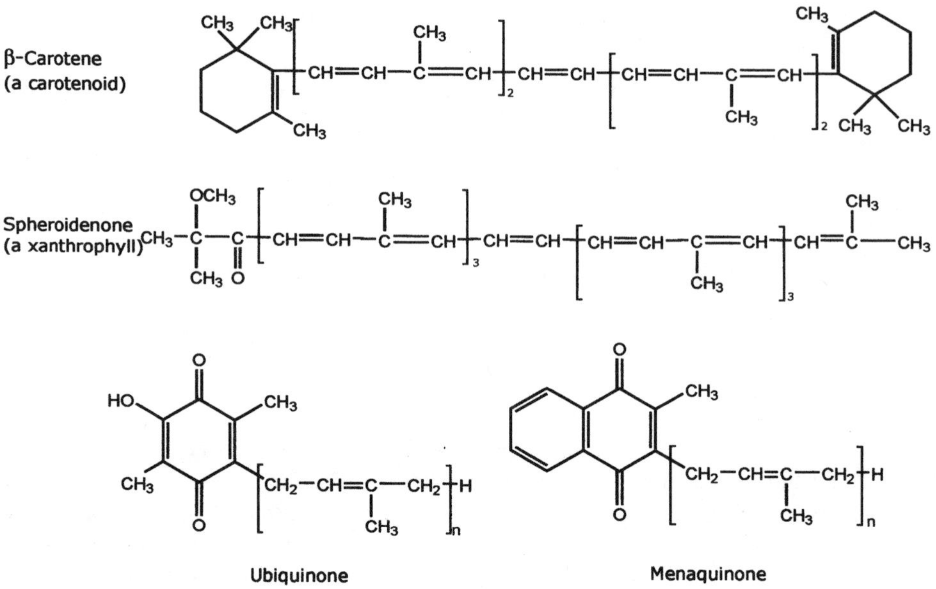

Figure 2.9. Structures of carotenoids and quinones in bacteria. With the hundreds of carotenes found in prokaryotes, β-carotene and spheroidenone are given as examples. The basic structures of ubiquinones and menaquinones are also given.

to the membrane can be removed by treatment with mild detergents or ion-exchange reactions using NaCl solutions. The membranes can be stabilized with the presence of $Ca^{2+}$ or $Mg^{2+}$. Certain enzymes or proteins are strongly attached to the membrane and can be used as markers; examples of these were given in the previous chapter. To solubilize membrane-bound enzymes, either chaotrophic agents or nonionic detergents are used. The chaotrophic reagents (e.g., NaSCN, $NaClO_4$, guanidine·HCl, and urea) modify the structure and lipophilicity of water so that nonelectrolytes and charged proteins containing polar groups are carried into the aqueous phase. Nonionic detergents (e.g., Tween-20, Triton X-100, Brij-58, and n-octyl-β-D-glucopyranoside) emulsify proteins in the membrane with relatively little modification of the solubilized enzymes. The use of 3-[(3-cholamidopropyl) dimethylammonio]-1-propanesulfonate (CHAPS), a zwitterionic detergent, is also effective in solubilization of membrane-bound enzymes. In general, cationic detergents are too aggressive to be used as solubilizing agents and their use frequently results in denaturation of enzymes. To prevent proteolytic digestion of proteins as they are being isolated, protease inhibitors must be added to all buffers.

## 2.3.2.2. Electron Transport Proteins

The dehydrogenases function to assist in the oxidation of substrates with the movement of electrons into the plasma membrane. Many bacteria contain succinic dehydrogenase, malate dehydrogenase, and NADH dehydrogenase, all of which reside in the plasma membrane. Cytochromes are electron-carrying molecules that are found as respiratory components in the plasma membrane. Cytochromes consist of protein plus heme, and the type of heme is used to identify the cytochrome. That is, cyto-chrome *a*, cytochrome *b,* and cytochrome *c* contain heme A, heme B, and heme C, respectively. Owing to the presence of several transmembrane helices, cytochromes *a* and *b* are stabilized within the membrane and are not found in the cytoplasm. On the other hand, there are two types of cytochrome *c*: those that have transmembrane segments and are found only in the plasma membrane and those that have no lipophilic regions and are found only in the periplasm or cytoplasm. Some respiratory enzymes contain heme and may be designated either as a cytochrome or as a heme-containing enzyme. There is considerable diversity in the molecular forms of cytochromes in prokaryotes, and the various types are discussed at length in subsequent chapters.

In terms of structure, several types of dehydrogenase proteins are associated with the plasma membrane; models of three types are given in Figure 2.10. In one type the dehydrogenase is positioned on the face of the plasma membrane where substrates have unlimited access to the enzyme and electrons can be shunted to the quinones in the membrane. The enzyme is loosely attracted to the membrane by charge interac-tions and the protein can be dislodged with ultrasonic or pressure treatments used in cell disruption. The dislodged protein can reassemble on both sides of the plasma membrane as well as being dispersed in the soluble fraction. A second type of de-hydrogenase has multisubunits with only one subunit being transmembrane while the other subunits are in the cytoplasmic region but attached to the transmembrane subunit.

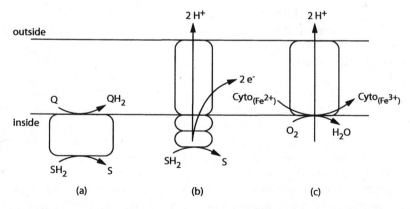

Figure 2.10. Models of dehydrogenase proteins associated with the plasma membrane. (**a**) Protein loosely associated with the inner side of the plasma membrane which is characteristic of numerous dehydrogenases. (**b**) Enzymes such as NADH dehydrogenase I have a protein with three subunits and only one is transmembrane. The subunits may contain Fe-S clusters, flavin, or Mo-cofactor. (**c**) Transmembrane protein consisting of several subunits and this con-figuration is characteristic of respiratory oxidases. (From Kadner, 1996.[2])

The protein subunits do not readily dissociate because ionic charges between the subunits as well as the presence of disulfide linkages hold the protein units together. Thus, multisubunit proteins will retain significant orientation throughout the isolation protocols for plasma membranes. In another type of dehydrogenase structure protein subunits are attached but all are within the plasma membrane as transmembrane subunits. Substrate combines preferentially at one side of the membrane and electrons are passed through the quinone lipid to the substrate.

### 2.3.2.3. ATP Synthase/ATPase[12,13]

The plasma membrane contains novel structures for the concomitant synthesis of ATP coupled to an inward $H^+$ translocation. Isolation of the $F_1$ complex resulted in a protein system that catalyzes the hydrolysis of ATP to ADP $+$ $P_i$, and thus the term ATPase was applied. In some cases, the extrusion of protons or other ions outward across the membrane was observed in bacteria, and this activity was demonstrated to be driven by energy from hydrolysis of ATP. In general terms, those activities that involve energy from hydrolysis of ATP are referred to as ATPase activities. Whereas ATPase formerly applied only to the respiratory driven component of oxidative phosphorylation, several membrane activities are included in this designation at present. The portion of the ATPase that is localized in the plasma membrane ($F_0$) contains six transmembrane domains. Three proteins make up the $F_0$ segment; one protein has three transmembrane segments, another has two transmembrane segments, and a third has only one transmembrane segment. There are several different types of transport ATPases in the prokaryotic membrane and a comprehensive treatment of ATPases and transport is given in Chapter 9.

### 2.3.2.4. Electron Carrier Complexes

The plasma membrane is the site for prokaryotic respiration where electrons are moved from soluble substrates in the cytoplasm to membrane-associated electron acceptors. Several segments of the respiratory chain are involved in expelling protons from the cytoplasm and they consist of several subunits that function as a unit. These segments are composed of a protein and/or a protein–lipid complex that responds to redox changes initiated by adjacent portions of the electron transport chain, forming a compact unit to transport electrons as well as to translocate protons. For example, a prokaryote that contains a complete electron transport chain with oxygen as the electron acceptor may be expected to have four of these electron carrier complexes. Assuming strong similarity between the aerobic respiratory system of bacteria and mitochondria, the following four complexes would be expected to occur. One would be the NADH dehydrogenase system, complex I, which contains as many as 26 different polypeptides, Fe-S centers, and FMN. The second has the succinate dehydrogenase arrangement, complex II, which consists of at least two and not more than four proteins, FAD, and Fe-S centers. The third is known as the cytochrome $bc_1$ system, complex III, which consists of about six proteins including Fe-S centers and heme B and heme C. The fourth contains the cytochrome oxidase components, complex IV, which may contain 13 proteins with prosthetic groups of heme A and copper.

Complexes I, III, and IV are presumed to be the proton pumps for aerobically grown organisms and although these proteins can be isolated, the mechanisms of these proton translocation systems are not resolved. If the prokaryotic cell is utilizing respiration with an electron acceptor other than oxygen, the organism would not be restricted to the four complexes of the aerobe, but the specific electron carrier complex would be appropriate for that substrate. Nevertheless, these electron carrier complexes are an important segment of the plasma membrane. A model of these complexes is given in Figure 2.11.

## 2.4. Special Systems

When separating outer membranes and plasma membranes of disrupted Gram-negative bacteria, sucrose gradients and ultracentrifugation are frequently used to isolate plasma membranes. The plasma membranes have a buoyant density of 1.14 to 1.16 g/ml, which is less than the density of outer membranes. However, some protocols for differential fractionation of membranes produce fractions that are enriched for membrane segments associated with other cellular constituents. For example, some fractions may contain plasma membrane where DNA is bound through the *oriC* site, or fractions could contain membrane adhesion structures (Bayer's connections) that extend from the plasma membrane to the outer membrane. If recombinant proteins form an aggregate with the plasma membrane, the density of this membrane is greatly increased and the plasma membrane will be isolated in the fraction with the outer membranes.

### 2.4.1. Membranes of Cell Wall–Less Prokaryotes[14]

Two major types of bacteria do not have cell walls and these are known as the L-forms and the cell wall–less organisms. The L-forms, also known as L-phase variants,

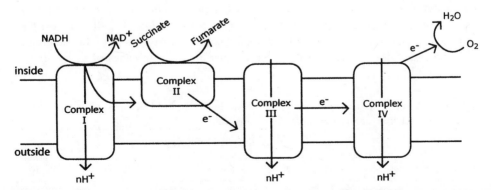

Figure 2.11. Organization of membrane into four electron transport complexes. Note that three of the complexes in aerobic bacteria containing a mitochondrial-like respiratory system are involved in proton transport outward across the plasma membrane.

(named for Lister Institute, where they were first observed in 1935) are derived from a distinct species such as *Streptobacillus moniliformis* and *Proteus mirabilis*. L-forms arise from either of two mechanisms: growth of cell wall–deficient mutants or induction of a nongenetic phenotypic change in which the cell has lost all existing cell wall so that new cell wall building units have no base to add onto. Under appropriate conditions, an L-form can revert to normal vegetative cells. Certain L-forms develop methods for strengthening the cell membrane by increasing the ratio of saturated to unsaturated fatty acids and are able to grow at the osmolarity of serum.

About 100 different species of bacteria and archaea have no cell walls. The membrane of cell wall–less bacteria contains about 30% lipid and 70% protein as in other membranes; however, important differences occur with respect to the type of lipid and proteins present. Membranes from most of the cell wall–less forms have 25% to 30% cholesterol, which is comparable to that of eukaryotic membranes while other bacteria have no cholesterol or sterols in their plasma membrane (see Table 2.6). The cell wall–less forms do not synthesize cholesterol nor do they esterify fatty acids onto the cholesterol structure but acquire both free cholesterol and esterified cholesterol from the growth medium. Even the *Acholeplasma* species, which do not require sterols, will acquire some cholesterol from the medium. To support growth, the sterol must have a planar nucleus, a hydrocarbon side chain, and a free hydroxyl group at the 3β-position, which are the same requirements for a sterol to influence the fluidity of eukaryotic biological membranes. Sterols satisfying these requirements for growth of cell wall–less forms include cholesterol, ergosterol, stigmasterol, and β-sitosterol but not coprostanol or epicholesterol.

As a result of the presence of cholesterol, the membranes of cell wall–less bacteria are susceptible to destabilization by various chemicals. Digitonin complexes with cholesterol in the membranes to cause extensive rearrangement of lipid, which results in the production of small pores that contribute to increased membrane permeability and cell lysis. Similarly, the polyene antibiotics of filipin and amphotericin B alter the cholesterol interaction with phospholipids in the membrane to produce a fragile membrane. The lytic thiol-activated bacterial toxins of streptolysin O (*Streptococcus pyogenes*), perfringolysin (*Clostridium perfringens*), tetanolysin (*Clostridium tetani*), and cereolysin (*Bacillus cereus*) will disrupt membrane stability in cell wall–less bacteria as a result of accumulation of the toxin–cholesterol aggregates in the membrane.

Some distinction in the type of phospholipids in the membranes of cell wall–less

Table 2.6. Presence of sterol and phospholipids in membranes of cell wall–less bacteria as a result of uptake from serum media by growing cultures.[a]

| Organism | Lipids (μmol/g of cell protein) | | | |
| --- | --- | --- | --- | --- |
| | Free cholesterol | Esterified cholesterol | Sphingomyelin | Phosphatidyl choline |
| *Acholeplasma laidlawii* | 10.2 | 0 | 0 | 0 |
| *Mycoplasma pneumoniae* | 85.9 | 58.0 | 40.3 | 16.6 |

[a]From Razin, 1981.[14]

forms of bacteria is also observed. While phosphatidyl glycerol and diphosphatidyl glycerol are synthesized in both wall-covered bacteria and cell wall–less bacteria, phosphatidyl ethanolamine is common in wall-covered bacteria but not in those lacking a cell wall. The mycoplasma do not synthesize phosphatidyl choline or sphingomyelin but readily accumulate these animal phospholipids in their membranes. For the mycoplasma lacking sterols, glycolipids represent a major segment of the membrane lipid and the glycolipid appears to provide structural stability to the membrane. The glycolipids are generally glycosyl diglycerides that contain up to five sugar residues. Because the different species have specific sugars along with unique linkages, these glycolipids provide important antigentic differences for the cell wall–less forms.

The cell wall–less bacteria have an abbreviated electron transport system that consists of a transfer of electrons from NADH to oxygen by an enzyme residing in the cytoplasm. The mycoplasma contain menaquinones in the membrane which are unusual because the structure present is MK-4, which has a very short side chain. F-type ATPases are present in *Acholeplasma laidlawii*, *Mycoplasma mycoides*, and *Mycoplasma gallisepticum* which are specific for $Na^+$ translocation. With the use of two-dimensional electrophoresis, it has been determined that there are about 140 major peptides in the membrane and approximately 180 major peptides in the cytoplasm of *Acholeplasma laidlawii*. The proteins in the membrane are not subject to as much change as the lipids in the membrane. It has been observed that in aged cultures of cell wall–less bacteria the membranes have an increased level of protein conferring increased resistance to osmotic lysis.

*Thermoplasma acidophilum*, a cell wall–less archaeon, grows as pleomorphic spheres at 60°C and pH 2.0. The membrane contains 60% protein, 25% lipid, and 10% carbohydrate. Of the 22 protein bands seen by sodium dodecyl sulfate (SDS)-polyacrylamide gel electrophoresis, the major band is a glycoprotein of 152 kDa. The carbohydrates consisting of mannose, glucose, and galactose (molar ratio: 31:9:4) are attached through an asparagine residue of the protein. This glycoprotein appears to contribute to the structural stability of the membrane in a hostile environment.

## 2.4.2. *Bacteriophage Membranes*[15]

Although it is not widely discussed, several of the bacterial viruses contain membranes with distinctive phospholipids and fatty acids. On a dry weight basis, the enveloped bacteriophages contain at least 12% lipid, and some viruses have been determined to have a membrane with a lipid bilayer. Lipid-containing viruses have been found in several different bacteria including *Altermonas espejiana*, *Escherichia coli*, *Pseudomonas aeruginosa*, *Bacillus anthracis*, *Bacillus acidocaldarius*, *Streptococcus pneumoniae*, *Acholeplasma laidlawii*, and *Mycoplasma smegmatis*. Although the type of lipid in the virus membrane is the same as found in the cell, the quantities of lipid are distinct for the virus. The lipid contents of virus membranes for *Altermonas espejiana* and *E. coli* are given in Table 2.7. In PRD1, a membrane-containing phage for *Enterobacteriaceae*, the virus membrane contains approximately 50% lipid and 50% protein. The lipid is from the plasma membrane and there are about 10 different

Table 2.7. Distribution of lipids in bacterial viruses as compared to bacterial cells.[a]

| Bacteria or virus | Percent of phospholipids present | | |
|---|---|---|---|
| | Phosphatidyl ethanola-mine | Phosphatidyl glycerol | Diphosphatidyl glycerol |
| *Altermonas espejiana* (PM) | 71.5 | 23.5 | — |
| *Altermonas espejiana* (OM) | 79.0 | 16.0 | — |
| PM2 bacteriophage[b] | 28.0 | 67.0 | — |
| *Escherichia coli*[c] | 74.0 | 10.0 | 10.0 |
| PR5 bacteriophage[d] | 43.0 | 44.0 | 3.0 |

PM, Plasma membrane; OM, outer membrane.
[a]Source of information is Steiner and Steiner, 1988.[9]
[b]Virus of *A. espejiana*.
[c]Values include phospholipids from both plasma membranes and outer membranes.
[d]Virus of *E. coli*.

proteins of viral origin in the viral membrane. The viral membrane is located inside the protein capsid and surrounds the dsDNA genome of the virus. The regulation for synthesis of virus membrane is unresolved, as are the details concerning the envelopment of the virus.

## 2.4.3. Membrane Vesicles[16,17]

In disrupted bacterial cells, the plasma membranes will naturally coalesce into vesicles owing to the strong hydrophobic interior of the membrane. Vesicles from the plasma membrane are usually 40 to 100 nm in diameter. Depending on the nature of the disruption, the vesicles will have different orientations. Disruption of cells and orientation of plasma membranes is given in Figure 2.12. In general, vesicles produced from the plasma membrane by ultrasonic treatment or osmotic disruption are everted or right side out where the outer side of the plasma membrane of the cell is generally the outer side of the vesicle. An electron micrograph of a vesicle produced from *Escherichia coli* by osmotic lysis is given in Figure 2.13. Membrane orientation is indicated from the closely packed randomly distributed proteins on the outer surface of the vesicle, which is characteristic of asymmetric distribution of proteins on the plasma membrane. Vesicles formed from disruption of bacterial cells by French pressure treatment are frequently inverted or inside out, where the outer side of the plasma membrane is generally on the inside of the vesicle. There is always some concern about the orientation of the membrane regardless of the disruption process so it is necessary to verify orientation by direct measurement.

In addition, membrane orientation is mixed and not 100% inverted or everted. For example, the membrane vesicles may be 80% inverted and 20% everted or 90% everted and 10% inverted. To determine orientation of plasma membrane vesicles, several biochemical tests can be conducted. Inverted vesicles will display ATPase activity, reaction of antibodies against ATPase, activity of D-lactate dehydrogenase, glycerol-3-phosphate dehydrogenase, and succinicate dehydrogenase. Everted or inside-out

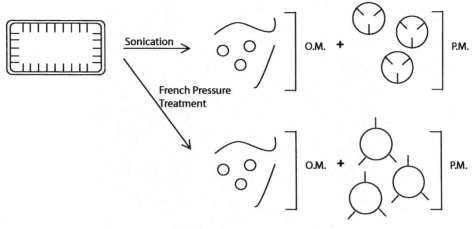

Figure 2.12. Disruption of Gram-negative bacteria is indicated in this figure. In Gram-positive bacteria, orientation of the plasma membrane would be the same as in Gram-negative bacteria.

vesicles will display active transport of sugars and amino acids but no activity with ATPase or NAD-coupled dehydrogenase reactions. However, the NAD-coupled reactions will occur with right-side-out vesicles if the membrane is permeabilized with Triton X-100 or similar surfactant to facilitate entry of the substrates.

Vesicles or blebs derived from plasma membrane are produced by *Neisseria gonorrhoeae* and vesicles are presumed to be produced by other bacteria. As cultures of *Neisseria gonorrhoeae* grow, vesicles are formed on the outer side of the plasma membrane by an unknown mechanism and these vesicles contain DNA. The size of DNA that is carried by a vesicle is unknown; however, the passage of R plasmids from cell to cell has been demonstrated to be mediated by these membrane structures. The form of DNA carried is either linear or circular and the acquisition of nucleic acid by vesicles is nonspecific, as some have been found to contain RNA. It is not known if the vesicle is produced before DNA is released from the bacterial cell or if the vesicle is formed around the DNA that has been translocated to that site. Although more studies need to be conducted, it is apparent that these plasma membrane vesicles are significant in the intercellular DNA transfer of certain bacteria.

The manner of distribution of proteins in a membrane has always been of interest and the distribution of proteins in membrane leaflets can be used to evaluate orientation of membrane vesicles. As seen in Figure 2.13, with freeze–fracture, more integral proteins extend from the inner face of the outer leaflet of the plasma membrane than from the inner face of the inner leaflet of the membrane. As we know, the types of proteins found on the outer face of the plasma membrane include cytochromes, respiratory enzymes, binding proteins for solute transport, signal receptors, structural lipoproteins, and chaperone protein systems. The inner face to the plasma membrane would be expected to have numerous proteins associated with it; however, when the bimolecular lipid leaflet is opened by freeze–fracture, the integral proteins remain associated with the outer leaflet. If one examines the outer surface of bacterial plasma

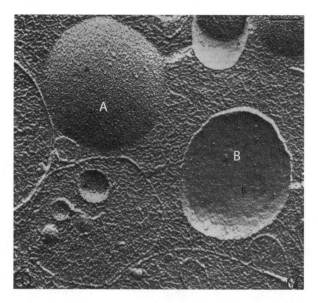

Figure 2.13. Using freeze–fracture electron microscopy to evaluate orientation of membrane vesicles of *E. coli* resulting from osmotic lysis. Freeze–fracture results in a split of the bi-molecular leaflet along a central hydrophobic plane to produce two segments: an inner leaflet or face and an outer leaflet. The intrinsic proteins in the membrane are asymmetrically distributed with protein aggregates associated with the inner side of the membrane leaflet. As a result the outer leaflet is smooth while the inner leaflet is rough. In Gram-positive bacteria, orientation of the plasma membrane is expected to have a similar appearance. The convex vesicle identified as **A** is the inner leaflet of a right-side-out vesicle while **B** indicates the outer leaflet of a concave vesicle. (From Altendorf and Staehelin, 1974.[16] Used with permission.)

membrane of intact cells the pattern of protein distribution is the same as seen with vesicles in Figure 2.13.

## 2.4.4. Adhesion Sites[18–20]

Membraneous structures that connect the outer membrane with the plasma membrane were initially described by M.E. Bayer. These connections, which suggested a fusion between the two membranes in Gram-negative bacteria, are known as adhesion sites or adhesion zones. Bayer suggests that these adhesion sites are comparable to the tight junctions formed by eukaryotic cells to form a seal against extracellular components and gap junctions that lie between adjacent eukaryotic cells to transfer chemicals between cells. It has been suggested that the adhesion sites may be important for translocation of cell wall precursor molecules to the region of wall growth and critical in transfer of double-stranded DNA from certain bacteriophage into the cell. Adhesion sites are demonstrated by electron microscopy when ultrathin sections of

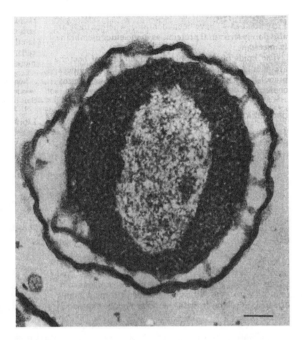

Figure 2.14. Thin sections of *E. coli* showing adhesion sites between plasma membrane and wall. For optimal visualization the cell was plasmolyzed in 20%. The cells were fixed in formaldehyde, stained with OsO$_4$, embedded, and sectioned. (From Bayer, 1968.[18] Used with permission.)

plasmolyzed cells are examined. The image in Figure 2.14 is a bacterial cell with adhesion sites where the plasma membrane is "attached" to the outer membrane. At the adhesion site, the outer membrane is continuous without any evidence of a break or hole being formed. The plasma membrane forms a contact zone with the outer membrane that is 200 to 300 Å wide and there is no evidence of fusion of the two membranes. At the site of adhesion, the plasma membrane becomes amorphous and loses its double track character, which is typical of bacterial membranes. Some are skeptical about these adhesion sites because they have been demonstrated in only a few Gram-negative bacteria: *E. coli, Salmonella typhimurium, Salmonella anatam, Shigella* spp., and *Caulobacter* sp. In addition, adhesion sites appear only in plasmolyzed cells following a chemical fixation process. It would be interesting to determine if adhesion sites are employed to "pull" the plasma membrane into the prosthecae of budding bacteria, *Hyphomicrobium*, or the stalk of *Caulobacter*. Clearly, additional research is required to determine if these membraneous contacts are induced on processing for electron microscopy or if they are indeed a valid structure.

## 2.4.5. Mesosomes[21]

A specific intracellular membrane in bacteria that was considered to be distinct from the plasma membrane was designated by P.C. Fitz-James in 1960 as the mesosome. From examination of thin sections of bacteria by electron microscopy, vesicular or lamellar membranes were observed in thin sections at approximately the region where cross-wall formation was expected to occur. These mesosomes were especially evident in thin sections of Gram-positive bacteria. The membranes of the mesosome were continuous with the plasma membrane and the chemical composition of the mesosome was not distinct from that of the plasma membrane. It has been observed that mesosomes may be produced during fixation of cells, as the volume of cells of *Enterococcus faecalis* may be reduced by 20% following fixation with glutaraldehyde where shrinking cell walls force the plasma membrane to fold into the cytoplasm. It is now considered by most microbial physiologists that mesosomes are a result of cellular perturbations and the invagination of the plasma membrane occurs at sites where the membrane–cell wall adhesion is compromised.

## 2.4.6. Proteoliposomes[22]

In vitro studies utilize proteosomes to examine a variety of membrane functions because the metabolic processes can be separated from transport activities. These membraneous vesicles contain protein and a phospholipid bilayer. To construct a proteoliposome, phospholipids are added to water and a lipid vesicle called liposome is produced. Phospholipids will coalesce when placed in water to form a layered unit with lipid on the interior and phosphate projected into the water phase. Liposomes have been examined by electron microscopy and they have the approximate thickness of a typical membrane. The phospholipid used for the liposome may be obtained from commercial vendors; however, phospholipids isolated from the organism of interest provided the best environment for the protein introduced. The protein is solubilized from the plasma membrane by detergents and this protein covered with detergent is added to the liposome. Although the protein will be slowly introduced into the lipid bilayer through the action of the detergent, it is frequently best to either subject the protein and liposome to freeze–thaw or to sonication. The resulting vesicle that contains protein inserted into the lipid bilayer is referred to as a proteoliposome. These proteoliposomes are about 100 nm in diameter and can be purified by ultracentrifugation procedures used to collect plasma membranes. If the protein inserted into the proteoliposome is associated with uptake transport, the ions or material transported will be concentrated inside the proteoliposome because the vesicle is a closed structure without pores. Proteoliposomes can be collected easily, and the amount of ions concentrated can be determined. If the protein introduced into the proteoliposome is an exporter, substrate and other components can be loaded inside the proteoliposome. The loading of proteins into a proteoliposome is accomplished by adding the chemicals to the buffer at the time the protein is added. Because no other proteins are present, synergistic and antagonistic activities are eliminated and clear results are produced. Proteoliposome studies are required to provide definitive evidence of nutrient transport coupled to movement of $H^+$, $Na^+$, $K^+$, or $Ca^{2+}$. In addition, the only way to demonstrate the proton-pumping activity of a specific respiratory complex is to employ proteoliposomes.

## 2.4.7. Signature Lipids[23]

Analysis of fatty acids extracted from cells have been used to identify bacteria. Because gas chromatography is used for this identification and fatty acids must be converted to their methyl esters, this method is frequently referred to as the FAME (fatty acid methyl ester) procedure. Because fatty acids content of bacteria may vary depending on the chemical and physical environment, it is necessary to grow the bacteria under prescribed conditions. For many bacteria that grow under only limited conditions, cultivation under specified conditions may not be achieved; however, the FAME procedure has been used with considerable success for the speciation of numerous bacteria.

Although total lipid present in a culture may not be used to follow growth or to estimate abundance of prokaryotes, the quantity of a specific lipid in the membranes has proven highly reliable in evaluating prokaryotes in various environments including bioremediation sites and terrestrial subsurface. D.C. White and colleagues developed chemical procedures to focus on signature molecules that function as indicators for specific physiological groups of bacteria and for species within a group. These signature lipid biomolecules come from the membrane, primarily plasma membrane, of bacteria. Total ester-linked phospholipid fatty acid (PLFA) gives a quantitative measure of viable or potentially viable biomass. Within minutes after a bacterial cell dies, phosphatases from the bacterial cell release phosphate from the phospholipid to yield diglycerides. When the PLFA is measured and the amount of diglyceride is quantitated, an estimation of living or potentially living bacteria and an estimation of dead bacteria is made. Other lipid analyses to quantitate specific physiological groups of bacteria include glycolipids for Gram-negative bacteria and phototrophs, lipid A or hydroxy fatty acids from the membranes of Gram-negative bacteria, and lipids containing ether bonds between glycerol and fatty acids in the plasma membranes of archaea.

## 2.5. Membrane Synthesis[24–26]

Membrane integrity is challenged whenever the environment changes, and replacement of membrane proteins or lipids is an ongoing process. Because the lipid and protein molecules making up the membrane structure are not covalently joined, newly synthesized proteins such as transport proteins and cytochromes can be readily inserted into the membrane. Clearly, the stabilization of a membrane by lipophilic and ionic interactions provides greater flexibility for the introduction of proteins than if the membrane structure was secured by covalent bonds. Cell division is not required for proteins to be inserted into plasma membranes.

Owing to the constitutive nature of bacterial plasma membranes, the lipid composition is markedly stable in a given species. *E. coli* grown in a variety of conditions produces 70% to 80% phosphatidyl ethanolamine, 20% to 25% phosphatidyl glycerol, and 5% phosphatidyl glycerol. With respect to fatty acid content of membranes from enterics, palmitic acid accounts for 45%, palmitoleic acid for 35%, *cis*-vaccenic acid makes up 18%, and myristic acid accounts for 2%. Some variation occurs with each phospholipid molecule because different fatty acids are attached to different carbon positions. This statement is supported by the analysis of phosphatidic acid from *E. coli* where the fatty acid at carbon position 1 is 89% palmitic acid, 1% palmitoleic acid, and 10% *cis*-vaccenic acid. At carbon position 2, the fatty acids are 10% palmitic acid, 52% palmitoleic acid, and 38% *cis*-vaccenic acid.  In Gram-positive and Gram-negative bacteria, palmitic acid ($C_{16:0}$) is usually found at position 1 of the phospholipid while the unsaturated fatty acids of palmitoleic acid ($C_{16:1}$) and *cis*-vaccenic acid ($C_{18:1}$) generally occupy position 2 on the glycerol moiety of the phospholipid.  This attachment change would give saturated fatty acids on position 1 a higher melting point while fatty acids in position 2 would have a lower melting point.  However, in *Mycoplasma* unsaturated fatty acids of lower melting points are generally found bound to position 1 while saturated fatty acids with higher melting points are bound to position 2.  Because membrane fluidity is based on rotational freedom of lipids, decreased fluidity occurs in the inner half of the membranes of *Mycoplasma*.  The rationale for this difference between *Mycoplasma* and other bacteria is unclear at this time.

## 2.5.1. Lipid Synthesis

Lipid synthesis is best characterized in *E. coli* and *S. typhimurium*, in which the phospholipids are the product of three separate activities: (1) synthesis of fatty acids, (2) binding of fatty acids onto *sn*-glycerol-3-phosphate, and (3) conversion of polar head groups to produce appropriate phospholipids.  Each of these activities is discussed in the following paragraphs.

### 2.5.1.1. Fatty Acid Biosynthesis

In general, enzymes for fatty acids are found in the cytoplasm while those for phospholipid are bound in the plasma membrane.  To initiate fatty acid synthesis, acetyl-coenzyme A (acetyl-CoA) is converted to malonyl-CoA by combined activities of carboxyl transferase and biotin carboxylase (see Figure 2.15).  The malonyl-CoA molecule is converted to malonyl-ACP by malonyl-CoA:ACP transcyclase.  For the elongation reaction, a small protein known as acyl carrier protein (ACP) carries the intermediates by binding through a sulfhydryl group.  The ACP is a 8.8-kDa protein that is present in *E. coli* at a level of $6 \times 10^4$ molecules/cell, which is about 0.25% of total soluble protein in the cell.  In comparison, CoA is present in the cell at $1 \times$

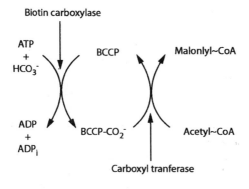

Figure 2.15. Reaction involving $CO_2$ assimilation with the synthesis of malonyl-CoA as a precursor for fatty acids. The reaction for conversion to malonyl-ACP is complex and can proceed by one of three different mechanisms. Carbon dioxide is converted to $HCO_3^-$ by carbonic anhydrase. BCCP, Product of *accβ* gene of lipid synthesis. (From Cronan and Rock, 1996.[24])

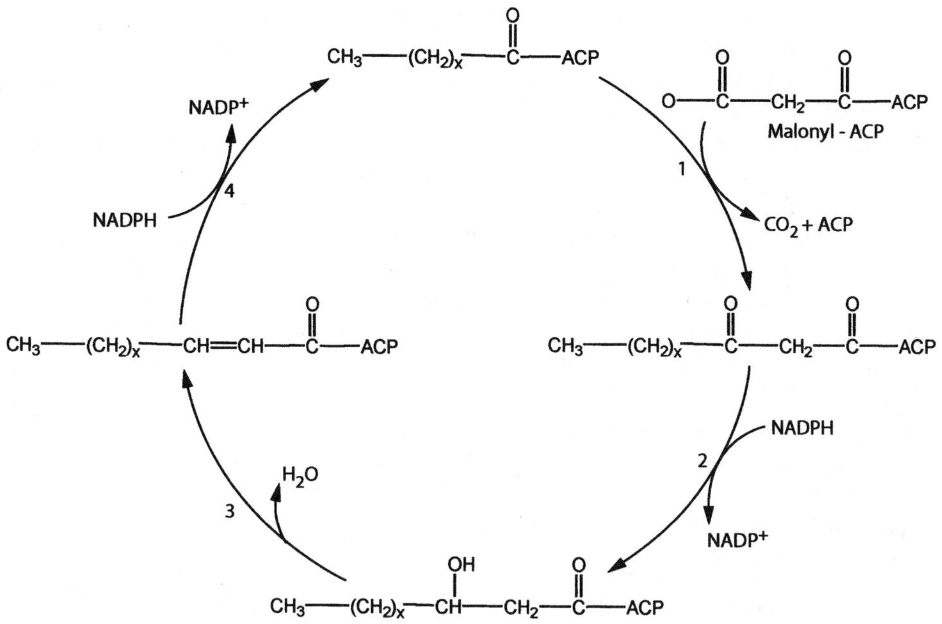

Figure 2.16. Reactions associated with elongation of fatty acids. ACP is attached to the carboxyl group of the fatty acid by a thioester bond. Enzymes are as follows: (1) β-ketoacyl-ACP synthase, (2) β-ketoacyl-ACP reductase, (3) β-ketoacyl-ACP dehydrase, (4) enoyl-ACP reductase. (From Cronan and Rock, 1996.[24])

$10^{12}$ molecules/cell and the synthesizing enzymes appear to be able to recognize either acyl-CoA or acyl-ACP. The elongation of fatty acids is extended by two carbon units at a time and these carbons originate from the three-carbon unit of the malonyl moiety. The reactions for elongation of fatty acids are given in Figure 2.16. There are three isofunctional enzymes (I, II, III) of β-ketoacyl-ACP synthase, with each enzyme form subject to its own regulation. The fatty acids produced are saturated and the length of fatty acid is regulated by the rates of synthesis of the acyl transferases. The apparent reason for the fatty acids in the membrane being $C_{16}$ to $C_{18}$ is the specificity of enzymes for recognizing lipids of this length. Thus, the enzymes for lipid synthesis play an important role in regulating synthesis. The formation of unsaturated fatty acids (e.g., palmitoleate and cis-vaccenate) in E. coli is attributed to a specific dehydratase acting on β-OH acyl-ACP. Odd-numbered fatty acids are produced when the initial condensing activity starts with a three-carbon unit and not a two-carbon moiety. In E. coli, the synthesis of unsaturated fatty acids is a result of β-hydroxydecanol-ACP dehydratase on Ho-10:0 β-hydroxydecanoyl-ACP. Inhibition of β-hydroxydecanol-ACP dehydratase occurs with 3-decanoyl-N-acetyl cysteamine. When this inhibitor is added to the cell, production of saturated fatty acids proceeds but synthesis of unsaturated fatty acids is inhibited. Cerulenin is a fungal product that inhibits fatty acid synthesis by interfering with enzymes involved in elongation of the fatty acid molecule.  Other inhibitors of fatty acid synthesis include thiolactomycin and diazorines.  In mycoplasma, fatty acid synthesis can be inhibited by avidin or *N,N*-dimethyl-4-oxo-2-*trans*-dodecanamide as well as by cerulenin.  These inhibitors are not specific for bacterial lipid synthesis but inhibit mammalian fatty acid synthesis as well.  As a result, these antibiotics are not employed to control bacterial infections in humans.  There is considerable interest in finding other antibiotics that will be capable of selectively inhibiting bacterial lipid synthesis.

The formation of cyclopropane fatty acids in bacteria is a highly specific process that involves modification of lipid after it is synthesized.  It is apparent that cyclopropane fatty acids are not abundant in rapidly growing cells but accumulate when cells enter the stationary phase.  The cyclopropane fatty acid synthase is a soluble enzyme of 80 to 90 kDa that interfaces with the fatty acids attached to phospholipids in the plasma membrane.  The methyl group from *S*-adenosylmethionine is enzymatically added across the *cis*-unsaturated acyl moiety to produce the *cis*-cyclopropane derivative.  Although the cyclopropane fatty acid synthase is highly conserved throughout bacterial species, the cyclopropane fatty acids are not essential for cellular activities and, in fact, have no known role in bacteria.

### 2.5.1.2. Phospholipid Biosynthesis

The first step in phospholipid synthesis is the attachment of fatty acids to positions 1 and 2 of glycerol-3-phosphate (G3P).  Glycerol-3-phosphate acyltransferase catalyzes the esterification of a fatty acid onto position 1 of G3P and 1-acyl-G3P acyltransferase adds another fatty acid onto position 2 of G3P with the production of phosphatidic acid.  This synthesis of phosphatidic acid and subsequent reactions involving synthesis of phospholipids in bacteria is given in Figure 2.17.  The synthesis of CDP-diacylglycerol proceeds from phosphatidic acid by CTP addition and the release of inorganic pyrophosphate by CDP-diacylglycerol synthase.  Phosphatidyl ethanolamine is an important phospholipid for *E. coli* and it is produced from phosphatidyl serine.  Serine is added to CDP-diacylglycerol to produce phosphatidyl serine by phosphatidyl serine synthase.  An interesting aspect of cellular fractionation is that phosphatidyl serine synthase is associated with ribosomes following cell disruption.  The reason is that because the attachment of the phosphatidyl serine synthase to ribosomes occurs through ionic bonding which is an artifact attributed to perturbations of the preparation process.  On addition of the substrate, CDP-diacylglycerol, the enzyme dissociates from the isolated ribosomes and associates with membranes.  Because phosphatidyl serine is an intermediate in production of phosphatidyl ethanolamine, phosphatidyl serine does not accumulate in the membrane.  The production of phosphatidyl glycerol phosphate results from the addition of G3P to CDP-diacyl glycerol.  Phosphatidyl glycerol is produced by phosphatase action on phosphatidyl glycerol phosphate, and condensation of two phosphatidyl glycerol molecules produces diphosphatidyl glycerol.

The synthesis of phospholipids or fatty acids in *E. coli* is not regulated by repression activities and this reflects the lack of clustering of the 26 genes for lipid synthesis.

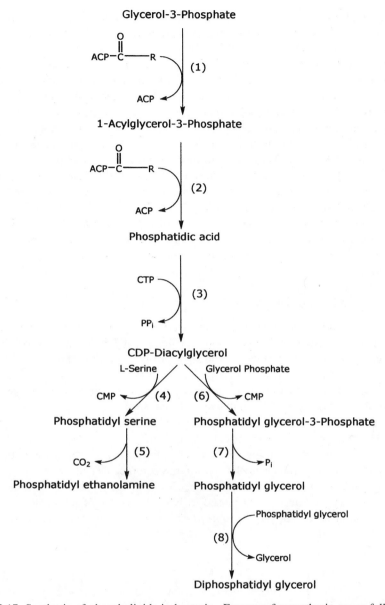

Figure 2.17. Synthesis of phospholipids in bacteria. Enzymes for synthesis are as follows: (1) *sn*-glycerol-3-phosphate acyltransferase (*sn* indicates stereospecifically numbered); (2) 1-acyl-*sn*-glycerol-3-phosphate acyltransferase; (3) phosphatidate cytidylyltransferase; (4) phosphatidyl serine synthase; (5) phosphatidyl serine decarboxylase; (6) glycerol phosphate phosphatidyl transferase; (7) phosphatidyl glycerol phosphate phosphatase; and (8) diphosphatidyl glycerol synthase. Other abbreviations are as follows: $R_1$ and $R_2$ are fatty acyl groups; ACP is acyl carrier protein. (From Cronan and Rock, 1987.[1])

The ratio of phospholipids to proteins in the plasma membrane is constant throughout the cell cycle. Fatty acid synthesis and phospholipid synthesis are coordinately regulated to prevent overproduction of free fatty acids or phospholipids.

A limited amount of information is available concerning the formation of the phospholipid layer on the outer boundary of the plasma membrane. Phospholipids are considered to move from the inner boundary of the plasma to the outer by some sort of a "flip-flop" system. For example, phosphatidyl ethanolamine is on the outer leaflet of the plasma membrane and other phospholipids are on the inner boundary. A specific enzyme known to be on the cytoplasmic side of the membrane decarboxylates the phosphatidyl serine as it is being formed to produce phosphatidyl ethanolamine. The phosphatidyl ethanolamine migrates to the outer leaflet where linear molecules such as MDO become attached to it. Approximately 5% of the head groups of phosphatidyl ethanolamine are attached to MDO which extends into the periplasm. Not all lipids are free to be reoriented across the plasma membrane because that would result in a homogeneous membrane and it is accepted that transmembrane lipid asymmetry exists. The regulatory mechanism for this membrane lipid asymmetry is unknown.

### 2.5.1.3. Biosynthesis of Quinones

Menaquinones and ubiquinones are synthesized from phosphoenol pyruvate and erythrose-4-phosphate, with shikimate being an important precursor. The origin of the various components of the menaquinone structure is given in Figure 2.18. The structure is derived from shikimate, 3-C comes from α-ketoglutarate, a methyl is from *S*-adenosylmethionine, and the prenyl groups are from prenyl pyrophosphate. In the biosynthesis the first part of the pathway is found in the cytoplasm and only the last enzyme before menaquinone production is bound to the membrane. In menaquinones, the prenyl group is the last component added to the structure. The pathway for ubiquinone formation is given in Figure 2.19. Ubiquinone is principally found in aerobes and of particular importance is the use of three molecules of $O_2$ required for the production of one molecule of ubiquinone. In those anaerobes that synthesize ubiquinone, the oxygen source is proposed to be water and unique enzymes would be employed for these reactions. The enzymes for ubiquinone synthesis have a small

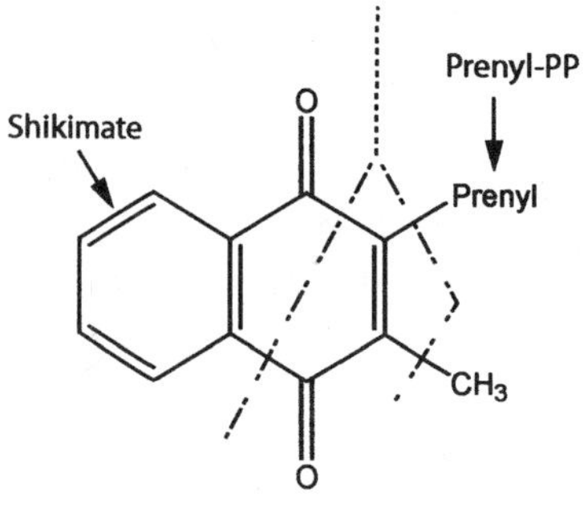

Figure 2.18. Origin of the various components of the menaquinone structure.

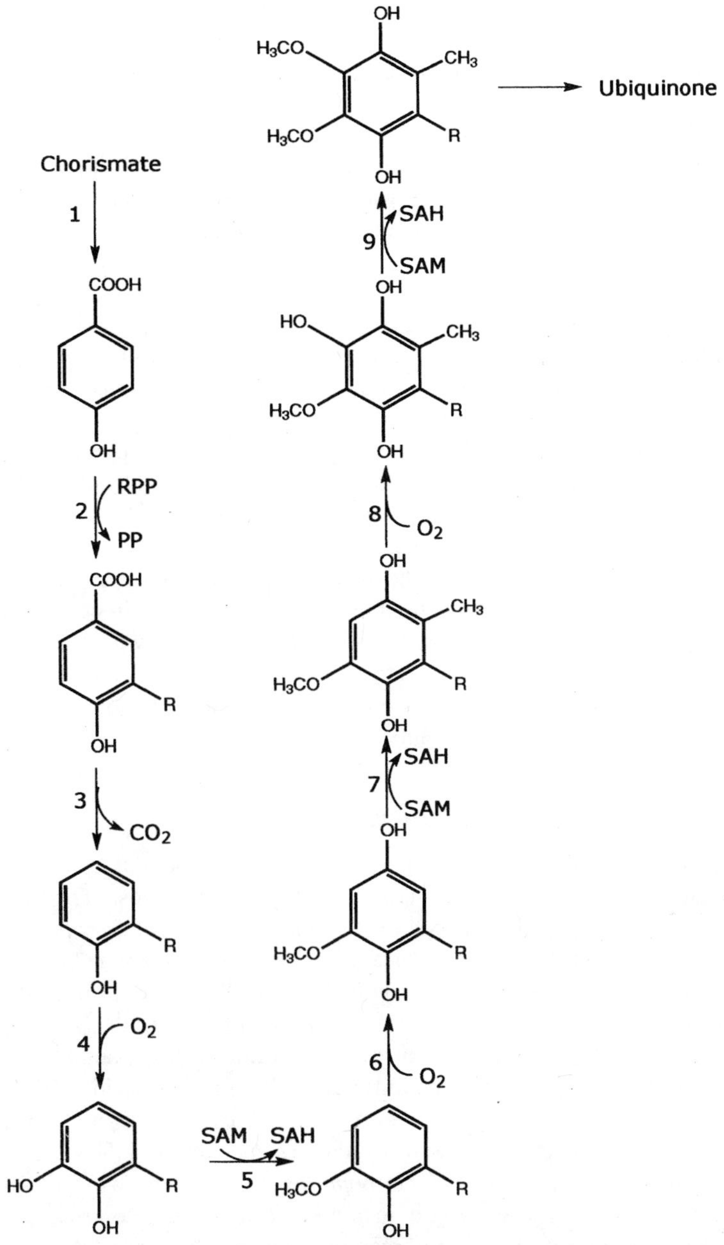

Figure 2.19. Metabolic pathway for the synthesis of ubiquinones. Structures: (A) 4-hydroxy-benzoate, (B) 3-octaprenyl-4-hydroxybenzoate, (C) 2-octaprenylphenol, (D) 2-octaprenyl-6-hy-droxyphenol, (E) 2-octaprenyl-6-methoxyphenol, (F) 2-octaprenyl-6-methoxy-1,4-benzoquinol, (G) 2-octaprenyl-3-methyl-6-methoxy-1,4-benzoquinol, (H) 2-octaprenyl-3-methyl-5-hydroxy-6-methoxy-1,4-benzoquinol, (I) ubiquinol. Enzymes: (1) Chorismate pyruvate lyase, (2) 4-hydroxybenzoate octaprenyltransferase, (3) 3-octaprenyl-4-hydroxybenzoate carboxylase, (4) an oxygenase, (5) a methyltransferase, (6) an oxygenase, (7) a methyltransferase, (8) an oxygenase, (9) 2-octaprenyl-3-methyl-5-hydroxy-6-methoxy-1,4-benzoquinol methyltransferase. SAM, S-Adenosylmethionine. (From Meganathan, 1996.[26])

mass of 20 to 50 kDa and only the first enzyme in the metabolic pathway is soluble while all others are membrane bound. In synthesis of quinones, the prenyl group is the first component added to the structure. There is a difference in synthesis of iso-prenyl phosphate in that the precursors for bacterial synthesis is sugars while for archaea and eukaryotes the precursor is acetate. Details of regulation for production of menaquinones and ubiquinones remain to be established.

## 2.5.2. Turnover of Lipid[1–3]

The phosphate of phosphatidylglycerol turns over with a half-time of 60 min for *E. coli* and 2 hours for *Bacillus subtilis*. For *Staphylococcus aureus*, the phosphatidyl glycerol content in membranes turns over three times in one generation. Liberation of phosphate from P-lipid may be attributed to the action of phosphatases that are present in the periplasm of cells for catabolism of organic-P molecules. However, the phosphate moiety of phosphatidyl ethanolamine does not readily turn over but phosphatidyl ethanolamine is degraded as cell lysis approaches. An important feature of the turnover of the head portion of phosphatidyl glycerol is the production of etha-nolamine phosphate, which is used for the synthesis of lipids important for the sta-bilization of oligosaccharides into the plasma membrane. Fatty acids, primarily from position 1 of glycerol, turnover at a rate of about 5% per generation and new fatty acids are inserted into the plasma membrane to maintain membrane fluidity as the cells undergo a change in the environment.

## 2.5.3. Response to Temperature

In general, bacteria that grow at high temperatures have saturated fatty acids while those that grow at low temperatures have unsaturated fatty acids. The effect of fatty acid content in phospholipids can be seen in Table 2.8. Cells of *E. coli* grown at 17°C have a ratio of saturated fatty acid to unsaturated fatty acid of about 1:2 while at 37°C the ratio of saturated to unsaturated fatty acid is 1:1. The thermal regulation of fatty acids is attributed to the increased rate at which palmitic acid ($C_{16:0}$) is enzymatically converted to *cis*-vaccenic acid ($C_{18:1}$) at low temperatures. For example, when *E. coli* is grown at 37°C it contains 45% palmitic acid, 2% myristic acid, 35% palmitoleic acid, and 28% *cis*-vaccenic acid. However, when *E. coli* is grown at 25°C, the pro-portions of *cis*-vaccenic acid and palmitic acid are changed to 45% and 28%, respec-tively. In addition, when *Thermus thermophilus* has been growing at 48°C and is shifted to 82°C the fatty acid length increases from *iso*-$C_{15}$ to *iso*-$C_{17}$. The fatty acids subject to change in the plasma membrane would be those fatty acids not sta-bilized by association with transmembrane proteins. The presence of unsaturated, branched, or cyclopropane fatty acids increases the fluidity of the hydrocarbon portion of the membrane by decreasing the critical temperature for the liquid–gel phase transition.

Table 2.8. Mid-transition temperatures for selected phospholipids in aqueous environments.[a]

| Fatty acids present | $T_m$ for specified phospholipid (°C) | | | |
|---|---|---|---|---|
| | PC | PE | PG | PA |
| Two molecules of $C_{14}$ | 24 | 51 | 23 | — |
| Two molecules of $C_{16}$ | 41 | 63 | 41 | 67 |
| Two molecules of $C_{18}$ | 58 | — | — | — |
| Two molecules of $C_{18:1}$ | 22 | — | — | — |
| One molecule of $C_{18:0}$ and One molecule of $C_{18:1}$ | 3 | — | — | — |

PE, Phosphatidyl ethanolamine; PA, phosphatidic acid; PC, phosphatidyl choline; PG, phosphatidyl glycerol.
[a]Modified from Jain and Wagner, 1980.[32]

## 2.6. Physical Considerations with Regard to Membranes

The rigid gel state occurs when acyl side chains (fatty acids on a glycerol backbone) are closely packed in an ordered array. With an increase in temperature, the acyl groups display an increase movement and individual phospholipid molecules can rotate around the axis perpendicular to the membrane. As lipid movement increases in the membrane, it is said that the lipids "melt" and display a liquid-crystalline structure. With an increase in fatty acyl movement, the membrane becomes thinner because lipid molecules are not fully extended. In the gel phase, the polar heads of the phospholipids are separated by 6.5 nm but in the fluid or liquid-crystalline phase, these heads of the phospholipids are more tightly packed and separated by only 5.5 nm. The temperature at which the lipid in the membrane shifts from a gel to a liquid-crystal is referred to as $T_m$. As shown in Table 2.8, the fatty acid and the polar head of the phospholipid account for different $T_m$. Saturated fatty acids produce a higher $T_m$ while branched, cyclopropane, or unsaturated fatty acids produce a lower $T_m$. It should be noted that much of the charge on the outer surface of the plasma membrane is neutralized by the binding of divalent cations which are principally magnesium and calcium ions. The charges on the inner leaflet of the plasma membrane attract charged amino acids, peptides, and other cytoplasmic molecules.

### 2.6.1. Membrane Fluidity[1-3]

The structures of the hydrocarbons of the lipids are the primary determinants of the fluid character and the gel-to-liquid-crystalline phase of the membrane. As the length of the fatty acids increases the interactions between adjacent hydrocarbons also increase, which together with the attractive forces between the phospholipids produces a highly organized lipid backbone in the membrane and low fluidity in the membrane. Under these conditions, the freedom of movement of the hydrocarbons is low and high temperatures are required to shift the lipids from the liquid to the crystalline phase. However, the attractive interactions between the hydrocarbons in the membrane is reduced if the fatty acids are short, unsaturated, or contain methyl side chains or cyclopropane groups. With a reduction of lipid–lipid interaction, there is an increase in membrane fluidity and a decrease in temperature for the conversion of the lipids in the transition from the liquid to the gel phase. If the membrane lipid is too fluid, the cohesive character is lost and the membrane dissolves owing to a lack of organization. If the membrane lipid is too rigid, the elastic character of the membrane is lost and membrane rupture occurs with cell growth and division. Optimally, cells maintain a balance between the gel and liquid phases of the membrane lipids. Thus, fatty acids in the plasma membrane are essential for maintaining a fluid character of the membrane at temperatures required for growth. Fatty acids vary considerably among species of bacteria with branched fatty acids being $C_{15}$ while linear fatty acids are $C_{16}$ or $C_{18}$ in length. The cell wall–less bacteria have served as excellent models to study the effect of fatty acids on membrane fluidity because membranes can be synthesized in vivo with different fatty acids. The cell wall–less forms of bacteria are unable to synthesize long-chain fatty acids and will incorporate fatty acids from the culture medium into the membrane structure. McElhaney and colleagues have provided excellent experimentation to expand our understanding of membrane fluidity and they suggest that *Acholeplasma* can grow with only one tenth of the membrane lipid in the fluid state.

Cholesterol has been implicated as a membrane stabilizer and in the cell wall–less bacteria, cholesterol is taken up from the media with distribution to both faces of the lipid bilayer. The hydroxyl at the 3β-position of cholesterol secures the sterol molecule to the polar surface of the membrane. The linear cholesterol molecule is oriented parallel to the hydrocarbon chains of the phospholipids in the membrane, thereby enabling greater freedom of movement by the fatty acids of the phospholipids. Cholesterol has a condensing effect on membranes at temperatures greater than the thermal phase transition and prevents cooperative crystallization of hydrocarbons at temperatures below that for phospholipid phase transition. While various bacteria use fatty acids to regulate membrane fluidity, the cell wall–less forms use cholesterol for this purpose. Cholesterol may play a secondary role in the membrane for activation of lipid-dependent enzymes.

The manner of distribution of proteins in a membrane has always been of interest. While the fluid mosaic model for membranes as proposed by Singer and Nicolson in 1972 applies to plasma membranes in prokaryotes, the plasma membrane may have proteins aggregated and proteins are not uniformly dispersed in the lipid. A highly stylized model indicating the topography of membrane proteins is given in Figure 2.20. Proteins appear as "islands" in the model where the "sea" of lipid is somewhat exaggerated. Perhaps this arrangement of proteins reflects some related functional activities in which these protein aggregates are needed for a specific process. For example, there are the cytochromes and dehydrogenase complexes associated with

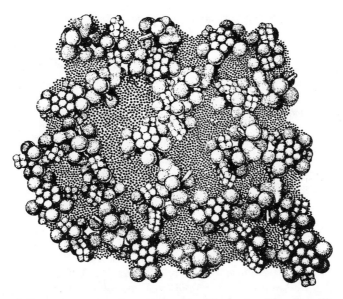

Figure 2.20. Model of membrane surface proposed by Sjöstrand and Barajas with protein islands interspersed in lipids. (Reprinted from Sjöstrand, F.S. 1973. The problems of preserving molecular structure of cellular components in connection with electron microscopic analysis. *Journal of Ultrastructure Research* 55:271–280. Copyright 1973, with permission from Elsevier.)

bacterial respiration, binding proteins associated with the ABC transport systems, transporters for various nutrient uptake systems, $F_0$ subunits of ATPase, chaperones and structural proteins for pili or flagella, and regions of membrane where only lipid is present. It has also been suggested that as membranes are shifted from an optimal temperature for growth to a lower temperature at which cells can no longer divide there may be a rearrangement of proteins in the membrane. That is, as lipids "freeze," the lipid molecules may coalesce and shift proteins into an abnormal position which results in a loss of function. Clearly, much remains to be understood concerning the membrane architecture and synthesis.

## 2.6.2. Membrane Asymmetry[14,27]

Initially, it was considered that lipids making up the bimolecular leaflet were the same on both faces of the membranes; however, this is not the case. From various observations, it is clear that the placement of lipids within the membrane is such that the inner and outer faces of the plasma membrane are not identical at the molecular level. This lack of uniformity of lipids in the plasma membrane has been designated by S. Razin as "transbilayer lipid asymmetry." Careful examination of thin sections made from the plasma membrane of *Mycobacterium phlei* reveal that the electron-dense outer layer was thicker than the electron-dense inner layer. This is assumed to reflect

Table 2.9. Lipid asymmetry in plasma membranes of prokaryotic cells.

| Organism | Preferential outer side | Preferential inner side | Equal distribution |
|---|---|---|---|
| *Escherichia coli*[a] | — | PE | — |
| *Bacillus megaterium*[a] | — | PE | — |
| *Micrococcus lysodeikticus*[a] | PG | PI | DPG |
| *Acholeplasma laidlawii*[b] | DGDG | GPMDGD | DPG |
| | MGDG | PG | GPDGDG |

PE, Phosphatidyl ethanolamine; PI, phosphatidyl inositol; PG, phosphatidyl glycerol; DPG, diphosphatidyl glycerol; DGDG, diglucosyl diglyceride; MGDG, monoglucosyl diglyceride; GPMDGD, glycerolphosphoryl monoglucosyl diglyceride; GPDGDG, glycerolphosphoryl diglucosyl diglyceride.
[a]Modified from Op den Kamp, 1979.[27]
[b]Modified from Razin, 1981.[14]

the presence of phosphatidyl inositol mannosides located on the outer surface of the plasma membrane. *Mycoplasma* and related cell wall–less forms have sugar-containing lipids referred to as glycolipids that contain one to five sugar residues of glucose and galactose. As determined by antibody reactions, these glycolipids appear preferentially located on the outer leaflet. In prokaryotes that contain phosphatidyl ethanolamine, these charged lipids are generally found on the inner leaflet of the plasma membrane (see Table 2.9). Proteins in the cytoplasm readily interact with phosphatidyl ethanolamine and other phospholipids on the inner side of the plasma membrane to provide a close union between the membrane and the cellular proteins for the various signaling and transfer activities. Additional research is needed to clarify the synthesis of lipid asymmetry in plasma membranes.

## 2.7. Destabilization of Membranes

When lipids are in the gel–liquid crystalline state, disruption of the plasma membrane can be accomplished by enzymatic degradation of lipids or disruption of the bimolecular leaflet. Proteases will digest surface peptide loops of integral proteins but will not attack the transmembrane peptide loops. Unsaturated fatty acids will react with hydrogen peroxide and oxygen free radicals to produce a lipid hydroperoxide and this process may lead to damage of membrane structure. A detoxification process of lipid peroxides is the regeneration of the lipid through reaction with alkyl hydroperoxidases. Without covalent bonds to hold the lipids in the bimolecular leaflet, disruption of the plasma membrane can be accomplished by various surface active physical agents that have a soap-like effect in disrupting membrane structure. Solubilization of proteins from membranes is discussed in Section 1.5.2.2. of Chapter 1. The forces holding the plasma membrane are distinguished from those with the outer membrane because addition of lipophilic reagents solubilize the plasma membrane while ethylenediamine-$N,N,N',N'$-tetraacetic acid (EDTA) treatment solublilizes the outer membrane.

## 2.7.1. Membrane-Active Chemicals

Cationic compounds are the most effective detergents with bactericidal activity. Non-ionic detergents have no charge and do not have any significant antibacterial activity, and some of these compounds serve as nutrients for cells. Anionic detergents carry a negative charge and lack antibacterial activity because the molecule is repelled by the negative charge on the cell. Sodium stearate or sodium palmate are examples of anionic detergents and they act to emulsify dirt particles including bacterial cells. The cationic detergents have a net positive charge on the molecule, as is the case with polymyxin B, and at only a few parts per million will distort the organization of the membrane structure. The most common of the cationic detergents are the quaternary ammonium compounds and examples are given in Figure 2.21. Both benzalkonium chloride and cetrimide have a long-chain alkyl group at one end of the molecule and this aliphatic portion of the molecule promotes the interaction of the positively charged segment of the molecule with lipids in the membrane. Because the quaternary ammonium compounds have bactericidal action against both Gram-negative and Gram-positive bacteria they are often used as disinfectants.

Polymyxin B is one of the chemicals widely used to control bacteria and it damages the integrity of the plasma membrane. *Bacillus polymxa* produces several different cyclic polypeptide antibiotics and the structure of polymyxin B is given in Figure 2.21. There is an unusual amino acid, $\alpha,\gamma$-diaminobutyric acid, in polymyxin B and this basic amino acid carries a positive charge at physiological pH levels. An aliphatic moiety of polymyxin B directs the charged portion of the molecule toward the lipid-rich membranes. The charged segment of polymyxin B has detergent-like activity and disrupts the membrane with the release of cytoplasmic constituents. Polymyxin B complexes with negatively charged or weakly charged phosphatides in the membrane but not with the highly positively charged phosphatidyl choline. Because the mode of action of polymyxin B is membrane solubilization, prokaryotic cells are sensitive to this antibiotic even under conditions in which cells are not growing. Another chemical that would disorganize membranes is amphotericin B. However, amphotericin B has an affinity for sterols but because few bacteria contain cholesterol, it is not important as a chemotheraputic agent to control prokaryotes.

## 2.7.2. Hydrolytic Enzymes

The hydrolysis of the phospholipids is accomplished by a series of enzymes that recognize specific bonds in the molecule. The four most common enzymes are lipase $A_1$ and $A_2$ plus phospholipases C and D. The sites of hydrolysis for these enzymes are as indicated in Figure 2.22. Phospholipase C is produced by *Clostridium perfringens*, the causative agent of gas gangrene, and is also known as $\alpha$-toxin, or lecithinase. Because this phospholipase can hydrolyze lecithin, sphingomyelin, and other phospholipids in the membranes of animal cells, it is a necrotizing toxin that produces considerable damage to the tissues. The production of phospholipase C may not be detrimental to *Clostridium perfringens* because it is synthesized at the time of endospore formation and endospores are highly resistant to enzyme hydrolysis. Phospho-

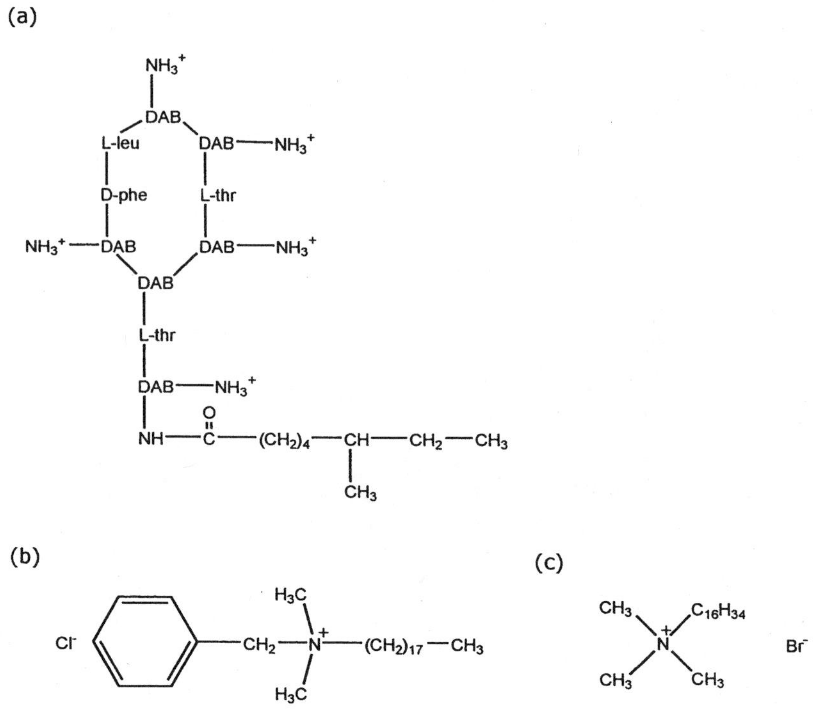

(a)

(b)

(c)

Figure 2.21. Structures of chemicals active against membranes. (a) Polymyxin B has a ring containing basic charges on DAB ($\alpha,\gamma$-diaminobutyric acid) and an aliphatic side chain. Examples of quaternary ammonium compounds with a positively charged region and a lipophilic side chain are (b) benzalkonium chloride or zephiran and (c) cetrimide, also known as CTAB.

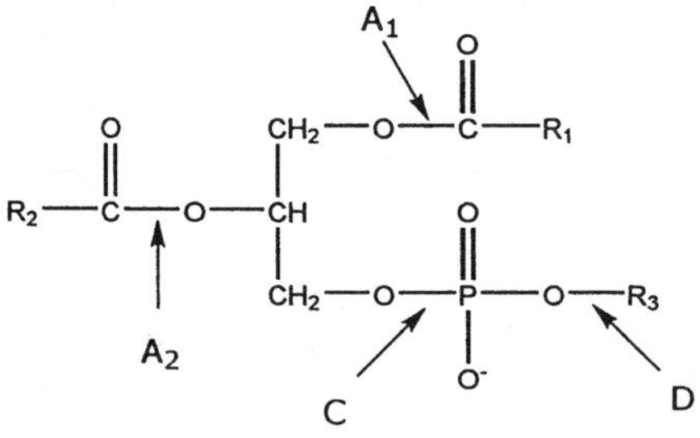

Figure 2.22. Specificity of enzymes attacking phospholipids. Lipases $A_1$ and $A_2$ and phospholipases C and D hydrolyze specific bonds at the *arrows*.

lipase D, like phospholipase C, attacks the phosphate ester bond and there is high specificity for the substrate. Lipases $A_1$ and $A_2$ are not highly specific in that they will hydrolyze fatty acids of various carbon lengths. Although various types of phosphatases reside in the periplasm, phosphate is not released from the outer plasma membrane when cells are growing. Hydrolysis of phospholipids with the destruction of the biomolecular leaflet occurs when bacterial viability is lost.

# References

1. Cronan, J.E., Jr. and C.O. Rock. 1987. Biosynthesis of membrane lipids. In Escherichia coli *and* Salmonella typhimurium: *Cellular and Molecular Biology*, 1st edit., Vol. I (F.C. Neidhardt, J.L. Ingraham, K.B. Low, B. Magasanik, M. Schaechter, and H.E. Umbarger, eds.). American Society for Microbiology Press, Washington, D.C., pp. 474–497.

2. Kadner, R.J. 1996. Cytoplasmic membrane. In Escherichia coli *and* Salmonella: *Cellular and Molecular Biology*, 2nd edit., Vol. I (F.C. Neidhardt, J.L. Ingraham, E.C.C. Lin, K.B. Low, B. Magasanik, W.S. Reznikoff, M. Riley, M. Schaechter, and H.E. Umbarger, eds.). American Society for Microbiology Press, Washington, D.C., pp. 58–87.

3. Engelman, D.M., A. Goldman, and T.A. Steitz. 1982. The identification of helical segments in the polypeptide chain of bacteriorhodopsin. In *Methods in Enzymology*, Vol. 88. Biomembranes, Part 1 (L. Packer, ed.). Academic Press, New York, pp. 81–88.

4. Dowhan, W. 1997. Molecular basis for membrane phospholipid diversity: Why are there so many lipids? *Annual Review of Biochemistry* 66:199–232.

5. Koga, Y., M. Nishihara, H. Morii, and M. Akagawanatsushita. 1993. Ether polar lipids of methanogenic bacteria: Structures, comparative aspects, and biosynthesis. *Microbiology Reviews* 57:164–182.

6. Langworthy, T.A. 1985. Lipids of archaebacteria. In *The Bacteria*, Vol. VIII: *Archaebacteria* (C.R. Woese and R.S. Wolfe, eds.). Academic Press, New York, pp. 459–497.

7. Sprott, G.D. 1989. Inorganic ion gradients in methanogenic archaebacteria: A comparative analysis to other prokaryotes. In *Metal Ions and Bacteria* (T.J. Beveridge and R.J. Doyle, eds.). John Wiley, New York, pp. 91–120.

8. Ratledge, C. and S.G. Wilkerson. 1988. *Microbial Lipids*, Vol. 1. Academic Press, New York.

9. Steiner, M. and S. Steiner. 1988. Distribution of lipids. In *Microbial Lipids*, Vol. 1 (C. Ratledge and S.G. Wilkerson, eds.). Academic Press, New York, pp. 83–116.

10. Sahm, H., M. Rohmer, S. Bringer-Meyer, G.A. Sprenger, and R. Welle. 1993. Biochemistry and physiology of hopanoids in bacteria. *Advances in Microbial Physiology* 35:247–273.

11. Armstrong, G.A. 1997. Genetics of eubacterial carotenoid biosynthesis. *Annual Review of Microbiology* 51:629–659.

12. Epstein, W. 1990. Bacterial transport ATPases. In *The Bacteria*, Vol. XII (T.A. Krulwich, ed.). Academic Press, New York, pp. 87–110.

13. Fillingame, R.H. 1990. Molecular mechanics of ATP synthesis by $F_1F_0$-type $H^+$-transporting ATP synthase. In *The Bacteria*, Vol. XII (T.A. Krulwich, ed.). Academic Press, New York, pp. 345–391.

14. Razin, S. 1981. The mycoplasma membrane. In *Organization of Prokaryotic Cell Membranes*, Vol. I (B.J. Ghosh, ed.). CRC Press, Boca Raton, FL, pp. 165–250.

15. Rydman, P.S. and D.H. Bamford. 2002. Phage enzyme digest peptidoglycan to deliver DNA. *ASM News* 68:330–335.

16. Altendorf, K.H. and L.A. Staehelin. 1974. Orientation of membrane vesicles from *Escherichia coli* as detected by freeze-cleave electron microscopy. *Journal of Bacteriology* 117:888–899.

17. Dreiseikelmann, B. 1994. Translocation of DNA across bacterial membranes. *Microbiological Reviews* 58:293–316.

18. Bayer M.E. 1979. The fusion sites between outer membrane and cytoplasmic membrane of bacteria: Their role in membrane assembly and virus infection. In *Bacterial Outer Membranes* (M. Inouye, ed.). John Wiley, New York, pp. 167–202.

19. Bayer, M.E. 1968. Areas of adhesion between wall and membrane of *Escherichia coli*. *Journal of General Microbiology* 53:395–404.

20. Oliver, D.B. 1996. Periplasm. In Escherichia coli *and* Salmonella: *Cellular and Molecular Biology*, 2nd edit., Vol. I (F.C. Neidhardt, J.L. Ingraham, E.C.C. Lin, K.B. Low, B. Magasanik, W.S. Reznikoff, M. Riley, M. Schaechter, and H.E. Umbarger, eds.). American Society for Microbiology Press, Washington, D.C., pp. 88–103.

21. Higgins, M.L., L.C. Parks, and L. Daneo-Moore. 1981. The mesosome. In *Organization of Prokaryotic Cell Membranes*, Vol. II (B.K. Ghosh, ed.). CRC Press, Boca Raton, FL, pp. 75–94.

22. Kaback, H.R. 1990. Active transport: Membrane vesicles, bioenergetics, molecules, and mechanisms. In *The Bacteria*, Vol. XII (T.A. Krulwich, ed.). Academic Press, New York, pp. 151–202.

23. White, D.C. and D.B. Ringelberg.1998. Signature lipid biomarker analysis. In *Techniques in Microbial Ecology* (R.S. Burlage, R. Atlas, D. Stahl, G. Geesey, and G. Sayler, eds.). Oxford University Press, New York, pp. 255–272.

24. Cronan, J.E., Jr. and C.O. Rock. 1996. Biosynthesis of membrane lipids. In Escherichia coli *and* Salmonella: *Cellular and Molecular Biology*, 2nd edit., Vol. I (F.C. Neidhardt, J.L. Ingraham, E.C.C. Lin, K.B. Low, B. Magasanik, W.S. Reznikoff, M. Riley, M. Schaechter, and H.E. Umbarger, eds.). American Society for Microbiology Press, Washington, D.C., pp. 612–636.

25. Mindich, L. 1973. Synthesis of bacterial membranes. In *Bacterial Membranes and Walls* (L. Leive, ed.). Marcel Dekker, New York, pp. 1–36.

26. Meganathan, R. 1996. Biosynthesis of the isoprenoid quinones menaquinone (vitamin $K_2$) and ubiquinone (coenzyme Q). In Escherichia coli *and* Salmonella: *Cellular and Molecular Biology*, 2nd edit., Vol. I (F.C. Neidhardt, J.L. Ingraham, E.C.C. Lin, K.B. Low, B. Magasanik, W.S. Reznikoff, M. Riley, M. Schaechter, and H.E. Umbarger, eds.). American Society for Microbiology Press, Washington, D.C., pp. 642–656.

27. Op den Kamp, J.A.F. 1979. Lipid asymmetry in membranes. *Annual Review of Biochemistry* 48:47–72.

28. Kyte, J. and R.F. Doolittle. 1982. A simple method for displaying the hydropathic character of a protein. *Journal of Molecular Biology* 157:105–132.

29. Engelman, D.M., T.A. Steitz, and A. Goldman. 1986. Identifying nonpolar transbilayer helices in amino acid sequences of membrane proteins. *Annual Review of Biophysics and Biophysical Chemistry* 15:321–353.

30. Gel'man, N.S., M.A. Lukoyanova, and D.N. Ostrovskii. 1967. *Respiration and Phosphorylation of Bacteria*. Plenum Press, New York.

31. Odom, J.M. and H.D. Peck, Jr. 1981. Localization of dehydrogenases, reductases and electron transfer components in the sulfate-reducing bacterium *Desulfovibrio gigas*. *Journal of Bacteriology* 147:161–169.

32. Jain, M.K. and R.C. Wagner. 1980. *Introduction to Biological Membranes*. John Wiley, New York.

33. Smith, P.F. 1988. Archaebacteria and other specialized bacteria. In *Microbial Lipids*, Vol. 1 (C. Ratledge and S.G. Wilkerson, eds.). Academic Press, New York, pp. 489–553.

34. Sjöstrand, F.S. 1973. The problems of preserving molecular structure of cellular components in connection with electron microscopic analysis. *Journal of Ultrastructure Research* 55:271–280.

35. Funnell, B.E. 1996. *The Role of the Bacterial Membrane in Chromosome Replication and Partition*. R.G. Landes, Austin, TX.

# Additional Reading

Luria, S.E. 1960. The bacterial protoplasm: composition and organization. In *The Bacteria*, Vol. I (I.C. Gunsalus and R.Y. Stanier, eds.). Academic Press, New York, pp. 1–34.

## *Lipids*

De Rosa, M., A. Gambacorta, and A. Glizzi. 1986. Structure, biosynthesis, and physiochemical properties of archaebacterial lipids. *Microbiology Reviews* 50:70–80.

## *Proteins*

Salton, M.R.J. 1974. Membrane associated enzymes in bacteria. *Advances in Microbial Physiology* 11:213–283.

Salton, M.R.J. 1971. Bacterial membranes. *CRC Reviews in Microbiology* 1:161–190.

Senior, A.E. 1990. The proton-ATPase of *Escherichia coli*. *Current Topics in Membranes and Transport* 23:135–151.

Suzuki, K., M. Goodfellow, and A.G. O'Donnell. 1993. Cell envelopes and classification. In *Handbook of New Bacterial Systematics* (M. Goodfellow and A.G. O'Donnell, eds.). Academic Press, New York, pp. 195–250.

# 3
# Cell Wall Structure

The modern microbiologist has come to accept "diversity" as one of the central features of microbial life. M.R.J. Salton, *The Bacteria*, Vol. I, 1960

## 3.1. Introduction

Most prokaryotic cells have a rigid structure external to the plasma membrane called the cell wall. A notable exception is the unique group collectively referred to as the cell wall–less bacteria, which are commonly given the suffix of "plasma" (i.e., *Acholeplasma*, *Mycoplasma*, *Spiroplasma*, *Ureaplasma*). In one cell wall–less genus (*Spiroplasma*), a 55-kDa protein is found associated with the plasma membrane to provide osmotic stability for the cell. The cell wall in prokaryotes is important to protect the cell against osmotic rupture, to contribute to cell form, and to participate in cell division. The cell wall composition accounts for the distinction of the two major staining groups: Gram-positive bacteria and Gram-negative bacteria. Evaluation of these cell walls by electron microscopy reveals that the cell wall of Gram-positive is a thick amorphous structure that appears as a single layer. In Gram-positive bacteria, approximately 80% of the cell wall consists of peptidoglycan while the cell wall of Gram-negative bacteria is comprised of a multilayered structure consisting of a thin layer of peptidoglycan and an outer membrane positioned exterior to the peptidoglycan. Members of the archaea also have a cell wall and according to recent literature, it is evident that these organisms do not have a typical peptidoglycan. However, the extent to which the archaeal cell walls contain complex chemical components influences the differential staining with subsequent separation of the *Archaea* into Gram-positive and Gram-negative cell types.

The focus of this chapter is to addresses the topic of the rigid cell wall structure in prokaryotes. Structural features of the peptidoglycan macromolecule are examined to reveal the chemical basis for the rigidity of the cell wall. Because the peptidoglycan composition of many different bacteria has been examined, the review of variable and essential units in the peptidoglycan of bacteria is approached from a comparative perspective. Included in this chapter is a presentation of the synthesis of peptidoglycan and the destabilizing activities of antibiotics. Although not much is known about the

---

**Box 3.1**
**The Prokaryotic Wall: A Structure or a Region**

Science requires the use of specific words and phrases for description; however, the outermost structure of prokaryotic cells remains to be designated by a term that is applicable to all cells. Early microscopic observation revealed a thick stained structure defining the cell perimeter and the designation of it as a wall seemed appropriate. Thin sections of bacteria observed through the electron microscope revealed structural differences of the bacterial cell "walls." Biochemical analysis reveals that at the molecular level there were numerous differences in the cell "walls" of Gram-positive bacteria, Gram-negative bacteria, and archaea. The requirement of a wall to provide cell stability and shape is not absolute because unique cellular shapes are observed in the cell wall–less bacteria and archaea. In terms of function, the presence of a cell wall is considered important to prevent the rupture of the protoplast, but to paraphrase Robert Frost in "Mending Wall," it has a role in keeping items out as well as keeping items inside the wall. The molecular diversity seen in the cell wall region of prokaryotes implies the multifaceted role of a "wall" and complicates the designation by a specific term. Some scientists have elected to refer to the cell surface as the *envelope*, which is a term that has relatively little specificity for prokaryotes. Until there is a collective acceptance of terms, students of microbiology will be exposed to numerous expressions to describe cell walls.

---

structures of archaeal cell walls, chemical features of the cell wall in archaea are presented. The chapter concludes with a discussion of the chemistry and distribution of S-layers that are found in the cell walls of many prokaryotes.

## 3.2. Peptidoglycan: An Essential Structure in Bacteria[1]

Bacteria have a cell wall that contains a covalent structure most commonly referred to as peptidoglycan, although an alternate name that is frequently used is murein. Other terms used to describe the structural segment of the cell wall include glycopeptide, muropeptide, mucopeptide, glucosamino-peptide, and mucopolymer. Two fundamental components of the peptidoglycan include peptides and glucans or sugars. The basic unit of the sugars in the peptidoglycan is a dimer of *N*-acetylglucosamine (NAG) and *N*-acetylmuramic acid (NAM) (see Figure 3.1). *N*-Acetyl muramic (from the Latin *murus*, meaning wall) acid is present throughout the groups of bacteria but is not found in archaea or in eukaryotic cellular systems. NAG is broadly distributed throughout biology with polymers of NAG making up chitin which is found in cell walls of fungi and the exoskeleton of crustaceans. The presence of the acetyl group

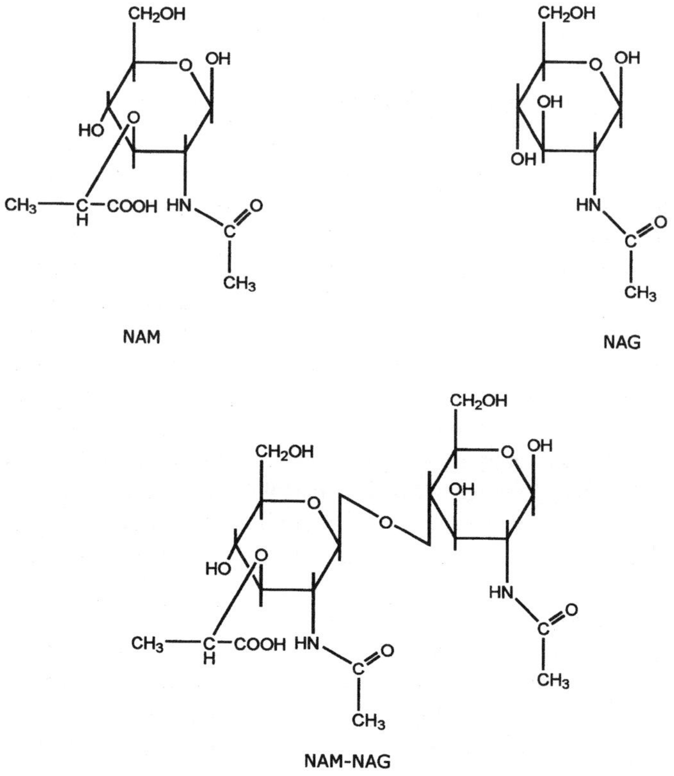

Figure 3.1. Structures of *N*-acetylmuramic acid (NAM), *N*-acetylglucosamine (NAG), and the NAM–NAG dimer. Between NAM and NAG in the dimer and between the linear dimers is a β-1,4 glucosidic bond.

on N of the amino sugars is a common occurrence in bacteria and has a role in balancing the charge on that region of the molecule and preventing the condensation of other molecules onto the amino moiety of these amino sugars. The rigidity of the peptidoglycan structure is apparent from the electron micrograph of the isolated peptidoglycan from *Bacillus stearothermophilus*, shown in Figure 3.2. The construction and stability of this structural girdle are some of the primary subjects of this chapter.

A few groups of bacteria have a cell wall but do not have peptidoglycan. The obligately intracellular *Chlamydia*, which stains Gram-negative, has an outer membrane but no peptidoglycan. Because *Chlamydia* grow only inside animal cells and are not exposed to dilute growth environments, there is less of a requirement to prevent cell lysis because the osmotic differences between the host cell and *Chlamydia* are not great. *Isosphera, Planctomyces,* and *Pirella* are genera of budding bacteria that lack muramic acid and therefore do not have a true peptidoglycan layer. These budding organisms have a protein structure that provides strength to the cell wall. The peptidoglycan as a structure providing stability against osmotic imbalance is found in

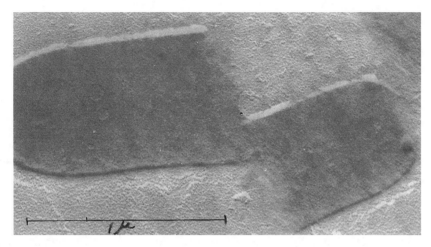

Figure 3.2. Peptidoglycan structure purified from *Bacillus stearothermophilus*. Specimen was shadowed with gold. Bar = 1 μm (Courtesy of Sandra Barton.)

nearly all bacteria. Perhaps the explanation for bacteria maintaining the genetic profile for peptidoglycan production is related to the efficiency in synthesis and to the unique molecular stability provided by the peptidoglycan structure.

## 3.3. Chemistry of Peptidoglycan[2]

The attachment of each sugar is via a β-1,4 glucosidic bond that results in a linear molecule containing 3 to 125 dimers of NAG and NAM. The linear NAG–NAM structure is sometimes called a glucan and there is no apparent difference in the length of the glucan in Gram-positive or Gram-negative bacteria. NAG functions as a spacer molecule while NAM provides the site for peptide attachment to the glycan chain. Peptides are linked by a peptide bond into the carboxyl of the three-carbon unit that extends from carbon number 3 of NAM (see Figure 3.3).

Peptidoglycan isolated from bacterial cell walls contains a small number of amino acids uniquely bonded to the NAM. There are four amino acids in this peptide and the first amino acid usually is L-alanine and the fourth is always D-alanine. A few variations have been found at amino acid position 2 and the greatest number of variations occur at amino acid position 3. The second amino acid is D-glutamic acid, with, in a few cases, addition of small molecules (i.e., amine or glycine) to the free carboxyl group adjacent to the amino group. The α-carboxyl of D-glutamic acid is available because the peptide bond to the third amino acid involves the γ-carboxyl of glutamic acid. In the four amino acid peptide, the D-alanine has a free carboxyl group. The third amino acid in the peptide from NAM must have the capability of bonding to the free carboxyl of an adjacent D-alanine molecule. A distinctive feature of the peptide is the unique amino–carboxyl bond formed. The conventional peptide bond

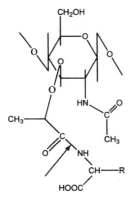

Figure 3.3. Attachment of amino acids onto the NAM unit. *Arrow points to the peptide bond between the three-carbon structure of NAM and L-alanine of the peptide. R indicates other amino acid residues attached to L-alanine.*

found throughout biology connects two amino acids of the L-form through an amino–carboxyl attachment (see Figure 3.4). However, the carboxyl–amino bonds between D- and L-forms of amino acids form linkages other than the common peptide bond. The participation of amino acids in the construction of the peptidoglycan is depicted in Figure 3.5. Different peptidoglycan structures are found as a result of substitutions of diamino acids found at the third position of the peptide. The most common diamino acids in the bacterial peptidoglycan are lysine and diaminopimelic acid.

## 3.3.1. Lysine-Containing Structure

In the peptidoglycan, lysine is present in the L-form with an epsilon (ε) amino group covalently bonded to the carboxyl group of D-alanine in an adjacent peptide. An amine group couples to the carboxyl group of D-glutamic acid. The L-alanine-D-glutamic acid-L-lysine-D-alanine type of peptide is found in *Staphylococcus aureus* and *Streptococcus pyogenes*. Both *Micrococcus lysodeikticus* and *Sarcina lutea* have a lysine-containing peptide; however, they have glycine substituted for amine added onto D-glutamic acid (see Figure 3.6).

## 3.3.2. Diaminopimelic Acid–Containing Structure

Diaminopimelic acid (DAP) is a seven-carbon amino acid that is a precursor for lysine formation and differs from lysine in that it has an additional carboxyl group. DAP is highly unusual with distribution primarily limited to prokaryotic organisms. Like

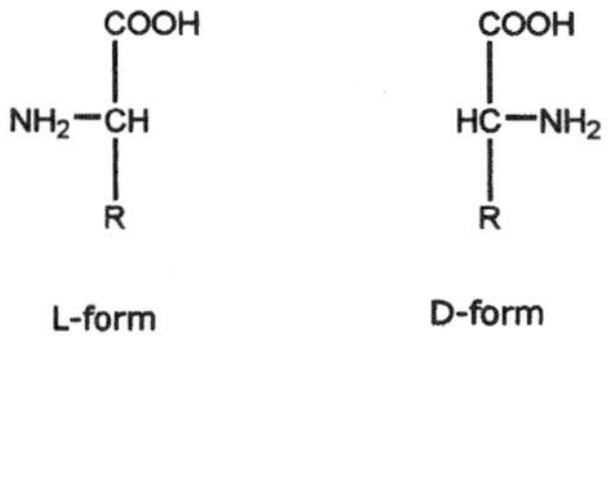

Figure 3.4. Structural distinctions between L- and D-forms of amino acids.

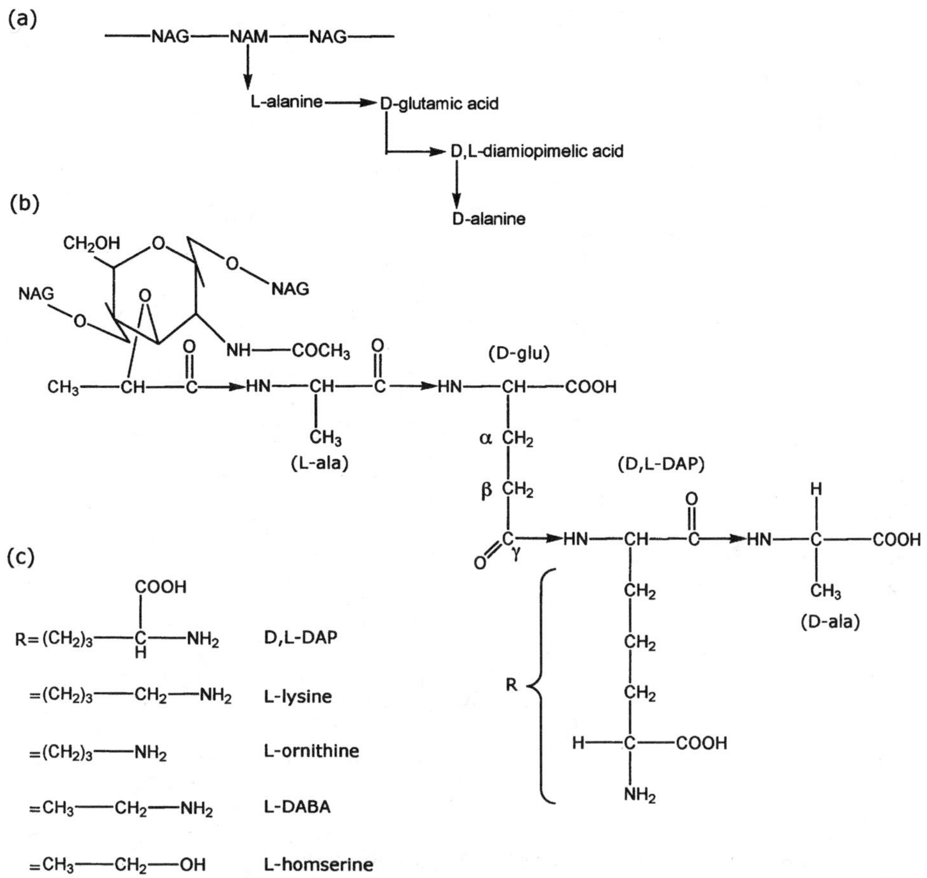

Figure 3.5. Peptides attached to glycan of peptidoglycan. The peptide in peptidoglycan structure of *E. coli* consists of four amino acids with model in (**a**) and molecular structure in (**b**). In (**c**) are the linkages between the amino acids and possible amino acids that may be found at position 3 of the peptide. Note the γ-linkage between D-glutamic acid and D,L-diaminopimelic acid (DAP). DABA, Diaminobutryic acid.

lysine in the peptidoglycan, DAP has an ε-amino group free for bonding to an adjacent peptide of the cell wall. Because DAP has two optically active carbons, it is found in peptidoglycan as either the L,L or D,L (*meso*) enantiomer. Bacteria containing DAP include *Escherichia coli*, *Bacillus megaterium*, *Bacillus subtilis*, *Prochloron* spp., many of the cyanobacteria, and the purple photosynthetic bacteria. Most of these bacteria with DAP-containing peptidoglycan have a direct bond between DAP and D-alanine of an adjacent peptidoglycan structure. In mycobacteria, the DAP can partic-ipate in either of two cross-bridge formations: (1) between the DAP and D-alanine and (2) between two adjacent DAP residues.

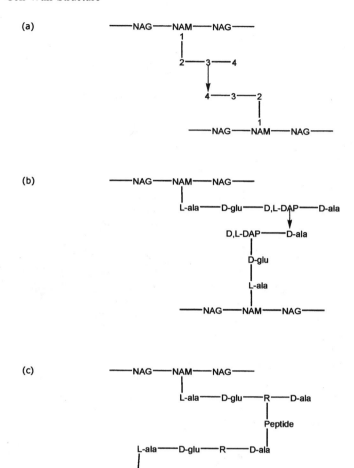

Figure 3.6. Cross-bridge linkages producing the polymeric peptidoglycan structure. (**a**) Model for linkage between the third amino acid and the fourth amino acid. (**b**) Example of cross-bridge linkage in peptidoglycan of *E. coli*. (**c**) Variations in amino acids making up the peptides of peptidoglycan. Major changes are at the third amino acid (R) and composition of the cross bridge.

## 3.3.3. Ornithine-Containing Structure

In *Butyribacterium rettgeri*, the third amino acid position is occupied by the unusual amino acid of L-ornithine. Ornithine is also found in the peptidoglycan of *Deinococcus radiodurans*, *Lactobacillus cellobiosus*, spirochetes, green nonsulfur photosynthetic bacteria, and all strains of *Thermus*. In *Deinococcus radiodurans*, a cross bridge of (Gly)₂ is attached to L-ornithine, whereas in the spirochetes and the green nonsulfur phototrophs, there is a direct linkage between L-ornithine and D-alanine in an adjacent peptidoglycan strand.

## 3.3.4. Diaminobutyric Acid–Containing Structure

L-Diaminobutyric acid (DABA) has been reported to be present in the third amino acid of the peptide making up the peptidoglycan of *Brevibacterium helvolum*, *Microbacterium liquefaciens*, *Corynebacterium sepedonicum*, *Corynebacterium tritici*, and *Corynebacterium insidiosum*. A rather curious feature of these cell walls is that D-DABA is also the amino acid involved in the cross-bridge formation between the third amino acid of one peptide to the fourth amino acid of an adjacent strand.

## 3.3.5. Homoserine-Containing Structure

In *Corynebacterium poinsettiae* (currently *Curthobacterium flaccumfaciens*), the third amino acid is homoserine, which has no free amino or carboxyl groups for bonding to an adjacent peptide chain. In this instance the cross-bridge involves the α-COOH of glutamic acid (the second amino acid) and the -COOH of D-alanine (the fourth amino acid). Direct linkages do not occur between these carboxyl groups but cross-bridge amino acids involve basic (polyamine) amino acids.

## 3.3.6. The Amino Acid Cross Bridge

Cross-bridge formation occurs between two different strands of peptidoglycan and is not an intramolecular feature. A cross bridge occurs between the amino acids in the peptidoglycan to provide structural stability. The most common system is the 3–4 cross bridge between the free amino moiety on amino acid 3 of one peptide and the carboxyl of D-alanine of an adjacent peptide (see Figure 3.6). There is considerable variation in how this bridge is produced in the various bacterial species. *E. coli* has a direct bonding without cross bridge amino acids while many bacteria have a specific set of amino acids providing the cross bridges (see Table 3.1). It appears that in some bacteria in which the cross bridge consists of diamino acids bonding may occur between the second amino acid in a peptide to the fourth amino acid in an adjacent peptide to form a 2–4 cross bridge.

Not all possible attachments on the four amino acids extending from NAM are used in linkage to an adjacent peptidoglycan structure. In several instances, stereochemistry limits the cross-bridge formation, and the number of cross linkages between the adjacent strands depends on the flexibility of the peptide chains. In *E. coli* and other bacteria that do not have an intervening peptide between the third amino acid and the amino acid for cross-bridge formation, relatively few cross bridges are formed. Some estimate that only 5% of all possible cross-bridge bonds are formed in the *E. coli* peptidoglycan. On the other hand, in *Staphylococcus aureus* more than 60% of possible bonds are formed because it has a flexible $(Gly)_5$ that provides transient interaction with the terminal D-alanine and the transpeptidase that forms the peptide bond.

Table 3.1. Examples of cross-bridge linkages in peptidoglycan of bacteria.[a]

| Third amino acid | Genera | Linkage |
|---|---|---|
| **L-Lysine** | *Staphylococcus* | -Gly-Gly-Gly-Gly-Gly- also known as $(Gly)_5$ |
| | | $-(Gly)_2$-Ser-$(Gly)_2$- |
| | | -L-ala-$(Gly)_2$- |
| | *Aerococcus, Gaffkya* | Direct peptide bond between L-lysine and D-Ala |
| | *Lactobacillus* | -D-Asp- |
| | | -(L-Ala)_2- |
| | | -L-Ser-(L-Ala)_2- |
| | *Leuconostoc* | -L-Ser-(L-Ala)_2- |
| | | -L-Ala-L-Ser- |
| | *Micrococcus* | -(L-Ala)_4- |
| | and *Sarcina* | -D-Ala-L-Ala-D-Glu-L-Ala- |
| | | -(L-Ala)_3- |
| | | -Gly-L-Glu- |
| | | -L-Ala- |
| | | -Gly- |
| | *Planococcus* | -D-Glu- |
| | *Pediococcus* | -D-Asp- |
| | *Streptococcus* | -L-Asp- |
| | | -Thr-L-Ala- |
| | | -(L-Ala)_3- |
| ***meso*-DAP** | *Bacillus, Arthrobacter, Clostridium, Escherichia coli*, and many Gram-negative bacteria | Direct peptide bond between *meso*-DAP and D-Ala |
| | Mycobacteria | -D-Ala-*meso*-DAP- |
| | | -*meso*-DAP-*meso*-DAP- |
| **L-DAP** | *Propionibacterium, Streptomyces, Clostridium* | $-(Gly)_{1-4}-$ |
| **L-Ornithine** | *Bifidobacterium* | -L-Ser-L-Ala-L-Thr-L-Ala- |
| | | -L-Lys-L-Ser-L-Ala-L-Ala- |
| | | -L-Lys-D-Asp- |
| | *Butyribacterium,* | -D-Orn- |
| **L-DABA** | *Corynebacterium tritici, Corynebacterium michiganense*[b] | Cross-linkage is through the $\alpha$-COOH of D-glutamic acid to D-ala and not through the diamino acid at position 3 |
| **L-Homoserine** | *Corynebacterium betae, Corynebacterium poinsettiae*[b] | -D-Orn- |

[a]Modified from Cummings, 1974.[2]
[b]In these organisms L-Ala is replaced in the peptide by Gly.
DAP, Diaminopimelic acid; DABA, diaminobutyric acid.

Thus, it appears that the quantity of peptidoglycan synthesized by the Gram-positive bacteria exceeds that required to provide structural form and osmotic stability. From a stability perspective, scientists have observed that it is more difficult to disrupt bacterial cells if they have a high quantity of peptidoglycan and it is extensively cross linked.

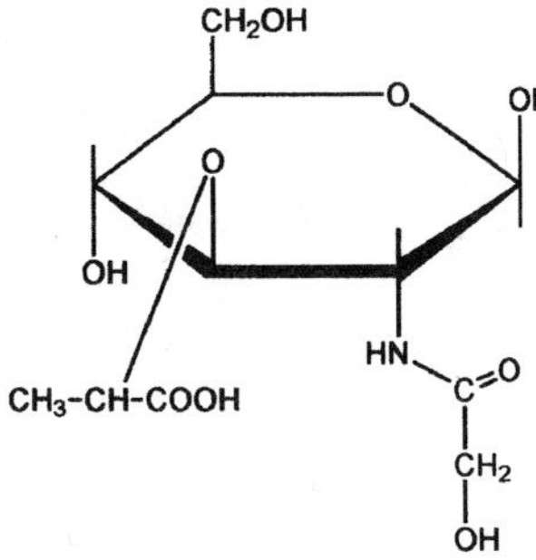

Figure 3.7. Structure of N-derived muramic acid found in mycobacteria.

### 3.3.7. Peptidoglycan of Mycobacteria and Nocardia

Unlike other bacteria, mycobacteria and nocardia do not have *N*-acetylated muramic acid. These bacteria have a cell wall with a peptidoglycan structure consisting of *N*-acetylglucosamine and *N*-glycolylmuramic acid. The structure of the *N*-modified muramic acid component in mycobacteria is given in Figure 3.7. In comparison to the NAM found in other bacteria, the *N*-glycolmuramic acid in mycobacteria and nocardia is only a slight change and can be used as an example of diversity in bacterial peptidoglycan. The benefit of this *N*-glycolylmuramic acid structure over the *N*-acetylmuramic acid moiety in the cell wall structure is unclear.

## 3.4. Molecular Events in Synthesis of Peptidoglycan[3]

The formation of peptidoglycan in eubacteria is a three-step process: (1) synthesis of the NAG and NAM in the cytoplasm, (2) transport of the dimer across the plasma membrane, and (3) insertion of the dimer into the growing point of the peptidoglycan. In both Gram-positive and Gram-negative bacteria, the synthesis of the NAG–NAM dimer proceeds by the same general mechanism by attachment to a lipid component in the plasma membrane. In biosynthesis, the initial step is the formation of NAM-lipid, which is often termed lipid I, and this is followed by the formation of NAG–NAM-lipid that is called lipid II. The formation of a polymeric peptidoglycan with the formation of a single unit often referred to as the sacculus is the activity of four enzymes known as PBP1A, PBP1B, PBP2, and PBP3. These enzymes all bind antibiotics and it is easier to quantitate these penicillin-binding proteins (PBP) because measuring the amount of protein attached to [14C]penicillin is easier than synthesizing the substrates for the enzymes.

### 3.4.1. Synthesis of NAM–NAG Dimers

The process for molecular construction of the glycan dimer used for the backbone of the peptidoglycan is given in Figure 3.8. Glucose, the initial molecule for the for-

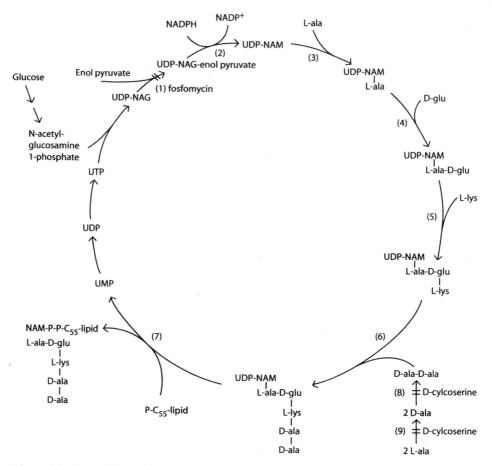

Figure 3.8. Biosynthesis of the UDP–NAM–pentapeptide precursor for peptidoglycan. Points of inhibition are indicated by a double line across the *arrow*. The $C_{55}$ lipid is found in the membrane as shown in Figure 3.9. Enzymes are as follows: (1) UDP–NAG–enolpyruvate transferase, (2) UDP–NAG–enolpyruvate reductase, (3) L-alanine ligase, (4) D-glutamic acid ligase, (5) L-lysine ligase, (6) D-alanyl-D-alanine ligase, (7) a transferase reaction, (8) D-alanyl-alanine synthase, (9) alanine racemase.

mation of NAG, is converted to glucosamine and the addition of an acyl moiety to the amino group prevents random additions to the amino sugar. Free glucosamine is not used for biosynthesis but uridine diphosphate (UDP) carries glucose-modified moieties including NAG and NAM for the production of the dimer. Apparently, enzyme recognition for dimer synthesis is more precise if a sugar nucleotide is involved instead of a single sugar molecule. Energy can be provided to the various reactions through the high-energy phosphates UTP and UDP. To initiate peptidoglycan synthesis, *N*-acetylglucosamine-1-phosphate is combined with UTP to produce UDP–NAG by a specific uridyltransferase. UDP–NAM is formed following sequential reactions of phosphoenolpyruvate (PEP) addition and reduction by NADPH. The addition of the amino acids through NAM occurs while NAM is linked to UDP and the first three amino acids are added in sequence. The final two amino acids (i.e., D-alanyl-D-alanine) are added as a unit using a ligase which is specific for D-alanyl-D-alanine. Other amino acids are also added by ligases that recognize specific amino acid structure. While the first two amino acids are generally L-alanine and D-glutamic acid, the third amino acid varies with the bacterial species. Enzyme specificity accounts for the construction of the peptide and it has been demonstrated that the ligase that adds the D,L-DAP moiety will not recognize L-lysine. It should be emphasized that the final two amino acids in the peptides extending from NAM are always D-alanine-D-alanine.

There is an apparent order of amino acid isomers for peptide synthesis and it follows the sequence: L-amino acid–D-amino acid–L-amino acid–D-amino acid–D-amino acid. Some scientists suggest the order of amino acids with the isomeric forms of L–D–L–D represents a remnant of early evolution in which mRNA and ribosomes were not involved in protein synthesis but amino acid addition to the peptide was regulated by enzyme fit or stereoproperties of the substrates. In addition to the peptidoglycan containing D-forms of amino acids, several antibiotics produced by prokaryotes also include D-amino acids. As a point of reference, the only D-amino acids in eukaryotic peptides have been reported in amphibian skin, snail ganglia, and spider venom. The significance of the D-forms is that a nontypical peptidoglycan bond is formed and these bonds are stable to proteolytic enzymes that attack the common peptide bond. The two D-alanine residues at the end of the pentapeptide provide a most unique stereo-orientation of a peptide which makes it difficult for any other amino acids to be attached through the terminal D-alanine.

The γ-linkage between D-glutamic acid and the third amino acid provides for a flexible peptide to enable enzymes for cross-bridge formation to interact with the peptide. In addition, the γ-linked peptide bond provides structural stability against common proteases in the environment.

## 3.4.2. Membrane-Associated Activities

A specific lipid in the plasma membrane participates in transmembrane movement of cell wall precursors. An isoprenoid lipid composed of 55 carbons (see Figure 3.9), with a terminal phosphate is known as P-undecaprenyl. The lipid segment of the molecule is extremely hydrophobic and resides only in the lipid domain of the plasma membrane while the phosphate end is hydrophilic and extends out from the membrane. The UDP–NAM–(pentapeptide) segment interacts with the P-undecaprenyl to produce NAM–(pentapeptide)–P–P–undecaprenyl which in turn combines with UDP–NAG to form NAG–NAM–(pentapeptide)–P–P–undecaprenyl. The model for generating pre-

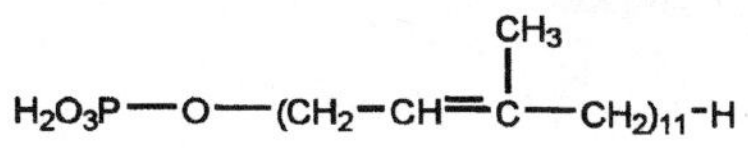

Figure 3.9. Structure of phosphoundecaprenyl: a phosphomonoester of $C_{55}$ polyisoprenoid alcohol.

$$H_2O_3P - O - (CH_2 - CH = \overset{\overset{\displaystyle CH_3}{\displaystyle |}}{C} - CH_2)_{11} - H$$

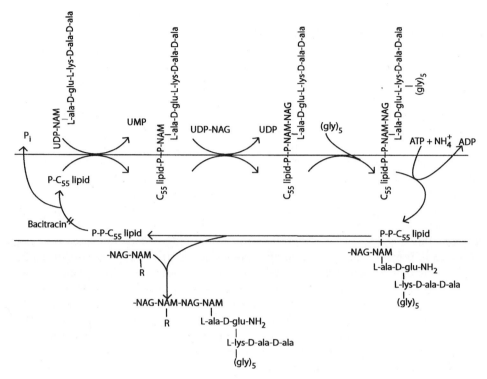

Figure 3.10. Membrane-associated system for peptidoglycan synthesis in *Staphylococcus aureus*. Inhibition by bacitracin is indicated by a double line across the *arrow*.

cursors to be used in peptidoglycan biosynthesis is given in Figure 3.10. If the cross bridge contains amino acids, they are added at this time. In those instances in which an amino or glycine is found on the glutamic acid, that addition occurs while the NAG–NAM (pentapeptide) is tethered to the membrane. By mechanisms that are not yet clear, the entire NAG–NAM dimer tethered to –P–P–undecaprenyl is moved from the inner face of the plasma membrane to the outer face and the peptidoglycan precursor is released from the –P–P–lipid. To recycle the –P–P–undecaprenyl, the terminal P is removed and released internally into the nutrient pool of the cell and the P-undecaprenyl is available to acquire another NAM. Approximately 4,000 molecules of P-undecaprenyl are committed to the synthesis of peptidoglycan in *E. coli*.

   Most bacteria studied have undecaprenol in the plasma membrane and this molecule is involved in transmembrane movement of precursors of peptidoglycan synthesis. A notable exception is mycobacteria, which lacks undecaprenol but has decaprenol and octahydroprenol in the plasma membrane. These polyprenols in mycobacteria are presumed to be involved in the reactions that undecaprenol mediates in other bacteria.

### 3.4.3. Dimers and Synthesis of Peptidoglycan

The insertion of the dimer of NAG–NAM (with affiliated pentapeptide) into the growing point of the peptidoglycan occurs in the cell wall. ATP is not directly required for this bond formation but energy may come from breaking the phosphate–phosphate linkage in the P–P–lipid molecule. The pentapeptide from the NAM contains the terminal two D-alanine residues and removal of one D-alanine is required to provide the necessary stereo-configuration for potential cross-bridge formation (see Figure 3.11). The terminal D-alanine is released by a specific D,D-carboxypeptidase and this amino acid may be transported into the cell for synthesis of other cell wall peptides.

### 3.4.4. Hydrolytic Enzymes and Peptidoglycan Turnover[4]

Enzymes displaying peptidoglycan hydrolase activities have been isolated from more than 25 species of bacteria and many of the genes for these enzymes has been cloned. The bonds attacked by autolyzing enzymes are indicated in Figure 3.12. These hydrolytic enzymes have a stereochemical requirement for substrate which accounts for high levels of specificity. It has been determined from X-ray crystallographic analysis that the peptidoglycan can fit into the lysozyme molecule in only one orientation and water can be added across the β-1,4-glucosidic bond following a NAG–NAM dimer. Thus, exhastive digestion of peptidoglycan by lysozyme will not produce NAG and NAM monomers. On the other hand, β-1,4-glucosidase will hydrolyze the bond between NAG and NAM following a dimer of NAM–NAG. Of special interest are the enzymes that hydrolyze bonds adjacent to D-alanine because these enzymes display penicillin-binding characteristics. Many of these autolytic enzymes are attached in the cell wall and participate in creating growing points for additions to peptidoglycan.

Hydrolytic enzymes are present in the expanding regions of the cell wall and contribute to the synthesis of the cell wall which is a combination of controlled autolysis and biosynthesis. In certain bacteria, it has been demonstrated that there is a loss of peptidoglycan from the assembled insoluble structure by autolytic enzymes and this is referred to as turnover of peptidoglycan. If autolytic enzymes release fragments of the wall into the culture fluid for Gram-positive bacteria or into the periplasm of Gram-negative bacteria during active growth, this material may be reutilized for

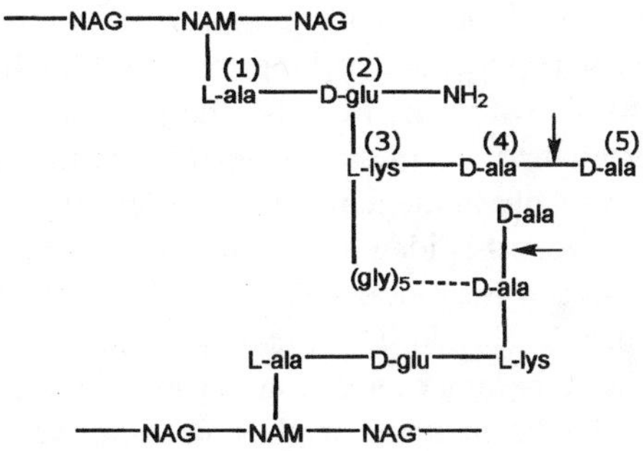

Figure 3.11. Cross-bridge formation in peptidoglycan following removal of terminal D-alanine in peptapeptide. *Arrows* indicate bonds attacked by carboxypeptidase. *Dotted line* indicates peptide bond of cross bridge.

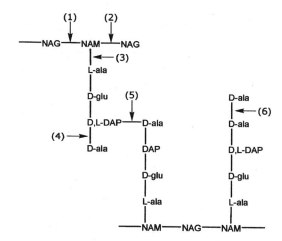

Figure 3.12. Enzyme specificity of bacterial peptidoglycan hydrolases. Structure shown is peptidoglycan found in *E. coli*. *Arrows* indicate points of hydrolysis and specific enzymes are as follows: (1) endo-β-*N*-acetylglucosaminidase (glucosamidase), (2) endo-β-*N*-acetylmuramidase [also known as lytic transglycosylase, muramidase, or lysozyme], (3) *N*-acetylmuramyl-L-alanine amidase, (4) L,D-carboxypeptidase, (5) D,D-endopeptidase, (6) D,D-carboxypeptidase.

peptidoglycan synthesis. Most of the components in the peptidoglycan are subject to autolytic release and are available for recycling into new structural material. The rates of turnover vary from 1% to 2% to 10% to 20% in a generation with differences attributed to the method of analysis. Most certainly it can be said that the cell wall of bacteria is highly dynamic with autolysis and biosynthesis delicately balanced. A possible role of teichoic acids and other sugar polymers in bacteria is the interaction with autolytic enzymes to reduce the degradation of peptidoglycan in a mature cell. In some bacterial cultures, the cells lyse on reaching stationary phase and this destruction of the cell wall can be attributed to inadequate control of enzymes used for peptidoglycan degradation.

## 3.4.5. Inhibitors of Cell Wall Synthesis

Important inhibitors of peptidoglycan synthesis are penicillins and cephalosporins. The 6-aminopenicillanic acid is the basic structure for penicillins while cephalosporic acid is the fundamental structure found in all cephalosporins. There is considerable similarity between the 6-aminopenicillanic acid and the cephalosporic acid structures (see Figure 3.13). The acid moieties of these core structures inhibit the action of the carboxypeptidase, thereby preventing release of the terminal D-alanine from the NAM-pentapeptide and making the fourth amino acid available for cross-bridge formation. As shown in Figure 3.14, the β-lactam ring of 6-amino penicillic acid is structurally similar to the D-alanyl-D-alanine moiety and penicillin can be considered to bind irreversibly in the active site of the D,D-carboxypeptidase. The mode of action of cephalosporins is identical to that of penicillins because of the similarity of cephalosporic acid and 6-aminopenicillanic acid. Other members of the β-lactam class of antibiotics include cephamycins, carbapenems, and monobactams. All of these β-lactams have the same mode of action because all have the β-lactam ring configuration.

The attachment of phosphoenolpyruvate (PEP) to UDP–NAM is inhibited by fos-

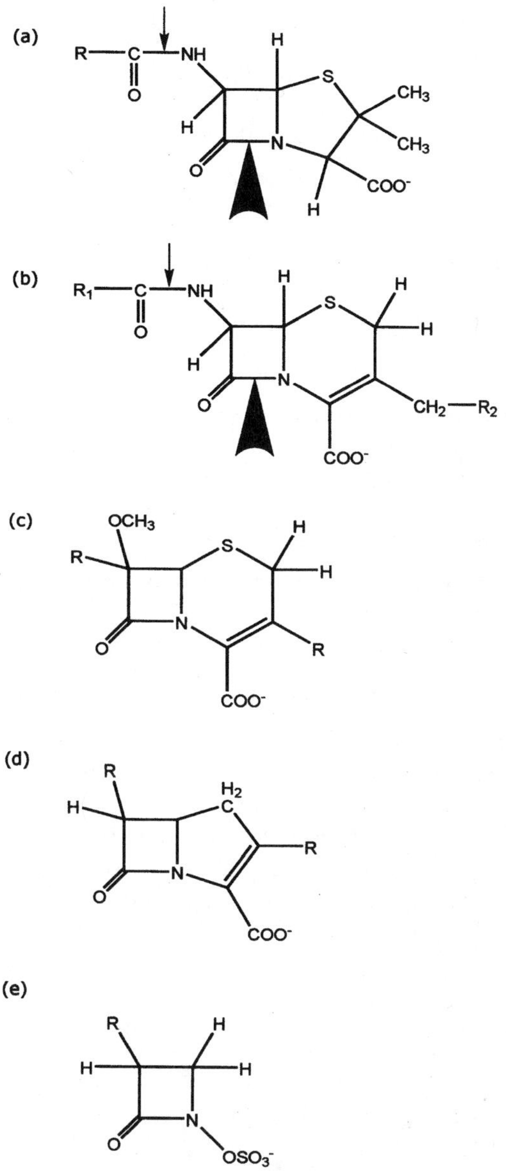

Figure 3.13. Essential structures of antibiotics that inhibit peptidoglycan synthesis. (**a**) 6-Aminopenicillinic acid and (**b**) cephalosporic acid. The *bold arrow* indicates bond disrupted in the β-lactam ring by penicillinase, also known as β-lactamase. *Small arrow* indicates bond hydrolyzed by enzymes belonging to the amidase class. The β-lactamase is associated with antibiotic resistance with transfer by plasmids or bacterial chromosome. Amidase would be released into the environment from various types of cell and would be the same enzyme involved in hydrolyzing any amide linkage. Structures of related antibiotics with a β-lactam ring include (**c**) cephamycin, (**d**) carbapenem, and (**e**) monobactam. The various side groups on the antibiotic structures are indicated as R. Specific examples of these antibiotics are given in Table 3.2.

fomycin (phosphonomycin) which binds irreversibly to sulfhydryl (-SH) group in the active site of the enzyme and prevents PEP binding. The Structural similarity between PEP and fosfomycin is shown in Figure 3.15. D-cycloserine, an antibiotic produced by *Streptomyces*, is a structural analog of D-alanine and inhibits the alanine racemase involved in the conversion of L-alanine to D-alanine. Because of toxicity reactions, fosfomycin and D-cycloserine are not used to control animal infections. The formation

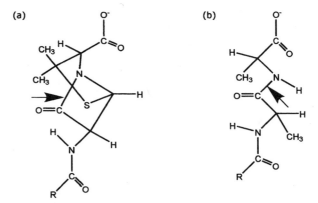

Figure 3.14. Structural similarity of substrates that bind to D,D-carboxypeptidase. (a) 6-Aminopenicillinic acid and (b) D-alanyl-D-alanine. *Arrow* indicates bond targeted for disruption by peptidoglycan D,D-carboxypeptidase.

of D-alanyl-D-alanine proceeds by the specific enzyme (D-alanyl-D-alanine synthase) that employs D-alanine and ATP for the formation of the diamino acid. The synthesis of D-alanyl-D-alanine is shown in the following reaction:

$$2 \text{ D-alanine} + \text{ATP} \rightarrow \text{D-alanyl-D-alanine} + \text{ADP} + \text{P}_i$$

D-cycloserine inhibits the above reaction by interfering with the addition of D-alanine to the enzyme. The enzyme responsible for the addition of D-alanyl-D-alanine onto the third amino acid extending from NAM is not inhibited by D-cycloserine.

Vancomycin and ristocetin are polyphenolic glycopeptides that have a mass of 3.5 kDa and are produced by *Nocardia*. Both of these antibiotics recognize the D-alanyl-D-alanine extending as the fourth and fifth amino acids from NAM and prevent cross-

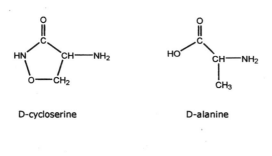

D-cycloserine

D-alanine

Phosphonomycin

Phosphoenolpyruvate

Figure 3.15. Structural analogs that function as inhibitors in synthesis of peptidoglycan. D-cycloserine is a structural analog of D-alanine and phosphonomycin is a structural analog of phosphoenol pyruvic acid.

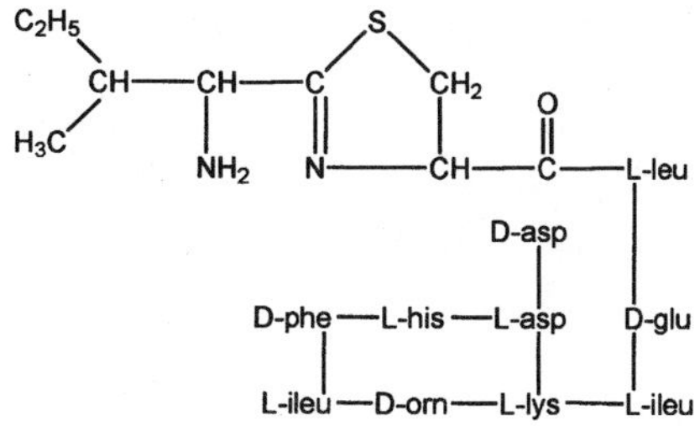

Figure 3.16. Structure of bacitracin A. Cysteine and isoleucine are the terminal amino acids attached to leucine and they are condensed into a thiazoline ring.

bridge formation in peptidoglycan synthesis. The action of vancomycin on peptidoglycan synthesis is distinct from the action of penicillins because vancomycin blocks the D,D-carboxypeptidase from gaining access to the substrate while penicillin binds to the enzyme. Tunicamycin, which is structurally similar to undecaprenyl phosphate, interferes with escorting of the NAG–NAM dimer across the plasma membrane. Bacitracin is a cyclic peptide produced by *B. subtilis* (see Figure 3.16) which complexes with the diphosphorylated undecaprenol in the membrane to block the release of inorganic phosphate and prevent the regeneration of P-undecaprenol. Thus, without the availability of a carrier for the NAG–NAM dimer, transport and synthesis of peptidoglycan stops. Owing to the extreme tissue toxicity of many of these antibiotics, only vancomycin and bacitracin are used as pharmacological agents in treatment of humans with bacterial infections. Vancomycin may be administered when there is an infection attributed to a penicillin-resistant Gram-positive bacterium or intestinal infections attributed to *Clostridium difficile*. The use of bacitracin is limited to topical application and is readily available in most pharmacies.

## 3.4.6. Antibiotic Resistance[5]

With the emergence of bacterial resistance to the antibiotics that prevent peptidoglycan synthesis, there has been considerable interest in the mechanisms of this resistance. Examples of resistance displayed by various antibiotics are given in Table 3.2. Owing to evolutionary pressures and the shuffling of genetic material between bacteria, considerable variation has resulted. Thus, it cannot be considered that there are separate strategies by the Gram-positive or Gram-negative bacteria for the resistance to antibiotics active against peptidoglycan synthesis. In general, it can be stated that there are three major strategies to achieve antibiotic resistance: (1) *Drug inactivation*: The production of β-lactamase (penicillinase) and other enzymes to hydrolyze the antibiotics. (2) *Altered uptake*: In Gram-negative bacteria the antibiotics diffuse through porins in the outer membrane and changes in the channel of the porin lead to a decrease of the influx of antibiotics. Although some bacteria actively export antibiotics by multifunctional transporters, antibiotics active in prevention of peptidoglycan biosynthesis are not required to enter the cytoplasm. (3) *Altered target site*: The resistance to penicillins commonly involves antibiotic-binding proteins and resistance to vancomycin may be the result of a change in the peptide attached to NAM.

Table 3.2. Resistance associated with antibiotics that inhibit peptidoglycan synthesis.[a]

| Antibiotic family | Patterns of resistance displayed by members of that type | Examples |
|---|---|---|
| β-Lactams | Altered target, or altered uptake, or drug inactivation | Penicillins (benzyl penicillin, cloxacillin, flucloxacillin, ampicillin, piperacillin, amoxycillin, ticarcillin, azlocillin) Cephalosporins (cephalexin, cefaclor, cefadroxil, cefuroxime, cefamandole, cefotaxime, ceftazidime, cefepime) Cephamycins (cefoxitin) Carbapenems (imipenem) Monobactams (aztreonam) |
| Glycopeptides | Altered target | Vancomycin and teicoplanin |
| Cyclic peptide | None known | Bacitracin |

[a]From Mims et al., 1998.[5]

Many of the Gram-negative bacteria produce enzymes to degrade penicillins and these genes are encoded on the *AmpC* gene of bacterial chromosome or encoded on *TEM-1* and related genes of resistant plasmids. At this time, there are more than 100 different molecular forms of β-lactamases and each has its own unique hydrolytic properties. Cloxacillin, ceftazidine, and imipenem are susceptable to hydrolysis of only a few of the many β-lactamases and as a result are often considered as β-lactamase-stable. Although widely used, ampicillin is readily hydrolyzed by action of the various β-lactamases.

In the case of vancomycin resistance, it has been observed that the fifth amino acid (D-alanine) of the peptide attached to NAM is replaced by D-lactate or D-serine and in some cases there are only four amino acids in the peptide attached to NAM. Vancomycin and related antibiotics will not bind to peptodoglycan segments unless the two terminal amino acids are D-alanyl-D-alanine. This modification of the peptidoglycan for vancomycin resistance requires that changes are made to at least nine genes. It is only after several generations that new strains of bacteria resistant to antibiotics emerge because several genetic alterations must occur to provide an appropriate phenotype.

In some instances, the resistance of bacteria to antibiotics may result from the development of alternate strategies for cell wall biosynthesis. One example would be the cross-bridge formation between amino acid number 2 and amino acid number 3 $(2 \rightarrow 3)$ instead of the standard $3 \rightarrow 4$ linkage between amino acid number 3 on one strand and amino acid on an adjacent strand.

A procedure has been developed for the in vivo treatment of bacteria producing β-lactamases using a penicillin that is commonly sensitive to the β-lactamase. The bond hydrolyzed by β-lactamase is indicated in Figure 3.13. Chemicals that have no antibiotic effect are administered to specifically bind the β-lactamases. Clavulanic acid, which is produced by *Streptomyces clavaligerus*, is considered to be a "suicide inhib-

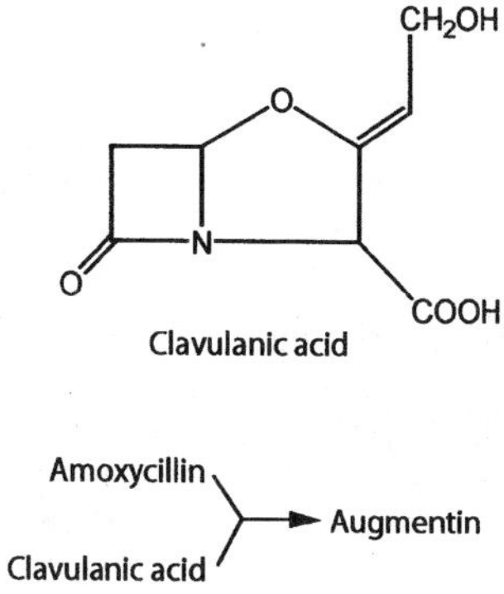

Clavulanic acid

Amoxycillin
Clavulanic acid → Augmentin

Figure 3.17. Structure of clavulanic acid and its role in augmentation administration.

itor" and its structure is given in Figure 3.17. Augmentin is used to treat bacteria producing a β-lactamase and it consists of two chemicals: amoxycillin plus clavulanic acid. Other combinations similar to augmentin contain ticarcillin plus clavulanic acid or piperacillin plus tazobactam. Because vancomycin inhibits peptidoglycan synthesis at a site adjacent to inhibition by penicillin, it would not be appropriate to administer a combination that contains vancomycin plus a penicillin.

## 3.4.7. Antibiotic-Binding Proteins[6]

Penicillin is an effective inhibitor of the peptidoglycan cross-linking system by inhibiting the enzymes involved. These enzymes that are sensitive to penicillins or cephalosporin bind to the active site and are also known as penicillin-binding proteins (PBPs). The PBPs are found principally associated with the penicilloyl serine transferase superfamily which comprises the D,D-transpeptidases, the D,D-carboxypeptidases, the β-lactamases, and the multifunctional high molecular weight PBPs. While the D,D-transpeptidase and the D,D-carboxypeptidase enzymes are involved in assembly of the peptidoglycan, the β-lactamases function as defense mechanisms against penicillin. The PBPs are broadly distributed throughout the Gram-positive and some Gram-negative bacteria with considerable research conducted on proteins found in *E. coli*, *Streptococcus pneumoniae*, *Staphylococcus aureus*, *Neiserria meningiditis*, and *Enterococcus hiriae*. In the penicilloyl serine transferase molecule, the active site for penicillin binding is created from three unique motifs: SXXK [serine-X-X-lysine], SXN [serine-X-asparagine] or YXN [tyrosine-X-asparagine], and K(H<R)T(S)G [lysine-(histidine<arginine)-threonine-(serine)-glycine]. As indicated in Figure 3.18, the multifunctional high molecular weight PBPs are anchored into the membrane at the N-terminus end by a series of hydrophobic amino acids. As the PBP extends from the outer face of the plasma membrane there are two identifiable domains. One is termed the N-terminal domain and the function of this segment is not obvious at this time. The other domain contains the unique motifs for penicillin binding and is near the carboxyl end of the protein. In *E. coli* there are over a half dozen major enzymes in the plasma membrane that bind penicillins or cephalosporins and examples

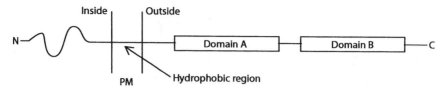

Figure 3.18. Model indicating essential motifs of multifunctional high molecular weight PBPs present in different species of bacteria. The protein is anchored into the plasma membrane by a hydrophobic region near the N-terminus and the PBP extends into the cell wall region. Approximately 250 amino acids are present in domain B, which is the penicillin binding segment. (Data from Ghuysen and Dive, 1994.[6])

of PBPs are given in Table 3.3. Based on these observations, it is apparent there are two or three enzymes in *E. coli* that are capable of cross-linking activity in the peptidoglycan and if one becomes inhibited the other can accomplish the needed biosynthetic activity. Redundant systems in bacterial cell wall synthesis reflects the importance of this process for cell viability.

## 3.4.8. Sites of Cell Wall Synthesis[7-10]

The synthesis of new peptidoglycan in cell walls is a discrete activity occurring at precise regions on the cell and not continuous across the cell wall. Because the peptidoglycan of cells is covalently linked and can be envisioned as a single "bag-shaped macromolecule" (see Figure 3.2), growth can occur only if autolytic enzymes (autolysins) are formed by the bacterial cell to produce a point for addition of the NAG–NAM dimer. The results of autolysin acting on cell wall would (1) expose free ends on the peptidoglycan for polymer growth and (2) hydrolyze peptidoglycan bonds for cell division. The old cell wall separates as new wall is introduced. The sequence of events involved in cell wall synthesis by cells having one plane of division was established by examining *Enterococcus* (*Streptococcus*) *faecalis* (Figure 3.19). At the

Table 3.3. Penicillin-binding proteins (PBPs) associated with *Escherichia coli*.[a]

| PBP | Molecular weight (KDa) | Molecules per cell | Activities of protein |
|---|---|---|---|
| 1A | 93.5 | 115 | Associated with cell elongation by synthesis of the side walls |
| 1B | 94.1 | 115 | Same as PBP1A |
| 2 | 70.8 | 20 | Contributes to rod shape of cell |
| 3 | 63.8 | 50 | Cell division attributed to septa formation and cell constriction |
| 4 | 49.5 | ? | Cell wall lysis and cell wall maturation |
| 5 | 41.3 | 2,400 | D-alanine carboxypeptidase |

[a]Modified from Matsuhashi, 1994.[9]

a

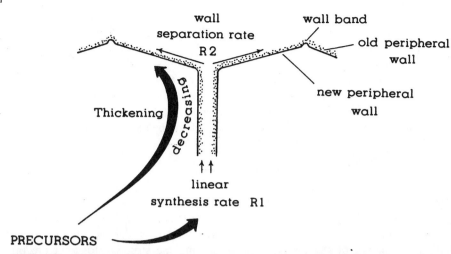

wall
separation rate
R2

wall band

old peripheral
wall

new peripheral
wall

Thickening

decreasing

linear
synthesis rate   R1

PRECURSORS

b

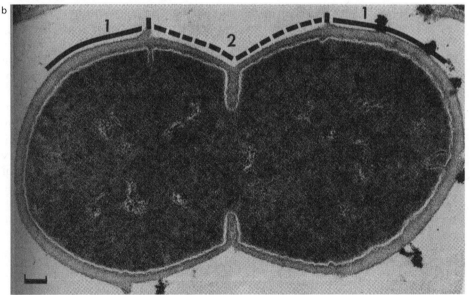

Figure 3.19. Site of peptidoglycan synthesis in *Enterococcus faecalis*. (**a**) Model indicating process of peptidoglycan synthesis in dividing cells. Precursors of cell walls are incorporated into the expanding exterior wall and into the cross wall at synthesis rates described as R2 and R1, respectively. A distinct band is present in which the old peripheral wall meets the new peripheral wall. (**b**) Electron micrograph indicating cell wall structure of dividing cells. The old peripheral cell wall is indicated as "1" while the newly synthesized cell wall is designated as "2." The juncture where 1 meets 2 is clearly obvious. Cells were stained with lead citrate and phosphomolybdate. (Reprinted with permission from Higgins, M.L. and G.D. Shockman. 1971. Prokaryotic cell division with respect to wall and membrane. *CRC Critical Reviews in Microbiology* 1:29–72. Copyright CRC Press, Boca Raton, Florida.)

periphery of the cell along the plane of septum formation, there is new peptidoglycan formation.

Research has revealed that there are 10–50 dimers of NAG–NAM as a unit in the peptidoglycan and the individual units are stacked either parallel or oblique to the plasma membrane. Cross-bridge formation between the peptidoglycan units provides structural stability. Curved rod shapes could result from cross-bridge formation between peptidoglycan units that are spatially separated and would reflect a rope-like twist on the peptidoglycan structure. Differences between coccus and bacillary forms may be attributed to the length and orientation of the peptidoglycan units.

### 3.4.9. Structural Stability[11]

There are several features of the molecular architecture of peptidoglycan that enable bacteria to retain a cell wall even in the presence of numerous hydrolytic enzymes. This structural uniqueness is the result of the presence of the *N*-acetylmuramic acid, the D-forms of amino acids, and diaminopimelic or other unique amino acids. In addition, the nature of the covalent bonds holding the structure together are often unusual with the β-glucosidic bond between the NAG and NAM residues and the γ-peptide bond in the peptide segment of peptidoglycan. The unique stereochemistry of the peptidoglycan requires specific autolytic enzymes.

The peptidoglycan isolated from *E. coli* and *Pseudomonas aeruginosa* has been determined to be the same size at the poles of the cell as along the linear channel of the cell. Beveridge and colleagues have found that the peptidoglycan structure for *E. coli* under aqueous conditions is 6.0 nm while the isolated peptidoglycan structure for *P. aeruginosa* is 3.0 nm. If these structures are dried, their thickness is reduced by 50%. The spacing between adjacent peptidoglycan strands is about 1.6 to 2.0 nm. Using an atomic force microscope under hydrated conditions, Beveridge's laboratory found that the intact cylinder of peptidoglycan from *E. coli* is not absolutely rigid but can be readily stretched and has an elastic modulus of $2.5 \times 10^7$ N/m$^3$. The elastic character of the peptidoglycan is attributed to the stretching of the short peptides that are held together by the cross bridges.

## 3.5. Cell Walls of Archaea[12]

The walls of archaea are biochemically distinct from those of bacteria in that the typical peptidoglycan structure is absent. D-forms of amino acids have not been reported for the cell walls of the archaea. Several different polymers make up the cell walls and their distribution in archaea is given in Table 3.4. Archaea that would be designated as Gram-positive have walls containing pseudomurein (pseudo-peptoglycan), methanochondroitin, and a complex acidic heteropolysaccharide.

### 3.5.1. Pseudomurein (Pseudopeptidoglycan)

The cell wall of *Methanobacterium* is made up of a glucan and a peptide, as is the case with bacterial peptidoglycan; however, muramic acid is replaced by talosamin-

Table 3.4. Surface and structural composition of archaea.[a]

| | Plasma membrane | S-layer (glycoproteins) | Protein sheath | Heteropolysaccharide | Methanochondroitin | Pseudomurein | Polyglutamic acid |
|---|---|---|---|---|---|---|---|
| *Thermoplasma* | Y | N | N | N | N | N | N |
| *Thermoproteus, Sulfolobus, Pyrodictium, Methanococcus, Halobacterium* | Y | Y | N | N | N | N | N |
| *Methanosarcina* | Y | Y | N | N | Y | N | N |
| *Methanothermus, Methanopyrus* | Y | Y | N | N | N | Y | N |
| *Methanobacterium, Halococcus* | Y | N | N | Y | N | Y | N |
| *Methanospirillum, Methanothrix* | Y | Y | Y | N | N | N | N |
| *Natronococcus* | Y | N | N | N | N | N | Y |

Y, Character is present; N, character is not present.
[a]Modified from König, 1988.[22]

uronic acid and the peptide has only the L-form of amino acids. The dimer glucan is composed of a dimer of D-*N*-acetylglucosamine (NAG) and L-*N*-acetyltalosaminuronic acid (NAT) with alternating $\alpha$-1,3 and $\beta$-1,3 glucosidic bonds. NAT is found only in these archaebacteria and the linkage between NAG and NAT is $\alpha$-1,3. There are four amino acids in the short peptide extending from *N*-acetyltalosaminuronic acid and the third amino acid is lysin, which is held by an $\varepsilon$-linkage. The cross-bridge peptide bond is from the terminal glutamic acid to lysine. This structure is termed pseudo-murein or pseudopeptidoglycan and a model is given in Figure 3.20.

## 3.5.2. Methanochondroitin

The rigid structure making up the cell wall of *Methanosarcina* spp. is an acidic poly-saccharide consisting of *N*-acetylgalactosamine and glucuronic acid. This structure is similar to chondroitin sulfate found in connective tissue of animals. The methano-chondroitin polymer is composed of a trimer consisting of [*N*-acetylgalactosamine-*N*-acetylgalactosamine-glucuronic acid] (see Figure 3.21). In some cases, small amounts of phosphate or sulfate have been found in isolated cell walls of the *Methanosarcina*. This chondroitin cell wall material has not been reported for other archaea.

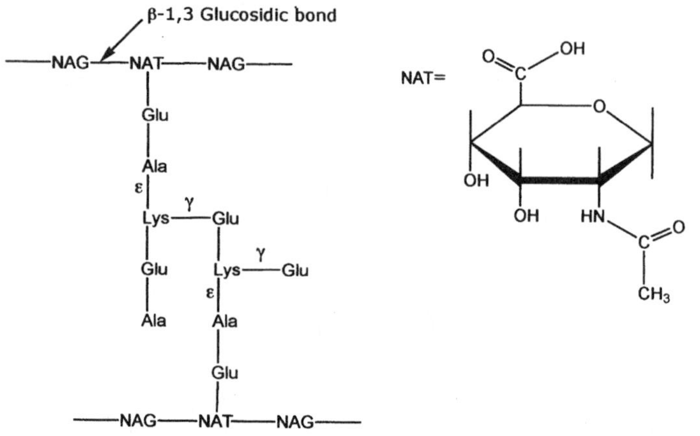

Figure 3.20. Pseudomurein of *Methanobacterium thermoautotrophicum*. T-NAT, *N*-acetyl-talosaminuronic acid. The linkage between NAG and NAT is a β-1,3 glucosidic bond. (Modified from Kandler and König, 1985.[12])

## 3.5.3. *Heteropolysaccharide*

*Halococcus morrhuae* has a thick cell wall composed of complex acidic polysaccharide polymer that contains several neutral sugar, uronic acid, and amino sugar residues. *N*-Glycylglucosamine and glucosaminuronic acid are two of the identified structures and both are highly unusual with limited distribution in biology. The proposed structure for this heteropolysaccharide wall polymer is given in Figure 3.22.

## 3.5.4. *Polyglutamine Glycoconjugate*[13]

*Natronococcus occultus* grows at pH 9.5 to 10.0 at a salt concentration of 3.5 *M*. As established by Niemetz and colleagues, the cell wall for this organism is a polymer containing γ-glutamine; D-forms of glucosamine, galactosamine, galacturonic acid, glucuronic acid; and glucose. A backbone consisting of about 60 residues of glutamate forms the polyglutamate and two different types of heterosaccharides are attached through the α-amide of the glucosamine polymer. One oligosaccharide structure contains a pentamer of glucosamine with an α-1,3 linkage at the reducing end and a pentamer of β-1,4-linked galacturonic acid at the nonreducing end. A second sugar

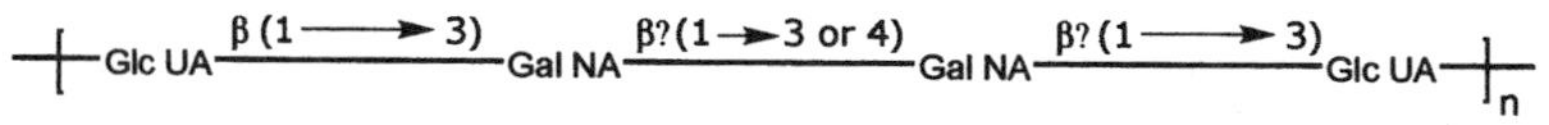

Figure 3.21. Proposed structure for the cell wall polymer of *Methanosarcina*. (Modified from Kandler and König, 1985.[12])

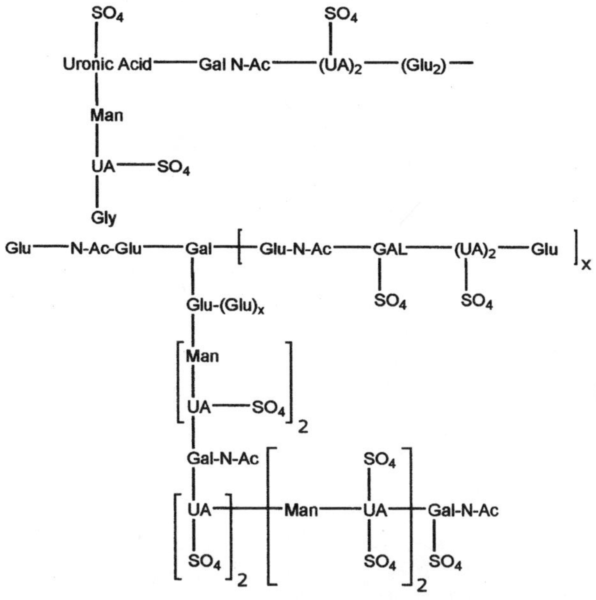

Figure 3.22. Proposed cell wall polymer in *Halococcus*. UA, Uronic acid. (Modified from Kandler and König, 1985.[12])

structure contained a galactosamine dimer in a β-1,3 linkage at the reducing end and maltose at the nonreducing end. The molecular mass of the polyglutamate glycocon-jugate is 54 kDa and a model of this molecular structure is given in Figure 3.23. It has been suggested that *Halococcus turkmenicus* also has a similar cell wall polymer consisting of glutamine and oligosaccharides.

## 3.6. Cell Walls of Endospores[14–18]

Bacterial production of endospores has been most extensively examined in *Bacillus*, and this activity is an example of cellular differentiation. Along with the numerous changes to produce an endospore there is the required formation of a spore wall that is chemically similar but not identical to the peptidoglycan of the mother cell. There are three phases of cell wall synthesis in sporulation and they include (1) formation of the septum or cross wall, (2) synthesis of the primordial germ-cell wall, and (3) synthesis of the cortex. To achieve this construction of the spore wall, various enzymes are required and many of these are identified as PBPs with specific functions. This section first addresses the physiological process of producing the spore pepti-doglycan and then discusses some of the activities associated with enzymes and PBPs specific for sporulation.

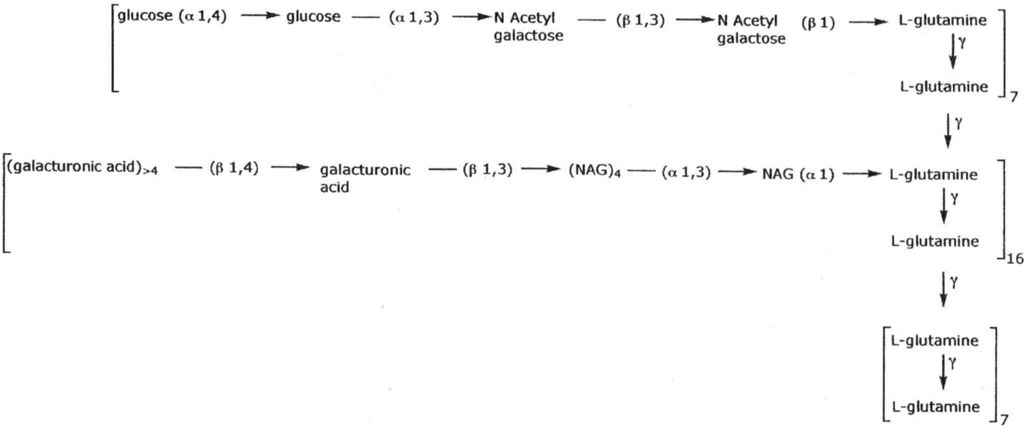

Figure 3.23. Proposed cell wall polymer in *Natronococcus*. (From Niemetz et al., 1997.[13] Used with permission.)

## 3.6.1. *Peptidoglycan Synthesis*

The production of cell walls in endospore-forming bacteria is a relatively unique process in that the peptidoglycan of the spore is unlike that of the vegetative cell. The first phase of peptidoglycan synthesis occurs when an asymmetric septation is produced in the cell. This peptidoglycan provides for compartmentalization of the cytoplasm in the mother cell where the prespore will be developed. Because this peptidoglycan divides the cytoplasm, there is no need to protect the developing prespore from osmotic imbalances. The function of this segment of peptidoglycan is not clear but it may provide a framework that is used to direct future development. Peptidoglycan synthesis associated with prespore development is similar if not identical to that in the vegetative peptidoglycan. As the prespore develops, peptidoglycan in the septation is partially hydrolyzed starting at the central core of the cell and progressing toward the sides of the vegetative cell. At this time, a flexible membrane extends from the walls of the cell and engulfs the prespore (see Figure 3.24).

The engulfed prespore inside the mother cell is separated by an inner forespore membrane (ifm) and an outer forespore membrane (ofm). It may be considered that this membrane configuration is an invagination of the plasma membrane and when the prespore is completely engulfed, the association of the ifm and ofm becomes independent from the plasma membrane. This double-membrane configuration has a space between the ifm and ofm and this space is likened to the extracellular space. Two types of peptidoglycans are produced with the precursors transported across each membrane (ifm and ofm). According to electron microscopic observations, a thin dense layer of peptidoglycan is formed on the outer face of the ifm and this is called the primordial germ-cell wall (PGCW). A thick and less dense layer is formed along the face of the ofm which becomes the cortex of the spore. In *Bacillus megaterium*, the PGCW has more cross linkages than the peptidoglycan in the cortex. Ultimately

**Bacterial Cell Structure**

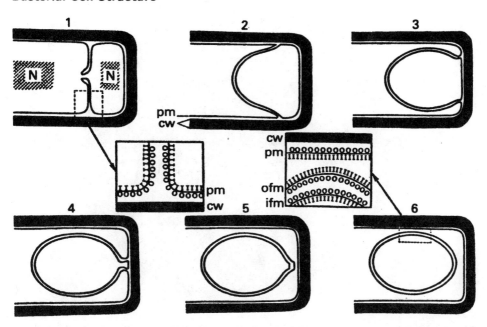

Figure 3.24. Model indicating early development of the bacterial endospore. Figures 1 through 6 indicate the orientation of the membranes of the forespore. pm, Plasma membrane; cw, cell wall; ofm, outer forespore membrane; ifm, inner forespore membrane. Note that in the final figure there are two sources of peptidoglycan and both originate from invagination of the plasma membrane. (From Rogers, 1983.[16])

with germination, the PGCW becomes the cell wall of the germinating spore while the cortex structure is degraded to enable water and nutrients to enter the germinating spore. Many of the enzymes in the PGCW are inactive in the mature spore but become functional in early spore germination.

The final phase of spore wall formation is the synthesis of the peptidoglycan in the cortex. Figure 3.25 shows an electron micrograph of a *Bacillus subtilis* spore with the various segments of the anatomy revealed. Repeating units of the cortex peptidoglycan are unique and contain muramyl δ-lactam, which is a modified muramic acid. In the muramyl δ-lactam, the carboxyl moiety from a lactic acid unit of muramic acid forms an internal amide (see Figure 3.26). There is no peptide unit attached to the muramyl lactam and in a substantial number of cases, *N*-acetylmuramic acid is linked to L-alanine without any additional amino acids. The structure given in Figure 3.26 is commonly found in the spore wall of about a half-dozen bacterial species examined. The average glucan chain length is 80 to 100 sugar residues and there are no teichoic acids in the peptidoglycan of spore walls. As only about 19% of the cross-bridge linkages through the tetraamino acid peptide are formed, the peptidoglycan of the cortex provides less structural strength than in cell walls of vegetative cells.

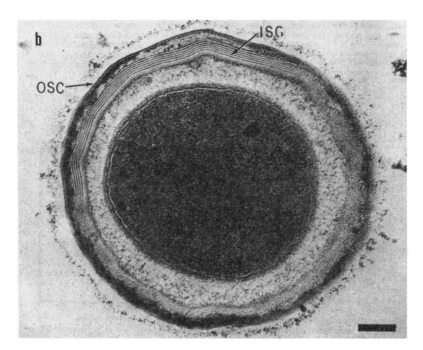

Figure 3.25. An electron micrograph of a section through a mature spore from *Bacillus subtilis*. The electron-transparent region of the spore is the cortex which contains $Ca^{2+}$-dipicolinic acid and spore-associated peptidoglycan. OSC, Outer spore coat; ISC, inner spore coat with several protein layers. (From Santo and Doi, 1974.[17] Used with permission.)

The small number of cross linkages in the peptidoglycan of the cortex produces a structure that is more susceptible to destruction than the cell wall peptidoglycan, and this degradation of the cortex structure is paramount for germination. The region of the cortex containing high concentrations of Ca-dipicolinic acid which accounts for thermal resistance.

## 3.6.2. Proteins and Enzymes

Genes for sporulation have been characterized and a partial listing is given in Table 3.5. These genes may be in clusters that are broadly distributed around the chromosome and many of these genes may be under regulation of specific sigma factors. The activities of various sigma factors in sporulation are discussed in Chapter 7. About 25 genes are needed for morophological changes dealing with prespore, forespore, and mother cell. In *Bacillus subtilis*, the synthesis of spore peptidoglycan is attributed to a series of membrane-bound enzymes that have penicillin binding activities and they are identified as PBP1, PBP2A, PBP2B, PBP3, PBP4, and PBP5. The amount of PBP1 decreases in the cells as the stationary phase is reached and from this it has been suggested that it has a function only in the vegetative spore. PBP2A

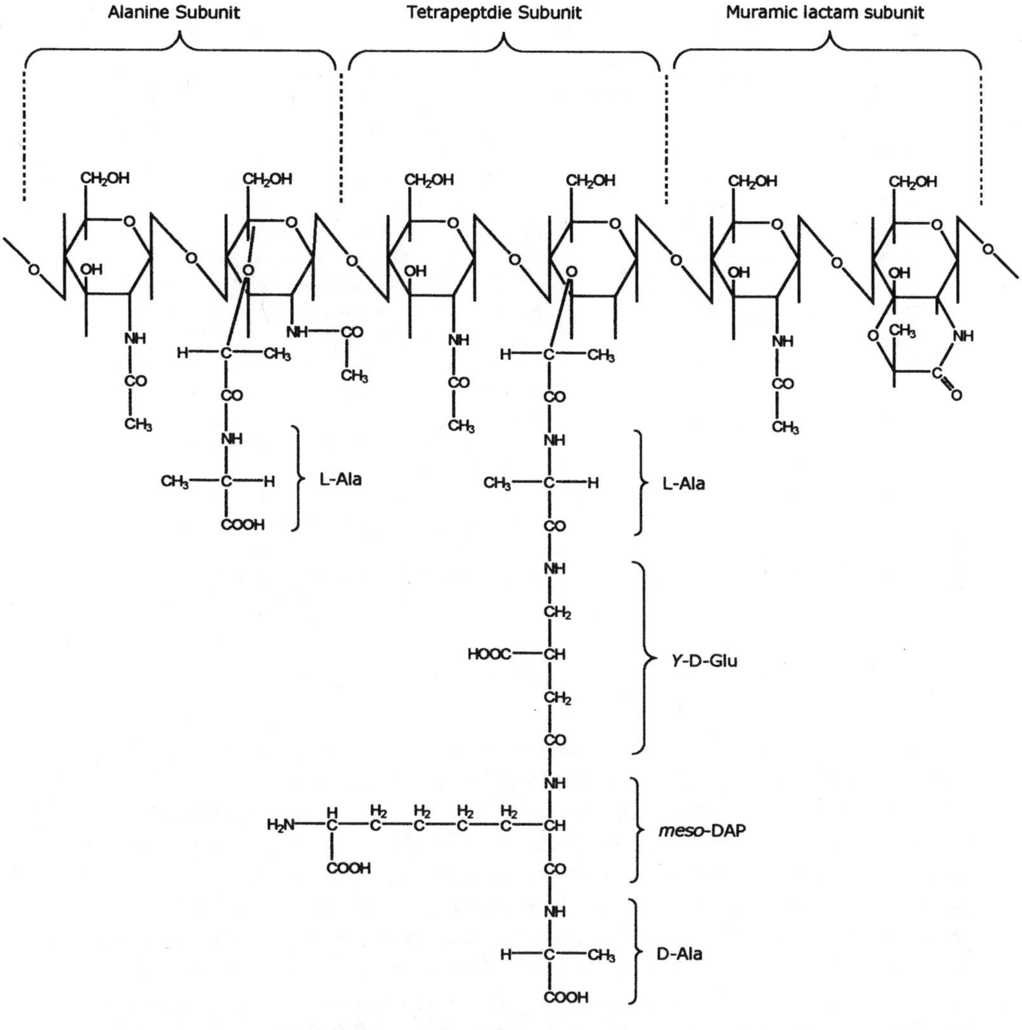

Figure 3.26. Repeating unit making up the cell wall polymer of endospore from *Bacillus*. The approximate abundance of the subunits in the endospore wall is as follows: Alanine subunit = 55%, tetrapeptide subunit = 15%, muramic lactam subunit = 30%. (From *The Microbial World*, 5/E by Stanier et al., © 1986. Reprinted by permission of Pearson Education, Inc., Upper Saddle River, NJ.)

has a function in synthesis of the sidewalls of vegetative cells of *Bacillus subtilis* and in terms of activity, may be comparable to PBP2 of *E. coli*. PBP2B may be required for septum formation in both vegetative cells and spores. There is a 28% similarity between PBP2B from *B. subtilis* and PBP3 from *E. coli*. During the phase of spore cortex development, synthesis of PBP3 increases, thereby suggesting it has a role in peptidoglycan synthesis in the cortex. Two PBPs produced with sporulation are PBP4 and PBP5 with PBP4 produced prior to PBP5. PBP4 has a sequence homology similar

Table 3.5. Genes of *Bacillus subtilis* associated with sporulation.[a]

| Gene | Size of protein encoded (kDa) | Proposed function of gene product |
|------|------|------|
| *pomA* | 100.5 | PBP1; vegetative transglycosylase/transpeptidase |
| *spoVD* | 71.1 | Similar to PBP2B of *E. coli* |
| *pbpA* | ? | PBP2A; cell elongation |
| *pbpE* | 51.4 | PBP4; unknown function |
| *dacA* | 45.5 | PBP5: a vegetative D-Ala carboxylase |
| *dacB* | 40.1 | PMP5*: D-Ala carboxylase in the cortex |
| *spoVE* | 40.1 | Similar to proteins FtsW and RodA of *E. coli* |
| *pbpB* | 77.6 | Septum formation during growth and sporulation |
| *pbpC* | 70.7 | Synthesis of side wall of cell and spore |
| *ftsA* | 48.1 | Septum formation during growth and sporulation |
| *ftsZ* | 40.3 | Septum formation during growth and sporulation |
| *mraY* | 35.6 | UDP-*N*-AcMur(pentapeptide):undecaprenyl phosphate phospho-*N*-Ac-Mur (pentapeptide) transferase |
| *murG* | 39.9 | UDP-*N*-AcGlc:*N*-Ac-Mur (pentapeptide) pyrophosphoryl-undecaprenol *N*-AcGlc transferase |
| *murE* | 54.3 | Enzyme for adding diaminopimelic acid |

[a]From Buchanan et al., 1994.[14]

to D-aminopeptidase of other bacteria but the role of this enzyme in sporulation is unknown. PBP5 is located in the ofm by a C-terminal amphiphilic helix and is synthesized only in the mother cell. There is 30% to 35% identity between PBP5 and D,D-carboxypeptidase from various bacteria. The PBP5 is required for synthesis of cortex peptidoglycan and it is indirectly associated in producing a heat resistant spore.

Studies with various *Bacillus* species revealed that the third amino acid residue in the peptidoglycan of the vegetative cell is L-lysine while the spore wall contains *meso*-diaminopimelic acid. In addition, the cross bridge in vegetative cells is D-alanyl-D-isoasparaginyl and its absence in spore walls suggests that the spore peptidoglycan has a direct cross bridge between the carboxyl group of D-alanine on one chain and the ε-amino of diaminopimelic acid of an adjacent peptide chain. These apparent differences in peptidoglycan between vegetative cells and spores reflect prokaryotic differentiation and account for preferential synthesis of endospore peptidoglycan.

Autolytic enzymes such as lysozymes are highly effective in digesting the structure of the peptidoglycan during sporulation and germination. One activity of peptidoglycan hydrolysis occurs after the septum is developed and a membraneous structure engulfs the prespore. The release of the mature spore from the mother cell is a result of cell wall autolysis of the mother cell. In the early stages of germination, the peptidoglycan structure in the cortex is hydrolyzed to allow nutrients to enter the spore more easily and also enables the germinating spore to extend the cell through the spore wall.

## 3.7. S-Layers of Prokaryotes[19]

Since the report by Houwink in 1956 indicating the presence of a unique structural array defining the surface layer of *Halobacterium halobium* cells, considerable attention has been given to the distribution, characterization, and function of surface layers (S-layers) in bacteria and archaea. While S-layers have been observed in hundreds of bacterial species and more than 50 species of archaea, there is minimal distribution of S-layers in eukaryotic cells and appears to be limited to a few species of algae. These crystalline surface layers are present on prokaryotes from various physiological and taxonomic groups. Recent isolates display the S-layer but cultures maintained in the laboratory for some time will grow without production of the S-layer. Models indicating the position of the S-layer with respect to other structures in cells from bacteria and archaea are given in Figure 3.27. The S-layer is composed of repeating subunits consisting of a carbohydrate moiety covalently attached to the protein unit to produce an outer surface that is highly hydrophobic.

### 3.7.1. Structure

The S-layers can be demonstrated by negative staining with phosphotungstic acid followed by observation with transmission electron microscope. Recently images have been obtained with atomic force microscopy and provide excellent information for mapping the structural detail of the S-layer lattice. There is a specific array or lattice

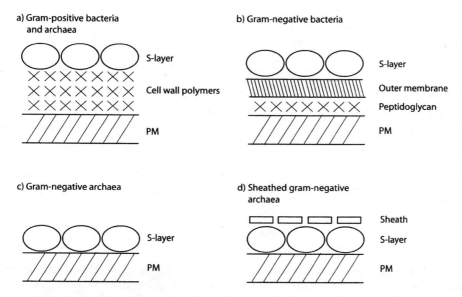

Figure 3.27. Location of S-layers on various prokaryotes. (From Sleytr et al., 1996.[20] Used with permission.)

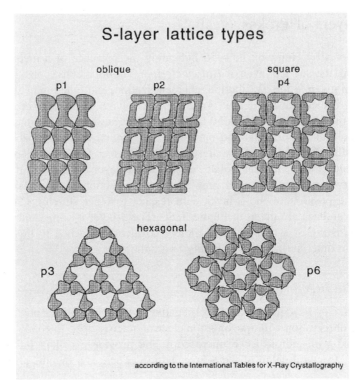

Figure 3.28. Models indicating different types of lattice in S-layers. The lattice types commonly found in prokaryotes are p2, p4, and p6. The protein subunits in each unit as well as the center-to-center spacing are highly reproducible and used to characterize the S-layers. (From Sleytr et al., 1996.[20] Used with permission.)

for the proteins of the S-layers and schematic representation of these are given in Figure 3.28. Hexagonal (p6) lattice is common in archaea while square (p4) and oblique (p2) lattices are associated with bacteria. Electron micrographs showing organisms with S-layers constructed with different lattices are in Figure 3.29. Between the protein aggregates that form the lattice of the S-layer are small pores that vary in size with the bacterial species. The size of the pores is expressed as the distances between their centers and range from 3 to 35 nm with specific examples given in Table 3.6.

## 3.7.2. Chemistry

Proteins in the S-layer can account for as much as 15% of the total cell protein and thus provide adequate material for examination. Characteristics of S-layers in various bacteria are found in Table 3.7. The number of hydrophobic amino acids in S-layer glycoproteins is 40% to 50% and generally the cysteine or methionine content is low.

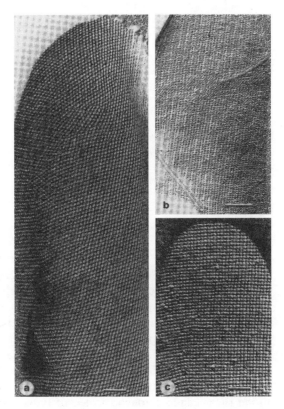

Figure 3.29. Electron micrographs of bacteria subjected to freeze-etched preparation to illustrate different lattice arrangements. (**a**) The p-6 lattice in *Thermoanaerobacter thermohydrosulfuricus* L111-69. (**b**) A p-2 lattice in *Bacillus stearothermophilus* NRS 2004. (**c**) The p-4 lattice in *Desulfotomaculum nigrificans* NCIB 8706. (From Sleytr et al., 1996.[20] Used with permission.)

Table 3.6. Crystalline surface layers on bacterial cells.[a]

| Organism | Lattice | Center to center spacing (nm) | $M_r$ (kDa) |
| --- | --- | --- | --- |
| **Bacteria** | | | |
| *Bacillus anthracis* | p6 | 8–10 | 87–95 |
| *Bacillus thermoaerophilus* | p4 | 10 | 116 |
| *Clostridium difficile* | p4 | 7.8 | 38–42 |
| *Clostridium xylanolyticum* | p2 | 6.6 | 180 |
| *Lactobacillus plantarum* | p2 | 8.3 | 56 |
| *Microcystis firma* | p6 | 14–16 | nr |
| *Propionibacterium jensenii* | p2 | 9.5 | 67 |
| *Treponema* sp. | p6 | 16.3 | 62 |
| **Archaea** | | | |
| *Methanocelleus marisnigri* | p6 | 13.5 | 138 |
| *Pyrobaculum organotrophum* | p6 | 20.6 | 80 |
| *Pyrodictium abyssi* | p6 | 15 | 126 |
| *Sulfolobus shibatae* | p3 | 20.3 | nr |
| *Thermococcus stetleri* | p6 | 18 | 210 |

[a]Modified from Messner and Sleytr, 1992.[21]
nr, Not reported.

Table 3.7. Characteristics of proteins in selected S-layers.[a]

| Organism | Signal sequence (number of amino acids) | $M_r$ (kDa) (from DNA sequence) | pI | α-Helix | β-Sheet | β-Turn | Random |
|---|---|---|---|---|---|---|---|
| | | | | Secondary structure (%) | | | |
| *Aeromonas salmonicida* | 21 | 50.8 | 4.8 | 25 | 26 | 7 | 43 |
| *Bacillus brevis* 47 (OWP) | 24 | 103.7 | 3.8 | 4 | 26 | 44 | 26 |
| *Deinococcus radiodurans* | nr‡ | 98.0 | nr | 7 | 31 | 11 | nr |
| *Halobacterium halobium* | 34 | 86.5 | nr | nr | nr | nr | nr |
| *Campylobacter fetus* | nr | 96.7 | 4.6 | 39 | 46 | 7 | 8 |
| *Rickettsia prowazekii* | nr | 169.8 | 5.8 | nr | nr | nr | nr |
| *Methanothermus sociabilis* | 22 | 65.0 | 8.4 | 7 | 44 | nr | nr |

[a]Modified from Messner and Sleytr, 1992.[21]
nr, Not reported.

The levels of lysine, glutamic acid, and aspartic acid are each approximately 10%. The subunits of the S-layer range from 50 to 220 kDa but owing to the presence of carbohydrates, the molecular mass of the proteins would be expected to be smaller. Most have an isoelectric point (pI) that is acidic and exceptions to this generalization are *Methanothermus* spp. which have an isoelectric point of pH 8.4 and *Lactobacillus* sp. which have a pI of 9.5. About 20% of the protein is in an α-helix format with about 40% as a β-sheet. Important posttranslation changes to the S-layer proteins include cleavage of C- or N-terminal fragments, phosphorylation of amino acids, and addition of sugar groups to amino acids. Amino acids residues participate in linkage to the sugar chains through oxygen of tyrosine or serine and through the amine of asparagine (see Table 3.8).

The carbohydrates are either linear or branched and may constitute 1% to 20% of the glycoprotein mass. The attachment of the polysaccharides into the S-layer proteins is through either asparagine or threonine. This presence of glycoproteins extremely unique and is one of the few instances of postribosomal processing of protein in eubacteria. Neutral hexoses, especially rhamnose which is a 6-deoxysugar, contribute to the hydrophobic character of the glycoprotein. A few examples of the different types of carbohydrate structures in S-layers are given in Table 3.8.

## 3.7.3. Biosynthesis

Proteins of the S-layer are one of the major proteins constituents in a cell, with more than $5 \times 10^5$ protein monomers required to cover an average bacterial cell. With a generation time of 20 min for this hypothetical organism, there would be a requirement for 500 copies of S-layer protein to be translocated to the cell surface and incorporated into the lattice every second. Currently, more than 25 genes that carry the information for the S-layer monomers have been cloned and sequenced. While the mRNA in prokaryotes has a 3- to 4-minute half-life, the intracellular half-life of the mRNA for S-layer protein monomers is 15 to 22 min. Stability of the mRNA is attributed to the 13 base inverted repeats that account for stem-loop structures that function as atten-

Table 3.8. Examples of oligosaccharides attached to protein molecules of the S-layer.[a]

| Organism | Structure of S-layer protein | Attached to amino acid |
|---|---|---|
| *Bacillus stearothermophilus* | $[\alpha\text{-L-Rha}(1\rightarrow2)\text{-}\alpha\text{-L-Rha-}(1\rightarrow3)\text{-}\beta\text{-L-Rha}]_{50}1\rightarrow N$-Asparagine | |
| *Clostridium symbiosum* | $\alpha$-D-ManNAc-$(1\text{—}4)$-$\beta$D—GalNAc$(1\rightarrow3)$-$\alpha$-D-BacNAc-$(1\rightarrow4)$-$\alpha$-D-GalNAc$[PO_4]_{15}$-**?????**[b] | |
| *Halobacterium halobium* | GluA-$(1\rightarrow4)$GluA$(1\rightarrow4)$GluA$(1\rightarrow4)\beta$-D-Glu $1\rightarrow$Asparagine<br>$\updownarrow$     $\updownarrow$     $\updownarrow$<br>$OSO_3^-$   $OSO_3^-$   $OSO_3^-$ | |
| *Lactobacillus buchneri* | $\alpha$-D-Glu$\rightarrow$6)$[\alpha$D-GluA$(1\rightarrow6)]_{4\text{-}6}\alpha$D-Glu-O-Serine | |
| *Thermoanaerobacter thermohydrosulfuricus* | 3-O-Methyl-L-Rha$(\alpha1\rightarrow4)$D-Man$(\alpha1\rightarrow3)$[-L-Rha-$(\alpha1\rightarrow4)$D-Man$]_{27}(\alpha1\rightarrow3)$-L-Rha-$(\alpha1\rightarrow4)$D-Man$(1\rightarrow3)$[L-Rha$]_3(\alpha1\rightarrow3)$D-Gal-O-Tyrosine | |

[a]From Messner, 1996.[23]
[b]Unknown aminoacid. Glu, Glucose; Rha, rhamnose; Man, mannose, GluA, glucuronic acid; GalNAc, galactose *N*-acetyl; ManNAc, mannose *N*-acetyl; BacNAc, *N*-acetylbacillosamine (2-acetamido-4-amino-2,4,6-trideoxyglucose).

uators and also protect the 5' end of the mRNA from exonuclease digestion. The exit of the S-layer monomer across the plasma membrane is controlled by a gene for S-protein secretion; however, little is known about this mechanism. Although yet unexplained, considerable homology was observed in the DNA content of genes for S-layer proteins and flagellin. It is not uncommon for bacteria to have several genes for S-layer monomers, and when this occurs only one gene is expressed at a time. Changes in protein synthesis can be regulated by the environment as is the case with *Bacillus stearothermophilus*, in which production of a given type of S-layer monomer is influenced by oxygen stress.

## 3.7.4. Function

Because bacteria of diverse phylogenetic groups representing unique physiological processes produce S-layers, it has often been suggested that S-layers have a broad spectrum of functions. Pathogenicity of *Aeromonas salmonicida*, *Campylobacter fetus*, and *Rickettsia* is enhanced by the presence of the S-layer. This increase in virulence results from resistance to phagocytosis, changing of antigenic surface of the cell to evade antibody reactions, or stimulation of the production of toxin-neutralizing antibodies. *Clostridium difficile* has two distinct types of S-layer proteins, both of which function in adhesion to intestinal cells. The S-layer in *Bacillus anthrasis* provides capsulation in those strains that do not produce capsules. The cells of *Lactobacillus* have an acid-resistant S-layer that enables them to attach to cells in regions of the human body where low pH conditions occur.

In addition, S-layers have been reported to function as adhesion sites for bacterial exoenzymes. Enzymes such as pullulanase, which hydrolyzes both $\alpha$-1,4 and $\alpha$-1,6 linkages in sugar polymers, have a catalytic and a noncatalytic domain. The noncatalytic region of the enzyme contains three repeats each of about 50 amino acid residues called SLH (S-layer homologous domains) that appear to be similar to the repeats found in proteins of the S-layer. A partial listing of the exoenzymes with SLH domains is given in Table 3.9.

The S-layer functions as a molecular sieve to prevent entry of hostile molecules that would be detrimental to underlying structures. Hydrolytic enzymes larger than 40 kDa would be excluded from the cell. In addition, bacteriophage would be prevented from attaching to the sites on the stratum beneath the S-layer and grazing by protozoa would be deterred. Because the glycoproteins of the S-layer are hydrophobic, the pores or channels would be hydrophobic also and would promote the uptake of hydrophobic antibiotics (i.e., streptinigrin and chloramphenicol) and porphyrins.

The S-layer from the cyanobacterium, *Synechococcus* sp. has been suggested to participate in the formation of $CaSO_4$. The unmineralized S-layer has diamond-shaped pores and when gypsum is formed, the pores become smaller and change shape. It has also been proposed that S-layers of aquatic prokaryotes may be involved in the formation of calcite, celestite, and strontionite. Further discussion on mineral formation by cyanobacteria is in Chapter 14.

Finally, it is important to recognize the S-layer in determining the shape of archaea and providing a rigid structure to protect the cell from osmotic stress. By careful examination with electron microscopy, it is evident that S-layers are free from lattice faults or cracks along the cylindrical part of the cell but some disruption of the lattice is apparent at the poles of the cells. In *Thermoproteus* and *Pyrobaculum*, the glucoprotein subunits of the S-layer are covalently united to provide a strong exoskeleton. *Halobacterium halobium* requires the presence of a high concentration of salt and linearity of the cell is dependent in part on the presence of the sulfate moieties on the charged glycan chains, see Table 3.8. The hexagonally arranged lattice (p6) of the S-layer in *Sulfolobus* provides a high degree of flexibility and accounts for pleomorphism.

In 1980, "square bacteria" with cells measuring 0.15 to 0.5 µm by 5 µm were found by A.E. Walsby in saline waters of the Sinai Peninsula. Since then, bacterial cells in the shape of irregular rectangle, square, triangle, or sphere have been reported in samples from hypersaline pools of Southern California and other regions. Examples

Table 3.9. Exocellular enzymes of bacteria containing S-layer homologous domains.[a]

| Organism | Enzyme |
| --- | --- |
| *Anaerocellum thermophilum* | Endoglucanase |
| *Bacillus* sp. KSM-635 | Alkaline cellulase |
| *Bacillus* sp. XAL601 | α-Amylase pullulanase |
| *Clostridium josui* | Endoglucanase |
| *Clostridium thermocellum* DSM | Lichenase |
| *Clostridium thermocellum* | Xylanase |
| *Thermoanaerobacterium thermosulfurigenes* EM1 | Pullulanase, polygalacturonate hydrolase, xylanase |

[a]Adapted from Beveridge et al., 1994.[24]

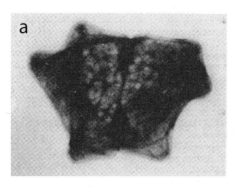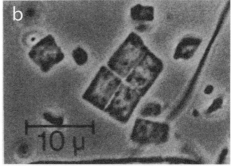

Figure 3.30. Prokaryotes with unusual shapes. (a) Star-shaped bacterial cell is known as *Stella vascolata* and was stained with 1% uranyl acetate. $\times$ 18,000. (From Vasilyeva, 1985.[25] Used with permission. (b) Archael cell with square corners. Phase contrast micrograph and gas vacuoles were collapsed in preparation. (From Stoeckenius, 1981.[26] Used with permission.)

of cells with star and rectangle shape are given in Figure 3.30. Although prokaryotes that display geometric shapes are not yet cultivated in the laboratory, it is known from electron microscopic studies that the surface of these cells display an hexagonal or tetragonal arrangement which is characteristic of S-layers.

## 3.7.5. Potential Applications[20]

The physical and chemical characteristics of the S-layers have prompted many to explore their possible application in biotechnology. With the uniform molecular lattices, these S-layers would make ideal ultrafiltration membranes and their stability can be enhanced by cross linking the lattices by using glutaraldehyde or other reagents. The S-layer from bacteria can be isolated and when reconstituted serves as an excellent support matrix for the immobilization of proteins, enzymes, or antibodies, and various diagnostic test systems have been constructed using them. Appropriate antigens have been fixed onto the S-layers for the immunization of fish against *Aeromonas salmonicida* and humans against *Streptococcus pneumoniae*. Bacteria producing S-layers have been cultivated in large quantities and the proteins from the S-layer have been disassociated from the bacterial cell wall and have been reformed on appropriate boards to make many different types of ampherometric and optical biosensors.

## 3.8. Distribution of Cell Wall Structures

A summary of cell wall structures is provided in Table 3.10. Although there is considerable difference between bacterial and archaeal cells, a pattern is apparent in the molecular structure of prokaryotic cell walls. The highly resistant S-layers are present

Table 3.10. A comparison of cell wall properties and structures in prokaryotes.

| Characteristic | Gram-positive bacteria | Gram-negative bacteria | Archaea[a] |
|---|---|---|---|
| Osmotic pressure in cytoplasm | 20 atm | < 10 atm | Probably < 10 atm |
| Amount of cell wall that is peptidoglycan | 40–95% | 10–20% | Not present |
| Diaminoacid common in peptidoglycan | Lysine or DAP | DAP | Not present |
| Amino acids present | L and D forms | L and D forms | Only L forms |
| Action of penicillin and bacitracin | Sensitive | Sensitive | Resistant |
| S-layer | Present in some | Present in some | Present in some |
| Pseudomurein | Present only in spores | Absent | Present in some |
| Methanochrodroitin | Absent | Absent | Present in some |
| Heteropolysaccharide | Absent | Absent | Present in some |
| Polyglutamine glyconjugate | Absent | Absent | Present in some |

[a]Although archaea have both Gram-positive and Gram-negative types based on the staining reaction, sufficient research has not yet been conducted on each of the cell types to enable one to make meaningful generalizations about cell wall properties.

in various species of prokaryotes, and this may suggest that this protein-containing structure was characteristic of early evolutionary forms of cells. The macromolecules of the cell wall contain repeating molecules that form a backbone from which side chains extend. In the case with peptidoglycan and pseudomurein, cross-bridge formations contribute to increased strength of the cell wall. Unusual bonds between monosaccharides and between amino acids produce a macromolecule highly resistant to many of the enzymes associated with cellular metabolism. The use of muramic acids and D-forms of amino acids in the structures contributes to the inert character of the cell wall. Penicillin binding proteins initially evolved to enable bacteria to grow in environments where these antibiotics were produced by different types of microorganisms.

# References

1. Salton, M.R.J. 1964. *The Bacterial Cell Wall.* Elsevier, New York.
2. Cummins, C.S. 1974. Bacterial cell wall structure. In *Handbook of Microbiology* (A.I. Laskin and H.A. Lechevalier, eds.). CRC Press, Cleveland OH, pp. 251–284.
3. Shockman, G.D. and A.J. Wicken (eds.). 1981. *Chemistry and Biological Activities of Bacterial Surface Amphiphiles.* Academic Press, New York.
4. Shockman, G.D. and J.-V. Hültje. 1994. Microbial peptidoglycan (murein) hydrolases. In *Bacterial Cell Wall* (J.-M. Ghuysen and R. Hakenbeck, eds.). Elsevier, New York, pp. 131–166.

5. Mims, C., J. Playfair, I. Roitt, D. Wakelin, and R. Williams. 1998. *Medical Microbiology.* Mosby, Philadelphia.

6. Ghuysen, J.-M. and G. Dive. 1994. Biochemistry of the penicilloyl serine transferases. In *Bacterial Cell Wall* (J.-M. Ghuysen and R. Hakenbeck, eds.). Elsevier, Amsterdam, pp. 103–129.

7. Ayala, J.A., T. Garrido, M.A. DePedro, and M. Vicente. 1994. Molecular biology of bacterial septation. In *Bacterial Cell Wall* (J.-M. Ghuysen and R. Hakenbeck, eds.). Elsevier, Amsterdam, pp. 73–99.

8. Higgins, M.L. and G.D. Shockman. 1971. Prokaryotic cell division with respect to wall and membrane. *CRC Critical Reviews in Microbiology* 1:29–72.

9. Matsuhashi, M. 1994. Utilization of lipid-linked precursors and the formation of peptidoglycan in the process of cell growth and division: Membrane enzymes involved in the final steps of peptidoglycan synthesis and the mechanism of their regulation. In *Bacterial Cell Wall* (J.-M. Ghuysen and R. Hakenbeck, eds.). Elsevier, Amsterdam, pp. 44–72.

10. Mirelman, D. 1979. Biosynthesis and assembly of cell wall peptidoglycan. In *Bacterial Outer Membranes* (M. Inouye, ed.). John Wiley, New York, pp. 116–166.

11. Yao, X., M. Jericho, D. Pink, and T. Beveridge. 1999. Thickness and elasticity of Gram-negative murein sacculi measured by atomic force microscopy. *Journal of Bacteriology* 181:6865–6875.

12. Kandler, O. and H. König. 1985. Cell envelopes of archaebacteria. In *The Bacteria*, Vol. VIII (C.R. Woese and R.S. Wolfe, eds.). Academic Press, New York, pp. 415–458.

13. Niemetz, R., U. Karcher, O. Kandler, B.J. Tindall, and H. Konig. 1997. The cell wall polymer of the extremely halophilic archaeon *Natronococcus occultus. European Journal of Biochemistry* 249:905–911.

14. Buchanan, C.E., A.O. Heneriques, and P.J. Poggpt. 1994. Cell wall changes during bacterial endospore formation. In *Bacterial Cell Wall* (J.-M. Ghuysen, and R. Hakenbeck, eds.). Elsevier, New York, pp. 167–184.

15. Gould, G.W., A.D. Russell, and D.E.S. Stewart-Tull. 1994. *Fundamental and Applied Aspects of Bacterial Spores.* Society for Applied Symposium Series Number 23. Blackwell, Boston.

16. Rogers, H.J. 1983. *Bacterial Cell Structure.* American Society for Microbiology Press, Washington, D.C.

17. Santo, L.Y. and R.H. Doi. 1974. Ultrastructure analysis during germination and outgrowth of *Bacillus subtilis* spores. *Journal of Bacteriology* 120:475–481.

18. Stanier, R.Y., E.A. Adelberg, and J.L. Ingraham. 1986. *The Microbial World*, 5th edit. Prentice-Hall, Englewood Cliffs, NJ.

19. Messner, P., G. Allmaier, C. Schäffer, T. Wugeditsch, S. Lortal, H. König, R. Niemetz, and M. Dorner. 1997. Biochemistry of S-layers. *FEMS Microbiology Reviews* 20:25–46.

20. Sleytr, U.B., P. Messner, D. Pum, and M. Sára. 1996. Occurrence, location, ultrastructure and morphogenesis of S-layers. In *Crystalline Bacterial Cell Surface Proteins* (U.B. Sleytr, P. Messner, D. Pum, and M. Sára, eds.). R.G. Landes/Academic Press, Austin, TX, pp. 5–34.

21. Messner, P. and U.B. Sleytr. 1992. Crystalline bacterial cell-surface layers. *Advances in Microbial Physiology* 33:213–275.

22. König, H. 1988. Comparative aspects on archaeobacterial proteinaceous cell envelopes. In *Crystalline Bacterial Cell Surface Layers* (U.B. Sleytr, P. Messner, D. Pum, and M. Sára, eds.). Springer-Verlag, New York, pp. 7–10.

23. Messner, P. 1996. Chemical composition and biosynthesis of S-layers. In *Crystalline Bac-*

*terial Cell Surface Proteins* (U.B. Sleytr, P. Messner, D. Pum, M. Sára, eds.). R.G. Landes/
Academic Press, Austin, TX, pp. 35–76.

24. Beveridge, T.J. and thirty-two co-authors. 1994. Functions of S-layers. *FEMS Microbiology Reviews* 20:99–149.

25. Vasilyeva, L.V. 1985. *Stella*, a new genus of soil prosthecobacterium, with proposals for *Stella humosa* sp. nov. and *Stella vacuolata* sp. nov. *International Journal of Systematic Bacteriology* 35:518–521.

26. Stoeckenius, W. 1981. Walsby's square bacterium: Fine structure of an orthogonal prokaryote. *Journal of Bacteriology* 148:352–360.

# Additional Reading

## *Cell Shape*

Harold, F.M. 1990. To shape a cell: An inquiry into the causes of morphogenesis of microorganism. *Microbiological Reviews* 54:381–431.

Höltje, J.-V. 1998. Growth of the stress-bearing and shape-maintaining murein sacculus of *Escherichia coli*. *Microbiological and Molecular Biology Reviews* 62:181–203.

Satta, G., R. Fontana, and P. Canepari. 1994. The two-competing site (TCS) model for cell shape regulation in bacteria. *Advances in Microbial Physiology* 36:182–246.

## *Cell Wall Structure*

Kandler, O. and H. König. 1993. Cell envelopes of archaea: Structure and chemistry. In *Biochemistry of Archaea (Archaebacteria)* (M. Kates, D.J. Kushner, and A.D. Matheson, eds.). Elsevier, Amsterdam, pp. 223–225.

Mohan, S., C. Dow, and J.A. Coles (eds.). 1992. *Prokaryotic Structure and Function*. Cambridge University Press, New York.

Salton, M.R.J. 1960. Surface layers of the bacterial cell. In *The Bacteria*, Vol. I (I.C. Gunsalus and R.Y. Stanier, eds.). Academic Press, New York, pp. 97–152.

Scherrer, R. 1984. Gram's staining reaction, Gram types and cell walls of bacteria. *Trends in Biochemical Science* 9:242–245.

Wientjes, F.B., L. Conrad, and N. Nanninga. 1991. Amount of peptidoglycan in cell walls of Gram-negative bacteria. *Journal of Bacteriology* 173:7684–7691.

## *Cell Wall Turnover*

Archibald, A.R., I.C. Hancock, and C.R. Harwood. 1993. Cell wall structure, synthesis, and turnover. In Bacillus subtilis *and Other Gram-Positive Bacteria* (A.L. Sonenshein, J.A. Hoch, and R. Losick, eds.). American Society for Microbiology Press, Washington, D.C., pp. 381–410.

Doyle, R.J., J. Chaloupka, and V. Vinter. 1988. Turnover of cell walls in microorganisms. *Microbiological Reviews* 52:554–567.

## *Antibiotic Resistance*

Grochs, P., L. Gutmann, R. Legrand, B. Schoot, and J.L. Mainardi. 2000. Vancomycin resistance is associated with serine containing peptidoglycan in *Enterococcus gallinarum*. *Journal of Bacteriology* 182:6228–6232.

Koch, A. 2000. Penicillin binding proteins, beta-lactams, and lactamases: Offenses, attacks, and defensive counter measures. *Critical Reviews in Microbiology* 26:205–220.

## S-Layers

Garduno, R.A., A.R. Moore, G. Olivier, A.L. Lizana, E. Garduno, and W.W. Kay. 2000. Host cell invasion and intracellular resistance by *Aeromonas salmonicida*: Role of the S-layer. *Canadian Journal of Microbiology* 46:660–668.
Sleytr, U.B. and T.J. Beveridge. 1999. Bacterial S-layers. *Trends in Microbiology* 7:253–260.
Sleytr, U.B., P. Messner, D. Pum, and M. Sára. 1988. *Crystalline Bacterial Cell Surface Layers*. Springer-Verlag, New York.

## Spore Cell Wall

Meader-Parton, J. and D.L. Popham. 2000. Structural analysis of *Bacillus subtilis* spore peptidoglycan during sporulation. *Journal of Bacteriology* 182:4491–4499.

# 4
# The Cell Wall Matrix

Chemical and morphological structures that comprise the envelope can no longer be dealt with as unrelated entities. The investigator must now ask how these structures interact dynamically—both temporally during their assembly and functionally during their operation. Loretta Leive, *Bacterial Membranes and Walls*, 1973

## 4.1. Introduction

Cell walls of prokaryotes are more than just structural frameworks for the cells. The cell wall in bacteria is a highly complex region that contains different molecular components in addition to the peptidoglycan and the S-layer. In Gram-negative bacteria, the cell wall encompasses an outer membrane that is unique in that it has a highly organized structure and functions as a permeability barrier for many compounds. Various types of ionized molecules are attached to the peptidoglycan of Gram-positive bacteria and they distribute charges within the cell wall as well as at the surface. These extra components have diverse but specific roles and may even enable bacteria to grow in hostile chemical environments. From various research reports concerning the cell walls of archaea, it can be inferred that these organisms also contain molecules in addition to those that are responsible for cell shape and cell stability.

This chapter examines the chemical structure and biological functions of polysaccharides, acidic polymers, and proteins in the cell walls of prokaryotes. On the basis of the Gram-staining reaction, two distinct types of prokaryotes are differentiated and analysis of the cell wall of these two groups of organisms indicates a fundamental difference in chemical composition. Besides providing structural rigidity for both Gram-positive and Gram-negative bacteria, the cell wall has additional functions that include the following: (1) transports nutrients from the environment to the cell membrane; (2) enhances virulence of pathogens by attaching microbial cells to host animal cells through specific binding activities; (3) protects the cell membrane against enzymes, bile salts, and immune proteins that would lead to destruction of the cell; and (4) influences synthesis of peptidoglycan biosynthesis through the modulation of biosynthetic enzymes and the binding of antibiotics that inhibit peptiglycan formation. To understand the different mechanisms for achieving these functions, it is crucial to examine the composition of lipids, proteins, and carbohydrates found in the cell walls of prokaryotes. Not only is it important to focus on the polymeric nature of these compounds but it is also necessary to examine the mechanisms that are used by microbes to localize these compounds in the cell wall region.

## 4.2. Cell Walls of Gram-Positive Bacteria

To detail fully all the structures in the cell wall of a prokaryote would be tedious, and in most instances all molecular components present have not been resolved. There is a continuum of cellular material from the plasma membrane to the surface of the cell, resulting in the cell wall structure being highly infiltrated with carbohydrates and proteins. Some investigators refer to this somewhat diffuse segment of the cell as the cell envelope, which would include the cell wall and the surface materials. As the designation *envelope* has specific meaning in virology, it may be best to avoid employing this term when discussing bacteria. While the cell wall region of Gram-positive bacteria may appear to be relatively simple on examination by electron microscopy, analytical biological and chemical studies reveal that a variety of materials exist in the cell wall region. Like those of bacteria, the cell walls of archaea may also harbor numerous compounds; however, generalizations about the nonstructural segments of archaeal cell walls cannot be made until additional research has been conducted.

### 4.2.1. Polysaccharides

Neutral polysaccharides are found in the cell walls of several Gram-positive bacteria and perhaps the polysaccharides of the streptococci are best characterized. In 1933, Rebecca Lancefield separated the streptococci into groups A through O based on neutral polysaccharides present in the cell walls. These polysaccharides are covalently bound into the cell wall but can be extracted by a mild acid hydrolysis, which is a treatment of cells with HCl (pH 2) at 100°C for 15 min. The group-specific carbohydrate antigens of β-hemolytic streptococci are, in general, a reflection of the type, amount, and sequence of hexose moieties present. For the β-hemolytic streptococci, the sugar determinants for group A are rhamnose and *N*-acetylglucosamine while for group C they are rhamnose, *N*-acetylglucosamine, and *N*-acetylgalactosamine. Although various polysaccharides have been found associated with the cell walls of Gram-positive bacteria (see Table 4.1), the general theme is that hexoses are more commonly present than pentoses.

### 4.2.2. Acidic Polymers[1–3]

In 1958, Baddiley and co-workers presented evidence for complex acidic polymers in the cell wall that they termed teichoic (from the Greek *teichos*, meaning "wall") acids.

Table 4.1. Monomeric units of polysaccharides found in cell walls of Gram-positive bacteria.[a]

| Genus or group | Sugars present |
| --- | --- |
| β-Hemolytic streptococci | |
| Group A | *N*-acetylglucosamine (30%) and rhamnose (60%) |
| Group A-intermediate | *N*-acetylglucosamine (17%) and rhamnose (70%) |
| Group A-variant | *N*-acetylglucosamine (4%) and rhamnose (85%) |
| Group C[a] | *N*-acetylgalactosamine (40%), rhamnose (42%) and *N*-acetylglucosamine (5%) |
| Group C-intermediate | *N*-acetylgalactosamine (22%), rhamnose (59%), and *N*-acetylglucosamine (4%) |
| Group C-variant | *N*-acetylgalactosamine (2%), rhamnose (88%), and *N*-acetylglucosamine (3%) |
| Actinomycetes (group 1) | Arabinose, galactose |
| (group 2) | Madurose (3-*O*-methyl-D-galactose) |
| (group 3) | Arabinose, xylose |
| *Actinomyces bovis* | Rhamnose, glucose, fucose, 2-deoxytalose |
| *Bacillus anthracis* | Galactose(50%), *N*-acetylglucosamine (33%), and *N*-acetylmannosamine (17%) |
| *Bacillus thuringiensis* | Glucose, glucosamine, galactosamine |
| *Corynebacterium diphtheriae* | Arabinose, galactose |
| *Lactobacillus casei* | Rhamnose, galactose, glucose |

[a]From Davis et al., 1980.[2]
[b]A terminal disaccharide of *N*-acetylgalactosamine is the determinant for group C.

Teichoic acids are linear polymers of ribitol phosphate or glycerol phosphate with D-alanine, monosaccharides, or amino sugars attached through the ribitol or glycerol residue. Because of the phosphate moieties in these wall polymers, teichoic acids are negatively charged under physiological conditions at neutral pH. Based on the site of attachment of the teichoic acids to the cell, there are two types of glycerol teichoic acids. One type is a "wall teichoic acid" in which the glycerol phosphate polymer is attached to the peptidoglycan. These polymers are readily extracted from cell walls of *Staphylococcus*, *Bacillus*, or *Lactobacillus* and may constitute up to 50% of the cell wall or 10% of cell dry mass. In the second type, the glycerol phosphate polymer is tethered to the plasma membrane and this type should be referred to as "membrane teichoic acid" or lipoteichoic acid.

### 4.2.2.1. Teichoic Acids

The ribitol phosphate polymers exist principally as wall-type teichoic acids and there is only a single report of their extension from the plasma membrane. Typically, the ribitol phosphate polymer are attached to a short "linkage unit" that consists of a short glycerol phosphate polymer of two or three units plus a monosaccharide phosphate or a disaccharide phosphate. The wall-type ribitol phosphate teichoic acid and glycerol phosphate polymers can be released from peptidoglycan by mild acid hydrolysis with dilute trichloroacetic acid. The general structures of wall teichoic acids associated with bacteria are given in Figure 4.1. It is not uncommon for a given bacterial species to have both types of wall teichoic acids (polyglycerol phosphate and polyribitol phosphate) as well as a lipoteichoic acid.

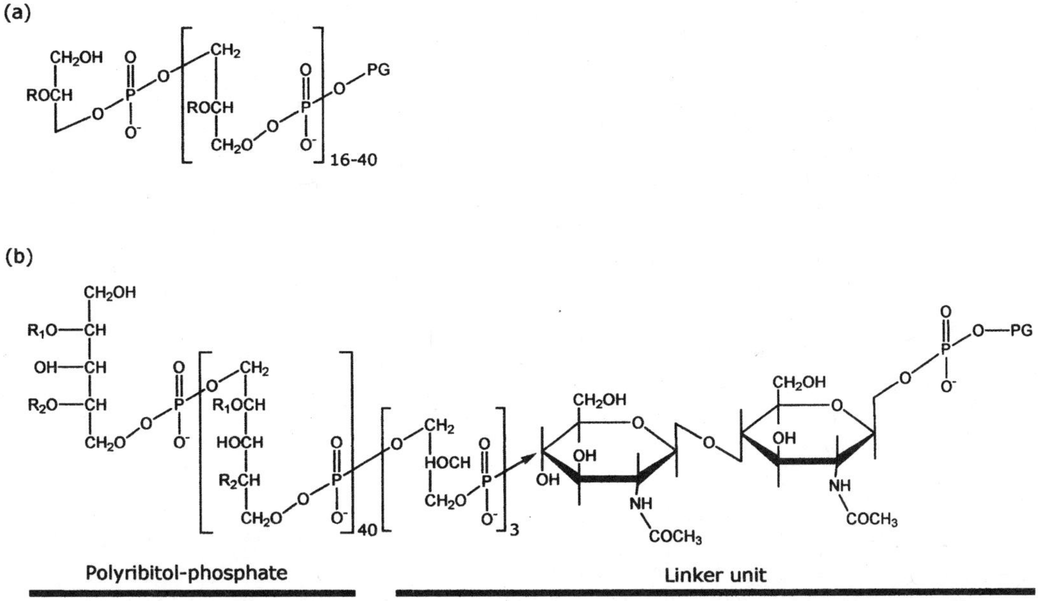

(a)

(b)

**Polyribitol-phosphate**                                    **Linker unit**

Figure 4.1. Structures of teichoic acids in cell walls of *Bacillus subtilis*. (**a**) Polyglycerol phosphate teichoic acids. R = H, D-alanine, or glucose. (**b**) Polyribitol phosphate. The linker unit is attached to the peptidoglycan (PG) with $R_1$ = D-alanine or H and $R_2$ = glucose or H.

Bound to the ribitol or glycerol units are neutral monosaccharides that do not detract from the linearity or the negative charge of the teichoic acid. In the ribitol phosphate polymer, D-glucose or a structure derived from glucose is attached to the 4-hydroxyl position of the ribitol residue. D-Alanine is covalently bound to the 2-hydroxyl position of the ribitol moiety. As seen from examples presented in Table 4.2, glucose or galactose units are associated with glycerol and ribitol teichoic acids. However, in the case of *Streptococcus pneumoniae* choline is present and contributes to the virulence of this organism. It is well accepted that wall teichoic acids are attached to the peptidoglycan although the exact bonding is not always fully resolved. In several instances, it is clear that the teichoic acids are covalently bonded by a phosphodiester linkage to the carbon at the number 6 position of *N*-acetylmuramic acid.

Synthesis of ribitol phosphate, a wall teichoic acid, employs the P-undecaprenyl ($C_{55}$) lipid in the plasma membrane by a mechanism similar to that used in peptidoglycan synthesis (Figure 4.2). With *Bacillus subtilis* as a specific example, there are four phases to the biosynthesis of wall teichoic acids: (1) Soluble nucleotide-monosaccharide precursors are synthesized in the cytoplasm. This involves the enzymatic formation of UDP-*N*-acetylglucosamine and UDP-*N*-acetylmannosamine. (2) Linkage units are attached to the P-undecaprenyl lipid in the plasma membrane to provide organization for subsequent synthesis. (3) The next phase is formation of the main chain onto the linkage units with the addition of the linear polymer of P-glycerol

Table 4.2. Teichoic acids found associated with various Gram-positive bacteria.[a]

| Organism | Type of polymer | Components attached to glycerol or ribitol moieties |
|---|---|---|
| *Bacillus licheniformis* | Glycerol-P | Glucose and galactose |
| *Bacillus subtilis* strain W23 | Ribitol-P | Glucose |
| *Bacillus subtilis* strain 168 | Glycerol-P | Glucose |
| *Lactobacillus plantarum* | Ribitol-P | Glucose |
| *Staphylococcus aureus* | Ribitol-P | *N*-acetylglucosamine |
| *Streptococcus mutans* | Glycerol-P | Galactose |
| *Streptococcus pneumoniae* | Glycerol-P | Choline |

[a]Adapted from Rogers, 1983.[25]

from the cytoplasm with glycerol and CDP interaction. Phosphate for the teichoic acid polymer comes from CDP. Concomitant is the addition of D-glucose to the central carbon of glycerol. (4) The final phase is the linkage of teichoic acid to the peptidoglycan. The linker unit plus the P-glycerol polymer is transferred from the cytoplasmic side of the membrane to the outer side of the plasma membrane at the region of cell wall synthesis. As in the case of peptidoglycan synthesis, this movement of the polymer attached to the P-undecaprenyl is by a proposed "flip-flop" mechanism. The linker unit along with the teichoic acid polymer is transferred from the P-undecaprenyl (P-$C_{55}$) lipid to carbon 6 on the *N*-acetylmuramic acid molecule in the peptidoglycan. This membrane shuttle activity and the insertion at the growth region of the peptidoglycan is indeed a busy traffic site. The addition of teichoic acid occurs at the time the NAG–NAM dimer is added to the peptidoglycan. Studies have concluded there is sufficient P-undecaprenyl lipid in the plasma membrane to account for both the carrying of the peptidoglycan dimer and teichoic acid as cell wall extension. It has been suggested, although not substantiated, that the distribution of teichoic acid around the cell may not be uniform but more teichoic acid occurs along the cylindrical walls of the bacilli than at the poles of the cells.

The addition of teichoic acid to the peptidoglycan is inhibited by bacitracin but not by vancomycin. Other bacteria would have the same general theme for synthesis with accommodations for different linkage groups and additions of monosaccharides and D-alanine. In ribitol-containing teichoic acids, glucose is usually added at the second carbon of ribitol while D-alanine is added at the fourth carbon of ribitol.

### 4.2.2.2. Teichuronic Acids

Cells deprived of phosphorus will discontinue production of phosphate-containing wall-type teichoic acids or lipoteichoic acids but instead use uronic (sugar) acids to

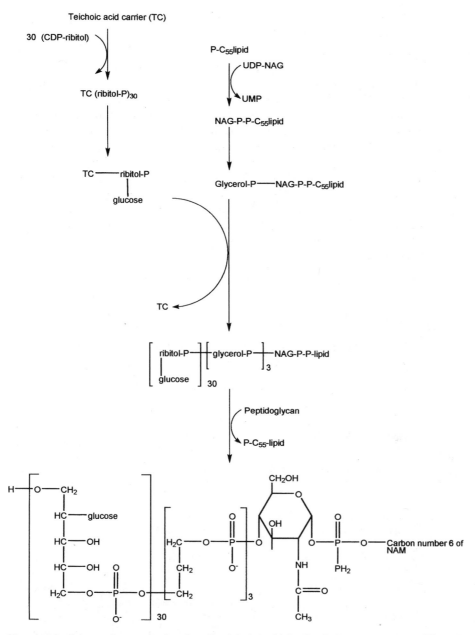

Figure 4.2. Scheme for synthesis of wall teichoic acid in *Staphylococcus aureus*. There are two parts to the biopolymer: polyribitol phosphate and a linkage unit consisting of glycerol phosphate-*N*-acetylglucosamine.

form an acidic polymer. These linear molecules containing uronic acids are termed teichuronic acids and are considered to be minor polymers in the cell walls of Gram-positive bacteria. Mild acid hydrolysis releases teichuronic acids from the peptido-glycan, indicating that a covalent bond is used to link teichuronic acids to the cell wall. In *Micrococcus luteus*, results suggest that teichuronic acids are covalently attached to the fourth carbon of *N*-acetylglucosamine in the peptidoglycan. Analysis of teichuronic acids reveals the presence of several monosaccharides in addition to the uronic acids, and as indicated in Table 4.3, the monosaccharides vary with the bacterial species. Commonly, the teichuronic acids contain acid forms of glucose and mannose. While chemical composition can be readily determined, the repeating unit of these linear anionic polymers is more difficult to obtain. In *Bacillus licheniformis*, the polymeric base of teichuronic acid has been proposed to consist of [→4)-D-glucuronic acid (α 1→3)-N-acetylgalactosamine-(1→] while in *Micrococcus luteus*, the repeating unit is suggested to be [→4)-*N*-acetyl-D-mannosaminuronic acid-(β1→6)-D-glucose-(1→]. In *B. subtilis* W23, the repeating subunit is [→3)-D-glucuronic acid-(1→3)-*N*-acetylglucosamine-(1→].

Table 4.3. Polymeric segment of teichuronic acids found associated with selected Gram-positive bacteria.[a]

| Organisms | Components |
| --- | --- |
| *Bacillus* sp. strain C-125 (Alkalophilic) | Glucuronic acid<br>L-glutamic acid |
| *Bacillus subtilis* var. *niger* | *N*-acetylgalactosamine<br>Glucuronic acid |
| *Bacillus megaterium* strain M46 | Glucose<br>Glucuronic acid<br>Rhamnose |
| *Curthobacterium flaccumfaciens* | Galactose<br>Glucuronic acid<br>Mannose<br>Pyruvic acid<br>Rhamnose |
| *Corynebacterium betae* | Fucose<br>Glucuronic acid<br>Mannose<br>Rhamnose |
| *Micrococcus lysodeikticus* | Glucose<br>*N*-acetyl-D-mannosaminuronic acid |
| *Staphylococcus aureus* T | *N*-acetyl-D-fucosamine<br>*N*-acetyl-D-mannosaminuronic acid |

[a]Adapted from Rogers, 1983.[25]

### 4.2.2.3. Lipoteichoic Acids[4,5]

Lipoteichoic acid was first reported by Wicken and Knox in 1970 as a result of their examination of the cell structures of *Lactobacillus fermentum*. Lipoteichoic acids appear to be more broadly distributed than teichoic acids in Gram-positive bacteria and may account for at least 1% to 2% of cell dry mass. It has been estimated that 12% to 20% of the lipid molecules in the outer leaflet of the plasma membrane of *Staphylococcus aureus* and *Lactobacillus lactis* is attributed to lipoteichoic acids.

The linear polyglycerol phosphate–containing lipoteichoic acids are found in species of *Bacillus, Lactobacillus, Staphylococcus*, and *Streptococcus*. As illustrated in Figure 4.3, there are two parts to the lipoteichoic acid: the linear teichoic acid or glycerol phosphate polymer segment and the glycolipid (glyco refers to sugar) unit. The teichoic acid segment is generally a 1,3 phosphodiester–linked glycerol unit with D-alanine bound to the central carbon of glycerol. In streptococci, the teichoic acid segment contains polyribitol phosphate. The hydrophobic end of the lipoteichoic acid is anchored into the plasma membrane while the other end extends through the cell

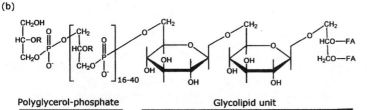

Figure 4.3. Lipid-containing linear polymers in cell walls of Gram-positive bacteria. (**a**) Lipoglycan from *Bifidobacterium bifidum* has R = L-alanine or H. (**b**) Lipoteichoic acid from *Bacillus subtilis* where R = D-alanine, glucose, or H. In both structures, FA = fatty acids.

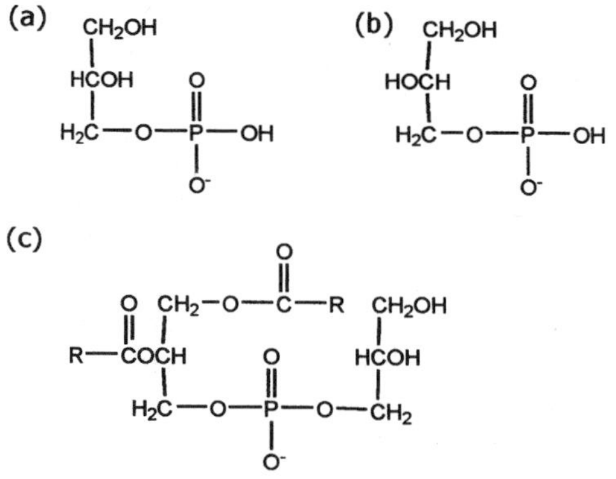

**(a)**

**(b)**

**(c)**

Figure 4.4. Configuration of glycerol phosphate molecules. (**a**) *sn*-Glycerol-1-phosphate as it occurs in lipoteichoic acids and glycerophosphoglycolipids. (**b**) CDP-glycerol and wall teichoic acids. (**c**) Both enantiomers are present in phosphatidyl glycerol.

wall. Lipoteichoic acids can be isolated from intact cells by extraction with hot 45% aqueous phenol.

It should be noted that an important stereochemical configuration is found in the glycerol phosphates. In Figure 4.4 it can be seen that the *sn*-glycero-1-phosphate present in lipoteichoic acids is an isomer of the molecule found in wall teichoic acids. The difference is the orientation at carbon position 2 of the glycerol structure, and phosphatidyl glycerol contains both of these isomers.

The glycolipid unit of the glycerol-type lipoteichoic acids varies with the bacterial genera. The simplest type of glycolipid consists of two fatty acids attached to glycerol as shown in Figure 4.3 for *Bacillus*, while more complex glycolipid structures are found in *Lactococcus* (see Table 4.4) in which an additional fatty acid is attached through glucose adjacent to the glycerol-fatty acid region. In *Listeria*, a supplemental anchor is phosphatidyl glycerol, which is attached to one of the monosaccharides in

Table 4.4. Glycolipid units of lipoteichoic acids indicating two or three fatty acids per unit.[a]

| Bacteria | Structure |
|---|---|
| *Streptococcus* | Glucose ($\alpha1\rightarrow2$) glucose ($\alpha1\rightarrow3$) glycerol—fatty acid<br>$\mid$<br>fatty acid |
| *Bacillus* | Glucose ($\beta1\rightarrow6$) glucose ($\beta1\rightarrow3$) glycerol—fatty acid<br>$\mid$<br>fatty acid |
| *Lactococcus* | Glucose ($\alpha1\rightarrow2$) glucose ($\alpha1\rightarrow3$) glycerol—fatty acid<br>$\mid$         $\mid$<br>6$\rightarrow$fatty acid    fatty acid |
| *Listeria* | Galactose ($\alpha1\rightarrow2$) glucose ($\alpha1\rightarrow3$) glycerol—fatty acid<br>$\mid$         $\mid$<br>6$\rightarrow$phosphatidyl    fatty acid<br>glycerol |

[a]From Fischer, 1994.[3]

Table 4.5. Fatty acids present in teichoic acids and plasma membranes of selected bacteria.[a]

| Microbial system | Bound fatty acids expressed as percent of total | | | | | | |
|---|---|---|---|---|---|---|---|
| | $C_{14:0}$ | $C_{14:1}$ | $C_{16:0}$ | $C_{16:1}$ | $C_{18:0}$ | $C_{18:1}$ | $C_{19:0}$ |
| *Enterococcus faecalis* NCIB 8191 | | | | | | | |
| Lipoteichoic acid | 5.7 | — | 37.7 | 8.7 | — | 40.0 | 7.9 |
| Membrane lipids | 3.4 | — | 37.7 | 7.5 | — | 35.8 | 15.5 |
| *Lactococcus garvieae* Kiel 42172 | | | | | | | |
| Lipoteichoic acid | 14.4 | 0.5 | 33.5 | 4.5 | 2.9 | 43.2 | 2.1 |
| Membrane lipids | 4.5 | 0.5 | 28.5 | 2.8 | 4.7 | 48.5 | 10.6 |

[a]Adapted from Fisher, 1988.[23]
The (−) indicates fatty acid is not present.

the glycolipid. Thus, the lipoteichoic acid is secured to the membrane by fatty acids and typically these are the same fatty acids found in plasma membranes. For example, considerable similarity is observed in fatty acid content of membranes and lipoteichoic acids of *Enterococcus faecalis* and *Lactococcus garvieae* (Table 4.5). With *Lactococcus garvieae*, the only major difference is the abundance of $C_{14:00}$ and $C_{19:00}$ fatty acids with greater amounts of $C_{14:00}$ found with lipoteichoic acid and greater amounts of $C_{19:00}$ present in membrane lipids.

A lipoteichoic acid containing polyribitol phosphate is produced by *Streptococcus pneumoniae*. Attached to the ribitol unit is a tetrasaccharide segment and one phosphocholine moiety is bonded to *N*-acetylgalactosamine in the tetrasaccharide (see Figure 4.5). This polymeric unit containing choline is similar, if not identical, to that found in teichoic acids. The presence of choline is indeed interesting because at physiological pH levels, there is a positive charge on the nitrogen atom of choline. Present in the glycolipid and in the polyribitol phosphate residue is 2-acetamido-4-amino-2,4,6-trideoxy-D-galactose. It is not clear why unique side groups are present on this specific galactose residue.

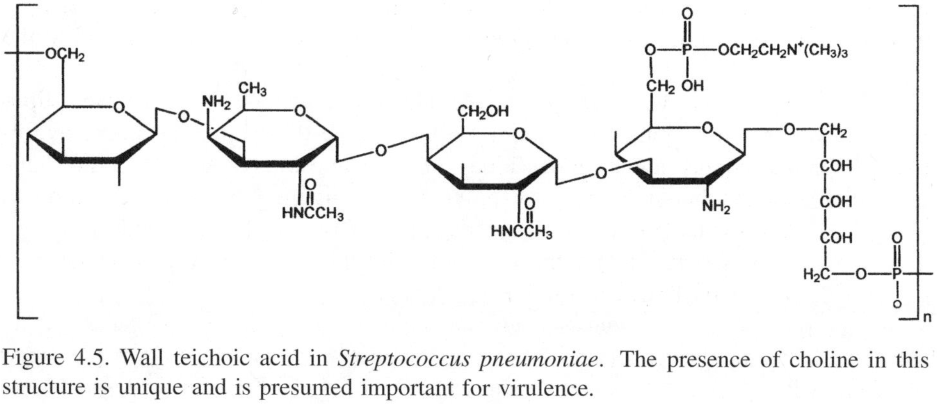

Figure 4.5. Wall teichoic acid in *Streptococcus pneumoniae*. The presence of choline in this structure is unique and is presumed important for virulence.

In biosynthesis of polyglycerol phosphate in lipoteichoic acid, a key molecule is phosphatidyl glycerol and not CDP-glycerol as used in synthesis of wall teichoic acids. Because phosphatidyl glycerol is the source for glycerol phosphate, there is a close relationship between lipoteichoic acid biosynthesis and lipid biosynthesis. As discussed in Section 2.9 of Chapter 2, lipid turnover occurs in the plasma membrane of growing cells and the turnover rate for phosphatidyl glycerol exceeds that of other phospholipids. Although S.A. Short and D.C. White demonstrated a high turnover of phosphatidyl glycerol in the early 1970s, it was not until about 20 years later that phosphatidyl glycerol was found to be a precursor for biosynthesis of lipoteichoic acids. This activity of phosphatidyl glycerol in bacteria is an excellent example of how scientific knowledge builds on results obtained from apparent unrelated pursuits.

D-alanine is added to lipoteichoic acid by a novel system involving the requirement of ATP for the formation of D-alanyl-AMP. A carrier protein is used to shuttle D-alanine between D-alanyl-AMP and the glycerol residue of the teichoic acid segment. The addition of D-alanine to ribitol occurs at the same time as teichoic acid is bound into the P-undecaprenyl lipid. Once D-alanine is bound into the lipoteichoic acid it is highly mobile and can be transferred within the cell wall region by inter- and intrachain translocation. From research reported by several laboratories, it has been proposed that D-alanyl moieties can be transferred from lipoteichoic acids to teichoic acids.

### 4.2.2.4. Lipoglycans[4–6]

Gram-positive bacteria with a guanine plus cytosine (G + C) content higher than 50% appear to have lipoglycans but not lipoteichoic acids. *Actinomyces, Actinoplanes, Arthrobacter, Bifidobacterium, Corynebacterium, Frankia, Micrococcus, Mycobacterium,* and *Streptomyces* are examples of bacteria with a high G + C% that produce lipoglycans. The general structure of a lipoglycan is a hydrophilic polysaccharide and a lipophilic segment that anchors the molecule into the plasma membrane. In *Bifidobacterium bifidum* the polysacccharide is two linear segments consisting of (1→5)β-D-galactofuranan and (1→6)β-D-glucopyranan. Alanine is bonded to the glycerol phosphate; however, this is L-alanine and not the D-isomer of alanine. Also, glycerol phosphate residues are attached to position 6 of the galactofuran moieties. The lipid unit is D-galactopyranoside (β1→3) glycerol containing two fatty acids bound to the glycerol.

*Micrococcus luteus, Micrococcus flavus,* and *Micrococcus sodenensis* have lipomannans which are polymers containing only one type of sugar. The linear hydrophilic region contains about 60 residues of D-mannose and about 15 succinate molecules attached into the mannose polymer, see Figure 4.6. The free carboxyl groups of this polymer provide a strong negative charge at physiological pH levels. The lipomannan is secured into the membrane by the lipophilic region which is a diglyceride covalently linked to the linear polymannose.

*Actinomyces viscosus* produces a unique molecule that consists of a fatty acid–substituted heteropolysaccharide. Mannose, glucose, and galactose form the sugar backbone with glycerol phosphate, lysine, and alanine linked into the polysaccharide

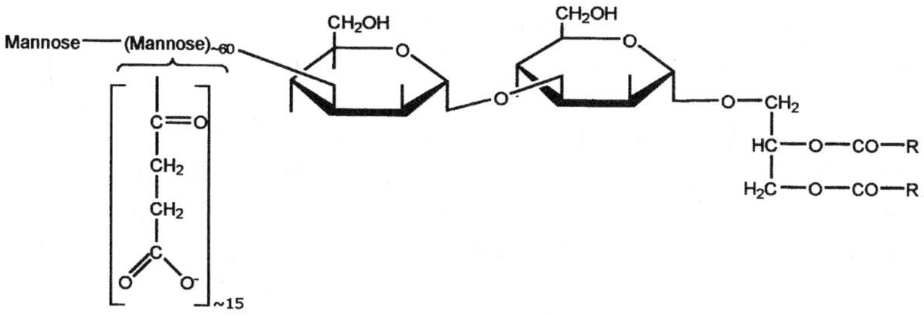

Figure 4.6. The lipomannan in *Micrococcus luteus* with succinic acid attached to mannose.

to provide an overall negative charge on the molecule. These complex arrangements are known as hexose–lysine amphiphile molecules.

### 4.2.2.5. Biological Activities

The wall teichoic acid polymers and lipoteichoic acids are long flexible molecules that have a negative charge attributable to the phosphodiester groups. These structures not only fill space around the peptidoglycan layers but also extend beyond the cell surface. Several functions have been ascribed to teichoic acids, including (1) binding $Mg^{2+}$ cations in the cell wall, (2) establishing a negative charge on the cell surface for interfacial cellular properties, (3) serving as a phosphate reserve, (4) influencing cell permeability by providing a charged barrier throughout the wall region, (5) participating in site recognition for bacteriophage, and (6) binding autolytic enzymes associated with peptidoglycan synthesis. Presumably, many of these functions would also be associated with lipoteichoic acids and certain glycolypids.

The teichoic acids and lipoteichoic acids function as a cation-exchange system between the surface of the cell and the plasma membrane. Because of charges on the exterior of the plasma membrane, 10 to 30 mM $Mg^{2+}$ are needed to promote membrane stability. From numerous studies, it has been determined that negatively charged wall polymers bind $Mg^{2+}$ with decreasing affinity: polyglycerol phosphate (wall teichoic acids) > succinylated lipomannan > polyribitol phosphate (wall teichoic acid) > polyglycerol phosphate (lipoteichoic acid).

Lipoteichoic acids, and presumably teichoic acids, prevent degradation of the peptidoglycan by autolytic enzymes as a result of binding muramidases, amidases, and the β-$N$-acetylglucosamidase in several species. The enzymes that attack the peptidoglycan are referred to as autolysins, and with sufficient degradation of the peptidoglycan, cell autolysis occurs. Because the autolysins are also needed for insertion of new units for peptidoglycan synthesis, these enzymes that attack the peptidoglycan must be highly regulated, and this task is attributed to the lipoteichoic acids. If the binding of peptidoglycan synthesizing activity is too great and sufficient enzymes are not allowed to react with the growing peptidoglycan to produce cell separation, multicellular filaments will result. If the binding by teichoic acids is not sufficient, cell lysis may result from overactive autolytic cell wall enzymes.

The lipoteichoic acids of bacteria are considered to be important as virulence-enhancing factors. Under conditions that are not fully resolved, lipoteichoic acids are released from bacterial cells. Either the released lipoteichoic acid or that attached to the bacterial cell will bind fibronectin and serum proteins including albumin. Lipoteichoic acids bind to mammalian cells and promote cell lysis by complement-fixation reactions and by the alternate complement reaction. Although lipoteichoic acids are neither pyrogenic nor cytotoxic, they are known to stimulate animal cells to release interleukin-1, interleukin-6, tumor necrosis factor, and superoxide radical. Pneumococcal lipoteichoic acid specifically activates the alternate complement pathway, presumably as a result of the presence of choline in the lipopolymer. The lipoarabinomannan from *Mycobacterium* suppresses T-lymphocyte action, inhibits release of tumor necrosis factor, and inhibits antigen presentation to appropriate antibody secreting cells.

## 4.2.3. Proteins[4,5]

In addition to "housekeeping" proteins needed for peptidoglycan synthesis, several different types of proteins are present in the cell walls of Gram-positive bacteria. Some of these proteins have an important function in binding collagen, antibodies, fibrinogen, laminin, plasmin, prothrombin, salivary glycoproteins, and complement proteins. In certain cells, hydrolytic enzymes including amylases, collagenases, and proteases may be associated with the cell wall. Some proteins at the surface of the cell walls may have specific functions, as is the case with *Streptococcus salivarius*, which uses proteins to attach to oral surfaces.

### 4.2.3.1. Attachment of Proteins

To prevent proteins from diffusing through the cell, Gram-positive bacteria retain proteins through the use of two distinctive molecular designs. One type attaches the protein by a specific amino acid sequence to lipids in the plasma membrane. At the N-terminal region of the protein there is a tetrapeptide sequence in which cysteine is the fourth amino acid. The consensus sequence of the peptide sequence at the N-terminus is leucine–X–Y–cysteine where X and Y are neutral amino acids. A specific signal peptidase cleaves the first three amino acids from the signal sequence, leaving cysteine as the terminal amino acid. Glycerol from phosphatidyl glycerol is transferred to cysteine and fatty acids from phospholipids are added to produce the diglyceride proteins. The diglyceride unit is localized at the outer leaflet of the membrane and thereby serves to tether the protein to the cell. Proteins that have such an N-terminal signal include penicillinases produced by *Bacillus licheniformis* and *Staphylococcus aureus* and adhesins produced by *Streptococcus gordonii* and *Streptococcus sanguis*.

Another mechanism of binding proteins into the cell wall is to employ specific C-terminal sequences that have a function in protein localization. A model indicating the production and anchoring of proteins into the cell wall is given in Figure 4.7.

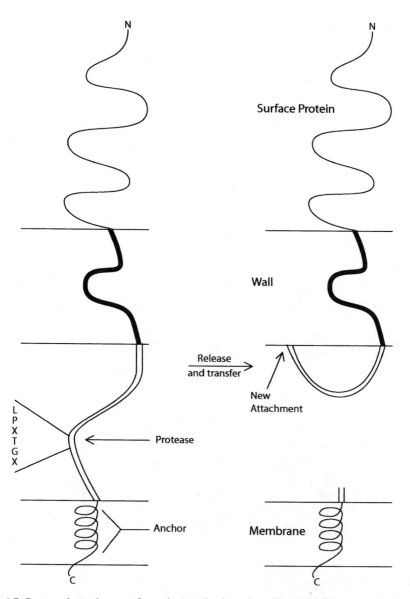

Figure 4.7. Proposed attachment of certain proteins into the cell wall of Gram-positive bacteria.

Table 4.6. Examples of cell wall proteins with unique C-terminal sequences that are found in Gram-positive bacteria.[a]

| Name of protein or gene for protein | Function of protein | Bacterial species |
|---|---|---|
| M 2 | Unknown function | *Streptococcus pyogenes* |
| ARP 4 | Binds immune globulin A | *Streptococcus pyogenes* |
| ProtH | Binds Fc segment of human immune globulin G | *Streptococcus pyogenes* |
| *bac* | Binds immune globulin A | *Streptococcus* Group B |
| *fnbA* | Binds fibrinonectin | *Streptococcus dysgalactiae* |
| DG 12 | Binds albumin | *Streptococcus* Group G |
| *asa1* | Aggregation substance | *Enterococcus faecalis* |
| *cna* | Binds collagen | *Staphylococcus aureus* |
| Fimbrae | Type 1 fimbrae | *Actinomyces viscosis* |
| Hyaluronidase | Hyaluronidase | *Streptococcus pneumoniae* |
| *nanA* | Neuraminidase | *Streptococcus pneumoniae* |

[a]From Fischetti, 1996.[6]

From analysis of more than 50 such proteins, it is evident that these proteins have three important structural regions. At the C-terminus, there is a series of 15 to 20 hydrophobic amino acids that span the plasma membrane and serve to anchor the protein into the membrane. Following the hydrophobic/charged-tail sequences is a peptide containing the X-glycine-threonine-X-proline-leucine sequence. This hexa-peptide contains the site for protease cleavage and release of the principal protein unit from the portion anchored into the membrane. Following protease action the newly established C-terminus is bound into the cell wall, thereby securing the protein to the cell surface. The streptococci and staphylococci produce a variety of cell wall proteins that have a unique C-terminal structure; examples of these and other surface proteins of Gram-positive bacteria are listed in Table 4.6. Several of these proteins bind to immune globulins or extracellular matrix of animal origin and a few are hydrolytic enzymes that promote bacterial invasion through tissues.

### 4.2.3.2. Protein M

When viewed by electron microscopy, some of these proteins provide a "fuzzy" texture to the surface of the cell. Proteins making up the fuzzy surface may be fibrils that are 200 to 400 nm long and may be either single molecules or aggregates. Early classification of proteins in the cell walls of group A hemolytic streptococci resulted in the designations of M, T, and R proteins which are associated with virulence enhancement. There are more than 100 different type-specific M antigens and more than 25 different serotypes of T protein. Complete sequences of the cloned M protein genes (*emm*) have been conducted and these genes contain typical C-terminal wall-associated signals. In addition, these genes have sizeable domains that may be repeated several times. Proteins from *Streptococcus, Corynebacterium, Staphylococcus,* and *Actinomyces* extend 0.5 to 4 μm from the cell as an α-helix coiled secondary structure. On occasion these proteins have been designated as "fimbrae"; however, these Gram-positive "fimbrae" are morphologically and structurally distinct from fimbrae found on cells of the Gram-negative enterics. Considerable attention is now being given by molecular biologists to the construction of gene fusions with the M protein gene for the purpose of developing vaccines for animals.

### 4.2.3.3. Streptococcal M Protein and Myosin

Rheumatic heart disease has long been recognized as an important consequence of recurring infection caused by group A streptococci. Some consider that damage to heart valves is attributed to the release of streptolysin O or another toxin from the bacteria. Many consider that the damage to the heart is a consequence of the bacteria releasing an antigen that is similar in structure to that of the heart. At one time it was thought that the capsule of *Streptococcus pyogenes* could be responsible because it consisted of hyaluronic acid and this polymer was involved in cementing eukaryotic cells. Recent information suggests that production of antibody to the streptococcal M protein may be responsible. Analysis of amino acids in the streptococcal M protein reveals sequences that are similar to human cardiac myosin (see Table 4.7). It would appear that M proteins are released from the streptococcal cells and these proteins would stimulate the humoral response. With the second streptococcal infection, the individual would be exposed to the M proteins and antibody development would be enhanced. The antibodies would cross react with the heart myosin and one of the responses could include complement activation to attack the heart cells. A thorough analysis of the genome of streptococci from group A may provide additional information concerning conserved sequences in other cell wall proteins and eukaryotic cell proteins.

### 4.2.3.4. Protein A

In 1963, Löfkvist and Sjöquist provided the initial characterization of protein A obtained from the cell walls of *Staphylococcus aureus*. Over the years an interest has

Table 4.7. Amino sequences in polypeptide segments from three different types of M proteins from group A streptococcal cell walls are compared to human cardiac myosin polypeptide.

| Protein | Sequences of amino acids in proteins |
| --- | --- |
| M5peptide B2 (bacterial protein) | TIG**TL K** K I**L** DE**TV** K**D**K**IA** |
| Cardiac myosin | LE**DL** K**R**QL EE**EV** K**A**K**NA** |
| NT4 peptide (bacterial protein) | G**L**K**T ENEG** LKTENEGLKTE |
| Cardiac myosin | K**L**Q**T ENGE** |
| M6 protein (bacterial protein) | **L**TD Q**NK**N**L** **TT**E**N** |
| Cardiac myosin | **L**TS Q**R**A**K**L QTE**N** |

[a]Modified from Salyers and Whitt, 2002.[5]
Letters in bold indicate similarities.

continued in this cell wall protein because it was found that protein A binds the constant portion (Fc) of immunoglobulin G (IgG). The complex resulting from protein A and bound immunoglobulins produces a variety of hypersensitivity reactions in humans. Protein A has a molecular mass of 42 kDa and is rapidly incorporated into the peptidoglycan through covalent bonds. Protein A can be isolated from the peptidoglycan following enzymatic digestion. In certain strains of methicillin-resistant *Staphylococcus aureus*, protein A can be isolated from the culture fluid because protein A is released from the cell and is not linked to the peptidoglycan. In clinical and research laboratories, protein A has become a powerful tool for cell biologists to conduct cytochemical and immunological studies. Various streptococci produce a Fc-binding protein that has been called protein G and the binding to Fc is at a site different than that of protein A. Like protein A, protein G binds the plasma protease inhibitors of kininogen and $\alpha_2$-macroglobulin.

## 4.3. Lipids in Cell Walls of Acid-Fast Bacteria[7–9]

When a smear of *Mycobacterium* is flooded with carbofuchsin dissolved in 5% aqueous phenol and the smear is heated to steaming, the cells become stained and retain the dye even when washed with acid alcohol (3% HCl in ethanol). This acid-fast staining characteristic is associated with bacteria belonging to the genus *Mycobacterium* and slight acid fast staining is observed with some *Nocardia* and *Corynebacterium*. Acid fastness in these bacteria is attributed to the high concentration of bound lipid in the cell walls and this lipid is of two classes: waxes and glycolipids. Waxes are composed of long-chain fatty acids linked to long-chain alcohols and in the mycobacteria mycolic acids esterified to an arabinogalactan that serves as the alcohol. The mycolic acids are $\alpha$-alkyl, $\beta$-hydroxy fatty acids and these $C_{70}$ to $C_{90}$ fatty acids contain either cyclopropane, methoxy, or keto groups. Examples of some of the mycolic acid structures are given in Figure 4.8. In the closely related *Corynebacterium* and *Nocardia*, mycolic acids are found that are $C_{32}$ to $C_{36}$ and $C_{48}$ to $C_{58}$, respectively. Future research will be necessary to determine the degree of similarity that exists between the cell walls of corynebacteria, nocardia, and mycobacteria.

The lipids in the cell wall contribute to the resistance of these organisms to acids, alkali, dyes, and other chemicals. Because *Mycobacterium tuberculosis* can survive exposure to 2% NaOH at 37°C, this rigorous treatment is used to recover viable bacteria from sputum and infected animal tissues. The Löwenstein–Jensen solid medium used to cultivate primary isolates of *M. tuberculosis* contains malachite green to inhibit other bacteria but the lipid in the cell wall prevents passage of the dye into the mycobacterial cells. The mycobacteria display remarkable resistance to phenol and for the cells to be killed, a 24-hr exposure is required when 5% phenol is used. In addition, the high lipid content in the cells of mycobacteria is responsible for the aggregation or clumping of cells, and to achieve greater dispersion in aqueous environments, Tween-80 is frequently incorporated into growth media.

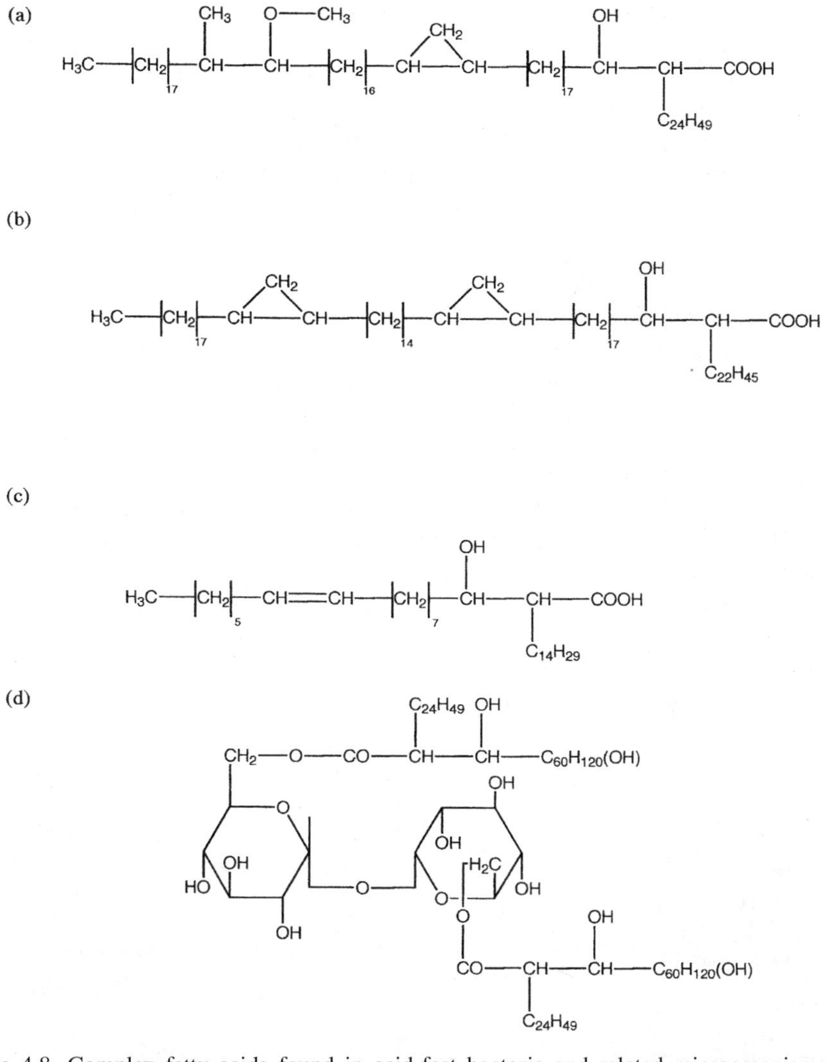

(a)

(b)

(c)

(d)

Figure 4.8. Complex fatty acids found in acid-fast bacteria and related microorganisms. (**a**) Methoxylated mycolic acid from *Mycobacterium tuberculosis*. (**b**) Kansamycolic acid from *Mycobacterium kansasii*. (**c**) Corynomycolenic acid from *Corynebacterium diphtheriae*. (**d**) 6,6'-Dimycolyltrehalose.

## 4.3.1. Firmly Attached Polymers

The cell wall of mycobacteria is extremely complex, and the molecular configurations in this region have attracted considerable interest over the years. However, the high lipid character makes it difficult to collect components in the cell wall because relatively few biochemical fractionations of cell structure are useful in the high lipid environment. There are two major means of attaching glycans to the cell: through the peptidoglycan and by fatty acyl interactions with the outer side of the plasma membrane. We will discuss the mycolic acid-arabinoglycan (MAG), which can be considered a unique type of teichoic acid because the attachment is similar to that found with glycan polymers attached to the peptidoglycan of Gram-positive bacteria. The second group of wall glycolipids includes the lipomannans and lipoarabinomannans that are attached through two fatty acyl groups into the plasma membrane.

### 4.3.1.1. Mycolyl-Arabinogalactan

The mycolyl-arabinogalactan (MAG) is a 30-kDa complex consisting of a galactose linear polymer with side branches consisting of arabinose polymers. About 30 mycolic acids are attached to the terminal ends of the arabinose polymers (see Figure 4.9), making this polymer highly hydrophobic. The MAG is covalently attached to the peptidoglycan in the cell wall. In mycobacteria, the peptidoglycan consists of repeating units that contain N-acetylglucosamine and N-glycol muramic acid. The MAG is attached through the C-6 of the N-glycolated muramic acid; however, only in about 10% of the N-glycolated muramic acid residues is the mycolic acid-containing arabinoglycan attached.

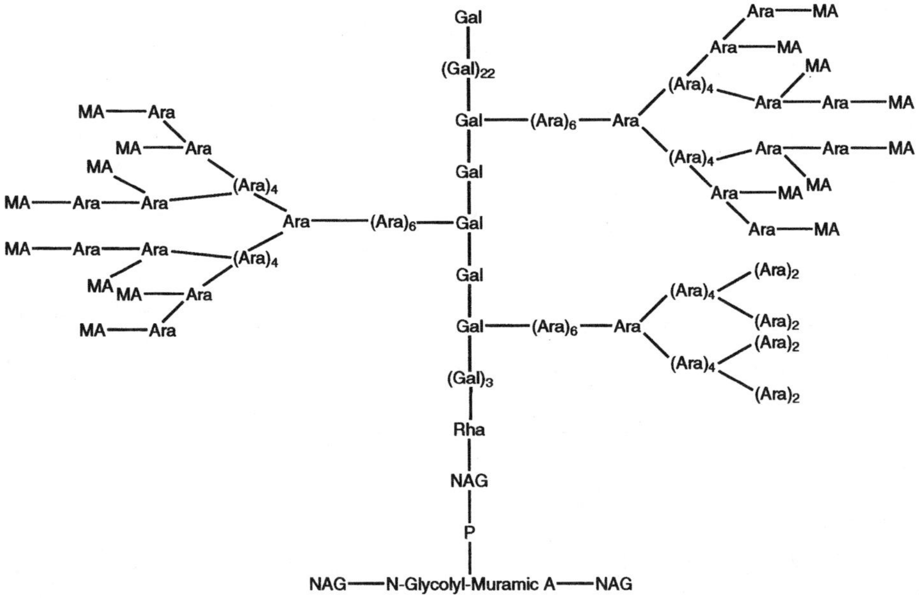

Figure 4.9. A model of the arabinoglyctan structure in the cell walls of *Mycobacteria*. (From Brennan and Nikaido, 1995.[7])

### 4.3.1.2. Lipoarabinomannan and Lipomannan

The lipoarabinomannan (LAM) consists of a mannose polymer as a backbone for the molecule with side chain polymers consisting of arabinose polymers. The LAM is tethered into the plasma membrane by two fatty acid residues of a phosphatidylinositol unit that is bonded to the mannose backbone. The fatty acids for this attachment in *Mycobacterium tuberculosis* are palmitate and tuberculostearate. The physical organization of LAM in the cell wall is given in a model depicting the cell wall of mycobacteria (see Figure 4.10). A related structure is lipomannan, which is similar to LAM but it is a mannose polymer without an arabinose-containing side chain. The synthesis of LAM and lipomannan appears to be limited to the acid-fast bacteria and the attachment of the sugar polymer to phosphatidyl inositol is extremely rare in bacteria. The function of LAM and lipomannan in the mycobacterial wall remains to be established.

## 4.3.2. Extractable Lipids

Lipids account for about 60% of the cell wall dry weight of mycobacteria and 25% to 30% of the cell wall is free lipids, which are readily extracted with an organic

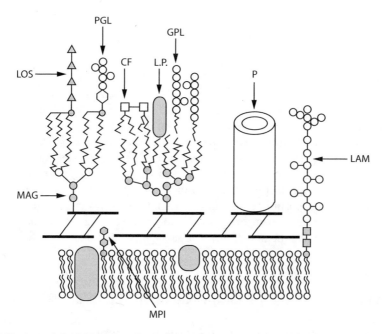

Figure 4.10. A model of the cell wall of mycobacteria indicating a complexity that is not found in cell walls of either Gram-positive or Gram-negative bacteria. MPI, Mycolyl phosphatidyl inositol; MAG, mycolyl-arabinogalactan; LOS, trehalose-containing lipooligosaccharides; PGL, phenolic glycolipids; CF, cord factor; LP, lipoprotein; GPL, glyceropeptidolipids; P, porin-like structure; LAM, lipoarabinomannan. (From Poxton, 1993[9] and related articles.)

solvent. The glycolypids in acid-fast organisms contain both lipid and carbohydrate moieties. There are several different types of extractable glycolipids in mycobacteria and a partial listing includes: glycopeptidolipids (GPL), trehalose-containing lipooligosaccharides (LOS), phenolic glycolipids (PGL), sulfolipids (SL), and phthiocerol dimycocerosate (PDM). Extractable lipids are on the outer surface of the mycobacterial cell and are considered by some to be attached into the mycolic acids of the cell wall and to produce a structure that is somewhat similar to a bimolecular lipid leaflet. An important glycolipid in the cell wall of *Mycobacterium tuberculosis* is the "cord factor," which consists of two mycolic acids attached to trehalose. The cord factor promotes bacterial grow as thread-like filaments in broth cultures and functions as a virulence enhancing factor in that it not only inhibits migration of polymorphonuclear leukocytes but also alters metabolism in animal cells. Another extractable component is wax D, which has long been used to enhance immune production in animals. Wax D contains mycolic acids and a glycopeptide and it is the active component in Freund's adjuvent. Sulfolipids are relatively rare in prokaryotes; however, the sulfolipid of mycobacteria is extremely important because sulfolipids may contribute to virulence in some mycobacteria. The abundance of mycolic acids in these extractable lipoglycans underscores the heavy energy commitment that the mycobacteria make to the formation of these very large fatty acids. The payoff for these bacteria is the great resistance of these cells to physical changes and chemical attack.

### 4.3.3. Porin-Like Structures

The mycolic acids are oriented to form a lipid layer parallel to the peptidoglycan structure, and porin-like structures provide entry of small molecules across the lipid barrier (see Figure 4.10). At this time porin-like components have been demonstrated only in *Mycobacterium chelonae* and *Mycobacterium smegmatis*; however, it is assumed that these porins are widely distributed in mycobacteria. Porin-like structures have been found in *Nocardia asteroides* and *Corynebacterium glutamicum*, which appear to be closely related to the mycobacteria. The low permeability of nutrients into the bacterial cell is considered to account for the slow growth rate of these organisms. The limiting factor for cell permeability and cell growth is the layers of lipid in the wall, and porin-like structures facilitate the transfer of nutrients through the cell wall. However, in comparison to Gram-negative enterics, mycobacteria have fewer porins per cell and the rate of nutrient passage through them is slower by several orders of magnitude.

## 4.4. Outer Membrane of Gram-Negative Bacteria[10,11]

Characteristic of Gram-negative bacteria is the presence of an outer membrane that contains phospholipids, lipopolysaccharides, and lipoproteins. This outer membrane has the same general thickness as the plasma membrane, with the structure of the outer membrane maintained by hydrophobic and hydrophilic boundaries. The outer membrane is structurally asymmetrical with the inner side of the outer membrane

Table 4.8. Composition of outer membrane of *Salmonella typhimurium*.[a]

| Component | Composition (wt%) | Average molecular mass | Fraction of surface area on outer face occupied by component (%) |
|---|---|---|---|
| Lipopolysaccharide | 19 | 10.3 kDa | 22 |
| Phospholipids | 19 | 0.7 kDa | 20 |
| Protein | 62 | 44 kDa | 58 |

[a]Modified from Nikaido, 1973.[10]

containing phospholipids while lipopolysaccharide is found exclusively on the outer side of the outer membrane. The composition of the outer membrane of a typical Gram-negative bacterium is given in Table 4.8. It is to be noted that phospholipids are important on the outer face of the outer membrane at about the same concentration as lipopolysaccharide but that about 60% of the surface of the outer membrane is protein. Porins constitute a major fraction of the protein in the outer surface of outer membrane.

## *4.4.1. Lipopolysaccharide*

The lipopolysaccharide structure that is found in the outer membrane can be divided into three fundamental segments: the O side chain, the core oligosaccharide, and lipid A. The best characterized lipopolysaccharide is from *Salmonella typhimurium* and a model showing these components is given in Figure 4.11. Similar regions have been found in other strains of *Salmonella* as well as for *E. coli* and *Shigella flexneri*. The lipopolysaccharide can be readily extracted from cells with 45% phenol at 65°C and isolated by ultracentrifugation of the aqueous fraction. Based on dry weight measurements conducted with enteric bacteria, Gram-negative bacteria contain about 3.4% lipopolysaccharide. Each strain of bacteria has only one molecular form of lipopolysaccharide, and changes in the lipopolysaccharide synthesis resulting from genetic transformations result in different antibody reactions.

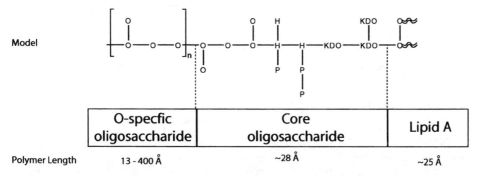

Figure 4.11. The structural organization of lipopolysaccharide in Gram-negative bacteria.

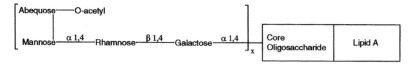

Figure 4.12. O-specific side chain of *Salmonella typhimurium*.

### 4.4.1.1. O-Side Chain

The O-side chain is projected outward from the cell and serves as the major antigen for the Gram-negative bacterial cell. The type of sugar present in the O-side chain, the sequence of the sugars and structural arrangement of these sugars in the side chain are influenced by the specific bonds between the individual sugars. A model of O-side chain in the outer membrane of *Salmonella* is given in Figure 4.12. Each strain of *Salmonella* has a unique O-chain composition of a few sugar residues that form a repeating unit of considerable size. Different strains of *Salmonella* have their own unique set of sugars making up the O-side chain, and therefore function as different serotypes that react with a specific antibody. A few of these differences are given in Table 4.9 and serve to illustrate the composition of the O-side chain in various

Table 4.9. Selected examples of O-side chains in different strains of bacteria.[a]

| Serotype | Repeating unit of O-side chain |
|---|---|
| | Paratose          Glucose-2-O-acetyl |
| | I(1,3)                 I(1,4) |
| *Salmonella typhi* | Mannose (1→4) Rhamnose (1→3) Galactose (1→2)→ |
| | Tyvelose             Glucose |
| | I(1,3)                 I(1,4) |
| *Salmonella parathyphi* A | Mannose (1→4) *O*-Acetyl (1→3) Galactose (1→2)→ |
| | Rhamnose |
| | Abequose-2-*O*-Acetyl          Glucose-2-*O*-Acetyl |
| | I(1,3)                      I(1,4) |
| *Salmonella typhimurium* | Mannose (1→4) Rhamnose (1→3) Galactose (1→2)→ |
| *Salmonella newington* | Mannose (1→4) Rhamnose (1→3) β-Galactose (1→6)→ |
| | Galactose                 *N*-Acetylglucosamine |
| | I(1,3)                      I(1,4) |
| *Salmonella minnesota* | β-Galactose(1→3) *N*-Acetyl-(1→3)*N*-Acetylgalactosamine→ |
| | Galactosamine |
| *Klebsiella pneumoniae* (type 03) | Mannose→mannose→mannose→mannose→ |
| | Fucose |
| | I |
| *Escherichia coli* (type 086) | Galactose→*N*-acetyl→*N*-acetylgalactosamine→ |
| | Galactosamine |
| | Glucose→Glucose |
| | I |
| *Escherichia coli* (type 04) | *N*-acetylglucosamine→rhamnose→*N*-acetylgucosamine→ |

[a]Structures from Rogers, 1983.[25]

serotypes of *Salmonella*. The linear carbohydrate repeating unit in *Salmonella* is mannose-rhamnose-galactose and the different serotypes of *Salmonella* reflect the different carbohydrate residues attached to galactose and to mannose. Further variation in O-side chain of Gram-negative bacteria is demonstrated by review of the structures found in *Escherichia coli* and *Klebsiella pneumoniae*. Only mannose residues are present in the O-side chain of *K. pneumoniae*. In *E. coli*, there are more than 160 different serological groups where the O-side chain is composed of different monomers in the repeating unit as well as different sugars that form the short branches.

### 4.4.1.2. Core Oligosaccharide

The structure of the core oligosaccharide has been well characterized in *Salmonella* spp. and appears as a basic unit. As illustrated in Figure 4.12, this oligosaccharide has a backbone that consists of eight residues with several short side chains. Unique to the core oligosaccharide are two molecules: L-glycero-D-manno-heptose, a heptose, and 2-keto-3-deoxyoctonic acid (KDO, also known as 3-deoxy-D-mannooctulosonic acid). Structures for KDO and heptose are given in Figure 4.12. The core oligosaccharide contains a cluster of phosphate moieties plus KDO units that at neutral pH anionic groups are highly ionized and hydrophilic. In native cell wall structures, divalent cations and especially $Mg^{2+}$ bind to the anionic (phosphate) groups of the core oligosaccharide. Under laboratory conditions, chelators such as EDTA can be used to remove the cations and instability of the outer membrane structure results.

The core oligosaccharide can be divided into two regions, the hexose oligosaccharide (outer core) and the heptose/KDO oligosaccharide (diheptose region) (see Figure 4.13). In both of these regions, there are considerable variations in sugar composition in both segments and perhaps the variation parallels what is observed for the O-side chain. Usually, the hexose oligosaccharide unit is composed of four sugars, which is remarkably similar to the unit observed in the O-side chain. The core region

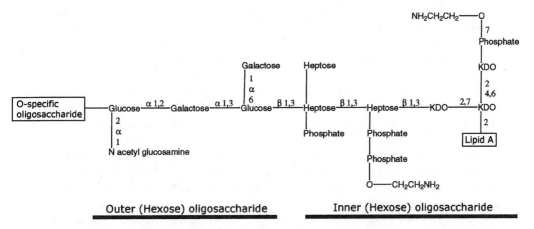

Figure 4.13. Core oligosaccharide of *Salmonella typhimurium* with outer and inner oligosaccharides.

Table 4.10. Sugars in the core region of the lipopolysaccharide of
*Escherichia coli*.[a]

| Organism | Hexose units present |
|---|---|
| | Galactose    Glucose |
| | ↕          ↕ |
| *Escherichia coli* R1 | →Galactose→Glucose→Glucose→Heptose→ |
| | *N*-Acetylglucosamine          Galactose |
| | ↕          ↕ |
| *Escherichia coli* R2 | →Glucose→Glucose→Glucose→Heptose→ |
| | Glucose    *N*-Acetylglucosamine |
| | ↕          ↕ |
| *Escherichia coli* R3 | →Glucose→Galactose→Glucose→Heptose→ |
| | Galactose    Galactose |
| | ↕          ↕ |
| *Escherichia coli* R4 | →Galactose→Glucose→Glucose→Heptose→ |

is less variable than the O-side chain with five different types in *E. coli* and only one in *Salmonella* (Table 4.10).

The outer core of the core polysaccharide and the O specific side chain are not essential for cell viability. In *Salmonella* it has been determined that mutants can grow without the production of the core oligosaccharide but do require the presence of lipid A plus the (KDO)$_3$ segment. In *Chlamydia*, the outer membrane contains lipopolysaccharide; however, it contains only lipid A and the three KDO residues but lacks the O-side chain and the core oligosaccharides.

### 4.4.1.3. Lipid A

A feature of the core oligosaccharide is to promote the extension of the O specific side chain outward from the lipid membrane while a function of lipid A is to tether the lipopolysaccharide into the hydrophobic region of the outer membrane. Under certain conditions, the lipopolysaccharide unit including the lipid A component may be stripped from the cell and in those cases, the phospholipids from the inner side of the outer membrane self-orients across the outer membrane to restore the biomolecular leaflet character. The structure of lipid A is given in Figure 4.14. In *Salmonella* spp., the central unit of the lipid A segment consists of two molecules of *N*-β-hydroxymyristroyl-D-glucosamine that are linked by a β-1,6 glucosidic bond. Additional β-hydroxymyristic acids (C$_{14}$) are attached as 3-myristoyl diesters to the β-hydroxymyristic acid moieties on the glucosamine to give four molecules of β-hydroxymyristic acid for each lipid A segment. Although α-hydroxymyristic acid residues are broadly distributed in animal systems, the presence of β-hydroxymyristic acid appears to be limited to bacteria and endotoxin activity has been attributed, in part, to β-hydroxymyristic acid. With the β-hydroxymyristic acid bonded to the sugar of lipid A, the hydroxyl of this myristic acid residue is a potential site for binding by phosphate or a KDO residue or another β-hydroxymyristic acid.

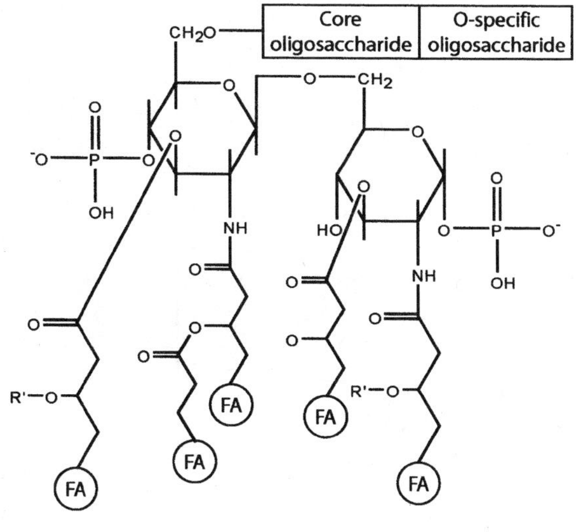

Figure 4.14. Lipid A of *Salmonella typhimurium*.

There is appreciable variability in fatty acid content of lipid A but saturated forms are prevalent. Hydrolysis of lipid A produces the following residues: *Salmonella* spp. yields one dodecanoic acid ($C_{12}$), four β-hydroxymyristic acids, and one hexadecanoic acid residues; *Bacteroides fragilis* yields two $C_{15}$, two $C_{16}$ and one $C_{17}$ residues; *Pseudomonas aeruginosa* yields one $C_{10}$ and four $C_{12}$ acid residues; and *Chromobacter violaceum* yields three $C_{10}$ and four $C_{12}$ acids. Thus, with five, six, or seven fatty acid residues for each lipid A there is considerable lipophilic action to secure the lipopolysaccharide into the outer membrane.

### 4.4.1.4. Lipopolysaccharide Synthesis[6]

The synthesis of the core oligosaccharides and the O-antigen of *Salmonella* spp. has been extensively studied (i.e., are the segments of the lipopolysaccharide that are the best characterized). The stepwise addition of the four hexose moieties is proposed to follow the sequence given in Figure 4.15. The mechanism for transport of lipid A plus the core polysaccharide to the outer membrane has not been resolved; however, it is considered that the sugars that are synthesized in the cytoplasm and transported across to plasma membrane to the exterior by the lipid carrier are used for the transport of the *N*-acetylglucosamine-*N*-acetylmuramic acid dimer in peptidoglycan synthesis. However, the free hexoses are not condensed into the oligosacccharide but the hexose conjugated to a nucleotide carrier is used to synthesize the unit of the O-side chain. The transmembrane movement of the synthesized oligosaccharides occurs by the same $C_{55}$ lipid carrier that is used for the synthesis of the bacterial peptidoglycan.

## 4.4.2. Wall (Braun) Lipoprotein[13]

To some extent, lipoproteins are associated with membranes and this is generally considered a mechanism of anchoring the protein into the membrane. In Gram-

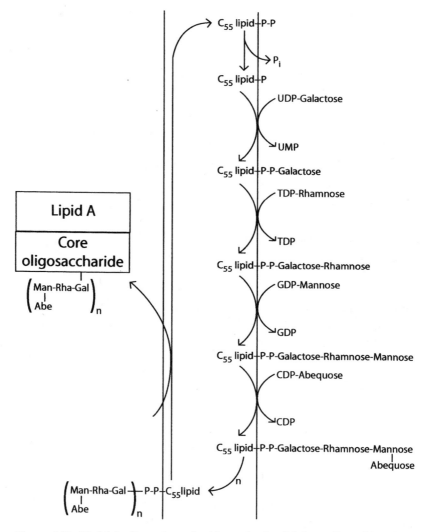

Figure 4.15. Model for lipopolysaccharide synthesis of *Salmonella typhimurium*.

positive bacteria and cell wall–less forms, the lipoproteins are associated with the plasma membrane. Proteins tethered to the outer face of the plasma membrane are used by Gram-positive bacteria for solute binding and are part of the ABC transport system. *Acholeplasma laidlawii* and *Mycoplasma capricolum*, cell wall–less forms of bacteria, are reported to each have more than 30 lipoproteins attached to their plasma membranes. In Gram-negative bacteria many of these lipoproteins are associated with the outer membrane and more than 75 lipoproteins from 34 different species of Gram-negative bacteria have been characterized. The outer membrane of Gram-negative bacteria contains a small number of low molecular weight lipoproteins in addition to porins.

# Box 4.1
# Bricks and Mortar in the Wall
A Perspective by Volkmar Braun

In 1966, I was a postdoc at the California Institute of Technology in Pasadena. There I attended a lecture series given by Max Delbrück, one of the founders of phage genetics, which became the basis of molecular biology and gene technology. Delbrück reported on biomembranes and I was so intrigued that I decided to enter this field, which was still in its infancy. When I returned to Germany, I chose *E. coli* as a model organism. My aim was to isolate succinate dehydrogenase by cleaving the enzyme from a presumed membrane anchor using proteases. Trypsin decreased most strongly the optical density of the cell envelope, suggesting a trypsin-sensitive protein with an important role in the integrity of the cell envelope. I then embarked on finding the trypsin-sensitive bond, which in light of millions of potential trypsin cleavage sites was like finding a "needle in a haystack." I found that isolated murein (peptodoglycan) could be freed from protein via cleavage of a trypsin-sensitive lysine bond that covalently fixes a defined lipoprotein to diaminopimelate of every tenth murein subunit. The structure of the lipid and its binding to the protein was determined, and hence became the first known protein with a covalently linked lipid. This murein lipoprotein ("Braun's lipoprotein") was also the first microbial membrane protein whose primary structure was determined. It became one of the model proteins with which secretions across the cytoplasmic membrane, chaperone-assisted transfer through the periplasm, and insertion into the outer membrane was and still is studied. Hundreds of lipoproteins in a large variety of Gram-negative and Gram-positive bacteria have since been identified which contribute to the Bricks and Mortar that make up bacterial membranes, serve as substrate transport proteins, contribute to protein secretion, display various enzyme activities, and elicit strong immune responses.

No function other than a structural role was assigned to the murein lipoprotein of *E. coli* when I turned my attention to a multifunctional protein in the outer membrane that plays a role in the infection of three different phages and a protein toxin called colicin M. It took several years of research to discover that this protein also serves as a transporter of an iron complex termed ferrichrome. It was fortuitous that I was at that time appointed to the chair of Microbiology at the University of Tübingen, Germany. This department had worked for more than a decade on iron-complexing compounds. The TonA (now called FhuA) protein became the first defined iron transport protein and the second purified protein of the outer membrane, besides the murein-lipoprotein. As it turned out, both FhuA-mediated transport and phage receptor activities consume energy, which is provided by the proton motive force of the cytoplasmic membrane. A complex consisting of three proteins (TonB, ExbB, and ExbD) transfers energy from the cytoplasm into the outer membrane. In addition, we found that transport across the cytoplasmic membrane occurs independently to transport across the outer membrane and is achieved by the ABC transporters that derive energy from ATP hydrolysis. Phage infection, export and import of protein toxins, and their mode of action became additional fields of active research that began with FhuA, but soon encompassed a larger variety of transporters, receptors and toxins, and novel mechanisms of transcription regulation by metal ions.

In 1969, Volkmar Braun discovered a structural lipoprotein in the cell wall of *E. coli* that contains 58 amino acids residues with lysine at the carboxyl end and cystine at the N-terminus. The configuration of the wall (Braun) lipoprotein at the cysteine region is given in Figure 4.16. A glycerol moiety is attached through the sulfur atom of cysteine and fatty acids are esterified through position 1 and position 2 on the glycerol molecule. A third fatty acid is attached by an amine linkage to cysteine. On the C-terminal end, the structure is attached through the ε-amino of lysine to the carboxylic group of diaminopimelic acid in the peptidoglycan. This lipoprotein contains most of the common amino acids; however, it lacks proline, phenylalanine, glycine, histidine, and tryptophan. It has been calculated that *E. coli* contains about $7.2 \times 10^5$ Braun molecules per cell and a Braun lipoprotein is attached to approximately

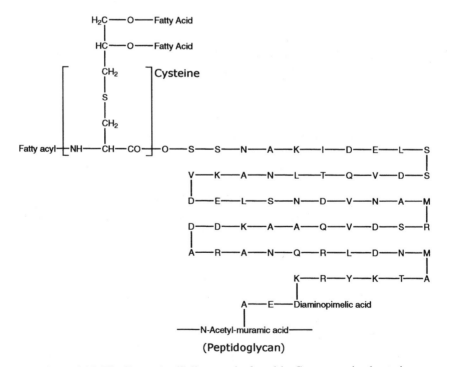

Figure 4.16. The Braun (wall) lipoprotein found in Gram-negative bacteria.

one tenth of available peptides containing diaminopimelic acid in the peptidoglycan. It is considered that similar types of lipoprotein molecules are present in cell walls of other Gram-negative bacteria.

Steps in the synthesis of wall lipoprotein by *E. coli* are given in Figure 4.17. The unmodified protein from the ribosomes contains a leader sequence of several amino acids followed by a chain of 58 amino acids with cysteine as the first amino acid and lysine at the terminal carboxyl end. The glycerol that is added to the cysteine moiety is from phosphatidyl glycerol and initiates the posttranslation processing. The fatty acids added to the glycerol moiety are from one of the phospholipids (phosphatidyl ethanolamine, phosphatidyl glycerol, or diglycerol phosphate). A specific signal peptidase cleaves the leader peptide with the establishment of cysteine as the N-terminus amino acid of the lipoproteins. A fatty acid, commonly palmatate, is added to the amino segment of cysteine from a phospholipid. The fatty acids making up the lipid unit of the lipoprotein become anchored at the inner leaflet of the outer membrane. Approximately one third of the lipoproteins are covalently linked to the peptidoglycan, thereby linking the outer membrane to the peptidoglycan and preventing the outer membrane from being released from the cell.

## *4.4.3. Porins*[14–16]

In 1976 Taiji Nakae used the term *porin* to describe the proteins that accounted for water-filled channels in the outer membranes of Gram-negative bacteria. Models for porins in the outer membrane are given in Figure 4.18. Genes for porins have been isolated from more than a dozen different species and sequence information indicates a phylogeny of the bacteria similar to that obtained by 16S RNA analysis. The most abundant protein in the outer membrane of *E. coli* is OmpA, with about $10^5$ molecules/cell. OmpA provides a nonspecific diffusion channel for solutes to traverse the outer membrane. According to projected structural analysis, the 35-kDa monomer of OmpA crosses the outer membrane eight times and forms antiparallel β-sheets typically found in porins. OmpC and OmpF are outer membrane porins of *E. coli*, which are present as trimers and regulate the passage of hydrophilic solute molecules with a mass of 300 to 5,000 daltons through the outer membrane. In addition to serving as selective channels for cations, OmpC and OmpF are responsible for structural maintenance of the cell and have a stable interaction with the peptidoglycan. Phosphoporin (PhoE) is also considered nonspecific in that it enables negatively charged solutes such as phosphate and chloride to be taken up by the cell. From three-dimensional structures obtained from X-ray crystallography of OmpF and OmpE, it is apparent that porins consist of 16-stranded antiparallel β-sheet proteins in a barrel topology. (Figure 4.19). There is an area on the outer surface of the barrel that is hydrophobic and interacts with the lipid center of the outer membrane, and the center of the barrel is an aqueous pore. The pore size changes with a progression through the central core. The pore entrance is narrowed by the protein loops at the exterior of the end of the barrel with a diameter of 11 to 19 Å. Approximately halfway through the membrane is the smallest pore region, resulting from the internal protein loops and at the bottom of the barrel adjacent to the peptidoglycan, the pore diameter increases owing to the lack

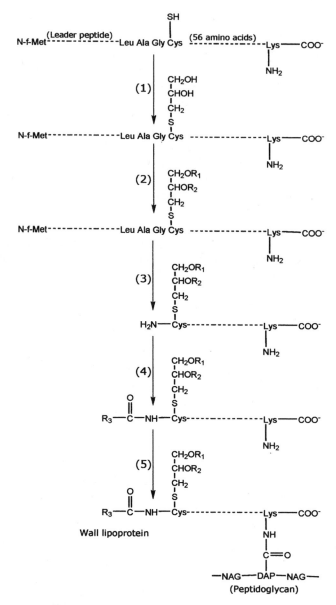

Figure 4.17. Scheme for biosynthesis of wall (Braun) lipoprotein. Enzymes are (1) glyceryl transferase, (2) *O*-acyltransferase, (3) signal peptidase II, (4) *N*-acetyltransferase, and (5) transpeptidase. Fatty acids are at $R_1$, $R_2$, $R_3$.

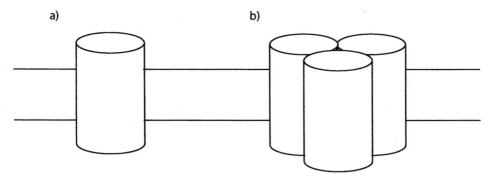

Figure 4.18. Model representing porins associated with the outer membrane of Gram-negative bacteria. In *Escherichia coli*, the outer membrane may contain the monerimic OmpA (**a**) as well as the trimers such as OmpC or OmpF (**b**).

of internal protein loops. The configuration of the internal loops regulating passage through the porins is apparent from the structures in Figure 4.19.

Solute selectivity is regulated, in part, by the peptide that extends across the central region of the pore and in the pore channels. Some examples of porins are given in Table 4.11. The porin proteins function both for uptake transport and as receptors for bacteriophages, antibiotics, or colicins. It is difficult to study porins in the native state in the outer membrane because of the other proteins that are present. The binding and permeation properties of porins are best determined by placement of isolated porins in artificially constructed lipid bilayer membranes. Owing to the unique mo-

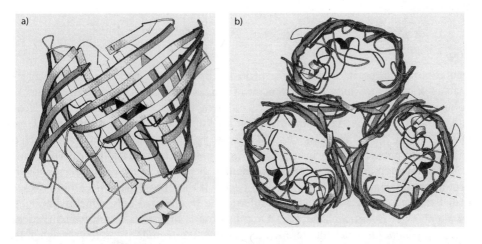

Figure 4.19. Ribbon models of porins. The OmpF monomer is shown in (**a**) as it would occur in the outer membrane and the view of the OmpF trimer (**b**) as viewed down the threefold axis. Used (From Cowan and Schirmer, 1994.[15] Used with permission.)

Table 4.11. List of some of the porins and transport proteins in the outer membrane of Gram-negative bacteria.[a]

| System | Protein | Porin structure | Solute transported | Bacteriophage, colicin, or antibiotic receptors |
|---|---|---|---|---|
| *Escherichia coli* | | | | |
| | OmpA (35 kDa) | No (?) | Nonspecific | |
| | OmpF (B)(38 kDa) | Yes | Cation selective | T$_2$, TP1, TP2, TP5, Colicin A |
| | OmpD (38 kDa) | Yes | Cation selective | |
| | OmpC (37 kDa) | Yes | Cation selective | PA2, 434, SS1, TP2, TP5, TP6 |
| | PhoE (36 kDa) | Yes | Anion selective | TC23, TC45 |
| | Txs (26 kDa) | No | Nucleosides and deoxynucleosides | Colicin K, T6 |
| | LamB (474 kDa) | Yes | Maltose, maltodextrin | λ, K10, TP1, TP5, SS1 |
| | Btu (66 kDa) | No | Vitamin B$_{12}$ | Colicins E1 and E2, BF23 |
| | FepA (81 kDa) | No | Fe(III)-enterochelin | Colicins B and D |
| | TonB | No | Fe(III)-ferrichrome | Albomycin, Colicin M, T1, T5, ø80 |
| | Cir (74 kDa) | No | Fe(III)-citrate | |
| *Salmonella typhimurium* | | | | |
| | OmpF (39 kDa) | Yes | Cation selective | |
| | OmpC (40 kDa) | Yes | Cation selective | PH42, PH105, PH221 |
| | OmpD (38 kDa) | Yes | Cation selective | PH31, PH42, PH51 |
| *Pseudomonas aeruginosa* | | | | |
| | F | Yes | Cation selective | |
| | P | Yes | Anion selective | |

[a]From Nikaido and Vaara, 1987.[16]

lecular structure, porins will self-orient in the lipid bilayer and this artificial matrix enables scientists to study the movement of ions or molecules through the porins.

Various strains of bacteria may have different porins. For example, *Escherichia coli* strain B produces OmpF under normal growth while *E. coli* K12 has Omp F and Omp C. Porin P from *Pseudomonas aeruginosa* is highly selective for anions such as chloride and phosphate. Porins such as Omp F exist as trimers and have a sufficiently stable interaction with the peptidoglycan that they can be isolated along with the peptidoglycan. Induction of the P porin occurs if the phosphate concentration is less than 1 m*M* phosphate. A structure similar to the P porin is PhoE, which is produced by members of *Enterobacteriaceae*. Omp F is found in *E. coli*, *Enterobacter cloacae*, and *K. pneumoniae* when the temperature is low, the osmolarity is low, or when the cellular concentrations of cAMP are high. Omp C is also cation selective; however, production occurs in media of high molarity.

## *4.4.4. Enzymes, Cytochromes, and Toxins*[17–19]

Various types of proteins that are structurally and functionally distinct from porins are found in the outer membranes of Gram-negative bacteria. These proteins are less abundant than porins and have a molecular mass of 60 to 80 kDa. A partial listing of some of these outer membrane proteins with mixed functions are listed in Table 4.12. Proteins attached to the outer membrane include hydrolytic enzymes, which are frequently called exoenzymes because they exist external to the plasma membrane. A few exoenzymes are held in the cell wall region by covalent attachment to the S-layer and these are not discussed here because they are covered in Chapter 3. Exoenzymes in the outer membrane have the following characteristics: Phenylalanine is the C-terminal amino acid residue, several putative amphiphatic β-strands are present, and the exoenzymes are similar at the C-terminal region to suggest secretion by an autotransporter system. Active proteins that are anchored into the outer membrane include an esterase produced by *Pseudomonas aeruginosa,* serine protease synthesized by *Serratia marcescens*, a vacuolating cytotoxin produced by *Helicobacter pylori*, and an esterase found in outer membranes of *Salmonella typhimurium*. The presence of toxins for host cells may be important to enhance the virulence of pathogenic bacteria and may occur in numerous other Gram-negative bacteria.

Historically, cytochromes have been associated with plasma membranes; however, recently it has been established that *c*-type cytochromes are present in the outer mem-

Table 4.12. Examples of cytochromes, enzymes, and toxins localized in outer membranes of Gram-negative bacteria.

| Proteins | Bacteria |
|---|---|
| **Electron transport** | |
| *c*-Type cytochromes | *Desulfovibrio vulgaris* |
| | *Shewanella oneidensis* |
| | *Geobacter sulfurreducens* |
| | |
| **Enzymes** | |
| Chitobiase | *Vibrio harveyi* |
| Endoglucanase | *Pseudomonas aeruginosa* |
| Esterases | *Pseudomonas aeruginosa* |
| | *Salmonella typhimurium* |
| Serine protease | *Serratia marcescens* |
| Pullulanase | *Bacteroides thetaiotaomicron* |
| | *Klebsiella oxytoca* |
| Phospholipase A | *Escherichia coli* |
| Protease | *Escherichia coli* |
| | |
| **Toxin** | |
| Cytotoxin | *Helicobacter pylori* |

branes of a few Gram-negative bacteria.  It has been suggested that these cytochromes may have a major role in electron transfer between the bacterial cell and minerals in the environment.  Lovley and colleagues have demonstrated that *Geobacter sulfurreducens* has a 41-kDa *c*-type cytochrome in the outer membrane and this molecule is proposed to have a role in the dissimilatory transfer of electrons from the cytoplasm to extracellular $Fe^{2+}$ which results in the formation of $Fe^{2+}$.  In this transfer of electrons to $Fe^{3+}$, there is a cascade of reactions with a 9-kDa *c*-type cytochrome in the periplasm transferring electrons from the plasma membrane to the 41-kDa cytochrome located at the surface of the cell.  In *Shewanella oneidensis* (formerly *putrefaciens*), a *c*-type cytochrome is present in the outer membrane and it is presumed to function in transfer of electrons from the cytoplasm to $Fe^{3+}$ or other oxidized metal ions.  Laishley and colleagues have demonstrated a cytochrome in the outer membrane of *Desulfovibrio vulgaris* and it has been proposed that this cytochrome has a role in transferring electrons from metallic iron in the oxidation phase of ferrous biocorrosion.  Although the presence of cytochromes in outer membranes in Gram-negative bacteria is considered to be relatively novel, it may be a widely distributed phenomenon that has eluded scientists.  Clearly, this is a fascinating area for future physiological studies.

## *4.4.5. Membrane Blebs and Vesicles*[20,21]

Many different species of Gram-negative eubacteria release small structures that appear to be an overexpression of the outer membrane.  These structures were initially called blebs by Elizabeth Work in 1966 and this designation would appear to be preferred over *vesicles*, as plasma membrane structures resulting from disruption of cells are known as vesicles.  Figure 4.20 reveals blebs on the surface of a bacterial cell.  In many instances, these blebs are released from the cell into the surrounding media.  These outer membrane blebs are not pelleted along with cells when using low-speed centrifugation of 10,000*g* for a few minutes.  However, if the centrifugate of the culture fluid is subjected to a force of 100,000*g* for 1 hr, the blebs will be pelleted.

The significance and the reason for production of these blebs is under investigation in several laboratories.  If the Gram-negative bacterium is a pathogen, the outer membrane blebs may contain lipopolysaccharide material that can be distributed into the environment with cell-free culture fluids.  Such activity could be important in intestinal infections and oral inflammation, especially gum disease.  The following activities have been associated with blebs:  endotoxin with blebs from *Neisseria meningitidis*, alkaline phosphatase with blebs of *Salmonella typhimurium*, and formate dehydrogenase with blebs from *Desulfovibrio gigas*.  Recently, some of the anaerobic Gram-negative bacteria including *Desulfovibrio*, *Shewanella*, and *Geobacter* have been found to have special *c*-type cytochromes in the outer membrane that react with minerals and metals in the environment.  Thus, the blebs released from certain species may play an important role in diseases, biocorrosion, mineral leaching, and geomicrobiology.

The utilization of water-insoluble hydrocarbons requires a special interface between the lipid and bacterial cells.  *Acinetobacter* sp. HO1-N is a Gram-negative bacterium

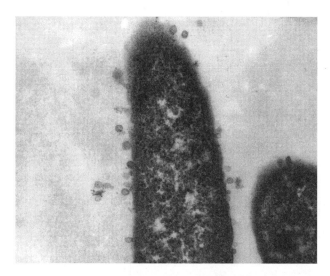

Figure 4.20. Electron micrograph of outer membrane blebs from the Gram-negative bacterium *Desulfovibrio gigas*. The cell was fixed in 2% glutaraldehyde, stained with 5% $OsO_4$, and sections were stained with 0.1% uranyl acetate and 1% lead citrate. $\times$ 40,000. (Courtesy of J. Ruocco and L. Barton.)

that was shown by Finnerty and colleagues to grow on hexadecane and release membranous vesicles that serve as surfactants to emulsify the hydrocarbon. To isolate the vesicles, the spent culture medium was subjected to centrifugation at 65,000$g$ for 5 hr. These extracellular vesicles are 20 to 50 nm in diameter and are similar, but not identical, to the outer membrane of the cell. The quantity of phospholipids present in the extracellular vesicle was about five times greater than that of the outer membrane and the lipopolysaccharide content was about 360 times greater than in the outer membrane. However, the numbers of polypeptides in the vesicle and outer membrane appeared to be relatively similar. The production of these vesicles in other bacteria remains to be established.

## 4.5. Surface Proteins of *Borrelia*[4,5]

Spirochetes are rigid spiral bacteria that have a unique cell structure with a plasma membrane and an outer covering designated by various terms including outer membrane, sheath, or surface structure. Owing to the unique anatomical structure of spirochetes and the potential diversity of the structure, the discussion here will use outer surface structure to designate the membrane-like structure exterior to the plasma membrane. Motility in the spirochetes is attributed to endoflagella located between the plasma membrane and the outer surface structure (see Chapter 6). Proteins on the outer surface structure have been examined in relatively few spirochetes, and this discussion will focus on *Borrelia*.

Table 4.13. Characteristics of lipoproteins in the outer surface structure of spirochetes.[a]

| Organism | Lipoprotein | Activities |
|---|---|---|
| *Borrelia burgdorferi* | OspA | Produced in tick larva at time of nymph development and quantity of protein decreases when the nymph consumes blood. |
| | OspB | Similar to OspA |
| | OspC | Concentration of protein per cell increases as bacteria move from midgut to salvary gland in tick. Production occurs when bacteria is in the mammal. |
| | OspD | Production occurs when bacteria infects mammal. Production is induced by temperature increases with greatest induction at 35°C. |
| | DbpA | A group of lipoproteins that bind to decorin which is a proteoglycan of the intracellular matrix. |
| | Vmp | Antigenic variable protein that is carried on a 28-kbp linear plasmid. |

[a]From Salyers and Whitt, 2002.[5]

*Borrelia burgdorferi* is the agent of Lyme disease, which is transmitted by *Ixodes* ticks and infects numerous mammals including humans. This bacterium is a rigid spiral of 4 to 20 coils and is 20 to 30 μm long but only 0.2 μm wide. Although *Borrelia burgdorferi* does not have lipopolysaccharides in the outer surface structure, it does produce several lipoproteins that are assumed to elicit the inflammatory response. Several of these outer surface proteins (Osp) appear to be encoded on plasmids. Some strains of *Borrelia burgdorferi* have 9 circular plasmids and 12 linear plasmids, which is an unusually high number of plasmids for a cell. These lipoproteins appear to be produced in specific environments with OspA and B produced in the tick larva under cool conditions. Following exposure to blood and a temperature of at least 32°C, the bacteria produces OspD, PbpA, and Vmp. The characteristics of these lipoproteins in the outer surface structure of *Borrelia burgdorferi* are given in Table 4.13.

# 4.6. Surface Amphiphiles[22,23]

Amphiphiles are a diverse polymeric set of molecules that have both a hydrophilic and a lipophilic region. The hydrophilic region is either a sugar polymer or a peptide while the lipophilic region contains fatty acids. The lipid portion of the molecule is anchored in the outer leaflet of the plasma membrane or the outer membrane and the hydrophilic region extends from the cell, providing an interface between the cell and physical environment. Many of the polymeric molecules in cell walls of Gram-positive or Gram-negative bacteria that were presented earlier in this chapter have amphiphilic characteristics. Virulence-enhancing activities may be attributed to amphiphilic proteins of pathogenic bacteria, and some of the structural features of am-

Table 4.14. Biological activities of amphiphiles.[a]

| | Lipopolysaccharide | Lipoteichoic acid | Enterobacterial common antigen | Lipoprotein | Lipomannan | Hexose-lysine amphiphile |
|---|---|---|---|---|---|---|
| Binds to eukaryotic cells | Y | Y | Y | Y | Y | Y |
| Immunogenic | Y | Y | Y | Y | Y | Y |
| Mitogenic | Y | Y | ? | Y | ? | Y |
| Stimulate bone resorption | Y | Y | ? | ? | ? | Y |
| Pyrogenic | Y | N | N | N | N | N |
| Cytotoxic | Y | N | N | N | ? | ? |

[a]Adapted from Wicken and Knox, 1981[32] and Fischer, 1994.[3]
Y, Yes; N, no; ?, unknown response.

phiphiles are summarized in Table 4.14. All of the molecules at the surface of bacterial cells interface with the host cells and are highly immunogenic. At least one of these amphiphilic molecules has a detrimental effect on host systems by being mitogenic, displaying cytoxicity, or promoting calcium resorption from bone.

## 4.6.1. Enterobacterial Common Antigens

The amphiphilic molecule present in most of the *Enterobacteriaceae* ia a hydrophilic linear polymer consisting of *N*-acetyl–D-glucosamine and *N*-acetyl-D-mannosamino-uronic acid. This structure is referred to as the enterobacterial common antigen and has an average molecular mass of 2.7 kDa.

## 4.6.2. Adhesions and Cell Clumping[24]

The various molecules at or near the surface of the cell contribute to the charge on the cell surface. The isoelectric point (pI), the pH at which there is no charge on the surface of the cell, reflects the specific type of charged groups on the surface of the cells. An evaluation of charge on many different bacteria indicates a difference between Gram-negative and Gram-positive bacteria (see Figure 4.21) but this difference in pI for the bacteria is not responsible for the Gram reaction. Some of the molecules contributing to the charge on the bacterial cells display either acidic or basic properties. As indicated in Table 4.15, an important feature of these molecules is the interaction with eukaryotic cells attributable to proteins on the bacterial cell surface that are referred to as adhesions. Considerable specificity is displayed by the adhesions for a type of molecule found in the interstitial extracellular matrix or the base-

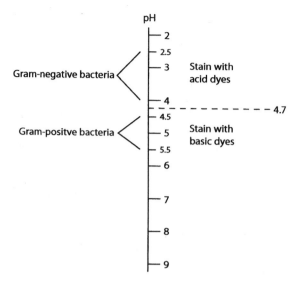

Figure 4.21. Surface charge of bacteria. Although bacteria have an acidic charge, Gram-negative bacteria have a cell surface that is more negative than the charge on Gram-positive bacterial cells. Since the pH of a cell must be at least pH 4.7 to be stained by a basic dye, bacteria from extremely acidic environments are optimally stained with acid dyes.

ment membrane network. Each polymer has a specific adhesion and *Staphylococcus aureus*, for example, is a potent pathogen in that it has adhesions that recognize nine different tissue polymers. The type of adhesion is important for pathogens and not the number of different types of adhesion molecules, because pathogens such as *Vibrio cholera* and *Mycobacterium tuberculosis* each have a single adhesion but display considerable pathogenicity. Adhesions on bacterial cells serve to target the pathogen within the host. In the case of *Mycobacterium leprae*, the adhesin binds to the laminin

Table 4.15. Interactions between selected bacterial adhesions and animal tissues.[a]

| Structure with animal cell | *Staphylococcus aureus* | *Escherichia coli* | *Streptococcus pyogenes* | *Vibrio cholera* | *Mycobacterium tuberculosis* |
|---|---|---|---|---|---|
| **Interstitial extracellular matrix** | | | | | |
| Collagens I, II, III | + | + | | | |
| Fibrinogen | + | + | + | | |
| Fibronectin network | + | + | + | | + |
| Elastin microfibrils | + | | | + | |
| Hyaluronan | + | | | | |
| Vitronectin | + | + | + | | |
| Bone sialoprotein | + | | | | |
| **Basement membrane** | | | | | |
| Collagen IV network | + | | + | | |
| Laminin network | + | | + | | |

[a]Modified from Preissner and Chhatwal, 2000.[33]
Adhesion production by bacteria is indicated by "+."

of the peripheral Schwann cells and this leads to the anesthetic character of infections with *M. leprae* and contributes to localization of the bacteria at the cooler regions of the body. The molecular characteristics of the bacterial cell–surface adhesions are relatively similar to adhesions on the tips of the bacterial pili, and some suggest that there is an evolutionary relationship between these two adhesive molecular forms.

Bacterial cells bind not only to animal cells but also to various organic molecules in the environment. If adhesions bind serum or antibodies, these bonding proteins are virulence enhancing and important for disease production. Examples of some virulence-enhancing compounds are presented in Table 4.16. Specific proteins that serve as adhesions have been characterized from pathogenic Gram-positive and Gram-negative bacteria. The binding affinity of adhesions on the surface of *E. coli* and *Helicobacter pylori* indicates that individuals with specific blood group antigens are at greater risk for infection. Binding of bacteria to host tissues is dependent on the target for the specific adhesions. The model in Figure 4.22 reveals the binding to the host cell surface as well as to extracellular matrix.

Table 4.16. Examples of proteins associated with adhesive properties or immune reactivity on the surface of Gram-positive and Gram-negative bacterial cells.[a]

| Activity of protein | Distribution |
|---|---|
| **Adhesions** | |
| Afa-I and Afa-III proteins recognize Dr[a] blood group antigens in intestine | *Escherichia coli* |
| Protein complexes with Lewis[b] antigen in gastrointestional tract | *Helicobacter pylori* |
| FHA (200 kDa protein) adheres to respiratory surfaces | *Bordetella pertussis* |
| Opacity protein (Opa) binds to DEA or $CD_{66}$ proteins of human cells | *Neisseria gonorrhea* |
| Specificity for individual CEA proteins accounts for tissue specificity | *Neisseria meningitidis* |
| Binding to fibronectin by proteins of the BCG85 complex. | *Mycobacterium tuberculosis* |
| Binding to laminin of Schwann cells of the peripheral nerves | *Mycobacterium leprae* |
| Binding to extracellular matrix of human cells | *Staphylococcus* sp. and *Streptococcus* sp. |
| **Immune related** | |
| Immunogenic membrane protein (Protein C) | *Treponema pallidum* |
| Lipoprotein binds immune globulin D (Protein D) | *Haemophilus influenzae* |
| Immunogenic membrane protein (H-8 antigen) | *Neisseria meningitidis* |
| Protein A binds the Fc region of immune globulins | *Staphylococcus aureus* |

[a]From Braun and Wu, 1994[13]; [b]From Finlay and Caparon, 2000.[28]

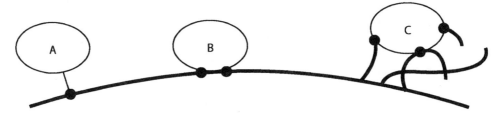

Figure 4.22. Model showing methods of attachment by adhesions to tissues in animal hosts. In **A** the bacterial cell is attached to the host cell by pilus (fimbrae), in **B** the bacterial cell is attached to the host cell by adhesions on the surface of the bacterial cell, and in **C** the bacterial cell is attached to exoskeleton matrix. (From Finlay and Caparon, 2000.[28])

A rather unique process of conjugation occurs with Gram-positive bacteria, and the best characterized system is found with *Enterococcus* (*Streptococcus*) *faecalis*. Cells have a directional transfer of DNA, with the donor cells having a conjugative plasmid and the recipient cell lacking this plasmid. The recipient cell releases an oligopeptide that attracts the donor cell, modifies the surface of this donor cell, and promotes clumping of donor and recipient cells. These oligopeptides are referred to as sex pheromones and are diffusible soluble peptides with molecular mass of less than 1 kDa. The nature of cell surface modification or the mechanism accounting for agglutination between donor and recipient cells remains to be resolved.

## 4.6.3. Resistance to Hydrolytic Enzymes

With the exposure of the microbial cells to enzymes in the environment, the resistance of molecules at the cell surface to these enzymes is extremely important. Molecules composed of different types of sugar residues and sugars that are linked together by different covalent bonds would require several different hydrolytic enzymes for digestion. Some of the unique sugars present in the lipopolysaccharide include rhamnose, heptose, and KDO, an acid consisting of eight carbon atoms per molecule. These unusual sugars are sparsely distributed in nature and as there is a requirement for specific enzymes for different substrates, the enzymes that would disrupt the bonds adjacent to these molecules would be expected to be rare. The bonds are not only $\alpha$ $1\rightarrow4$ or $\alpha$ $1\rightarrow6$ linkages, commonly found in starch and glycogen, but may be in a $\beta$ as well as $\alpha$ configuration. All the carbon atoms on the sugars may be involved in the bonding, producing some bonds that will require unique enzymes for autolysis. The presence of acetyl groups on the amino moiety of the amino sugars prevents random addition of amino acids. In addition, a few of the bacterial structures are glycoproteins and the sugar components would serve to protect the protein against enzyme attack. The unique hydrophobic molecules found in the cell walls of acid-fast bacteria are highly resistant to enzymatic hydrolysis, as evidenced from the inability of phagocytes to digest the mycobacterial cells.

## 4.7. Distribution of Hydrolases on the Cell Surface

Complex carbon materials of biological origin are important substrates for prokaryotes, and various systems have evolved for optimal utilization of cellulose, starch, and lipids. Interfacial reactions are important with these substrates because their rigidity, bulky size, or hydrophobic character present problems for enzyme attack. If a bacterium uses cellulose, starch, or lipids as a source of both carbon and energy, the cell has multiple enzymes for the efficient utilization of the substrate. However, in several instances, bacteria use these substrates to assist their other metabolic activities and only one or two enzymes are produced by a given cell. A steady progression in our understanding of these enzymes is attributed to the interest that they are generating in the commercial sector.

### 4.7.1. Cellulosome[25–27]

Some clostridia and rumen bacteria produce cellulases and xylanases that result in complete digestion of cellulosic plant material. To achieve this, enzymes are highly organized on the surface of the cell to bind cellulose and utilize the sugars released by various exoenzymes. These bacteria would have a "complete" cellulosic conversion system and are in contrast to "incomplete" cellulose utilizing systems that have only one or two of the cellulose digesting enzymes. Bacteria with incomplete cellulose utilizing systems include plant pathogens (e.g., *Erwinia chrysanthemi* and *Pseudomonas solanacearum*) and nitrogen-fixing symbionts (e.g., *Rhizobium leguminosarum*). The complete cellulose digesters have a bulbous structure on the cell surface which are the cellulose-binding domains (CBD) that contain binding proteins and enzymes. Not all bulbous-like structures on the surface of bacteria are cellulosomes because noncellulolytic bacteria degrade starch and are proposed to have hydrolytic enzymes localized in structures that could be referred to as amylosomes.

R. Lamed and collaborators, in 1984, designated a structure on the surface of *Clostridium thermocellum* as the cellulosome and established that it was responsible for the hydrolysis of cellulose. Isolated cellulosomes display considerable stability with no loss of activity when stored in 50% ethanol for several months at room temperature. The multicomponent surface structure of cellulosomes varies somewhat with bacterial strains but in general they are about 18 nm in diameter with a mass of 2 to 6.5 × $10^6$ Da. Polycellulosomes have a mass of 50 to 80 × $10^6$ Da. A model for the cellulosome of *Clostridium thermocellum* is given in Figure 4.23. In young cells, the cellulosomes are tightly packed in a highly ordered array but as the cells age, the cellulosomes become loosely packed and disintegrate to release individual enzymes. At least 20 different genera of clostridia hydrolyze cellulose, and cellulosome-like structures have been reported on the cell surface of *Clostridum cellulovorans*, *Acetivibrio cellulolyticus*, *Bacteroides cellulosolvens*, *Ruminococcus albus*, and *Clostridium cellobioparum*.

Each cellulosomal unit of *Clostridium thermocellum* may contain as many as 14 to 26 polypeptides and a large polypeptide that serves as a scaffolding protein. The

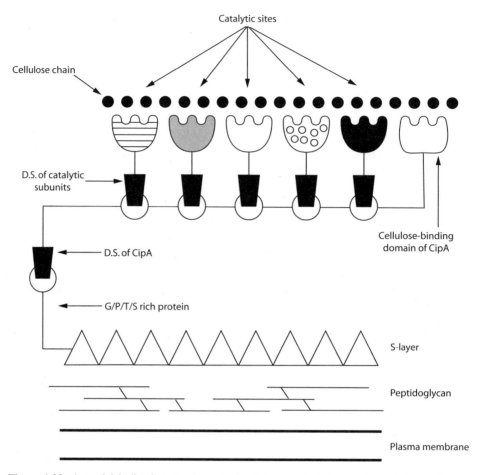

Figure 4.23. A model indicating attachment of cellulosome to S-layer of *Clostridium thermocellum*. DS is a duplicated peptide segment consisting of a 22 amino acid unit. Different shading of the enzymes that interact with the cellulose chain reflects the different endocellulases present in the cellulosome. (Modified from Felix and Lungdahl, 1993.[26])

enzymes are organized around a noncatalytic protein of 210 kDa, termed CipA for *c*ellulosome-*i*ntegrating *p*rotein, that recognizes cellulose. In *Clostridium thermocellum*, a carotenoid-like pigment is produced that coats the cellulose fibers and enhances the binding of CipA in the cellulosome to cellulose. Also, CipA appears to have SLH (S-layer homologous) domains that may be important in binding the proteins into the underlying S-layer. The major hydrolytic enzymes in the cellulosome have three types of activities: endo-1,4-β-glucanase, exo-1,4-β-glucosidase, and cellobiase activity. Enzymes in the cellulosome of *Clostridium thermocellum* have been characterized and are summarized in Table 4.17. There are numerous endoglucanase genes in *Clostridium thermocellum*, and eight of these genes (*celA, celB, celC, celD, celE*,

Table 4.17. Characteristics of enzymes in the cellulosome of *Clostridium thermocellum*.[a]

| Enzyme | Amino acids | Size | Substrate |
|---|---|---|---|
| **Endogluconase** | | | |
| CelA | 488 aa | 52,503 Da | Carboxycellulose |
| CelB | 536 aa | 63,857 Da | Carboxycellulose, cellote-traose, cellopentaose |
| CelC | | 40,439 Da | Cellotriose, cellotetraose, cellopentaose |
| CelD | 649 aa | 72,334 Da | Carboxycellulose |
| CelE | 780 aa | 84,016 Da | Carboxycellulose |
| CelF | 730 aa | — | ? |
| CelH | 900 aa | 102,301 Da | Carboxycellulose |
| CelX | 673 aa | — | Carboxycellulose (?) |
| **Exogluconase or cellobiohydrase** | | | |
| $S_8$-tr | | 68,000 Da | Amorphous cellulose, xylan |
| **Cellobiase or β-glucosidase** | | | |
| Enzyme | | — | Cellobiose |

[a]From Felix and Ljungdahl, 1993.[26]

*celF*, *celH*, and *celX*) have been cloned and sequenced. At least one of these endoglucanase (product of the *celA* gene) is a glycoprotein with 11% to 12% of the molecule as carbohydrate, making it one of the few glycoproteins in bacteria. Many of these endonucleases reflect a similar structure: a signal peptide of 30-plus amino acids, a catalytic site at the N-terminal end, a cellulose binding region in the central domain of the enzyme, and a reiterated segment of about 20 amino acids at the C-terminus. The endoglucanases randomly hydrolyze the linear cellulose into short chain lengths, exoglucanases release cellobiose from the nonreducing ends of the cellulose chains, and cellobiase hydrolyzes cellobiose to glucose. Purified enzymes do not hydrolyze crystalline cellulose but the mixture of enzymes from solubilized cellulosomes synergistically degrades cellulose. Some organisms, such as *Clostridium thermocellum*, do not convert cellobiose to glucose in the cell wall region but actively transport cellobiose across the plasma membrane, and the disaccharide is metabolized by a cellobiose phosphorylase.

## 4.7.2. Starch Degradation[28–30]

Starch is a storage polysaccharide in plants that consists of amylose and amylopectin. Amylose is made up of several hundred glucose units attached by α 1→4 glucosidic bonds to produce a linear polymer. Amylopectin is made up of glucose units with a linear segment of α 1→4 bonds and branch points attributed to an α 1→6 glucosidic linkage. Bacteria produce several different extracellular enzymes that hydrolyze starch and some of these amylases remain attached to the cell. There are four major classes of starch degrading enzymes and these include α-amylase, β-amylase, glucoamylase, and pullulanase. The α-amylase is a liquifying or endoamylase because it hydrolyzes the α 1→4 bond of the amylopectin or amylose molecule. Bacteria that produce α-amylase include *Bacillus stearothermophilus*, *Bacillus subtilis*, *Clostridium acetobutylicum*, *Pseudomonas stutzeri*, and *Pseudomonas saccharophilia*. The enzyme that hydrolyzes the α 1→4 linkage at the end of the molecule with the release of maltose is β-amylase. Bacteria that produce β-amylase include *Bacillus cerus*, *Bacillus polymyxa*, *Bacillus circulans*, and *Clostridium thermosulfurogens*. Glucoamylase hydrolyzes both α 1→4 and α 1→6 glucosidic linkages with the conversion of starch to glucose. *Clostridium acetobutylicum* is one of the few bacteria that produces glucoamylase. The fourth class of amylase includes pullulanase, which preferentially hydrolyzes the α 1→6 bond but some enzymes also attack the α 1→4 linkage. Bacteria that produce pullulanase include *Klebsiella aerogenes*, *Bacillus stearothermophilus*, and *Bacteroides thetaiotaomicron*.

Polysaccharide utilization by Gram-negative bacteria appears to be associated with the substrate binding to the outer membrane and this is followed enzymic hydrolysis of the bound substrates. The best example of this bacterial activity is the starch utilization system by *Bacteroides thetaiotaomicron* developed by A.A. Salyers and collaborators. *Bacteroides thetaiotaomicron* has eight genes for starch utilization and these are designated as *sus* genes. The *susR* regulates the promoters of *susA* and the operon for *susB*, *C*, *D*, *E*, *F*, *G*. Products of these genes include SusA, a periplasmic pullanase; SusG, a pullanase bound onto the exterior of the outer membrane; and SusB, an α-glucosidase in the cytoplasm. SusC is a porin that facilitates the transport of maltodextrins that range in size from maltose, a glucose dimer, to maltoheptose, a polymer of seven glucose moieties. About 70% of the starch binding to the outer membrane is by the proteins SusC and SusD with the remaining starch binding attributed to SusE or SusF. The organizational pattern of these proteins in the outer membrane is not yet established but perhaps there is a specific complex for starch utilization similar to the cellulosome for cellulose-utilizing bacteria. It remains to be determined if protein complexes are on cell surfaces for utilization of other biopolymers.

## *4.7.3. Lipases*[18,31]

The hydrolysis of lipids by bacteria is a unique activity involving a hydrophilic substrate and an enzyme in an aqueous environment. The cleavage of esters between glycerol and long-chain fatty acids is an interfacial phenomenon with enzyme kinetics distinct from those for catabolic reactions. Initially, the binding of the enzyme to an emulsified substrate occurs with an apparent change in structural conformation of the enzyme to expose the active site. A second activity is the release of the product from the enzyme–substrate complex. With the substrate consisting of three fatty acids esterified to a glycerol backbone, there are two groups of bacterial lipases. Enzymes produced by *Staphylococcus aureus*, *Corynebacterium acnes*, and *Chromobacterium viscosum* have no specificity for esterified fatty acids attached onto the glycerol moiety.

**Box 4.2**
**A Research Path Is Rarely a Straight One**
A Perspective by Abigail A. Salyers

I entered the field of bacteriology through the back door. My Ph.D. was in theoretical physics and I had worked as a physics professor for several years when I decided to switch fields. At the time, both jobs and research funding in theoretical physics were scarce, and many young physicists were jumping ship. I began taking courses in biochemistry at the Virginia Polytechnic Institute in Blacksburg, Virginia. At a department party, I met my future postdoctoral mentor, Tracy Wilkins. On the basis of that conversation, Tracy offered me a position in his laboratory working on the obligate anaerobic bacteria that are the numerically predominant bacteria in the human colon. At the time, I did not realize how unfashionable this area was. There was little metabolic information about the human colonic anaerobes and there was no genetic system in sight.

Nonetheless, these poorly understood bacteria seemed likely to be important. After all, we carry a large number of them in our intestines throughout life. They have been suspected, but still not proven, to have a role in colon cancer and inflammatory bowel disease. They are known to have a nutritional role; they ferment polysaccharides that are indigestible to us, producing acetate, propionate, and butyrate, which are taken up and used as energy sources by intestinal cells. Finally, some of them, such as *Bacteroides* species my laboratory studies, are opportunistic human pathogens.

When virtually nothing is known about a group of bacteria, it can be difficult to decide where to start. I took as a starting point the fact that *Bacteroides* species can ferment a variety of dietary and host-derived polysaccharides. Over the years, scientists working in my laboratory have developed a model for how *Bacteroides* species utilize polysaccharides. Rather than excreting degradative enzymes into the external environment, they bind the polysaccharide to their outer membranes and translocate fragments of the polysaccharide into the periplasmic space as they are degrading it. This strategy is probably necessary to prevent competing bacteria from stealing the mono- and disaccharides released when *Bacteroides* species degrade polysaccharides.

My research group had not gotten far into the study of polysaccharide utilization by *Bacteroides* species when we realized that it was going to be essential to have some way of genetically manipulating the bacteria. We could purify polysaccharide-degrading enzymes and characterize them, but without the ability to disrupt the gene encoding a particular enzyme, we could not be sure how important that enzyme was in the intact cell. Also, genetic techniques such as transposon mutagenesis can help to identify genes-encoding proteins for which there is no assay. It was only when we were able to take this approach that we found the outer membrane proteins that bind polysaccharides to the bacterial surface.

After much labor and numerous failures, we finally managed to develop a genetic system for *Bacteroides*. As often happens in science, we started out in one direction and ended up in taking a detour that took us to a completely unexpected place. To make genetic tools for *Bacteroides*, we needed selectable markers. This led us to try cloning various antibiotic resistance genes. One of these genes proved to be located on a new type of transmissible element. At that time, plasmids were considered to be the only genetic elements capable of transfer by conjugation. Yet, groups working with Gram-positive bacteria had reported finding some conjugally transferred elements that were located in the chromosome. These elements were called conjugative transposons. Our resistance gene cloning experiments led us to the *Bacteroides* version of conjugative transposons. Conjugative transposons are now being found in many bacteria, including *Salmonella* and *Vibrio*, and may be as important as plasmids in transferring antibiotic resistance genes.

The *Bacteroides* conjugative transposons have now become a major focus of my research group. Over the past 10 years, we have learned a lot about how conjugative transposons excise from the chromosome, transfer to a recipient cell, and integrate into the recipient's genome. We have also become interested in the ecology of resistance genes—how genes move from bacterium to bacterium in real environments like the colon. Our work on the ecology of resistance genes has had the effect of drawing me out of the academic ivory tower into the real world of economics and politics. The unique blend of expertise I had acquired in such diverse areas as the normal microbiota of the human colon, the normal microbiota of the colons of livestock animals, and the transfer of antibiotic resistance genes was suddenly in demand. I was first called upon to testify before regulatory agencies on possible safety problems associated with the antibiotic resistance marker genes found in transgenic plants. Next, I was drawn into the debate over agricultural use of antibiotics. Talking to regulators, farmers, and scientists in biotechnology companies has been time consuming but has provided a useful reminder that what scientists like myself do in the laboratory has considerable relevance in the real world.

Most recently, my laboratory has taken yet another leap into the unknown. We are working with a geologist, Bruce Fouke, on the geomicrobiology of some Yellowstone hot springs and on a disease of corals called black band disease. This experience, together with my experience as a co-director of the Microbial Diversity summer course at the Marine Biological Laboratory in Woods Hole, Massachusetts—a course that focuses on marine prokaryotes—has made me more aware than ever of the enormous diversity of microbial life. It is now clear that most of the biodiversity on planet Earth resides in the microbial world. So many microbes, so little time!

In 1998, the magazine *Business Week* proclaimed that whereas the 20th century had been the century of the computer, the 21st century would be the century of biology. This prediction is proving to be an accurate one. The amazing advances in biochemical and genetic technology that have occurred in the last two decades have opened up vast new areas of biological research, especially in prokaryotic biology. This is truly a wonderful time to be a microbiologist!

Lipases produced by *Pseudomonas fragi, Pseudomonas fluorescens*, and *Pseudomonas glumae* are specific in that they hydrolyze only the ester bonds at C-1 or C-2 of glycerol. The secretion of lipases does not follow a single system in bacteria, with the most common systems including the Sec system with a leader signal peptide or translocation systems involving from 3 to 12 different proteins. An organism is not limited to producing a single lipase; *Pseudomonas aeruginosa* is known to secrete at least five different enzymes.

Based on structural analysis, there is little similarity in amino acid sequence between the various lipases, and the molecular masses for these enzymes generally range from 29 kDa to 40 kDa. At this time, the lipases are grouped into different classes based on size and a few molecular characteristics but continued research is needed to provide a complete characterization of these bacterial lipases. From three-dimensional structural analysis, some bacterial lipases resemble human lipases in that they have a conserved folding region known as the $\alpha/\beta$-hydrolase fold. The active site contains three amino acids: serine, histidine, and aspartate or glutamate. With respect to this three amino acid catalytic site, lipases resemble serine proteases. Lipases are held in the outer membrane by a series of transmembrane segments with the active site accessible from the outer side of the cell surface. The active site is inside the enzyme and an $\alpha$-helix peptide structure forms a cover for the lipase. With interaction between the lipase from *Pseudomonas glumae* and an appropriate substrate, the cover is removed to expose the enzyme active site by a process known as interfacial activation. Although it is not yet confirmed, the lipases may be constructed by antiparallel $\beta$-sheet segments that form a barrel-like unit somewhat similar to porins. In this case, the catalytic site of the lipases would regulate molecular passage through the channel of the transmembrane lipase. An example of a lipolytic enzyme bound to the outer membrane is the 55-kDa esterase of *Pseudomonas aeruginosa*, which is proposed to form 11 transmembrane $\beta$-strands with 10 to 12 amino acids in each strand (see Figure 4.24). Lipases are considered to function as virulence-enhancing factors because bacteria with surface lipases can compromise cells of the human immune defense system. In addition, lipases participate in colonization of *Propionibacterium acnes* and *Staphylococcus epidermidis* on human skin.

## 4.8. Distribution of Cell Wall Components in Prokaryotes

It would be simplistic to state that cell walls of Gram-positive bacteria are composed only of peptidoglycan and teichoic acids while cell walls of Gram-negative bacteria consist only of peptidoglycan, lipoprotein, and lipopolysaccharide. Many additional biopolymers are present in microbial cell walls and these are summarized in Table 4.18. The molecular forms can be summarized as belonging to three different types: (1) proteins found on the surface of both Gram-positive and Gram-negative bacteria, (2) anionic polymers found on Gram-positive bacteria, and (3) lipopolysaccharides and lipoproteins found in the outer membranes of Gram-negative bacteria.

Various hydrolytic enzymes are present on the outer surface of prokaryotic cells that function to degrade molecular complexes consisting of carbohydrates, nucleic acids, proteins, or lipids. In addition, numerous proteins are present on the surfaces

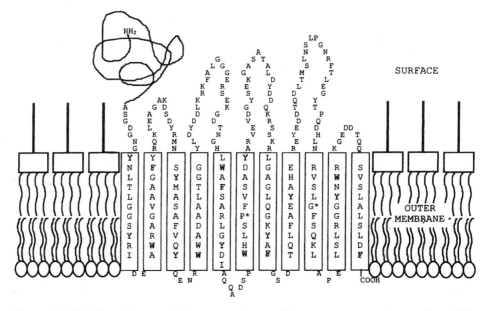

Figure 4.24. Binding of esterase in outer membrane of *Pseusomonas aeruginosa*. (From Wilhelm et al., 1999.[31] Used with permission.)

Table 4.18. Molecular forms present in prokaryotic cell walls in addition to peptidoglycan or other structural polymers.

| Components | Gram-positive bacteria | Gram-negative bacteria |
|---|---|---|
| **Proteins** | | |
| Hydrolytic enzymes | Present | Present |
| Proteins with adhesion properties | Present | Present |
| **Anionic polymers** | | |
| Teichoic acids | Present | Absent |
| Teichuronic acids | Present | Absent |
| Lipoteichoic acids | Present | Absent |
| Lipoglycans | Present | Absent |
| **Outer membrane components** | | |
| Lipopolysaccharides | Absent | Present |
| Porins | Absent[a] | Present |
| Braun (wall) lipoproteins | Absent | Present |

[a]Although mycobacteria stain as Gram-positive, the chemistry of the cell wall is distinct from that of typical Gram-positive cell types.

of bacteria that function to bind the cell to animal or plant surfaces. These adhesive proteins are distinct from pili found on Gram-negative bacteria. Proteins that bind antibiotics active against peptidoglycan synthesis are found primarily in cell walls of Gram-positive bacteria; however, a few penicillin-binding proteins are found in Gram-negative bacteria and this may be attributed to lateral gene transfer.

The anionic polymers derive their charge from phosphate and carboxyl groups that are supported on carbohydrate polymers. Thus, while Gram-positive organisms lack an outer membrane for regulation of chemical passage, they have different types of charged polymers interspersed between the peptidoglycan units to provide a sieving effect that limits chemical migration through the cell wall region. These acidic charges attributed to teichuronic acid, teichuronic acid, lipoteichuronic acid, and lipoglycans on the surface of Gram-positive bacteria are responsible for the isoelectric point of pH 4.5 to 5.5 (see Figure 4.21).

The structure of the outer membrane in Gram-negative bacteria has several unique components. Lipopolysaccharides have highly charged phosphate groups at the external surface of the cell and abundance of these anionic charges accounts for the low isoelectric point of pH 2.5 to 4.0 found with Gram-negative bacteria. The porins and Braun proteins are lipoproteins that have functional roles in Gram-negative bacteria but are absent in Gram-positive bacteria.

Another way of examining the components in the cell walls of prokaryotes is to focus on the mechanism of attaching the protein or glycan into the cell wall. The major types of attachment of the structures into the wall are summarized in the models presented in Figure 4.25. The two major mechanisms for attachment of the protein or glycan polymers into the cell wall are covalent attachment or lipophilic interaction. Several of the structures are attached into the carbon 6 of muramic acid or into the cross bridging peptides extending from the peptidoglycan. The individual enzymes comprising the cellulosomes are attached into the S-layer by covalent bonds. Lipid interactions are found in transmembrane protein loops of enzymes and porins in the outer membrane. In several instances, the carbohydrate polymers or proteins use the lipophilic action of fatty acids to secure the molecule to the plasma membrane or the outer membrane.

| Type | Model |
|------|-------|

**Transmembrane protein loops**

    Enzymes and porins in outer membranes of Gram-negative bacteria

**Lipophilic attachment**

    Lipotiechoic acid and lipoglycans of Gram-positive bacteria

    Lipopolysaccharides of Gram-negative bacteria

**Covalent bonds into the cell wall**

    Teichoic acids and teichuronic acids in Gram-postive bacteria attached to peptidoglycan

    Proteins in Gram-positive bacteria attached to wall structure

    Mycolic acids in Acid-fast bacteria attached to peptidoglycan

    Polysaccharides in Gram-positve bacteria attached to wall structure

    Cellulosomes attached to S-layer

**Mixed attachment: Lipophilic and covalent**

    Wall (Braun) protein of Gram-negative bacteria

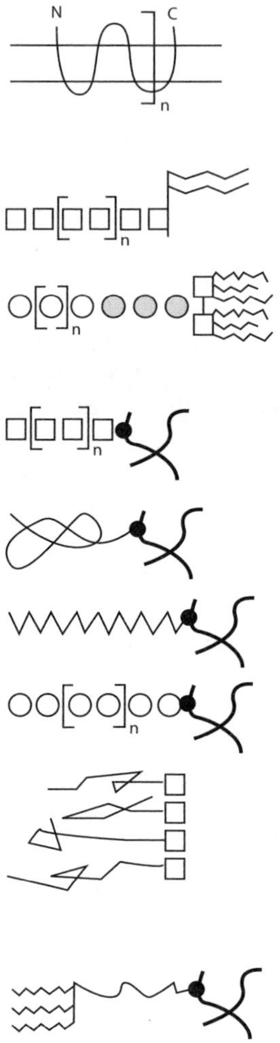

Figure 4.25. Summary of the attachment of polysaccharides, acidic polymers, and proteins into the cell walls of bacteria.

# References

1. Archibald, A.R. 1994. The structure, biosynthesis and function of teichoic acid. *Advances in Microbial Physiology* 11:53–97.
2. Davis, B.D., R. Dulbecco, H.N. Eisen, and H.S. Ginsberg. 1980. *Microbiology*. Harper & Row, Philadelphia.
3. Fischer, W. 1994. Lipoteichoic acids and lipoglycans. In *Bacterial Cell Wall* (J.M. Ghuysen and R. Hakenbeck, eds.). Elsevier, New York, pp. 199–216.
4. Kehoe, M.A. 1994. Cell-wall associated proteins in Gram-positive bacteria. In *Bacterial Cell Wall* (J.M. Ghuysen and R. Hakenbeck, eds.). Elsevier, New York, pp. 217–262.
5. Salyers, A.A. and D.D. Whitt. 2002. *Bacterial Pathogenesis*. American Society for Microbiology Press, Washington, D.C.
6. Fischetti, V.A. 1996. Gram-positive commensal bacteria deliver antigens to elicit mucosal and systematic immunity. *ASM News* 62:405–410.
7. Brennan, P.J. and H. Nikaido. 1995. The envelope of mycobacteria. *Annual Review of Biochemistry* 64:29–63.
8. Daffé, M. and P. Draper. 1997. The envelope layers of mycobacteria with reference to their pathogenicity. *Advances in Microbial Physiology* 39:132–205.
9. Poxton, I.R. 1993. Prokaryote envelope diversity. *Journal of Applied Bacteriology Symposium Supplement* 74:1S–11S.
10. Nikaido, H. 1973. Biosynthesis and assembly of lipopolysaccharide and the outer membrane layer of Gram-negative cell wall. In *Bacterial Membranes and Walls* (L. Leive, ed.). Marcel Dekker, New York, pp. 132–208.
11. Rietschel, E.T., L. Brade, O. Holst, V.A. Kulshin, B. Lindner, A.P. Moran, U.F. Schade, U. Zahringer, and H. Brade. 1990. Molecular structure of bacterial endotoxin in relation to bioreactivity. In *Endotoxin Research Series*, Vol. 1. *Cellular and Molecular Aspects of Entotoxin Reactions* (A. Nowotny, J.J. Spitzer, and E.J. Ziegler, eds.). Excerpta Medica, Amsterdam, pp. 15–32.
12. DiRienzo, J.M., K. Nakamura, and M. Inoye. 1978. The outer membrane proteins of Gram-negative bacteria: Biosynthesis, assembly and functions. *Annual Review of Biochemistry* 47:481–532.
13. Braun, V. and H.C. Wu. 1994. Lipoproteins, structure, function, biosynthesis and model for protein export. In *Bacterial Cell Wall* (J.-M. Ghuysen and R. Hakenbeck, eds.). Elsevier, New York, pp. 319–343.
14. Benz, R. 1988. Structure and function of porins from Gram-negative bacteria. *Annual Review of Microbiology* 42:359–393.
15. Cowan, S.W. and T. Schirmer. 1994. Structures of non-specific diffusion pores from *Escherichia coli*. In *Bacterial Cell Wall* (J.-M. Ghuysen and R. Hakenbeck, eds.). Elsevier, New York, pp. 353–362.
16. Nikaido, H. and M. Vaara. 1987. Outer membrane. In Escherichia coli *and* Salmonella: *Cellular and Molecular Biology*, Vol. I (F.C. Neidhardt, J.L. Ingraham, K. Brooks Low, B. Magasanik, M. Schaechter, and E.H. Umbarger, eds.). American Society for Microbiology Press, Washington, D.C., pp. 7–22.
17. Barton, L.L., R.M. Plunkett, and B.M. Thomson. 2003. Reduction of metals and non-essential elements by anaerobes. In *Biochemistry and Physiology of Anaerobic Bacteria* (L.G. Ljungdahl, M.W. Adams, L.L. Barton, J.G. Ferry, and M.K. Johnson, eds.). Springer-Verlag, New York, pp. 220–234.

18. Jaeger, K.-E., S. Ransac, B.W. Dijkstra, C. Colson, M. van Heuvel, and O. Misset. 1994. Bacterial lipases. *FEMS Microbiological Reviews* 15:29–63.

19. Osborne, M.J. and H.C.P. Wu. 1980. Proteins of the outer membranes of Gram-negative bacteria. *Annual Review of Microbiology* 34:349–422.

20. Haynes, T.S., D.J. Klemm, J.J. Rucco, and L.L. Barton. 1995. Formate dehydrogenase activity in cells and outer membrane blebs of *Desulfovibrio gigas. Anaerobe* 1:175–182.

21. Rosenberg, E. and N. Kaplan. 1987. Surface-active properties of *Acinetobacter* exopolysaccharides. In *Bacterial Outer Membranes as Model Systems* (M. Inouye, ed.). John Wiley, New York, pp. 311–342.

22. Shockman, G.D. and A.J. Wicken (eds.). 1981. *Chemistry and Biological Activities of Bacterial Surface Amphiphiles.* Academic Press, New York.

23. Fischer, W. 1988. Physiology of lipoteichoic acids in bacteria. *Advances in Microbial Physiology* 29:233–302.

24. Saier, M.H. and G.R. Jacobson. 1984. *The Molecular Basis of Sex and Differentiation.* Springer-Verlag, New York.

25. Rogers, H.J. 1983. *Bacterial Cell Structure.* American Society for Microbiology Press, Washington, D.C.

26. Felix, C.R. and L.G. Ljungdahl. 1993. The cellulosome: The exocellular organelle of *Clostridium. Annual Review of Microbiology* 47:791–819.

27. Tomme, P.R., A.J. Warren, and N.R. Gilkes. 1995. Cellulose hydrolysis by bacteria and fungi. *Advances in Microbial Physiology* 37:2–82.

28. Finlay, B.B. and M. Caparon, 2000. Bacterial adherence to cell surfaces and extracellular matrix. In *Cellular Microbiology* (P. Cossart, P. Boquet, S. Normark, and R. Rappuoli, eds.). American Society for Microbiology Press, Washington, D.C., pp. 67–80.

29. Shipman, J.A., J.E. Berleman, and A.A. Salyers. 2000. Characterization of four outer membrane proteins involved in binding starch to the cell surface of *Bacteroides theaiotao-micron. Journal of Bacteriology* 182:5365–5372.

30. Ferrari, E., A.S. Jarnagin, and B.F. Schmidt. 1993. Commercial production of extracellular enzymes. In Bacillus subtilis *and Other Gram-Positive Bacteria* (A.L. Sonenshein, J.A. Hoch, and R. Losick, eds.). American Society for Microbiology Press, Washington, D.C., pp. 917–937.

31. Wilhelm, S., J. Tommassen, and K.-L. Jaeger. 1999. A novel lipolytic enzyme located in the outer membrane of *Pseudomonas aeruginosa. Journal of Bacteriology* 181:6977–6986.

32. Wicken, A.J. and W. Knox, 1981. Chemical composition and properties of amphiphiles. In *Chemistry and Biological Activities of Bacterial Surface Amphiphiles* (G.D. Shockman and A.J. Wicken, eds.). Academic Press, New York, pp. 1–9.

33. Preissner, K.T. and G.S. Chhatwal, 2000. Extracellular matrix and host cell surfaces: Potential sites of pathogen interaction. In *Cellular Microbiology* (P. Cossart, P. Boquet, S. Normark, and R. Rappuoli, eds.). American Society for Microbiology Press, Washington, D.C., pp. 49–66.

# Additional Reading

## Outer Membrane Structure

Beverage, T.J. 1999. Structure of Gram-negative cell walls and their derived membrane vesicles. *Journal of Bacteriology* 181:4725–4733.

Nelson, D.E., and K.D. Young. 2000. Penicillin binding protein 5 affects cell diameter, contour, and morphology of *Escherichia coli*. *Journal of Bacteriology* 182:1714–1721.

## Outer Membrane Proteins

Brinkman, F.S.L., M. Bains, and R.E.W. Hancock. 2000. The amino acid terminus of *Pseudomonas aeruginosa* outer membrane protein OprF forms channels in the lipid bilayer membranes: Correlation with three dimensional model. *Journal of Bacteriology* 182:5251–5255.

Maier, E., G. Polleichtner, B. Boeck, R. Schinzel, and R. Benz. 2000. Identification of the outer membrane porin of *Thermus thermophilus* HB8: The channel-forming complex has an unusually high molecular mass and an extremely large single channel conductance. *Journal of Bacteriology* 183:800–803.

Winder, C.L., I.S.I. Al-Adham, S.M.A. Abdel Malek, T.E.J. Buaitjens, A.J. Horrocks, and P.J. Collier. 2000. Outer membrane proteins shifts in bioside-resistant *Pseudomonas aeruginosa* PA0. *Journal of Applied Microbiology* 89:289–295.

Wong, K.K., F.S. Brinkman, R.S. Benz, and R.R.E.W. Hancock. 2001. Evaluation of a structural model of *Pseudomonas aeruginosa* outer membrane protein PprM, an efflux component involved in intrinsic antibiotic resistance. *Journal of Bacteriology* 183:367–374.

## Porins

Benz, R. 2001. Porins—Structure and function. In *Microbial Transport Systems* (G. Winkelmann, ed.). Wiley-VCH, New York, pp. 227–246.

Rieß, F.G., T. Lichtinger, A.F. Yassin, K.P. Schaal, and R. Benz. 1999. The cell wall porin of the Gram-positive bacterium *Nocardia asteroides* forms cation-selective channels that exhibit asymmetric voltage dependence. *Archives for Microbiology* 171:173–182.

Van der Helm, D.R. and R. Chakraborty. 2001. Structures of siderophore receptors. In *Microbial Transport Systems* (G. Winkelmann, ed.). Wiley-VCH, New York, pp. 261–288.

Zhang, H.H., D.R. Blanca, M.M. Exner, E.S. Shang, C.I. Champion, M.L. Phillips, J.N. Miller, and M.A. Lovett. 1999. Renaturation of recombinant *Treponema pallidum* rare outer membrane protein I into a trimeric, hydrophobic, and porin-active conformation. *Journal of Bacteriology* 181:7168–7175.

## Cellulosome

Fuhino, T., P. Beguin, and J.-P. Aubert. 1993. Organization of a *Clostridium thermocellum* gene cluster encoding the cellulosomal scaffolding protein CipA and protein possibly involved in attachment of the cellulosome to the cell surface. *Journal of Bacteriology* 175:1891–1899.

# 5
# Capsules, Pili, and Internal Structures

Scientific knowledge does not grow evenly within any subject, partly because of the different degrees of difficulty involved in understanding problems and partly because of the different degrees of effort that they have attracted. This frequently leads to paradoxes that may be difficult to understand. Frustrating though these paradoxes may seem, they nevertheless are the very stuff of the excitement and fun of biological research. To watch and contribute to the piecemeal growth of answers to these problems is the joy of those involved in research. H.J. Rogers, *Bacterial Cell Structure*, 1983

## 5.1. Introduction

Prokaryotes possess a fine structure at the cell surface and display an elaborate detail at the molecular level. The chemical nature of the surface structures of the cell is of great importance because the cell surface interfaces with the environment. The anatomical term glycocalyx is used to describe the network of polymeric material at the surface of microbial cells that contains biopolymeric debris from dead cells or inorganic particulate material. This glycocalyx or irregular form and varying composition surrounds bacterial cells in natural environments. At the surface of most prokaryotes is an amorphous extracellular (or exocellular) structure that is of a single chemical type and is termed the capsule. In essence, it can be said that the capsule containing other materials makes up the glycocalyx. Care should be used in reference to the glycocalyx because historical precedence supports the use of capsule as the chemical structure exterior to bacterial cell wall and glycocalyx is employed in description of animal cell anatomy. It should be noted that capsule formation is characteristic of prokaryotes and that in eukaryotes only a few strains of yeast produce capsules.

This chapter examines the basic characteristics of capsules and pili as examples of surface features of bacteria and the various internal bodies associated with bacteria. Information in this chapter reflects the extensive chemical evaluation that has been directed toward capsule characterization and summarizes the basics of capsular biosynthesis, transport, and attachment to the cell surface. An example of bacterial attachment through the capsule is provided in Figure 5.1. The surface appendages with Gram-negative bacteria may be separated according to function into categories of pili

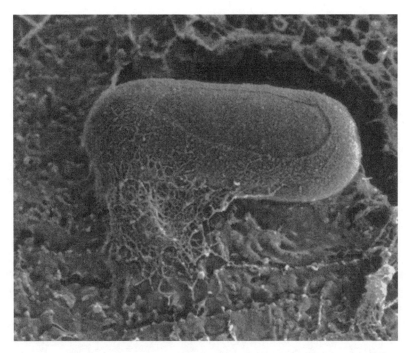

Figure 5.1. The adhesive character of a bacterial capsule in the binding of bacterium DOE SMCC BO693 to the surface of bytownite plagioclase. The cell is 1 μm long and visualization is with a high-resolution scanning electron microscope. (From Ribbe, 1997.[1] Used with permission.)

and fimbrae; however, the term *pili* is being used increasingly and also includes fimbrae. In this text, *pili* is used collectively to include fimbrae along with the pili. Currently there is great interest in pili assembly, and some of these advances are reviewed. Finally, the chapter examines the various specialized internal structures and the activities associated with these bodies found inside prokaryotes. The capsules, pili, and internal components are not generally considered to be obligatory structures for bacterial cell viability; however, their presence provides special advantages that enable bacteria to compete successfully in natural environments.

## 5.2. Capsules[1-3]

The bacterial capsule [from the German *Kapsel*] is referred to as the K antigen with specificity attributed to the monomers making up the structure and the three-dimensional character of the molecule. This K antigen is to be distinguished from the O- or somatic antigen which is attributed to the lipopolysaccharide on the outer membrane of Gram-negative bacteria. Capsules may be demonstrated by cytological methods to be several times thicker than the diameter of the cell that it surrounds or they may be detected by immunological procedures but not readily seen using microscopic procedures. In some strains of bacteria, the production of polysaccharides is not associated with the cell but is secreted into the medium. This is referred to as slime and characteristic of slime-producing organisms is the viscous characteristic of fluid cultures and poor visualization by capsule stains. Copious production of polysaccharide capsule has been called gum and with some bacteria (i.e., *Xanthomonas*) gum produced by a colony is of sufficient quantity to flow onto the lid of an inverted Petri dish.

It is considered that all bacteria produce a capsule when grown in the natural environment. This capsule is not required for cell viability as is the plasma membrane or cell wall but its presence provides additional advantages to the bacteria. Capsules are present on both Gram-positive and Gram-negative cells and various activities associated with capsules are summarized in Table 5.1. Pathogenic bacteria may produce smooth colonies of high virulence as a result of capsule formation while non-capsule producing strains would produce rough colonies and have reduced virulence. Present on certain pathogenic Gram-negative is an intermittent thin polysaccharide layer surrounding the cell which is not visible by cytological observation but is demonstrated by immune reaction. This polysaccharide was initially considered to account for virulence and was designated as the Vi antigen. The Vi microcapsule in *Salmonella typhi* is a homopolymer consisting of *N*-acetylgalactosamine while in *Escherichia coli* it is a polymer of acetylaminohexuronic acid.

While tremendous variation exists in the chemistry of bacterial capsules, it has been determined that in Gram-negative bacteria the capsule genes are found as a cluster at a single locus on the chromosome. As a result of genes for capsule formation localized at a single site, an exchange of genetic information between bacteria could account for a loss of existing capsule genes and acquisition of new genes for capsule production. Such genetic shuffling between bacteria does appear to have occurred

Table 5.1. Functions and applications of bacterial capsules.

1. Capsules prevent phagocytosis of bacteria by protozoa in aquatic environments.
2. Phagocytosis of bacteria by neutrophiles and related animal cells is prevented by capsules.
3. The release of cyanobacterial capsules contributes to moisture retention in desert microbial crusts and capsule production by heterotrophic bacteria increases the water holding capacity in soil.
4. Capsules increase bacterial survival by preventing desiccation.
5. Polysaccharides released from cyanobacteria in water precipitate clay particles and clarify the photosynthetic zone.
6. In aquatic alkaline environments, capsules enhance cell nutrition by accumulating iron, manganese, and other metal ions.
7. The presence of a capsule on a bacterial cell prevents bacteriophage from attacking the cell.
8. Promote adhesion of bacterial cells to the environment which may include plant roots, oral surfaces, indwelling catheters, surfaces of metals, and soil particles.
9. The adhesive character of capsules is important for nonspecific attachment of bacteria in biofilm formation.
10. Secondary recovery of oil from subsurface deposits is accomplished by surfactant properties of extracellular polysaccharides.

because identical capsule structures have been found in the K1 polysaccharide of *E. coli* and the group B polysaccharide of *Neisseria menningitidis*. In addition, a great similarity exists in the chemistry of capsules produced by serotype b of *Haemophilus influenza* and the K18 polymer of *Escherichia coli*. Also, the cloned *cps* genes from *Escherichia coli* have a higher G + C value than is characteristic of *Escherichia coli* DNA and this suggests that *Escherichia coli* acquired the capsular genes from an organism with a high G + C DNA content. Molecular biologists are intrigued by the driving force that accounts for genetic changes resulting in capsule diversity and are seeking to understand the factors that maintain capsule diversity within a given bacterial species.

## 5.2.1. Heteropolymers[4,5]

Carbohydrate capsules are very large molecules that have molecular weights of 100 to 1,000 kDa. Their composition is polymeric and most commonly contains hexoses, amino sugars, and uronic acids. In terms of chemical composition, capsules may be divided into two types: homopolymers (one monomer) or heteropolymers (two or more monomers).

### 5.2.1.1. Diversity of Capsules

Heteropolysaccharide capsules are composed of more than one type of sugar and are distinguished by the type of molecules present, the sequence in which the sugars make up the polymer and the type of bonds between the sugar monomers. With these distinctions, it may be proposed that each species of bacteria may have more than one type of capsule, and multiple antigenic surfaces are indeed found on the bacteria. For example, streptococci are distinguished by the type of heteropolysaccharide capsules produced with more than 80 different serologic types of pneumococci. More than 72 immunological distinct capsular structures are produced by *Klebsiella pneumoniae* and more than 80 different antigenic capsules are produced by *Escherichia coli*.

The molecular configuration of heteropolysaccharide residues may be linear or highly branched. Specific examples of polysaccharide capsules constructed from linear molecules are given in Table 5.2. The importance of different bond formations between monomers are given for *Neisseria meningitidis* (see Table 5.2) and these bonds account for distinctions between the serogroups B and C.

Specific strains of *Escherichia coli* produce a linear heteropolysaccharide capsule of sialic acid, which is a polymer of *N*- and *O*-acetyl derivatives of *N*-neuraminic acid. The polysialic acids contains glucosidic linkages of α-2→9 and α-2→8 which are alternating in the polymer. An example of a highly branched polysaccharide moiety making up the capsule is colianic acid found in several enteric bacteria (Table 5.3). The presence of pyruvic and glucuronic acid residues in colianic acid account for the acidic character of this capsule.

Some bacteria that have capsules composed of heteropolymers where the monomers may include residues in addition to sugars. *Haemophilus influenza* has six serological types of capsular carbohydrates designated "a" through "f." Of the various strains of

Table 5.2. Structures of some linear capsular polysaccharides.

| Bacteria | Name | Repeating unit | Polymer bond |
|---|---|---|---|
| *Salmonella* (Some strains) | Vi antigen | -α-D-*N*-Acetylgalactosamine uronic acid-<br>↕<br>*O*-acetyl | α1→4 |
| *Klebsiella* sp. (Group I-like) | K54 | -α-D-Glucuronic acid 1→3α-L-Fucose 1→6β-D-Glucose-<br>↕ (β1→4)<br>D-Glucose | α1→4 |
| *Streptococcus* (group A β-hemolytic) | Hyaluronic acid | –[D-Glucuronic acid β1→3 *N*-Acetylglucosamine]– | β1→3 |
| *Neisseria meningitidis* | Serogroup[a] A, Serogroup B, | –[2-Acetaminido-2-deoxy-D-mannopyranosyl phosphate]–<br>–[D-*N*-Acetylneuraminic acid]– | α1→6 |
| | Serogroup C | –[D-*N*-Acetylneuraminic acid]–<br>↕<br>*O*-acetyl | α2→8<br>α2→9 |
| *Enterococcus aerogenes* | Mixed hexose | –[Rhamnose α1→3 Rhamnose α1→4 Glucuronic acid-<br>-β1→2 Rhamnose α1→3 Rhamnose α1→3-Galactose]– | β1→2 |

[a]Serogroup is based on capsular polysaccharide and serotype is based on outer membrane proteins.

*Haemophilus influenza*, the most important pathogen for humans is serological type "b" which has a molecular weight of 550 kDa and is a polymer of ribose, ribitol, and phosphate.

Various species of cyanobacteria including *Anabaena, Nostoc, Cyanothece, Oscillatoria, Lymgbya, Phormidium,* and *Spirulina* have been examined for capsule production. The number of monosaccharides present in the cyanobacterial capsular polysaccharides may be 10 or fewer and the molecular mass ranges from 100 to 2,000 kDa. Sugars present in these capsules include hexoses (glucose, galactose, and mannose), pentoses (ribose, xylose, and arabinose), deoxyhexose (fucose and rhamnose),

Table 5.3. Structure of colonic acid, a branched capsular heteropolysaccharide, found in enteric bacteria.[a]

| Bacteria | Repeating unit | Structure |
|---|---|---|
| *Escherichia coli, Salmonella* spp., *Enterobacter cloacae* | –[> 4 α-L-Fucose (1→3) β-D-Glucose (1→4) α-L-Fucose-1]–<br>↕ (1→4)          *O*-acetyl<br>β-Galactose<br>↕ (1→3)<br>β-Glucuronic acid<br>↕ (1→4)<br>β-Galactose<br>↕<br>Pyruvic acid | Colonic acid |

[a]From Roberts, 1996.[10]

Colonic acid is also known as M (mucoid) antigen.

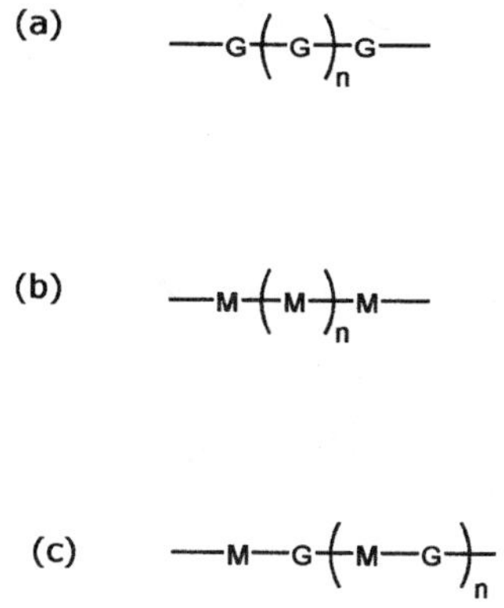

(a)

$$-G\left(G\right)_n G-$$

(b)

$$-M\left(M\right)_n M-$$

Figure 5.2. Block structure of alginate capsule in bacteria. (M = D-manuronic acid; G = L-glucuronic acid). (a) Mixed block polymer = M-G-M-G-; (b) polymanuronic acid = -M-M-; (c) polyglucuronic acid = -G-G-G-.

(c)

$$-M-G\left(M-G\right)_n$$

and acidic hexoses (glucuronic acid and galacturonic acids). Sulfate is associated with some of the capsular polysaccharides in trace levels; however, the exact distribution of sulfate molecules is not currently resolved.

Alginate is a mixture of copolymers containing D-manuronic acid and L-glucuronic acid in which there is no repeating unit. The basic structure of alginate, as produced by *Azotobacter vinelandii* and *Pseudomonas aeruginosa*, is composed of of a block structure seen in Figure 5.2. The exact content of the blocks of alginate appear to vary with the species and as a result the properties of alginate may vary. The higher the content of L-glucuronic acid block, the greater the binding of $Ca^{2+}$ and thus the more rigid the capsule. Another variable in alginate capsules is the amount of acetate attached through hydroxyl groups to produce *O*-acetyl groups on carbon 2 and on carbon 3 of the uronic acids.

### 5.2.1.2. Genetic Organization in Gram-Negative Bacteria[6]

As a group, the most extensively studied capsules are the series found with the Gram-negative bacteria. Properties of the three classes of K antigens of *Escherichia coli* are summarized in Table 5.4. The best characterized antigens are of group I and II and relatively little is known about group III. The polysaccharides of group I are attached to the outer membrane through the lipid A moiety while some examples of group II are by phosphatidic acid. Several molecular characteristics are used to distinguish capsules in group I from those in group II and these are indicated in Table 5.4. The similarity of the structure between group I and lipopolysaccharide (both have lipid A) is interesting and reflects the exploitation of the lipid A residue for anchoring large polymers to the cell.

The cloning of genes has helped to define the genetic organization of capsule formation in Gram-negative bacteria. In *Escherichia coli*, the genes for group II capsule formation have one site on the chromosome and they have been mapped near *serA*, which encodes for 3-phosphoglycerate dehydrogenase associated with biosynthesis of serine. The genetic organization for K5 antigen is diagrammatically presented in Figure 5.3, which shows that the genes for capsule formation are arranged in three

Table 5.4. Comparison of Group I and Group II capsular (K) antigens of
*E. coli.*[a]

| Characteristic | Group I | Group II |
|---|---|---|
| Antigen characteristic of group | K30, K40 | K1, K5, K7, K22, K23, K92 |
| Attachment of polysaccharide to cell surface | By lipid A | By phosphatidic acid |
| Site on chromosome | near *rfb*[b] | Near *serA*[c] |
| Molecular weight of polysaccharide | <Group II | >Group I |
| Charge density of molecule | >Group II | <Group I |
| Temperature of production | Not below | Produced at all temperatures |
| Amino sugars present | No | NeuNAc or KDO |
| Mechanism of export across plasma membrane | Unknown | By ABC transporter |

[a]From Roberts, 1996.[10] KDO, 2-keto-3-deoxymannooctonic acid; NeuNAc, *N*-acetylneuraminic acid.
[b]Gene for synthesis of lipid A.
[c]Gene for serine biosynthesis.

regions: transport, biosynthesis, and transport. The *ksps* genes encode for transport activities of *Escherichia coli* while the *kfi* genes encode for biosynthesis. Although functions of several of these genes remain to be established, the activities of a few genes are proposed to include the following: (1) Products of *kpsD* and *kpsE* genes account for transport of the polysaccharide outward across the periplasmic region. (2) Proteins from *kpsC* and *kpsS* attach KDO to the nonreducing end of the polysaccharide and the polysaccharide is linked through the KDO to a phospholipid. (3) UDP-glucose dehydrogenase is a product of *kfiD* and catalyzes the conversion of UDP-glucose to UDP-gluconic acid. (4) A glycosyltransferase is produced from *kfiC* and this enzymes

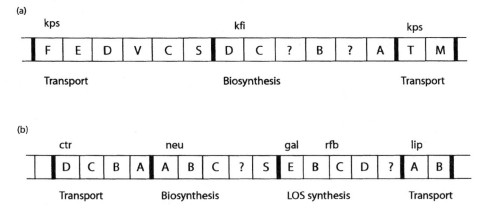

Figure 5.3. Gene cluster for capsules of Gram-negative bacteria. (**a**) *Escherichia coli* K5 and (**b**) *Neisseria meningitidis* group B. (From Roberts, 1996.[10])

adds glucuronic acid and *N*-glucosamine in an alternate fashion to the nonreducing end of the polysaccharide. (5) The products of *kpsT* and *kpsM* are involved in export of the polysaccharide across the plasma membrane. Organisms such as *Neisseria meningitidis* with type b antigen, *Haemophilus influenza* with group B antigen, and *Salmonella typhi* with Vi antigen which have capsule genes existing as a cluster in a manner similar to that of *E. coli* with group II capsules. The model of gene organization in *Neisseria meningitidis* group B is given in Figure 5.3 and a notable difference between it and *Escherichia coli* is the presence of lipooligosaccharide synthesis (LOS) involved in synthesis of lipid A core. It has been proposed that the LOS genes were inserted into the capsule gene cluster as a chromosomal rearrangement process.

### 5.2.1.3. Genetic Organization in Gram-Positive Bacteria[7-9]

The study of capsules associated with the Gram-positive *Streptococcus pneumoniae* has been associated with major contributions in biology. In the description of *Streptococcus pneumoniae* by Pasteur in 1880, the initial reference was made to a structure now recognized as the polysaccharide capsule. In 1925, the capsule of the pneumonoccus was demonstrated by a research group led by Avery to be the first nonprotein antigen to react with antibodies. The transformation of nonvirulent to virulent pneumococci by Griffith in 1928 lead to the demonstration by Avery, MacLeod, and McCarty in 1944 that DNA is the carrier of genetic information. Even today *Streptococcus pneumoniae* is one of the principle Gram-positive bacteria studied with respect to genetics of capsule production. The other organism receiving considerable attention with regard to the genetics of polysaccharide capsule formation is *Staphylococcus aureus*. The organization of capsule genes of these two pathogenic bacteria provides excellent model systems for polysaccharide production in Gram-positive bacteria. The capsules for these Gram-positive bacteria appear to be attached to the peptidoglycan; however, the nature of the bond remains unresolved. The synthesis of the sugar unit making up the repeating unit is in the cytoplasm and the P-undecaprenyl lipid is suggested to be involved in transmembrane movement in a manner similar to lipopolysaccharide synthesis in Gram-negative bacteria.

About 90 different types of polysaccharides make up the capsules of *Streptococcus pneumoniae* and 23 of these serotypes are blended to form the basis of a polyvalent vaccine used in the United States. While the genes for more than 10 different types of *Streptococcus pneumoniae* capsules have been cloned and sequenced, most attention has been given to type 3 and type 19 polysaccharides. Bacteria with the type 3 capsule are historically important because they were used by Griffith in the transformation experiment and the type 3 capsule consists of only three genes. The type 3 capsule is composed of a repeating unit of glucose and glucuronic acid which produces an innert barrier that is about 200 to 400 nm thick. The genes for this polysaccharide production form a small cluster on the chromosome between *dexB* and *aliA* (Figure 5.4). There are two functional aspects of the capsular gene system for *cps3*: (1) conserved genes found in all polysaccharide capsules and (2) specific structural genes. The conserved genes include *cps3A*, *B*, *C*, and *P* which function in regulation and transport of the capsule repeating units across the plasma membrane. Enzymes for

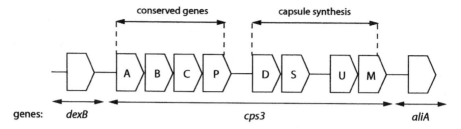

Figure 5.4. Gene system for capsular formation in serogroup 3 of *Streptococcus pneumoniae*. (Data from Paton and Morona, 2000.[8])

capsular synthesis are produced from three serotype specific genes and these are as follows: *cps3D*, which encodes for UDP-glucose dehydrogenase; *cps3S*, which encodes for a β-glucosyltransferase; and *cps3U*, which encodes for glucose-1-phosphate uridylyltransferase. The *cps3M* is an open reading frame and is not considered essential for capsule formation.

Unlike the type 3 capsule which has only seven genes, most of the other serotypes are much larger with 12–18 genes. Type 19F of *Streptococcus pneumoniae* has 15 genes that make up a transcriptional unit. There are four different serotypes making up the type 19 polysaccharide group (i.e., 19F, 19A, 19B, and 19C) and these structures are given in Figure 5.5. The only difference between the 19F and the 19A serotypes is the nature of the bond connecting the repeating trisaccharide. For 19F it is an α 1→2 bond while with 19A it is an α 1→3 linkage. The genes responsible for production of these four serotypes are given in Figure 5.5 and there are 13 genes conserved in these serotypes. Although definitive results have not yet been obtained for all the Type 19 polysaccharide genes, the functions are proposed as follows: *cps19A*—regulatory function for expression of *cps* genes; *cps19B*—function unknown; *cps19C & D*—proteins with two hydrophobic membrane spanning structures that may be involved in chain length and export; *cps19E*—a glucosyl transferase that transfers glucose-1-phosphate to undecaprenyl-P in the membrane; *cps19F*—N-acetylmannosamine transferase; *cps19G*—unknown function; *cps19H*—rhamnosyl transferase; *cps19I*—a protein with 12 membrane-spanning domains and the function is polysaccharide polynmerase; *cps19J*—transfer of repeating unit for chain elongation; *cps19K*—UDP-N-acetylglucosamine-2-epimerase. For the synthesis of dTDP-rhamnose there are the following genes: *cps19L*—glucose-1-phosphate thymidylyl transferase; *cps19M*—dTDP-4-keto-6-deoxyglucose-3,5-epimerase; *cps19N*—dTDP-glucose-4,6-dehydratase; and *cps19O*—dTDP-L-rhamnose synthase. The three genes of *cps19P, Q, R* are found only in the serotypes of Cps19Fb and Cps19Fc and contribute to formation of the rhamnose-ribose side chain. The *cps19S* gene is found only in Cps19Fc serotype and is a glucosyltransferase that is used to produce a second side chain onto the core repeating unit. The genes responsible for the type 19 polysaccharide group appear to have a common ancestry and may have originated from type 19F.

More than 90% of the clinical isolates of *Staphylococcus aureus* have capsules and

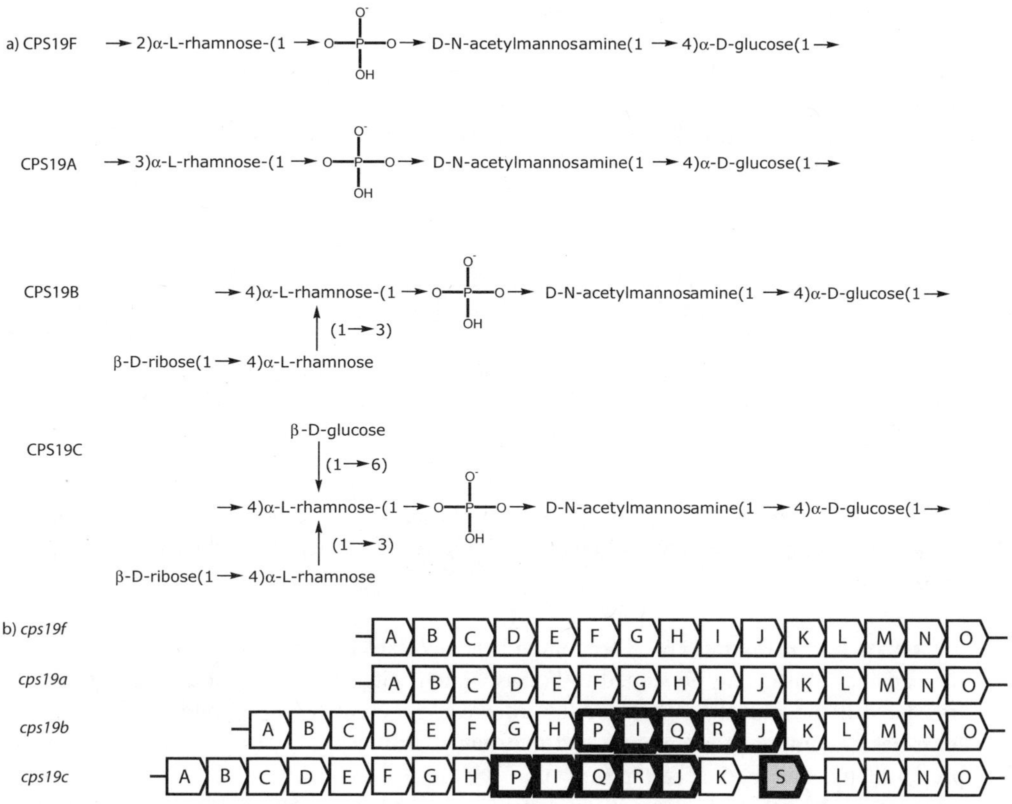

a) CPS19F    → 2)α-L-rhamnose-(1 → O—P—O → D-N-acetylmannosamine(1 → 4)α-D-glucose(1 →

CPS19A    → 3)α-L-rhamnose-(1 → O—P—O → D-N-acetylmannosamine(1 → 4)α-D-glucose(1 →

CPS19B    → 4)α-L-rhamnose-(1 → O—P—O → D-N-acetylmannosamine(1 → 4)α-D-glucose(1 →
          (1→3)
β-D-ribose(1 → 4)α-L-rhamnose

CPS19C
β-D-glucose
(1→6)
          → 4)α-L-rhamnose-(1 → O—P—O → D-N-acetylmannosamine(1 → 4)α-D-glucose(1 →
          (1→3)
β-D-ribose(1 → 4)α-L-rhamnose

b) cps19f    A B C D E F G H I J K L M N O

cps19a    A B C D E F G H I J K L M N O

cps19b    A B C D E F G H P I Q R J K L M N O

cps19c    A B C D E F G H P I Q R J K S L M N O

Figure 5.5. Comparison of structures and genes in serogroup 19 of *Streptococcus pneumoniae*. Capsular polysaccharide (*cps*) genes for types 19F and 19A have considerable similarity with genes A-F, I, J, M-O having only 70% to 90% identity while other genes have >90% identity. The genes with *heavy borders* are specific for types 19b and 19c while S gene is specific for type 19c. Genes A–H and K–O for the types 19b and 19c have >90% identity to the corresponding genes in 19f. (From Paton, J.C. and J.K. Morona. 2000. *Streptococcus pneumoniae* capsular polysaccharide. In *Gram-Positive Pathogens* (V.A. Fischetti, R.P. Novick, J.J. Ferretti, D.A. Portnoy, and J.I. Rood, eds.). American Society for Microbiology Press, Washington, D.C., pp. 201–212.)

these are divided into 11 serotypes. Approximately 25% of the clinical isolates produce a polysaccharide attributed to type 5 capsular antigen. Only serotypes of 1 and 2 produce large capsules and the remaining types produce microcapsules which cannot be seen by routine laboratory microscopy. The chemical compositions of types 1, 2, 5, and 8 are given in Figure 5.6. The genes for the production of these capsules are designated as *cap* and they are flanked on the chromosome by two regions of genes. These flanking genes are common for capsule producing genes in the staphylococci. Based on gene analysis of *cap5* and *cap8*, only one fourth of the genes are specific for a given serotype (see Figure 5.6). Functions of the genes specific for a serotype

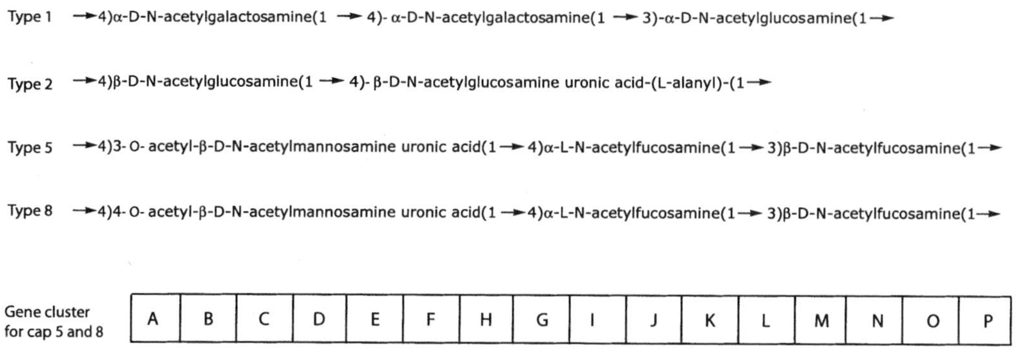

Type 1    →4)α-D-N-acetylgalactosamine(1 → 4)- α-D-N-acetylgalactosamine(1 → 3)-α-D-N-acetylglucosamine(1→

Type 2    →4)β-D-N-acetylglucosamine(1 → 4)- β-D-N-acetylglucosamine uronic acid-(L-alanyl)-(1→

Type 5    →4)3- O- acetyl-β-D-N-acetylmannosamine uronic acid(1→ 4)α-L-N-acetylfucosamine(1→ 3)β-D-N-acetylfucosamine(1→

Type 8    →4)4- O- acetyl-β-D-N-acetylmannosamine uronic acid(1→ 4)α-L-N-acetylfucosamine(1→ 3)β-D-N-acetylfucosamine(1→

| Gene cluster for cap 5 and 8 | A | B | C | D | E | F | H | G | I | J | K | L | M | N | O | P |
|---|---|---|---|---|---|---|---|---|---|---|---|---|---|---|---|---|
| | | | | | | | | | | | | | | | | |

Figure 5.6. Specific serotype of staphylococcal capsules. The repeating units in type 1 and type 2 are distinct. The difference between type 5 and type 8 is not in the trisaccharide repeating unit but the position of the *O*-acetylation. Genes A–G and L–P are for regions common to type 5 and type 8 with 97% to 99% identity. Genes H–K are specific regions for type 5 and type 8 with very little similarity. (From Lee, C.Y. and J.C. Lee. 2000. Staphylococcal capsule. In *Gram-Positive Pathogens* (V.A. Fischetti, R.P. Novick, J.J. Ferretti, D.A. Portnoy, and J.I. Rood, eds.). American Society for Microbiology Press, Washington, D.C., pp. 361–366.)

are also given in Figure 5.6. The common genes flanking specific serotype genes in the capsular cluster are presumed to be associated with synthesis and transmembrane transport of the repeating unit. A few of these genes in the *cap5* and *cap8* cluster have been examined for function and they are as follows: (1) *H* in the *cap5* gene is proposed to add an *O*-acetyl group onto the third carbon of *N*-acetylmannosamine uronic acid; however, this activity is considered to be attributed to *J* in *capJ*. (2) The function of *cap5O* and *cap5P* is proposed to be UDP-*N*-acetylglucosamine-2-epimerase and UDP-*N*-acetylmannosamine dehydrogenase, respectively. (3) The chain length of the expolysaccharide is proposed to be regulated by *cap5B* and *cap8B*. Other functions of these genes await further evaluations.

## 5.2.2. Homopolymers

The presence of a single type of sugar are found in capsules produced by several different bacteria and examples are given in Table 5.5. Polymers of D-galacturonic acid make up the capsule of cyanobacteria. Oral streptococci and *Leuconostoc mesenteroides*, a soil bacterium involved in fermentation of cabbage that produces sauerkraut, have a dextran capsule composed of polymers of D-glucose. Strains of *Pseudomonas* and *Bacillus* have a levan capsule consisting of polymeric D-fructose. *Streptococcus mutans* produces a highly branched polymer of D-glucose which is a variation of dextran and some have designated this polymer as a mutan. *Acetobacter xylinum* produces a capsule that is a polymer of D-glucose connected by a β-glucosidic bond which results in both amorphous and crystalline cellulose. As seen in Figure 5.7, crystalline cellulose fibrils occur outside of the cell and a gelatinous material covers the fibrils on the cell surface. The bacterial cell is highly committed to cel-

Table 5.5. Structures of representative extracellular homopolysaccharides.[a]

| Structure | Repeating unit | Monomer joined by |
|---|---|---|
| Dextran | –[D-Glucose $\alpha1\rightarrow6$ D-Glucose $\alpha1\rightarrow6$ D-Glucose]– $\updownarrow (\alpha1\rightarrow3)$ D-Glucose | $\alpha1\rightarrow6$ bond |
| Mutan | –[D-Glucose $\alpha1\rightarrow6$ D-Glucose]– $\updownarrow (\alpha1\rightarrow3)$ D-Glucose $\updownarrow (\alpha1\rightarrow3)$ D-Glucose | $\alpha1\rightarrow6$ bond |
| Levan | –[Fructose $\alpha1\rightarrow5$-Fructose]– | $\alpha1\rightarrow5$ bond |

[a]From Berkeley, 1979.[2]

lulose production, with about 19% of glucose converted to cellulose. Several commercial efforts have focused on using the cellulose material produced by bacteria. In the World War I era, there was an effort to use tannic acid to precipitate the bacterial cellulose and related capsular material to produce a "bacterial leather." More recently there has been interest in using the cellulosic material of bacterial origin for color printing because of the desired properties. Although the formation of crystalline cel-

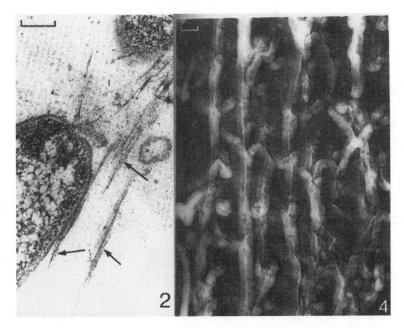

Figure 5.7. Cellulose production by *Acetobacter xylinum*. (**2**) Thin section of stained cell with rigid fibrils of cellulose (arrows) outside of the cell. $\times$ 50,000. (**4**) Parallel cellulose fibrils 100 nm wide on the surface of the cell. (From Colvin and Leppard, 1977.[12] Used with permission.)

lulose is rare among prokaryotes, several bacteria in addition to *Acetobacter xylinum* have been reported to produce cellulose.

## 5.2.3. Biosynthesis of Capsules[10-12]

Over the years a considerable amount of research has been focused on the biosynthesis of capsules; however, most of the attention has been directed to polysaccharide capsule formation and very little is known about synthesis of peptide capsules. There are two distinct types of synthesis used for polysaccharide capsules and these reflect the location of enzymes for capsule production. Polysaccharide synthesizing enzymes are found either at the surface of the cell or inside the cell at the cytoplasm–plasma membrane interface.

With dextran and levan production, the capsular polymers are synthesized at the surface of the cell because this is where the enzyme is localized. The production of dextran is attributed to dextran sucrase while the formation of levan is by levan sucrase. In both cases, the substrate is sucrose and the reactions involved are indicated in Figure 5.8. Dextran is a linear polymer that is produced by addition of each glucose by an $\alpha$ 1→6 glucosidic bond. For *Streptococcus mutans* to produce a highly branched dextran containing $\alpha$ 1→3 glucosidic linkages, a unique enzyme must be present. Because both the enzymes and the substrates are at the surface of the cell, polymer synthesis is regulated by the amount of sucrose present in the environment outside the cell. As the homopolysaccharide accumulates, the glucose or fructose released from sucrose is available to be used by bacteria. Dextran production by oral *Streptococcus* is an important factor in tooth decay because the dextran will adhere to the tooth and bacteria associated with solubilization of the enamel of the tooth become attached to the capsule material. To prevent formation of the oral dextran, carbohydrates that have a sweet taste are substituted for sucrose in candies and gums.

The production of heteropolymeric capsules is relatively complex and is the product of several specific activities. First is the intracellular synthesis involving nucleotide-monosaccharides as precursor molecules and not free carbohydrate molecules. As presented in Table 5.6, each monosaccharide has a specific nucleotide that functions as a carrier to provide enzyme specificity for synthesis in the cytoplasm. In most

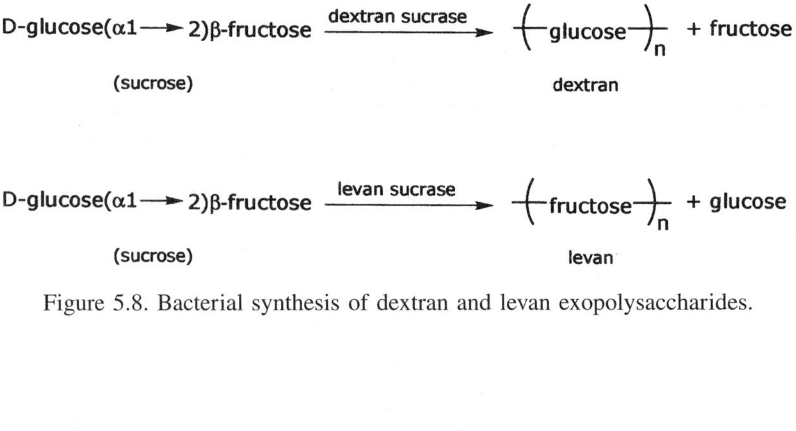

Figure 5.8. Bacterial synthesis of dextran and levan exopolysaccharides.

Table 5.6. Nucleotides used as carriers for carbohydrates in the formation of polysaccharides.[a]

| Monosaccharides | Nucleotide carrier | Principal polymer of bacteria |
|---|---|---|
| Abequose (3,6-Dideoxy D-galactose) | CDP- | Lipopolysaccharide |
| Fucose (6-Deoxy-L-galactose) | GDP- | Lipopolysaccharide and polysaccharide capsule |
| Galactose | UDP- | Lipopolysaccharide and polysaccharide capsule |
| Galacturonic acid | UDP- | Teichuronic acids |
| Glucose | UDP- | Lipopolysaccharide and polysaccharide capsule |
| N-acetylglucosamine | UDP- | Peptidoglycan |
| Glucuronic acid | UDP- | Teichuronic acids |
| Heptose (L-glycero-D-mannoheptose) | NDP- | Lipopolysaccharide |
| Mannose | GDP- | Lipopolysaccharide |
| Aminomannuronic acid | GDP- | Teichuronic acids |
| N-acetylmuramic acid | UDP- | Peptidoglycan |
| Rhamnose (6-Deoxy-L-mannose) | TDP- | Lipopolysaccharide and polysaccharide capsule |
| Tyvelose (3,6-Dideoxy D-mannose) | CDP- | Lipopolysaccharide |

[a]From Mandelstam et al., 1982.[11]

cases, the precursor carbohydrates are assembled into a unit attached to P-P-undecaprenyl ($C_{55}$) lipid by membrane-associated enzymes. Once the polymeric unit is formed, it is transported across the plasma membrane by a manner similar to synthesis of peptidoglycan and lipopolysaccharide biosynthesis.

In Gram-negative bacteria, translocation of the polymeric unit must be through the periplasm and across the outer membrane before the capsule is assembled at the surface of the cell. In Gram-positive bacteria, the unit must be transported through the cell wall. These mechanism of translocation through the plasma membrane is not well understood but in some bacteria, such as with *Escherichia coli* with group II capsule, an exit ABC transporter system functions. The ABC system is part of the active transport processes of bacteria and energy from hydrolysis of ATP is used to transport the polysaccharide across the plasma membrane. Additional research is needed to provide an understanding of how the polymeric unit or the assembled polysaccharide migrates through the cell wall.

## 5.2.4. *Exopeptides as Capsules*[13]

Polyglutamate peptides are found in a few species of organism belonging to the *Bacteria* and *Archaea* domains. There are two forms of glutamic acid that may be involved and the resulting exopolymer depends on the configuration of the amino acid. Polymers constructed from the D-form of glutamic acid contain the γ-linkage and polymers constructed from the L-form of glutamic acid contain the α-linkage. Organisms that produce the α-linked polyglutamate include: *Mycobacterium tuberculosis*, *Flexithrix dorotheae*, *Xanthobacter autrophicus*, *Xanthobacter flavus*, *Natronococcus occultus*, and *Halococcus turkmenicus*. The γ-linked polyglutamate is produced by *Bacillus anthrasis*, *Bacillus licheniformis*, *Bacillus megaterium*, *Bacillus subtilis*, *Planococcus halophilus*, and *Sporosarcina halophila*.

Generally these exopeptides occur as capsular components of the cell; however, the cell wall of *Mycobacterium tuberculosis* contains a polyglutamyl polymer with a molecular mass of 35 kDa and it makes up 8% of the cell wall. In the cell wall of *Natronococcus occultus*, the polyglutamate forms a linear "backbone" structure which contains about 60 residues with side chains composed of glucose, *N*-acetylglucosamine, *N*-acetylgalactoseamine, and galacturonic acid.

The biosynthesis of polyglutamyl polymers has not been extensively described. While the polypeptide capsule for *Bacillus licheniformis* is 173 kDa, the mass of the native polymer will approach $1.5 \times 10^6$ Da, which is comparable in size to the polysaccharide capsule. The genetic code for peptide capsule production by *Bacillus licheniformis* is carried on the chromosome while for *Bacillus anthrasis* it is on the pXO2 plasmid as genes *capA*, *capB*, and *capC* which produce proteins of 46 kDa, 44 kDa, and 16 kDa, respectively. In *Bacillus licheniformis*, the polyglutamyl synthetase for capsule biosynthesis is localized in the plasma membrane and it requires ATP and $Mg^{2+}$, but synthesis is independent of ribosomes and an RNA template.

## 5.2.5. *Visualization*

Demonstration of capsules by electron microscopy is limited to a few staining procedures. Generally, the dehydration procedure used in preparation of specimens for electron microscopy results in the collapse of the capsule to produce irregular bundles on the surface of bacterial cells. The only chemical that provides a positive stain of polysaccharide capsule is ruthenium red, which prevents collapse of the capsule and infiltrates the hydrated polysaccharide with an electron-dense material. Some capsules can be treated with polycationic ferritin which attaches to the polysaccharide and provides an electron-dense region resulting from the high number of iron atoms (approximately 4,500 iron atoms) carried by the ferritin molecule. It is also possible to introduce antibodies specific for the capsule prior to fixation and stain the proteinaceous antibody with appropriate heavy metals. Using antibodies specific for the polysaccharide in the capsule, an uncollapsed capsule of *Escherichia coli* is seen in Figure 5.9A. Bayer and Thurow have observed that the production of capsule by *Escherichia coli* occurs at discrete regions of the cell, see Figure 5.9B. By analysis of serial

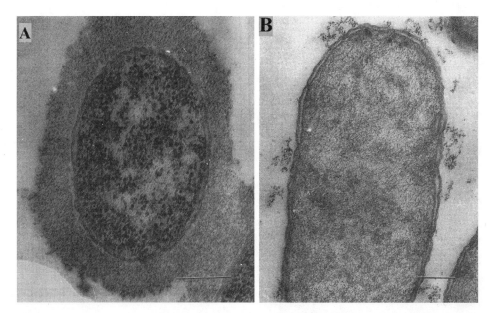

Figure 5.9. Electron micrograph of capsule produced by *Escherichia coli*. (**A**) Cells were pretreated with anticapsular immunoglobulin before fixation, dehydration, embedding, and sectioning. Capsule is 150 nm to 350 nm thick. (**B**) Sites of polysaccharide transport demonstrated by adding ferritin-conjugated antibodies specific for capsule to cells prior to electron microscopy procedures. (From Bayer and Thurow, 1977.[9] Used with permission.)

sections made of an *Escherichia coli* cell producing capsule, it has been proposed that there are only 10 to 40 sites on the cell where new capsular material is released from the cell.

## 5.3. Sheaths[14–16]

Some aquatic bacteria produce a sheath that has the appearance of a small cylindrical tube and bacterial cells are distributed inside this sheath. The cells and sheath of *Sphaerotilus natans* are given in Figure 5.10. Filamentous sheaths serve to concentrate bacteria in appropriate ecological niches where essential nutrients are available as well as to protect cells against viruses, ultraviolet irradiation, oxygen-containing free radicals, and grazing protozoa. Although little information is available on the chemical content of sheaths associated with various prokaryotes, the sheath of *Sphaerotilus natans* has been found to contain a carbohydrate–protein–lipid complex that consists of 36% reducing sugars, 11% hexosamine, 27% protein, and 5% lipid. A polysaccharide layer consisting of fucose, glucose, galactose, and glucuronic acid has been found to cover the sheath of *Sphaerotilus natans*. The sheath produced by *Leptothrix*

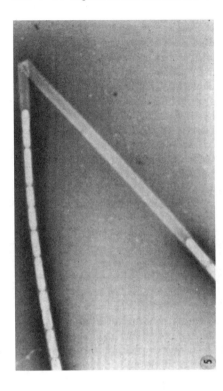

Figure 5.10. Straw-like sheaths containing cells of *Sphaerotilus natans*. Stained with nigrosin. × 1,700. (From Stokes, 1954.[15] Used with permission.)

*discophora* consists of 20% protein, 8% lipid, and 40% carbohydrate. The sheath is composed of fibrals stabilized by disulfide bonds. In *Leptothrix discophora*, the sheath may be a matrix to support the manganese-oxidizing protein. The sheath has considerable rigidity, as the "straw-like" structure is maintained even when it contains no cells; however, muramic acid, teichoic acids, or other molecules characteristic of cell wall structure have not been detected in the sheath. Empty sheaths that do not contain bacterial cells will collect at the water surface owing to buoyancy from trapped air. Environmental problems have been associated with the floating masses of sheath which collect metal ions and metal oxides.

## 5.4. Pili[17–21]

Following independent studies, nonflagellar appendages on bacterial cells were reported by T.F. Anderson and by A.L. Houwink in 1949. Over the years these structures have been referred to as fimbriae (Latin for "thread" or "fiber") and as pili (Latin for "hair-like structure"). With the observation that certain of the surface appendages participated in genetic exchange between bacteria, J.C.G. Ottow suggested that pili should be used to designate conjugative appendages while fimbriae would refer to all other nonflagellar appendages. In current usage, pili is applied to all these surface

Table 5.7. Characteristics of pili.

| Microscopic appearance: | Very thin, flexible (3–4 nm diameter) |
|---|---|
| | Thin, flexible (6–7 nm diameter) |
| | Thick, flexible (8–10 nm diameter) |
| | Thick, rigid (8–11 nm diameter) |
| | Tapered tip at distal end of pilus |
| | Swollen region or knobs at base of pilus |
| | 1–6 per cell, polar location |
| | 100–500 per cell, peritrichously arranged |
| **Function:** | Conjugation with unidirectional DNA exchange from cell to cell |
| | Specific sites for bacteriophage adsorption |
| | Adhesive property with attachment to other cells with pili |
| | Adhesion to mammalian cells of mucosal surfaces, intestinal epithelium, or urinary tract |
| | Adhesion to inanimate environmental objects |
| | Twitching mobility of cells due to polar located pili |
| | Enhance nutrient acquisition of cells |
| **Biochemical characteristics:** | Mannose-sensitive (MS) and mannose-resistant (MR) types based on inhibition of binding to erythrocytes by D-mannose. |
| | Pap type binds to the $\alpha$-D-Gal-(1→4)-$\beta$-D-Gal of P-blood group substance. |
| | N-terminal region of pilin contains hydrophobic amino acids and in one type $N$-methylphenylalanine (NMePhe) is the initial residue. |

structures and use of the term *fimbriae* has declined. Pili have been primarily reported for various Gram-negative bacteria including *Pseudomonas aeruginosa, Moraxella* sp. *Neisseria gonorrhea, Bordatella pertussis, Vibrio cholera,* and *Dichelobacter (Bacteroides) nodosus.* The electron micrograph of *Escherichia coli,* in Figure 1.5 of Chapter 1, reveals the presence of pili on cells. In Table 5.7, are listed the structural characteristics and some of the different activities associated with bacterial pili.

## 5.4.1. Types of Pili

Pili have been examined from various bacteria, and attempts to organize them into groups have not been successful because of the great number of differences in protein structure, agglutination reactions, antigenic characteristics, and genetic features. Although there have been numerous attempts to use morphology, function, and biochemical properties to classify pili, few scientists are in agreement. The following names have been used to identify pili over the years and currently they have received some acceptance: type 1 (I), type 4 (IV), type P, and type F. Some of the characteristics of these types of pili are given in Table 5.8. It would seem unlikely that all the various pili produced by bacteria would fit within this list of only four types; however, this classification serves to focus on the major characteristics of pili. It is not uncommon for a given strain of bacteria to be able to produce several types of pili. For example, enterotoxigenic strains of *Escherichia coli* produce over 11 different pili

Table 5.8. Features of some of the major types of pili.

| Type | Pilin | Size | | Characteristics |
|------|-------|------|------|-----------------|
| | | Outer diameter | Inner diameter | |
| Type 1 | 14 kDa | 7 nm | 2 nm | Found in pathogenic and nonpathogenic *Escherichia coli*. D-Mannose inhibits aggregation with erythrocytes. 100–500 pili peritrichously arranged on the cell. The operon consists of 6 genes on a 9.4-kb segment of DNA at 98 min on *Escherichia coli* linkage map. |
| Type 4 | 16–19 kDa | 5–6 nm | 1.2 nm | Polar located on cells and associated with twitching. Most have a 32 amino acid sequence which starts with *N*-ethylphenylanaline. There is a conserved sequence for most from amino acid 33 to glycine at residue 54 or 55 with differences between species but not within species. There is considerable variability within species and strains from amino acid 54/55 to end of segment which is residue number 145 or 160. |
| Type F | 7.2 kDa | 8 nm | 2 nm | This is the conjugative pilus and bacteriophage may specifically attach. There are 14 contiguous genes within the *traYZ* operon of 33 kb. One gene is required for pilin synthesis and others are associated with the outer membrane, transport across the plasma membrane, and protein folding. |
| Type P | 20 kDa | 7 nm | 2–3 nm | Virulence of uropathogenic strains of *Escherichia coli* is due to these pili that bind to α-D-Gal-(1→4)-β-D-Gal unit of P blood group substance. These pili have an adhesin found on the tip. |

while uropathogenic strains of *Escherichia coli* are known to produce at least 7 different types of pili.

## 5.4.2. Pili Structure

From structural analysis, it appears that the pili are composed of repeating protein subunits termed pilin. In general, pilin consists of 140 to 200 amino acids with two cystine residues and are acidic proteins with isoelectric points (pI) of pH 4 to 5, and

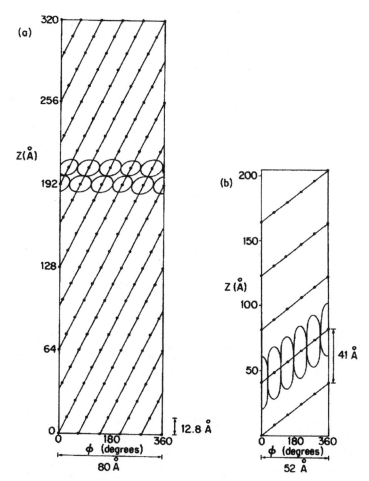

Figure 5.11. Models of bacterial pili. **(a)** The F pilus of *Escherichia coli* with 25 units in two turns of the helix and **(b)** the PAK pilus of *Pseudomonas aeruginosa* with 5 units per turn. (From Paranchych and Frost, 1988.[21] Used with permission.)

the N-terminus of pilin is highly hydrophobic. From X-ray analysis of the cylindrical body of type 1 pilus, it was observed that several thousand copies of pilin are arranged in a right handed helix with 3.14 units per turn of helix with a pitch of 2.54 nm. Models of the molecular organization of protein in bacterial pili are given in Figure 5.11.

In addition to the central body or core of the pilus, a portion of the pilus anchors into the bacterial cell surface and specific extensions at the pilus tip terminate in the adhesion protein. The proteins that attach the adhesion to the pilus body are desig-

Table 5.9. Proposed operon in *Escherichia coli* for Pap (pyleonephritis-associated pili).[a]

| Genes | Function | Function of gene products |
|---|---|---|
| *papI* | Regulation | Associated with regulation |
| *papB* | Regulation | Contains promoter region |
| *papA* | Structure | Pilin subunit, a major protein |
| *papH* | Structure | Production of minor protein that serves to anchor pilin to outer membrane. |
| *papC* | Assembly | 88-kDa product is associated with the outer membrane translocating |
| *papD* | Assembly region | 28.5-kDa chaperone protein involved with transport in periplasmic |
| *papJ* | Assembly | Chaperone function and integrity for pilin assembly |
| *papK* | Tip structure | Inactivation and production of minor pilin protein |
| *papE* | Tip structure | Functions as adopter to couple product of G gene to tip of pilin, a minor pilin protein |
| *papF* | Tip structure | Initiates assembly of pilus filament |
| *papG* | Adhesion | 35 kDa protein that is the adhesin found at the tip of the pilus that specifically recognizes galactose $\alpha$ (1→4) galactose |

[a]From Finlay and Caparon, 2000.[17]
Genes are found on the chromosome according to the sequence given below.

nated as the tip fibrillum. Different pili will adhere to specific molecular conformations on structures in the immediate environment. P type pili bind to structures containing a galactose-galactose motif while type 1 bind to mannose-containing structures. In the case of the sex pilus, type F, the tip of the pilus has a receptor site that recognizes the surface of the recipient cell. In addition, the sex pilus contains receptor sites for male-specific bacteriophage. The f2 phage attaches at a site along the side of the cylindrical core of the F pilus and the f1 phage is adsorbed at the end of the pilus near the site where the recipient cell may attach.

## 5.4.3. Gene Clusters

The genes for pili formation are clustered on plasmids or chromosomes and the size of the gene cluster varies with the specific pili. The type 1 operon and the *pap* (type P) operon of *Escherichia coli* are relatively small with 9 genes and 11 genes, respectively. The genes in the *pap* operon are given in Table 5.9 along with their function. The operon for type 4 pilus of *Pseudomonas aeruginosa* is markedly larger with more than 30 genes for biosynthesis and these genes are located in 6 distinct chromosomal clusters. In the cases examined, the number of genes for pilus structure are small while numerous genes are involved in transport and assembly of pili into the outer membrane. There are different types of exit transport for the pilin proteins, with the more complex systems requiring numerous genes. The greatest quantity of gene product is that making up the core of the pilus and this is found at the leading segment of the structural genes of the pili operons. Conversely, only one adhesion protein is required for each pilus and the gene for this is at the terminal end of the operon. The operon for pili formation would be expected to be highly regulated. It is known that the *pap* operon of *Escherichia coli* is regulated by cAMP and the CRP (catabolite repressor protein) owing to the presence of appropriate binding sites. In some, there is antigenic variation in the pilin protein production that apparently is attributed to the activation of silent genes. Also, bacteria change rapidly from pilated to nonpilated state.

## 5.4.4. Biosynthesis

The biogenesis of pili on bacteria indicates there are relatively complex levels of movement of the proteins across the plasma membrane and through the periplasm to the outer membrane where the pilus is assembled. No doubt there are many different schemes for the transport of pilus protein across the plasma membrane but at this time it is known that type 1 pilus and type 4 pilus use different methods. The process for type 1 and also for type P involves protein secretion into the periplasm where chaperone proteins transfer the pilin proteins to the outer membrane where it is assembled. Type 1 pilus can be completely synthesized from pilin subunits in only about 3 min. The model in Figure 5.12(a) presents a scheme for assembly of a pilus using nomenclature for the type P structure. The pilin protein has a short leader segment in type 1 of 23 amino acids and in type P of six or seven amino acids to direct the pilin protein outward across the plasma membrane into the periplasmic space. The N-terminus of the protein extends into the periplasm and the pilin protein is released from the leader sequence. The pilin protein in the periplasm forms a complex with the chaperone which is a product of *papD* of the *pap* operon. The products of *papC* are termed usher proteins and these ushers must be inserted into the outer membrane to enable pilus proteins to traverse the outer membrane. The PapC products are β-stranded proteins that occur as hexamer complexes with a circular configuration that has a central pore of 2 to 3 nm. Proteins making up the tip fibrillae readily pass through the central pore of the PapC protein. Because the pilus shaft or core is 7 nm in diameter, the individual pilin subunits must pass through the outer membrane before they are assembled into the helical structure. When there is a mutation in the *papC* gene, there is no production of the usher proteins and the porin proteins collect in the periplasm. This process of secretion using the N-terminal secretory proteins is referred to as the general secretory process which uses the Sec-system (also known as type I secretion system).

The process for secretion of type 4 pilus involves the construction of the pilus core on the exterior of the plasma membrane and this core extends through the periplasm into the outer membrane. A model of type 4 synthesis is given in Figure 5.12(b) and this assembly has some characteristics found in type II secretion system of bacteria. The carboxyl end of the pilin protein is released into the periplasm while the N-terminus is anchored into the plasma membrane by a hydrophobic domain. With *Neisseria gonorrhea*, the major pilin protein is PilE and the model for assembly of pilus by this organism is given in Figure 5.12(b). The pre-pilin peptidase is encoded on *pilD* and a protein with an ATP-binding motif is produced from *pilT*. The *pilQ* gene produces a protein that forms a pore in the outer membrane with a pore diameter of about 55 Å. This pore-forming protein in the outer membrane is called a "secretin"

a)

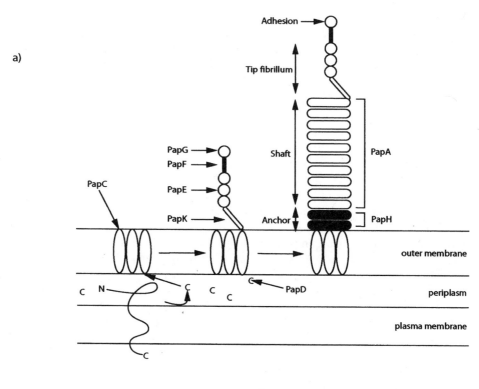

b)

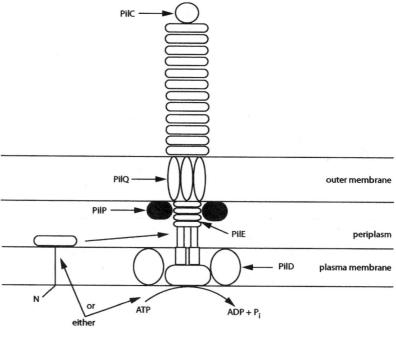

by some and the pilus extends through the pore because it is only about 52 Å. The tip of the type 4 pilus does not contain a tip fibrillum but does have an adhesion. The "twitching motility" attributed to type 4 pili may reflect the release and formation of molecular bonds in the synthesis of the pilus. Thus, the twitching of type 4 pili is a consequence of pilus biosynthesis and not associated with sensory activity.

Slightly different terminology is seen with other pili. In *Pseudomonas aeruginosa*, the pilin protein is a product of the *pilA* gene which is in the *pilA-D* locus. The product of *pilB* is a cytoplasmic protein with an ATP-binding sequence, the product of *pilC* is a protein in the plasma membrane, and *pilD* produces a pilin peptidase which releases the protein from the lipophilic N-terminus attached to the plasma membrane.

## 5.5. Internal Structures

Bacterial cells lack internal organelles; however, several structures are readily observed in the cytoplasm of specific bacterial cells. These structures are unique because enzymes important for granule formation are present at the interface of the granule and the cytoplasm; however, the surface of the granule does not have a true membrane with the presence of phospholipids and structural proteins.

### 5.5.1. Polyhydroxybutyrate[22]

Although polyhydroxybutyrate (PHB) bodies were discovered in bacteria in 1926, little interest was shown until these polymers were found to be of industrial importance. The production of PHB as discrete granules is widely distributed in various bacteria and archaea. The PHB granules are produced under conditions where nitrogen and phosphorus is limiting but carbon source is abundant. While the number and size of the PHB granules inside cells vary with the bacterial species, polymers of $2 \times 10^5$ to $3 \times 10^6$ monomeric units have been reported. The lipid polymer is amorphous; however, on heating, dehydrating, or treatment with solvent, the PHB material will crystallize. Thin sections through cells containing PHB granules reveal that there

◄──────────────────────────────────────

Figure 5.12. Synthesis of pili. (**a**) Scheme for assembly of type I pilus in *Escherichia* coli. A chaperone protein in the periplasm prevents premature aggregation of pilin protein in the periplasm. The proteins at the tip of the pilus (PapK, E, F, G) are initially assembled on the outer membrane and the PapA proteins are inserted through the outer membrane complex and attached to the proteins at the tip. Finally, the PapH is added at the interface of the pilus and the outer membrane to terminate further extension of pilus. (**b**) Model for type 4 pilus synthesis and assembly in *Neisseria gonorrhoeae*. The pilin protein (PilE) traverses the plasma membrane by either specific secretory N-sequences or ATP binding motifs. The pilE extends from the plasma membrane through the outer membrane and terminates with the adhesion (PilC) at the tip of the filament. PilQ is a specific secretory protein that forms a pore in outer membrane through which the pilin emerges. (Modified from Fernández and Berenguer, 2000.[18])

is thin structure delineating the electron transparent PHB material; however, this is not a true membrane because it contains only protein and is only about half as thick as a typical membrane. It is assumed that the protein associated with the PHB granule is involved with PHB metabolism.

The synthesis of PHB is attributed to three enzymes that are subject to tight regulation in bacteria. A pathway for the biosynthesis and degradation of PHB is given in Figure 5.13. The first enzyme is 3-ketothiolase, which catalyzes a reversible reaction in which the equilibrium lies in favor of acetyl-CoA formation. The ketothiolase is strongly inhibited by free CoA in the cell and under conditions of nutrient limitation, the concentration of CoA is low, which enables acetoacetyl-CoA to form. The conversion of acetoacetyl-CoA to 3-hydroxybutyryl-CoA is attributed to acetoacetyl-CoA reductase. *Ralstonia eutrophus* has two acetoacetyl-CoA reductases, one uses NADH as a cofactor and the other uses NADPH as the cofactor. The third enzyme is PHA polymerase or PHA synthase which adds only the D(-) enantiomers of monomers to the PHA granule. In *Ralstonia eutrophus*, the genes are adjacent to each other on a 5-kb DNA fragment in the following order: polymerase (*pha*C), ketothiolase (*pha*A), and reductase (*pha*B).

Copolymers consisting of PHB and hydroxyvaleric acid (HV) are formed by *Ralstonia eutrophus* using the metabolic pathway given in Figure 5.13. The abundance of co-polymer production in bacteria is given in Table 5.10. In other bacteria, polymers of referred to as polyhydroxylalkanoates (PHA) are found and they consist of various monomers including 1, 4-butanediol, butyrolactone, 4-chlorobutyric acid, and 3-hydroxypropionate. Several thousand tons of PHA are produced annually for production of biodegradable bottles and food trays.

While high molecular weight PHB are exclusively associated with prokaryotes, low

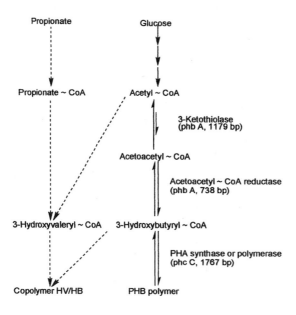

Figure 5.13. Pathway for biosynthesis and degradation of polyhydroxybutyrate granules. The synthesis of copolymer consisting of hydroxyvaleric acid (HV) and hydroxybutyric acid (HB) is also indicated. Genes and base pairs are from *Ralstonia eutropha*. (Data from Byrom, 1994.[22])

Table 5.10. Composition of polyhydroxylalkanoates produced by bacteria.[a]

| Organism | Media | PHA (% w/w) | Composition of PHA (mol %) | |
|---|---|---|---|---|
| | | | 3HB | 3HV |
| *Rhodococcus* NICMB 40126 | Glucose | 31 | 24 | 76 |
| | Fructose | 16 | 13 | 87 |
| | Fumarate | 8 | 47 | 53 |
| *Alcaligenes* SH-69 | Glucose-ammonium | 29 | 93 | 6 |
| | Glucose-urea | 15 | 86 | 14 |
| | Glucose-yeast extract | 45 | 98 | 2 |

[a]From Byrom, 1994.[22]

molecular weight PHB consisting of 100 to 200 monomers are found widely distributed in bacteria, plants, and animals. Calcium polyphosphate is found along with PHB in the plasma membrane of various Gram-positive and Gram-negative bacteria. It has been proposed that DNA transport into some bacterial cells is attributed to a membrane-spanning pore that consists of polyphosphate, calcium, and small polymers of PHB.

## 5.5.2. Polyglucose or Glycogen[23,24]

The accumulation of polymeric glucose in the cytoplasm as glycogen is widely distributed in over 40 species of organisms including both *Bacteria* and *Archaea* domains. The role of glycogen is not clear but is considered by many microbiologists to have a role in energy reserve. The polyglucose is a linear molecule with some branching, with more than 90% of the structure as an $\alpha$-1,4 glucosidic linkage and 10% as an $\alpha$-1,6 glucosidic bond. The polyglucose material aggregates in the cytoplasm of the cell to produce a discrete granule-like structure. When bacterial cells producing polyglucose are prepared for electron microscopy and sectioned, the polyglucose granule is electron transparent and with a single protein boundary between the cytoplasm and the granule. The presence of polyglucose granules in cells of *Clostridium pasteuranum* is revealed in Figure 5.14. Many cells accumulate glycogen during the log phase; however, most strains have greatest production of glycogen in stationary phase. Accumulation of glycogen is triggered in many bacteria when the carbon source is high but there is a limiting amount of nitrogen, sulfur, or phosphorus in the media. Most dramatic for synthesis of glycogen is the effect of nitrogen starvation and research is needed to assess the role of the sigma factor ($\sigma^{54}$) produced with nitrogen starvation. In some instances a shifting of the pH to a unfavorable level for growth may stimulate glycogen production. The glucose polymer is not sequestered in a discrete membranous structure but appears as an amorphous structure in the cytoplasm that may account for 50% of the cell dry weight.

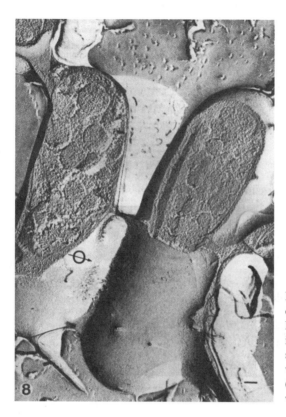

Figure 5.14. Polyglucose granules produced by *Clostridium pasteurianum*. Freeze–fracture procedure is used to indicate that a true membrane does not surround the polyglucose granules which are 12 to 16 nm in diameter. (From Laishley et al., 1973.[23] Used with permission.)

The system used by most organisms in the production of glycogen involves the ADP glucose pathway. In *Escherichia coli* the genes have been identified and are listed in Table 5.11. The ADP glucose pyrophosphorylase is an initial enzyme in the polymerization of glucose and catalyzes the following reaction:

$$\text{ATP} + \alpha\text{-glucose-1-phosphate} \leftrightarrow \text{ADP-glucose} + \text{PP}_i$$

Although some variation occurs with the various microbial strains, the ADP glucose pyrophosphorylase is activated by pyruvate, 3-phosphoglycerate, fructose 6-phosphate, fructose 1,6-bisphosphate, and NADPH. The glycogen synthase functions as a condensing enzyme and accomplishes the following reaction:

$$\text{ADP-glucose} + \alpha\text{-glucan} \rightarrow \alpha\text{-1,4 glucosyl-glucan} + \text{ADP}$$

The above two reactions will result in the synthesis of the polyglucose without any limitation on the type of carbon source. A system for synthesis of glycogen is found in a few bacteria growing on sucrose or maltose and these disaccharides accumulate in the cytoplasm. Certain strains of *Neisseria* have an amylosucrase that accomplishes the following reaction:

$$\text{sucrose} + \alpha\text{-glucan} \rightarrow \text{D-fructose} + \alpha\text{-1,4 glucosyl-glucan}$$

Table 5.11. Genetic organization for glycogen synthesis in *Escherichia coli*.[a]

| Genes | Enzyme or function |
|-------|--------------------|
| *glgC* | ADP glucose pyrophosphorylase for synthesis of glycogen |
| *glgA* | Glycogen synthase for synthesis of glycogen |
| *glgB* | Branching enzyme for synthesis of glycogen |
| *glgY* | Glycogen phosphorylase for utilization of glycogen |
| *glgX* | Glycogen hydrolase for utilization of glycogen |
| *glgQ* | Regulation of above genes with overproduction of proteins from genes *glgC*, *glgB*, and *glgA*. Acts at *trans*-site on DNA. |
| *glgR* | Promoter or operator leads to overproduction of proteins from *glgC*, *glgB*, and *glgA*. Acts at *cis*-site on DNA. |

[a]From Preiss and Romeo, 1989.[24]

Some of the enteric bacteria have an amylomaltase that catalyzes the following synthesis of glycogen:

$$\text{maltose} + \alpha\text{-glucan} \rightarrow \text{D-glucose} + \alpha\ 1,4\text{-glucosyl glucan}$$

The synthesis of internal storage of glycogen by both the amylosucrase and the amylomaltase are similar to the two enzymes that synthesize the homopolymer capsule in certain bacteria where the starting substrates are maltose and sucrose.

## 5.5.3. *Polyphosphate Granules*[25,26]

The presence of condensed inorganic phosphates was reported by Liebermann in 1888; however, the study of inorganic phosphate granules in bacteria was not started until the late 1950s. Inorganic polyphosphate has the general formula shown in Figure 5.15 and the polyphosphates will accumulate in many bacteria to account for up to 20% of the cell dry weight. The polyphosphates as granules are not an obligatory component of cells because many bacteria will not accumulate them except under specific nutritional conditions. A procedure that favors the formation of polyphosphate granules is to initially starve the cells for phosphate and then provide phosphate to the culture. The formation of inorganic polyphosphate granules is considered to result from an uptake of phosphate that is far greater than needed for cell growth. In a single granule there are thousands of phosphate residues making up the energy-rich phospho-anhydride bonds of polyphosphate granules.

There are various methods of detecting polyphosphates in bacterial cells. Perhaps the oldest procedure is to stain bacterial cells with basic dyes such as methylene blue or toluidine blue and to look for metachromatic granules. These aggregates have long been designated as volutin granules and may stain other material associated with the polyphosphate aggregates. An alternate staining method is to use 4',6'-diamino-2-phenylindole (DAPI) as a fluorochrome with examination for localized fluorescence in the cells by the use of fluorescence microscopy.

Another *in situ* procedure with high reliability is the use of high resolution $^{31}$P nuclear magnetic resonance to detect polyphosphates. Polyphosphates can be assayed

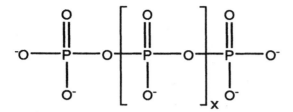

Figure 5.15. Polymeric structure of polyphosphate granule in bacteria.

by providing $K_2HPO_4$ labeled with $^{32}P$ to the growth medium and the amount of isotope released from the isolated polyphosphate granule following degradation by exopolyphosphatase. A nonradioactive assay can be conducted that measures the amount of ATP produced from the reaction catalyzed by polyphosphate kinase which uses ADP and polyphosphate as substrates. Polyphosphate granules are obtained from cells by extracted with a solution containing EDTA, urea, sodium dodecyl sulfate (SDS), and formic acid.

A highly acceptable method of identifying polyphosphate granules is to examine thin sections of bacteria with the electron microscope. The uniformly spherical polyphosphate granules are electron dense and without an enveloping membrane. Results obtained with X-ray dispersion analysis combined with transmission electron microscopy (TEM) gives definitive identification because phosphorus and oxygen will be present in a ratio consistent with polyphosphate. It has been determined that in several different bacteria the polyphosphate granules contain $Mg^{2+}$ in addition to polyphosphate and that in low $Mg^{2+}$ environments $Ca^{2+}$ is found along with $Mg^{2+}$. There are several enzymes in prokaryotes that have been implicated in the formation and utilization of polyphosphates (see Table 5.12); however, the enzymes that are the most important have not been determined.

Arthur Kornberg and other scientists consider that polyphosphates may have had a prebiotic role since polyphosphates are polyanionic molecules similar to RNA and

Table 5.12. Polyphosphate metabolizing enzymes in bacteria.[a]

| Enzyme | Reaction catalyzed |
| --- | --- |
| **Formation of polyphosphate** | |
| 1. Polyphosphate: ADP phosphotransferase | $ATP + (polyphosphate)_n \rightarrow ADP + (polyphosphate)_{n+1}$ |
| 2. 1,3-Diphosphoglycerate:polyphosphate phosphotransferase | 1,3-Diphosphoglycerate $+ (polyphosphate)_n \rightarrow$ 3-phosphoglycerate $+ (polyphosphate)_{n+1}$ |
| **Utilization of polyphosphate** | |
| 1. Polyphosphate: AMP phosphotransferase | $AMP + (polyphosphate)_n \rightarrow ADP + (polyphosphate)_{n+1}$ |
| 2. Polyphosphate-Dependent $NAD^+$ kinase | $NAD^+ + (polyphosphate)_n \rightarrow NADP^+ + (polyphosphate)_{n+1}$ |
| 3. Polyphosphate: D-glucose-6-phosphate transferase | D-glucose $+ (polyphosphate)_n \rightarrow$ D-glucose-6-phosphate $+ (polyphosphate)_{n+1}$ |
| 4. Polyphosphohydrolase (exopolyphophatase) | $(polyphosphate)_n + H_2O \rightarrow P_i + (polyphosphate)_{n+1}$ |

[a]Adapted from Kulaev and Vegabov, 1983.[26]

DNA; however, the exact activity of polyphosphates in early evolution is unknown. Recently attention has been given to physiological activites of polyphosphates and some of their activities include the following. Polyphosphates (1) are considered as an energy source and ATP substitute since phosphate from polyphosphate can drive many reactions, (2) can serve as a reservoir for phosphate to support growth when environmental levels of phosphate are insufficient, (3) serve as chelators of various metal ions including $Ca^{2+}$ and $Mg^{2+}$ which are essential for cellular activity or for binding of toxic metal cations, (4) may serve as internal buffers against cations in alkaline environments, (5) form channels for DNA entry by competent bacteria, (6) serve as regulators in response to a variety of environmental stresses, and (7) function in fruiting body formation, spore formation, and other developmental changes in microorganisms. Clearly, elucidation of definitive roles for polyphosphates in prokaryotic cells awaits further discoveries.

A key enzyme in polyphosphate synthesis is polyphosphate: ADP phosphotransferase, also known as polyphosphate kinase (PPK). This enzyme, encoded by the gene *ppk*, is a homotetramer of 80-kDa subunits which is present in more than 25 different species of *Bacteria* and *Archaea*. In *Escherichia coli*, the PPK is located on the inner side of the plasma membrane and in addition to catalyzing the formation of ATP from ADP and polyphosphate it will also promote the conversion of GDP to GTP in the presence of polyphosphate, although at a greatly reduced rate. The PPK from *Escherichia coli* will also promote the formation of linear ppppG according to the following reaction:

$$(\text{Polyphosphate})_n + \text{GDT} \rightarrow (\text{Polyphosphate})_{n-2} + \text{ppppG}$$

The interaction of ppppG with the formation of polyphosphate is indicated in Figure 5.16.

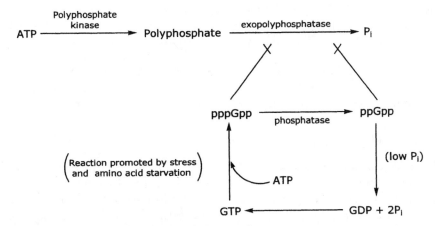

Figure 5.16. Interaction of pppG with polyphosphate formation. (Reprinted, with permission, from the *Annual Review of Biochemistry*, Volume 68. © 1999 by Annual Reviews, www. annualreviews.org.

## 5.5.4. Magnetic Crystals[27,28]

In 1975, R. Blakemore isolated a bacterium, *Magnetospirillum (Aquaspirillum) magnetotactium*, that contained numerous magnetic crystals in the cytoplasm. The magnetic crystals, also known as magnetosomes, are intracellular inclusions that are responsible for the migration of the cell in a magnetic field. The alignment of these magnetosomes in the cell accounts for bacterial cells as north-seeking in the northern hemisphere but the bacterial cells are south-seeking in the Southern Hemisphere. The function of the magnetosomes is to orient bacteria toward the water–mud interface in aquatic environments, thereby enabling the bacteria to select the appropriate oxygen concentration for growth.

As revealed by Schüler and Bäeuerlein in 1997, bacteria display considerable variability in the shape, size, and arrangement of magnetosomes. The magnetosomes in bacteria range from cubo-octrahedral crystals at 50 nm $\times$ 50 nm to hexagonal crystals at 140 nm $\times$ 70 nm. Bacteria commonly have approximately 20 magnetosomes per cell, although one strain produces up to 1000 magnetic crystals per cell. There is also considerable variability in magnetosome arrangement in the cell with distribution in the cytoplasm with one or more chains of magnetosomes or magnetosomes distributed in an appparent random array. In bacterial cells that belong to the $\alpha$-subclass of *Proteobacteria* the magnetic crystal is magnetite ($Fe_3O_4$) and microaerophilic conditions for growth are required for this formation. An isolate assigned to the $\delta$-subclass of *Proteobacteria* produces magnetic iron-sulfide crystals of gregrite ($Fe_3S_4$) and nonmagnetic iron pyrite ($FeS_2$) deposits.

Unique metabolic activities may be expected to be present in magnetotactic bacteria. Because of the presence of magnetosomes, magnetotactic bacteria may contain as much as 2% iron while in contrast, only 0.055% of heterotrophic bacteria is iron. Both the mechanism for the bacteria to acquire large quantities of iron and for coupled deposition of iron as magnetite crystals are not well understood at this time. The biomineralization of magnetite by magnotactic bacteria is dependent on growth conditions and physiological activity of these bacterial cells. The magnetosomes of *Magnetospirillum gryphiswaldense* (see Figure 5.17B) are covered by a membrane which is 6 to 8 nm thick and this membrane differs from other cellular membranes by lipid and protein composition. Presumedly proteins in the membrane that envelopes the magnitite crystals are involved in crystal nucleation. Future studies are needed to establish the mechanisms for magnetite formation in magnetotactic bacteria.

## 5.5.5. Carboxysomes

Polyhedral bodies in the sulfur-oxidizing autotrophic bacteria *Thiobacillus neopolitanus* were reported by J. Shivley to contain ribulose 1,5-bisphosphate carboxylase/oxygenase (RuBisCO), the enzyme for autotrophic fixation of $CO_2$. These inclusion bodies are called carboxysomes and are widely distributed in cyanobacteria, in some but not all species of chemoautotrophic bacteria, and a few nonoxygenic, photosynthetic bacteria. The best characterized RuBisCO granules are from *Thiobacillus neopolitanus*, which serves as the basis for the model. The distribution of carboxysomes

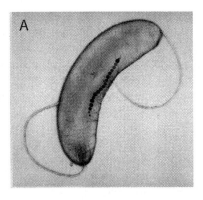 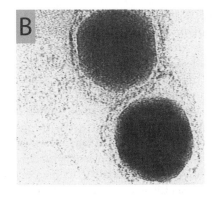

Figure 5.17. Electron micrograph of magnetosomes in bacteria. (**A**) A magnetotactic bacterium containing magnetic crystals. (From Blakemore, 1982.[27] Used with permission.) (**B**) Magnetosomes from *Magnetospirillum gryphiswaldense*. There is a 6-to-8 nm thick membrane covering the particles. (Used with permission from D. Schüler.)

in *Thiobacillus neopolitanus* is apparent from Figure 5.18. Carboxysomes in *Thiobacillus neapolitanus* vary from 70 to 500 nm in diameter with a mean of about 120 nm. Around the carboxysomes is a 3.5-mm shell and inside the granule are 11-nm particles arranged in rows or circles consisting of RuBisCO. Eleven polypeptides have been isolated from the carboxysomes and the 51-kDa and 9-kDa forms are the L (large) and S (small) subunits of RuBisCO, respectively. Polypeptides presumed to be associated with the carboxysome shell have molecular masses of 130 kDa, 80 kDa, 44 kDa, 15 kDa, and 5 kDa. A relative unique feature of these polypeptides of the

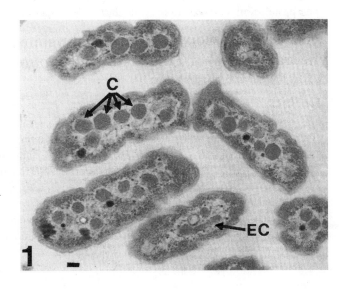

Figure 5.18. Carboxysomes in thin sections of *Thiobacillus neapolitanus*. C, Carboxysomes; EC, elongated carboxysome. Bar = 100 nm. (From Shively et al., 1973.[31] Used with permission.)

shell is that they may be glucosylated. A polypeptide of 38 kDa was isolated from the carboxysome and the function is not yet established. Evidence for the other two polypeptides comes from inference made concerning the putative carboxysome operon. There is no membrane surrounding the granule, and freeze–etch (fracture) microscopy provides evidence that the shell is not a membrane. The surface of the polyhedral bodies is smooth and does not display protein globules that are found when the freeze–fracture occurs along the hydrophobic center of a membrane.

Several structural differences are observed in the various bacteria that have been examined. The RuBisCo enzymes from most autotrophic bacteria have a molecular mass of 500 to 600 kDa and a quaternary structure that consists of eight large (L) subunits with eight small (S) subunits. This configuration is referred to as a 8L8S structure and it is designated as form I of RuBisCo. In three species of *Rhodopseudomonas*, there is an additional RuBisCo with a molecular mass of 360 kDa and a quaternary structure of 6L in that it contains only six large subunits. *Rhodospirillum rubrum* contains a RuBisCo that is a 2L molecule. This presence of a RuBisCo enzyme with a quaternary structure containing only the large subunit has been designated as form II. The large subunits from form I and form II RuBisCo enzymes have been analyzed and their phylogenetic relationship is shown in Figure 5.19. There are two distinct clusters of form I; one is referred to as "green-like" in that it is found in green plastids and the other "red-like" because it is found in the red and brown plastids. The form II enzymes are proteins distinct from those of form I and the genes of form II are unrelated to either the green-like or the red-like form I RuBisCo.

The carboxysomes genes and the Calvin cycle genes are frequently found clustered in large operons. The Calvin cycle genes are designated *cbb* (in honor of Calvin, Benson, and Bassham) while the carboxysome genes are referred to as *cso*. The genes for form I (e.g., *cbbL* and *cbbS*) are adjacent to each other in the operon and for form II (*cbbM*) there is no gene for a small subunit. Some of the unresolved issues of carboxysomes deals with the genetic organization and regulation of genes for individual proteins in carboxysomes. It should be mentioned that not all bacteria with RuBisCo activity have carboxysomes. For example, *Ralstonia (Alcaligenes) eutrophus* has the 8L8S (type I) enzyme that is found as particles of 10 to 12 nm with a fourfold symmetry. Why this type I enzyme does not occur as a carboxysome is unresolved.

## 5.5.6. Gas Vesicles[32,33]

In 1888, Winogradski reported the observation of spaces or hollow cavities in bacterial cells and some years later these were called "gasvakuolen" by Klebahn. These structures are not a discrete membranous unit term vesicle, and in his extended studies A.E. Walsby suggests that *vesicle* is an appropriate term. Gas vacuoles are found predominately in aquatic bacteria and are important to provide cell buoyancy. About 150 different species of bacteria produce gas vesicles, and these include cyanobacteria, halophilic bacteria, methanogens, gliding bacteria, sheathed bacteria, spore-forming rods, chemolithotrophic bacteria, budding bacteria, and Gram-negative aerobic rods. Both longitudinal and cross section of gas vesicles in a cyanobacterial cell is given

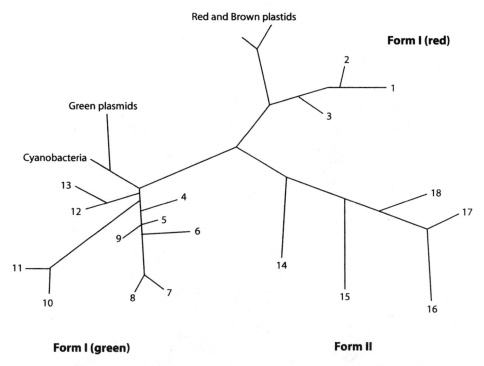

Red and Brown plastids

**Form I (red)**

Green plasmids

Cyanobacteria

**Form I (green)**

**Form II**

Figure 5.19. Unrooted phylogenetic tree indicating relationships of form II and large subunit of form I ribulose-1,5-bisphosphate carboxylase/oxygenase (RuBisCo). Organisms are as follows: 1. *Ralstonia eutropha* (chromosomal), 2. *Ralstonia eutropha* (plasmid), 3. *Rhodobacter sphaeroides*, 4. *Hydrogenovibrio marinus* 1, 5. *Thiobacillus neopolitanus*, 6. *Nitrobacter vulgaris*, 7. *Thiobacillus denitrificans*, 8. *Chromatium vinosum* 1, 9. *Acidithiobacillus ferroxidans*, 10. *Pseudomonas hydrogenothermophila*, 11. *Rhodobacter capsulatus*, 12. *Chromatium vinosum* 2, 13. *Hydrogenovibrio marinus* 2, 14. *Thiobacillus denitrificans*, 15. *Hydrogenovibrio marinus*, 16. *Rhodobacter sphaeroides*, 17. *Rhodobacter capsulatus*, 18. *Rhodospirillum rubrum*. (From Shively, J.M., G. Van Keulen, and W.G. Meijer. 1998. Something from almost nothing: Carbon dioxide fixation in chemoautotrophs. With permission, from the *Annual Review of Microbiology*, Volume 52, © 1998 by Annual Reviews, *www.annualreviews.org*.)

in Figure 5.20. The gas vesicles are cylinders with cones at each end (bipyrimidal) and although there appear to be size differences with the physiological types of bacteria, the mean length is 200 to 400 nm, width is 65 nm to 75 nm, and the wall thickness is 1.5 to 2.0 nm. Vesicles contain proteins that have both hydrophobic and hydrophilic regions containing α-helix, β-sheet, and random coil structures. The wall is composed of two layers of β-sheet proteins with an orientation at 35° to the vesicle long axis. As indicated in Figure 5.21, the interaction between proteins could account for structural stability. The gas vesicles of most cyanobacteria can be collapsed with pressures of 4 to 8 bar that sufficiently damage the vesicle so that it cannot be inflated.

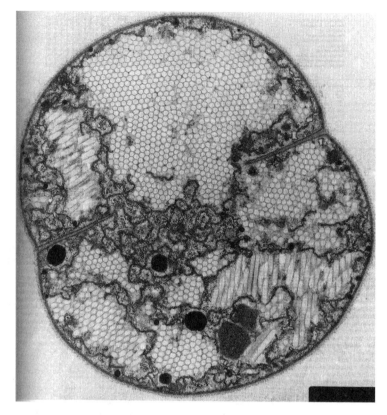

Figure 5.20. Gas vacuoles highly packed in the cell of *Microcystis* sp. $\times$ 31,500. (From Walsby, 1994.[33] Used with permission.)

Gas vesicles can be readily released from cells by mild chemical or physical procedures that do not collapse vesicles. The separation of gas vesicles from carefully lysed cells is accomplished by floatation.

The function of gas vesicles is indeed curious because the vesicle is not a balloon that entraps gas but is highly permeable to gas but not water vapor. The hydrophilic region is on the surface interacting with the cytoplasm and the hydrophobic region is on the inner wall of the vesicle. The gas vesicles are permeable to air, $O_2$, $N_2$, $H_2$, Ar, CO, $CO_2$, and $CH_4$. The pores through which the gases diffuse are 0.3 to 0.6 nm in diameter and are considered to be between the molecules of the structural proteins. In most cases, it is incidental that gas is present inside the vesicle.

Gas vesicles first appear in cells as tiny biconal structures that increase in width and length until the diameter is reached, which is characteristic of that species and then the length increases. A model indicating this is given in Figure 5.22. Every 1.7 sec there are additions to the vesicle equivalent to a 15-kDa protein. It is assumed

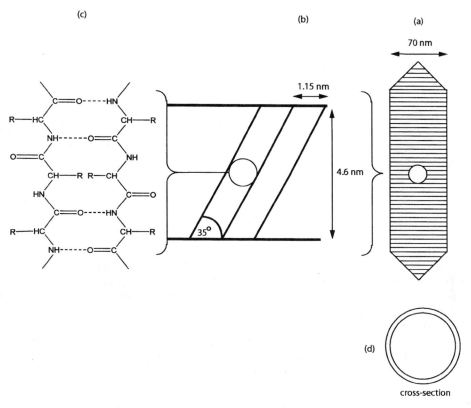

Figure 5.21. Proposed structure of gas vesicles consisting of layers of β-sheet proteins in the cell wall. The vesicle (**a**) is constructed of protein layers or ribs that are 4.6 nm wide. An enlarged rib (**b**) indicates that there is a series of proteins layers that are 1.15 nm thick and positioned at a 35° angle. The proposed β-sheet protein is shown in (**c**). The center of the cylindrical vesicle is hollow (**d**) and has side walls that are 2 nm thick. (Reprinted with permission from Armstrong, R.E. and A.E. Walsby. 1981. The gas vesicle: A rigid membrane enclosing a hollow space. In *Organization of Prokaryotic Cell Membranes* (B.K. Ghosh, ed.). CRC Press, Boca Raton, FL, p. 116. Copyright CRC Press, Boca Raton, Florida.)

that energy must be required for assembly because the vesicle is growing against internal cell pressures.

Molecular biology studies have provided considerable information about gas vacuoles in cyanobacteria and halobacteria. One of the principal proteins found in cyanobacteria is a small hydrophobic protein (GvpA) of about 7.4 kDa. In *Pseudoanabaena* sp., a single gene, *gvpA*, exists but in *Calothrix* sp. the tandem repeat of *gvpA* occurs while in *Anabaena flos-aquae* there are at least five copies of *gvpA* as a tandem repeat. This GvpA protein with 70 amino acids is one of the most hydrophobic proteins characterized with a 15 aliphatic amino acid segment consisting of

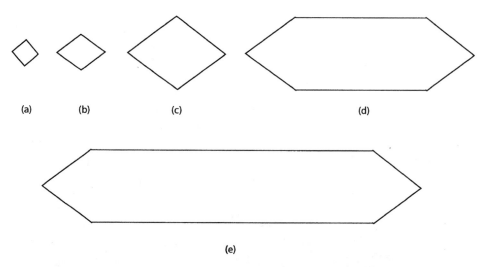

Figure 5.22. Proposed process for growth of gas vesicles. Starting with the initially small biconal structure (**a**), the vesicle increases in size (**b**) until the width is reached which is characteristic of that species (**c**). The side walls of the vesicle lengthen (**d**) until the length for that species is produced (**e**). (Reprinted with permission from Armstrong, R.E. and A.E. Walsby. 1981. The gas vesicle: A rigid membrane enclosing a hollow space. In *Organization of Prokaryotic Cell Membranes* (B.K. Ghosh, ed.). CRC Press, Boca Raton, FL, p. 117. Copyright CRC Press, Boca Raton, Florida.)

six valines, four leucines, three isoleucines, and two alanines. Unlike most proteins, the GvpA protein from cyanobacteria does not contain phenylalanine, histidine, methionine, or cysteine. There is a large surface protein, GvpC, that consists of 193 amino acids in *Anabaena flos-aquae* including histidine, methionine, and phenylalanine with glutamine, alanine, and glutamic acid accounting for more than 40% of the residues. It has been estimated that GvpC accounts for about 9% of the gas vesicle mass. Perhaps one of the most unique features of GvpC is the highly conserved segment of 33 amino acid residues which is repeated five times in protein from *Anabaena*. It is speculated that the GvpA protein makes up the two layers of the β-sheet and the 33 amino acid repeats in GvpC protein are in α-helical conformation. The GvpC protein would provide a scaffold to support the sheets of GvpA proteins.

## 5.5.7. Sulfur Globules[34]

Several different microorganisms oxidize sulfide to inorganic sulfur and the sulfur collects inside the cell. The sulfur aggregate is referred to as a globule and not as a granule because sulfur does not exist only in the elemental form with eight atoms of sulfur ($S_8$) forming a circular crown-like configuration. Rather the sulfur aggregate consists of long chains of polythionates with only traces of $S°$ rings. In some it has been found that the inner core of the globule consists of sulfur in chains consisting of seven to nine atoms and the outer layer consists of polythionates of 19 or more sulfur atoms. This entire globule is covered with a hydration shell and is surrounded by a non-unit membrane that lacks lipid. Sulfur globules are found in numerous anaerobic phototrophic organisms and sulfur-oxidizing bacteria including *Leucothrix*, *Macromonas*, *Thiobacterium*, *Thiospira*, and *Thiovulum*.

## 5.5.8. Cyanophycin Granules[35]

Granules of cyanophycin are highly refractile bodies in cyanobacteria that were first described by Borzi in 1886. These inclusion bodies are found almost universally in both nitrogen-fixing and non–nitrogen-fixing species of cyanobacteria but are not found in other prokaryotic or in eukaryotic cells. Electron microscopic observations of these granules indicates a dense structure that varies in size and shape but has no membrane surrounding it. Chemical evaluation of these granules reveals that it consists of 25 to 100-kDa peptides containing equal quantities of two amino acids: L-aspartic acid and L-arginine. Each polypeptide consists of polyaspartic acid attached by α-linked peptide bonds which constitute the core and an asparagine molecule is attached to the free carboxyl group of the aspartic acid polymer. A model of the cyanophycin structure is given in Figure 5.23. This amino acid polymer is insoluble and is highly resistant to attack by commonly available proteases. The cyanophycin granule accumulates in the cells as the culture enters stationary phase and synthesis does not require ribosomal activity. Synthesis of this polymeric structure involves the addition of an aspartic acid residue to an existing polyAsp molecule followed by the addition of an arginyl residue to the aspartic acid. Enzymes responsible for the synthesis of this polymer have not been purified but it known that ATP is required. In addition, enzymatic activity leading to the initial formation of the aspartic acid polymeric unit serving as the primer have not been elucidated.

The function of this Arg-poly(Asp) aggregate is nitrogen storage and under conditions of excess nitrogen, the cyanophycin granule may account for 10% to 12% of cell dry weight. In many aquatic environments nitrogen is frequently a limiting nutrient and storage of nitrogenous compounds would favor these photosynthetic bacteria. In addition, nitrogen storage in nitrogen-fixing cyanobacteria would be an important energy conservation mechanism because 16 to 24 ATP molecules are required for each $N_2$ molecule reduced to ammonia. Under conditions of nitrogen lim-

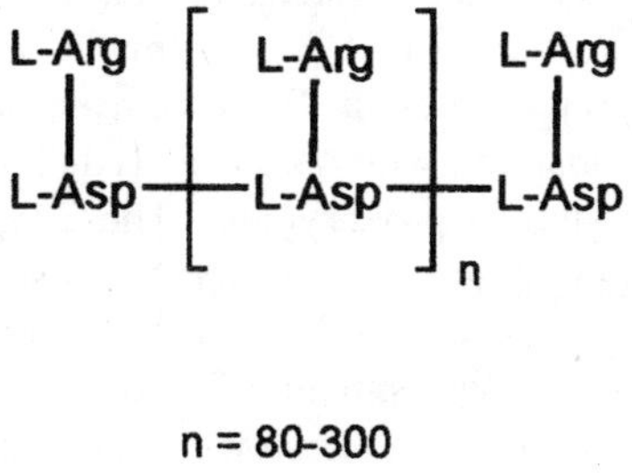

Figure 5.23. A model of cyanophycin structure.                 n = 80-300

itation, Arg-poly(Asp) is degraded by a specific exopeptidase that releases the dipeptide of Arg-Asp from the polymer and the amino groups from these amino acids would then enter the nitrogen pool of the cells. As a nitrogen storage molecule the cyanophycin system is highly effective in that it contains five atoms of nitrogen for every dipeptide of Arg-Asp. Although other polymers, such as polyarginine, could also serve as an effective nitrogen storage molecule, polyarginine would not be desirable because it would be soluble under the conditions in the cytoplasm and would interfere with numerous cell activities including DNA metabolism.

## 5.5.9. Proteasomes[37–38]

Proteasomes are a unique form of molecular compartmentalization in which proteases aggregate to produce a structure with a cylinder shape. These macromolecular structures have a hollow core with proteases forming the outer structure and active sites of the proteases are exposed only to the inner region. Attached to the proteases are accessory components that bind appropriate substrates and have ATPase subunits to utilize ATP for the energy-dependent hydrolysis of proteins. The high molecular weight proteinases, approximately 500 kDa, in prokaryotes participate in cytoplasmic-protein turnover by degradation of metabolic enzymes, various types of regulatory proteins, and nonfunctional or damaged proteins. Proteasomes generally are multi-catalytic and display hydrolysis similar to trypsin, chymotrypsin, or postglutamyl proteinases. There are approximately 14 active sites for protein degradation associated with the 20S proteasome and the size of the peptides released ranges from 4 to 25 amino acid residues. It appears that a protein molecule is degraded and released before a second protein molecule enters the proteosome to be hydrolyzed.

The proteasome from the archaeon *Thermoplasma acidophilum* has been extensively examined and has been found to have a structure composed of four stacked rings which consist of 28 subunits in an $\alpha_7\beta_7\beta_7\alpha_7$ configuration. The $\alpha$-type subunits are related but distinct from the $\beta$-type subunits with the two outer rings containing the $\beta$-type subunits while the two inner rings containing the $\beta$-type subunits. Proteolytic activity is associated with 6 to 14 active sites in the central core containing the $\beta$-type subunits. The dimension of the 20S proteasome is 15 nm high and 11 nm in diameter (see Figure 5.24) with near identical size found for prokaryotic and eukaryotic structures. Access to the region where proteolysis occurs is through openings that are 13 Å in diameter which is about the diameter of protein in an $\alpha$-helix but is too small to allow entry of folded proteins.

Although proteasomes were first reported in eukaryotic cells in 1968, it was not until 1989 that B. Dahlmann, F. Kopp, L. Kuehm, B. Niedel, and G. Pfeifer found proteasomes in *Thermoplasma acidophilum*. Eukaryotic cells contain a 26S proteasome unit with a 20S catalytic core and this core appears similar to the 20S proteasome found in prokaryotes. The 20S proteasome is considered to be broadly distributed throughout the *Archaea* but absent in the *Bacteria*. However, it is now considered that proteasomes in bacteria may not have the highly characteristic architecture of $\alpha$-type and $\beta$-type subunits found in archaea. Genes encoding for the $\beta$-type subunits have been found by reviewing the published genomes of *Bacillus subtilis*, *Escherichia coli*,

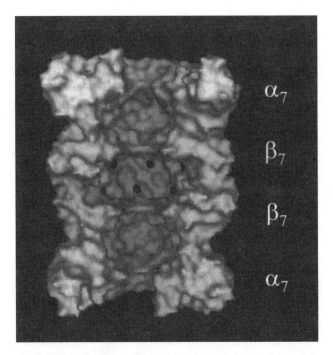

Figure 5.24. Model of 20S proteosome proposed for *Methanosarcina thermophila*. Active sites for hydrolysis of peptide bonds are indicated as *dark circles* in this model. (Reprinted from Maupin-Furlow, J.A., S.J. Kaczowka, M.S. Ou, and H.L. Wilson. 2001. Archaeal proteasomes: Proteolytic nanocompartments of the cell. *Advances in Applied Microbiology* 50:279–338. Copyright 2001, with permission from Elsevier.)

*Haemophilus influenzae, Mycobacterium leprae,* and *Pasteurella haemolyticus. Rhodococcus,* an actinomycete, has a cylinder-like structure with β-type subunits but other proteins are only distantly related to the α-type subunits found in archaeal proteosomes. In *Escherichia coli, Pasteurella haemolytica,* and *Haemophilus influenzae,* the α-type subunits are absent but instead there is *hslU,* an ATPase gene. In fact, the *hslU* gene in *Escherichia coli* is found at the position in the operon at the location of the α-subunit protein in *Rhodococcus* and *Mycobacterium leprae.*

The proteasomes from both prokaryotic and eukaryotic cells degrade proteins by an energy-dependent process and contain ATPases in association with the α-type proteins. The ATPases form a regulatory complex that translocates the unfolded protein substrate to the central core of the barrel-like structure where it is unfolded in a manner that could be described as a reverse chaperone. *Methanococcus jannaschii* has a "PAN" (proteasome-activating nucleotidase) ATPase where protein degradation is stimulated by concomitant hydrolysis of ATP, CTP, GTP, or UTP. *Rhodococcus erythropolis* has an "ARC" ATPase that is not closely related to that of the archeal ATPases and produces a hexameric arrangement in the proteasome that is the source of its name (ARC = *A*TPase forming *r*ing-shaped *c*omplexes). Other prokaryotes produce an ATP-dependent protease such as is formed as part of the heat-shock response to degrade the aberrant heat shock proteins. ClpP, a serine protease, is produced by bacteria and is a structure of about 10 nm × 10 nm. Covering each end of the ClpP chamber are ATPase complexes which energize an unfoldase, a reverse chaperone, to permit the protein to enter the proteolytic chamber. Clearly, additional research is needed to expand current information concerning proteasome and controlled protease activity in prokaryotes.

# References

1. Ribbe, P.H. 1997. Cover picture. *Reviews in Mineralogy*, Vol. 35. Mineralogical Society of America, Washington, D.C.
2. Berkeley, R.C.W., G.W. Gooday, and D.C. Ellwood. 1979. *Microbial Polysaccharides and Polysaccharases*. Academic Press, New York.
3. Sutherland, I.W. 1988. Bacterial surface polysaccharides: Structure and function. *International Review of Cytology* 113:187–233.
4. De Philippis, R. and M. Vincenzini. 1998. Exopolysaccharides from cyanobacteria and their possible applications. *FEMS Microbiology Reviews* 22:151–175.
5. Whitfield, C. and I.S. Roberts. 1999. Structure, assembly and regulation of expression of capsules in *Escherichia coli*. *Molecular Microbiology* 31:1307–1319.
6. Whitfield, C. and M.A. Valvano. 1993. Biosynthesis and expression of cell-surface polysaccharides in Gram-negative bacteria. *Advances in Microbial Physiology* 35:136–246.
7. Lee, C.Y. and J.C. Lee. 2000. Staphylococcal capsule. In *Gram-Positive Pathogens* (V.A. Fischetti, R.P. Novick, J.J. Ferretti, D.A. Portnoy, and J.I. Rood, eds.). American Society for Microbiology Press, Washington, D.C., pp. 361–366.
8. Paton, J.C. and J.K. Morona. 2000. *Streptococcus pneumoniae* capsular polysaccharide. In *Gram-Positive Pathogens* (V.A. Fischetti, R.P. Novick, J.J. Ferretti, D.A. Portnoy, and J.I. Rood, eds.). American Society for Microbiology Press, Washington, D.C., pp. 201–212.
9. Bayer, M.F. and H. Thurow. 1977. Polysaccharide capsule of *Escherichia coli*: Microscopic study of its size, structure, and sites of synthesis. *Journal of Bacteriology* 130:911–936.
10. Roberts, I.S. 1996. The biochemistry and genetics of capsular polysaccharide production in bacteria. *Annual Review of Biochemistry* 50:285–315.
11. Mandelstam, J., K. McQuillen, and I. Dawes. 1982. *Biochemistry of Bacterial Growth*. Blackwell, London.
12. Colvin, J.R. and G.G. Leppard. 1977. The biosynthesis of cellulose by *Acetobacter xylinum* and *Acetobacter acetigenus*. *Canadian Journal of Microbiology* 23:701–709.
13. Birrer, G.A., A.-M. Cromwick, and R.A. Gross. 1994. γ-Poly(glutamic acid) formation by *Bacillus licheniformis*. Activation, racemation, and polymerization of glutamic acid by a membraneous polyglutamyl synthesis complex. *Journal of Biological Chemistry* 254: 6262–6269.
14. Emerson, D. 2000. Microbial oxidation of Fe(II) and Mn(II) at circumneutral pH. In *Environmental Microbe–Metal Interactions* (D.R. Lovley, ed.). American Society for Microbiology Press, Washington, D.C., pp. 31–52.
15. Stokes, J.L. 1954. Studies on the filamentous sheathed iron bacterium *Sphaerotilus natans*. *Journal of Bacteriology* 67:278–291.

16. vanVeen, W.L., E.G. Mulder, and M.H. Deinema. 1978. The sphaerotilis-leptothrix group of bacteria. *Microbiological Reviews* 42:329–356.

17. Finlay, B.B. and M. Caparon. 2000. Bacterial adherence to cell surfaces and extracellular matrix. In *Cellular Microbiology* (P. Cossart, P. Boquet, S. Normark, and R. Rappuoli, eds.). American Society for Microbiology Press, Washington, D.C., pp. 68–80.

18. Fernández, L.A. and J. Berenguer. 2000. Secretion and assembly of regular surface structures in Gram-negative bacteria. *FEMS Microbiology Reviews* 24:21–44.

19. Hultgren, S.J., S. Normark, and S.N. Abraham. 1991. Chaperone-assisted assembly and molecular architecture of adhesive pili. *Annual Review of Microbiology* 45:383–415.

20. Krueger, R.G., N.W. Gillham, and J.H. Coggin, Jr. 1973. *Introduction to Microbiology*. Macmillian, New York.

21. Paranchych, W. and L.S. Frost. 1988. The physiology and biochemistry of pili. *Advances in Microbial Physiology* 29:53–114.

22. Byrom, D. 1994. Polyhydrocyalkanoates. In *Plastics for Microbes* (D.P. Mobley, ed.). Hanser, New York, pp. 5–34.

23. Laishley, E.J., T.J. MacAlister, I. Clements, and C. Young. 1973. Isolation and morphology of native intracellular polyglucose granules from *Clostridium pasteurianum*. *Canadian Journal of Microbiology* 19:991–994.

24. Preiss, J. and T. Romeo. 1989. Physiology, biochemistry and genetics of bacterial glycogen synthesis. *Advances in Microbial Physiology* 30:183–239.

25. Kornberg, A., N.N. Rao, and D. Ault-Riché. 1999. Inorganic polyphosphate: A molecule of many functions. *Annual Review of Biochemistry* 68:89–125.

26. Kulaev, I.S. and V.M. Vegabov. 1983. Polyphosphate metabolism in micro-organisms. *Advances in Microbial Physiology* 24:83–171.

27. Blakemore, R.P. 1982. Magnetotactic bacteria. *Annual Reviews of Microbiology* 36:217–238.

28. Schüler, D. and E. Bäeuerlein. 1997. Iron uptake and magnetite crystal formation of the magnetic bacterium *Magnetospirillum gryphiswaldense*. In *Transition Metals in Microbial Metabolism* (G. Winkelmann and C.J. Carrano, eds.). Harwood, Amsterdam, pp. 159–186.

29. Codd, G.A. 1988. Carboxysomes and ribulose bisphosphate carboxylase/oxygenase. *Advances in Microbial Physiology* 29:115–166.

30. Shivley, J.M., F.L. Ball, and B.W. Kline. 1973. Electron microscopy of the carboxysomes (polyhedral bodies) of *Thiobacillus neopolitanus*. *Journal of Bacteriology* 116:1405–1511.

31. Shively, J.M., G. Van Keulen, and W.G. Meijer. 1998. Something from almost nothing: Carbon dioxide fixation in chemoautotrophs. *Annual Review of Microbiology* 52:191–230.

32. Armstrong, R.E. and A.E. Walsby. 1981. The gas vesicle: A rigid membrane enclosing a hollow space. In *Organization of Prokaryotic Cell Membranes* (B.K. Ghosh, ed.). CRC Press, Boca Raton, FL, pp. 95–130.

33. Walsby, A.E. 1994. Gas vesicles. *Microbiological Reviews* 58:94–144.

34. Steudel, R. 1989. On the nature of the "elemental sulfur" ($S^0$) produced by sulfur-oxidizing bacteria—a model for $S^0$ globules. In *Autotrophic Bacteria* (H.G. Schlegel and B. Bowien, eds.). Science Tech. Publishers, Madison, WI and Springer-Verlag, New York, pp. 289–304.

35. Simon, R.D. 1987. Inclusion bodies in the cyanobacteria: Cyanophycin, polyphosphate, polyhedral bodies. In *The Cyanobacteria* (P. Fay and C. Van Baalen, eds.). Elsevier, Amsterdam, pp. 199–225.

36. Coux, O., K. Tanaka, and A.L. Goldberg. 1996. Structure and functions of the 20S and 26S proteosomes. *Annual Review of Biochemistry* 65:801–847.

37. Maupin-Furlow, J.A., H.C. Aldrich, and J.G. Ferry. 1998. Biochemical characterization of the 20S proteasome from the Methanoarchaeon *Methanosarcina thermophila*. *Journal of Bacteriology* 180:1480–1487.
38. Maupin-Furlow, J.A., S.J. Kaczowka, M.S. Ou, and H.L. Wilson. 2001. Archaeal proteasomes: Proteolytic nanocompartments of the cell. *Advances in Applied Microbiology* 50: 279–338.

# Additional Reading

Rogers, H.J. 1983. *Bacterial Cell Structure*. American Society for Microbiology Press, Washington, D.C.

## *Capsules*

Leigh, J.A. and D.L. Coplin. 1992. Exopolysaccharides in plant-bacterial interactons. *Annual Review of Microbiology* 46:307–346.
Pigeon, R.P. and R.R. Silver. 1997. Analysis of the G93E mutant allele of KpsM, the membrane component of an ABC transporter involved in polysialic acid translocation in *Escherichia coli* K1. *Microbiology Letters* 156:217–222.
Ross, P., R. Mayer, and M. Benziman. 1991. Cellulose biosynthesis and function in bacteria. *Microbiology Reviews* 55:35–58.
Whitfield, C. and M.A. Valvano. 1993. Biosynthesis and expression of cell-surface polysaccharides in Gram-negative bacteria. *Advances in Microbial Physiology* 35:136–246.

## *Proteasomes*

Voges, D., P. Zwickl, and W. Baumeister. 1999. The 26S proteasome: A molecular machine designed for controlled proteolysis. *Annual Review Biochemistry* 68:1015–1068.
Maupin-Furlow, J.A., S.J. Kaczowka, M.S. Ou, and H.L. Wilson. 2001. Archaeal proteasomes: Proteolytic nanocompartments of the cell. *Advances in Applied Microbiology* 50:279–338.

## *Internal Structures*

Bazylinski, D.A. and R.B. Frankel. 2000. Biologically controlled mineralization of magnetic iron minerals by magnetotactic bacteria. In *Environmental Microbe–Metal Interactions* (D.R. Lovley, ed.). American Society for Microbiology Press, Washington, D.C., pp. 109–114.
Byrom, D. (ed.). 1991. *Biomaterials. Novel Materials from Biological Sources*. Macmillan, London.
Offner, S., U. Ziese, G. Wanner, D. Typke, and F. Pfeifer. 1998. Structural characteristics of halobacterial gas vesicles. *Microbiology* 144:1331–1342.
Shively, J.M., D.A. Bryant, R.C. Fuller, A.E. Konopka, S.E. Stevens, Jr. and W.R. Strohl. 1988. Functional inclusions in prokaryotic cells. *International Review of Cytology* 113:35–101.

## Fimbrae/Pili

DeGraaf, F.K. and F.R. Mooi. 1986. The fimbrial adhesins of *Escherichia coli. Advances in Microbial Physiology* 28:65–143.

Jann, K. and B. Jann. 1989. Bacterial adhesions. In *Current Topics in Microbiology and Immunology.* Springer-Verlag, New York, pp. 205–225.

Mattick, J.S., M. Hobbs, P.T. Cox, and B.P. Dalrymple. 1992. Molecular biology of the fimbrae of *Dichelobacter* (previously *Bacteroides*) *nodosus.* In *Genetics and Molecular Biology of Anaerobic Bacteria* (M. Sebald, ed.). Springer-Verlag, New York, pp. 517–545.

# III
## Profiles of Bacterial Growth

# 6
# Cell Motion, Sensing, and Communication

The motion of most of these animalcules in the water was so swift and so various, upwards, downwards and roundabout, that t'was wonderful to see.... their belly is flat and provided with diverse incredibly tiny feet or little legs which were moved very nimbly. *Anton van Leeuwenhoek, Philosophical Transactions of the Royal Society of London*, 1684

## 6.1. Introduction

One of Leeuwenhoek's observations in viewing bacteria under the microscope was that cells displayed movement. Over the years, cell motion has been used to identify prokaryotic cells, and flagellar movement has received considerable study. In nature, microbial movement is not random but is directional, with the objective of moving to a region for optimum growth. To achieve directionality in movement, cells have developed a system of chemotaxis that couples cell movement to chemical sensing of the environment.

The mechanism for chemotaxis is part of a larger cell sensing process that bacteria use for communicating external stimuli across the membrane to the cytoplasm. Cells have a primitive, but effective system for sensing changes in the environment and their response to changes enables them to persist throughout time. In some cases, the changes in the environment are independent of the organisms but in other cases a chemical released from one cell elicits a response in another cell. Signal response systems are chemical reactions that require cells to process external stimuli across the plasma membrane. Although there is considerable variation in the types of chemicals that stimulate the different species of bacteria, this response commonly involves histidine kinases with the transfer of a phosphate group from a signal receptor protein to a response protein. Communication between bacterial cells is also a chemical-based system with one cell responding to a specific chemical released by another cell. In some bacteria, chemicals serve as cues for developmental processes when bacteria undergo morphological changes as they execute a life cycle. This chapter explores some of the sensory processes used by bacteria for response to chemical stimulation.

## 6.2. Systems of Cell Motion[1,2]

Movement by bacteria is an important activity enabling cells to relocate for the purpose of growing under optimal conditions. Bacteria have developed a variety of procedures for lateral movement through water columns and in water films on a solid substrate. Vertical distribution of bacteria in a water column is a result of gas vesicles with buoyancy directly related to the amount of gas in the cell. This physiological activity has interested scientists for many years, but only recently has research proposed mechanisms to explain why cells move and to suggest mechanisms for bacterial movement.

In an effort to characterize fully all types of movement, Henrichsen described bacterial motion as swarming, swimming, gliding, twitching, sliding, or darting (see Table 6.1). The most common mechanism of bacterial motility is swimming by individual cells that requires flagella for the polar-driven activity displayed in aqueous environments. A great deal is known about the swimming process and it will be discussed in some detail in this chapter. Swarming is a two-dimensional movement that also uses flagella; however, movement appears attributable to lateral flagella. The function of swimming is to propel a cell through the water environment while the function of swarming may be to spread aggregates of bacterial cells. For example, *Bacillus alvei* swarming in raft-like clusters or microcolonies over a solid surface travels in counterclockwise curves and produces colonies on an agar plate some distance from the point of inoculation. The movement of cells in colonies on agar plates is recorded in Figure 6.1. Cells in the motile microcolonies appear to rotate and are not tightly bound together but can be dispersed with slight physical agitation. Swarming is displayed by many different genera of bacteria and would appear to be important in bacterial interactions with surfaces such as biocorrosion of metals, biodecomposition of solid materials, rhizospheric bacterial interaction with plant roots, and colonization of various animal tissues. The production of lateral flagella used for swarming may be induced, as is the case with *Vibrio parahaemolyticus*, when the bacterial cell con-

Table 6.1. Characteristics of major systems of bacterial surface translocation.[a]

| Type | Rate at which edge of colony spreads (μm/min) | Movement attributed to | Cells move together | Cells move separately |
|------|------|------|------|------|
| Swarming | 20–75 | Flagella | + | |
| Swimming | 50 | Flagella | | + |
| Gliding | 3.5–15 | Cellular activity | + | |
| Twitching | 4–5 | Cellular activity | | + |
| Sliding | 2–10 | Centrifugal force in expanding colonies | + | |
| Darting | 6 | Tension force in cell aggregates | + | |

[a]From Henrichsen, 1972.[2]

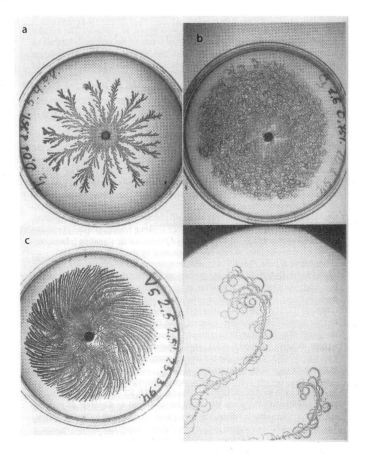

Figure 6.1. Examples of motility seen in bacterial colonies as a cooperative organization of cells. Three different morphotypes are represented: (a) The tip-splitting (T-morphotype) or branching is observed with colony growth of *Paenibacillus dentritiformus*. (b) The chiral (C-morphotype) has extensions from the central colony that have a left-handed twist and this is observed with a specific strain of *Paenibacillus dentritiformus*. The whole colony in an 8 cm diameter petri plate is observed in the top picture and a magnified segment of an extension from the colony is seen in the lower picture. (c) The V-morphotype is seen in a colony of an unidentified soil bacterium where the colony branch moves from a central point of inoculation in a clockwise motion with a droplet consisting of many cells at the tip of the branch. (From Ben-Jacob, E., I. Cohen, and D.L. Gutnick. Cooperative organization of bacterial colonies: From genotype to morphotype. With permission from the *Annual Review of Microbiology*, Volume 52, © 1998 by Annual Reviews, www.annualreviews.org.)

tacts an appropriate surface or when the nodule-forming bacterium initiates contact with a plant root. At this time, there is considerable interest in the propulsive force of flagella and cell–cell communication that occurs with the swarming phenomenon.

Other systems for bacterial movement are less common but are included here because they address an important environmental issue, namely, surface translocation phenomenon. While twitching movement may be relatively similar to gliding in many respects, the fundamental difference is that cells exhibiting twitching motion are not highly organized in a unit as opposed to gliding cells, which move as a single body. Cells displaying twitching and gliding move by nonflagellated means. Recent observations suggest that surface translocation by sliding motion is characterized by cells moving as a monolayer from a colony via forces created by the expanding colony. This sliding movement has been used to describe the spreading of colonies of nonflagellated bacteria such as *Streptococcus* and *Neisseria gonorrhea*. Darting is the phenomenon used to describe the ejection of individual cells from a cluster as is the case often found with *Staphylococcus*. Often these activities described as twitching, sliding, or darting are incorporated into the gliding category. Movement of archea has been largely unexplored, and because many members of the *Archaea* domain do not have flagellar mechanisms, surface translocation would be an interesting research subject.

## 6.2.1. Gliding Motility[3,4]

Gliding is another mechanism of movement in bacteria and it has been known since the late 1800s. The gliding phenomena has been found widely distributed throughout the bacteria and examples of genera displaying this feature are listed in Table 6.2. Gliding bacteria can move over a solid surface at a rate of 1 to 10 µm/sec, which is about one cell length per minute. Colonies of gliding bacteria do not have a discrete boundary but gradually slope from the center as cells move over each other and onto the agar surface. In liquid environments, the cell surface of gliding bacteria displays movement. It has been proposed that more than 20 genes are required for gliding in certain prokaryotic cells and that there is considerable regulation in gliding activity

Table 6.2. Examples of gliding bacteria.[a]

| Physiological groups | Selected genera of bacteria |
|---|---|
| Bacteria with low G+C values | *Filibacter, Helicobacter, Mycoplasma pneumoniae, Mycoplasma mobile* |
| Green nonsulfur photosynthetic bacteria | *Chloroflexus, Heliothrix* |
| Green sulfur photosynthetic bacteria | *Chloroherpetom* |
| Cyanobacteria | *Oscillatoria, Anabaena, Spirulina* |
| *Cytophaga–Flavobacterium* group | *Flavobacterium, Flexibacterium, Cytophaga* |
| β-Proteobacteria | *Vitroscilla* |
| γ-Proteobacteria | *Beggiatoa, Thiothrix, Leucothrix* |
| δ-Proteobacteria | *Myxococcus, Stigmatella* |

[a]From McBride, 2000.[3]

as cells do not glide for several minutes prior to cell division. Several different mechanisms have been proposed to explain gliding in bacteria and these include the following: (1) There is a forceful release of extracellular polysaccharide from the cell. (2) There is a movement of structures in the cytoplasm or periplasm that account for gliding. (3) There are gliding motors that move in a rotary direction to disperse the polysaccharide over the surface of the cell or to move the outer membrane of the cell.

When cells of *Flavobacterium* (formerly *Cytophaga*) *johnsoniae* are labeled with latex spheres, the spheres will rotate along the axis of the cell or across the cell and even change direction of movement. This type of surface movement of cells is found only in gliding bacteria. The gliding in *Flavobacterium johnsoniae* is driven by a proton-motive force and accounts for the movement of the proteins in the outer membrane. Apparently sulfonolipids are important for gliding activity because mutants that lack sulfonolipids do not display gliding activity.

There are two types of gliding systems in *Myxococcus xanthus*: "S" system to identify social gliding and "A" for adventurous gliding. In the "S" system, small motile microcolonies are formed similar to that of the swarming bacteria; however, gliding bacteria do not have flagella, so the direction of microcolony movement is random. Social gliding has been described for cell aggregates of myxococci that result in the formation of fruiting bodies. Relatively large aggregates consisting of hundreds to thousands of cells of *Myxococcus xanthus* have been described as moving like "wolf packs" to surround and digest bacteria by enzymes secreted by cells making up the loosely formed aggregate. In some bacteria, type VI pili are involved in S-type gliding and in mutations in which pili are not produced there is a loss of gliding activity. The mechanism for gliding of single cells by the A-type system is unresolved; however, it is known that pili are not required. Apparently, *Myxococcus xanthus* has a signal transduction system that is somewhat similar to the chemotactic or "*che*-like" system found in swimming bacteria.

## 6.2.2. Swimming Motility in Bacteria[5]

Flagellar activity is required for the three-dimensional swimming of bacteria. As observed by electron microscopy, flagella (from the Latin, meaning "whip") have a helical structure of a uniform diameter and a length that is 10 to 20 times greater than that of the cell. The number of flagella per cell is genetically controlled, with some species having a single flagellum per cell while other species are peritrichous with flagella distributed over the entire cell surface. Amphitrichous or bipolar bacteria have flagella at both ends (poles) of the cell while lophotrichous cells have a tuft of several flagella. Peritrichous cells have a variable number of flagella per cell. In *Escherichia coli* or *Salmonella typhimurium*, each cell has 5 to 10 flagella while in *Proteus vulgaris* each cell may have several hundred flagella. Flagella are found in both Gram-negative and Gram-positive bacteria, and are often found in bacillary cells and only rarely in cells of coccus form. Swimming bacteria may appear to move at only a few micrometers per second (see Table 6.3); however, movement is relative to size of the individual cell and when expressed as the number of cell lengths traveled per second, bacteria are remarkably fast. In comparison, a human running at world

Table 6.3. Motility characteristics of selected prokaryotes.

| Microorganism | Type of flagellation | Velocity | Cell length/sec |
|---|---|---|---|
| **Intracellular movement** | | | |
| *Listeria, Rickettsia, Shigella* | Movement is due to actin filaments of host | 6–90 µm/min | |
| **Gliding** | | | |
| *Myxococcus xanthus* | None | 2.2–3.7 µm/min | — |
| **Spirochetes** | | | |
| *Leptospira interrogans* | Axial filament | 30 µm/sec | — |
| **Swimming** | | | |
| Archaea | | | |
| *Halobacterium salinarum* | Lophotrichous, polar | 2–3 µm/sec | — |
| Bacteria | | | |
| *Pseudomonas aeruginosa* | Monotrichous, polar | 55.8 µm/sec | 37 |
| *Chromatium okenii* | Lophotrichous | 45.9 µm/sec | 5 |
| *Thiospirillum jenense* | Lophotrichous | 86.2 µm/sec | 2 |
| *Escherichia coli* | Peritrichous | 16.5 µm/sec | 8 |
| *Bacillus licheniformus* | Peritrichous | 21.4 µm/sec | 7 |
| *Streptococcus* sp. | Peritrichous | 28.1 µm/sec | 7 |

record speed travels at about 5.5 body lengths/sec, which is roughly equivalent to that of *Escherichia coli* and many multiflagellated bacteria but slower than that of polar flagellates such as *Bdellovibrio* or *Pseudomonas*. Bacteria can swim at approximately the rate of yellow tuna, one of the fastest fish, which moves at 5 to 20 body lengths/sec. Viscosity is an important parameter for movement of flagellated bacteria. Organisms are unable to move if the viscosity of the suspending medium exceeds 60 centipoise (cp). As a point of reference, cytoplasmic gels of cells, intestinal contents, and mucous secretions that line respiratory systems in animals have viscosity values of several hundred centipoise.

## 6.2.2.1. Characteristics of Flagella

The flagellar antigen is termed the "H antigen" because flagellated cells produce a film of growth on an agar surface. Cells without flagella display a somatic antigen or "O antigen." The designation of "H" comes from the German word *Hauch*, which describes a watery film on a solid surface, and "O" comes from the German term *ohne Hauch* describing the absence of a watery film on a solid surface. Cells with flagella frequently produce spreading colonies on agar surfaces that resemble a watery film. One of the difficulties in using anti-H (flagellar) antibodies to identify bacteria is that bacteria have been observed to change the filament to produce an immunologically different antigen. *Salmonella* will shift from production of flagellin protein from gene *H1* to production of flagellin protein from gene *H2*, and this is termed phase variation. This shift is reversible and is attributed to the inversion of an 800-nucleotide segment.

Genetic analysis of *E. coli* and *Salmonella typhimurium* indicates that about 60 genes are organized in 14 transcriptional units that are clustered in three different regions of the chromosome. The transfer of DNA between bacteria cannot result in conversion of a nonflagellated cell to one with flagella because of the large number of genes required for flagellar synthesis and the great distribution of flagellar genes throughout the chromosome. A unique bacteriophage, phage χ, can infect cells by attaching to the flagellar filaments of *E. coli* or *S. typhimurium* but not to nonflagellated cells. The hollow core of the filament enables the nucleic acid of phage χ to penetrate the host cells. This flagella-specific virus is useful in selecting cells that have a mutation for filament production.

It is generally assumed that in peritrichous bacteria all the filaments are the same; however, variations in filament assembly could result in flagellar filaments of different length or altered wave character. A strain of *Proteus* displays biplicity, which is the simultaneous presence of two different types of flagella on a cell. Motility is attributed to the flagellar structures, which have filament segments in a left-hand helix and rotation is counter-clockwise. The mutant filaments are right-handed helices with half the normal helical pitch and are nonmotile.

A few bacteria have sheaths that cover the filament and the function could protect the filament from digestion by proteases in the environment. The flagellar sheath of *Vibrio cholerae* appears to have the same lipopolysaccharides in sheath as in the outer membrane while the sheath of *Bdellovibrio bacteriovorus* appears distinct from the outer membrane. Other bacteria with sheathed flagella include *Helicobacter pylori* and spirochetes.

## 6.2.2.2. Fine Structure and Synthesis of the Flagellar Unit[6–10]

The anatomical features of the bacterial flagellum are based on research with *E. coli* and *S. typhimurium*. A model of the fine structure of a bacterial flagellum is given in Figure 6.2. The flagellar unit consists of three major structural segments: filament, hook, and basal body. The basal body consists of the L-ring, the P-ring, the MS-ring, the C-ring, the central rod, and multiprotein components associated with the plasma membrane that make up the motor and switch mechanism. The M-ring is within the plasma membrane and the thinner S-ring is stacked on top of the M-ring. The L-ring interfaces with the lipopolysaccharide in the outer membrane and the P-ring is attached into the peptidoglycan layer. The C-ring is attached at the base of the flagellum but extends into the cytoplasm where it forms the motor/signal system for chemotaxis.

*E. coli* and *S. typhimurium* are nearly identical in that they have the motility/chemotaxis genes clustered in about a dozen operons organized in a three-tiered hierarchical system as indicated in Figure 6.2. At the first level is the master operon, which contains the *flhC* and *flhD* genes. The protein products of these two genes are transcriptional activators, such as sigma factors, for operons in the second level of orga-

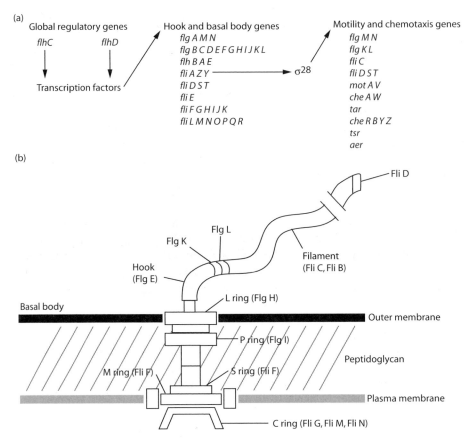

Figure 6.2. Genes and flagellar apparatus for bacterial motility and chemotaxis. (**a**) Genes for motility are organized into three tiers of expression. (**b**) Structural components of the bacterial flagellum and the interface of the flagellum with the cell surface. Precise location and orientation of many of the motor/switch proteins cannot be made until additional experimentation is conducted. (Data from Macnab, 1995[9]; Chilcott and Hughes, 2000[7]; and Aizawa, 2001.[6])

nization. The third level of organization is activated by the product of the *fliA* gene of the second level. As with many catabolic enzymes, the expression of the *flhD* operon is dependent on the presence of cAMP and the catabolite activator protein. The operon at the second level contains the *fliA* gene and structural genes for basal flagellar body. The *fliA* gene produces an alternate sigma factor, $\sigma^{23}$, needed for synthesis of mRNA from the operons in the third level of regulation. In this final organizational tier, the *fliC* operon codes for the flagellar filament while the *motA* and *tar* operons code for signal transduction proteins, flagellar motor proteins, and transmembrane receptor proteins. Many of the genes needed for flagellar synthesis encode for chaperone proteins needed in folding and translocation of flagellar proteins.

At the end of the proximal end of the filament is a flexible hook, which is about

50 nm long and is genetically determined. The hook is constructed from about 120 subunit monomers and is about 100 times more flexible than the filament. Between the hook and the filament are two hook-associated proteins (HAP1 and HAP3) that are important in polymerization of the filament proteins and in transfer of the rotational motion from the hook to the filament. In *E. coli*, HAP1 and HAP3 are polymers of 10 to 20 protein subunits that provide for only about two turns around the hook-filament structures. A third protein (HAP2) regulates the flagellin insertion at the tip of the filament and also prevents the release of the flagellin into the environment. Some consider that HAP2 functions as a "cap" at the tip of the filament and it prevents foreign material from entering the cell via the flagellum. In de novo synthesis, the first structure formed is the rod with the SM-ring attached. Next, additions are made in the following order: P-ring, L-ring, hook, HAP1 and HAP3, and finally filament.

The flagellar filament is hollow with a 30 Å channel and consists of a globular protein subunit termed flagellin. Filaments are generally 5 to 10 μm in length, consist of a homopolymer of about 10,000 copies of flagellin, and have a molecular mass of 25 to 60 kDa. For synthesis, flagellin moves through the hollow core of the filament to self-assemble a hollow filament. The length of the filament is controlled by environmental factors such as the forces in the core of the filament that diminish the rate of elongation with increasing filament length. Following breakage of filament owing to environmental perturbations, reformation of a filament occurs at a rate of 1 μm every 2 min, and this assembly of the filament is not affected by inhibitors of protein synthesis because flagellin is stored at the base of the flagellum.

The filament of the flagellum can be readily broken by subjecting the flagellated cell to conditions of agitation. Staining of flagella for viewing by light microscopy is frequently disappointing because the flagella are broken in the transfer and staining manipulations. For harvest of large quantities of filament, shearing forces of a laboratory blender are effective in breaking the filaments from the cell. The intact flagellum can be released from protoplasts or spheroplasts by treatment with Triton X-100 and EDTA.

### 6.2.2.3. Mechanism of Motility[11]

Discoveries by H. Berg and by M. Silverman and M. Simon revealed that the mechanism of bacterial motility is attributed to flagellar rotation driven by proton-motive force. An inward current of protons applied at the outer edge of the M-ring drives the flagella, with approximately 300 to 1,000 protons required for a single rotation. The rotation of the flagella is initiated by about 25 mV across the plasma membrane and reaches speeds up to 3,000 revolutions per minute. This rotation of the flagellum accounts for about $10^{-16}$ W of power for each flagellar motor. Under normal growth conditions, about 0.1% of the cell energy is for flagellar rotation while about 2% of cell energy is for flagellar biosynthesis. The rod, which is secured to the SM-ring, rotates and the L-ring/P-ring complex serves as a bushing to facilitate the rotation of the massive filament. Bacteria with left-handed helical filaments will rotate the filaments in a counter-clockwise (CCW) rotation.

Direct observation of flagellar rotation was obtained by tethering *Escherichia coli*

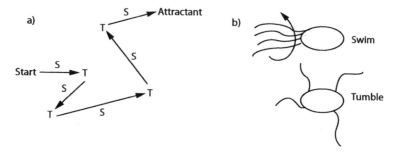

Figure 6.3. Movement by flagellated cells. (**a**) Random-walk activity with positive chemotaxis. S, Swim mode; T, tumble mode. (**b**) Flagellar arrangement of peritrichous bacterium in swim and tumble modes.

cells to a microscope slide by the use of antibody specific for the flagellar filament and following the rotation of the bacterial cell. The sequence of events was that the flagellum rotated CCW, stopped, rotated CW, stopped, rotated CCW, and repeated the sequence. Generally, the filament rotation is from 0.1 to 1.0 sec and the time interval for rotation varies with the metabolic state of the cell. Specific factors that function to influence movement are given in the section concerning chemotaxis.

If bacterial cells are observed in a hanging drop slide preparation, the movement of the cell would be a series of movement, stop, movement, stop, movement, and so forth. The movement period is referred to as swim or run and the stop period is referred to as tumble. During the swim phase, the flagella are rotating counterclockwise (CCW) and the cell is propelled in a straight line and if the cell is peritrichous the filaments are collected in a bundle. In the tumble phase, the rotation of the flagellum is reversed and turns in a clockwise (CW) direction. During the tumble phase the flagella are released from the bundle if the cell is peritrichous or if the cell is monotrichous forward movement is slowed. Thus, the movement of bacterial cells will be a series of swim–tumble–swim–tumble events described as "random walk." The process for movement of *Salmonella* or *Escherichia coli* is illustrated in Figure 6.3. During the tumble phase, the cell becomes reoriented for either positive or negative chemotaxis.

### 6.2.2.4. Variations in Bacterial Motility[12]

Some of the structural parts of the flagella are markedly different from what is found with *Escherichia coli* and *Salmonella typhimurium*. The flagellar unit of the Gram-positive *Bacillus* has been examined by electron microscopy and would appear to lack the L-ring or the P-ring and the flagellar hook goes through a sleeve in the thick peptidoglycan. Thus, the L-ring and the P-ring would not be essential for flagellar motility. It should be noted that the sizes of the substructures of flagella in both *Escherichia coli* and *Bacillus* are about the same. Considerable complexity of the filament occurs with *Rhizobium meliloti*, which has two types of flagellin making up the filament, and for *Caulobacter crescentus*, which has three types of flagellins distributed in three regions along the axis of the filament.

Variations also exist with respect to chemotaxic and motor activity in bacteria. Flagellar motility in some alkalophilic bacteria is attributed to $Na^+$-driven rotation and not to proton-motive force. When attached to a surface, *Vibrio haemolyticus* has a proton-driven flagella but when the cells are swimming, the flagella are driven by sodium uptake. In some bacteria, the rotation of flagella is only in the CW direction and these cells would not display a typical tumble mode. *Rhodobacter sphaeroides* is a purple nonsulfur phototroph that has a single polar flagellum that rotates only in a CW direction. Cells change direction in the stop mode. The flagellum moves very slowly to reorient the cell so that when flagellar rotation is restarted the cell moves in a new direction. With *Rhizobium meliloti*, which has 5 to 10 peritrichous flagella, the flagella rotate continuously to move the cell by short straight runs with intermittent quick turns. It appears that rotation of all flagella may not be highly coordinated and different flagellar motors on the same cell may operate at different rates. Uncoordinated motors will result in the disruption of the flagella cluster when cells are in the swim mode while engagement of lateral flagella will influence cells to move in a different direction. Flagellar rotation in *Rhodobacter sphaeroides* is also in a single direction and changes in direction are attributed to changes in speeed of rotation. Bacteria with bipolar flagella, such as the magnotactic spirillum and the archaenon *Halobacterium halobium*, have flagella at both ends of the cell so that when the cell is in a swim mode one flagellum rotates CCW while the other moves CW. When the rotation of the flagella is reversed, the cell does not tumble but instead it reverses. Owing to this change in direction of the flagellar rotation the direction of the cell is random. While this does not include all possible types of variations, it does reflect the numerous possibilities that may be present in motile bacteria.

## 6.2.3. Spirochete Motility[13,14]

Spirochetes are Gram-negative bacteria that have a helical shape and do not have external flagella but employ an axial filament for motility. As indicated in Table 6.4, these bacteria are broadly distributed in the environment. The helical cells of these bacteria have distinctive cellular characteristics which include cell length, cell diameter, wavelength, and wave amplitude.

The axial filament contains a flagellum-like structure that resides in a region between the protoplasmic cylinder and the outer sheath. The protoplasmic cylinder consists of the cytoplasm of the cell, plasma membrane, and the cell wall. The axial filament extends along the longitudinal axis of the cylindrical cell and has flagella-like structures that originate at the subpole location of the spirochete where they are folded back along the axis of the cell. With many of the spirochetes, the filaments extending from each pole overlap near the median region; however, filaments from leptospiral species are short and do not overlap. A model of a spirochete with axial filament is given in Figure 6.4. The outer sheath covers the entire cell and this sheath appears to be a three-layered membrane consisting of protein, lipid, and carbohydrate.

Table 6.4. Examples of bacteria that have axial filaments.

| Genus | Number of endoflagella | Diameter of endoflagella | Discs number (size) | Characteristics |
|---|---|---|---|---|
| *Borrelia* | 7–20 | 13 nm | 2 (30–35 nm) | Anaerobic or microaerophilic, pathogens for humans |
| *Cristispira* | >100 | 18 nm | ? | Present in digestive tract of marine and fresh water mollusks |
| *Leptospira* | 2 | 13–18 nm | 3–5 (20–26 nm) | Aerobic. Present in soil, water, and some associated with animals. Ends of cells may have hook-like shape. |
| *Spirochaeta* | 2–40 | 15–21 nm | 3 (40 nm) | Anaerobic and facultative anaerobes. Present in water and mud. |
| *Treponema* | 2–15 | 15–21 nm | 1–2 (30 nm) | Anaerobic, associated with mouth, genital, and intestinal tracts of animals and humans. |

The outer sheath interfaces with the environment and to some extent serves as a permeability barrier to protect the endoflagella. The number of endoflagella per cell is commonly two; however, some spirochetes have been reported to have more than 100 and these endoflagella are positioned along the longitudinal axis of the cell.

Although a genetic system has not been developed for axial filaments, considerable insight has been gained from examination of the fine structure and chemical composition of endoflagella of spirochetes. The endoflagella consist of a shaft or filament and an apparatus that attaches the filament to the cell. The filaments consist of two parts: an inner core and an outer covering or sheath. The diameters of several filaments are given in Table 6.4. The inner core of the filaments have been isolated from several different species and it has been determined that this structure consists of protein that contains an abundance of aspartic acid, glutamic acid, alanine, leucine, and glycine but lacks cysteine, tyrosine, or tryptophan. The filament core can be dissociated into subunits, which in the case of *Treponema phagedenis* is 32 to 36 kDa. At the proximal end of the filament is a hook that links the endoflagella to the layers of the protoplasmic cylinder through a series of discs. Based on the similarity of amino acid composition of endoflagella and flagellar filaments, the presence of discs in spirochetes and rings in bacterial flagella, and the dissociation of filaments from flagella and endoflagella into subunits it is generally assumed that the endoflagella are structurally and functionally similar to the bacterial flagella. In addition, it is assumed that endoflagella rotate by mechanisms similar to those found for bacterial flagellar rotation.

The spirochetes move in liquid by two different processes: rotation about the longitudinal axis and flexing. Cell rotation is one of the most commonly observed activities for movement of spirochetes. The internalized flagella rotate by mechanisms similar to extracellular flagella in that motility is driven by proton-motive force across

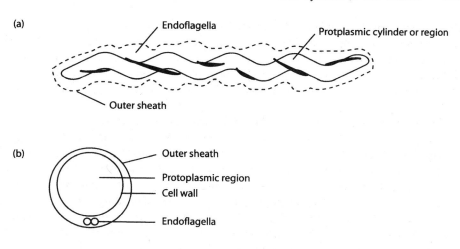

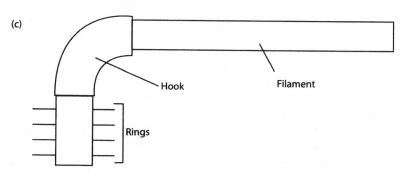

Figure 6.4. Axial filament structure as found in spirochete bacteria. (**a**) Position of filament along helical cell with origin of filaments at each pole of the cell. (**b**) Cross section of spirochete showing location of filaments in cell. (**c**) Structure of endoflagella as it would be found at one end of the spirochete.

the plasma membrane. The rotation of the spirochete cell is counter to the rotation of the flagella in the axial filament and this cellular rotation propels the cell forward in a cork screw-like fashion. The movement of *Spirochaeta aurantia* in buffer has some activities that are reminiscent of bacteria with typical flagella. The spirochete cells rotate in a straight line (analogous to swim activity), periodically stop with cell flexing that contributes to reorientation (analogous to tumble activity), and resume movement in a new direction. Flexing of the cell is likened to the bending of a spring and even though the spirochetes have a cell wall of peptidoglycan, flexing may result in acute bending of the cell. Cells of *Borrelia hermsii* have been noted to move by continuous rotation; however, this movement ceases when the density reaches several million cells. Although this may suggest quorum sensing in *Borrelia hermsii*, definitive evidence of this needs to be presented. Cells of *Leptospira icterohaemorrhagiae* have a hook at one end and this hook has been suggested to rotate and propel the cell as it moves through a liquid medium. The spirochetes have been described to "creep" or "crawl" on solid surfaces by mechanisms involving rotational movement similar to that in liquid media or nonrotational movement by a process that is similar to that of gliding bacteria.

Unlike flagellated prokaryotes, spirochetes with axial filaments can move through media with high viscosity and through a gel such as 1% to 2% agar. While movement of flagellated bacteria ceases at about 60 cp, spirochetes display translational motion from 1 to 1,000 cp. For example, *Leptospira interrogans* moves at 30 μm/sec at a viscosity of 300 cp, which is in the general range for movement of flagellated bacteria which are usually present in aqueous media where the suspending medium is only 1 to 2 cp (see Table 6.3). The axial filament enables spirochetes to move through rumen fluids, genital fluids, mucosal fluids, viscous muds, and environments with a viscosity equivalent to that found in cell membranes of animal cells. The axial filament may have evolved to enable spirochetes to move through environments where bacteria with exposed flagella were immobile. The movement of spirochetes through gels enables them to penetrate gingival crevices and various matrices found in the genital tract. The movement through a gel has been exploited for the isolation of spirochetes where a sample containing a mixture of organisms is placed in a well on an agar plate and after several days of incubation, spirochetes are isolated in the agar some distance from the region where bacteria were deposited.

A mechanism for axial filament-driven movement of spiral bacteria has been proposed by Berg and it assumes the following: (1) the cell is rigid, (2) the flexible outer sheath is not attached to the cell, and (3) the endoflagella rotate. When the endoflagella from each pole of the cell rotate in opposite direction, the outer sheath rotates about the cell and the cell is propelled forward. It is proposed that if both endoflagella reverse the direction of rotation, the spirochete moves in reverse. The leading end of the spiral cell moves through the medium and the trailing end may display a "hook" form due to flexing of the spiral. When the endoflagella rotate in the same direction, both ends of the cell display a "hooked" form and no translational movement occurs. The shape of the ends of the cells, that is, hooked or nonhooked, is a response of the rotational dynamics of the endoflagella. Future work with motility mutants with different spirochetes are needed to illuminate further the mechanism of movements by the rigid spiral bacteria.

## 6.2.4. Motility of Archaea[15]

Several members of the *Archaea* domain have flagella for motility and some of these organisms are identified in Table 6.5. By a manner similar to bacterial flagellar movement, the flagella of archaea propel the cells as a result of flagellar rotation. However, there are several striking differences in filament structure and flagellar biosynthesis that distinguished archaeal and bacterial flagellation. These distinctions are as follows: (1) Flagellar structure—the archaeal filaments are only 10 to 14 nm in diameter while those of bacteria are generally 20 nm. The filaments in bacteria consist of a single

Table 6.5. Distribution of flagella in the *Archaea*.

| Genera with flagella and flagellin genes | Genera that lack flagella |
| --- | --- |
| *Aeropyrum* | *Halococcus* |
| *Archaeoglobus* | *Methanobacterium* |
| *Desulfurococcus* | *Methanosarcina* |
| *Halobacterium* | *Natronococcus* |
| *Methanococcus* | *Thermofilum* |
| *Methanoplanus* | |
| *Methanospirillum* | |
| *Methanopyrus* | |
| *Methanothermus* | |
| *Natronomonas* | |
| *Natrialba* | |
| *Pyrobaculum* | |
| *Pyrodictium* | |
| *Pyrococcus* | |
| *Sulfolobus* | |
| *Thermoplasma* | |
| *Thermoproteus* | |

type of protein while in archaea, the filament contains several different protein units. Observations by electron microscopy indicates that the flagellar hooks in many of the archaeal cells are about twice as long as in bacteria. (2) Flagellar genes—from examination of genome sequences, flagellar gene families in the archaea have no homologues with flagellar genes found in bacteria. (3) Leader peptides—unlike bacterial filament proteins, archaeal flagellins have a highly conserved hydrophobic N-terminal amino acid sequence and a short leader sequence that is hydrolyzed from the flagellin protein by membrane-bound peptidases. (4) Flagellar synthesis—the synthesis of archaeal flagella is proposed to be similar to type IV pili and not involving the system used for synthesis of bacterial flagella.

### 6.2.4.1. Structure

There are several unresolved issues pertaining to structure of archaeal flagella. For example, the mechanism for anchoring the archaeal flagellum to the cell remains to be established. Generally, the isolation of flagella from archaeal cells occurs following treatment with Triton X-100 and this treatment produces flagella without rings. However, two rings have been found associated with the flagella isolated from *Methanococcus thermolithotrophicus* and *Methanospirillum hungatei*. It may be that when flagella are associated with the cell, rings are present but the absence of rings in isolated flagella is a technical problem. The structural location of these rings in the cell wall of archaea would be different than that of bacteria because the cell wall of archaea does not contain the peptidoglycan structure. Also, the presence of flagella on cells *Thermoplasma* species presents the question of how the flagella are stabilized in these cell because *Thermoplasma* do not contain a cell wall. The mechanism of

Table 6.6. Amino acid sequences at the N-terminus of archaeal flagella.[a]

| Organism | Protein | Sequence | | | | | | | |
|---|---|---|---|---|---|---|---|---|---|
| *A. fulgidus* | FlaB1 | MGMRFLKNEKG | FTGLE | AAIVL | **IAFVT** | VAAVF | SYVLL | GAGFF | |
| | FlaB2 | MRVGSRKLRRDEKG | FTGLE | AAIVL | **IAFVV** | VAAVF | SYVML | GAGFY | TTQK |
| *H. salinarum* | FlgA1 | MFEFITDEDERG | QVGIG | TLIVF | **IAMVL** | VAAIA | AGVLI | NTAGF | LQSK |
| | FlaA2 | MFEFITDEDERG | QVGIG | TLIVF | **IAMVL** | VAAIA | AGVLI | NTAGF | LQSK |
| | FlgB1 | MFEFITDEDERG | QVGIG | TLIVF | **IAMVL** | VAAIA | AGVLI | NTAGY | LQSK |
| | FlgB2 | MFEFITDEDERG | QVGIG | TLIVL | **IAMVL** | VAAIA | AGVLI | NTAGY | LQSK |
| | FlgB3 | MFEFITDEDERG | QVGIG | TLIVF | **IAMVL** | VAAIA | AGVLI | NTAGY | LQSK |
| *M. jannaschii* | FlaB1 | MKVFEFLKGKRG | AMGIG | TLIIF | **IAMVL** | VAAVA | AAVLI | NTSGF | |
| | FlaB2 | MKVFEFLKGKRG | AMGIG | TLIIF | **IAMVL** | VAAVA | AAVLI | NTSGF | |
| | FlaB3 | MLLDYIKSRRG | AIGIG | TLIIF | **IALVL** | VAAVA | AAVLI | NTAAN | |
| consensus | | KKG | AVGIG | TLIVF | **IAMVL** | VAAVA | ASVLI | NTSGY | LQQK |
| sequence[b] | | RR | S | | | | G | A   F | |
| | | | | | | | A | | |

[a]From Thomas et al., 2001.[15]

[b]Established from review of sequences of more than 40 archaeal flagella and is found in more than 70% of archaeal flagella. Amino acids that do not vary are bold and major alternate amino acids are listed beneath the consensus amino acid.

flagellar rotation in archaea may be quite unique especially if rings are absent from flagella.

Based on the amino acid content from cloned filament genes, the protein content of archaeal filaments is 20.5 to 61.1 kDa. Unlike bacterial flagella, the filament proteins of archaea contain cysteine and are glycoproteins with carbohydrate segments added by posttranslational processes. The prefilament in a variety of archaea contains a signal sequence that is generally 12 amino acids in length and this peptide is released from the filament by a membrane-bound peptidase. Based on site-directed mutagenesis studies, it has been determined that the glycine residue must be present for action by this protease. On cleavage of the signal peptide from the prefilament, the membrane peptidase methylate the amino acid at the N-terminus of this filament. The N-terminus of the filament protein is highly hydrophobic and a 34 amino acid sequence is highly conserved. The sequences of amino acids in the signal peptide and filament of several archaea are given in Table 6.6.

### 6.2.4.2. Synthesis

Further distinctions between archaeal and bacterial flagella are observed following evaluation of the molecular genetics. Analysis of genes for archaeal motility reveals little similarity to genes found in bacteria. The archaea have clusters of genes for flagellar and accessory proteins. Examples of flagellar genes are given in Figure 6.5. In *Halobacterium salinarum*, the principle filament genes A and B are on separate regions of the chromosome. With *Methanococcus jannaschii* and *Pyrococcus kodakarensis*, the multiple filament genes are present in a linear array within the cluster of

*Halobacterium salinarum*

*Methanococcus jannaschii*

*Pyrococcus kodakarensis*

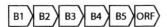

Figure 6.5. Flagellar genes in selected archaea. Multiple genes for filament in *Halobacterium salinarum* are B1, B2, and A1, A2, which are located on two separate genetic segments. Genes for *Methanococcus jannaschii* are B1–B4 and for *Pyrococcus kodakarensis* are B1–B5, which are located at a single site on the chromosome of these archaea. (From Thomas et al., 2001.[15])

flagellar genes. No homologues of bacterial flagellins, rod proteins, hook proteins, or ring proteins present in archaea. However, the *cheA*, *cheW*, *cheR*, *cheB*, and *cheY* genes for bacterial chemotaxis have strong homologues in *Halobacterium salinarum*.

The transport and assembly of the flagellar proteins resembles synthesis of type IV pili and appears to be distinct from the process observed with bacterial flagella. The presence of signal peptides on the flagellar proteins suggests that the proteins are not moved across the plasma membrane by the type III secretion system used in synthesis of bacterial flagella. The synthesis of the flagella does not appear to be at the tip because there is no evidence that has been presented for hollow flagella in archaea. In addition, if a pore would be present but escape detection, the diameter of a possible pore would be too small to accommodate passage of protein to the distal end of the flagella. Further support for the suggestion that the archaeal flagella is assembled exterior to the plasma membrane is the observation that glycoprotein formation with addition of carbohydrate residues occurs after flagella proteins have been transported across the plasma membrane.

## 6.3. Cell Sensing and Signaling Pathways[16]

Bacteria respond to various changes in the chemical environment by systems involving cell sensing which involves a complex system of interactions. The stimulus for physiological change most likely goes from the environment external to the cell and in this case the information must be relayed across the plasma membrane. A soluble receptor stimulates DNA to initiate gene expression to accommodate the environmental change. In most instances, the system for signal transduction involves two distinct protein activities; however, numerous variations of this two-component theme is observed. We shall first examine the structure of a two-component signal transfer system and then provide description of alternate sensing systems.

## 6.3.1. Two-Component Signal Transduction[17–21]

Broadly distributed throughout the bacteria is a two-component sensing system and the general model summarizing these activities is presented in Figure 6.6. A receptor protein binds the small signaling molecule and prompts the phosphorylation of a histidine residue in the receptor protein from the $\gamma$-phosphate of ATP. In some of the systems, the structure phosphorylated is not the one that interacts with the signal but it is an adjacent receptor molecule. Because the receptor molecule is involved in phosphorylation, the process has been termed "auto phosphorylation." The second component is a response regulator protein and the high-energy phosphate is transferred from the receptor protein to an aspartic acid moiety of the response regulator protein. The receptor regulator~phosphate molecule is responsible for a variety of physiological activities including chemotaxis and DNA regulation. This new physiological expression may be discontinued by a phosphatase that removes the ~phosphate from the activated receptor regulator.

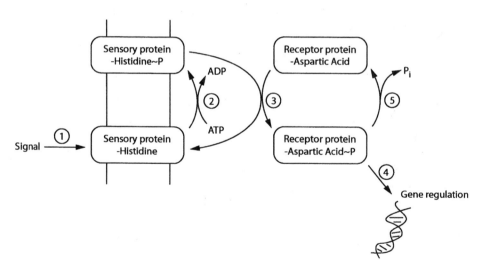

Figure 6.6. Generalized model of the two-system sensor system in bacteria. The principal areas of activity are as follows: (1) signal recognition by a transmembrane sensory protein, (2) autophosphorylation of histidine-containing sensory protein, (3) phosphotransferase activity with phosphate transfer from protein–histidine–phosphate to aspartate in receptor protein, (4) direct interaction of receptor protein aspartate–phosphate with DNA to initiate transcription, and (5) phosphatase activity to reduce level of protein–phosphate in the cytoplasm that reduces transcription induced by protein–aspartate–phosphate.

Table 6.7. The generalized phosphotransferase reactions involved in signal transduction.

| Proteins involved | Reactions |
| --- | --- |
| Histidine kinase (HK) | ATP + Protein$_{HK}$ − Histidine ↔ ADP + Protein$_{HK}$ − Histidine-Phosphate |
| Response regulator (RR) | Protein$_{HK}$ − Histidine-Phosphate + Protein$_{RR}$ − Aspartate ↔ Protein$_{HK}$ − Histidine + Protein$_{RR}$ − Aspartate~Phosphate |
| Phosphatase | Protein$_{RR}$ − Aspartate~Phosphate + H$_2$O ↔ Protein$_{RR}$ − Aspartate + Phosphate |

The chemical reactions associated with the two-component transduction system are given in Table 6.7. The histidine protein kinase and the aspartic acid kinase system have been extensively studied in eukaryotic systems and it is considered that these two activities are associated with bacterial ATPases. Many of the bacterial sensory proteins do not have ~P attached to aspartic acid in the receptor protein but another amino acid is involved. In addition, in some cases, there may be three or more components in the phospho-relay system.

Structural analysis of various sensory proteins reveals that there are several instances in which the amino acid sequences are highly conserved. These regions of protein with about 70% similarity are referred to as "blocks." A model indicating the presence of five blocks of consensus sequence and their respective amino acid composition are given in Figure 6.7. Each block is named for a key amino acid present in that sequence. The H block is in the N-terminus of the molecule and contains the histidine residue which becomes phosphorylated by autophosphorylation. The remaining blocks are found associated with the C-terminus of the protein. The N and D blocks derive their name from the presence of asparagine and aspartate, respectively. The F block is part of a short amino acid sequence containing phenylalanine and it links the D and the G blocks. Since the G block and the D block contain several glycine residues, it is inferred that they function as the binding site for Mg-ATP. At the N-terminus and generally prior to the H block is a hydrophobic block which accounts

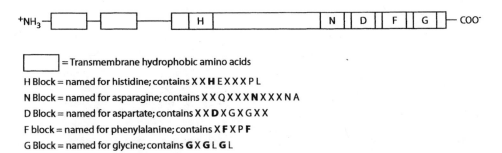

□ = Transmembrane hydrophobic amino acids

H Block = named for histidine; contains X X **H** E X X X P L

N Block = named for asparagine; contains X X Q X X X X **N** X X X N A

D Block = named for aspartate; contains X X **D** X G X G X X

F block = named for phenylalanine; contains X **F** X P **F**

G Block = named for glycine; contains **G** X **G** L **G** L

Figure 6.7. Motifs that are characteristic of classic sensory transmitters found in bacteria. (From Parkinson and Kofoid, 1992.[19])

for transmembrane distribution. There is some variation in the positioning of these blocks in the sensory proteins and in some instances the block may be absent. Structural analysis indicates that a DNA binding segment is near the C-terminus of the receptor protein molecule.

The bacterial systems of sensing and signal transfer have been extensively studied and some of the two-component sensing systems found in bacteria are listed in Table 6.8. Additional bacterial activities that are considered to be regulated by sensing-response systems include heterocyst production in cyanobacteria, bacteroid development in rhizobia, and cell division by the parasitic *Bdellovibrio*. Some characteristics of a few of the two-component systems are given in the following paragraphs and are found in other areas of this book.

Table 6.8. Examples of two-component sensory systems found in bacteria.[a]

| Systems | Bacteria employing this system | Stimulus receptor | Regulator |
|---|---|---|---|
| **Developmental and structural features** | | | |
| Chemotaxis | Various Gram-negative bacteria | CheA | CheY, CheB |
| Sporulation | *Bacillus subtilis* | KinA, KinB | SpoOA, SpoOB |
| Polar morphogenesis | *Caulobacter crescentus* | DivJ, DivK | CtrA |
| Fruiting body formation | *Myxococcus xanthus* | FrzE | FrzZ |
| Symbiosis | *Bradyrhizobium japonicum* | NodV | NodW |
| Gene transfer | *Bacillus subtilis* | ComP | ComA |
| Capsule synthesis | *Pseudomonas aeruginosa* | AlgR2 | AlgR1 |
| **Metabolic activities** | | | |
| Phosphorus assimilation | Various bacteria | PhoR | PhoB(PhoP) |
| Nitrogen assimilation | Various bacteria | NtrB | NtrC |
| $N_2$ fixation | *Klebsiella pneumoniae, Rhizobium meliloti*, and *Bradyrhizobium japonicum* | FixL | FixJ |
| Uptake of carboxylic acids | *Rhizobium meliloti* and *Bradyrhizobium japonicum* | DctB | DctD |
| Antibiotic synthesis | *Enterococcus faecium* | VanS | VanR |
| **Stress response** | | | |
| Redox changes | *Escherichia coli* | NarQ, NarX | NarL |
| Osmolarity | Enteric bacteria | EnvZ | OmpR |
| Heavy metal resistance | *Streptomyces lividans* | CutS | CutR |
| Starvation | *Bacillus subtilis* | DegS | DegU |
| Turgor pressure | *Escherichia coli* | KdpD | KdpE |

[a]From Parkinson and Kofoid, 1992.[19]

## 6.3.1.1. Porin Regulation[22]

There are two distinct porins present in *Escherichia coli* that enable hydrophilic molecules to diffuse across the outer membrane into the periplasm. One is the OmpC porin which is a trimer that forms a channel of 1.08 nm and the other is OmpF, which also is a trimer and has an opening with a diameter of 1.16 nm. Small molecules of <600 Da can diffuse through the OmpF pore at a faster rate than through the OmpC pore as OmpC is more restrictive in the size of molecules that can get across the outer membrane. Other bacteria have porins with pore diameters significantly greater than that found with *Escherichia coli*. Porins with a small channel or pore enable *Escherichia coli* to live in an intestinal environment where nutrients are abundant and toxic bile salts are present. In fact, owing to the elevated temperature which increases the rate of diffusion and the nature of the amino acids lining the pore, negatively charged bile salts diffuse through OmpF porins at a greater rate than through OmpC. As a defense against bile salts and other harmful salts, the high osmolarity of the intestine signals *Escherichia coli* to produce only OmpC porins. Conversely, in aquatic environments where nutrients are dilute, *Escherichia coli* produces only the larger OmpF porin as a means of acquiring nutrients at the greatest rate of diffusion.

This regulation of the porins is attributed to a two-component sensory system in which the stimulus is osmotic pressure. The exact type of extracellular molecules accounting for this regulation remain to be established. A model of the regulation of the porin genes is given in Figure 6.8. The sensing protein is EnvZ and a single cell is believed to contain only 10 EnvZ molecules. The 450 amino acid protein contains two transmembrane segments and as a result of histidine protein kinase activity, the histidine moiety at position 243 becomes phosphorylated. The response regulator protein is OmpR and there are about 1,000 copies of this molecule per cell. OmpR consists of 239 amino acids and aspartic acid at position number 55 is phosphorylated owing the aspartic kinase activity of OmpR. The C-terminus of OmpR contains a DNA-binding region which interacts with the *ompF* and *ompC* genes. Distinct molecular conformations are observed with the phosphorylated and nonphosphorylated forms of OmpR. It may be that each of these forms of OmpR recognizes different regions of DNA so that at high osmolarity OmpR represses the formation of OmpF but promotes the synthesis of OmpC and at low osmolarity, the major porin present is OmpF.

## 6.3.1.2. Bacterial Virulence[23–26]

Bacteria that are capable of causing diseases in humans produce virulence factors that enable the bacteria to overcome the defense systems of the host. Growth of pathogens in living tissues is a consequence of bacteria producing capsules, toxins, adherence factors, enzymes, and numerous other components promoting invasion of host cells. It is apparent that different virulence factors are required as the bacteria progress through the various stages of infection and for efficient energy allocation, the production of virulence factors must be closely regulated. One of the systems for the pathogens to adapt to the host is to employ the two-component signal transduction system. Extracellular signals such as low pH, elevated temperature, and osmolarity

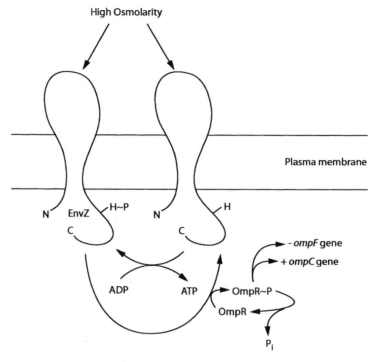

Figure 6.8. Sensory system accounting for porin production in *Escherichia coli*. EnvE has two transmembrane segments with both N and C terminating in the cytoplasm. EnvE senses the osmolarity and transmits information across the plasma membrane to the cytoplasm where EnvE becomes phosphorylated. The phosphate moiety is passed on to OmpR, the response regulator, and this phosphorylated molecule has an influence on transcription. (From Pratt and Silhavy, 1995.[22])

can stimulate regulatory proteins to promote desired gene expression. Some of the activities associated with virulence production by pathogenic bacteria is given in Table 6.9. It should be emphasized that not all virulence factors are produced as a result of two-component signal transduction systems because there are several other mechanisms employed by pathogenic bacteria to regulate production of virulence factors.

The system that controls production of exotoxin by *Vibrio cholera* is a variant of two-component signal transduction regulation and this system serves as an excellent example of adaptation by a pathogenic agent. The model for production of virulence enhancing factors by *Vibrio cholera* is given in Figure 6.9. The external stimulus is considered to include a slight alkaline environment of pH 6.5, an elevated temperature, or increased osmotic pressures. These stimuli are received by ToxS, which is a membrane protein that extends into the periplasm and serves as the sensory protein. The sensitized ToxS molecule interacts with a ToxR protein and a region of the activated ToxR molecule interacts with specific sites on DNA to account for the induction of an outer membrane protein (OmpU), the production of cholera toxin (CtxAB),

Table 6.9. Two-component signal transduction systems associated with virulence in bacteria.[a]

| Pathogen | Extracellular signal | Phenotypic change | Regulatory proteins |
|---|---|---|---|
| *Bacteroides fragilis* | Tetracycline | Porin synthesis | RprX/RprY |
| *Bordatella pertussis* | Temperature | Toxin | BvgA/BvgS |
| *Citrobacter fruendii* | ? | Vi antigen | ViaA/ViaB |
| *Clostridium perfringens* | ? | Collagenase synthesis | VirR |
| *Enterococcus faecium* | Vancomycin | Vancomycin resistance | VanR/VanS |
| *Neisseria gonorrhea* | Stress | Pili synthesis | PilA/PilB |
| *Pseudomonas aeruginosa* | Osmolarity | Alginate synthesis | AlgR1/AlgR2 |
| *Shigella flexneri* | Osmolarity | Porin synthesis | OmpR/OmpZ |
| *Streptococcus* spp. | Carbon dioxide | M-protein synthesis | Mry |
| *Vibrio cholera* | pH, temperature, osmolarity | Toxin and pili synthesis | ToxR/ToxS |

[a]From Dziejman and Mekalanos, 1995.[24]

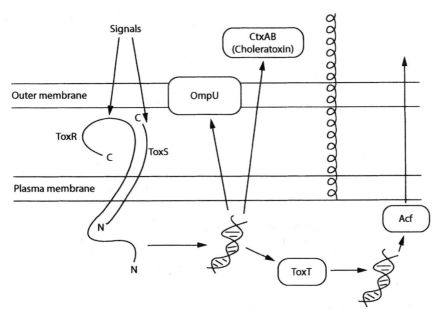

Figure 6.9. Virulence expression in *Vibrio cholera* is attributed to regulation of cholera toxin operon. The regulation involves three protein components: ToxR, a response regulator; ToxS, an effector protein; and ToxT, a transcriptional activator. ToxR interacts with ToxS to produce an active form of ToxR designated here as ToxR*. The ToxR* interacts with DNA for the production of ToxT, a porin and cholera. The ToxT results in the production of toxin-coregulated pilus (Tcp) by stimulation of major protein subunits and proteins for pilus assembly. ToxT also stimulates the production of accessory colonization factor (AFC). Both pilus and ACF account for localization of the bacteria in the intestine. There is a slight difference in signal control with production of classical cholera toxin and El Tor cholera toxin. (From DiRita, 1995.[23])

and production of ToxT which serves as an additional regulatory protein. The ToxT molecule is responsible for localization of *Vibrio cholera* in the intestine because it regulates the production of pili and an accessory colony formation (AcfA) factor. It can be considered that the three-component system includes ToxS, ToxR, and ToxT.

### 6.3.1.3. Regulation of Sugar Phosphate Transport[27]

Many bacteria acquire phosphate from phosphorylated sugars in the environment through the use of an alkaline phosphatase to release inorganic phosphate with the subsequent uptake of phosphate by a phosphate transport system. A more efficient process of acquiring sugar phosphates is direct import of the phosphorylated organic molecule. The transport of sugar phosphates in several Gram-positive and Gram-negative bacteria is by the Uhp (*u*ptake of *h*exose *p*hosphate) system. Although the only inducer for the Uhp system is the presence of glucose-6-phosphate in the extra-

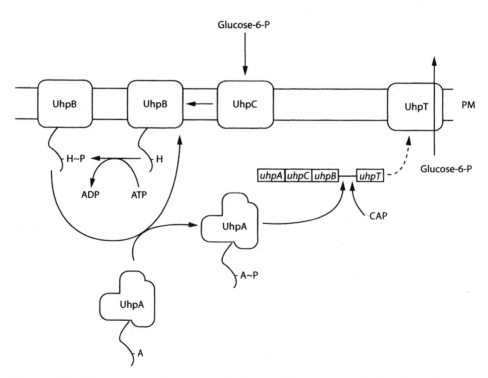

Figure 6.10. Signal system for uptake of glucose-6-phosphate in *Escherichia coli*. Two membrane-associated signaling proteins initiate uptake with the binding of glucose-6-phosphate to UhpC initiating the formation of a UhpC–UhpB complex. Phosphate is transferred from UphB~P to UmpA which activates transcription of the UhpT promoter in the presence of cAMP and catabolite activator protein (CAP). UhpT transports glucose-6-phosphate and other substrates. In contrast, GlpT is specific for glycerol-3-phosphate and PgtP transports phosphoenol pyruvate and phosphoglycerates. (From Kadner, 1995.[27])

cellular environment, the Uhp system indiscriminately imports phosphate esters of various hexoses (galactose, glucose, fructose, mannose, and mannitol), a sugar with seven carbons (sedoheptulose), and pentoses (arabinose and ribose).

The Uhp genetic system consists of four genes: *uhpA*, *B*, *C*, and *T*. A model of the *uhp* system is given in Figure 6.10. While UhpT can independently transport sugar phosphates, products of the *uhpA*, *B*, and *C* genes are required for synthesis of *uhpT*. The induction of UhpT is a two-component system where UhpC and UhpB are the sensory proteins and UhpA is the response regulator protein. The UhpC is stimulated by extracellular levels of glucose-6-phosphate and in cooperation with UhpB the sensory proteins are autophosphorylated at one of the histidine residues. The UhpA has an aspartic acid moiety that receives phosphate from UhpB~P and binds to the DNA upstream from the *uhpT* gene. In the presence of the catabolite activating protein (CAP) and the UhpA protein, induction of the *uhpT* gene occurs. Thus, controls on the expression of the hexose-phosphate transport system is attributed to both an extracellular signal that glucose-6-phosphate is present in the environment and to the high concentration of cAMP which occurs when glucose is depleted inside the cell.

### 6.3.1.4. Degradative System of *Bacillus*[28]

As *Bacillus subtilis* shifts from the log phase to stationary phase of growth or as it adjusts to nutrient limitation, new phenotypes are displayed. Some of these changes include synthesis of hydrolytic enzymes that degrade macromolecules, promotion of competence in cells for DNA transformation, increased cell movement and chemotaxis, production of antibiotics, and sporulation. The regulatory system for these activities has characteristics similar to a two-component signal transduction system in that they contain a histidine protein kinase and a response regulator.

When *Bacillus subtilis* is in an environment deficient in readily metabolizable sugars, one of the responses displayed may include the synthesis of enzymes that degrade macromolecules and a model for this activity is given in Figure 6.11. The induction of several different hydrolytic enzymes is controlled by the two-component system consisting of DegS and DegU. The DegS is assumed to be a soluble enzyme because it lacks the hydrophobic amino acids typically found in transmembrane loops. The DegS protein interacts with an unknown stimulus and functions as a histidine protein kinase by interacting with ATP to autophosphorylate DegS. The phosphate is transferred from DegS~phosphate to DegU and the resulting DegU~phosphate interacts with specific regions on DNA to produce the following hydrolytic enzymes: (1) proteases, which are encoded by *aprE*, *nprE*, and *isp*; (2) α-amylase, which is produced from the *amyE* gene; (3) levansucrase, which is encoded by *sacB*; and (4) β-glucanase, which is produced from the *licS* gene. DegU is a unique response regulator in that in phosphorylated state it induces hydrolytic enzymes but in the nonphosphorylated state it stimulates the production of competence factors. The DegU system is highly complex because DegU~phosphate is responsible for the production of DegQ which regulates the production of additional proteases. Thus, *Bacillus subtilis* has a highly efficient mechanism for the simultaneous induction of numerous carbohydrases and proteases to attack macromolecules and release carbon units for metabolism.

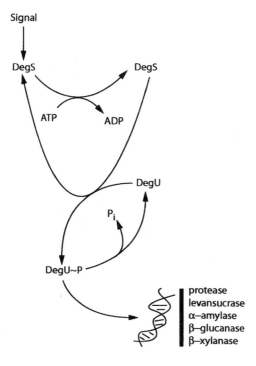

Figure 6.11. Signals for the synthesis of degradative enzymes in *Bacillus subtilis*. In the post-exponential phase, the DegS/DegU system controls production of an intracellular protease and several secreted carbohydrases. DegU/DeU~P also interfaces with other signal systems involved with development of competence, sporulation, and motility. (From Msadek et al., 1995.[28])

## 6.3.2. Chemotaxis[29,30]

The overall goal of cell motility is not for random activity but for directional movement toward an environment that is favorable for growth. Although the initial experiments employing capillary tubes to quantitate chemotaxis were conducted by W. Pfeffer in 1884, attention was not directed to chemotaxis until the last few decades. The directionality of cell movement is a relatively complex process that involves sensing of chemicals in the environment, transfer of a signal across the plasma membrane, processing of the stimulus inside the cell, and regulation of flagellar rotation. As 2% to 5% of the bacterial genome is allocated for chemotaxis, bacteria have a considerable commitment to swim toward optimal growth environments. There are several influences on cellular movement and these are summarized in Figure 6.12. It has been demonstrated that amino acids, sugars, oxygen, high-intensity blue light, external pH, and temperature may act as either attractants or repellents for microbial chemotaxis.

The best studied systems for chemotaxis involve *Escherichia coli* and *Salmonella typhimurium*. The membrane receptor proteins involved in chemotaxis are referred to as chemotactic signal transducers. In *Escherichia coli*, there are four transducer proteins and they are Tsr, Tar, Tap, and Trg while *Salmonella typhimurium* has the Tsr, Tar, Tap, and Tcp. The chemoattractants and repellents for these signal transducers are listed in Table 6.10. The number of Tsr molecules per cell is 2,000 to 2,500 while

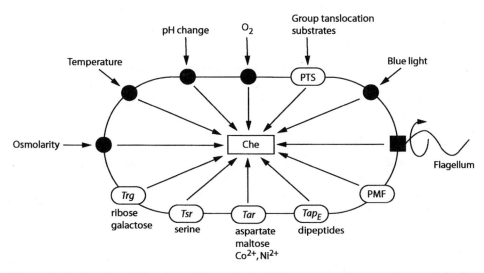

Figure 6.12. Example of the physical and chemical factors that influence chemotaxis in *Escherichia coli*. This model indicates that the Che apparatus has a central role in processing the environmental signals and interacting with the flagellar motor (FliM, FliN, FliG) to regulate rotation of the flagellum. (From Macnab, 1987.[9])

the number of Tar proteins is 1,000 to 1,500 per cell. Only a few hundred molecules of Trg and Tap are present in each cell. It is interesting that with enteric bacteria the only amino acids that are strong attractants are L-serine and L-aspartate. Presumedly these two amino acids function as indicators for the other amino acids that would be found in protein hydrolysate. Soil bacteria live in an environment in which amino

Table 6.10. Chemoeffectors for chemotaxis by enteric bacteria.

| Membrane receptors | Bacteria | Attractants | Repellents |
|---|---|---|---|
| **Tsr protein** | *Escherichia coli* and *Salmonella typhimurium* | Alanine Glycine Serine | Leucine Acetate Indole Benzoate |
| **Tar protein** | *Escherichia coli* and *Salmonella typhimurium* | Aspartate Glutamate Maltose | Cobalt Nickel |
| **Trg protein** | *Escherichia coli* and *Salmonella typhimurium* | Galactose Ribose | ? |
| **Tap protein** | *Escherichia coli* | Glycyl-L-leucine L-leucylglycine | ? |
| **Tcp protein** | *Salmonella typhimurium* | Citrate | ? |

acids are available only transiently and, therefore, in *Bacillus subtilis* all amino acids serve as attractants for chemotaxis.

Observations by J. Adler, R. Macnab, and D. Koshland have revealed that bacteria are capable of detecting chemical attractants by specific receptors and that the mechanism for signal transduction involves either a methylation-dependent or nonmethylation processes. The transducer proteins have also been called methyl-accepting chemotaxis proteins (MCPs) because they become methylated when appropriate effector molecules bind to them. Methylation of transducer molecules promotes cellular memory and promotes CCW rotation of the flagellum. Transducer proteins span the membrane and have four to six sites of methylation in the cytoplasmic region near the carboxyl end while the binding sites for affectors are on the outer side of the membrane. The transducer molecules generally have a mass of 60 kDa and a model depicting the structural configuration is given in Figure 6.13. At this time there are five known chemoreceptors with respective chemoattractants and they include: (1) the aspartate receptor, which recognizes both aspartate and maltose-binding protein; (2) the serine receptor, which recognizes serine; (3) the galactose and ribose receptor, which recognizes ribose-binding protein and glucose or galactose-binding protein; (4) the dipeptide receptor, which binds the dipeptide-binding protein; and (5) the citrate receptor, which recognizes either free citrate or citrate-binding protein. Transducer molecules are located at the poles of the cells and have two segments that span the membrane. It is considered that with the binding of chemoattractant molecules to the transducer molecules induces a molecular change in shape that increases activity in the methylation region. Methylation provides the cell with a memory of chemical exposure with the extent of methylation of the transducer molecule causing cells to sense concentration gradients. The binding of an attractant to the transducer molecule stimulates methylation with promotion of a swim phase while the binding of a repellent accounts for reduced methylation with the result of a tumble phase, which would enable a cell to seek a new and possibly more productive environment. Methylation is promoted by CheR, which functions as a methyltransferase by transferring the methyl group from *S*-adenosylmethionine to the transducer molecule.

As shown in Figure 6.14, there is another protein, CheB, that acquires phosphate from CheA-P to produce CheB-P. The phosphorylated CheB, CheB-P, functions as a methyl esterase to release the methyl group from the transducer molecule. Unlike flagellar rotation, chemotaxis requires ATP to establish phosphorylated proteins that function in cell sensing. The role of these proteins are summarized in Table 6.11. CheA$_l$, which has the phosphorylation site at histidine 48 near the N end of the protein, along with CheA$_S$, which lacks the phosphorylation site, and CheW, form a stable complex. With the formation of CheA-P, phosphate is transferred to aspartate number 57 of CheY. The CheY molecule is a protein of five β-sheets and 5 α-helices that interact with the flagellar switch (FliM, FliN, and FliG) to increase tumbling (CW). To prevent accumulation of CheY-P, CheZ functions as a phosphatase to release phosphate from CheY-P and CheZ-P may also interact with the flagellar switch proteins. Thus, CheA, CheY, CheZ, and CheB function as signaling proteins to modulate the activity of the attractants or repellents. As the various facets of this chemotaxis pro-

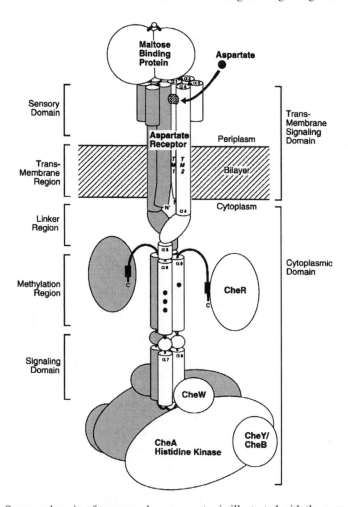

Figure 6.13. Sensory domain of transmembrane receptor is illustrated with the aspartate receptor system that binds either aspartate or maltose-binding protein. The complex occurs as a dimer that transcends the plasma membrane from the periplasm to the cytoplasm. CheW is the protein that complexes the receptor protein to the histidine kinase, CheA. CheR is the methyltransferase that promotes methylation and CheB is the methylesterase that hydrolyzes the methyl ester. CheY is the motor response regulator. (Model reproduced from Falke, J.J., R.B. Bass, S.L. Butler, S.A. Chervitz, and M.A. Danielson. The two-component signaling pathway of bacterial chemotaxis: A molecular view of signal transduction by receptors, kinases, and adaptation enzymes. With permission from the *Annual Review of Cellular and Developmental Biology*, Volume 13, © 1997 by Annual Reviews, www.annualreviews.org.)

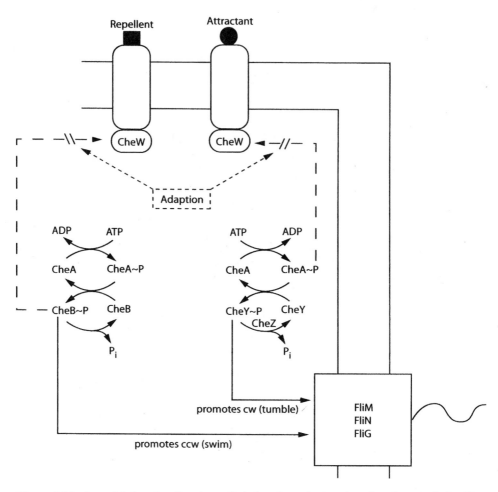

Figure 6.14. A model for signaling in methylation-dependent system for chemotaxis by *Escherichia coli* and *Salmonella typhimurium*. Binding of attractants to the transmembrane sensors results in relaying information to CheW, which promotes methylation of the sensory receptor and reduces autophosphorylation of CheA. See Figure 6.12 for site of methylation. The CheA~P relays phosphate to CheY and CheY promotes CW rotation of the flagellum (tumble mode). In *Escherichia coli*, CheZ is present and it will dephosphorylate CheY~P. Repellents increase the production of the CheA~P with more CheB~P formed as a consequence. CheB~P promotes demethylation of the sensory receptor with the formation of methanol and the motor is stimulated for CCW flagellar rotation (swim mode). In adaption, both methylation and demethylation activity is reduced. (Modified from Stock et al., 1989[20] and Falke et al., 1997.[17])

Table 6.11. Signaling proteins for chemotaxis in *Escherichia coli* and *Salmonella typhimurium*.

| Protein | Proposed function |
|---------|-------------------|
| CheA$_L$ | Carries phosphate, interacts with MCPs and phosphorylates CheY and CheP proteins |
| CheA$_S$ | Similar to CheA$_L$ but smaller and does not have a phosphorylation site |
| CheW | Important in CheA phosphorylation and interaction between CheA and MCPs |
| CheY | Becomes phosphorylated and CheY-P interacts with flagellar switch to increase probability of tumbles (CW rotation) |
| CheZ | Counteracts the tumble signal by dephosphorylating the CheY-P |
| CheR | Promotes methylation of MCP |
| CheB-P | Promotes demethylation of MCP |

cess are understood, the delicate features of bacterial physiology at the molecular level are revealed.

One of the mechanisms employed by prokaryotes for directed cell movement involves methylation. Methylation activity provides the cell with a memory for exposure to molecules that function as attractants. In several bacteria, demethylation results in release of the C-1 group as methanol. Adaption occurs when the stimulus, positive or negative, has been received for some time and methylation/demethylation no longer is produced as a response. The molecules that function as attractants and repellents may vary considerably even with closely related species. As seen in Table 6.12, *Escherichia coli* and *Salmonella typhimurium* have considerable difference in the types of molecules that serve as stimulants for chemotaxis.

There are four known methyl-accepting chemotaxis proteins (MCP) in *Escherichia coli*—Tsr, Tar, Trg, and Tap—but from genome sequencing five genes for MCP were

Table 6.12. Variability of regulation concerning chemotaxis in bacteria.

| Chemotaxis receptor | Location of receptor | Chemical for induction in *Escherichia coli* | Chemical for induction in *Salmonella typhimurium* |
|---------------------|----------------------|----------------------------------------------|----------------------------------------------------|
| Galactose | Periplasm | Galactose, fucose | Fucose |
| Maltose | Periplasm | Maltose, trehalose | No maltose chemotaxis |
| Ribose | Periplasm | Ribose | Ribose |
| Aspartate | Periplasm and membrane | Constitutive | Constitutive |
| Serine | Membrane | Constitutive | Constitutive |
| Glucose | Membrane | Constitutive | Glucose, mannose |
| *N*-acetylglucosamine | Membrane | *N*-Acetylglucosamine | *N*-Acetylglucosamine |
| Fructose | Membrane | Fructose | Constitutive |
| Mannitol | Membrane | Mannitol | Mannitol |

Modified from Koshland, 1980.[8]

found. The number of MCP in bacteria varies markedly with the species and some examples of this variation are given in Table 6.13. From 10 to 15 different MCP are found in each of the following prokaryotes: *Bacillus subtilis, Desulfovibrio vulgaris, Halobacterium salinarum, Rhodobacter sphaeroides*, and *Caulobacter crescentus*. Many bacteria have the full complement of the Che motility signal proteins; however, CheZ is not an obligatory requirement for the control of chemotaxis because there are several examples where bacteria do not have it. CheZ, the protein for dephosphorylation of the CheY~P, is present in *Escherichia coli*, but in other species this dephosphorylation may be attributed to other Che proteins in the system.

There are several instances where microbial motility is not attributed to MCP as chemosensory proteins. Evaluation of the completed genome sequence for *Methanococcus jannaschii* reveals that there are no DNA sequences resembling *che* genes present in this archeon. Phosphotransferase systems (PTS) for uptake of sugars is methylation independent and is highly diversified in bacteria. With PTS, enzyme I and HPr are associated with phospho-relay to enzyme II which interacts with CheA and CheY to produce a chemotaxis signal. *Escherichia coli* has at least 15 PTS receptors that function with the Che system for motility. In some bacteria, fumarate is part of a sensory system that controls motor switching with increase in CW rotation of flagella. The fumarate system is independent of the Che system and is a metabolic sensing system that is as important as environmental sensing. The rational for methylation versus nonmethylation is not apparent in most systems.

Several other stimuli have been important in contributing to microbial chemotaxis. Blue light exposure to nonphotosynthetic bacteria interrupt the swim mode and produces tumbling. Apparently, this action is due to the photooxidation of porphyrins in the cells and this product functions as a repellent in the classical MCP system. *Halobacterium salinarum* has four light-sensitive pigments which include a retinol-based proton pump (bacteriorhodopsin); halorhodopsin (HP) which uses light to pump chloride out of the cell. Two additional sensory rhodopsin pigments (SRI and SRII) are constitutive and are part of the methylation system which interacts with the Che proteins. SRI provides a positive signal in the orange light and a negative signal in blue light while SRII absorbs blue light and in high light intensity is a repellent. The phototactic systems may not be able to position prokaryotic cells within a gradient light field but would be expected to function in directing microbes into a light region. Two other stimuli for chemotaxis are temperature and oxygen concentration. The

Table 6.13. Distribution of proteins in bacteria as determined by genome sequencing.[a]

| Organism | MCP | CheA | CheB | CheW | CheY | CheR | CheZ | W-Y |
|----------|-----|------|------|------|------|------|------|-----|
| *Escherichia coli* | 5 | 1 | 1 | 1 | 1 | 1 | 1 | — |
| *Bacillus subtilis* | 10 | 1 | 1 | 1 | 1 | 1 | — | — |
| *Helicobacter pylori* | 3 | 1 | — | 1 | 1 | — | — | 3 |
| *Treponema pallidum* | 4 | 1 | 1 | 2 | 1 | 1 | — | — |

[a]From Armitage, 1999.[12]

temperature response is to have cells swim toward the environment where the temperature would be that of previous growth. While one may consider that cells would move to the temperature of optimum growth that is not what occurs. With aerotaxis, it is important for facultative and anaerobes to sense and respond to molecular oxygen or to redox changes. Oxygen sensing may be either direct or through a sensory protein containing a cytochrome. Coupled with the loss of molecular oxygen there is a shift in metabolism to nitrate respiration and nitrate becomes an attractant for chemotaxis. Thus, nitrate as a positive chemoattractant is a result of metabolic or energy sensing.

## 6.3.3. Cross Talk

With the multicomponent system for chemotaxis, there is great opportunity for these proteins to receive signals from other cellular processes. Some of the following interactions are the ones readily apparent. (1) The phosphotransferase system for sugar transport interacts with the signaling process for chemotaxis. Enzyme II, the phosphorylated transport protein in the plasma membrane, replaces the MCP proteins as the signal transducers and chemotaxis proceeds without the methylation activity. (2) The requirement of cAMP and catabolite activator protein for expression of all flagellar genes and for chemotaxis activity suggests that bacterial motility is product of the cAMP regulated global control system. (3) Expression of the *flhD* gene, which is part of the master operon, is influenced by the presence of heat-shock response proteins from *DnaK*, *DnaJ*, and *GrpE*. (4) The signaling proteins of CheA, CheY, and CheB also regulate a variety of responses to environmental stress. (5) A sudden increase of intracellular levels of $Ca^{2+}$ promotes CW rotation of the flagellum and interaction by CheA, CheW, and CheY. (6) Proton movement associated with protonmotive force increases CCW rotation of the flagellum and CW rotation almost disappears.

## 6.4. Quorum Sensing[31,32]

Bacterial communication between cells assumes many different avenues, and one of these unique systems is the response of cells to chemicals produced by organisms of the same species. The process of autoinduction, which has subsequently been referred to as quorum sensing, was introduced by K. Nealson and by J. Hastings as a result of their studies with bioluminescent bacteria. Basically, the bacterial cell produces a chemical at a concentration too low to stimulate a cell response; however, when cells reach a critical mass, or quorum, the concentration of chemical produced exceeds the threshold needed to induce a physiological change. A listing of several examples of quorum sensing is given in Table 6.14 and, at this time, the known signal compounds are either peptides, homoserine lactones, or γ-butyrolactones. As indicated in Table 6.14, the phenomenon of quorum sensing is practiced by various genera of bacteria and there are considerable differences in the system stimulated. To illustrate the basic principles of quorum sensing, it is useful to examine the bacterial systems responsible for bioluminescence and for pheromone activities.

Table 6.14. Examples of cell–cell communication in bacteria.[a]

| Signal molecules | System | Genus |
|---|---|---|
| Homoserine lactones | Bioluminescence | *Vibrio* |
| | Plant symbiosis or pathogenesis | *Rhizobium, Erwinia* |
| | Human pathogenesis | *Pseudomonas* |
| γ-Butyrolactone | Antibiotic production | *Streptomyces* |
| Peptides | Competence and virulence | *Streptococcus* |
| | Development of fruiting body | *Myxococcus* |
| | Conjugation and pathogenesis | *Enterococcus* |
| | Pathogenesis | *Staphylococcus* |
| | Bacteriocin production | *Lactococcus* |

[a]From Dunny and Winans, 1999.[32]

## 6.4.1. *Chemiluminescence and Homoserine Lactones*[33–36]

Examination of *Vibrio* (*Photobacterium*) *fischeri*, a marine symbiont, and *Vibrio harveyi*, a free-living bacterium, revealed that light production was attributed to luciferase action. The substrate for luciferase is reduced FMN, molecular oxygen, and a long-chain aliphatic aldehyde. The reaction resulting in light formation is given below:

$$FMNH_2 + O_2 + RCHO \xrightarrow{\text{Luciferase}} FMN + RCOOH + H_2O + \text{light}$$

NADH from the tricarboxylic acid cycle provides the reducing power to reduce FMN with the biological formation of $FMNH_2$. The long-chain aliphatic aldehyde used as the substrate may be hexadecanal (palmitaldehyde) or dodecanal (lauraldehyde) and

---

**Box 6.1**
**"Quorum Sensing": How Did the Process Get This Name?**

At a Thanksgiving dinner Stephen C. Winans challenged his guests to suggest a new name for the autoinducer-mediated process that was observed in bacterial cell–cell communication. "Pheromone" and "autoinducer" had been used; however, these terms implied that the process was restricted and highly academic. During the course of the dinner, the terms "communulin" and "gridlockin" were suggested; however, the term "quorum sensing," which was proposed by the nonscientist Rob Johnson, seemed suitable. As it turns out, "quorum sensing" is not only descriptive of the process but also has a general appeal to audiences. This illustrates that the appropriate naming of a process appears to be important in communicating between scientists and may even stimulate research in the area. (From Dunny and Winans, 1999.[32])

the product of the reaction is the corresponding organic acid. Light production is attributed to the flow of electrons through bacterial luciferase.

The genes for chemiluminescence in *Vibrio fischeri* are found clustered in a unit referred to as the *lux* operon. A model of this gene cluster consisting of about eight kilobase pairs is given in Figure 6.15. One transcriptional unit consists of *luxR* which produces the LuxR protein and a second unit consists of *luxI, A, B, C, D, E*, and *G*. These two units are transcribed in opposite directions with a binding site for cAMP receptor protein (*crp* box) and a "*lux* box" which serves as a binding site for LuxR located on the DNA segment connecting the two units. Near the *luxR* gene is the *crp* box and concomitant binding of cAMP receptor protein and cAMP is required for expression of *luxR*. The LuxR molecule contains 250 amino acids with two distinct domains. The transmembrane segment of the LuxR protein is associated with the N-terminus segment and contains the site for binding by the autoinducer. The C-terminal domain of LuxR is found in the cytoplasm and contains the H–T–H (helix–turn–helix) motif which interacts with DNA and functions as the transcription activator. The *luxI* gene encodes for the protein associated with the synthesis of the autoinducer. In *Vibrio harveyi*, the autoinducer is *N*-(3-hydroxybutonoyl)-L-homoserine lactone (3-hydroxy-C4-HSL) and it is found to stimulate expression of *lux* genes at nmolar levels. The genes of *luxA* and *luxB* encode for the production of the two subunits of luciferase while *luxC, luxD*, and *luxE* encode for enzymes involved with the synthesis of the aldehyde which is required as a substrate for the luciferase.

The bioluminescent activity in *Vibrio harveyi* is attributed to quorum sensing and the involvement of multiple signal receptors. The two-component signaling system regulating the luminescence operon is given in Figure 6.16. Based on analysis of mutants, it is considered that *Vibrio harveyi* has separate sensory systems for the two autoinducer molecules: HA1 and HA2. HA1 is *N*-(3-hydroxybutonoyl)-L-homoserine lactone (3-hydroxy-C4-HSL) and its synthesis is attributed to products of the *luxLM* genes. The LuxN protein is the sensory receptor for HA1 and this molecule has both histidine kinase and aspartic acid kinase activity. By the phospho-relay system, the LuxO protein is activated and through the H–T–H (helix–turn–helix) domain, interacts with DNA to induce expression of structural genes in the lux operon for the production of luciferase and the long-chain aldehyde used as a substrate for luciferase. A second autoinducer molecule is assumed to be produced; however, the exact structure and the genes for its production have not been determined. The HA2 would interact with the

Figure 6.15. Genetic organization of the lux operon in *Vibrio fischeri*. The gene *luxI* produces the autoinducer and *luxR* produces the protein that combines with the autoinducer to promote transcription of the *lux* operon. Also, regulation of this operon is by cAMP and CAP. (From Nealson, 1999.[35])

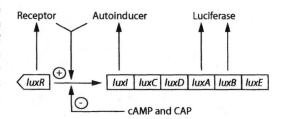

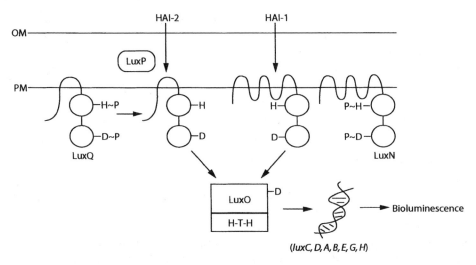

Figure 6.16. A model of the dual signaling system for bioluminescence in *Virbrio harveyi*. System 1 has 3-hydroxy-C4 HSL and enzymes for its production are encoded on *luxL* and *luxM* genes. The nature of the autoinducer in system 2 is unresolved. Without accumulation of HAI-1 or HAI-2, LuxQ and LuxN autophosphorylate at H (histidine) and transfer the phosphate to D (aspartate) of the respective proteins. Finally, the phosphate is transferred to LuxO. LuxO~P inhibits transcription of the *lux operon*. In the presence of HAI-1 or HAI-2, phosphorylation of Lux Q and Lux N is reduced with LuxO becoming nonphosphorylated. The nonphosphorylated LuxO interacts with DNA by the hinge–turn–hinge (H–T–H) motif to induce genes for luminescence. (From Bassler, 1999.[33])

LuxQ protein and in a manner similar to that of LuxN, the sensory LuxQ would activate LuxO. The necessity for this dual sensory system in *Vibrio harveyi* is un- known; however, the identification of HA2 may provide a needed clue for this activity.

While autoinducer molecules were initially considered to be found in only a limited physiological group of bacteria, that is not the case because many bacteria are now known to produce HSL and related molecules, the γ-butyrolactones, for signaling activities. Examples of some of these HSL structures and the bacteria that produce them are given in Figure 6.17. The γ-butyrolactones are molecules similar to HSLs in that it has autoinducer activity in streptomyces. These small autoinducer molecules are well suited for quorum sensing because they are neither intermediates nor are they products of standard cellular metabolism.

Streptomyces are filamentous soil bacteria that produce antibiotics and chains of spores at the tips of aerial mycelium. In the 1960s Khokhlov determined that the regulatory factor, termed A-factor, induced the production of aerial mycelium and the synthesis of streptomycin. In a general sense, the A-factor in streptomyces is similar to sex hormones produced by fungi and yeast. In *Streptomyces griseus* the A-factor is 2-iso-capryl-3R-hydroxymethyl-γ-butyrolactone and this structure is given in Figure 6.17. The A-factor receptor protein (ArpA) has high affinity binding for the A-factor

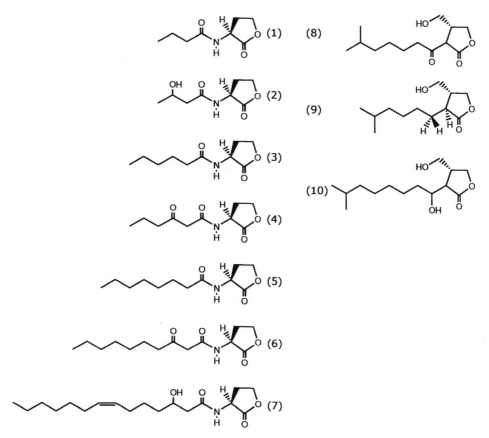

Figure 6.17. Examples of chemical signaling molecules produced by bacteria. Acyl homoserine lactone (HSL) molecules and the bacteria that produce them: (1) C4-HSL; *Serratia liquefaciens*, *Pseudomonas aeruginosa*, and *Aeromonas salmonicida*. (2) 3-Hydroxy-C4-HSL; *Vibrio harveyi*. (3) C6-HSL; *Serratia liquefaciens*, *Pseudomonas aureofaciens*, and *Chromobacterium violaceum*. (4) 3-Oxo-C6-HSL; *Vibrio fischeri*, *Erwinia stewarii*, *Enterobacter agglomerans*, and *Yersinia enterocolitica*. (5) C8-HSL; *Vibrio fischeri*. (6) 3-oxo-C8-HSL; *Agrobacterium tumefaciens*. (7) △7-3-Hydroxy-C14-HSL; *Rhizobium leguminosarum*. Several molecules having the γ-butyrolactone structure are produced by *Streptomyces griseus* (8); *Streptomyces virginiae* (9); and *Streptomyces cyaneofuscatus* (10). (From Horinouchi, 1999[34]; Fuqua and Eberhard, 1999.[37])

with a $K_D$ of 0.7 n*M* and this specificity is needed because other streptomyces produce additional types of γ-butyrolactone compounds for their specific activities. The A-factor receptor is activated with the binding of the A-factor and as a consequence of the H–T–H motif, it transmits a signal to DNA for the synthesis of streptomycin. Another protein stimulated by γ-butyrolactone is considered to serve as a signal for cell diffraction with activation of sporulation genes. While the divergence of *Bacillus subtilis* and *Streptomyces* occurred at least 700 million years ago, it may be that the signal systems used for sporulation in *Streptomyces* is as complex as that found in *Bacillus subtilis*.

## 6.4.2. *Pheromone and Competence Activities*[34,37–38]

Research with the streptococci has a notable position in science because it was with this organism that Griffith in 1928 first demonstrated transformation. It is now understood that the uptake of DNA by *Streptococcus* is in response to a signal initiated by an extracellular peptide or competence activator secreted by the streptococcal cell. Although not all details are available concerning the mechanisms involved, a model for signal transduction leading to cell competence is given in Figure 6.18. With Gram-

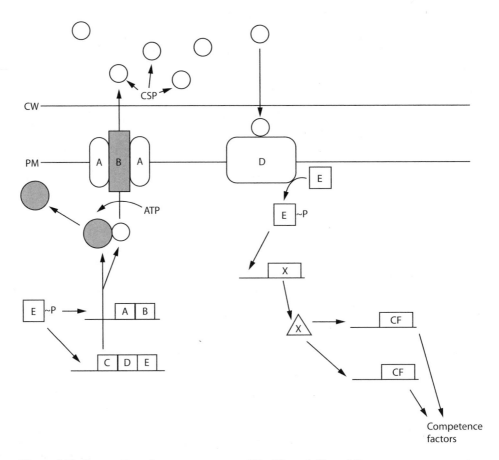

Figure 6.18. Competence pheromone operons of *Bacillus subtilis* and *Streptococcus pneumoniae*. The protein produced by the precursor gene is systematically degraded to produce the peptide pheromone. The receptor kinase has seven membrane-spanning loops and it requires the response regulator for action. (From Lazazzera et al., 1999[39]; Håvarstein and Morrison, 1999.[38])

Table 6.15. Examples of peptide signals in bacterial communication.[a]

| Genus | Signal peptide | Size of precursor signal peptide | System regulated by peptide |
|---|---|---|---|
| *Streptococcus pneumoniae* | L/F/WXXL/ FFKR[b] | 14–23 Amino acids | Competence |
| *Bacillus subtilis* | ERGMT | 40 Amino acids | Competence and sporulation |
| | ADPITRQWGD | 55 Amino acids | Transformation competence |
| | ARNQT | 44 Amino acids | Sporulation |
| *Enterococcus faecalis* | LFSLVLAG | 21–23 Amino acids | Conjugation |
| | LFSVVTLVG[c] | | Inhibitor of conjugation signal |

[a]From Håvarstein and Morrison, 1999[38]; Lazazzera et al., 1999[39]; Clewell, 1999.[42]
[b]Motif found in all signal peptides.
[c]Antisignal peptide.

positive bacteria, signals for communication between cells are frequently attributed to short peptides and several examples of these peptides with signal activity are given in Table 6.15. In *Streptococcus pneumoniae*, there are four sets of competence (*com*) genes: two gene clusters are well characterized; however, the other two are proposed to occur as a result of mutation selections.

One of the best characterized gene clusters is the *comCDE* operon of *Streptococcus pneumoniae*, shown in Figure 6.19. The precursor peptide is encoded by the *comC* gene and the uptake transported for the pheromone peptide is produced from *comD*. The ComD protein has two domains: a membrane-associated portion with seven trans-

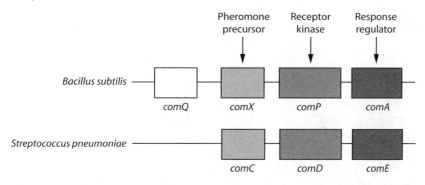

Figure 6.19. A model for signal transduction and information processing for development of competence in *Streptococcus pneumoniae*. All proteins are encoded on the *com* genes. A and P proteins are an ABC transporter with a proteolytic region. The *comC* gene produces the competence-stimulating peptide (CSP) precursor and CSP is released from the precursor by the protease in the ABC transporter. The CSP binds onto the CSP receptor (D) and this information is passed on to the transcription factor (E) which activates the *comAB* (AB) operon, *comCDE* (CDE) operon, and a putative *comX* (X) operon. The comX produces a protein that promotes the expression of competence factors from the various competence factor (CD) operons. (From Håvarstein and Morrison, 1999.[38])

membrane hydrophobic helices and the second portion of the molecule extends into the cytoplasm and it has a kinase region at the C-terminus. In addition to comC and comD on the operon, there is comE which encodes for a response regulator. The ComE detects intracellular concentrations of pheromone peptide and signals increased transcription of competence genes. The comAB operon encodes for the ABC transporter (ComA) which contains a proteolytic domain to release the pheromone peptide from the precursor peptide before the peptide is exported. ComB encodes for an accessory protein that assists the ComA in its organization and/or function. A *comX* operon is considered to produce a protein that interacts with another gene system to produce the proteins for development of competence.

The pheromone peptide is also known as the competence stimulating peptide (CSP) and it is synthesized as a precursor molecule from the *comC* gene. In *Streptococcus pneumoniae*, the comCDE operon is located between the genes for Arg-tRNA and Glu-tRNA with a very low level of CSP produced on a regular basis in the cells. At the N-terminus of the molecule is a leader sequence that may serve to prevent premature signaling by the CSP. The leader sequence is attached to the signal peptide through a G–G dimer, and this dimer is cleaved by a protease at the time of CSP export. This action of a protease is commonly employed in other bacterial systems and is not unique for rlease of pheromone peptide from the precursor. From the examination of several strains of streptococci it is found that the pheromone peptide contains only 14–23 amino acids and a positive charge occurs at the C-terminus with a negative charge at the N-terminus. Most peptides contain an amino acid sequence of –L/F/WXXL/FFXR– which provides the needed conformation for interaction with the signal receptor molecule. Specificity for the induction on transcription for competence proteins is the presence of a base sequence referred to as the "com box" which contains the following sequence: –TACGAATA–. This "com box" would also be important in activities of *Staphylococcus*, *Bacillus* and other Gram-positive bacteria.

Competence in *Bacillus subtilis* has been demonstrated to be attributed to the *comP* gene which produces a protein of 55 amino acids. In the export of this molecule, 45 amino acids at the N-terminus are removed and the pheromone peptide of 10 amino acids is released from the cell. This small peptide is the major competence factor for transformation of *Bacillus subtilis* and will accumulate to 50 n*M* in the culture fluid. The pheromone peptide is ADPITRQWGD with a farnesyl group added to the tryptophan (W) residue by posttranslation processing. Farnesyl modification is common in eukaryotic cells through cysteine but this modification of tryptophan is unique in biology and details remain to be established. With an increasing cell density, the concentration of this 10 amino peptide increases and competence is developed as a result of a signal system consisting of a membrane protein with histidine kinase activity and a response regulator with an aspartate residue available for phosphorylation.

## 6.4.3. Sporulation[39,40]

In addition to the 10 amino acid peptide functioning for competence, *Bacillus subtilis* produces a peptide that has been designated as the competence and sporulation factor (CSF). A model of activity of CSF is given in Figure 6.20. The CSF can be isolated

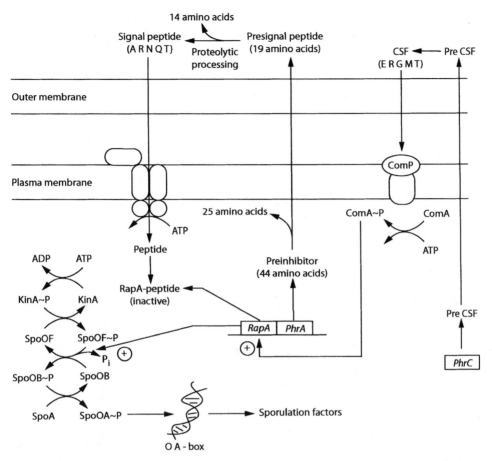

Figure 6.20. Pentapeptides as signals for sporulation in *Bacillus subtilis*. The competence and sporulation factors (CFS) is a pentapeptide produced from protein encoded on the *PhrC* gene. The CSF promotes ComA~P formation which in turn stimulates the expression of the *RapA* and *RhrA* gene. The PhrA protein is processed as it exits the cell and ultimately results in formation of a pentapeptide that serves as a signal peptide. On transport into the cell, the signal pentapeptide binds to the RapA and inactivates it. RapA protein activates the dephosphorylation of SpoOF~P but with inactivation of RapA, the cascade of phosphate transfer proceeds to SpoOA, with the induction of sporulation factors. (From Perego, 1999.[40])

from culture media and is a peptide of five amino acids with a sequence of ERGMT (Glu-Arg-Gly-Met-Thr). A precursor peptide of 40 amino acids is encoded on *phrC* and transcription is controlled by σ H-containing RNA polymerase. At the N-terminus of this peptide is a signal sequence involved in peptide export and coupled to the transport is proteolytic processing that releases the CSF from the cell. At extracellular concentrations of 1–5 n$M$, CSF signals production of SrfA (protein for surfactant production) but at concentrations greater than 20 n$M$, it inhibits production of SrfA and enhances sporulation.

CSF enters the cell through a specific peptide transporter and influences the concentration of ComA~P. High concentration of ComA~P stimulates the production of RapA which is a phosphatase that inactivates SpoOF~P and prevents sporulation events. However, the ComA~P also initiates the production of PhrA which is a peptide of 44 amino acids. After proteolytic digestion in the cytoplasmic, 25 amino acids are removed and the 14 amino acids are reacted with extracellular proteases to yield the signal peptide (ARNQT). The peptide enters the cell by the ABC transport system and binds to RapA, the phosphatase. This enables the SpoOA~P to accumulate and promote expression of sporulation factors by binding to "OA-box" regions on the DNA. These "OA-boxes" are found in the promoter regions of relavent genes and contain the sequence of 5'NGNCGAA3' where N is any base but it usually is A or T. It is not understood why peptide export followed by peptide import is required in this signaling process.

## 6.5. Diversity in Cell Signaling[41]

Previously in this chapter we have examined communication between morphologically identical cells or internal signaling of cells to achieve optimum physiological expression. Other forms of bacterial signaling occurs with cells that display developmental stages where cells may assume different forms. Some attention has been given to cell signaling in bacteria that have holdfasts and in bacteria that produce fruiting bodies. Although it has not yet been demonstrated, the heterocyst and akinete formation by cyanobacteria may be expected to be regulated by cell signaling. *Anabaena flos-aquae* produces heterocysts for fixation of nitrogen and akinetes for resistance. These distinct cells produced by *Anabaena* are given in Figure 6.21. A system of complex signaling occurs with the establishment of nodules in legumes by nitrogen-fixing bacteria and it may be expected that molecular signaling occurs with the parasitic *Bdellovibrio*. In addition to the establishment of signal processes for the various developmental features of bacteria, the emerging field of cellular microbiology evaluates the interactions between bacterial pathogens and animal host cells. This section examines the signal systems in stalked bacteria, fruiting-body bacteria, and pathogens to illustrate some of the signal processes used by bacteria.

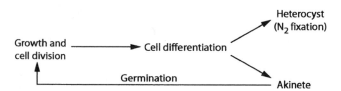

Figure 6.21. The developmental stages of *Anabaena* sp. cells resulting in production of akinetes and heterocysts.

## 6.5.1. Conjugation Signal[42]

*Enterococcus faecalis* is an opportunistic pathogen that contains numerous plasmids for virulence enhancement, drug resistance, and conjugative activities. This Gram-positive bacterium is involved in various processes of genetic transfer and one of the conjugative systems requires a peptide that serves as a signal between the two partners. The producer cell does not contain a conjugative plasmid but releases a pheromone peptide that communicates with a responding cell of the same or related species. The sex peptide is encoded on the chromosome of the producer cell and it stimulates the cell which contains a specific conjugative plasmid. The best conjugative plasmids studies are pADI, pPD1, pCF10, pAM373, and OB1 and there is a specific pheromone peptide for each plasmid system. In general, the sex peptide consists of seven or eight amino acids which is released from a 21- to 23-amino precursor peptide at the time of export. The enterococci have great signaling activities for pheromone peptides because levels of $5 \times 10^{-11}$ $M$ are sufficient to initiate a cell response. The cell stimulated by the pheromone peptide produces a lipoteichoic acid or related cellular component that promotes aggregation between the donor and recipient cell. In addition, the cell stimulated by the pheromone produces an anti-pheromone peptide that is encoded on the plasmid and released to cells in the immediate vicinity including the pheromone-producing cell. Presumedly, the function of the anti-pheromone peptide is to prevent self-induction and ensure the directional flow of DNA in the conjugative activity. Considerable similarity in peptide size occurs between the pheromone peptide and the anti-pheromone peptide. For example, the pheromone produced to stimulate cells with the pAD1 plasmid is LFSLVLAG and the antipeptide produced is LFVVTLVG. This system of signaling conjugation by sex peptides is not restricted to *Enterococcus* but is also found in *Staphylococcus* and *Streptococcus*.

## 6.5.2. Signal Transduction in Caulobacter[43,44]

*Caulobacter crescentus* is a highly unusual prokaryote in that the motile cell will become an attached cell following the release of the polar flagellum and formation of a stalk. To understand the signal transduction activity responsible for this developmental activity in *Caulobacter crescentus*, it is useful to examine the cell cycle as given in Figure 6.22. The stalked cell becomes attached to the substratum by a holdfast and on division produces a flagellated cell which is referred to as the swarmer cell. In addition to a single flagellum at the end of the cell distal to cell division, the swarmer cell also has pili at the same pole as the flagellum and is unable to divide. The swarmer cell becomes a stalked cell; however, prior to the formation of the stalk, the flagellum with all chemotaxis proteins and pili are released from the cell. The stalk is an extension of the cell in that it contains plasma membrane, peptidoglycan, and outer membrane. The stages of the cell cycle have been divided into three segments: (1) the S phase represents chromosome replication and biogenesis of the stalk, (2) the $G_2$ phase is the time of cell division, and (3) the $G_1$ phase includes cell motility and transition of flagellated cell to stalked cell.

The control of flagellar development is the developmental stage of the cell cycle of

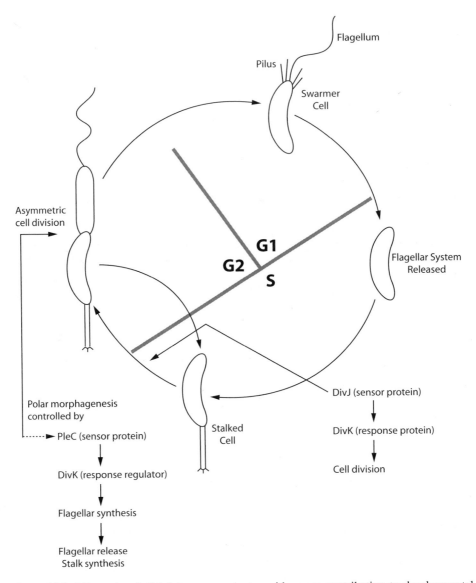

Figure 6.22. Life cycle of *Calulobacter crescentus* with genes contributing to developmental changes. (From Jenal, 2000,[43] Lane et al., 1995,[44] and Plamann and Kaplan, 1999.[45])

*Caulobacter crescentus* that has attracted the greatest amount of research. It is known that formation of the flagellum is attributed to four different classes of genes and these are summarized in Table 6.16. There is a cascade of activities that follows the sequence where a product of class I genes regulates expression of genes in class II, a product of class II regulates expression of class III genes, and a class III product

Table 6.16. Genes involved in development of flagella in *Caulobacter crescentus*.

| Gene class | Activity attributed to genes |
| --- | --- |
| I | Contains regulatory genes including *ctrA*, a proposed global regulator, which controls transcription of class II genes. Expression of class I genes is linked to the cell cycle. |
| II | Encodes synthesis of M and S ring proteins and produces proteins for export of flagella. Expression of class II genes are required for transcription of class III and IV genes. |
| III | Encodes for synthesis of rods, outer rings, hook, and hook assembly for flagella. Accounts for attachment of flagella. Transcription of class III genes by unique sigma factor ($\sigma^{54}$). |
| IV | Encodes for flagellar filament biosynthesis and uses $\sigma^{54}$. |

regulates genes in class IV. Encoded in the genes of class I is CtrA that has a C-terminal region that binds to DNA to initiate transcription of genes in class II. Thus, an important regulator in polar morphogenesis of this organism is the CtrA protein which is part of the signal sensing system. The molecule stimulating flagellar formation is not established but the initial sensor protein is DivJ which has kinase activity and becomes phosphorylated. A second protein in this phospho-relay system is DivK and ultimately CtrA serves as the response regulator. It remains to be established if a multiple component for signal sensing is involved in flagellar ejection and pili formation in *Caulobacter crescentus*.

## 6.5.3. *Signaling in Fruiting Body Bacteria*[45–48]

The fruiting body bacteria consists of organisms that produce aerial spores under conditions of starvation and desiccation. An example of sporulation structures of fruiting body bacteria is given in Figure 6.23. One of the model organisms for examination of the developmental activities of these spore producing bacteria is *Myxococcus xanthus* and the sporulation cycle is given in Figure 6.24. This Gram-negative bacterium grows by binary fission and individual cells move by gliding. Under starvation, about $10^5$ cells collect and form a fruiting structure which consists of about $5 \times 10^4$ cells. Considerable interest has been given to the social behavior that results in the multicellular development of *Myxococcus xanthus*. The movement of cells to form the fruiting structure is attributed to social gliding which is regulated by the *frz* genes. Although there is little known about the *frz* system, it is known that there are at least seven genes (*frzA, B, CD, E, F, G,* and *Z*) making up the *frz* operon. Several of these *frz* genes have considerable sequence similarity to the chemotaxis (*che*) genes of enteric bacteria.

The social movement of *Myxococcus xanthus* is to provide an environment for cooperative uptake of nutrients. Because the individual bacterial cells secrete enzymes for digestion of complex polymers, the concentration of these enzymes, and also the resulting soluble nutrients, is greatest where cells aggregate. To produce a fruiting

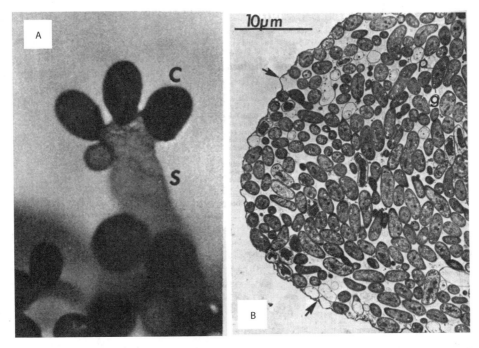

Figure 6.23. Fruiting body of *Stigmatella aurantiaca*. (**A**) Aerial structure showing the stalk (S) and cysts (C), × 310. (**B**) Section through a cyst showing the cyst wall (*arrows*), electron-dense granules (p), and electron-transparent inclusions (g). The cyst was stained with osmium tetroxide. × 3,500. (From Voelz and Reichenbach.[48] Used with permission.)

structure, cells respond to cell-density signaling which is initiated by nutrient depletion resulting from dense growth. Under starvation stress, guanosine tetra- and penta-phosphate [(p)ppGpp] accumulates and is considered to be a stimulus for signal trans-duction using a phosphorelay pathway similar to the two-component system discussed earlier. A model of the cell-density sensing of *Myxococcus xanthus* is given in Figure 6.25. The AsgA is a histidine protein kinase that autophosphorylates and interacts with a series of proteins. Finally, AsgB is phosphorylated to produce AsgB~phosphate which interacts with DNA and results in the production of $\sigma^{54}$. This sigma factor in *Myxococcus xanthus* is key for the production of numerous proteases that release amino acids and short peptides from complex proteins. In the presence of high cell density, some of the proteins shed from the vegetative cells as they differentiate to produce spherical spores that may serve as substrates for the proteases. The stimulus for the expression of differentiation genes is referred to as the "A-signal amino acids." The extracellular concentration of A-signal amino acids reaches a maximum after 2 hr of development and will produce a concentration of 10 $\mu M$ in the periplasmic region of the cells. The A-signal amino acids stimulate the two-component signal transducing system, which is referred to as the A-signaling system and consists of

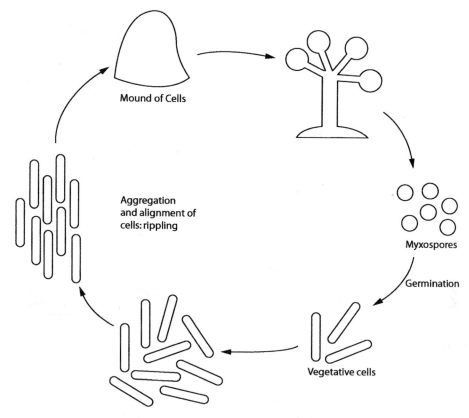

Figure 6.24. Development of fruiting body with spores in *Myxococcus xanthus*. (From Shi and Zusman, 1995[46]; Shimkets and Kaiser, 1999.[47])

SasS, SasN, and SasR proteins. The SasS sensory proteins are situated in the plasma membrane, and after autophosphorylation they relay the phosphate group to produce phosphorylated SasR which promotes elements essential for the formation of fruiting structure and spore development.

In *Myxococcus xanthus*, there is the stimulation of a unique cell contact-dependent activity referred to as the C-signaling system. The movement of the bacterial cells along a solid surface follows three morphological stages: rippling, aggregation, and sporulation. The gliding cells move in rhythmic waves with a wavelength of 45 to 70 µm and a period of about 20 min. The formation of waves is in response to a tactile signaling system that senses cell alignment, regulation of cell directional movement, and expressions of proteins encoded on *csgA*, the C-signaling gene. Aggregation reflects the formation of mounds of bacteria consisting of tens of thousands of cells and sporulation is the development of spores from cells in the fruiting structure. The product of *csgA* is a 24.6-kDa protein that has some structural similarity to the short-chain alcohol dehydrogenases including possible binding of NAD(P). Because

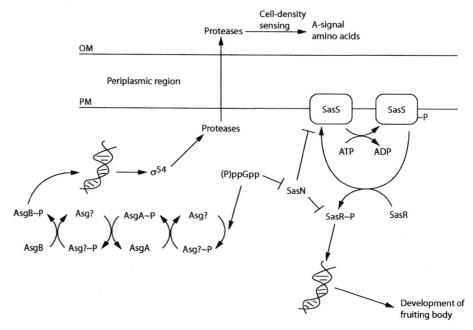

Figure 6.25. Role of A-signal in development of fruiting body in *Myxococcus xanthus*. A starvation signal (pppGpp) initiates a phosphorelay involving Asg proteins. AspB~P interacts with DNA to induce the production of a transcriptional factor that results in production of secreted proteases. The A-signal, consisting of amino acids, is bound by SasS and becomes phosphorylated. The SasS~P relays the phospho moiety to SasR and the resulting complex induces the production of fruiting body proteins. Under vegetative growth conditions, the SasN blocks the development of fruiting bodies but with increased concentration of A-signal, the threshold for inhibition by SasN is exceeded. (From Plamann and Kaplan, 1999.[45])

antibodies against CsgA prevent rippling and other developmental stages, it is considered that CsgA is an extracellular signal protein. Future studies will clarify the role of CsgA in the various developmental features of *Myxococcus xanthus*.

## 6.5.4. Actin Rearrangements in Host Cells[45,49]

There is a complex communication between pathogenic bacteria and the host animal cell that involves the internalization of the pathogen. Depending on the bacteria, phagocytosis may be induced by either extracellular or intracellular stimulation of the actin cytoskelton. There are two basic mechanisms for internalization of bacterial cells by eukaryotic cells and these are the "zipper" and "trigger" mechanisms (see Figure 6.26). *Listeria monocytogenes* is the pathogen often used as a model for zipper-based phagocytosis. On the surface of *Listeria monocytogenes* are two distinct adhesions: Internalin A, an 88-kDa protein and Internalin B, a 65-kDa protein. Following the binding of these internalins to the cytoskeleton at the surface of the host

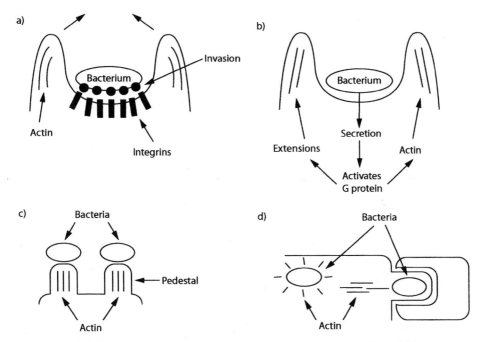

Figure 6.26. Signals from bacteria influence localization of bacteria inside or on the surface of host cells. (**a**) Species of *Listeria* and *Yersinia* have high affinities between invasins on the bacterial surface and integrins on the host cell surface which results in driving internalization of the bacterial cell. This invasin is attributed to reorganization of actin and it is termed a "zippering" process. (**b**) *Salmonella* and *Shigella* are internalized by a "trigger" process in which the bacteria inject chemicals into the host cell and there is a global reorganization of the actin filaments involving G-proteins. Chemicals are injected into the host cell from the bacterium by a type III secretion mechanism. (**c**) *Escherichia coli* forms colonies on the surface of intestinal epithelial cells. The intimin, a bacterial surface protein, and bundle-forming pili produce a firm binding to the surface of the epithelial cell. *Escherichia coli* injects by type III secretion into the host cell chemicals to retract the microvilli and promote actin filaments under the bacteria to form a pedestal. (**d**) Intracellular spreading of *Listeria*, *Shigella*, and *Rickettsia* is attributed to actin induced "tails" that propel bacteria into adjacent cells. (From McCallum and Theriot, 2000.[49])

cell, the pathogenic bacteria activate tyrosine kinases and lipid kinases in the host eukaryotic cell. These kinases are responsible for the polymerization and rearrangement of actins for incorporating the parasite in an intracellular compartment that does not fuse with the lysosomes (see Figure 6.26). Not all bacteria that bind onto the surface of the host cell are engulfed by the host cell. The enteropathogenic *Escherichia coli* (EPEC) establishes microcolonies on the surface of the intestinal brush border cells in actin-containing segments referred to as "pedestals." Phagocytosis is prevented as a result of EPEC injecting factors by type III secretion into the host cell to promote actin-stabilized pedestals. Another system for preventing phagocytosis is displayed by *Yersinia* which injects a series of factors into macrophages by a type III system. The Yops (*Yersinia* outer membrane proteins) are also soluble proteins that are encoded on a 70-kb plasmid. YopA, YopD, and YopN are proteins used in delivery of YopH, YopM, and YopE into host cells. YopH is a 55-kDa protein with strong protein tyrosine phosphatase activity. The dephosphorylation of protein tyrosine phosphates is considered to disrupt the actin-rich structures and account for release of bacterial cells from the phagocyte.

A second mechanism of phagocytosis involves a trigger model that has been called "macropinocytic-like" mechanism. This is the avenue of phagocytosis for *Shigella* and *Salmonella* species. In *Shigella*, the proteins for trigger-mediated uptake is carried on a 220-kb plasmid. The type III secretory system is encoded by the *Mxi-spa* genes while the IpaB, IpaC, and IpaD are proteins released from the *Shigella* into the phagocyte. The result is the activation of members of GTPases that serve as signals in eukaryotic cells and promote formation of actin fibers. For both *Salmonella* and *Shigella*, the uptake of microorganisms is an energy-dependent process that involves actin molecules inside the host cells.

The spread of bacteria from cell to cell by a direct route without exiting the cell is an important advantage for the pathogen because it can avoid interaction with antibodies and nonspecific defense chemicals outside of the host cells. Although *Shigella* and *Rickettsia* can pass directly from cell to cell, the model system for this capability is *Listeria*. On entry into the cell by a zipper mechanism, *Listeria* releases listeriolysis O which attacks the internal vacuole housing the *Listeria* and releases the bacteria into the cytoplasm. Short actin filaments collect around the *Listeria* cells producing a tail-like structure of several microns in length. Critical for intracellular movement is the rearrangement and cross linking of actin molecules at one pole of the bacterium to produce bacterial polarity. The bacterial cell is moved through the cytoplasm by contractile activities with energy supplied by the host cell. The result is the propulsion of the bacterial cell at a rate of 6 to 90 μm/min which is considerably less than observed with other motile bacteria (see Table 6.3). Entry into adjacent cells occurs via protrusion structures that extend from the infected cell into adjacent healthy cells. This mechanism of actin-dependent movement of bacteria is necessary because bacteria cannot use flagella to propel them through the highly viscous cytoplasm.

# References

1. Ben-Jacob, E., I. Cohen, and D.L. Gutnick. 1998. Cooperative organization of bacterial colonies: From genotype to morphotype. *Annual Review of Microbiology* 52:779–806.
2. Henrichsen, J. 1972. Bacterial surface translocation: A survey and a classification. *Bacteriology Review* 36:478–503.
3. McBride, M.J. 2000. Bacterial gliding motility: Mechanisms and mysteries. *ASM News* 66:203–210.
4. Spormann, A.M. 1999. Gliding motility in bacteria: Insights from studies of *Myxococcus xanthus*. *Microbiology and Molecular Biology Review* 63:621–641.
5. Manson, M.D. 1992. Bacterial motility and chemotaxis. *Advances in Microbial Physiology* 33:277–346.

6. Aizawa, S-I. 2001. Bacterial flagella and type III secretion systems. *FEMS Microbiology Letters* 202:157–164.

7. Chilcott, G.S. and K.T. Hughes. 2000. Coupling of flagellar gene expression to flagellar assembly in *Salmonella enterica* serovar typhimurium and *Escherichia coli*. *Microbiology and Molecular Biology Reviews* 64:694–708.

8. Koshland, D.E., Jr. 1980. *Bacterial Chemotaxis as a Model Behavioral System*. Raven Press, New York.

9. Macnab, R.M. 1987. Motility and chemotaxis. In Escherichia coli *and* Salmonella typhimurium (J.L. Ingraham, K.B. Low, B. Magasanik, M. Schaechter, and E. Umbarger, eds.). American Society for Microbiology Press, Washington, D.C., pp. 732–759.

10. Macnab, R.M. 1995. Flagellar switch. In *Two-Component Signal Transduction* (J.A. Hoch and T.J. Silhavy, eds.). American Society for Microbiology Press, Washington, D.C., pp. 181–202.

11. Berry, R.M. and J.P. Armitage. 1999. The bacterial flagella motor. *Advances in Microbial Physiology* 41:292–338.

12. Armitage, J.P. 1999. Bacterial tactic response. *Advances in Microbial Physiology* 41:229–291.

13. Canale-Parola, E. 1978. Motility and chemotaxis of spirochetes. *Annual Review of Microbiology* 32:69–99.

14. Holt, S.C. 1978. Anatomy and chemistry of spirochetes. *Microbiology Reviews* 42:114–160.

15. Thomas, N.A., S.L. Bardy, and K.F. Jarrell. 2001. The archaeal flagellum: A different kind of prokaryotic motility structure. *FEMS Microbiology Reviews* 25:147–174.

16. Parkinson, J.S. 1993. Signal transduction schemes of bacteria. *Cell* 73:857–871.

17. Falke, J.J., R.B. Bass, S.L. Butler, S.A. Chervitz, and M.A. Danielson. 1997. The two-component signaling pathway of bacterial chemotaxis: A molecular view of signal transduction by receptors, kinases, and adaptation enzymes. *Annual Review of Cellular and Developmental Biology* 13:457–512.

18. Hoch, J.A. and T.J. Silhavy (eds.). 1995. *Two-Component Signal Transduction*. American Society for Microbiology Press, Washington, D.C.

19. Parkinson, J.S. and E.C. Kofoid. 1992. Communication modules in bacterial signaling proteins. *Annual Review of Genetics* 26:71–112.

20. Stock, J.B., A.J. Ninfa, and A.M. Stock. 1989. Protein phosphorylation and regulation of adaptive response in bacteria. *Microbiology Reviews*. 53:450–490.

21. Stock, J.B. M.G. Surette, M. Levit, and P. Park. 1995. Two-component signal transduction systems: Structure-function relationships and mechanisms of catalysis. In *Two-Component Signal Transduction* (J.A. Hoch and T.J. Silhavy, eds.). American Society for Microbiology Press, Washington, D.C., pp. 25–51.

22. Pratt, L.A. and T.J. Silhavy. 1995. Porin regulon of *Escherichia coli*. In *Two-Component Signal Transduction* (J.A. Hoch and T.J. Silhavy, eds.). American Society for Microbiology Press, Washington, D.C., pp. 105–127.

23. DiRita, V.J. 1995. Three-component regulatory system controlling virulence in *Vibrio cholera*. In *Two-Component Signal Transduction* (J.A. Hoch and T.J. Silhavy, eds.). American Society for Microbiology Press, Washington, D.C., pp. 351–365.

24. Dziejman, M. and J.J. Mekalanos. 1995. Two component signal transduction and its role in the expression of bacterial virulence factors. In *Two-Component Signal Transduction* (J.A. Hoch and T.J. Silhavy, eds.). American Society for Microbiology Press, Washington, D.C., pp. 305–317.

25. Grosman, E.A. and F. Heffron. 1995. Regulation of *Salmonella* virulence by two-component regulatory system. In *Two-Component Signal Transduction* (J.A. Hoch and

T.J. Silhavy, eds.). American Society for Microbiology Press, Washington, D.C., pp. 319–332.

26. Uhl, M.A. and J.F. Miller. 1995. *Bordatella pertussis* BvgAS virulence control system. In *Two-Component Signal Transduction* (J.A. Hoch and T.J. Silhavy, eds.). American Society for Microbiology Press, Washington, D.C., pp. 333–350.

27. Kadner, R.J. 1995. Expression of the Uhp sugar-phosphate transport system of *Escherichia coli*. In *Two-Component Signal Transduction* (J.A. Hoch and T.J. Silhavy, eds.). American Society for Microbiology Press, Washington, D.C., pp. 263–274.

28. Msadek, T., F. Kunst, and G. Rapoport. 1995. A signal transduction network in *Bacillus subtilis* includes the DegS/DegU and ComP/ComA two-component systems. In *Two-Component Signal Transduction* (J.A. Hoch and T.J. Silhavy, eds.). American Society for Microbiology Press, Washington, D.C., pp. 447–471.

29. Armitage, J.P. 1999. Bacterial tactic responses. *Advances in Microbial Physiology* 41:231–291.

30. Berry, R.M. and J.P. Armitage. 1999. The bacterial flagella motor. *Advances in Microbial Physiology* 41:292–338.

31. Dworkin, M. (ed.). 1991. *Microbial Cell–Cell Interactions*. American Society for Microbiology Press, Washington, D.C.

32. Dunny, G.M. and S.C. Winans. 1999. Bacterial life: Neither lonely nor boring. In *Cell–Cell Signaling in Bacteria* (G.M. Dunny and S.C. Winans, eds.). American Society for Microbiology Press, Washington, D.C., pp. 1–8.

33. Bassler, B.L. 1999. A multichannel two-component signaling relay controls quorum sensing in *Vibrio harveyi*. In *Cell–Cell Signaling in Bacteria* (G.M. Dunny and S.C. Winans, eds.). American Society for Microbiology Press, Washington, D.C., pp. 259–273.

34. Horinouchi, S. 1999. γ-Butyrolactones that control secondary metabolism and cell differentiation in *Streptomyces*. In *Cell–Cell Signaling in Bacteria* (G.M. Dunny and S.C. Winans, eds.). American Society for Microbiology Press, Washington, D.C., pp. 193–210.

35. Nealson, K.H. 1999. Early observations defining quorum-dependent gene expression. In *Cell–Cell Signaling in Bacteria* (G.M. Dunny and S.C. Winans, eds.). American Society for Microbiology Press, Washington, D.C., pp. 277–290.

36. Stevens, A.M. and E.P. Greenberg. 1999. Transcriptional activation by LuxR. In *Cell–Cell Signaling in Bacteria* (G.M. Dunny and S.C. Winans, eds.). American Society for Microbiology Press, Washington, D.C., pp. 231–242.

37. Fuqua, C. and A. Eberhard. 1999. Signal generation in autoinduction systems: Synthesis of acylated homoserine lactones by *LuxI-type* proteins. In *Cell–Cell Signaling in Bacteria* (G.M. Dunny and S.C. Winans, eds.). American Society for Microbiology Press, Washington, D.C., pp. 211–230.

38. Håvarstein, L.S. and D.A. Morrison. 1999. Quorum sensing and peptide pheromones in streptococcal competence for genetic transformation. In *Cell–Cell Signaling in Bacteria* (G.M. Dunny and S.C. Winans, eds.). American Society for Microbiology Press, Washington, D.C., pp. 9–26.

39. Lazazzera, B.A., T. Palymer, J. Quisel, and A.D. Grossman. 1999. Cell-density control of gene expression and development in *Bacillus subtilis*. In *Cell–Cell Signaling in Bacteria* (G.M. Dunny and S.C. Winans, eds.). American Society for Microbiology Press, Washington, D.C., pp. 27–46.

40. Perego, M. 1999. Self-signaling by Phr peptides modulates *Bacillus subtilis* development. In *Cell–Cell Signaling in Bacteria* (G.M. Dunny and S.C. Winans, eds.). American Society for Microbiology Press, Washington, D.C., pp. 243–258.

41. Cossart, P., P. Boquet, S. Normark, and R. Rappuoli (eds.). 2000. *Cellular Microbiology*. American Society for Microbiology Press, Washington, D.C.

42. Clewell, D.B. 1999. Sex pheromone systems in enterococci. In *Cell–Cell Signaling in Bacteria* (G.M. Dunny and S.C. Winans, eds.). American Society for Microbiology Press, Washington, D.C., pp. 47–66.

43. Jenal, U. 2000. Signal transduction mechanisms in *Caulobacter crescentus* development and cell cycle control. *FEMS Microbiology Reviews* 24:177–191.

44. Lane, T., A. Benson, G.B. Hecht, G.J. Burton, and A. Newton. 1995. Switches and signal transduction networks in the *Caulobacter crescentus* cell cycle. In *Two-Component Signal Transduction* (J.A. Hoch and T.J. Silhavey, eds.). American Society for Microbiology Press, Washington, D.C., pp. 403–417.

45. Plamann, L. and H.B. Kaplan. 1999. Cell-density sensing during early development in *Myxococcus xanthus*. In *Cell–Cell Signaling in Bacteria* (G.M. Dunny and S.C. Winans, eds.). American Society for Microbiology Press, Washington, D.C., pp. 67–82.

46. Shi, W. and D.R. Zusman 1995. The *frz* transduction system controls multicellular behavior in *Myxococcus xanthus*. In *Two-Component Signal Transduction* (J.A. Hoch and T.J. Silhavy, eds.). American Society for Microbiology Press, Washington, D.C., pp. 419–430.

47. Shimkets, L.J. and D. Kaiser. 1999. Cell contact-dependent C signaling in *Myxococcus xanthus*. In *Cell–Cell Signaling in Bacteria* (G.M. Dunny and S.C. Winans, eds.). American Society for Microbiology Press, Washington, D.C., pp. 83–97.

48. Voelz, H. and H. Reichenbach. 1969. Fine structure of fruiting bodies of *Stigmatella aurantiaca*. *Journal of Bacteriology* 99:856–866.

49. McCallum, S.J. and J.A. Theriot. 2000. Bacterial manipulation of the host cell cytoskeleton. In *Cellular Microbiology* (P. Cossart, P. Bosquet, S. Normark, and R. Rappuoli, eds.). American Society for Microbiology Press, Washington, D.C., pp. 171–192.

# Additional Reading

## *Chemotaxis*

Blair, D.F. 1995. How bacteria sense and swim. *Annual Review of Microbiology* 49:487–522.

Ordal, G.W., L. Màrquez-Magaña, and M.J. Chamberlin. 1993. Motility and chemotaxis. In *Bacillus subtilis and Other Gram-Positive Bacteria* (A.L. Sonenshein, J.A. Hoch, and R. Losick, eds.). American Society for Microbiology Press, Washington, D.C., pp. 765–784.

## *Communication, Cell Signals, and Differentiation*

Bassler, B.L. and M.R. Silverman. 1995. Intercellular communication in marine *Vibrio* species: density-dependent regulation of bioluminescence. In *Two-Component Signal Transduction* (J.A. Hoch and T.J. Silhavey, eds.). American Society for Microbiology Press, Washington, D.C., pp. 431–445.

Bourret, R.B., K.A. Borkovich, and M.I. Simon. 1991. Signal transduction pathways involving protein phosphorylation in prokaryotes. *Annual Review of Biochemistry* 60:401–441.

Dworkin, M. 1996. Recent advances in the social and developmental biology of myxobacteria. *Microbiology Reviews* 60:70–102.

Groisman, E.A. and F. Heffron. 1995. Regulation of *Salmonella* virulence by two-component regulatory system. In *Two-Component Signal Transduction* (J.A. Hoch and T.J. Silhavey, eds.). American Society for Microbiology Press, Washington, D.C., pp. 319–333.

Huang, T.-C. and N. Grobbelaar. 1995. The circadian clock in the prokaryote *Synechococcus* RF-1. *Microbiology* 141:535–540.

Losick, R. and D. Kaiser. 1997. Why and how bacteria communicate. *Scientific American* XX: 68–73.

Parkinson, J.S. and E.C. Kofoid. 1992. Communication modules in bacterial signaling proteins. *Annual Review of Genetics* 26:71–112.

Saier, M.H., Jr. and G.R. Jacobson. 1984. The molecular basis of sex and differentiation. Springer-Verlag, New York.

Uhl, M.A. and J.F. Miller. 1995. *Bordatella pertussis* BvgAS virulence control system. In *Two-component Signal Transduction* (J.A. Hoch and T.J. Silhavey, eds.). American Society for Microbiology Press, Washington, D.C., pp. 431–445.

Viollier, P.H., K.T. Nguyen, W. Minas, M. Folcher, G.E. Dale, and C.J. Thompson. 2001. Roles of aconitase in growth, metabolism, and morphological differentiation of *Streptomyces coelicolor*. *Journal of Bacteriology* 183:3193–3203.

Withers, H., S. Swift, and P. Williams. 2001. Quorum sensing as an integral component of gene regulatory networks in Gram-negative bacteria. *Current Opinion in Microbiology* 4: 186–193.

## Flagellar Structure

Chunhao, L., L. Corum, D. Morgan, E.L. Rosey, T.B. Stanton, and N.W. Charon. 2000. The spirochete FlaA periplasmic flagella sheath protein imparts flagellar helicity. *Journal of Bacteriology* 182:6698–6706.

Kim, K.K. and L.L. McCarter. 2000. Analysis of the polar flagellar gene system of *Vibrio parahaemophilus*. *Journal of Bacteriology* 182:3695–3704.

Tokuo, S., T. Ueno, T. Kubori, S. Yamaguchi, T. Iino, and S. Aizawa. 1998. Flagellar filament elongation can be impaired by mutations in the hook protein FlgE of *Salmonella typhimurium*: A possible role of the hook as a passage for the anti-sigma factor FlgM. *Molecular Microbiology* 27:1129–1139.

## Motility

Berg, H.C. 1993. *Random Walks in Biology*, 2nd edit. Princeton University Press, Princeton, NJ.

Deziel, E., Y. Comeaau, and R. Villemur. 2001. Initiation of biofilm formation by *Pseudomonas aeruginosa* 57RP correlates with emergence of hyperpiliated and highly adherent phenotype variants defiient in swimming, swarming, and twitching motilities. *Journal of Bacteriology* 183:1195–1204.

Dibb-full, M.P., E. Allen-Vercoe, C.J. Thorns, and M.J. Woodward. 1999. Fimbriae and flagella mediated association with and invasion of cultured epithelial cells by *Salmonella enteritidis*. *Microbiology* 145:1023–1031.

Gosink, K.K. and C.C. Häse. 2000. Requirements for conversion of the $Na^+$-driven flagella motor of *Vibrio cholera* to the $H^+$-driven motor of *Escherichia coli*. *Journal of Bacteriology* 182:4234–4240.

Hoiczyk, E. and W. Baumeister. 1995. Envelope structure of four gliding filamentous cyano-bacteria. *Journal of Bacteriology* 177:2378–2395.

Kohler, T., L.K. Curty, F. Barja, C. Vandelden, and J.C. Pechere. 2000. Swarming of *Pseudomonas aeruginosa* is dependent on cell-to-cell signaling and requires flagella and pili. *Journal of Bacteriology* 182:5990–5996.

Kojima, S., K. Yamamoto, I. Kawagishi, and M. Homma. 1999. The polar flagellar motor of *Vibrio cholera* is driven by an $Na^+$ motive force. *Journal of Bacteriology* 181:1927–1930.

Leeuwenhoek, A.V. 1684. Microscopic observations about animals in the scruf of the teeth. *Philosophical Transactions of the Royal Society of London* 14:568–574.

Lünsdorf, H. and H.U. Schairer. 2001. Frozen motion of gliding bacteria outlines inherent features of the motility apparatus. *Microbiology* 147:939–947.

Martinez, A., S. Torello, and R. Kolter. 1999. Sliding motility in mycobacteria. *Journal of Bacteriology* 181:7331–7338.

Rashid, M.H., N.N. Rao, and A. Kornberg. 2000. Inorganic polyphosphate is required for motility of bacterial pathogens. *Journal of Bacteriology* 182:225–227.

Roebler, M., G. Wanner, and V. Muller. 2000. Motility and flagellum synthesis in *Halobacillus halophilus* are chloride dependent. *Journal of Bacteriology* 182:532–535.

Semmler, A.B.T., C.B. Whitchurch, A.J. Leech, and J.S. Mattick. 2000. Identification of a novel gene, *fimV*, involved in twitching motility in *Pseudomonas aeruginosa*. *Microbiology* 146:1321–1332.

Yang, Z., X. Ma, L. Tong, H.B. Kaplan, L.J. Shimkets, and W. Shi. 2000. *Myxococcus xanthus dif* genes are required for biogenesis of the cell surface fibrils essential for social gliding motility. *Journal of Bacteriology* 182:5793–5798.

# 7
# Cellular Growth and Reproduction

[T]he growth of bacterial cultures, despite the immense complexity of the phenomena to which it testifies, generally obeys relatively simple laws, which make it possible to define certain quantitative characteristics of the growth cycle.
J. Monod, *Annual Review of Microbiology*, 1949

## 7.1. Introduction

Growth is an orderly increase of all the chemical constituents in an organism and associated with this increase is cellular multiplication. Thus, growth in prokaryotes encompasses both a population increase and an increase in cell size. Through the study of bacterial growth as a model system, considerable insight into biology at the cellular level has been achieved. Prokaryotes display asexual growth which is a result of cell division that is frequently referred to as binary fission. The mother cell divides to produce two nearly identical daughter cells, with each containing genomic information consisting of double-stranded DNA. Because of semiconservative DNA replication, each new cell contains a newly replicated strand of DNA and an original unmodified strand of DNA. There is close coordination between DNA replication and cell division. To ensure that each cell contains genomic information, cell division does not occur until replication had been completed. Too often growth by organisms is dismissed as a simple event of cell division; however, on close examination, it is seen that growth is an expression of several hundred different cellular activities. Not only is science capable of enumerating some of these critical regulations in cell division, but considerable attention has been given to establishing mechanisms for cell response to a variety of stress-related events.

This chapter summarizes the dynamic activities that prokaryotes display in association with cell growth. Because cell growth follows a definite scheme or chain of events, it is highly predictable and this chapter characterizes growth by a series of mathematical expressions. Not only will the events be characterized that lead to an increase in the microbial population but also attention is given to the changes that occur within an individual cell as it passes through the growth cycle. Starvation and nutrient stress are examined because it is well recognized that relatively few species of prokaryotes produce spores, so cells must continuously grow to maintain the species. This chapter also examines activities of secondary metabolism because in nature, microbial populations are generally in a growth pattern that could be characterized as similar, if not identical, to the stationary phase. Sporulation in bacteria is examined because it is a response to starvation and is also an expression of secondary metabolism.

## 7.2. Measurement of Populations

In multicellular organisms, cellular multiplication leads to an increase in size of the individual; however, in unicellular prokaryotic organisms growth most commonly refers to an increase in the number of individuals. Over the years, several parameters have been used to evaluate growth of a bacterial population and all are dependent on a valid sampling procedure. Some of the parameters used for measuring growth are summarized in Table 7.1. If microorganisms grow as single dispersed cells and not as cell aggregates in culture media, a small volume can be taken of the broth culture and highly reproducible measurements can be obtained. However, if cells form cell aggregates as is the case with *Staphylococcus, Mycobacterium*, or *Streptococcus*, special techniques must be employed to assess growth.

### 7.2.1. Direct Measurements

The measurement of a population can be accomplished by counting the number of cells that are present. Viable cells can be determined using standard plate count

Table 7.1. Parameters used to measure bacterial growth.

| Method | Application | Features of method |
|---|---|---|
| **I. Direct measurements** | | |
| A. Microscope: cell counts | Calculation of generation time | Difficult with cells that clump. Fluorescent dyes can be used to distinguish between living and dead cells. |
| B. Colony counts | Used to determine generation time | Excellent for determining viable cells. It is difficult if cells clump. |
| **II. Indirect methods** | | |
| A. Dry weight | Calculation of yield values | Requires 50–100 ml if growth is not dense. |
| B. Protein content | Follow growth response Basis for enzyme measurements | A rapid test that can be performed with 1 ml of culture. |
| C. Turbidity | Follow growth response | A rapid measurement but difficult to determine if cells clump |
| D. Amount of an endproduct | Best with homofermentative lactic acid bacteria | Served as the basis for bioassays of vitamins and amino acids |
| E. Reduction of tetrazolium salts | To determine respiration rates in complex systems such as soil | Aerobic or anaerobic method for accepting electrons from diverse electron transport chains |

procedures with the assumption that each colony results from a single cell. However, it is not uncommon that a cell is in the process of division when deposited on the nutrient agar surface. Thus, it is reliable to refer to the number of viable units producing a colony as "colony forming units" and not number of cells. The common heterotrophic bacteria readily produce colonies on agar plates. Some prokaryotes, most notably the chemolithotrophs, have relatively few colonies resulting when cells are placed on solid support media and these organisms are referred to as having "low plate count efficiency."

Growth in broth has frequently been used to estimate the number of viable units per volume. Prokaryotes that do not readily produce colonies may be enumerated by diluting cells to extinction, placing a fraction of each dilution in five identical broth tubes, and determining the highest dilution that produces growth. This method is termed "most probable number" and through the use of statistical tables growth in broth tubes is obtained. Of the various methods used for measuring growth, the most probable number procedure is the least desired because it results in hundreds to thousands of tubes for a single growth curve.

The total number of bacteria present in a given volume of culture may be determined by counting the number of cells present. Counting devices such as the Petroff Houser chamber have a small sample depth suitable for use with oil immersion objective lenses that have a small focal length. In counting chambers, the number of cells counted includes both dead and living cells. However, the use of commercially available dyes can be used to differentiate between dead and live cells.

Another means of detecting bacteria is to label cells with fluorescent probes and employ fluorescence-activated cell cytometry (FACS). Bacterial cells can be labeled by the addition of nonlethal fluorescent dyes or fluorescent molecules attached to lectins. Cultures can be evaluated rapidly because FACS instruments can analyze a few thousand fluorescent cells in a fraction of a minute. The application of FACS to analyze the interaction of antibiotics with microbial cells and the potential for evaluation of environmental samples ensures future development of this electronic instrumentation.

## 7.2.2. Indirect Measurements

Several procedures are used to follow microbial growth that reflect cell abundance without actually counting cells and these are indirect measurements. One of the more common procedures is to measure the optical density (OD) because turbidity of a culture is directly related to cell density. The instrument most commonly used is a spectrophotometer with the wavelength set between 500 nm and 600 nm where there is no absorption by the cellular pigments or the culture media. Spectrophotometers are fairly inefficient in measuring culture turbidity because these instruments measure only the amount of light that passes through the sample. A more sensitive instrument to measure bacterial density is the nephlometer, which measures the amount of light deflected by the cells in the light beam. Because nephlometers have a limited application in the laboratory, the instrument usually purchased for measurement of cell optical density (OD) is the spectrophotometer because it also can be used to measure the absorbancy of colored solutions. Spectrophotometers are dependent on the presence of sufficient bacteria to produce enough turbidity for detection and this usually requires $10^6$ cells/ml.

Measuring the cell mass is widely used to assess growth; however, the sample size needs to be adequate because of the small size of the cells. For example, a living cell of *Escherichia coli* has a mass of $9.5 \times 10^{-13}$ g and if 70% of the cell is water, the dry weight of a single cell would be $2.8 \times 10^{-13}$ g. Measurement of cell mass reflects the total biosynthetic activity of a culture, and therefore, is an excellent parameter to assess conversion of nutrients into cell material. Mass measurements not only are useful for bacteria that grow as dispersed single cells but are also perhaps the best indicator of growth for cells that clump.

Various chemicals making up the bacterial cells have been used to assess growth and composition of bacterial cells is given in Table 7.2. While the chemical content may vary with bacteria of different species, the chemical measurements are highly reliable if growth is under carefully controlled conditions. Also, these chemical analyses include both living and dead cells. Protein content is readily assessed by treating cells with an alkaline solution to solubilize protein. Other chemicals that have been used to follow growth include total nitrogen, total carbon, and to a lesser extent, ATP, DNA, and total phosphorus. The carbohydrate level has not been useful for following population growth because capsular polymers or internal granules may be produced as a function of available carbon source. In cyanobacteria and other photosynthetic bacteria, the amount of chlorophyll extracted from a culture has often been used to assess the quantity of growth. By following the amount of unique compounds asso-

Table 7.2. Composition of *Escherichia coli* B/r cells based on a division time of 40 min.[a]

| Constituent | Percentage of total dry weight | Weight per cell ($10^{15} \times$ wt, g) | Molecular weight | Number of molecules per cell | Different kinds of molecules |
|---|---|---|---|---|---|
| Protein | 55 | 156 | $4.0 \times 10^4$ | 2,360,000 | 1,850 |
| RNA | 20.5 | 58 | | | |
| 23S rRNA | | 31.0 | $1.0 \times 10^6$ | 18,700 | 3,000 |
| 16S rRNA | | 15.5 | $5.0 \times 10^5$ | 18,700 | 1 |
| 5S rRNA | | 1.2 | $3.9 \times 10^4$ | 18,700 | |
| tRNA (4S) | | 8.2 | $2.5 \times 10^4$ | 198,000 | 60 |
| mRNA (6–50S) | | 2.3 | $1.0 \times 10^6$ | 1,380 | 600 |
| DNA | 3.1 | 9.0 | $2.5 \times 10^9$ | 2.1 | 1 |
| Lipid | 9.1 | 26.0 | 705 | 22,000,000 | 4[b] |
| Lipopolysaccharide | 3.4 | 9.7 | 4,070 | 1,430,000 | 1 |
| Peptidoglycan | 2.5 | 7.1 | $(904)_n$[c] | 1 | 1 |
| Glycogen | 2.5 | 7.1 | $1.0 \times 10^6$ | 4,300 | 1 |
| Soluble (pool) | 2.9 | 8.0 | | | 800 |
| Inorganic | 1.0 | 3.0 | | | |

Dry weight of a single cell is $2.84 \times 10^{-13}$ g. Cell moisture is 70%.
[a]Source is Ingraham et al., 1983.[5]
[b]Based on four major types of phospholipids.
[c]The number of N-acetyl glucosamine-N-acetyl muramic acid dimers with the associated peptide is not estimated.

ciated only with prokaryotes, such as *N*-acetylmuramic acid and lipid A, the growth of bacteria in eukaryotic systems can be followed.

In many instances growth can be estimated by quantitation of respiratory activity of an environmental sample. By using control cultures, the rate of oxygen utilized can be used to reflect aerobic respiring bacteria and the production of hydrogen sulfide from sulfate has been used to indicate the amount of anaerobic, dissimilatory sulfate-reducing bacteria present. The reduction of methylene blue, tetrazolium, and other dyes that function as electron acceptors has been used to estimate bacterial growth in soil and related environments. These dyes pick up electrons from the electron transport chain and color change is linked to reduction of the molecule. The basis of a commercial system for identification of bacteria is to evaluate growth through the reduction of a soluble tetrazolium to give a dark blue color.

Metabolic activity may be used to indicate growth, especially if the activity is tightly coupled to cell density. With bacteria that display fermentation, the amount of acid produced as an end product can be determined thorough titration and can be used to estimate bacterial growth. Although seldom used, bacterial growth can be determined by quantitatively determining the presence of succinic dehydrogenase or other constitutive enzymes.

With such a large of range of methods to follow bacterial growth, the selection of appropriate parameters is important and cannot be made without conducting preliminary measurements. To ensure that a method employed is growth dependent, graphs can be constructed to compare the parameter tested and the number of cells present.

## 7.3. Mathematics of Population Growth

Cell division in prokaryotes occurs by binary fission, which is an asexual process that produces two cells nearly identical in chemical composition and size. In exponential growth the population can be characterized by a log expression (Figure 7.1). As indicated in Table 7.3, growth is a function of log base 2 which can also be expressed as log base 10. Logarithmic growth is readily demonstrated in the laboratory; how-

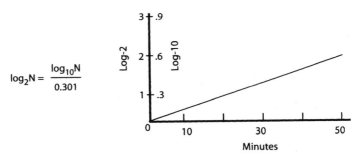

$$\log_2 N = \frac{\log_{10} N}{0.301}$$

Figure 7.1. Expression of bacterial growth by converting from log base 2 to log base 10.

Table 7.3. Binary fission in prokaryotes is a log function.

| Incubation time (minutes) | Number of cells | Number of cell divisions | Log 2 | Log 10 |
|---|---|---|---|---|
| 0 | 1 | 0 | 0 | 0 |
| 20 | 2 | 1 | 1 | 0.301 |
| 40 | 4 | 2 | 2 | 0.602 |
| 80 | 8 | 3 | 3 | 0.903 |

ever, growth under natural conditions is more readily characterized as arithmetic. Distinction between arithmetic and logarithmic growth can be demonstrated by graphic expression (see Figure 7.2).

A way to express growth rate of bacteria or archaea is to state the time required for cell division, which is also known as generation time. While many bacteria are described as having cell division every 20 min, this is the time for *Escherichia coli* and similar bacteria. Rates of growth are characteristic of various microorganisms and various examples are given in Table 7.4.

The calculations for growth rate can be either a closed system as in a flask or a continuous culture system as shown in Figure 7.2. In a closed growth system, the exponential growth rate, $k$, is defined as follows:

$$k = \frac{\log 10 \, N_t - \log 10 \, N_0}{0.301 \times t}$$

where $N_t$ is the number of microbial cells at the designated time, $N_0$ is the initial number of cells, and $t$ is the time for the measurements. The value for $k$ = generations/hr. The generation time in minutes is calculated by the following expression:

$$\frac{1}{k} \times 60 \text{ min/hr} = \text{number of minutes for a generation}$$

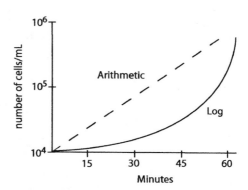

Figure 7.2. Graphic comparison of arithmetic and log growth.

Table 7.4. Generation times for several species of bacteria.

| Organisms | Time for one generation |
|---|---|
| *Bacillus stearothermophilus* | 8.4 min |
| *Vibrio natriegens* | 9.6 min |
| *Escherichia coli* | 17 min |
| *Methanococcus jannaschii* | 26 min |
| *Staphylococcus aureus* | 28 min |
| *Thermus aquaticus* | 1 hr |
| *Bacteroides fragilis* (with haemin) | 2 hr |
| *Mycobacterium tuberculosis* | 6 hr |
| *Nitrosomonas europaea* | 8 hr |
| *Anabaena* (phototrophic) | 9 hr |
| *Nitrobacter* agilis | 20 hr |
| *Treponema pallidum* (in rabbit testis) | 33 hr |
| *Mycobacterium lepraemurium* | 7 days |
| *Mycobacterium leprae* (in mouse foot pad) | 21 days |

---

## Box 7.1
## Calculation of Generation Time of a Bacterial Culture

*Problem*: If a bacterial population increases from $1 \times 10^2$ cells/ml to $1 \times 10^9$ cells/ml in 10 hr, what is the growth rate?  What is the generation time?

$$\text{Growth rate: } k = \frac{\log_{10} \cdot 10^9 \text{ cells} - \log_{10} \cdot 10^2 \text{ cells}}{0.301 \times 10 \text{ hr}}$$

$$= \frac{9\text{–}2 \text{ generations}}{3.01 \text{ hr}}$$

$$= 2.33 \text{ generations/hr}$$

$$\text{Generation time: } Gt = 1/k \text{ generations/hr}$$

$$= \frac{1}{2.33 \text{ gen/hr}} = 0.429 \text{ hr/generation}$$

$$\text{To convert to } Gt \text{ in minutes} = 0.429 \text{ hr/generation} \times 60 \text{ min/hr}$$
$$= 25.74 \text{ min/generation}$$

## 7.4. Cell Division and the Z Ring[1–3]

Cell division in prokaryotics is described as binary fission, in which the dividing cell produces two identical daughter cells. Successful division involves separation of microbial chromosomes followed by the biosynthesis of cell wall along the division plane. As the microbial chromosome is replicated, the two macromolecules of DNA are separated and the chromosomes are localized at the two poles of the cell. The mechanism of this segregation of chromosomes is unresolved but some have suggested that the plasma membrane may participate by transiently binding the DNA macromolecule. Prior to separation of the cells a septum forms that is the new cell wall and a plasma membrane is constructed between the dividing cells.

The construction of a septum is an extremely critical event because peptidoglycan synthesis is redirected to be perpendicular to the peripheral wall without cell rupture even though the cell is under several atmospheres of turgor pressure. In addition, in many cells there is a slight constriction of the cell wall as the cells divide to produce cells with a tapered end. Some species of Gram-negative bacteria have highly tapered ends of cells and these are frequently referred to as "lemon shaped." The formation of a zone or ring structure that is at the site where septum development is observed in dividing cells. To engineer this division process, prokaryotes employ a series of proteins that are products of the *Fts* genes. The term "Fts" was used to describe the *f*ilamentous *t*emperature-*s*ensitive mutant of *E. coli* that did not display cell division but instead cells grew as long filaments. A major Fts protein found in most bacteria and archaea, as well as many eukaryotic cells, is the FtsZ. The FtsZ protein is tubulin-like with GTPase activity and polymerizes to produce the Z ring. Because organisms such as *E. coli* contain enough FtsZ to form about 20 Z rings, the limiting factor for cell division would appear to be initiation of Z ring formation. To prevent premature formation of the Z ring, the aggregation of the FtsZ proteins is negatively regulated by the Min system. Cell division occurs at a specific stage of the cell cycle and an unknown stimulus triggers the construction of the Z ring which mediates cytokinesis.

With *E. coli* as the model system, the following series of events have been proposed by W. Margolin and others in the field. Establishment of the midpoint of the cell is determined by three proteins encoded on the *min* operon. The cytosol of vegetative cells contain FtsZ proteins and aggregation is prevented by the cooperative action of MinC and MinD. At the site of future septa formation, MinE protein serves as a nucleation site on the membrane and initiates the localization of the FtsZ with the development of the Z ring. The FitsA and ZipA, FtsW and FtsK, FtsQ, FtsL, FtsI, and FtsN are added in sequence to the division apparatus and this scheme is given in Figure 7.3.

Although the roles of the various proteins in the Z ring are not fully understood, it is clear that these proteins are critical in stabilizing the Z ring and in closure of the septum. FtsA and FtsZ are cytoplasmic, with FtsA functioning as an ATPase and binds to monomers of FtsZ which produce the Z ring. The N-terminus of ZipA is bound to the plasma membrane while the C-terminus is in the cytoplasmic region. It

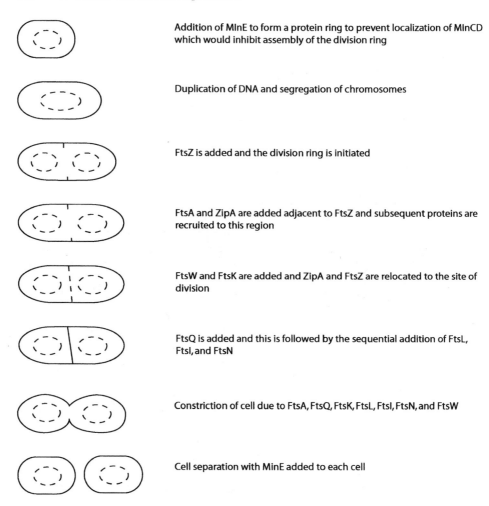

Addition of MinE to form a protein ring to prevent localization of MinCD which would inhibit assembly of the division ring

Duplication of DNA and segregation of chromosomes

FtsZ is added and the division ring is initiated

FtsA and ZipA are added adjacent to FtsZ and subsequent proteins are recruited to this region

FtsW and FtsK are added and ZipA and FtsZ are relocated to the site of division

FtsQ is added and this is followed by the sequential addition of FtsL, FtsI, and FtsN

Constriction of cell due to FtsA, FtsQ, FtsK, FtsL, FtsI, FtsN, and FtsW

Cell separation with MinE added to each cell

Figure 7.3. Model for assembly of cell division apparatus leading to cell constriction and cell division. (Modified from Margloin, 2000.[3])

has been proposed that FtsK and FtsW are transmembrane proteins where FtsW may serve as a sensory protein for communication between the periplasmic and the cytoplasmic regions of the cell. The N-terminal region of the FtsK interacts with the Z ring and the C-terminal domain may participate in the separation of chromosomes. The FtsQ, FtsL, FtsI, and FtsN proteins are attached to the plasma membrane by short transmembrane segments with most of the C-terminal domain residing free in the cytoplasm. FtsI is a transpeptidase enzyme that functions in peptidoglycan synthesis at the region of the septa. Although FtsQ, FtsL, and FtsN are important for cell division, their specific roles remain unclear.

Although the most extensive analysis of proteins in cell division has been conducted in *E. coli*, interesting variations are observed with other prokaryotes. In *Bacillus*, cell division is accomplished with no constriction of the cell wall and the formation of the Z ring at midcell occurs only in vegetative cells. In sporulation, the assemply of the Z ring is at the pole where the spore septum is generated and this shift in location is attributed to the action of SpoOA, which is a master regulator of sporulation. In *Caulobacter crescentus*, there is a tight regulation of the FtsZ levels with swarm cells prevented from producing FtsZ by inhibition of the CtrA regulator. When stalked cells do divide, the FtsZ is digested on completion of cell division. FtsZ is widely distributed throughout the bacteria and the only organisms known to lack this cell division system are two species of host-dependent bacteria (*Chlamydia trachomatis* and *Chlamydia pneumoniae*) and a mycoplasma species (*Ureaplasma urealyticum*). The FtsZ system of cell division has been found in the archaea, except for the crenarchaeon *Aeropyrum pernix*.

In a mutant strain of *E. coli* cell division occurs near one of the poles and the product of this asymmetrical cell division is a small cell and a larger cell. Presumably, this mutant had a defect in locating the Z ring. This error in cell division was initially described by H. Adler and he termed the newly formed unit a minicell. An electron micrograph of a minicell is given in Figure 7.4. Because the minicell does not contain DNA, there can be no further growth or cell division by the minicell. Not only are these minicell mutants smaller than the other cells but they are also less dense because they lack DNA. Minicells can be separated from other cells in the culture by density centrifugation techniques and have been proposed to be useful for biotechnology applications such as vaccine administration. The surface of a minicell can be programmed to carry the antigen for pathogenic bacteria but disease production can be dispelled because the minicells cannot divide.

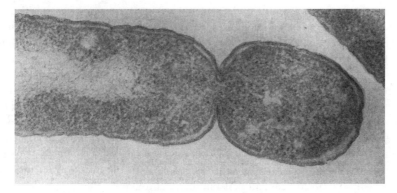

Figure 7.4. Electron micrograph of a thin section of *Escherichia coli* showing the formation of a minicell devoid of DNA. From Adler, 1967.[1]

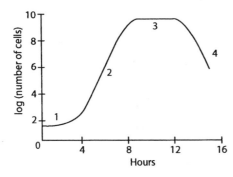

Figure 7.5. Idealized growth response of a bacterial culture in a semilogarithmic plot. (1) lag phase, (2) exponential or log phase, (3) stationary phase, and (4) death or decline phase.

## 7.5. Phases of Growth[4-7]

Bacterial growth is readily evaluated by following the increase in cell numbers. With cultivation in a batch culture system such as a tube or flask, growth of bacteria is highly predictable. When the number of viable bacteria are expressed as a log function, the growth over time follows an expression that has been designated as the growth curve (see Figure 7.5). There are four major phases making up the growth curve and these have been described as lag phase, logarithmic (log) or exponential phase, stationary phase, and decline or death phase. Between the lag phase and the log phase as well as between the log phase and the stationary phase are two important transitional phases. Activities of cells in each of these growth phases will be discussed.

### 7.5.1. Lag Phase

Under most conditions with the inoculation of bacteriological media, there is a period during which growth is not observed and this is termed the lag phase. If growth is followed by counting the number of cells under a microscope, assessing the number of viable cells through colony counts, or measuring OD, there is no increase in the parameter followed for several minutes. Following inoculation, there is an adjustment of cells to the carbon source, nitrogen source, pH, and oxygen tension. Nutrients that are new to the cells will be taken into the cells by diffusion, especially if active transport systems for a specific chemical are not constitutive in that organism. In the early part of the lag phase, cellular changes in regulation of protein synthesis through induction and repression occur.

As cells start to metabolize, there is a gradual increase in the rate of respiration and utilization of electron donor. There is also a progressive increase in cellular constituents displayed by an increase in cell mass (see Figure 7.6), as cells increase in size but without an increase in cell number. Throughout the lag phase there is no cell division because cells require a signal to divide after replication of DNA is completed. More precisely, there is a lag in cell division in this phase and cell division does not occur until the transition phase when cells enter the log phase. Thus, to

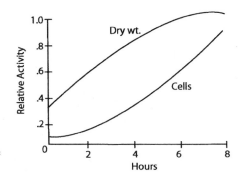

Figure 7.6. Cell mass increases in late lag phase as cells elongate prior to division.

display the idealized growth curve, it is necessary to follow growth with those parameters that measure individual cells.

The lag phase is not fixed for a culture but is influenced by cellular characteristics of the bacteria at the time of transfer. The length of lag is generally influenced by the age of the culture. If actively growing cells are used as the inoculum, there may be no lag observed; however, when cells for the inoculum are taken from the stationary phase where cells are not growing, there is an appreciable lag phase. Actively growing cells with high levels of metabolism are referred to as physiologically active or in terms of cell division are designated as physiologically young cells. Inoculum from the log phase would be comprised of physiologically young cells while inoculum from the stationary phase would consist of physiologically older cells.

In certain instances the size of inoculum is critical to the length of the lag. In aerobic eubacteria, an inoculum of 0.1% may be used whereas in anaerobic bacteria the size of inoculum may increase to as much as 5%. This large inoculum may be needed because methods used for cell transfer may not adequately protect anaerobes to air. A large inoculum with sulfate-reducing bacteria will introduce hydrogen sulfide into media that promotes an appropriate environment for growth. In some instances metabolic products transferred with the cells promote bacterial growth. For example, cyanobacteria have a shorter lag phase if a large inoculum is used because vitamins and cofactors produced by these bacteria are present in the inoculum and promote bacterial growth. Thus, in certain cultures, the larger the inoculum the shorter the lag phase. It is of course important to know the limitation of the method as a means to follow growth. As it generally takes $>10^6$ cells/ml to produce an OD reading, the addition of only a few cells may not indicate visible growth because the cell density is not sufficient for detection by this method.

The type of nutrients present in the culture media will have a significant impact of the length of the lag phase. If the inoculum consists of physiologically young cells from a medium identical to that used for growth, the lag will not be observed. However, if the carbon source is different in the growth medium than for the inoculum, then a lag is expected. For example, an inoculum of *E. coli* grown on glucose and added to the growth medium with a sugar other than glucose would produce a lag (see Figure 7.7). With *E. coli*, glucose is the preferred carbon substrate in metabolism

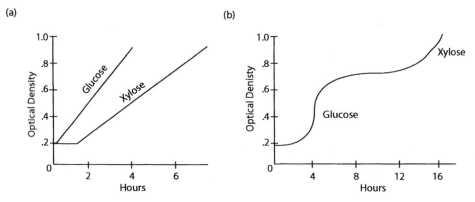

Figure 7.7. Lag phase is influenced by the carbon source. (**a**) The transfer of glucose-grown cells into media containing glucose shows growth before cells transferred into media containing xylose. (**b**) Diauxic growth where glucose is preferred over xylose as the carbon source.

and induction is not required because the enzyme pathways are always present. If *E. coli* from the stationary phase is inoculated into a medium that contains a small quantity of glucose and xylose, a multiple growth response is observed. A lag is observed because physiologically young cells were not used and glucose is the first sugar used because it is the preferred substrate. When glucose is consumed in the medium, a period of induction is required before xylose is metabolized. This unique response is termed diauxic growth and was extensively studied by Monod.

The length of lag phase extends to chemicals other than the sugar used as the carbon and energy source. In minimal media, bacteria produce enzymes to synthesize all vitamins and amino acids required for growth whereas in complex media, cells are provided with vitamins and amino acids and the production of these enzymes is repressed. When *E. coli* is grown in complex media but transferred to minimal media, a considerable lag is observed while enzymes for the synthesis of vitamins and amino acids are produced. However, when *E. coli* is grown in minimal medium and phys-

Figure 7.8. Culture grown in minimal salts medium containing only inorganic compounds and glucose as the carbon and energy source. Growth occurs immediately with transfer from minimal medium if inoculum is from the log phase. A lag in growth does occur even though the cells are from log phase if the inoculum is from complete medium containing amino acids, vitamins, and nucleic acid precursors.

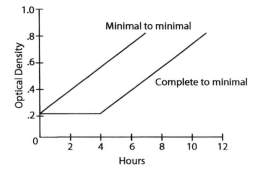

iologically young cells are transferred to minimal medium, no lag is observed because the bacterial cells already have the enzymes for biosynthesis of vitamins and amino acids (see Figure 7.8). Sometimes inhibitors such as heavy metals are present in the growth media and account for a longer lag phase. In general, when physiologically young cells are inoculated into media that has chemicals optimal for growth, there is no lag but if the chemicals are suboptimal for growth, then a lag is observed.

## 7.5.2. Log Phase

Cells in the log phase display the fastest rate of growth with cell division following a log function. The rate of synthesis of protein, DNA, and RNA is comparable to the division rate. This rate of cell division is not fixed but is influenced by the chemical environment with the amount of biosynthetic energy required inversely related to the rate of growth. With respect to nitrogen source in a medium, the greatest rate of growth is seen with amino acids and slower rates of growth are seen with ammonium and nitrate (see Figure 7.9). With bacteria such as *Salmonella typhimurium*, growing in lysine salts medium, glucose salts medium, or nutrient broth, the generation time is shortest with nutrient broth and slowest with lysine as the carbon and energy source (see Table 7.5). If the energy source or one of the essential nutrients is limiting then the growth is reduced because there is insufficient chemical for saturation of enzymes. Another factor that greatly influences growth rate of bacteria is temperature. A good example is the growth of *E. coli* at temperatures of $10°$ to $47°C$. Each organism has an optimal temperature for growth and at this temperature the generation time is shortest. Also, bacteria will display a pH optimum and growth is greatest at that pH value.

The growth rate is not influenced by the lag phase. That is, if there is a long lag due to history of inoculum, once growth has been initiated the rate of growth is influenced only if the chemical composition of the growth medium is changed. As is shown in Figure 7.7 (a), the lag of *E. coli* is lengthened when the inoculum is transferred from glucose into xylose-containing medium. Once growth is initiated the rate is the same as when the inoculum used came from cells grown in xylose-containing

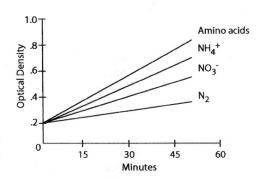

Figure 7.9. The influence of nitrogen source on rate of growth. The only variable in the media is the nitrogen added.

Table 7.5. Impact of growth medium on chemical composition of *Salmonella typhimurium*.

| Medium | Generation time (min) | Dry weight/ cell (g $\times 10^{15}$) | Nuclei/ cell | Ribosomes/ cell | Protein ($10^{-19}$ g) synthesized/minute/ ribosome |
|---|---|---|---|---|---|
| Lysine salts | 96 | 240 | 1.1 | 4,000 | 3.3 |
| Glucose salts | 50 | 360 | 1.5 | 10,000 | 4.6 |
| Nutrient broth | 25 | 840 | 2.4 | 40,000 | 5.3 |
| Heart infusion both | 21 | 1,090 | 2.9 | 70,000 | 4.6 |

From Maaløe and Kjeldgaard, 1966.[10]

medium. The metabolic rate of bacterial cells changes as the culture progresses through the various growth phases and the greatest rate of respiration, utilization of carbon and energy sources, and formation of end products occurs in the log phase.

## 7.5.3. Stationary Phase

The stationary phase is characterized by a period during which the number of viable cells is constant. Some cells are dividing; however, the number of cells produced from division is equal to the number of cells dying. There is relatively little protein synthesis, DNA replication, RNA production, utilization of energy sources, or respiration. Some biosynthetic activities of cells continue in the stationary phase and the best example is that of synthesis of peptidoglycan in Gram-positive bacteria with a marked increase in the thickness of the cell wall. Values reflecting cell composition are usually based on measurements at the stationary phase. The length of cells is based on measurements taken at the stationary phase and the chemical composition of cells generally is from the stationary phase.

The length of the stationary phase can range from a few minutes to several hours or days depending on the environment. Of the various factors that may influence length of stationary phase, most important are concentration of acidic end products released and temperature of incubation of the culture. Because acidic products from glycolysis are important, stock cultures can be maintained on amino acid–containing medium without carbohydrate to eliminate the production of organic acids. Alternately, high concentration of buffers or calcium carbonate can be added to the growth medium to prevent acidic conditions from being produced. In most bacteriological media, the concentration of phosphate greatly exceeds that required for growth and the high level of phosphate provides buffering to the media. The shifting of a culture grown at 37 to 4°C would reduce the rate of cell death, and thereby result in an increased length of stationary phase.

## 7.5.4. Death Phase[8–10]

When cell death occurs at a rate that is faster than cell division, the culture is in the death or decline phase. Each culture appears to have a death rate that is characteristic for that bacterial strain under a given set of culture conditions. The mathematics of cell death was described by H. Chick in 1908 and many of his expressions are still used today. When evaluating the effectiveness of chemical and physical treatments in controlling bacterial viability, it is important to predict the time necessary for all cells to be killed. By plotting the logarithm of the number of bacteria alive against time, a straight line is obtained and the time required for death of all cells can be determined. Alternately, the following formula can be used:

$$kt = \ln (X_0 / X_0 - X_d)$$

where ln is the natural logarithm, $X_0$ is the initial number of live cells, $X_d$ is the number of dead cells, and $t$ is the time. The death of cells and spores as a result of exposure to heat is often expressed as a $D_x$ value which reflects the time for 90% of the population to die at the temperature of $x$ which is measured in °C. For example, cells of a certain microbial species may have a $D_{50°C}$ of 10 min and spores of *Bacillus stearothermophilus* may have a $D_{121°C}$ of 16 min. In addition to the $D$ value, other designations frequently used include thermal death time (TDT) and thermal death point (TDP). The TDT refers to the time required for heat sterilization with a specific concentration of organisms using a prescribed volume and medium. The TDP is a growth parameter and it is the lowest temperature at which an organism can grow. For example, the TDP for *E. coli* is 4°C.

Over the years, considerable interest has been given to *programmed cell death* (PCD) or apoptosis in eukaryotic cells and recently attention has focused on factors controlling death in prokaryotes. In the case of sporulating cells, death to the mother cell is an obvious consequence of endospore production; however, to consider this as an example of PCD may be extremely generous. While several provocative observations concerning cell death are listed in Table 7.6, definitive examples of PCD remain to be presented. Cultures most likely to display PCD would be those that undergo differentiation to produce two or more different cell types such as stalked bacteria, endospore producing bacteria, and cyanobacteria.

# 7.6. Dynamics of the Single Cell[4–7]

The size of cells is variable and subject to change as the culture passes through the various phases of growth. The greatest cell length occurs at the transition where lag goes into log phase and cells become shorter as the culture progresses toward the stationary phase (see Figure 7.10). The longer cells develop because the initiation of cell division lags behind the increase in cell biosynthesis. The lag reflects the time required for replication of DNA. Using *E. coli* as a point of reference, replication of DNA requires 60 min; however, with DNA replication initiated every 20 min from the point of origin on the DNA, cell division at 20 min after the first replication of DNA would provide each cell with DNA. Thus, as a culture progresses from early log to late log, cell division gradually catches up with the rate of DNA replication so that at the stationary phase cells have the minimal amount of DNA per cell. Cells in early log phase will have three to four times the DNA content of cells in stationary phase.

Table 7.6. Activities resulting in prokaryotic cell death.[a]

| Organism | Activity |
|---|---|
| *Bacillus subtilis* | *cwlC* and *cwlH* are transcribed by a spore specific $\sigma^K$ RNA polymerase to produce two separate autolysins that lyse the mother cell. |
| *Fibrobacter succinogenes* | Significant lysis of cells during log phase occurs due to autolysin; however, if protease is induced by transfer of culture to a medium deficient in energy source, the autolysin is degraded and no lysis occurs. |
| *Myxococcus xanthus* | Dense cultures have lysis of vegetative cells where nutrients from cells are used by fruiting body and developing spores. |
| *Streptococcus pneumoniae* | The *lytA* gene is activated by antibiotics and other factors to lyse cells and prevent expression of damaged DNA that has been acquired by homologous recombination. |
| *Escherichia coli* | DNA damaged by mutagens promotes activation of RecA which subsequently activates *sulA* to produce a protein that inhibits cell division and is continuously degraded by the Lon protease. With repair of DNA, the level of RecA decreases as well as the production of SulA and residual SulA is degraded by the Lon protease. This ensures that cells with defective DNA do not divide. |

[a]Data from Lewis, 2000.[9]

In the log phase, cells will have a macromolecular composition that reflects the growth rate. As seen from Table 7.5, cells in the log phase with the highest growth rate have the greatest DNA and RNA content per cell. Cells in the log phase will have several thousand ribosomes per cell while cells in the stationary phase will have only a few hundred ribosomes per cell. The number of haploid genome copies varies

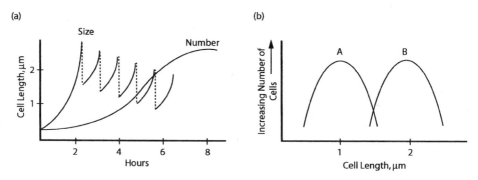

Figure 7.10. Cell length of a growing culture is reduced as the cells progress through the log phase. (**a**) Cells are longest at the transition from lag to log phase and get progressively shorter until the stationary phase. (**b**) The range of cell length is shorter in the stationary phase (A) and longer in early log phase (B).

with the prokaryotic species and the phase of growth. *E. coli* has four copies of the genome in the log phase but only a single copy in the stationary phase. The number of genome copies varies with the bacterial species, with eight to ten copies present in the log phase with *Deinococcccus radiodurans* and four copies in the stationary phase.

## 7.7. Balanced Growth

In balanced growth all biosynthetic activities occur at the same rate as cell division. During the log phase of growth, protein, DNA, RNA, and cell division all proceed at the same rate. In nature, the length of exponential growth by bacteria is relatively short; however, bacteria can be maintained in balanced growth if a chemostat is used (see Figure 7.11). Cultures can be maintained in continuous growth by regulating the

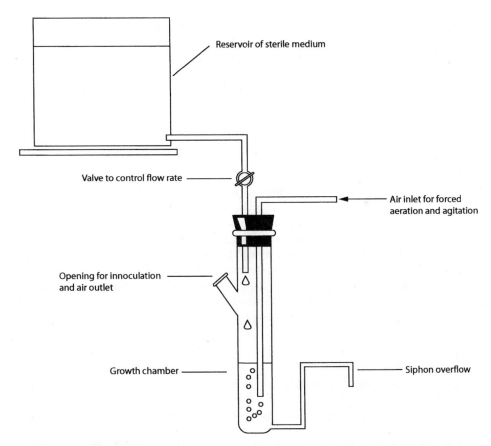

Figure 7.11. A chemostat in which residence time in the growth chamber is regulated by the rate of medium flowing into the reaction vessel. Because the chemostat is sealed, the rate of overflow containing bacterial cells is equal to the rate of nutrients flowing into the growth chamber.

rate at which essential chemicals are fed into the growth chamber. The growth of bacteria in a chemostat is expressed as follows:

$$\frac{dX}{dt} = DX = \mu X$$

where $X$ is the number of bacteria, $t$ is time, $D$ is the dilution rate, and $\mu$ is the instantaneous growth-rate constant. While the growth-rate constants of $\mu$ and $k$ reflect activities in continuous and batch culture, respectively, the two constants are related and characteristic of that bacterial culture. Chemostat cultures are especially useful if small quantities of cells of a given growth rate are needed over a period of several weeks for different experiments. The growth rate of bacteria in a chemostat is controlled by regulating the dilution rate.

In a continuous culture system, the instantaneous growth rate, $\mu$, is calculated by the following formula:

$$\mu = \frac{dN}{dt} = \frac{d(\ln x)}{dt} = \frac{\ln 2}{t_d}$$

where $x$ = dry weight cells/ml and $t_d$ = doubling time.

## 7.8. Unbalanced Growth

In nature, bacterial cells have a relatively short exponential phase during which growth is balanced. Cells are commonly responding to changes in the nutrient state of the environment and this can be studied in the laboratory by shifting bacteria to an environment that has more nutrients or fewer nutrients. As the culture passes from the log to the stationary phase there is a change in the composition of cells that can be characterized as a downshift in metabolism, as shown in Figure 7.12(a). Following the reduction in rate of cell division, there is a reduction in the rate of synthesizing DNA, protein, and RNA, respectively. The composition of cells also changes as the culture passes from lag to log phase and can be characterized as displaying an upshift in metabolism [see Figure 7.12(b)]. The first macromolecule to increase is RNA for production of mRNA, tRNA, and rRNA. The rate of protein synthesis increase precedes the increase of DNA replication and the final activity in upshift is cell division.

## 7.9. Synchronous Growth[4–7]

One of the more difficult features of studying bacterial growth is to try to assess the activity of a single cell. Bacteria are dividing randomly throughout the growth phase and analytical methods cannot measure the composition of a single cell. In an effort to characterize physiological activity with respect to nuclear division, methods were developed for the synchronization of cell division. One of the more successful pro-

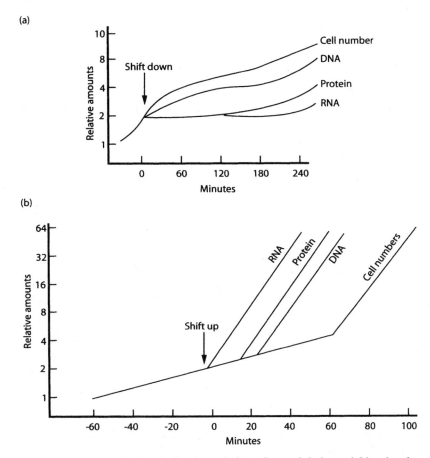

Figure 7.12. Semilogarithmic plots indicating unbalanced growth in bacterial batch culture. (**a**) The culture showing logarithmic growth in a rich nutrient medium is transferred to an inorganic medium with a result in growth occurring at a slower rate. (**b**) The shift of a culture growing in an inorganic medium to a rich nutrient medium with the result of a greater rate of cell division.

cedures to synchronize bacterial cell division is to employ cyclic temperature changes. The culture of *Salmonella typhimurium* can be synchronized by exposing it to 25°C for 28 min followed by a shift to 37°C for 8 min. Cell division and DNA synthesis occur predominately at the time of exposure of the culture to 37°C and perhaps this is the effect of increased thermal energies on macromolecular configurations and the migration of the proteins for setting up the ring for division. The rate of RNA production and protein synthesis is constant throughout the period of cell division and is not influenced by the temperature of incubation. This cycle can be repeated three to four times before cell division is no longer synchronous but becomes random. These experiments with synchronous cells emphatically demonstrated that biosynthetic ac-

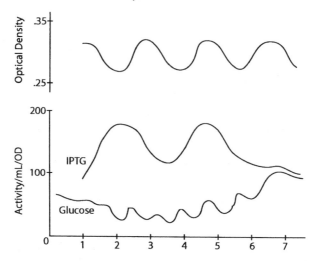

Figure 7.13. Production of β-galactosidase by *Escherichia coli* growing in a chemostat using addition of phosphate to synchronize the culture. The cyclic optical density of the culture reflects the growth response of the culture. The addition of IPTG, the gratuitous inducer, produces β-galactosidase that is dependent on cell density. The addition of glucose to a culture growing on lactose produces a cyclic response of uniformly low periodic formation of β-galactosidase that reflects CAP repression. With the depletion of IPTG or glucose, a minimal level of β-galactosidase is produced as observed in the final growth of cells. (From Goodwin, 1969.[4])

tivities in bacteria could be spatially separated even though there is no structural localization of activities.

Another means of inducing synchrony is to grow bacteria in a chemostat and add an essential nutrient at defined intervals. In what is now considered a hallmark growth experiment, Goodwin examined the dynamics of enzyme production in a culture of *E. coli* that had been synchronized by periodic phosphate addition. As presented in Figure 7.13, synchrony with a 100-min period was maintained as long as the phosphate addition was pulsed into the chemostat. When the carbon source was 0.2% glycerol and 0.1 m*M* isopropylthiogalactoside (ITPG) was added, there was an induction of β-galactosidase with a single enzyme oscillation per generation time. ITPG is a gratuitous inducer because it is not metabolized by the cell. This indicates that production of β-galactosidase is dependent on the quantity of cells present and not a specific segment of the cell cycle. When 0.02% glucose is added to the culture, a perturbation of the culture results with several irregular oscillations without a clear periodicity. After four generations the influence of glucose or ITPG is lost and a minimal level of β-galactosidase is produced. This experiment indicates that DNA replication, septation, cross wall formation, and synthesis of β-galactosidase is temporally organized and periodic. The variation in β-galactosidase production with glucose indicates that numerous variations in concentrations of inducer and cAMP occur in the cell. Clearly, the concentrations of metabolites in the cell are not static but their level fluctuates as a result of contributing factors. Although only the oscillation of β-galactosidase is shown here, other enzymes have been demonstrated to oscillate and it has been proposed that several hundred oscillations are occurring in a growing cell. Research with synchronous cells reveals some of the dynamic activities found in cells.

## 7.10. Energetics of Growth

The chemical composition of bacterial cells reflects the various biosynthetic processes that are required for cell growth. As listed in Table 7.7, an organism such as *E. coli* consists primarily of protein and the biosynthesis of proteins consumes a major frac-

Table 7.7. ATP Requirement for production of *E. coli* growing on pyruvate as the carbon and energy source.[a]

| Biomolecules | Percentage of total[c] | ATP required ($10^4 \times$ mol/g cells formed) | |
| --- | --- | --- | --- |
| | | Medium A[b] | Medium B[b] |
| Protein | 52.4 | | |
|   Amino acid formation | | 0 | 148 |
|   Polymerization | | 191.4 | 191.4 |
| Polysaccharide | 16.6 | | |
|   Glucose-6-phosphate formation | | 61.5 | 61.5 |
|   Polymerization | | 10.3 | 10.3 |
| RNA (total) | 15.7 | | |
|   Nucleoside monophosphate formation | | 37 | 62 |
|   Polymerization | | 9.2 | 9.2 |
| Lipid | 9.4 | 27 | 27 |
| DNA | 3.2 | | |
|   Deoxynucleoside monophosphate formation | | 8 | 14 |
|   Polymerization | | 1.9 | 1.9 |
| Turnover mRNA | | 13.9 | 13.9 |
| Transport of | | | |
| Pyruvate | | 58.1 | 148 |
|   Amino acids | | 47.8 | 0 |
|   Ammonium ions | | 0 | 42.4 |
|   Potassium ions | | 1.9 | 1.9 |
|   Phosphate | | 7.7 | 7.7 |
| Total ATP required | | 475.7 | 740.2 |
| Gram cells per mole of ATP | | 21.0 | 13.5 |
| Minerals, soluble nutrients | 2.7 | | |

[a]From Stouthamer, 1979.[12]
[b]Medium A is a complex medium containing amino acids and medium B is an inorganic salts medium with ammonium as the nitrogen source.
[c]Based on dry weight measurements.

Table 7.8. Biosynthetic activities of a typical bacterial cell with a cell division of 20 min.

| Biological molecule | Number of molecules to be synthesized per second[a] | Number of ATP molecules required per second to support bacterial growth | Percent of total biosynthetic energy required |
|---|---|---|---|
| Protein | 20,000 | 2,120,000 | 51.8 |
| Lipo- and polysaccharide | 12,000 | 65,000 | 15.0 |
| RNA | 2,125 | 75,000 | 13.0 |
| Lipid | 183,000 | 87,500 | 3.7 |
| DNA | 0.017 | 60,000 | 2.5 |
| Solute transport | — | — | 15.0 |

[a]Number of molecules is from Neidhardt and Umbarger, 1996[11] and ATP required per molecule is according to Stouthamer, 1979.[12]

tion of biosynthetic energy. If the culture is growing on pyruvate, energy expended on protein synthesis ranges from 40% to 45%. There is a greater demand for energy when ammonium is the nitrogen source as compared to amino acids provided in the medium. Although the quantity of energy needed for ion transport and synthesis of polysaccharide, RNA, DNA, and lipids is significant, no one segment of metabolism requires as much energy as protein synthesis. Based on energy demands for protein synthesis, the regulation of amino acid synthesis and protein formation enables cells to optimize energy expression. For bacterial cells to divide every 20 min, the synthesis of the various biopolymers proceeds at an extremely high rate. Because an average bacterial cell contains about 26 million molecules, the number of molecules synthesized per second is extremely high (see Table 7.8). Even though the greatest number of molecules produced per second are lipids, the production of lipids requires little energy in comparison to protein synthesis. Approximately 2.5 million molecules of ATP are used every second to support the biosynthetic activities of cells.

In 1942, Monod found that bacterial growth measured as cell mass was proportional to the amount of glucose consumed and this was one of the contributions that earned him the Nobel Prize. At low levels of glucose, the amount of growth of *E. coli* in a glucose–mineral salts medium is directly related to the amount of glucose added to the medium. When glucose is supplied in excess in a medium and glucose is not the carbon source but is only the energy source, the amount of cell growth is directly related to the amount of glucose utilized. One of the contributions of Monod to microbiology was his description of bacterial growth through mathematical equations which led to the establishment of yield coefficients to characterize bacterial growth.

## 7.10.1. Growth Yield Coefficient[11–13]

To evaluate microbial growth of different substrates, many have found it useful to determine the amount of cell mass produced for a given quantity of energy source consumed. One expression is $Y_g$ = g dry weight of cells produced for per gram of sugar (or other energy source) used. The $Y_g$ is calculated according to the following formula:

$$Y_g = X - X_0 / C_0 - C$$

where $X$ is the g dry weight/ml of cells at the end of log phase, $X_0$ is the weight/ml of cells near the beginning of the log phase, $C$ is the g of substrate present/ml at the time when $X$ is determined, and $C_0$ is the amount of substrate present in g/ml at the time when $X_0$ is measured. For the calculation of $Y_g$, it is not necessary to know the metabolic pathways used by the bacteria for a given substrate. In addition, it is a means of evaluating aerobic or anaerobic growth using simple or complex energy sources and to establish the economics of cell growth. To evaluate the growth response with energy sources of considerable difference in molecular weight, it may be useful to determine growth based on the consumption of molecules of sugar or amino acids. For these evaluations, the coefficient of $Y_g$ is used and this is the amount of g dry weight of cells produced for each mole of substrate used as an energy source.

## 7.10.2. $Y_{ATP}$ and $Y_{kcal}$ Yield Coefficients[11–13]

A more meaningful expression of growth is frequently employed in which growth is equated to the amount of energy produced by the substrate. Because the catabolic pathways in bacteria are well established, it is possible to establish the growth yield values for many bacteria. In 1960, T. Bauchop and S.R. Elsden observed that bacterial growth was directly related to the amount of ATP obtained from the energy source. Anaerobic bacteria use fixed fermentation pathways, and ATP yield from substrate-level phosphorylation is highly predictable. In aerobic bacteria, however, the ATP yield for a given substrate varies with the electron transport system used by the organism and growth is not easily predicted. The following yield coefficient was used to describe the energetics of growth for fermentative bacteria:

$Y_{ATP}$ = g dry weight of cells/mol of ATP generated from the energy source

Using continuous flow to cultivate heterotrophic bacteria, the $Y_{ATP}$ yield coefficient for many bacteria was determined and when all the measurements were taken into consideration, it was found that the $Y_{ATP}$ yield coefficient for many heterotrophic bacteria growing on glucose was 10.5 ± 0.5. A few examples of $Y_{ATP}$ yield with bacteria are given in Table 7.9. It appears that the energy required to synthesize building blocks

Table 7.9. Growth yields for several different organisms involving different metabolic pathways.[a]

| Culture | Energy source | Fermentation pathway | ATP yield (mole energy source) | $Y$ | $Y_{ATP}$ |
|---|---|---|---|---|---|
| *Streptococcus faecalis* | Glucose | Glycolysis | 2–3 | 23–32 | 10.8 |
| | Gluconate | | 1 (2?) | 17.6 | 9.8 |
| | Arginine | | 1 | 10.5 | 10.2 |
| *Lactobacillus plantarum* | Glucose | Glycolysis | 2 | 18.8 | 9.4 |
| *Zymomonas mobilis* | Glucose | Entner–Douderoff | 1 | 8–9.3 | 8.5 |

[a]From Stouthamer, 1969.[4]

of cell material is relatively constant for the different bacteria and one may assume also for the archaea. By understanding the metabolic pathways of prokaryotes it is possible to predict the cell growth with different substrates.

The $Y_{ATP}$ value of 10.5 $\pm$ 0.5 is valuable in predicting the relationship of growth and energy production for most bacteria. A few organisms have a $Y_{ATP}$ value that is either larger or smaller than 10.5 $\pm$ 0.5 and this may be the result of the presence of storage products in the cells or erroneous assumptions about the catabolic pathways used. Generally it has been found that when cultures deviate from the $Y_{ATP}$ value of 10.5 $\pm$ 0.5 it is because growth is influenced by (1) the amount of energy required for cell maintenance, (2) the specific growth rate, (3) the pathways for assimilation of carbon and nitrogen sources into organic compounds, or (4) an energy requiring transport or respiratory activity. If the $Y_{ATP}$ value is low, it could be the result of an uncoupling of growth from energy production and this occurs when (1) minimal medium is used, (2) chemicals are present that inhibit cell growth, or (3) the temperature for incubation exceeds that needed for optimal growth. A limitation of the $Y_{ATP}$ value is the inability to predict the growth yield of aerobic, heterotrophic cultures and the metabolic pathway used may vary with the stage of growth and with the concentration of glucose in the media.

In an effort to evaluate both anaerobic and aerobic bacterial growth and to express it in a yield coefficient, W.J. Payne proposed that growth be expressed as $Y_{kcal}$. By measuring the heat of combustion of dry cells produced under specific growth conditions, it was determined that 5.3 kcal is required for the biosynthesis of each g dry weight of cells. The $Y_{kcal}$ was calculated according to the following equation:

$$Y_{kcal} = Y_{(s)} / E_a + E_d$$

where $Y_{(s)}$ is the g dry weight of cells, $E_a$ is 5.3 kcal/mol substrate used multiplied by $Y_{(s)}$, and $E_d$ is the difference obtained after subtracting the heat of combustion of end products from the heat of combustion of the substrate. The values for $Y_{kcal}$ are estimated to be 0.119 g dry weight cells/kcal of substrate used. Unlike the $Y_{ATP}$ value, the $Y_{kcal}$ value is the same for growth of both aerobic and anaerobic bacteria.

## 7.10.3. Growth Yield and Efficiency[11–13]

The yield coefficient frequently used to characterize growth is $Y_S$, which is calculated according to the following formula:

$$Y_S = X - X_0 / C_0 - C$$

where $X$ is g dry weight of cells at the end of log phase, $X_0$ is the g dry weight of cells near the beginning of the log phase, $C$ is the number of moles of substrate present/ml at the time when $X$ is determined, and $C_0$ is the number of moles of substrate present at the time when $X_0$ is determined. By definition, $Y_S$ is the g dry weight of cells produced per mole of substrate consumed.

The efficiency for converting free energy from the substrate into ATP for cellular biosynthesis can be determined by comparing actual growth yield to that which could be theoretically produced from the specific reaction. The calculations are as follows:

Table 7.10. Energy yield for anaerobic reactions.

| Characteristic | Alcoholic Glycolytic pathway | Homolactic Glycolytic pathway | Phosphoroclastic pathway |
|---|---|---|---|
| Reactions | Glucose → 2 ethanol + 2 formate | Glucose → 2 lactate + 2 H⁺ | Pyruvate + 2 H₂O→ acetate + formate |
| Moles ATP produced per mole substrate used | 2 | 2 | 1 |
| $Y_s$ (grams cells/mole of substrate) | 21 | 21.5 | 10 |
| $Y_{ATP}$ | 10.5 | 10.5 | 10 |
| $\Delta G'_0$ (kcal/reaction) | −53.9 | −46.8 | −12.1 |
| Efficiency[a] | 29.7% | 35.0% | 62.9% |

[a]See text for calculation.

$$Y_{(ts)} = \frac{\Delta G'_0 \text{ (kcal/reaction)} \times 10.5 \text{ g cell dry weight per mole of ATP}}{7.5 \text{ kcal/mol (energy from hydrolysis of ATP)}}$$

The efficiency of growth is calculated from the following formula:

$$X\% = \frac{Y_s \times 100}{Y_{(ts)}}$$

$Y_s$ is obtained from experimental measurements. The energy yields for various anaerobic reactions have been calculated and they are expressed in Table 7.10.

## 7.11. Maintenance Energy[14]

Energy is required to maintain cellular homeostasis and research on maintenance energy in bacteria was pioneered by L.J. Pirt. As a result of his efforts, it is now recognized that maintenance energy contributes to the following: (1) turnover of cellular materials such as lipids in the membrane, peptidoglycan in the cell wall, mRNA, and cytoplasmic proteins; (2) maintaining intracellular pH and appropriate ionic character of the cytoplasm; and (3) maintaining a cytoplasmic concentration or pool of amino acids, nucleic acids, vitamins, and other metabolites. Perhaps the greatest energy demands for cell maintenance are associated with membrane transport–related activities. Maintenance energies are not constant but may vary considerably with the species, phase of growth, and environmental conditions. Approximately 5% to 10% of cell energy may be used for maintenance energies when cells grow on complex or rich media but as much as 18% to 20% of cell energy used for maintenance when cells are growing in minimal media. In rapidly metabolizing cells, the energy needs for maintenance are easily met and cellular growth proceeds; however, when the rate of metabolism is minimal there may be insufficient energy for cell maintenance and the cells dies. With a reduction in growth rate due to a change in pH, temperature,

## Box 7.2
## Cellular Energy and ATP

ATP is the universal energy source in biology and Fritz Lipmann, in 1941, described the intracellular activity of this "high-energy compound." Because the hydrolysis of ATP produces ADP plus inorganic phosphate ion, there is a common perception that a specific bond in ATP is responsible for holding this energy. However, at neutral pH, energy in ATP is attributed to resonance forces of the phosphate moieties, the release of a proton in the hydrolysis, and the disordering of water molecules around the phosphates. At pH 7.0, ATP has a −4 charge while ADP has a −3 charge. Energy released from ATP hydrolysis is also dependent on the concentration of $Mg^{2+}$ in the system. The hydrolysis of ATP in the absence of $Mg^{2+}$ yields −8.4 kcal/mol; however, under physiological conditions in the presence of physiological levels of 10 m$M$ $Mg^{2+}$, the value of −7.6 kcal/mol most accurately expresses the quantity of energy released from ATP. Some prefer to express the energy as kJ/mol of ATP rather than kcal/mol and for these expressions one can use a conversion of 1 cal = 4.184 J. As listed below, there are several "high-energy compounds" in bacterial cells; however, ATP is most important for cellular energy because it participates in the greatest number of reactions.

| Energy-rich compounds | kJ/mol | kcal/mol |
|---|---|---|
| Adenylyl sulfate (APS) | −88.0 | −21.0 |
| Phosphoenol pyruvate (PEP) | −51.6 | −12.3 |
| Acetyl ~ phosphate | −44.8 | −10.7 |
| Acetyl ~ CoA | −35.7 | −8.5 |
| **Adenosine triphosphate (ATP)** | **−31.8** | **−7.6** |
| Adenosine diphosphate (ADP) | −31.8 | −7.6 |

or other parameters of growth, there is a decrease in observed growth yields which is attributed to an increase in maintenance energy.

Maintenance energy in bacteria can be expressed by the following equations:

$$1/Y_{glu} = m_s/\mu + 1/Y_{glu}^{Max}$$

where $Y_{glu}$ is the molar growth yield for glucose, $m_s$ is the maintenance coefficient (mol glucose/g dry wt cells/hr), $\mu$ is the specific growth rate, and $Y_{glu}^{Max}$ is the growth yield for glucose corrected for maintenance energy.

In terms of maintenance energy using the ATP growth yield, the following equation can be used:

$$q_{ATP} = \mu/Y_{ATP} = \mu/Y_{ATP}^{Max} + m_e$$

where $m_e$ is the maintenance coefficient ($mol_{ATP}$/g dry wt cells/hr) and $q_{ATP}$ is the specific rate of ATP production (mol ATP/g dry wt cells/hr). For *Enterobacter aerogenes* grown under conditions in which glucose is the limiting factor, values for $m_e$ range from 7 to 9 mol ATP/g dry wt cells/hr while values for $Y_{glu}^{Max}$ vary from 14 to 20 mol ATP/g dry wt cells/hr.

## 7.12. Starvation Response[15]

There are many reasons for cells to cease dividing and enter the stationary phase. Perhaps the most clearly recognized event for cells occurs when a specific nutrient becomes limiting for a microbial culture. As listed in Table 7.11, a series of changes occur as a culture enters stationary phase as a result gene induction in response to starvation. When cells respond to a nutrient deficiency, a series of genes are expressed to produce proteins that enable the organism to persist in a nongrowing state. Some of the deficiencies to elicit a response in prokaryotes include the Pho system for phosphate starvation, the SSI response for sulfate starvation, and the siderophore response for $Fe^{3+}$ starvation. In a general sense, the limitation for a nutrient generally causes the cell to produce solute binding proteins which are located in the periplasm of Gram-negative cells or attached to the outer face of the plasma membrane of Gram-positive cells. In addition to induction of important genes, nutrient starvation will also account for various physiological responses as the cells conserve energy.

### *7.12.1. Physiological Changes*[16–18]

The physiological response of bacteria in stationary phase can produce several cellular changes. As seen with Gram-negative bacteria responding to prolonged starvation of

Table 7.11. Genes induced with starvation as cells enter the stationary phase.[a]

| Genes | Activity of gene product |
| --- | --- |
| *katE* | Production of HPII catalase |
| *xthA* | Production of exonuclease III |
| *cstaA* | Encodes for a peptide transport system |
| *appA* | Production of acid phosphatase |
| *dnaK, groEL,* and *htpG* | Production of heat shock proteins |
| *bilA* | Contributes to cell shape |
| *osmB* | Produces a lipoprotein |
| *treA* | Produces periplasmic trehalase |
| *glgS* | Produces the primer for synthesis of glycogen |
| *dps* | Produces proteins for DNA protection |
| *csgA* | Produces fibrocectin-binding fibers |

[a]From Hochman, 1997.[8]

several weeks, there is a dramatic change in morphology with cells becoming markedly smaller and spherical. These smaller starved cells or ultramicrobacteria (UMB) are a result of cell division without an increase in cell mass and are considered to give cells an advantage as a result of increase in cell numbers, increase in surface-to-volume ratio of cells, and condensation of the cytoplasm. Although the formation of UMB is readily demonstrated in marine bacteria, these morphological changes are expected also in terrestrial bacteria. Owing to the small size of UMB and potential migration in subsurface environments, UMB have been proposed to augment native bacteria for remediation of contaminated terrestrial environments. However, UMB may have reduced viability which would limit the effectiveness of planned introduction. When UMB are introduced into adequate nutrients, cells grow and return to the morphology that is characteristic for that species.

Several structural changes occur as cells become starved and enter the stationary phase. Hydrophobic molecules are more abundant on the surface of the bacteria to promote adhesion to structures in the environment and cell aggregation. Cell wall alterations are observed to increase resistance to autolysis. In *E. coli* and *Virbio* spp., it has been observed that cell surfaces will produce new structures including pili and fibronectin-binding filaments. Capsule formation generally increases dramatically in stationary phase which would enable bacteria to resist phagocytosis in natural environments. It is proposed that histone-like proteins are produced as a result of cells entering the stationary phase and these proteins would account for changes in negative superhelical density of plasmids and condensation of genomic DNA. In the photosynthetic bacterium *Rhodobacter capsulatus*, the nucleus disintegrates as cells reach the stationary phase with the formation of nucleoprotein particles called "gene transfer agent." These DNA-containing protein aggregates have molecular masses of about $3 \times 10^6$ kDa and all segments of the genome may be taken up by other strains of *Rhodobacter capsulatus*.

The second physiological change in bacteria is maintenance of cell viability while in stationary phase. One of the important characteristics is the alteration in proteins synthesis. With cells growing in minimal medium with glucose as the carbon/energy source, protein synthesis is reduced 20% with the onset of starvation and this reduction in protein synthesis continues so that at 11 days the rate is about 0.05% of a growing cell. In *E. coli*, 15 to 30 new proteins appear to be produced as a result of starvation and these have been designated as the *post-ex*ponential (Pex) proteins. Associated with macromolecular synthesis in the stationary phase would be proteins responsible for protection of bacteria against different environmental insults. Gram-negative cells in starvation are more resistant to heat shock, oxidative stress, antibiotic challenges, and osmotic perturbations than cells in the log phase. It is seen commonly that flagellar activity is greatly diminished in the stationary phase as compared to rapidly growing cells. Cell viability in the stationary phase is generally reduced when a minimal medium is employed but cells in complex media have fewer deficiencies and may retain cell viability at very high levels for many weeks. In the stationary phase, cryptic (i.e., hidden) growth occurs as a result of nutrients released from the lysis of dead cells. Often, the cells that grow in stationary phase are mutants that normally occur in extremely small numbers in the population but will not compete with the wild-type. Starvation signals expression of the *rpoS* gene which produces the sigma factor designated as $\sigma^{32}$ or $\sigma^s$ and is used for the expression of a large number of related genes. The mechanisms accounting for the induction of the *rpoS* gene and the physiological effects of these gene products attributed to $\sigma^{32}$ are unclear at this time.

Finally, some important changes are associated with cells exiting the stationary phase to initiate growth. With the introduction of nutrients to alleviate starvation, RNA synthesis can be observed within 20 sec and protein synthesis is observed within 2 min. As determined by the use of antibiotics, not all proteins synthesis is attributable to de novo RNA synthesis but long-life transcripts must already be present in the starved cell. Once growth is initiated, the production of the unique sigma factor, $\sigma^{32}$, ceases and the starvation-related genes cease to be expressed. In *E. coli*, there is the induction of the *fis* gene which encodes for a protein that binds to DNA and enables the bacterium to respond quickly to nutrient availability with an upshift in growth.

## 7.12.2. The Pho System[16–18]

The preferred form of phosphorus for cell metabolism is inorganic phosphate and when it is at growth-limiting concentrations in the environment, bacteria display a phosphate-starvation response. An inorganic phosphate concentration of 100 $\mu M$ enables *E. coli* to grow unrestricted; however, at 0.16 $\mu M$, the Pit transport system for inorganic phosphate is not saturated and the uptake of inorganic phosphate occurs at a minimal level. Under conditions of phosphate starvation, the *pho* regulon system is induced and cell division is extended from 80 min to 16 hr. In *E. coli*, inorganic phosphate interacts with more than 20 different promoters which influences the production of about 100 genes. One of the activities controlled by the concentration of inorganic phosphate is the *pho* regulon which consists of at least 30 genes for the utilization of alternate phosphorus sources. The regulatory genes for phosphorus utilization are *phoR* and *phoB* which are representatives of the two-component regulatory system in bacteria. A model indicating expression of the *pho* regulon is given in Figure 7.14. The concentration of inorganic phosphate is sensed by PhoR, a transmembrane protein, and a conserved histidine moiety of PhoR is autophosphorylated. The phosphate group is transferred from PhoR~P to a conserved aspartyl residue on PhoB, a cytoplasmic protein. The PhoB~P molecule is a transcriptional activator that recognizes an 18 nucleotide sequence known as the "*pho* box" which is located at the −10 region of the genes. As part of the promoter, genes in the *pho* regulon have a *pho* box that contains a consensus nucleotide sequence of CT(G/T)TCATA(A/T)A(T/A)CTGTCA(C/T). PhoB consists of 229 amino acids that have three functional segments: domain I contains the phosphorylation site, domain II binds to the *pho* box, and domain III interacts with the RNA polymerase. In the *pho* regulon, there are seven operons all with a *pho* box and these genes are associated with utilization of phosphate and phosponate compounds. Compounds with a phosphorus atom bound directly to a carbon atom are phosphonates.

## PHO REGULON

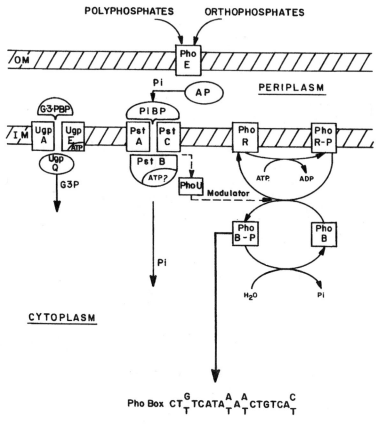

Figure 7.14. Model of the *pho* regulon with induction by phosphate starvation. PhoE, porin E; G3P, glycerol-3-phosphate; G3pBP, G3P binding protein; UgpA and UgpE, transmembrane proteins that form a channel for G3P uptake; UgpQ, glycerophosphororyldiesterphosphodies-terase; AP, alkaline phosphatase; PiBP, periplasmic binding protein for inorganic phosphate; PstA and PstC, transmembrane proteins that form a channel for inorganic phosphate; PstB, protein for active transport of $P_i$; PhoR, the $P_i$ sensory protein kinase; PhoB, inducer for the *pho* regulon; and PhoU, protein associated with the membrane that modulates $P_i$ transduction. (From Makino, K., M. Amemura, S.-K. Kim, A. Nakata, and H. Shinagawa. 1994. Mechanism of transcriptional activation of the phosphate regulon in *Escherichia coli*. In *Phosphate in Microorganisms* (A. Torriani-Gorini, E. Yagil, and S. Silver, eds.). American Society for Microbiology Press, Washington, D.C. Used with permission.)

## 7.12.3. The SSI Response[19,20]

From proteomic studies, it was determined that several *s*ulfate *s*tarvation-*i*nduced (SSI) proteins can be found in various species of Gram-negative and Gram-positive bacteria. The SSI response is specific for utilization of sulfur compounds just as the *pho* response is for phosphates and phosphonates. Among proteins produced as part of the SSI response are transport proteins including periplasmic binding proteins specific for sulfate, cysteine, and sulfonates. The production of arylsulfatase, and perhaps other sulfatases, is stimulated in the absence of sulfate. In the various bacteria tested, available sulfur prevents synthesis of the enzymes of the cysteine biosynthetic pathway by repressing the *cys* genes. With a reduction of sulfate, cyanobacteria produce an alternate set of phycobiliproteins that have reduced quantities of cysteine and methionine.

## 7.13. Activities in the Post-Exponential Phase

The onset of the stationary phase is not a shutting down of metabolic activity but frequently cells display new characteristics. Typically, microbial cultures respond to stationary phase with three activities: (1) At the early sensing of a nutrient limitation, an induction of genes occurs to correct the deficiency. (2) While cells are in the stationary phase, the metabolism changes to accommodate post-exponential phase stresses. (3) It is thought that in preparation for cells reentering the rapid phase of growth, there will be a series of molecular preparations; however, this adjustment is poorly understood. Rapidly growing cultures display an increase in cellular synthesis attributed to primary metabolism while slowly growing cells have activities resulting from secondary metabolism. Some of the activities of cells displaying secondary metabolism include development of motility in attached cells, establishment of competence for acquisition of DNA, developmental changes in cellular morphology, synthesis of antibiotics, production and release of extracellular products of metabolism, and production of spores. Cells in stationary phase produce many different types of secondary metabolites including heterocyclic compounds that come from aromatic amino acid biosynthesis. Cells in the post-exponential phase remain dynamic and responsive to the chemical and physical environment.

## 7.13.1. Antibiotic Production[21,22]

Antibiotic production by most *Streptomyces* does not occur during the phase of rapid growth but during a period when growth slows and cells enter the stationary phase. With the growth of *Actinomycetes* on solid surfaces, an additional secondary event includes the formation of aerial hyphae. The reason for the initiation of secondary metabolism is not known but it is considered to result from the expression of global physiological controls which may vary with the various species. In some *Streptomyces*, an increase in levels of ppGpp have been considered important for induction of secondary metabolism. In *Streptomyces coelicolor* A3, genes important in secondary metabolism include *abs*, *aba*, *afs*, and *bld*. Although the products of all these genes are not known, the function of *bldA* is to code for the tRNA that recognizes TTA, the leucine codon rarely used in most systems. This suggests that TTA codons are absent in genes for vegetative growth and perhaps regulate secondary metabolism. Strains of *Streptomyces griseus* use factor A (a γ-butyrolactone), which is a product of primary metabolism. Although factor A is produced is very small amounts during primary metabolism, it will accumulate in older cultures to sufficient levels and binds to critical cell proteins. The result of modifying certain proteins is the switching on of antibiotic production and formation of aerial hyphae. The reason for the emergence of these two phenotypes may appear unrelated; however, many scientists now consider that the production of antibiotics may function to exclude other organisms from a microcolony, thereby enabling *Streptomyces* to utilize nutrients released from lysed cells for the biosynthesis of aerial hyphae with formation of spores.

As a result of secondary metabolism, many *Bacillus* and other Gram-positive bacteria produce hundreds of different molecules and some of these compounds have been found to be highly effective antibiotics. The function of these antibiotics appears to enable the producing bacteria to compete more effectively for limited nutrient resources. The formation of these antibiotics can take place by either of the following mechanisms: ribosomal-mediated production or nonribosomal-mediated production. These antibiotics are commonly cyclic peptides consisting of fewer than 24 amino acids. The peptides from the ribosomes are extensively processed by posttranscription mechanisms in the formation of the active antibiotic and these activities are discussed in Chapter 12. An alternate mechanism of production of antibiotics involves multienzyme complexes in which amino acids are added to the growing peptide following stereospecific binding. Commonly present in antibiotics that display nonribosomal synthesis are D-forms of amino acids and some of the best known cyclic peptides in this category include Gramicidin S, Tyrocidine, and bacitracin (see Chapter 2 for the structure of bacitracin). D-forms of amino acids are not present in ribosomal-mediated antibiotic production because there are no codons specific for the D-amino acids. Synthesis of peptide antibiotics is enzyme directed and is similar to peptidoglycan synthesis in bacterial cell walls. In some instances, lipopeptide antibiotics are produced in which the cyclic molecules consist of amino acids and a fatty acid. Surfactin, produced by *Bacillus subtilis*, is one of the most powerful biosurfactants known and its structure is given in Figure 7.15.

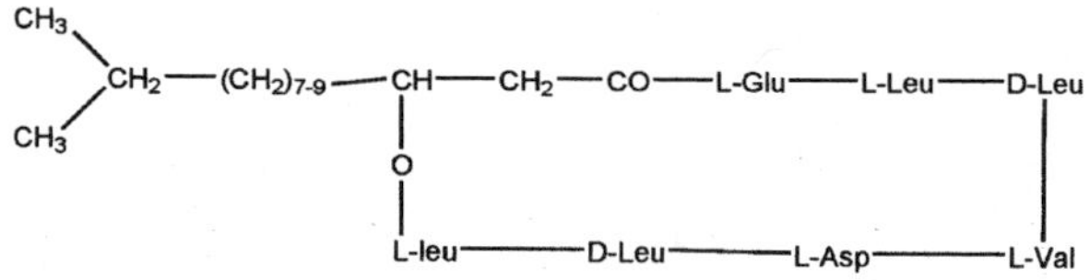

Figure 7.15. Structure of a chemical produced by bacteria. This molecule is known as surfactin.

## 7.13.2.  Toxin Production[23]

Some potent animal toxins are produced by clostridia as cells sporulate; however, most of the time the toxins are released as cells lyse or toxins are released from precursor molecules by proteases.  Using mutants blocked at a definite stage of sporulation, it was found that *Clostridium perfringens* produces the enterotoxin associated with food poisoning as a product of secondary metabolism.  The enterotoxin is encoded on a sporulation-specific gene with production initiated at stage III of sporulation, and synthesis of enterotoxin continues until the spore is released from the sporangium.  Enterotoxin-like proteins have been detected in the spore coat of *Clostridium pergringens* where these proteins are assumed to have a role in maintaining spore integrity.  Some investigators have speculated that non-enterotoxin strains of *Clostridium perfringens* would efficiently regulate the formation of these enterotoxin-like proteins but that enterotoxin-producing strains would be defective in regulation and overproduce this protein used in formation of the spore coat.

Some of the protein toxins produced by *Bacillus* are products of secondary metabolism when the cells have shifted to a slower metabolic rate.  At the time of sporulation, certain *Bacillus* are known to produce parasporal inclusions.  As shown in Figure 7.16, these parasporal crystals appear as a compact body in the cytoplasm of the cell.  Physiologically, these protein crystals are important because they are toxic when ingested by various insect larvae.  These parasporal inclusions are called protoxins and in *Bacillus thuringiensis* these proteins are referred to as δ-endotoxins. More than 50 different protoxins are known to be produced by *Bacillus* and usually the genes for their production are carried on megaplasmids.  The estimated mass of each of these protoxins is about 130 kDa and a list of some of these parasporal inclusions is given in Table 7.12.  In *Bacillus*, the synthesis of RNA polymerase specific for parasporal inclusions has been found to require the sigma factors of $\sigma^E$ and $\sigma^K$.  These are the same sigma factors involved in sporulation in that $\sigma^E$ is active over stages II to IV and $\sigma^K$ is involved with stages IV and V.  It appears that there

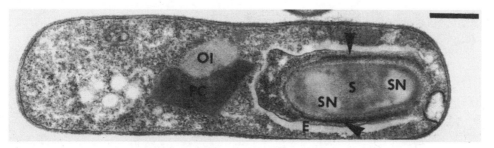

Figure 7.16. Parasporal body in *Bacillus thuringiensis*. On the lateral sides of the spore (S) is the spore coat (arrowheads). SN is the spore nucleoid, E is the exosporium, OI is ovoid inclusions, and PC is the parasporal crystal. Bar is 0.5 μm. (From Bechtel and Bulla, 1976.[43] Used with permission.)

Table 7.12. Insecticidal toxins produced as products of secondary metabolism.[a]

| Gene designation | Subspecies of *B. thuringiensis* where gene was found | Susceptible insects |
|---|---|---|
| *cryIA(a)* | *kurstaki* HD1 | (L) *Bombyx mori* |
| cryIA(c) | *kurstaki* HD1 | (L) *Heliothis virescens* |
| *cryIB* | *thuringeniensis* | (L) *Pieris brassicae* |
| crylG | *galleria* | (L) *Galleria mellonella* |
| cryIIIA | *san diego* | (C) Colorado potato beetle |
| cryIVA | *israelensis* | (D) *Aedes aegypti* and black flies |

[a]From Aronson, 1993.[23]
L, Lepidoptera; C, Coleoptera; D, Diptera.

may be a competition for $\sigma^E$ and $\sigma^K$ in the formation of sporulation and protoxin. It should be noted that a compensation for protoxin production is that it occurs over a broad period of sporulation. In addition, the half-life of mRNA for protoxin formation is about 12 min, which is in contrast to the half-life of 4 min for the sporulation mRNA. It has been suggested that this stability of protoxin mRNA is attributed to a stable stem-and-loop structure that is present at the 3' end of the molecule.

Cyanobacteria are the organisms that are commonly associated with nuisance blooms on lakes and in the 1970s as a result of studing eutrophication activities, cyanobacteria were determined to produce potent toxins. In certain strains of *Anabaena flos-aquae*, *Microcystis aeruginosa*, and *Aphanizomenon flos-aquae*, toxins are produced only in the late phase of growth, with the quantity of toxin produced influenced by light intensity, temperature of incubation, and presence of heterotrophic bacteria. Anatoxin-a is an eight-membered bicyclic ring with a molecular formula of $C_{10}H_{15}NO$ [see Figure 7.17(a)] and is produced by toxin-producing strains of *Ana-*

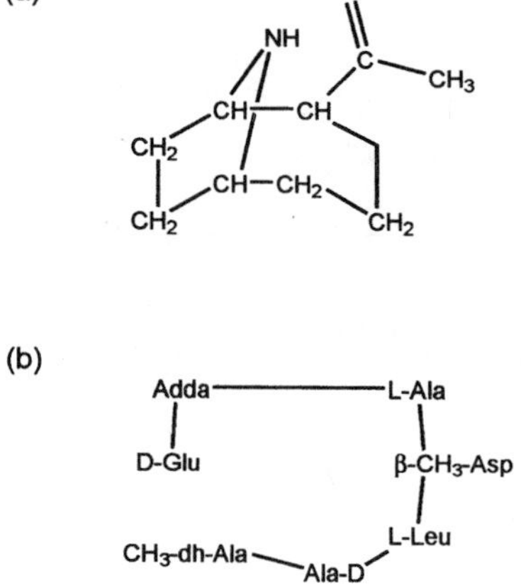

(a)

(b)

Figure 7.17. Toxins produced cyanobacteria. (**a**) anatoxin-a and (**b**) cyanoginosin-LA.

*baena flos-aquae*. Anatoxin-a is a neuromuscular inhibitor that is lethal to mammals, birds, and fish when injected intraperitonally. The anatoxin-a has attracted considerable interest owing to its structural similarity to cocaine. Strains of *Microcystis aeruginosa* produce heptapeptide toxins that are lethal to mammals and birds as they cause extensive bleeding. The best characterized toxin from *Microcystis aeruginosa* is cyanoginosin-LA which is a cyclic peptide containing unusual amino acid residues and some amino acids of the D-form [see Figure 7.17(b)]. A few strains of *Aphanizomenon flos-aquae* produce toxins that are also lethal when injected into animals and these toxins are related to saxitoxins produced by dinoflagellate, *Gonyaulax*, and concentrated by shellfish.

## 7.13.3.  Sporulation[24–28]

As a result of nutrient deprivation or exposure to hostile environments, some bacteria adjust with the formation of spores. Regulation of gene expression for sporulation occurs in the stationary phase by a process known as signal transduction which involves the activity of two proteins. A model showing this regulation is given in Chapter 6. One protein, in the plasma membrane, receives a signal and becomes phosphorylated through the reaction with ATP. The second is a regulator protein that interacts with DNA after it becomes phosphorylated following an encounter with the phosphorylated membrane protein. This regulation by a two-component signal transduction system is highly interdependent on other cellular activities and resolving reaction intricacies has contributed to a slow resolution of activities in the stationary phase of growth.

The formation of the bacterial spore occurs as the growing cell is subjected to conditions of starvation or some chemical signal and the resulting dormancy enables the species to persist under hostile conditions. Thus far, spore formation has been associated with members of the Domain *Bacteria* but not of the Domain *Archaea*. As can be seen in Table 7.13, there is considerable diversity in the bacterial types that produce resting cells. The spores produced by these various bacteria do not share the same level of resistance to physical and chemical changes but it varies with the species. With all systems studied, the development of sporulation has been found to be a highly ordered event. Because the molecular biology has been aggressively pursued, *Bacillus* has become the model for spore research and the following generalizations result from this research.

### 7.13.3.1. Morphological Development

The morphological changes of spore development in *Bacillus* spp. follow a specific pattern and can be divided into seven stages, with each stage persisting for about an hour. Electron micrographs of a cell producing a spore throughout the spore cycle are given in Figure 7.18. Phenotypic characteristics of spores in each of these stages are given in Table 7.14. Initially condensation of DNA occurs in the mother cell with the DNA forming a fibril that extends from one pole of the cell to the other pole. The segment of the genome destined for the spore is segregated by a membrane that appears to extend from the plasma membrane and between the two layers of this

Table 7.13. Examples of bacteria that produce spores.

| Endospore production | | Spores less resistant than endospores | |
|---|---|---|---|
| Genus | Characteristic of genus | Genus | Structure |
| *Alicyclobacillus* | Formerly *Bacillus acidocaldarius*, optimum growth at pH 3 | *Azotobacter* | Azocyst |
| | | *Coxiella* | Small resting cell |
| *Amphilobacillus* | Facultative anaerobe, optimum growth at pH 9 | Cyanobacteria: | |
| | | *Anabaena* | Akinete |
| | | *Chamaesiphon* | Exospore |
| *Anaerobacter* | Cells fix $N_2$, 1–5 spores/cell | *Pleurocapsales* | Baeocytes |
| | | *Westiella* | Hormocytes |
| *Bacillus* | More than 650 mesophilic strains, aerobic or facultative growth | Streptomyces | Conidia or hyphal spore |
| | | Fruiting body bacteria | Myxospores |
| *Clostridium* | Heterotrophic fermentative anaerobes | *Methylocystis* | Exospore |
| | | *Methylomonas* | Cyst-like body |
| *Desulfitobacterium* | Reductive dechlorination of chlorophenols | | |
| *Desulfotomaculum* | Sulfate-respiring anaerobes | | |
| *Heliobacterium* | Anaerobic photosynthetic | | |
| *Pasteuria* | Parasite for nematodes, host-dependent growth | | |
| *Sporohalobacter* | Halophile | | |
| *Sporolactobacillus* | Microaerophilic, lactic acid producer | | |
| *Sporomusa* | Gram-negative, grows on $H_2 + CO_2$ | | |
| *Sporosarcina* | Coccus, aerobe, spores in tetrads | | |

membrane there is extensive spore-related activity. The membranes for endospore development are described in Figure 3.24. The development of the endospore is reversible until the forespore is produced and at this time the cell becomes committed to sporulation. The genomes of both the mother cell and the forespore are active during sporulation, with about 20% of the genome associated with sporulation. Early in the forespore development intracellular proteases degrade various proteins to amino acids that support the biosynthesis of the endospore.

As the forespore develops in the mother cell of *Bacillus subtilis*, the spore wall develops between the layers of the double membrane that surrounds the developing spore. The spore peptidoglycan consists of two layers: (1) the germ wall, which is a relatively thin layer that is next to the spore plasma membrane and (2) the cortex wall, which is thicker and forms the outermost layer. The germ wall appears to contain *N*-acetylglucosamine (NAG) and *N*-acetylmuramic acid (NAM) while the cortex wall is suggested to contain NAG, NAM, and muramic-δ-lactam which is a modified NAM. The structure of the modified NAMs in the spore cell wall is given in Chapter 3. There is an equal distribution of NAM and muramic-δ-lactam in the peptidoglycan of the cortex wall with muramic-δ-lactam occurring in every second NAM position.

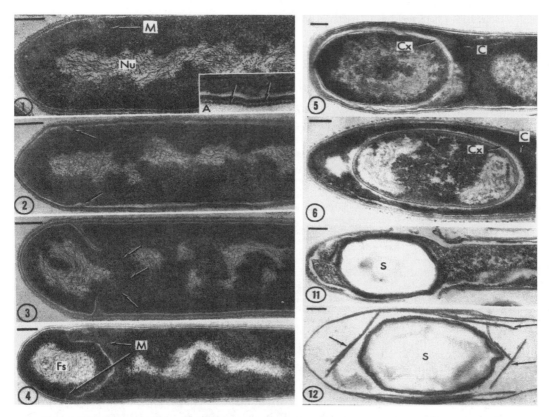

Figure 7.18. Sequential stages in bacterial sporulation. Nu = nuclear material, M = membranes, Fs = forespore, Cx = cortex, C = coat, S = spore, A = inset showing S-layer. Age of cells is as follows: 2 = 6 hr, 3 = 8 hr, 5 = 10 hr, 6 = 16 hr, 11 = 25 hr, and 12 = 33 hr. Bar is 0.2 μm. (From Santo et al., 1969.[25] Used with permission.)

Because the muramic-δ-lactam residues are without the standard peptide chain, they have no potential for stabilizing the structure through cross linking. Approximately one fourth of the NAM have been subjected to enzymatic digestion to leave L-alanine as the only residue and without a peptide chain. Thus, although 40% of the peptidoglycan structure is cross linked in vegetative bacteria, only about 4% of the NAM residues in the bacterial spore wall are secured by cross-bridge linkages. The low extent of crosslinking in peptidoglycan in the cortex wall is considered to be important for spore germination and outgrowth of the new cell.

The germ cell wall is formed first and it remains to be determined if the precursors for the cell wall are derived from the developing endospore or from the mother cell. The germ wall is analogous to the cell wall peptidoglycan in that it is highly cross linked. The peptidoglycan of the germ wall serves to stabilize the germinating spore and to serve as a site for synthesis of vegetative cell wall during germination. Al-

Table 7.14. Characteristics of bacterial sporulation.[a]

| Stage | Developmental activity | Enzymes present in highest levels |
|---|---|---|
| I | DNA forms an axial filament between the poles of the cells. | Serine protease, ribonuclease, α-amylase |
| II | DNA for spore is localized and asymmetric septation is initiated. | NADH oxidase, cytochrome synthesis, arginase |
| III | UV resistance of spore is observed. DNA for spore is totally engulfed by membranes from mother cells and the forespore is suspended in the cytoplasm. Coat proteins synthesized. | Alkaline phosphatase, DNase, aconitase, alanine dehydrogenase |
| IV | Prespores appear gray under light microscopy. Spore cell wall is laid down and cortex synthesis occurs. | Catalase, glucose dehydrogenase, diaminopimelate ligase |
| V | Prespores are refractile (bright) under light microscopy. Coat proteins assembled and Resistance to octanol, butanol, ethanol, and methanol is observed. | Adenosine deaminase, dipicolinate synthase, ribosidase, dipicolinic acid accumulation, $Ca^{2+}$ transport, alanine racemase |
| VI | Mature spore is produced along with heat resistance. Resistance to chloroform and phenol is observed. | |
| VII | Lysis of mother cell occurs and spore is released. | |

[a]Data from Mandelstam et al., 1982.[27]

though the spore walls do not have an active role in dehydration of the developing spore, they do have an important activity in maintaining spore wall stability while dehydration does occur. Some have suggested that $Ca^{2+}$ in the cortex complexes with the polyanionic cortex wall which results in contraction of the cortex and exclusion of water from that region of the developing spore.

Although heat resistance is a hallmark feature of endospores, the mechanism of resistance is not definitively established but may be a reflection of the dehydrated character of the spore. Some have suggested that the presence of $Ca^{2+}$ chelated to dipicolinic acid provides heat resistance as $Ca^{2+}$ and dipicolonic acid are found only in the cortex of the highly heat-resistant bacterial endospore. The structure of dipicolinic acid is given in Figure 7.19. It is not that chelated $Ca^{2+}$ serves as a thermoinsulator but it has a role in desiccation and maintaining the spore in the dehydrated state. Free water is not present in spores, and as a consequence steam is not generated inside the spore structure as the temperature is increased. Other bacterial spores that are really not endospores may employ other mechanisms to dehydrate the developing spore; however, the removal of water is not always complete and these non-endospores are sensitive to elevated heat.

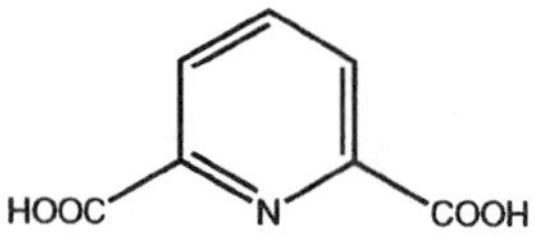

Figure 7.19. Structure of dipicolinic acid.

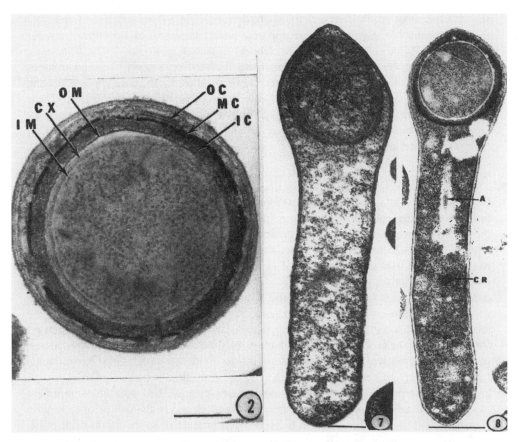

Figure 7.20. Fine structure of the endospore of *Clostridium cochlearium.* **(2)** Endospore showing inner coat (IC), outer membrane (OM), outer coat (OC), middle coat (MC), inner membrane (IM), and cortex (CX). **(7 and 8)** Stages in development of the endospore in the sporangium showing spore appendage (A) and crystalline inclusions (CR). (From Pope and Rode, 1969.[24] Used with permission.)

Exterior to the cortex wall is a dense protein layer that is the spore coat. An endospore consists of a keratin-like material with a high distribution of cysteine–cysteine (S-S) and dityrosine bonds. The stability of the spore coat is attributed to the posttranslation processing in proteins and this is discussed in Chapter 12. The mother cell and not the developing endospore is involved in the synthesis of the coat. By electron microscopic examination, the spore coat of *Bacillus cerus* appears as a single layer but in *Bacillus subtilis* it appears as a multilayered structure. The function of the spore coat is to protect the spore against free radicals, biocides, toxic metals, and enzymes present in the environment.

The outermost region of the spore is the exosporium, which consists of a lipid–protein membrane consisting of 20% carbohydrate. As seen in Figure 7.20, there are several layers that are produced as the spore develops. In a few cases extensions from the exosporium of the endospore are apparent and examples of exosporium structures are given in Table 7.15 with surface details presented in Figure 7.21. The spore appendages vary in appearance ranging from feather-like, ribbon-like, and pilus or tube-like structures. No active role has been proposed for these spore appendages and initial designations of these structures as organelles of motility are erroneous. These surface structures are not found in all spores but when present they may retard the rate of downward migration in a water column and contribute to dispersal in air currents and the attachment of spores to various surfaces in the environment.

### 7.13.3.2. Sigma Factors[29]

Important determinants for spore development are the specific sigma factors present in the developing endospore as well as in the mother cell. The sigma factors enable the RNA polymerase to interact specifically with the DNA by binding to the $-10$ region and the $-35$ region of the promotor. The production of specific sigma factors results in about 100 genes transcribed in *Bacillus* for cell differentiation resulting in endospore formation (see Table 7.16). The production of sigma factors is highly ordered and appears to follow a specific sequence: $\sigma^H$, $\sigma^F$, $\sigma^E$, $\sigma^G$, and $\sigma^K$. There is a spatial separation of activities with some sigma factor activities occurring in the mother cell and some in the developing spore. Functioning in the mother cell are $\sigma^A$ for routine cellular processes and $\sigma^H$ which is required to initiate sporulation processes. In stage II, during which the cell is preparing to divide asymmetrically to produce a forespore, the $\sigma^A$ that is produced by the mother cell initiates transcription of *spoIIE* and *spoIIG* operons. The gene *sigE* is one of the first transcribed with the commitment to sporulation and produces $\sigma^G$. The factor $\sigma^F$ is produced in the cell prior to septum formation with the synthesis of the forespore and is inactive in the mother cell but is activated in the prospore. The $\sigma^F$ factor is indispensable for sporulation because it directs the transcription of *spoIIIG* which results in the formation of $\sigma^G$. This factor, $\sigma^G$, is produced in the forespore and is responsible for the transcription of numerous genes that produce enzymes critical for spore formation. The $\sigma^H$ is produced in veg-

Table 7.15. Distribution of appendages on endospores from *Clostridium*.[a]

| Organism | Characteristic | Number per spore | Length of appendage (μM) |
|---|---|---|---|
| *C. bifermentans* 1A-SDH | Feather-like shaft with many filaments | 24–30 | 4–7 |
| *C. botulinum* type E | Tubular appendage with a distal hemispherical cap | 160–240 | 0.5 |
| *C. cochlearium* 6B | Tubular with a distal fibrillar region | 9–12 | 3–4 |
| *C. sordellii* | Tubular and smooth | 2–6 | 4 |
| *C. taeniosporum* | Ribbon-like and flat | 15–20 | 4–5 |

[a]From Rode, 1971.[32]

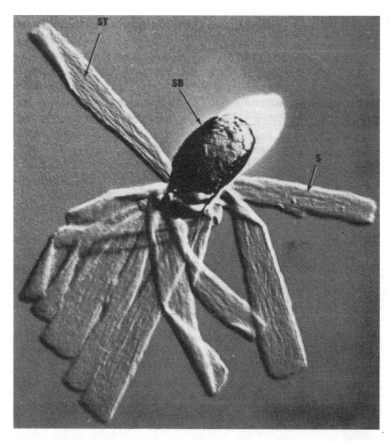

Figure 7.21. Appendages present of on spore produced by *Clostridium taeniosporum*. Extending from the spore body (SB) are ribbon-like smooth (S) or striated (ST) appendages. Carbon replica of spores was used to produce images. × 17,000. (Reprinted with permission from Rode, L.J. 1971. Bacterial spore appendages. *Critical Reviews in Microbiology* 1:7. Copyright CRC Press, Boca Raton, Florida.)

etative cells where it directs transcription of nonsporulation genes such as *citG*, the structural gene for fumarase in the tricarboxylic acid (TCA) cycle. The concentration of $\sigma^H$ increases dramatically with the onset of sporulation and $\sigma^H$ activates the promoter for the *spoIIA* operon which encodes $\sigma^F$. A similar cascade of regulation is used to produce $\sigma^A$, $\sigma^C$, $\sigma^D$, and $\sigma^L$.

### 7.13.3.3. Proteins[30–34]

Up until the time the cell has undergone differentiation to produce the forespore, the activity is reversible and the cell can revert to a vegetative cell. The formation of spore coats is dependent on the mother cell and the activity of $\sigma^K$. The spore coats

Table 7.16. Sigma factor expression during sporulation in *Bacillus* spp.[a]

| Sigma factor | −35 sequence | −10 sequence | Product of sigma factor expression |
|---|---|---|---|
| $\sigma^A$ | TTGAGA | TATAAT | Housekeeping activities of cell |
| $\sigma^B$ | AGGTTT | GGGTAT | Directs transcription of several promoters |
| $\sigma^C$ | AAATC | TANTGNTTNTA | Unknown |
| $\sigma^D$ | CTAAA | CCCATAT | Synthesis of flagella |
| $\sigma^E$ | KMATATT | CATACA-T | Stage II sporulation |
| $\sigma^F$ | X[a] | X[a] | Stage II sporulation |
| $\sigma^G$ | YGHATR | CAHWHTAH | Activity in forespore |
| $\sigma^H$ | AGGA—T | GAATT | Stage 0 sporulation |
| $\sigma^K$ | GKMACA | CATANNNT | Specific for mother cell |
| $\sigma^L$ | CTGGA[b] | TTGCA[c] | Levanase operon |

[a]From Moran, 1993.[29]
[b]Similar to that found with $\sigma^G$; [b]−24 region; [c]−12 region.

may account for about 50% of spore protein in *Bacillus subtilis*. These spore coats have been suggested to be important for spore resistance to mechanical rupture as well as resistance to chemicals and enzymes. To extract these proteins from the spore coat, it is necessary to treat them with 0.1 *M* NaOH for 15 min at 4°C before solubilizing the proteins with 1% sodium dodecyl sulfate (SDS) plus 50 m*M* dithiothreitol for 30 min at 65°C. From 10 to 20 individual proteins making up the multilayered spore and some of the properties of the spore coat proteins are given in Table 7.17. These spore coat proteins are produced by the mother cell at 3 to 4 hr after sporulation has been initiated by sigma factor $\sigma^E$.

DNA in the spore is highly condensed and there are several proteins that bind to double-stranded DNA. Approximately 8% to 10% of proteins in the spore are attributed to two different types of proteins which are referred to as *s*mall *a*cid-soluble *s*pore *p*roteins (SASP). One type of protein is the $\alpha/\beta$ type SASP which is associated with the nucleoid of the spore. In *Bacillus*, *Clostridium*, *Sporosarcina*, and *Thermoactinomyces* these small proteins of 60 to 72 amino acid residues display strong conservation of sequence, with about half of the molecule being identical. In *B. subtilis*, these proteins are encoded on *ssp* genes (*sspABCDF*) that are distributed throughout the chromosome. Production of the $\alpha/\beta$ type SASP molecules result from transcription due to the specific $\sigma^G$ at about 3 to 4 hr into sporulation. These DNA binding proteins are considered to account for stability to UV light because UV resistance is also observed at this time. On binding of the $\alpha/\beta$ type SASP to the double-stranded DNA, there is a marked conformational change of DNA and a structure similar to an A-like DNA helix is produced. DNA that has no protein associated with it has a structure markedly distinct from DNA with protein and the macromolecule of DNA that has no protein has the B-like helix. In addition to $\alpha/\beta$ type SASP, *Bacillus subtilis* spores contain a $\gamma$-type SASP which is the product of a single gene that is not associated with the spore DNA but may be localized within the spore. The production of the $\gamma$-type SASP is approximately at the time when $\alpha/\beta$ type SASP are

Table. 7.17. Characteristics of spore coat proteins in *Bacillus subtilis*.[a]

| Protein | Size (kDa) | Genetic control | Location in coat | Properties |
|---------|-----------|-----------------|------------------|------------|
| CotA | 65 | $\sigma^K$ | Outer region | Minor protein, controls brown pigment |
| CotB | 59 | $\sigma^K$ + GerE | Outer region | Minor protein |
| CotC | 12 | $\sigma^K$ + GerE | Outer region | Major protein |
| CotD | 11 | $\sigma^K$ | Outer region | Major protein |
| CotE | 24 | $\sigma^E$ | Outer region | Assembly of outer coat |
| CotF | 8 and 5+ | $\sigma^K$ | ? | Unknown function |
| CotG | 24 | $\sigma$ + GerE | Outer region | Assembles CotB, has 9 tandem copies of a 13 amino acid repeat. |
| CotH | 42.8 | $\sigma^K$ | Inner region | Assembly of outer coat |
| CotJA | 9.7 | $\sigma^E$ | Inner region | Assembly of outer coat |
| CotJC | 21.7 | $\sigma^E$ | ? | Association with CotJA |
| CotK | 6 | ? | ? | ? |
| CotL | 5.4 | ? | ? | ? |
| CotM | 14 | $\sigma^K$ | Outer coat | Control coat assembly |
| CotS | 41 | $\sigma^K$ + GerE | ? | ? |
| CotT | 7.8 | $\sigma^K$ | Inner region | Processed from a 14-kDa protein |
| CotV | 14 | $\sigma^K$, $\sigma^K$ + GerE | ? | ? |
| CotW | 12 | $\sigma^K$, $\sigma^K$ + GerE | ? | ? |
| CotX | 18.6 | $\sigma^K$, $\sigma^K$ + GerE | Outer region | Controls coat assembly |
| CotY | 17.9 | $\sigma^K$, $\sigma^K$ + GerE | Outer region | Controls coat assembly |
| CotZ | 16.5 | $\sigma^K$, $\sigma^K$ + GerE | Outer region | Controls coat assembly |

[a]From Setlow, 1993.[34]

[b]Processed by proteolytic action on a 14-kDa protein encoded on the *cotF* gene.

synthesized. On germination, all of these SASP molecules are rapidly degraded and the amino acids are used for cellular biosynthesis.

The enzyme for the hydolysis of both $\alpha/\beta$ type and $\gamma$-type SASP has been designated as the *germination protease* (Gpr). This endoprotease hydrolyzes at the conserved sequence of X-E-X-A-S-E-X to produce oligopeptides that are subsequently hydrolyzed by other peptidases. The GPR is produced from the *gpr* gene at approximately the time of SASP production; however, it is inactive because of the presence of several amino acids at the N-terminus of the GRP molecule. At the time when the cortex is being assembled and the spore is being dehydrated, 16 amino acids are removed from the GPR of *B. subtilis*. Even though these amino acids are removed, the GPR is still inactive because of the presence of one additional amino acid at the N-terminus of the GPR. One of the first activities of germination is the activation of an exoprotease that removes the terminal amino acid from GPR, and thereby converts it to an active endoprotease. The peptides are rapidly degraded to free amino acids by other proteases.

## 7.13.4. Bacteriocins[35–38]

Bacteriocins are proteins that are produced in the late log or early stationary phase of growth and have no effect on the cells that produce them but inhibit the growth of sensitive bacteria. The molecular stimulus triggering the synthesis of bacteriocins remains to be established. Since their discovery by Gratia in 1925, bacteriocin production has been demonstrated in many different types of bacteria. Collectively, bacteriocins are a heterogeneous group of proteins that have considerable range in size, structure, and activity. Bacteriocins produced by Gram-negative bacteria are 60- to 80-kDa proteins and are generally referred to as colicins because they were first demonstrated in *E. coli*. Colicins are plasmid encoded and generally affect bacteria by one of the following processes: (1) disruption of DNA in sensitive cells, (2) inhibition of protein synthesis, or (3) alteration of cell energetics by disruption of the plasma membrane.

Bacteriocins of only 20 to 40 amino acids are produced primarily by Gram-positive bacteria but a few are synthesized also by Gram-negative bacteria. These antibiotic-like molecules are divided into two groups: (1) class I, lantibiotics from *lant*hionine-containing ant*ibiotic* peptides and (2) class II, small heat-stable non–lanthionine-containing peptides. The primary mode of action of these class I and class II bacteriocins is the permeabilization of the plasma membrane through the formation of small pores with subsequent cell death. Examples of organisms producing class I and class II bacteriocins are given in Table 7.18. Class I bacteriocins have considerable posttranslation and examples of these activities are discussed in Chapter

Table 7.18. Examples of various types of bacteriocins produced by bacteria.[a]

| Type | Examples | Producers | Size (daltons) | Action |
|---|---|---|---|---|
| **Colicin** | E1 | *Escherichia coli* | 57,279 | Membrane depolarization |
| | E2 | *Escherichia coli* | 61,561 | DNA endonuclease |
| | DF13 | *Escherichia coli* | 59,283 | Ribosome inactivation |
| **Class I—type A** | | | | |
| | Nisin A | *Lactococcus lactis* | 3,353 | Permeabilize membrane |
| | Subtilin | *Bacillus subtilis* | 3,317 | Permeabilize membrane |
| | Salivaricin A | *Streptococcus salivaricus* | 2,321 | Permeabilize membrane |
| **Class I—type B** | | | | |
| | Cinnamycin | *Streptomyces cinnamoneus* | 2,043 | Permeabilize membrane |
| | Mersacidin | *Bacillus* ssp. | 1,825 | Permeabilize membrane |
| | Variacin | *Micrococcus varians* | 2,658 | Permeabilize membrane |
| **Class IIa** | | | | |
| | Leucocin A | *Leuconostoc carnosum* | 3,390 | Permeabilize membrane |
| | Enterocin A | *Enterococcus faecium* | 4,829 | Permeabilize membrane |
| | Sakacin A | *Lactobacillus sake* | 4,306 | Permeabilize membrane |
| **Class IIb** | | | | |
| | Lactacin F | *Lactobacillus johnsonii* | 6,300 | Permeabilize membrane |
| | Lactococcin G | *Lactobacillus lactis* | 4,346 and 4,110 | Both proteins are required for membrane permeabilization. |

[a]From Ennahar et al., 2000.[35]

12. Class I bacteriocins are classified according to structural and functional features, with type A consisting of linear molecules with one to four rings per peptide while type B is a globular peptide that is conformationally well defined with three or four rings per molecule. Class II bacteriocins are frequently divided into two major subgroups based on molecular composition: IIa, which contains YGNGVX as a consensus sequence at the N-terminus and IIb, which involves peptides that assemble a barrel-like structure in the plasma membrane that serves as a lethal pore. These class IIa compounds have been found to contain a molecular form consisting of four parts: the β-loop, the β-sheet, an α-helix, and a second amphiphilic α-helix unit. The β-loop has a consensus sequence of YGNDV while the β-sheet contains about 17 amino acids, the initial α-helix contains about 8 amino acids, and the final α-helix contains about 18 amino acids. A model indicating the attachment of the β-loop to sensitive bacterial cells and subsequet disruption of the plasma membrane by the amphiphilic α-helic structural at the carboxyl end of the peptide of class IIa bacteriocins is given in Figure 7.22. These class IIa bacteriocins are receiving attention from food processors because there is considerable potential to use them in foods to prevent spoilage due to *Listeria*, *Brochotrix*, *Clostridium*, *Bacillus*, and *Staphylococcus*.

## 7.14. Germination[39,40]

The spore is described as metabolically inert in part because it lacks many of the molecules needed for cellular activity. Some of these chemical constituents and macromolecules that are either absent or in low quantity in the spore are listed in Table 7.19. The chemically unstable ribonucleoside triphosphates are not stored in the spore but their stable precursors are present. Spores either lack or have greatly reduced levels of mRNA, ATP, citric acid cycle enzymes, and cytochromes but have minimal quantities of ribosomes, tRNA, and associated factors to initiate protein synthesis.

The overall process associated with the production of a vegetative cell from a spore is characterized as germination and this activity can take less than 90 min. The first of three stages in germination is activation. When placed in a favorable chemical and physical environment, spores generally will germinate; however, some require extreme conditions such as low pH, short exposure to heat, or exposure to sulfhydryl molecules for activation. The criteria used to designate germination of bacterial spores is heat inactivation, staining with bacteriological dyes, and loss of light refractivity. Germination, like sporulation, occurs as a sequence of reactions.

The regulatory signals responsible for outgrowth of a cell from the spore are difficult to assess through metabolic determinations. It is apparent that spores have signals for commitment to germination and that growth occurs when these signals are activated. Like sporulation, germination is an ordered process requiring the activity of numerous genes. Genetic analysis has revealed that the expression of *ger* genes is required for germination. One locus, *gerA*, consists of three genes (*gerAA*, *gerAB*, and *gerAC*) which are associated with utilization of L-alanine. The GerAA protein is 53.5 kDa with at least five putative membrane-spanning helices with a C-terminal hydrophilic region. The GerAB protein is 41.2 kDa with 10 potentially membrane-spanning hel-

(a)

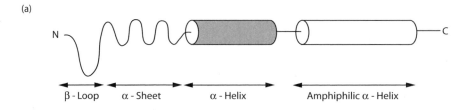

(b)

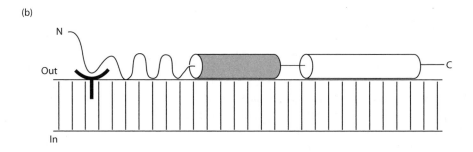

(c)

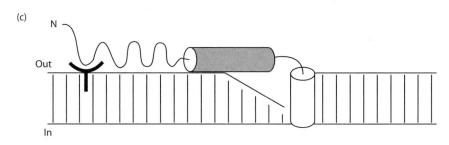

Figure 7.22. Model depicting mechanism of bacteriocin disruption of plasma membrane. (**a**) A model of the bacteriocin structure indicating four discrete regions of the molecule. (**b**) The β-loop recognizes a binding site on the plasma membrane and (**c**) once the bacteriocin is tethered, the amphiphilic "tail" of the bacteriocin penetrates the membrane and with several bacteriocin molecules attached to the membrane, a pore is created that destroys the cell. (From Ennahar et al., 2000.[35])

ices. The GerAC protein is 42.3 kDa, strongly hydrophilic, is transferred across the forespore membrane by a signal sequence, and is attached to the membrane by a cysteine–glyceride bond. The *gerB* locus also contains three genes (*gerBA*, *gerBB*, and *gerBC*) which are associated with the metabolism of asparagine, glucose, and fructose. The GerA and GerB proteins appear to be of low copy number and are produced by the sporulation sigma factor ($\sigma^G$) in the forespore at the time of sporulation. The proteins from *gerA* and *gerB* may be part of a two-component sensory and regulatory system.

Table 7.19. Abundance of metabolites in endospores and cells of *Bacillus megaterium*.[a]

| Chemical | Metabolite as μmol/g dry weight | |
| --- | --- | --- |
| | Growing cells | Spores |
| Amino acids | 1,400 | 150 |
| 3-Phosphoglyceric acid | <0.2 | 5–18 |
| ATP | 3.6 | <0.005 |
| ADP | 1 | 0.2 |
| AMP | 1 | 1.2 |
| NAD | 0.35 | 0.11 |
| NADH | 1.95 | 0.002 |
| NADP | 0.44 | 0.018 |
| NADPH | 0.52 | 0.001 |
| Glutamic acid | 38 | 26 |
| Dipicolinic acid | <0.1 | 440 |
| GTP + CTP + UTP | 680 | <10 |
| Nucleotides of G + C + U | 860 | 530 |
| Deoxyribonucleotides | 181 | <1.5 |
| $Ca^{2+}$ | — | 380–916 |
| $Mg^{2+}$ | — | 86–120 |
| $Mn^{2+}$ | — | 27–56 |
| pH | 7.5–8.2 | 6.3–6.5 |
| Ribosomes and tRNA | High levels | Very low levels |
| Cytochromes and respiratory components | High levels | Very low levels |
| TCA cycle enzymes | High levels | Absent |
| mRNA | High levels | Absent |

[a]From Leive and Davis, 1980[26] and Setlow, 1994.[28]

The second stage of germination is associated with elevated metabolic activity. Early in this stage is the formation of various hydrolytic enzymes that are associated with digesting the spore coat and other surface layers of the spore. A lytic enzyme with a molecular mass of 29 kDa has been purified from the cortex of endospores and it appears to be activated from a 68-kDa protein found in the cortex. In certain strains of *Bacillus*, an enzyme is produced that releases muramic acid δ-lactam from the spore wall. In other strains of *Bacillus*, there appears to be an activation of glucosamidase, endopeptidase or transpeptidase activity during germination. For elongation of the cell as it emerges from the spore, there would be a requirement for controlled autolysis of the spore cell wall structure. In addition, this enzymatic activity could contribute to the increased permeability of the germinating spore. As germination progresses, there is an uptake of water, increase in oxygen respiration, increase in glucose metabolism, a swelling of the spore, and release of about 30% of the spore material in the form of calcium dipicolinate, proteins, and glycopeptides occurs within minutes of initiation of germination. As the various surface layers are damaged by autolytic action, nutrients are rapidly taken into the spore. It has been determined that either $Mn^{2+}$, alanine, or valine is required for germination while a combination of asparagines, glucose, fructose, and $K^+$ stimulates germination. The formation of mRNA is rapid and provides the message for de novo enzyme formation. The energy storage molecule is 3-phosphoglyceric acid which is quickly converted to phosphoenol pyruvate and ultimately to ATP. While the pool of amino acids is absent in spores, one of the earliest stages of germination activates preoteases that hydrolyze SASPs, which serve as storage proteins to provide the needed amino acids.

The final stage is associated with the outgrowth of the vegetative cell from the spore. Two mechanistic models are proposed for outgrowth from a spore and these may function independently or in a cooperative process. One model involves the activation of a cortex enzyme that promotes hydration and nutrients to gain access to the spore membrane. The other is an allosteric alteration of membrane protein with the result of increasing water and ionic flow into the spore cytoplasm. Regardless of the mechanism, the various outer layers of the spore are disrupted to form a germination groove where the vegetative cell emerges from the spore. The form that is produced with the emergence of the cell from the spore is characteristic of that species.

## 7.15. Activities in the Stationary Phase

As our understanding of the molecular and biochemical activities of bacteria in the log phase of growth has increased, emphasis is shifting to understand activities in the stationary phase. In nature, where a prokaryote is in constant competition with numerous other prokaryotes, cells would be best characterized as being in the stationary stage of growth and persistence of these cells is attributed to their unique metabolic activities. From information described earlier in this chapter, it is apparent that metabolic activities change as cells leave the log phase and enter the stationary phase. Many of the activities in the stationary phase are to ensure the persistence of that species, and the production of spores is an obvious event for maintaining that specific bacterial strain. The formation of bacteriocins, toxins, antibiotics, and other compounds must have had an important function in the recent history of that organism or one would expect that the genes for these activities would have been lost. Bacteria living in different environments produce a secondary metabolite that assists that organism to be more successful in competing with bacteria in the immediate area. As research progresses, the chemical signals for production of these compounds and the mechanism for production of these secondary metabolites will be better understood. Considerable interest has been given to the synthesis of antibiotics and the general scheme for biosynthesis of several antibiotics is given in Figure 7.23. As indicated in the figure, a single bacterial species produces a specific antibiotic and through the employment of different types of bacteria, industry has perfected the production of a complete suite of useful pharmaceuticals. With the sequencing of prokaryotic genomes, a large number of genes with no known function are present and it is possible that some of undesignated genes are associated with stationary phase activity.

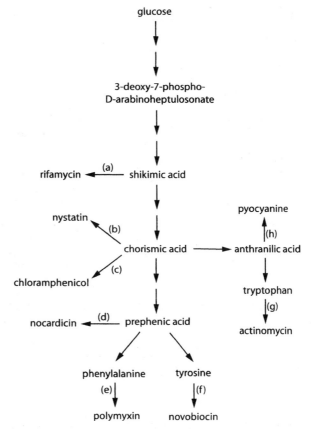

Figure 7.23. Metabolic scheme for synthesis of numerous antibiotics. Bacteria associated with these biosynthetic activities are as follows: (**a**) *Nocardia mediterranei*, (**b**) *Streptomyces noursei*, (**c**) *Streptomyces venezuelae*, (**d**) *Nocardia* sp., (**e**) *Bacillus polymyxa*, (**f**) *Streptomyces griseus*, (**g**) *Streptomyces* sp., and (**h**) *Streptomyces* sp. The synthesis of these antibiotics requires several metabolic steps in each case. (Data from Madigan et al., 2002.[7])

## 7.16.  Global Regulation[41,42]

Prokaryotes respond quickly to external environmental changes by a series of adjustments involving global regulation. From a systems approach, there are two processes for regulating transcription: (1) operons and regulons are controlled by a common regulatory protein and (2) several unrelated operons and regulons are controlled by a single stimulus. In an operon, genes are under a common regulatory control and are transcribed into a single mRNA. Regulons consist of separate but related operons that are controlled by the same regulatory protein. While an operon controls the expression of a small number of genes, a regulon controls several operons. Commonly, a regulon consists of enzymes for catabolism of a given compound and solute transport proteins for uptake of the compound. For example, the *pho* regulon controls operons for glycerol phosphate transport, production of the PhoE porin in the outer membrane, and the ABC (ATP binding cassette) transport system for inorganic phosphate (see Figure 7.14). A modulon consists of a series of operons and regulators that are controlled by another regulator in addition to the regulatory protein already

Table 7.20. Examples of modulons and global regulation in bacteria.[a]

| System | Characteristics |
|---|---|
| **Nutrients** | |
| 1. Carbon source influenced by cAMP | Carbohydrate transport by phosphotransferase system and operons for catabolism of carbohydrates |
| 2. Phosphate limitation | Pho regulon involving operons of organic and inorganic phosphate |
| 3. Anaerobic respiration | Nar Q, Nir, and Fnr systems involving proteins for anaerobic electron transport |
| **Stress response** | |
| 4. SOS response | Approximately 20 genes are regulated for repair of damaged DNA |
| 5. Oxidation response | OxyR and SoxR systems to metabolize oxygen free radicals |
| 6. Heat shock | Chaperones and related proteins that are produced with exposure to elevated temperatures |
| 7. Starvation response | Survival of bacteria in stationary phase due to CreC, RpoS, and other operons |
| **Differentiation** | |
| 8. Endospore formation | Because of energy starvation many genes including Spo, Div and Kin operons are activated |
| 9. Production of virulence factors | Following a shift in temperature, pH, or Fe concentration, enzymes and toxins are produced by pathogens |

[a]From Lengeler and Postma, 1999.[42]

controlling them. The operons and regulons making up the modulons may encode for unrelated processes but collectively these genes enable the organism to respond quickly to an external stimulus. Examples of global regulation that is initiated by a single stimulus are given in Table 7.20. In addition, modulons would be used for quorum sensing, chemotaxis, conjugation, secondary metabolism in antibiotic production, sporulation in streptomyces and fruiting body bacteria, cold shock, and osmotic shock.

## 7.16.1. The crp Modulon

As cells metabolize, low molecular weight compounds referred to as alarmones may accumulate and function as regulators for the modulons. One of these alarmones is cAMP, which is a catabolite repressor that influences transcription of various operons. cAMP is produced by adenylate cyclase from ATP and in enteric bacteria, the intracellular concentration of cAMP may approach 10 $\mu M$. Regulation of transcription is attributed to cAMP plus the cAMP receptor protein (crp) which binds to the promoter and enhances the interaction between RNA polymerase and DNA. Glucose is an effective catabolite repressor in that little cAMP is produced while glycerol promotes considerable cAMP levels. In addition to the *lac*-operon, other systems that are influenced by intracellular levels of cAMP include uptake of sugars by the phospho-transferase system, *glp* regulon for glycerol phosphate metabolism, and numerous catabolite operons and regulons.

## 7.16.2. The Stringent Response

In 1961, Stent and Brenner described the cellular activities following amino acid starvation and this has become known as the stringent response in bacteria. When an amino acid is limiting, the corresponding tRNA is not loaded with an amino acid and protein synthesis stops. The absence of a tRNA with an amino acid on the entry or A-site on the ribosome results in a stall of the overall ribosomal movement. A model indicating the activities associated with amino acid starvation and ribosomal stall is given in Figure 7.24. Because GTP and ATP are present along with appropriate enzymes, pppGpp (guanosine 5'-triphosphate, 3'-diphosphate) is formed. The enzyme ppGpp synthetase I is a product of *relA* gene and produces pppGpp with the pyro-phosphate from ATP added at the 3' hydroxyl group of GTP. The ppGpp synthetase I molecule is bound to the 50S robosomal subunit and when the unloaded tRNA binds and releases from the ribosome, GTP for peptide elongation is phosphorylated. The conversion of pppGpp to ppGpp is attributed to a phosphohydrolase (*gpp*) and the intracellular accumulation of ppGpp results in the stringent response. A second means of producing pppGpp from GTP involves the action of *spoT* which is activated by carbon starvation. Once produced, ppGpp has primary effects that not only inhibit peptide elongation but also prevent production of stable RNA molecules. By inhib-iting transcription of the relevant genes, the production of RNA polymerase is reduced and protein synthesis declines accordingly. There are several secondary effects fol-lowing inhibition of protein synthesis, including the following: inhibition of tRNA processing, reduction in phospholipid synthesis, inhibition of nucleotide synthesis, reduction of synthesis of peptidoglycan precursors, inhibition of DNA replication, and reduction of the rate of cell division. Thus, ppGpp serves as a signal for global regulation and has far reaching ramifications in cell metabolism.

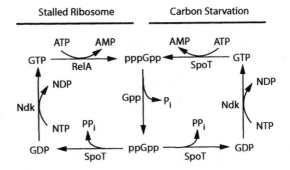

Figure 7.24. Production of ppGpp by the ribosome and the starvation path-ways. *SpoT*, (p)ppGpp3'-pyrophospho-hydrolase; *ndk*, nucleoside 5'-diphos-phate kinase; *relA*, ppGpp synthetase. (From Lengeler and Postma, 1999.[42])

# References

1. Adler, H.I., W.D. Fisher, A. Cohen, and A.A. Hardigree. 1967. Miniature *Escherichia coli* cells dificient in DNA. *Proceedings of National Academy of Sciences USA* 57:321–326.

2. Lutkenhaus, J. and S.G. Addinall. 1997. Bacterial cell division and the Z ring. *Annual Review of Biochemistry* 66:93–116.

3. Margloin, W. 2000. Themes and variations in prokaryotic cell division. *FEMS Microbiology Reviews* 24:531–548.

4. Goodwin, B.C. 1969. Control dynamics of β-galactosidase in relation to the bacterial cell cycle. *European Journal of Biochemistry* 10:515–522.

5. Ingraham, J.L., O. Maaløe, and F.C. Neidhardt. 1983. *Growth of the Bacterial Cell.* Sinauer Associates, Sunderland, MA.

6. Maaløe, O. 1962. Synchronous growth. In *The Bacteria*, Vol. IV: *Physiology of Growth* (I.C. Gunsalus and R.Y. Stanier, eds.). Academic Press, New York, pp. 1–32.

7. Madigan, M.T., J.M. Martinko, and J. Parker. 2002. *Brock Biology of Microorganisms.* Prentice-Hall, Upper Saddle River, NJ.

8. Hochman, A. 1997. Programmed cell death in prokaryotes. *Critical Reviews in Microbiology* 23:207–214.

9. Lewis, K. 2000. Programmed death in bacteria. *Microbiology and Molecular Biology Reviews* 64:503–514.

10. Maaløe, O. and N.O. Kjeldgaard. 1966. *Control of Macromolecular Synthesis.* W.A. Benjamin, New York.

11. Neidhardt, F.C. and H.E. Umbarger. 1996. Chemical composition of *Escherichia coli*. In Escherichia coli *and Salmonella Cellular and* Molecular Biology (F.C. Neidhardt, R. Curtiss, III, J.L. Ingraham, E.C.C. Lin, K. Brooks Low, B. Magasanik, W.S. Reznikoff, M. Riley, M. Schaechter, and H.E. Umbarger, eds.). American Society for Microbiology Press, Washington, D.C., pp. 13–16.

12. Stouthamer, A.H. 1979. The search for correlation between theoretical and experimental growth yields. In *International Review of Biochemistry, Vol. 21: Microbial Biochemistry* (J.R. Quayle, ed.). University Park Press, Baltimore, MD, pp. 1–48.

13. Stouthamer, A.H. 1969. Determination and significance of molar growth yields. *Methods in Microbiology* 1:629–664.

14. Pirt, S.J. 1965. The maintenance energy of bacteria in growing cultures. *Proceedings of Royal Society B* 163:224–239.

15. Kolter, R., D.A. Siegele, and A. Tormo. 1993. The stationary phase of the bacterial life cycle. *Annual Review of Microbiology* 47:855–874.

16. Makino, K., M. Amemura, S.-K. Kim, A. Nakata, and H. Shinagawa. 1994. Mechanism of transcriptional activation of the phosphate regulon in *Escherichia coli*. In *Phosphate in Microorganisms* (A. Torriani-Gorini, E. Yagil, and S. Silver, eds.). American Society for Microbiology Press, Washington, D.C., pp. 5–12.

17. Torriani-Gorini, A. 1994. The pho regulon of *Escherichia coli*. In *Phosphate in Microorganisms* (A. Torriani-Gorini, E. Yagil, and S. Silver, eds.). American Society for Microbiology Press, Washington, D.C., pp. 1–4.

18. Wanner, B.L. 1994. Multiple controls of the *Escherichia coli* pho regulon by the $P_i$ sensor phoR, the catabolite regulatory sensor CreC, and acetyl phosphate. In *Phosphate in Microorganisms* (A. Torriani-Gorini, E. Yagil, and S. Silver, eds.). American Society for Microbiology Press, Washington, D.C., pp. 13–21.

19. Kertesz, M.A. 1999. Riding the sulfur cycle—metabolism of sulfonates and sulfate esters in Gram-negative bacteria. *FEMS Microbiology Reviews* 24:135–175.

20. Gleason, F.A. and J.M. Wood. 1987. Secondary metabolism in the cyanobacteria. In *The Cyanobacteria* (P. Fay and C. Van Baalen, eds.). Elsevier, Amsterdam, pp. 437–452.

21. Chater, K.F. and D.A. Hopwood. 1993. Streptomyces. In Bacillus subtilis *and Other Gram-Positive Bacteria* (A.L. Sonenshein, J.A. Hoch, and R. Losick, eds.). American Society for Microbiology Press, Washington, D.C., pp. 83–99.

22. Zuber, P., M.M. Nakano, and M.A. Marahiel. 1993. Peptide antibiotics. In Bacillus subtilis *and Other Gram-Positive Bacteria* (A.L. Sonenshein, J.A. Hoch, and R. Losick, eds.). American Society for Microbiology Press, Washington, D.C., pp. 897–916.

23. Aronson, A.I. 1993. Insecticidal toxins. In Bacillus subtilis *and Other Gram-Positive Bacteria* (A.L. Sonenshein, J.A. Hoch, and R. Losick, eds.). American Society for Microbiology Press, Washington, D.C., pp. 953–963.

24. Pope, L. and L.J. Rode. 1969. Spore fine structure in *Clostridium cochlearium*. *Journal of Bacteriology* 100:994–1001.

25. Santo, L.M., H.R. Hohl, and H.A. Frank. 1969. Ultrastructure of putrefactive anaerobe 3679h during sporulation. *Journal of Bacteriology* 99:824–833.

26. Leive, L.L. and B.D. Davis. 1980. Cell envelope; spores. In *Microbiology* (B.D. Davis, R. Dulbecco, H.N. Eisen, and H.S. Ginsberg, eds.). Harper & Row, New York, pp. 71–110.

27. Manelstam, J., K. McQuillen, and I. Dawes. 1982. *Biochemistry of Bacterial Growth*. Blackwell, London.

28. Setlow, P. 1994. Mechanisms that contribute to the long-term survival of spores of *Bacillus* species. *Journal of Applied Bacteriology Symposium Supplement* 76:49S–60S.

29. Moran, C.P., Jr. 1993. RNA polymerase and transcription factors. In Bacillus subtilis *and Other Gram-Positive Bacteria* (A.L. Sonenshein, J.A. Hoch, and R. Losick, eds.). American Society for Microbiology Press, Washington, D.C., pp. 653–667.

30. Cano, R. and J.S. Colomé, 1986. *Microbiology*. West, St. Paul, MN.

31. Dirks, A. 1999. *Bacillus subtilis* spore coats. *Microbiology and Molecular Biology Reviews* 63:1–20.

32. Rode, L.J. 1971. Bacterial spore appendages. *Critical Reviews in Microbiology* 1:1–27.

33. Rode, L.J., M.A. Crawford, and M. Glenn Williams. 1967. *Clostridium* with ribbon-like spore appendages. *Journal of Bacteriology* 93:1160–1173.

34. Setlow, P. 1993. Spore structural proteins. In Bacillus subtilis *and Other Gram-Positive Bacteria* (A.L. Sonenshein, J.A. Hoch, and R. Losick, eds.). American Society for Microbiology Press, Washington, D.C., pp. 801–812.

35. Ennahar, S., T. Sashihara, K. Sonomoto, and A. Ishizahi. 2000. Class IIa bacteriocins: biosynthesis, structure and activity. *FEMS Microbiology Reviews* 24:85–106.

36. Klaenhammer, T.R. 1993. Genetics of bacteriocins produced by lactic acid bacteria. *FEMS Microbiology Reviews* 12:39–86.

37. Luria, S.E. and J.I. Suit. 1987. Colicins and col plasmids. In Escherichia coli *and* Salmonella typhimurium (F.C. Neidhardt, J.L. Ingraham, K.B. Low, B. Magasanik, M. Schaechter, and H.E. Umbarger, eds.). American Society for Microbiology Press, Washington, D.C., pp. 1615–1626.

38. H.-G. Sahl and G. Bierbaum. 1998. Lantibiotics: Biosyntheis and biological activities of uniquely modified peptides from Gram-positive bacteria. *Annual Review of Microbiology* 52:41–79.

39. Johnson, K. 1994. The trigger mechanism of spore germination: Current concepts. *Journal of Applied Bacteriology* (Symposium Supplement) 76:17S–24S.

40. Moir, A., E.H. Kemp, C. Robinson, and B.M. Corte. 1994. The genetic analysis of bacterial spore germination. *Journal of Applied Bacteriology* (Symposium Supplement) 76:9S–16S.
41. Cashel, M. and K.E. Rudd. 1987. The stringent response. In Escherichia coli *and* Salmonella typhimurium (F.C. Neidhardt, J.L. Ingraham, K.B. Low, B. Magasanik, M. Schaechter, and H.E. Umbarger, eds.). American Society for Microbiology Press, Washington, D.C., pp. 1410–1438.
42. Lengeler, J.W. and P.W. Postma. 1999. Global regulatory networks and signal transduction pathways. In *Biology of the Prokaryotes* (J.W. Lengeler, G. Drews, and H.G. Schlegel, eds.). Georg Thieme Verlag, Stuttgart, Germany, pp. 491–523.
43. Bechtel, D.B. and L.E. Bulla, Jr. 1976. Electron microscope study of sporulation and parasporal crystal formation in *Bacillus thuringiensis. Journal of Bacteriology* 127:1472–1481.

# Additional Reading

## *Growth Characteristics*

Bramhill, D. 1997. Bacterial cell division. *Annual Review of Cellular and Developmental Biology* 13:395–424.

Davey, H.Z. and D.B. Kell. 1996. Flow cytometry and cell sorting of heterogenous microbial populations: The importance of single-cell analysis. *Microbiology Review* 60:641–696.

Godfrey, A., N.D.H. Raven, and R.J. Sharp. 2000. Physiology and continuous culture of the hyperthermophilic deep-sea vent archaeon *Pyrococcus abyssi* ST549. *FEBS Microbiology Letters* 186:127–132.

Grover, N.B. and C.L. Woldringh. 2001. Dimensional regulation of cell-cycle events in *Escherichia coli* during steady state growth. *Microbiology* 147:171–181.

Hjort, K. and R. Bernander. 1999. Changes in cell size and DNA content in *Sulfolobus* cultures during dilution and temperature shift experiments. *Journal of Bacteriology* 181:5669–5675.

Kacena, M.A., G.A. Merrell, B. Manfredi, E.E. Smith, D.M. Klaus, and P. Todd. 1999. Bacterial growth in space flight: Logistic growth curve parameter for *Escherichia coli* and *Bacillus subtilis. Applied Microbiology and Biotechnology* 51:229–234.

Monod, J. 1949. The growth of bacterial cultures. *Annual Review of Microbiology* 3:371–394.

Neidhardt, F.C. 1999. Bacterial growth: Constant obsession with dN/dT. *Journal of Bacteriology* 181:7405–7408.

Page, W.J., A. Tindale, M. Chandra, and E. Kwon. 2001. Alginate formation in *Azotobacter vinelandii* UWD during stationary phase and the turnover of poly-beta-hydroxybutyrate. *Microbiology* 147:483–490.

Rajeshwari, R., J. Yashitola, A.P.K. Reddy, and R.V. Sonti. 1997. Characteristics of stationary-phase variation affecting virulence in *Xanthomonas oryzae* pv. *Oryzae. Canadian Journal of Microbiology* 43:862–867.

Ross, R.A., L.C. Madoff, and L.C. Paoletti. 1999. Regulation of cell component production by growth rate in the group B *Streptococcus. Journal of Bacteriology* 181:5389–5394.

## *Sporulation*

Caltalano, F.A., J. Meador-Parton, D.L. Popham, and A. Driks. 2001. Amino acids in the *Bacillus subtilis* morphogenetic protein SpoIVA with roles in spore coat and cortex formation. *Journal of Bacteriology* 183:1645–1654.

Gould, G.W., A.D. Russell, and D.E.S. Stewart-Tull (eds.). 1994. Fundamental and applied aspects of bacterial spores. *Journal of Applied Bacteriology* 76 (Suppl).

Jiang, M., W. Shao, M. Peregro, and J.A. Hoch. 2000. Multiple histidine kinases regulate entry into stationary phase and sporulation in *Bacillus subtilis. Molecular Microbiology* 38:535–542.

Meador-Parton, J. and D.L. Popham. 2000. Structural analysis of *Bacillus subtilis* spore peptidoglycan during sporulation. *Journal of Bacteriology* 182:4491–4499.

Paidhungat, M. and P. Setlow. 2000. Role of Ger proteins in nutrient and nonnutrient triggering of spore germination in *Bacillus subtilis. Journal of Bacteriology* 182:2513–2519.

Woods, D.R. and D.T. Jones. 1986. Physiological responses of *Bacteroides* and *Clostridium* strains to environmental stress factors. *Advances in Microbial Physiology* 28:1–64.

## *Energetics*

Dols, M., W. Chaibi, M. Remaud-Simeon, N.D. Lindely, and P.F. Monsan. 1997. Growth and energetics of *Leuconostoc mesenteroides* NRRL B-1299 during metabolism of various sugars and their consequences for dextransucrase production. *Applied and Environmental Microbiolgy* 63:2159–2165.

Schicho, R.N., K. Ma, M.W. Adams, and R.M. Kelly. 1993. Bioenergetics of sulfur reduction in the hyperthermophilic archaeon *Pyrococcus furiosus. Journal of Bacteriology* 175:1823–1830.

Scholten, J.C. and R. Conrad. 2000. Energetics of syntrophic propionate oxidation in defined batch and chemostat cocultures. *Applied and Environmental Microbiology* 66:2934–2942.

# 8

# Physiological Basis for Growth in Extreme Environments

Pasteur adopted what may be called the physiological point of view. He regarded the yeast cell or the lactic bacterium as a living organism with a life to lead and an axe to grind. Scientific discoveries made in advance of their time are like long salients in an army by which a position may be captured; if, however, the main army is too far behind to bring up the necessary supplies the successful advance is cut off and may be forgotten and the position has then to be recaptured at a later date. This explains why Pasteur's physiological approach to problems of microbiology was so successful. Marjory Stephenson, *Bacterial Metabolism*, 1949.

## 8.1. Introduction

Bacteria and archaea inhabit virtually every area on the surface of the planet, including those areas where the environment is extremely hostile. Not only do microorganisms survive but they also thrive under extreme conditions of temperature, drought, irradiation, pH, high osmotic pressure including salinity, and hydrostatic pressure. The basis for microbial growth at these environments is attracting considerable attention. With respect to temperature, organisms are generally assigned to one of four categories: psychrophiles, which grow optimally at temperatures less than 20°C; mesophiles, which grow at temperatures between 20°C and 45°C; moderate thermophiles, which grow at temperatures between 45°C and 70°C; and hyperthermophiles, which grow between 70°C and 110°C. The mesophiles are often subdivided into the terrestrial group, which has a temperature optimum of 25° to 30°C, and a group that includes animal pathogens has a temperature optimum of 35° to 45°C. *Archaea* are present as mesophiles, moderate thermophiles, and hyperthermophiles while genera of the domain *Bacteria* are found in all four temperature zones. As indicated in Figure 8.1, anaerobic archaea are the most prevalent at pH levels of 5–7 and temperatures of 80° to 100°C.

As a group, bacteria are less capable than fungi of growing in high osmotic environments. Extreme osmotic pressures generated by sugars are not tolerated by most bacteria; however, *Acetobacter diazotrophicus* is unique in that it will grow in media

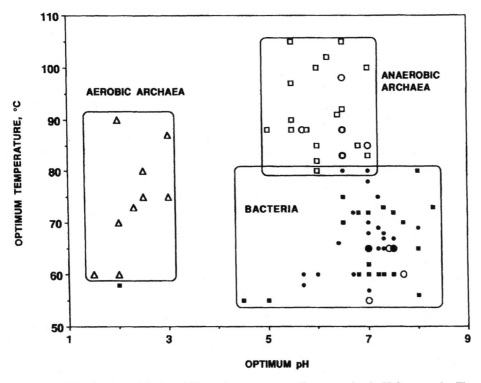

Figure 8.1. Distribution of thermophilic prokaryotes according to optimal pH for growth. The symbols are as follows: open circles, methanogens; open triangles, aerobic archaea; open squares, anaerobic archaea; solid circles, anaerobic bacteria; and solid squares aerobic bacteria. (Reprinted with permission from Kristjansson, J.K. and K.O. Stetter. 1992. Thermophilic bacteria. In *Thermophilic Bacteria* (J.K. Kristjansson, ed.). CRC Press, Boca Raton, FL. Copyright CRC Press, Boca Raton, Florida.)

with 30% sucrose. The mechanism accounting for this growth of *A. diazotrophicus* in the intracellular spaces of sugarcane stems remains to be established. Halophilic archaea include those organisms that can grow optimally only in the presence of >20% salt. Numerous bacterial strains, including *Staphylococcus*, are halotolerant and can exist in salt concentrations from 4.5% to 10% NaCl.

Prokaryotic tolerance to hydrogen ion concentration in the growth medium has resulted in designation of three groups: acidophilic, neutralophilic, and alkalinophilic. Acidophiles can grow optimally at pH 1.0 to 5.0, neutralophilic organisms grow optimally at 6.0 to pH 7.0, while alkalinophilic organisms grow from pH 9.0 to pH 11. Organisms that grow at these three different pH levels are found in the *Archaea* as well as in the *Bacteria*. The most extreme of the environments is low pH and high temperature and as indicated in Figure 8.1, several species of the aerobic archaea grow optimally at pH 2 to 3 and 60° to 90°C.

## 8.2. Moderate Thermophiles and Hyperthermophiles[1,2]

Moderate thermophiles include photosynthetic bacteria, spore-producing bacteria, sulfate-reducing bacteria, lactic acid bacteria, and numerous Gram-negative aerobes and Gram-negative anaerobes. Although some prokaryotes grow at elevated temperatures there is some compromise in the rate of growth, cellular maintenance energy, and enzyme activity. When comparing mesophilic, thermophilic, and extreme thermophiles it is apparent that the efficiency of physiological processes is reduced as the temperature increases. Specific examples pertaining to growth and enzyme activities with temperature changes are given in Table 8.1. As seen for malate dehydrogenase and isocitrate dehydrogenase, enzyme activity is dependent on temperature of bacterial cultivation, with the greatest activity at the lowest temperature. While enzymes from thermophiles display thermal resistance, enzymes from mesophiles are more temperature responsive than those from thermophiles. Lars Ljungdahl and collaborators have suggested, and there appears to be supportive evidence, that thermophilic enzymes have sacrificed catalytic activities for thermal stability. Thermophiles are capable of utilizing thermal energies to overcome some of the molecular rigidity that accounts for enzyme stability. Generally, the growth yield is greater at a low temperature (see Table 8.1) and maintenance energy increases with an increase in temperature.

Thermophiles are not a distinct physiological form of life with unique metabolic schemes but they have many of the same catabolic pathways, regulatory systems for DNA expression, cell wall chemistry, and cytochrome-mediated energetics as those found in mesophiles. Prokaryotes that grow at temperatures above 70°C include *Bacteria* of the genus *Thermotoga*, and numerous genera of sulfur oxidizing *Archaea*. Similarity, hyperthermophiles are not distinct forms of life because the mechanisms that account for cell energetics, processing of information for DNA replication, tran-

Table 8.1. Influence of elevated temperature on growth and enzyme activity.[a]

| Classification | Organism | Temperature[a] parameter measured (°C) | $Y_{glucose}$[b] | $M_{glucose}$[c] | Activity[d] MDH | IDH |
|---|---|---|---|---|---|---|
| Mesophile | *Bacillus subtilis* | 30 | | — | 322 | 41 |
| Thermophile | *Bacillus stearo-* | 55 | 65 | — | 187 | 21 |
| | *thermophilus* | 45 | 86 | — | — | — |
| Extreme thermophile | *Bacillus caldo-* | 70 | 89 | 4.1 | 83 | 4 |
| | *tenax* | 65 | 89 | 3.8 | — | — |
| | | 60 | 79 | 2.2 | — | — |
| Extreme thermophile | *Thermus* | 78 | 55 | 0.38 | — | — |
| | *thermophilus* | 70 | 64 | 0.28 | — | — |
| | | 60 | 64 | 0.18 | — | — |

[a]From Sundaram, 1986.[2]

[b]$Y_{glucose}$ = glucose yield in grams dry weight of cells per mole of glucose used.

[c]$M_{glucose}$ = maintenance energy coefficient which is defined as millimoles of glucose required per gram dry weight of cells per hour.

[d]Specific activities: MDH, malate dehydrogenase; IDH, isocitrate dehydrogenase.

scription and translation, and cell division are similar to those of mesophiles. After many years of studying moderate thermophiles and hyperthermophiles, many scientists agree that thermostability is attributable to the following: (1) intrinsic character of the protein molecules; (2) materials in the cytoplasm to protect nucleic acids, many of which include amines; and (3) lipid composition to maintain fluid character of the membrane.

## 8.2.1. Thermostability of Proteins[3–5]

A well known principle of biochemistry is that heat denatures plant and animal proteins. Therefore, it is of considerable importance to examine the proteins of thermophilic bacteria to determine the key to thermostability. Many different proposals have been put forth to account for protein stability in thermophilic bacteria but to date there has not been one single feature that appears to account for this activity. Enzymes from thermophiles and mesophiles (1) are about the same molecular weight with nearly identical three-dimensional structures, (2) use the same cofactors with similar catalytic mechanisms, and (3) have an amino acid composition that is similar to that found in organisms that grow at cooler temperatures. An important feature for distinction of thermophilic proteins is that they have about twice the number of subunits as compared to mesophilic enzymes. An enzyme with small subunits is more thermostable than a larger molecule without subunits.

It is well accepted that protein resistance to thermodenaturation is also accompanied by an increased stability to surfactants, pH extremes, or organic solvents. For an enzyme to withstand a high temperature, the protein must be rigid but also must be flexible to interface with the substrate. The enzymes from thermophilic bacteria are called thermozymes and the reason these enzymes are active at elevated temperatures has been examined for more than 50 years. Thermostability of proteins is not attributed to a single factor but can be promoted by different molecular activities, including: (1) amino acid sequences that promote hydrophobicity, (2) presence of disulfide bonds, (3) interactions between adjacent folds in the protein, (4) presence of metal cations as stabilizers, and (5) stabilization resulting from the presence of chaperone proteins.

The primary structure of thermophilic enzymes is one of the factors that provides thermostability. An analysis of the amino acid sequence of several enzymes revealed that mesophilic organisms have proteins containing peptides with the sequence of glycine, serine, serine, lysine, and asparagine which promote flexibility to the backbone of the protein. Thermophilic bacteria replace this amino acid sequence with alanine, alanine, threonine, arginine, and glutamic acid residues to reduce the free energy during folding and increase internal hydrophobicity. These replacement amino acids in thermophiles enhance stabilization of protein helices, which prevent complete loss of structural conformation at the high temperature.

In some thermozymes, there is a higher abundance of cysteine residues than what is found in mesophiles and as a result greater stability is obtained with S–S bond formation. In addition, some enzymes from thermophilic organisms have disulfide bonds that are deeply buried within the molecules, making the S–S bridges highly inaccessible to destabilizing agents. It has been determined that each disulfide bond contributes 2 to 5 kcal/mol in thermal stability. In a few enzymes, the ratio of arginine to lysine is greater in proteins from thermophilic bacteria than from mesophiles. The presence of arginine on the surface of a protein promotes a stable water shell that stabilizes the protein against heat as well as solvents.

Thermal stabilization of proteins in the secondary structure is promoted by both hydrophobic and charge interactions. These interactions are attributed to amino acids of the primary structure through hydrogen bonding, salt bridges, or ring–ring interaction. It has been proposed that some enzymes from thermophilic bacteria have a high number of H-bonds, which increases free energy of molecular stability including heat stability. Hydrogen bonds may increase the free energy of stability by as much as 5 kcal/mol. In some enzymes from thermophilic bacteria, it has been suggested that the electron cloud on the face of phenylalanine, tryptophan, or tyrosine may interact with hydrogen atoms on the face of a second aromatic ring. In addition, it is proposed that histidine, lysine, or arginine residues may interact with aromatic amino acids because of electronic charges about the aromatic rings. Ion pairs between segments of folded proteins can account for about 4 kcal/mol of stability for each pair. With respect to hydrophobic interactions, methyl groups of amino acids that are projected into the folded protein structure stabilize the protein at the level of 1.3 kcal/mol.

In some enzymes, it has been demonstrated that metal binding by thermophilic proteins may contribute to structural stability. For example, it is known that amylases and proteases from thermophiles have a requirement for $Ca^{2+}$ ions for catalytic activity and for heat stability. For structural stability, $Ca^{2+}$ was found buried deep within the molecule and not accessible to chelating agents such as EDTA. Another example of metal-induced thermostability is the presence of a zinc atom in ferredoxin from *Sulfolobus*, and mesophilic homologs that are sensitive to temperature lack the zinc binding site.

Recent investigations have demonstrated that chaperons may be important in providing thermal stability to proteins. *Thermus* sp. contains *cpn60* and *cpn10* genes that produce proteins of 58 kDa and 10 kDa, respectively. These chaperon proteins are involved in protein folding and promote heat stability when $Mg^{2+}$ and ATP are present.

It should be considered that all these approaches effecting protein activity are employed in thermal stability. There is no single mechanism accounting for thermostability of proteins in moderate thermophilic and hyperthermophiles, but rather different mechanisms are employed by different proteins. The extent to which enzymes from thermophiles are stable at elevated temperatures is given in Table 8.2.

## 8.2.2. Cytoplasmic Protectants[6]

Thermal protection of macromolecules and metabolites in the cytoplasm is required in thermophiles. Currently, the highest temperature for cell growth is displayed by the archaeon *Pyrolobus fumarii*, which grows at 113°C. At temperatures above 110°C, ATP, NADH, amino acids, and other molecules essential for living systems are unstable and spontaneously degrade in aqueous solution. However, the presence of certain chemicals will markedly increase the thermal stability of these molecules. Enzymes are less thermostable when purified than when there is a substrate–enzyme

Table 8.2. Thermostability of enzymes from thermophiles.[a]

| Domain | Enzyme | Function | Organism | Optimum temperature for reaction | Thermostability[b] ($t_{50}$), hr/°C |
|--------|--------|----------|----------|-------------------|---------------------|
| *Archaea* | Protease | Peptide hydrolysis | *Pyrococcus furiosus* | 115°C | 33/98 |
| | pfu polymerase | DNA polymerase | *Pyrococcus furiosus* | 75°C | 1/95 |
| | Ferredoxin | Electron transport | *Thermococcus litoralis* | >95°C | 24/95 |
| | Protease | Peptide hydrolysis | *Desulfurococcus mucosus* | 100°C | 1.5/95 |
| | GAPDH | Glycolysis | *Thermotoga maritima* | >90°C | 2/95 |
| *Bacteria* | L-asparaginase | Deaminase of L-asparagine | *Thermus aquaticus* | — | 0.3/85 |
| | *Taq* polymerase | DNA polymerase | *Thermus aquaticus* | 75°C | 1/95[c] |
| | Pyruvate carboxylase | Pyruvate metabolism | *Bacillus stearo-thermophilus* | — | 0.5/55 |

[a]Adapted from Adams, 1993.[3]
[b]Temperature at which 50% of catalytic activity is lost.
[c]Activity is 75%, not 50%.

complex or when organic components are present. Ribosomes in thermophiles are resistant to high temperatures owing to the character of the proteins present rather than to changes in RNA. Components that are present in the cytoplasm of these bacteria to protect nucleic acids, cofactors, and proteins include amines and polyamines. There is a distinction in the distribution of these stabilizing agents in that moderate thermophiles produce spermine, spermidine, and putrescine while hyperthermophiles have polyamines of thermine, caldine, and *sym*-homospermidine. Additional information about stabilization of DNA by proteins is covered in the section on histones in Chapter 12.

## 8.2.3. *Membrane Lipids*[7]

As cells are shifted to a different temperature of growth, several changes are observed and these include: (1) length of carbon chains, (2) degree of saturation, (3) extent of branching, and (4) type of head groups on phospholipids. These changes are responsible for lipids with higher melting temperatures. In thermophiles, as the temperature increases, C17 fatty acids replace C15 fatty acids, saturated fatty acids replace the unsaturated ones, and the extent of branching in fatty acids increases. Examples of branched lipids would include the *iso* and *anteiso* types in which a methyl group is present at the end or near the end of the fatty acids. In bacteria that produce glycolipids, these complex lipids containing multiple residues of glucose and galactose may constitute 70% of total lipids as bacteria grow at elevated temperatures.

Changes in membrane lipids are also observed with mesophilic bacteria, and as an example, the fatty acid composition in *Escherichia coli* growing at different temperatures is given in Table 8.3. In the outer membrane of *E. coli*, myristic and β-hydroxy myristic acids are in equal abundance at 43°C but at 10°C β-hydroxymyristic acid

Table 8.3. Major fatty acids in *Escherichia coli* as influenced by temperature.[a]

| Constituent | Percent of total fatty acid at specified temperature for growth | | |
| --- | --- | --- | --- |
| | 10°C | 30°C | 43°C |
| **Fatty acids in lipopolysaccharides** | | | |
| Myristic acid (S) | 3.9 | 4.1 | 7.7 |
| β-Hydroxymyristic acid (S) | 12.6 | 10.1 | 7.5 |
| **Fatty acids in phospholipids** | | | |
| Palmitic acid (S) | 18.2 | 28.9 | 48.0 |
| Palmitoleic acid[b] (U) | 26.0 | 23.3 | 9.7 |
| Methylene hexadecanoic acid (U) | 1.3 | 3.4 | 11.6 |
| *cis*-Vaccinic acid[c] (U) | 37.9 | 30.3 | 12.2 |

[a]Modified from Ingraham, 1987.[7]
[b]Hexadecanoic acid.
[c]Octadecanoic acid.
S, Saturated fatty acid; U, unsaturated fatty acid.

is at a greater concentration. With a shift to a higher growth temperature, the concentration of palmitic acid increases while the concentration of palmitoleic acid and *cis*-vaccinic acid decreases in phospholipids of *E. coli*.

## 8.3. Bacteria in Extreme Hydrogen Ion Concentrations[8–10]

Prokaryotes are found throughout the world including those areas that have a pH level that is hostile to most eukaryotic systems. While the greatest number of bacterial and archaeal species grow in environments near neutrality, a relatively small number of species are found to grow at acidic levels or under basic conditions. The microorganisms that grow in extremely acidic or extremely alkaline environments may be considered as specialists and they have special physiological activities to support growth under these extreme conditions. The acidity or alkalinity of the environment is one of the parameters selective for prokaryotic growth. Microorganisms have considerable variability in their capability of growing in environments of low pH, and some examples of this are given in Figure 8.2. Growth of aerobic and anaerobic prokaryotes at different concentrations of proton and hydroxyl ions indicates that there are both generalists and specialists when it comes to acid tolerance.

### 8.3.1. pH Homeostasis and Membrane Charge[8–14]

The prokaryotic plasma membrane carries a charge resulting from the segregation of proton and hydroxyl ions. Most of the bacteria and archaea have a plasma membrane with an exterior that is acidic and positively charged while the interior of the membrane is alkaline and negatively charged. A membrane charge is required for

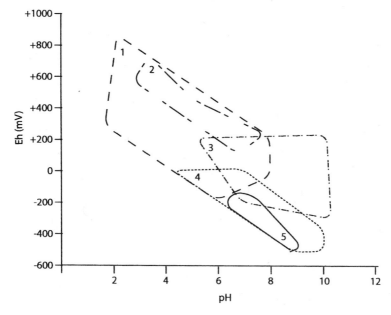

Figure 8.2. pH range for different physiological groups of prokaryotic organisms. 1. Sulfur oxidizing bacteria, 2. iron bacteria, 3. purple sulfur bacteria, 4. sulfate-reducing bacteria, 5. green sulfur bacteria.

cellular viability, and prokaryotes that grow in extremely acidic or alkaline environments must adjust their metabolism to maintain this charge. A large number of the bacteria that have been isolated display a mechanism for maintaining the cytoplasmic pH at a level that is considered to be slightly alkaline. With exposure of the organisms to either slightly alkaline or acid environments, there is a cellular response. While hydroxyl ions are prevented from passing across the membrane to make the cytoplasm alkaline, weak acids or weak bases can readily diffuse through the membrane to effect a pH change in the cytoplasm.

Scientists have been attracted to the bacteria that grow at extremes of pH because there is an interest to determine if the cytoplasmic pH is influenced by the environment and to determine the effect of extreme pH on membrane charge. While hydrogen ion concentrations external to the cell can be readily measured with a pH electrode, cytoplasmic pH is measured by following pH indicators; however, the use of pH indicators is limited owing to metabolism and selective membrane permeability. The distribution of a weak base, for example, methylamine, or a weak acid, such 5,5-dimethyl-2,4-oxazolidinedione, across the plasma membrane is preferred to the use of dyes. Considerable sensitivity in measurements with weak acids or weak bases is obtained when they are labeled with [14]C. Another method for measuring internal pH is to use [31]P-labeled phosphate and methylphosphonate as probes with the distribution of phosphorus measured by nuclear magnetic resonance. From an assessment

Table 8.4. Characteristics of bacteria growing at extremes of pH.[a]

| Character | Acidophiles | Neutralophiles | Moderate alkaliphiles | Extreme alkaliphiles |
|---|---|---|---|---|
| Principal organisms investigated | *Bacillus acidocaldarius* *Sulfolobus acidocaldarius* *Thermoplasma acidophila* *Acidithiobacillus thiooxidans* | *Enterococcus faecalis* *Escherichia coli* Various other bacteria | *Propionigenium modestum* *Vibrio alginolyticus* | *Bacillus alcalophilus* *Bacillus firmus* *Exiguobacterium aurantiacum* |
| pH for growth | 1–4 | 6–8 | 8–9 | 10–12 |
| Cytoplasmic pH | 5.5–6.9 | 7.2–8.4 | 8.0–8.6 | 7.6–9.2 |
| $\Delta\psi$ (Membrane charge) | 0 to small (Positive inside) | −110 mV (Positive outside) | −150 mV (Positive outside) | −170 mV (Positive outside) |
| $Na^+$ (outside) | 10–100 m$M$ | 10–100 m$M$ | 300 m$M$ | 400–500 m$M$ |
| Solute transport | $H^+$/solute symport | $H^+$/solute symport and $Na^+$/solute symport | $Na^+$/solute symport | $Na^+$/solute symport |
| $Na^+$/$H^+$ anteporter | Not common | Common | Yes | Yes |
| $K^+$/$H^+$ anteporter | ? | Yes | ? | Yes |
| ATPase | $H^+$ translocating and ($K^+$ driven ATP synthase) | $H^+$ translocating and $Na^+$ translocating | $H^+$ translocating and $Na^+$ translocating | $Na^+$ translocating |

[a]From Krulwich and Ivey, 1990[9]; Horikoshi and Akiba, 1982.[15]

of various bacteria it was determined that the cytoplasmic pH of neutralophilic bacteria is alkaline and not greatly affected by extracellular pH (see Table 8.4).

In the Gram-negative enteric bacteria, it has been established that bacteria use membrane transport activities to maintain the cytoplasmic pH at 7.8. With a shift to an acidic environment, the $K^+$/$H^+$ antiporter functions to pump in $K^+$, which makes the cytoplasm alkaline and neutralizes acidic compounds. If there is a shift in the environment to make it alkaline, the cytoplasm becomes acidified by the action of the $Na^+$/$H^+$ antiporter. These proton antiporters will be effective in transient pH shifts but metabolic changes are required for prolonged exposures to acid or alkaline environments.

In the streptococci and lactobacilli, shifts in cytoplasmic pH are permitted; however, there must be a $\Delta$pH across the plasma membrane of about 1 unit. In some lactobacilli, the intracellular pH is less than 5.5 and the organisms are more resistant to acetate and volatile fatty acids. With a slightly acidic pH in the cytoplasm, the lactobacilli have an increased action of the $F_1F_0$ ATPase to expel protons from the cytoplasm. The acidification of the cytoplasm promotes the stability of the ATPase subunits with complete assembly into the plasma membrane while a neutral environment promotes degradation of these subunits.

The production of volatile fatty acids of acetic, propionic, and butyric are important contributors to the acidification of the cytoplasm. In slightly acidic environments, these weak acids will be in the nonionized form and the free acid readily traverses the plasma membrane to acidify the cytoplasm. When the system for pH homeostasis is overtaxed, the cell dies.

## 8.3.2. Acidophilic and Acid-Tolerant Bacteria[8–10]

Prokaryotes are found in environments in which the pH is limiting for higher forms of life. While most life forms prefer an environment in which the pH is near neutral, some bacteria will grow at pH levels approaching 1.0. The most acidophilic bacterium is *Acidothiobacillus* (formerly *Thiobacillus*) *thiooxidans*, which grows from pH −1.0 to pH 5.0. Acidophilic bacteria grow on sulfur compounds in acid-mine streams and produce sufficient acidification to inhibit eukaryotic growth on the bank of a stream as well as in the water. The mechanism to explain why thiobacilli require a high concentration of protons in the environment is not clear. The general consideration is that a concentration of protons on the outer side of the plasma membrane is needed to assist the cell to carry out certain membrane functions. It is not simply that these protons serve as a source for ATP synthase through F-type ATPase, because these organisms require an external energy source for growth. Perhaps the thiobacilli are similar to the streptococci in that a high concentration of protons is required to stabilize the ATPase structure. Although considerable progress has been made in understanding acid tolerance by microorganisms, the mechanisms to explain why certain bacteria will grow only at pH levels <5 remain to be established.

The patterns of proton-motive force and ion currents of bacteria growing in extreme pH environments is an area of research that includes many dimensions. A comparison of bioenergetic characteristics of cell membranes with physiological features for bacteria growing in extreme hydrogen ion concentrations is given in Table 8.4. Unfortunately, the generalizations on acidophiles are based on research from relatively few aerobic bacteria, and many critical concepts including global cellular regulation, starvation response, and synthesis of structural architecture need to be addressed.

### 8.3.2.1. Metabolic Response

It is known that neutralophilic bacteria display two general types of responses with an increase in environmental acid concentration. One response is termed "the Gram-positive response" and the other is known as "the Gram-negative response." In the Gram-negative response, there is an effort by the cell to maintain a cytoplasmic pH at 7.5 to 8.0. In the Gram-positive system of acid tolerance, the cytoplasmic pH is permitted to acidify but it is kept more than 1 pH unit more alkaline than the external pH.

Many of the heterotrophic organisms will actively change the environment through specific cell metabolism. The homolactic fermentative bacteria produce copious quantities of lactic acid, which reduces the pH level in the environment, and thereby inhibit many of the neutralophilic organisms. If the environment becomes highly acidic, *E. coli* and *Streptococcus bovis* will stop producing the strong acids formic acid and acetic acid and shift fermentation to increase the formation of lactic acid. When the pH of the environment declines owing to the production of acetic and butyric acids, clostridia will shift the end products of fermentation to solvent (butanol, acetone, and isobutanol) production. Strains of *Helicobacter* are capable of growing in the highly acidic environment of the stomach, in part owing to the neutralization effect of ammonia that is produced from urea by the bacterial urease. The cholera agent increases the alkalinity of the media through the release of ammonia from deamination of amino acids. In these instances, changes in pH reflect a sensing of chemicals in the environment and the responses of the bacteria would serve to be selective for their presence.

Neutralophilic bacteria are killed by extracellular acid levels of pH 5.0 or less because the cytoplasm will become acidified. It has been demonstrated that if the internal pH for *E. coli* drops from pH 7.8 to pH 7.2 growth is decreased by a factor of two and that at pH 6.6 to 6.8 growth stops. The cytoplasm is poorly buffered and as a consequence marked changes in pH occur with the influx of protons or volatile fatty acids (VFA). The early food preservation process of pickling employed fermentative bacteria to acidify dairy products to produce yogurt and cabbage fermentation to produce sauerkraut. High acidity in foods is an effective food preservative that prevents the growth of common environmental bacteria.

### 8.3.2.2. Decarboxylase Reactions

Numerous acid survival mechanisms exist in bacteria and the two addressed here involve amino acid metabolism. The first system for acid resistance (AR) to be discussed involves glutamine decarboxylase. In *E. coli*, *Lactobacillus*, and *Shigella*, a cytoplasmic glutamate decarboxylase consumes a proton as γ-aminobutyric acid (GABA) is produced from L-glutamate. GABA is transported out of the cell by an exchange system that imports glutamate (see Figure 8.3). The protein responsible for the export of GABA is the product of the *gadC* gene. A glutamate/GABA anteporter system provides for the export of a proton from the cytoplasm.

A second system for surviving an acidic cytoplasm is an L-arginine decarboxylase reaction. This reaction is observed with *E. coli* and it is similar to the glutamate decarboxylase system. The cytoplasmic enzyme converts arginine to agmatine with the consumption of a proton. Presumably there is an anteport system for the uptake of L-arginine and the export of agmatine; however, evidence for this has not yet been presented. Lysine and ornithine decarboxylation reactions do not function in detoxifying the cytoplasm from protons even though these enzymes are present in *E. coli*. One explanation is that glutamate decarboxylase and arginine decarboxylase have an optimum pH of 4.5 and 5.0, respectively. The pH optima for lysine and ornithine decarboxylases are 5.7 and 6.9, respectively. Thus, the inability of these decarboxylases to serve in acid resistance could be due to lack of sufficient enzyme activity or it could be that there is not an appropriate anteporter system to remove protons from the cytoplasm.

In addition, proton-motive metabolic cycles may involve dicarboxylic acids and numerous amino acids in exchange reactions. In these reactions, the divalent anion

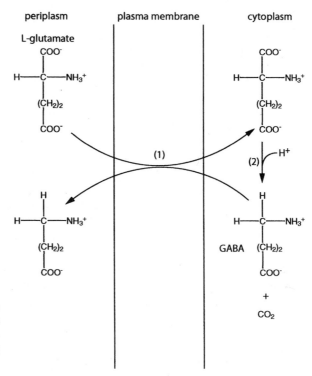

Figure 8.3. Model of glutamate system involved in acid tolerance by enteric bacteria. (1) glutamate/GABA antiporter, and (2) glutamate decarboxylase.

enters as $^-OOC-R-COO^-$ and with metabolism, the product would exit the cytoplasm carrying a proton. The decarboxylation reactions that may function in this capacity include the following precursor/metabolic products in an exchange reaction include the following: aspartate/alanine, histidine/histamine, phenylalanine/phenethylamine, tyrosine/tyramine, malate/lactate, and oxylate/formate. These reactions with protons removed from the cytoplasm could be important where proton pumps are limited by thermodynamics.

## 8.3.3. Alkaliphilic Microorganisms[15]

Although the first bacteria capable of growing in an extreme alkaline environment were reported in 1934 by T. Gibson and by A. Vedder, relatively little was known about these unusual organisms until the research of K. Horikoshi in the 1970s revitalized the area. Originally these alkaline-loving bacteria were called alkalophilic; however, the term "alkaliphilic" is increasing in popularity. Most of the bacteria that have been isolated as alkaliphiles belong to the genera of *Bacillus*, *Corynebacterium*, *Pseudomonas*, *Flavobacterium*, *Streptomyces*, and *Micrococcus*. Although the obligate alkaliphilic bacteria grow optimally at pH of 9.0 to 12 but not at pH 7.0, many strains of bacteria will grow optimally at near neutral pH level but will tolerate exposure to strong alkaline conditions. When the pH of the environment is not sufficiently alkaline, some of the alkaliphiles will release ammonia from the deamination of amino acids to enhance the alkalinity.

### 8.3.3.1. Physiological Characteristics

Some important issues for alkaliphiles include motility, proton-motive force, and pH homeostasis. In most cases, the natural alkaline environment for alkaliphiles has appreciable levels of $Na^+$ and as a result $Na^+$ assumes an important role in energetics. Alkaliphiles would not have a $\Delta pH$ across the plasma membrane with an acid exterior and an alkaline interior but, in fact, it would be reversed. Thus, the low electrochemical gradient of $H^+$ ($\Delta\bar{\mu}_{H+}$) is replaced by the electrochemical gradient of $Na^+$ ($\Delta\bar{\mu}_{Na+}$) and solute transport is coupled to transmembrane movement of $Na^+$ (see Table 8.4). In neutralophiles, some solute transport systems are driven by $Na^+$ with the majority driven by $H^+$ but alkaliphiles appear to have only $Na^+$ solute transport. The facultative alkaliphilic *Bacillus firmus* OF4 uses the $H^+$-driven $F_1F_0$-ATP synthase with a fivefold increase in the succinate: quinone oxidoreductase (complex II) as the cells are shifted from pH 7 to 10. In the anaerobic alkaliphile, *Propionigenium modestum*, $\Delta\bar{\mu}_{Na+}$-driven ATP synthesis is by an $F_1F_0$-ATPase while with *Vibrio alginolyticus* ATP synthesis appears to be by a $Na^+$ pump. It has been determined that motility of bacteria attributed to rotation of the flagella is driven by $Na^+$ entering at the flagellar rings in the plasma membrane.

From the studies on cell metabolism, alkaliphiles appear as other bacteria in that they have similar processing of DNA for replication and protein synthesis, metabolic schemes, cytochromes, and virus infection. The relatively few number of microorganisms that have been isolated as alkaliphiles should not be taken to indicate that these are the only ones. To date no archaea have been isolated as alkaliphiles.

### 8.3.3.2. Biotechnology Applications

Recent interest in alkaliphilic bacteria is attributed to the potential biotechnology applications. Of interest are the alkaline proteolytic enzymes that are produced by various strains of alkaliphilic *Bacillus* sp. Alkaline proteases are used in detergents because of the broad pH activity range, stability of the enzyme under alkaline conditions, high level of activity in the presence of surfactants, activity over a broad temperature range, and excellent storage stability. The use of detergents containing alkaline proteases reached its peak in sales in the early 1970s but recent concern about widespread allergic reactions to the enzymes has diminished sales. The use of proteases in detergents will increase as the public accepts the reports of nonallergenic responses to these proteases. Alkaline proteolytic enzymes are used in tanneries where unhairing of hides is conducted at pH 12. Other industrial applications for use of thermozymes include the degradation of cellulose coupled to the production of ethanol and the bleaching of wood fibers in the paper pulping process.

The production of cyclodextrins by cyclodextrin glycosyltransferase (CGTase) from alkaliphilic bacteria has considerable potential for industrial production because the CGTase from alkaliphilic *Bacillus* has several properties that makes it preferred over enzymes from neutralophilic bacteria. Cyclodextrins, frequently called Schardinger dextrins, are produced by *Bacillus macerans* and were first identified by Schardinger. The cyclodextrins are cyclic molecules consisting of six or more molecules of D-glucose covalently held by $\alpha$ 1,4 glucosidic bonds. These cyclic molecules are not readily decomposed by hot alkaline solutions or by strarch-splitting amylases and form inorganic complexes with various salts. Interest in cyclodextrins is attributed to the numerous commercial applications, including the following: emulsification of oils in foods with an increase in foaming power for cake mixes, whipping cream, and powdered egg whites; stabilizing of flavors in chewing gums and seasoning powders; masking colors and stabilizing fragrances in cosmetics; and increasing solubility and masking taste of pharmaceuticals.

## 8.4. Halophilic Microorganisms[16–18]

Halophilic bacteria require salt concentrations of 12% to 30% NaCl and some bacteria reside inside salt crystals. There are numerous examples of aerobic and anaerobic halophilic prokaryotes, and perhaps the best studied is *Halobacterium halobium*. Oxygen is frequently limiting for bacterial respiration in saline environments because the solubility of $O_2$ in saturated salt water may be only 20% of that found in fresh water. Under conditions of low oxygen, *Halobacterium halobium* produces a unique purple protein, bacteriorhodopsin, that is in sufficient quantity to give the cells a bright red color. The plasma membrane containing bacteriorhodopsin is termed the "purple protein," and this purple plasma membrane contains 25% phospholipid and 75% protein. Bacteriorhodopsin is a 26-kDa protein with seven helices that span the plasma membrane. The chromophore is a retinol molecule which is a $C_{20}$ carotenoid and is similar to the structure of vitamin A.

In the dark, carbon 18 of the light-sensitive retinol is in the *trans* form but with exposure to light it assumes a *cis* conformation (see Figure 8.4). One proton is pumped out of the cytoplasm with this shift from *trans* to *cis*. To achieve this $H^+$ translocation action, the protein uses an aspartate residue at position 96 which is near the inner side of the protein and aspartate 85 near the outer side of the plasma membrane. This is proton transfer without electron transfer and the complete proton translocation process takes only $\sim$50 $\mu$sec. The bacteriorhodopsin is attached through the $N^+$ to amino acid residue 216, which is lysine. The Schiff base that is formed is an essential component in proton export. When light at 568 nm strikes the bacteriorhodopsin in the plasma membrane, a proton is ejected through the central channel from the cytoplasm into the periplasm. The proton exported across the plasma membrane can be used to drive ATP synthase or the $H^+$ can exchange with a $Na^+$ ion in a proton symport process to remove $Na^+$ from the cytoplasm of the cell.

A second photosensitive pigment in the plasma membrane of *Halobacterium halobium* is halorhodopsin. When light activates halorhodopsin, $Cl^-$ is pumped into the cytoplasm of the prokaryotic organism. By a proton symport system, $K^+$ is transported into the cytoplasm by a process unrelated to the photosensitive reactions. Thus, an internal concentration of KCl is used to counterbalance the osmotic pressure gen-

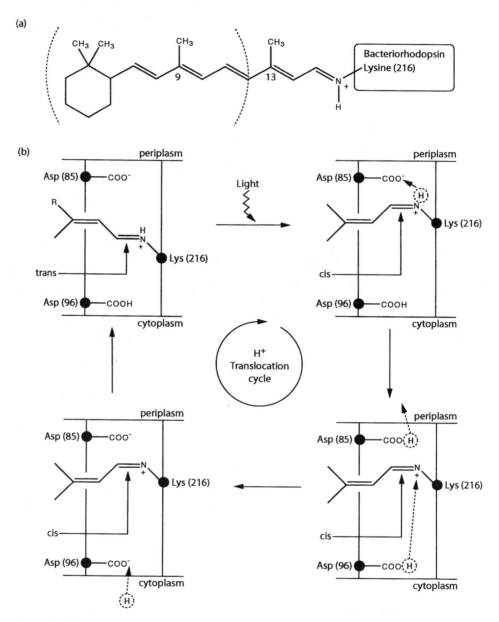

Figure 8.4. Light-driven proton transport in *Halobacterium halobium*. (**a**) Light-sensitive retinol structure is attached through lysine 216 of bacteriorhodopsin. (**b**) Conformational change of retinol moiety with *trans* in the dark and *cis* in the light. The R group is the segment of the retinol molecule in parentheses in (a). Protons move from the cytoplasm to the carboxyl of aspartate residue number 96 to the N of the retinol, to aspartate residue number 85, from where it is released to the periplasm. (From Nicholls and Ferguson, 1992.[18])

erated from the external environment that contains high levels of NaCl. In the dark, these photoreactions may be driven by other mechanisms including special ATP-driven reactions.

## 8.5. Psychrophilic Organisms[19]

The examination of bacterial growth at near-freezing temperatures has attracted attention owing to spoilage of foods at low temperatures and the activities of bacteria in Arctic and Antarctic environments. In 1902, Schmidt-Nelson coined the term "psychrophile" to describe bacteria that were capable of growing at 0°C, and although there have been efforts to change this designation to psychrotrophic or eurythermal mesophile, the term psychrophile remains the most broadly used. The cardinal growth temperatures that are used to characterize psychrophilic microorganisms were established by R. Morita in 1975 and are as follows: minimum, <0°C; optimum, <15°C; and maximum, <20°C. It has been proposed by Ingraham that the lowest possible temperature at which bacteria can grow is −12°C because at that temperature ice crystals will form in the cytoplasm. Therefore, it is recommended that frozen food should be stored at <−18°C to ensure that bacterial growth cannot occur.

There is considerable diversity in the psychrophilic group, which is composed of autotrophs, heterotrophs, aerobes, anaerobes, phototrophic cyanobacteria, attached populations, and free-swimming cellular forms. Abundant microbial flora is present in marine environments where the average temperature of the sea at the poles is −1.8°C and polar oceanic water is <5°C. Psychrophilic bacteria can be readily isolated from Arctic and Antarctic soils where the temperature rarely exceeds 5° to 10°C. Examples of psychrophilic bacteria isolated from various environments are given in Table 8.5. Although Gram-negative bacteria are considered to be most common in these environments, there are examples where the dominant psychrophilic bacteria present are of the Gram-positive type.

It is considered that physiological adjustments of membrane lipids and protein function are required to enable bacteria to grow at temperatures approaching freezing. To provide an appropriate liquid crystalline lipid phase for membrane structure and function of integral proteins, the fatty acids in psychrophilic bacteria are different from those growing at mesophilic conditions. For example, a comparison of lipids from *E. coli* growing at temperature extremes of 10° to 43°C is given in Table 8.3.

Table 8.5. Examples of psychrophilic bacteria isolated from cold environments.[a]

| Environment | Major genera of isolated organisms |
| --- | --- |
| Beaufort Sea in the sub-Arctic | *Flavobacterium, Microcyclus*, and *Vibrio* |
| Signy Island soil in Antarctic peninsula | *Pseudomonas, Achromobacter*, and *Alcaligenes* |
| Antarctic dry valley soil | *Planococcus* and *Micrococcus* |
| Mats in antarctic freshwater streams | *Oscillatoria, Lyngbya*, and *Microcoleus* |

[a]From Russell, 1992.[19]

Compared to mesophilic bacteria, psychrophilic ones frequently have a higher percentage of unsaturated fatty acids, shorter fatty acids, and reduced branching of the lipid components. The change of saturation in fatty acids may be attributed to activation of the desaturase mechanism, which converts the saturated fatty acid to an unsaturated form and accounts for an instantaneous change of the lipid. A somewhat slower method for changing the saturation of the fatty acids with a shift to a colder temperature would be the production of a specific fatty acid synthetase which selects a different type of fatty acid. With thermal shifts to a colder environment, shorter fatty acids with lower melting points are favored, and straight-chain fatty acids are preferred to methyl-branched fatty acids.

The stability of the proteins or protein complexes at psychrophilic temperatures appears to be attributable to ionic interactions involving salt linkages or hydrogen bonding and not hydrophobic linkages. Structural complexes containing ionic linkages may be somewhat sensitive to cold temperatures. Complexes can be envisioned as being dynamic with dissociation of subunits to produce a nonfunctional system as well as association of subunits to give a functional unit. When a cell is shifted to an environment too cold for growth to occur it may be that the dissociation of the complex occurs but the association into a functional unit does not. Under mesophilic conditions, thermal energies are responsible for the molecular movements that account for random interactions and subsequent complex formation by ionic interactions. For example, *E. coli* and other mesophiles cannot grow at <4°C because initiation of the complex for protein synthesis involving ribosomal assembly on mRNA does not occur. This may also explain the cold sensitivity of isolated large protein complexes such as nitrogenase and ATPase structures. The ability of psychrophiles to have a cold-stable translational system appears to be a special adaption. From in vitro experimentation with ribosome subunits and soluble factors needed for protein synthesis, it has been suggested that the ribosomal proteins of psychrophiles are unique. Ribosomes from psychrophiles have a greater requirement for $Mg^{2+}$ and a protein enabling translation to occur. Certain bacteria growing at cold temperatures will produce small acidic proteins that are termed cold acclimation proteins (CAPs) and these are discussed later in this chapter.

## 8.6. Irradiation-Resistant Organisms

Prokaryotic cells that are exposed to sunlight would be exposed to a broad spectrum of light, and some wavelengths are highly detrimental to the organisms. Visible light (400 to 700 nm) is not harmful to most cells; however, if protoporphyrin IX accumulates in the cells, it can be excited and generate superoxide radicals. Because protoporphyrin IX is an intermediate in heme synthesis, this indicates a benefit for tight regulation on this pathway. Bacterial resistance to visible light includes the production of yellow to red carotenoids which are localized in the plasma membrane. Near-ultraviolet (UV) light (290 to 400 nm) will destroy ubiquinones and napthoquinones in the plasma membrane. Because the carotenoids are buried within the membrane and a diversity of UV absorbing molecules are in the outer regions of the bacterial cell, damage to bacteria is rarely critical. For cell exposure to far-UV (200 to 290 nm) the principal UV absorbing structures are the bacterial nucleic acids at 260 nm.

Defense against far UV irradiation in prokaryotes is attributed to either the production of chemicals to screen the cell or to processes that repair damaged DNA. Owing to the growth of prokaryotes on various surfaces of the earth, the exposure of these organisms to UV light is a relatively common event. Because UV light is highly mutagenic, many bacteria have developed systems to protect against UV damage. A few organisms are notable in that they have unique molecules for photoprotection. The cyanobacterium *Nostoc commune* produces a yellow to red lipid-soluble pigment that is localized in the sheath. This pigment is scytoremin and it is considered to be synthesized from aromatic amino acids. Other cyanobacteria produce UV-absorbing pigments attached to carbohydrates that contain amino acids. Bacteria isolated from soil and antarctic sediments produce melanin-like pigments that provide some protection against UV light. In addition, special cellular structures consisting of cysts and spores provide some protection against irradiation. The unusually thick cell walls of the acid-fact bacteria and of the deinococci are considered to provide some protection against UV light. Natural geological formations and radioactive springs producing ionizing radiation are limited in the environment, and as a result the number of bacterial types displaying resistance to ionizing radiation is relatively small. In addition to the passive shielding of bacterial DNA to irradiation damage, it is well documented that resistant organisms display inducible repair systems as a result of DNA damage.

## 8.6.1. Ultraviolet Light[20,21]

Following UV exposure, thymine dimers (T–T) form that alter the conformation of the helix and prevent DNA replication past the replicating fork. Bacteria can repair the damaged DNA by removing thymine dimers by one of four mechanisms. A model indicating the four repair mechanisms is given in Figure 8.5. One mechanism is light induced and three mechanisms are distinct from light-induced repair. Systems of repair that are not induced by light include the following: (1) excision repair with the removal of thymine dimers in the single strand of DNA; (2) recombination repair, in which DNA is reconstructed from undamaged segments; and (3) SOS repair, which promotes the DNA polymerase to proceed past the T–T dimer and not stop at the site of damaged DNA.

The removal of damaged DNA may be accomplished by photoactivation, which is a DNA repair process that occurs following exposure of cells to visible light. An enzyme termed photolyase hydrolyzes the double bond between the two thymines and restores the DNA to a replicating form. Photolyase does not function in the dark and apparently light is used to provide the energy for the cleavage of the bond between the two thymines moieties. The photosensitive segment of the enzyme is unknown even though the purified photolyase has been available for more than 30 years.

In *E. coli*, damaged DNA is excised by a repair endonuclease that removes a 14- to 15-base segment containing the T–T dimer. The DNA strand is hydrolyzed eight nucleotides to the 5' side of the T–T dimer and four or five nucleotides to the 3' side of the T–T dimer. In *Micrococcus luteus*, the endonuclease hydrolyzes the *N*-glucosidic bond at the 5' end of the T–T dimer, and after the new DNA is synthesized,

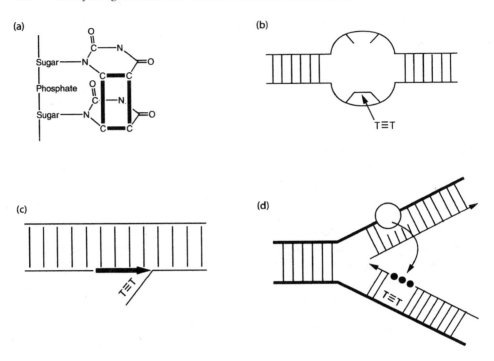

Figure 8.5. Mechanisms for repair of damaged DNA with thymine dimers. (a) Cytobutylthymine dimer produced from two adjacent thymine residues, *heavy line*, following UV light irradiation of DNA. (b) *Escherichia coli* type of excision repair with the segment of DNA removed at *arrows* and gap is replaced. (c) Recombination repair in which new DNA strand, *light line*, has a gap opposite the thymine dimer and gap is filled from parental DNA. (d) Excision repair by the *Micrococcus luteus* mechanism in which the thymine dimer is removed and the gap is filled by polymerase action.

a second excision removes the T–T dimer. DNA repair is by polymerase I followed by ligase action to covalently attach new DNA to the existing DNA strand. In *E. coli*, this replacement involves the Uvr system where *uvrA* and *uvrB* are genes for the subunits of the excision endonuclease and *uvrC* encodes for excision activity.

In recombination repair of DNA damaged by UV light, DNA replication produces gaps in the daughter strand as a result of thymine dimers in the parental strand. While DNA polymerase III stalls at the replication fork where thymine dimers occur, there will be an unprimed reinitiation of DNA replication past the site of the thymine dimer. The gap is repaired by copying the nucleotide sequence from a sister strand of DNA and the mechanism of this process remains to be established. This process of repair is important because there is no lag in DNA replication as is the case in photoactivation where the bonds attaching thymines require hydrolysis before replication can occur.

The final process to be discussed in repair of damaged DNA by UV light is referred to as SOS repair. SOS is not a biological acronym but an expression of alarm, with

SOS (initially a maritime alert of Save Our Ship) used as the universal signal for help. The gene product from *recA* binds to the thymidine dimer to influence the polymerase III action so that replication proceeds and is not stopped at the thymine dimer. In reading through the damaged segment of DNA, bases are randomly inserted into the damaged region, resulting in nonlethal mutations. RecA is constitutively produced at a minimal level that is insufficient to have an impact on genetic repair until the level of RecA increased by induction following UV damage. RecA is an inducer of the SOS response which involves the inactivation of LexA. LexA is a repressor that controls the SOS regulon including RecA production. In the presence of thymine dimers, RecA binds to thymine dimers and in the process activates a proteolytic action specific for LexA. In the absence of damaged DNA, LexA functions to repress protein production important for DNA repair.

## 8.6.2. Ionizing Radiation[22–24]

Members of the genera *Deinococcus* and *Deinobacter* are distinguished from other bacteria by their resistance to extremely high levels of far-UV and ionizing radiations. The unique cell structure of *Deinococcus* has been suggested to protection against far-UV radiation but not against ionizing radiation. As presented in Figure 8.6, the cells of *Deinococcus radiodurans* have a multilayered surface with several unique structures. The plasma membrane contains large quantities of phosphoglycolipids and not the phospholipids commonly found in other bacteria. A unique phosphoglycolypid, 2'-*O*-(1,2,-diacyl-*sn*-glycero-3-phospho)-3'-*O*-(galactosyl)-*N*-D-glyceroyl  alkylamine, has been found in the plasma membrane of *Deinococcus radiodurans* and this molecule is the major lipid in the plasma membrane. The cell wall has a peptidoglycan layer that is 14 to 20 nm thick. In addition to the unique amino acid (L-ornithine) in the cross bridge formation (see Chapter 3), it has been determined that "tunnels" with a diameter of about 10 nm permeate this layer. Exterior to the peptidoglycan layer

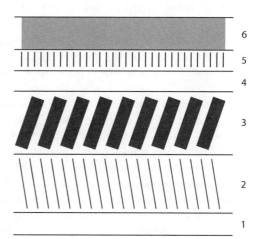

Figure 8.6. Model indicating cell wall of *Deinococcus radiodurans*. 1. Plasma membrane, 2. Peptidoglycan, 3. Compartmentalized layer, 4. Outer membrane, 5. S-layer, and 6. Capsule.

is an uncharacterized compartmentalized layer that consists of several different components. This compartmentalized layer is almost twice as thick as the peptidoglycan layer and has an unknown physiological application. It is known that a significant loss of chemicals occurs following treatment with high concentrations of NaCl or polar solvents. Exterior to the compartmentalized layer is the outer membrane in *D. radiodurans* and it is distinct from that found in Gram-negative bacteria because it does not contain lipopolysaccharides consisting of lipid A, heptoses, and hydroxyl-containing fatty acids. Exterior to the outer membrane is the S-layer which is composed of a 104-kDa protein highly organized in a hexagonally lattice as seen by electron microscopy. The S-layer consists of a glycoprotein with six monosaccharide residues per polypeptide chain and the sugars are attached into the protein through either serine or threonine structures. Approximately two fatty acids are bound into the N-terminus of the polypeptide. The outermost structure of the *D. radiodurans* cell is the carbohydrate layer or coat. These cells generally stain as Gram-positive, although they contain an outer membrane that is characteristic of Gram-negative cells. It is unclear why these organisms have a cell wall that is distinct from that found in other prokaryotes; however, it does not appear that the cell wall composition contributes to the resistance against ionizing radiation.

The deinococci can withstand a radiation dose that is several thousand–fold greater than the lethal dose for humans. An exposure of 0.6 megarads of ionizing radiation is sufficient to kill *E. coli* but not *D. radiodurans*. Survival of *D. radiodurans* is difficult to comprehend because an exposure of 0.6 megarads is expected to produce 200 double-strand breaks of DNA, 3,000 single-strand breaks, and 1,000 damaged nucleotide bases. Double-strand breaks of DNA are especially difficult for most bacterial cells to repair because there is no single strand to serve as a template for accurate repair and *E. coli* can survive only 10 to 15 double-strand breaks. However, *D. radiodurans* will survive irradiation of 1.75 megarads, which produces 1,000 to 2,000 double-strand breaks in the genomic DNA. The processing that would be required for repair of double-strand breaks is incredible, as each cell of *D. radiodurans* in stationary phase contains four genome copies and in logarithmic phase has 8 to 10 genome copies. The radiation resistance in *D. radiodurans* is attributed, in part, to the unique DNA repair system for processing the double-stranded breaks in DNA. The sorting of DNA strands to provide correct splicing of the double-strand breaks would be an astronomical puzzle, and a feature that may enable correct splicing of DNA double-strand breaks is possibly attributed to chromosomes in a highly ordered state.

The repair of damaged DNA in *D. radiodurans* is a highly regulated event. The first requirement is that the cell be in a nutrient media because a considerable amount of energy and biosynthesis is required for the DNA repair mechanisms. Damaged DNA is exported from the cell in segments of about 2,000 basepairs which prevents mutagenesis by incorporation of damaged DNA into the genome. Precursors for DNA synthesis are produced but cell division does not proceed until DNA repair has occurred. Nine proteins are stimulated by ionizing radiation and one of these proteins is considered to be RecA. The mechanism for repair of double-strand breaks in *Deinococcus radiodurans* remains to be established.

An additional feature of damage produced in cells exposed to high levels of ionizing radiation is the production of an oxygen radical ($O_2^-$) and hydrogen peroxide. *D. radiodurans* has a highly active superoxide reductase that enables the destruction of toxic oxygen compounds.

## 8.7. Osmotic Balance and Osmotic Stress[22–24]

When confronted with high osmotic environments, prokaryotic organisms have developed two strategies. One approach is to transport $K^+$ into the cell and this response, which occurs in halophilic bacteria, has been discussed in a previous section. A second approach is to concentrate compatable solutes inside the cell to provide osmotic balance. The use of compatable solutes is widely distributed throughout the prokaryotic groups. Frequently, cellular resistance to osmotic pressures is not attributed to any one given compound but is the combined effect of several molecules. Chloride accumulation is not an appropriate ion for osmoprotection in most bacteria because chloride readily inhibits ribosomal activity as well as enzymes in the tricarboxylic acid cycle.

### 8.7.1. Compatible Solutes

Molecules suitable for osmoprotection are polar, highly water soluble, generally non-charged at physiological pH, and stabilizers of proteins. The concentration of salt required to produce an osmotic response varies with the bacterial species. As little as 0.2 to 0.3 *M* NaCl will promote changes in *E. coli* while 0.7 to 0.8 *M* produces changes in *Staphylococcus aureus*, *Listeria monocytogenes*, or *Corynebacterium glutamicum*. Some of the chemicals that bacteria use to balance osmotic pressures are given in Table 8.6. Of the various chemicals to provide osmotic protection, some of the most commonly employed compounds include glycine betaine, ectoine, and hydroxy-ectoine. Some organisms use trehalose, proline betaine, choline, or proline to provide osmoprotection while a few organisms use glutamate, alanine, *N*-acetyllysine, or *N*-acetylornithine. Structures of some of these osmotic solutes are given in Figure 8.7. These osmotic solutes either may be synthesized by the bacteria or may be assimilated from the growth media.

### 8.7.2. Osmotic Stress Response

When there is an osmotic upshift, water is quickly lost from the cell into the extracellular environment. The efflux of water is considered to produce a physical change in the cytoplasm that initiates the osmosensing process. Water is lost through the plasma membrane by specific channels called aquaporins. The aquaporin is encoded on the *aqpZ* gene and present in both bacteria and archaea. Molecular characteristics of aquaporins are given in Chapter 10.

The reactions associated with exposing *E. coli* to a high osmotic environment involve several transporter systems and these are illustrated in Figure 8.8. It is known

Table 8.6. Examples of osmotic solutes used as osmoprotectants in moderately halophilic/haloduric bacteria.[a]

| Species | GB | E | OH-E | P | C | T | Other |
|---|---|---|---|---|---|---|---|
| *Actinopolyspora halophila* | + | | | | | + | |
| *Bacillus subtilis* | + | + | | + | + | | Carnitine and proline betaine |
| *Brevibacterium casei* | | + | | | | + | |
| *Chromatium purpuratum* | + | | | | | | N-Acetylglutaminyl glutamine amide and sucrose |
| *Corynebacterium glutamicum* | + | + | | + | | | |
| *Escherichia coli* | + | | | | + | | |
| *Halomonas halophila* | | + | + | | | | β-Glutamate |
| *Halomonas varibilis* | + | + | + | | | + | |
| *Micrococcus halobius* | | + | + | | | + | |
| *Planococcus citreus* | | | | + | | | N-Acetylornithine and N-acetyllysine |
| *Paracoccus halodenitrificans* | | + | | | | | |
| *Rhodobacter sulfidophilus* | + | | | | | | Glucosyl glycerol |
| *Salinicoccus roseus* | + | | | + | | | |
| *Sinorhizobium meliloti* | + | | | | | | |
| *Streptomyces griseolus* | + | + | | | | + | |
| *Acidithiobacillus ferroxidans* | + | | | + | | | |

[a]Adapted from Ventosa et al., 1998[17]; Csonka, 1989[23]; Galinski and Trüper, 1994.[24]
GB, glycine betaine; E, ectoine; OH-E, hydroxyectoine; P, proline; C, choline; T, trehalose.

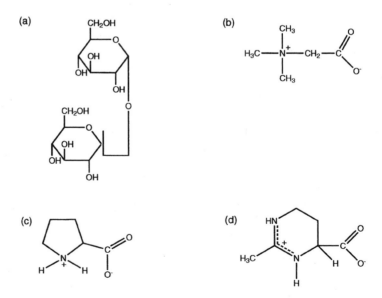

Figure 8.7. Structures of molecules commonly used by bacteria in response to osmotic stress. (a) Trehalose, a diglucose molecule; (b) glycine betaine; (c) proline; and (d) ectoine.

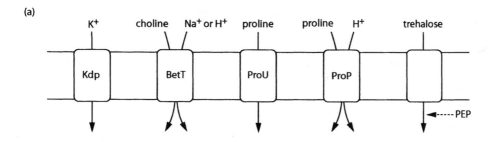

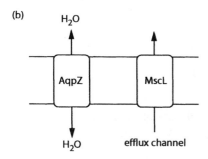

Figure 8.8. Transmembrane movement of chemicals in *Escherichia coli* in response to osmotic stress. Positioned in the plasma membrane are the following: (**a**) Kdp, a P-type ATPase; BetT, a choline transporter driven by $Na^+$ or $H^+$, exact ion is unknown; ProU, an ABC proline transporter; ProP, a proline transporter driven by $H^+$ Symport; and trehalose transport by the phosphotransferase with phosphoenol pyruvate as the energy source. (**b**) AqpZ, the aquaporin; and MscL, a gated channel to release cytoplasmic components when the cell is under a hypoosmotic state.

that $K^+$ accumulation is triggered by the exposure of cells to hyperosmotic media and the change of the cytoplasmic concentration of $K^+$ occurs within a few minutes. The first reaction is the dramatic uptake of $K^+$ attributed primarily to the Kdp transporter and to a smaller extent to the low velocity Trk transporter. While the Trk system is constitutive, the Kdp system is under transcriptional regulation with induction of the *kpd* operon linked to intracellular reduction of $K^+$ levels. The KdpD kinase is located on the inner side of the plasma membrane, where it receives a signal and autophosphorylates. The phosphate group is transferred to KdpE, the response regulator, which results in activating *kdp* transcription. It is proposed that this osmotic response follows the classic two-component signal transduction system discussed in Chapter 6.

Following the rapid uptake of $K^+$, there is an efflux of putrescine followed by an uptake of proline or glycine betaine. The osmoregulated uptake of proline and glycine betaine in *E. coli* and *Salmonella typhimurium* takes place via ProP and ProU transporters. The ProP system is a $H^+$ symporter that has a C-terminal extension that is proposed to respond with a structural change (helical coil) under osmotic increase.

The *proP* gene is controlled by two promoters: *proP*-P1 by a mechanism involving cAMP–CRP (catabolite repressor protein) and the *proP*-P2 which requires an alternate sigma factor that has been designated RpoS. The RpoS-dependent genes are induced in high osmolarity and in stationary phase. The ProU transporter is a member of the ABC transport family and the *proU* operon is induced several hundred–fold by hyperosmotic stress. It has been proposed that the binding of the RNA polymerase for expression of the ProU transporter is influenced by the ionic concentration of the cytoplasm. There is a third system in *E. coli* for transport of osmoregulatory compounds and it is the BetT transporter for choline which is an ion symport but at this time it is not known if it is a $Na^+$- or a $H^+$-coupled transporter. The osmotolerant protectant molecules accumulate in the cell after a few minutes and replace turgor activities attributed to $K^+$. With the uptake of $K^+$ the levels may reach amounts inhibitory for growth, and to alleviate this event, Kef functions as an exporter of $K^+$.

Other metabolic adjustments that may be observed in cells include the accumulation of glutamate or trehalose. Glutamate may be transported into the cell or it is synthesized from appropriate precursors. In some instances, cells will synthesize proline from glutamate and glycine betaine from choline. In those organisms that use trehalose for osmotic protection, the transport into the cytoplasm may occur by a phosphotransferase system specific for trehalose. In the case of *E. coli*, it must synthesize trehalose because it does not have a mechanism for transporting trehalose into the cell from the extracellular environment.

Another change observed in Gram-negative bacteria is in the porin proteins, OmpF and OmpC. These porin proteins provide channels for the movement of small molecules through the outer membrane. The total number of OmpF and OmpC proteins is fairly constant in cells; however, with low osmolarity the quantity of OmpF protein increases and the level of OmpC protein decreases, but with high osmolarity the number of OmpC molecules increases and that of OmpF decreases. The OmpC porin is more restrictive and less specific than the OmpF porin. Transcription of both of these porins is regulated by an OmpR protein that is modified by a membrane protein in response to osmotic changes. Porin synthesis is under control of a two-component sensory system that is discussed in Chapter 6.

In *Bacillus subtilis*, there is a highly developed system for transporting molecules that provide protection against osmotic stress. The specific transporter systems and substrates transported by *B. subtilis* are listed in Table 8.7. As in *E. coli*, *B. subtilis* has numerous avenues for acquiring solutes that are osmotic protectants. The redundancy of systems for uptake of osmoregulatory protectant molecules underscores the importance of this osmotic stress response.

## 8.7.3. Hypoosmotic Shock

When a cell has successfully defended against a high osmotic environment, solutes important in affording osmotic protection reach high concentrations in the cytoplasm. A rapid dilution of osmotic-protected cells will produce hyposmotic shock. Extreme turgor exists inside the cell owing to the presence of osmotic solutes plus the influx of water through aquaporins, and no metabolic energy is required for this activity.

Table 8.7. Osmoregulatory systems in a Gram-positive and a Gram-negative bacterium.[a]

| Bacteria | Plasma membrane protein | Transport mechanism | Substrates transported by specified protein |
|---|---|---|---|
| *Escherichia coli* | ProP | H⁺ symport | Carnitine, ectoine, glycine betaine, proline betaine |
| | ProU | ABC transporter | Glycine betaine, proline, proline betaine |
| | BetT | Na⁺ or H⁺ symport | Choline |
| *Bacillus subtilis* | OpuA | ABC transporter | Glycine betaine, proline betaine |
| | OpuB | ABC transporter | Choline |
| | OpuC | ABC transporter | Carnitine, choline, glycine betaine |
| | OpuD | Na⁺ symporter | Glycine betaine |
| | OpuE | Na⁺ symporter | Proline |

[a]Adapted from Bremer and Krämer, 2000.[22]

Release of cytoplasmic components through gated channels in the plasma membranes reduces the osmotic pressure. These avenues of release of trehalose are operated by mechanosensitive activities. *E. coli* has three gated efflux channels (MscL, MscS, and MscM) that can open for 0.1 to 3.0 nsec. Because the efflux channels are nonspecific, $K^+$, proline, and other molecules are lost from the cytoplasm. The efflux of proline and glycine betaine from *Lactobacillus plantarum* reveals that there are two different pathways: one is the mechanosensitive channel while the other is a typical membrane transporter. These two systems can be differentiated because the gated channel accounts for immediate, nonspecific release of cytoplasmic materials while the transporter is solute specific and kinetically slower. It is proposed that in the absence of gated channels, bacteria could rupture. However, with the sudden release of osmotic pressure, the cells can return to a state of moderate imbalance between the cell and their environment.

## 8.8. Barophilic Bacteria[25]

Prokaryotes living in deep marine environments are subjected to intense pressures; organisms growing under these extreme pressures were termed "barophilic" by C. Zobell in 1949. Barophilic bacteria either require hydrostatic pressures for growth or display greater growth rates at pressures exceeding atmospheric conditions. Not only do deep-sea bacteria withstand great pressures but the temperature of the deep ocean is 2°C, which categorizes these marine organisms as psychrophilic barophiles. The barophilic *Spirillum* sp. isolated at a depth of 5,700 m had a generation time of 4 to 13 hr at 500 atmospheres and 2°C; however, the generation time was 3 to 4 days if the culture was adjusted to atmospheric pressures. The obligate barophilic character is displayed by the strains of *Shewanella benthica*, isolated from the Mariana Trench at a depth of 11,000 m, which had a requirement of 80 MPa for growth. In marine organisms the temperature for growth is dependent on hydrostatic pressure. The hyperthermophilic archaeon *Thermococcus peptonophilus* displayed an optimum growth of 85°C when the pressure was 30 MPa but the optimum temperature of growth was found to be 90° to 95°C when the pressure was changed to 45 MPa.

Examination of the lipid content of the membranes from *Spirillum* sp. revealed that fatty acid composition was influenced by hydrostatic pressures. At high pressure, the $C_{16:1}$ and $C_{18:1}$ fatty acids increased over that observed at lower pressures but in comparison, the $C_{14}$, $C_{14:1}$, and $C_{16}$ fatty acids decreased with an increase of pressure. Other bacterial isolates obtained from depths of 1,200 m to 10,476 m contained $C_{22}$ polyunsaturated fatty acids, which is noteworthy because polyunsaturated fatty acids are rarely found in prokaryotes.

In one barophile, the transcription of a membrane protein is regulated by barophilic pressure. A Gram-negative bacterium, Strain SS9, isolated from a depth of 2,500 m in the Sulu Sea, had an outer membrane protein of 37 kDa that was induced with high pressure. It has been speculated that this protein, OmpH (H is for high pressure), may function as a porin because traditional outer membrane porins produced at low pressures may be damaged and collapse at high pressures. The *OmpH* gene from this organism has been cloned in *E. coli* where the induction of transcription by pressure has been confirmed. This raises the possibility that the barophiles may sense the increased pressures and respond with appropriate physiological inductions.

In facultative barophiles, there is a change in the cell physiology as the culture is moved from high pressure to low pressure. *Thermococcus barophilus* produces a 60-kDa protein when cultivated at atmospheric pressures; however, this protein is not present when cultured at deep-sea pressures of 40 MPa. Thus, it is obvious that prokaryotes not only adapt to environments where the pressure is increased but also will display changes when the pressure is decreased.

## 8.9. Water Activity and Desiccation Tolerance[26,27]

Water has a special role in support of cell metabolism, and with reduction of available water, microbial growth is markedly changed. To designate the level of water available to organisms in various environments the term "water activity" ($a_w$) is used. Water activity is calculated by the following expression:

$$a_w = P/P_0 = n_2/(n_1 - n_2)$$

In the equation $P$ = vapor pressure of the solution, $P_0$ = vapor pressure of water, $n_1$ = number of moles of solvent, and $n_2$ = number of moles of solute. In soil environments, water potential ($\psi$) is used and it is related to water activity by the following formula: $a_w$:c = $(1,000\ RT/W)$ where $R$ = the gas constant for water, $T$ = is the absolute temperature, and $W$ is the molecular weight of water. While microorganisms can grow from $a_w$ = 1.0 to 0.62, individual species can tolerate only a narrow range of water activity. Examples of bacterial tolerance to limiting levels of water are given in Table 8.8. Organisms growing in high salt concentration are actually growing in

Table 8.8. Growth of prokaryotes at different levels of water activity.[a]

| $a_w$ | Environment | Prokaryotes |
|---|---|---|
| 1.000 | Water | Aquatic prokaryotes |
| 0.995 | Blood | Numerous organisms including pathogenic bacteria |
| 0.980 | Sea water | Diverse types of bacteria and archaea |
| 0.950 | Bread | Gram-positive bacteria |
| 0.900 | Ham | *Staphylococcus aureus* and other Gram-positive bacteria |
| 0.850 | Salami | *Staphylococcus aureus* |
| 0.830 | Cortex of endospore | *Bacillus stearothermophilus* |
| 0.800 | Jams, fruitcakes | No bacteria but yeast and fungi |
| 0.750 | Salt lake, salted fish | Halophiles: *Halobacterium* and *Halococcus* |
| 0.700 | Cereals, dried fruit | None |

[a]From Brown, 1976.[26]

reduced levels of water because the salt ions consume space that water occupies in a dilute solution.

As a cell dries, water is lost from the cell surface and from the cytoplasm, resulting in a concentration of salts and macromolecules. Approximately 99% of a bacterial capsule is water and with drying this structure shrinks. Additional drying results in promoting damage to pili and flagella. Typically Gram-negative bacteria have about 80% water and 20% solids with a water-accessible volume of approximately 4.0. When the water-accessible volume reaches 0.3, the cell dies. Water inside the cell is not all free but much is bound as a monolayer on membranes, proteins, ribosomes, DNA, and other biostructures. Bound water is not static but in fact is highly mobile and changes with protein associations as evidenced by the release of approximately 65 water molecules when hexokinase binds glucose as a substrate. The binding of water to molecules is significant in that about one fourth of the weight of a soluble protein is water. When a cell dries, water is not removed randomly but there is an order with initial losses of water from –N, =N–H, and –OH moieties. With continued drying, water is removed from the =C=O and =P=O groups. One explanation of how desiccation-tolerant cells can survive is the idea that polyhydroxyl groups such as trehalose replace water in the hydrogen bonding activity. Proteins drying in the presence of trehalose will form intramolecular hydrogen bonds between the carbohydrate and the protein. Complete stripping of water molecules from the surface of proteins results in denaturation of proteins and the protein will never regain its structure when placed in water. Thus, carbohydrates stabilize the protein in the absence of water and when water is added the formation of a functional protein with water bonded on the surface is reestablished. DNA damage occurs after proteins become desiccated, and in *E. coli*, subjecting a cell to drying for 12 min produces 10 single-stranded breaks per genome of DNA. In phospholipids, 10 to 12 water molecules are attached to each phosphate moiety in phospholipids by hydrogen bonding, and with water removal the phospholipids become very tightly packed. When the desiccated membrane is rehydrated, "leaky" membranes result. This alteration of the membrane can be prevented by drying in the presence of trehalose. Stationary phase cells are more resistant to drying pressures than log phase cells, in part because trehalose is one of the components that accumulates in stationary phase cells.

In addition to a change in water activity attributable to the presence of high concentrations of salt or another solute, $a_w$ is also reduced in dry environments. Microorganisms that can remain viable or grow in low water environments are known as xerotolerant organisms. Prokaryotes that produce spores, akinetes, cysts, and resting stages are resistant to desiccation. Endospores of *Bacillus anthrasis* and akinetes of cyanobacteria have been demonstrated to be viable for a century. It is not known how long spores can remain viable but spore-forming bacteria and conidia-producing actinomycetes have been isolated from core samples at a 114 m depth at the Vostok station in Antarctic and are considered to be about 3,000 years old. Non–spore-forming *Listeria monocytogenes* will remain viable for about 20 years when dried on straw, and dry cells of *Thermoplasma acidophilum* remain viable for more than 10 years. When suspended in dust, cells of *Mycobacterium tuberculosis* remain viable for as long as 20 days. Bacteria suspended in foods or body fluids remain viable when dried for a longer time than when the cells are dried individually. *Staphylococcus* is a bacterium highly resistant to drying and this may be because the outer cells in the cell aggregate slowly dry, thereby causing the central cells in the clump to produce necessary measures for xeritolerance. A key for prokaryotic viability is slow drying to enable the cell to either produce a spore or accumulate trehalose which is used to replace water in formation of hydrogen bonds. Oxygen, UV light, and elevated temperatures have detrimental effects on dry microbial cells because these additional stresses produce damages that cannot be corrected without cell metabolism.

Several methods have been used for preservation of prokaryotic cultures, and these relate to water management. Microbiologists observed that it was possible to maintain spore-forming bacterial cultures by adding a few drops of broth culture to sterile soil and allowing the soil to dry for several days during which time spores were produced. Non–spore-forming cultures can be maintained for several years by covering an agar slant of the culture with sterile mineral water and refrigerating the tube. Freezing of cultures is not desirable because ice crystals produced in the cells or media will mechanically damage microbial membranes. To alleviate the problem of ice crystal formation, cells are suspended in 10% to 30% glycerol or 5% to 8% dimethyl sulfoxide before freezing at $-70°C$. For long-term storage of various types of bacteria, freeze-drying procedures are used in which the cells are suspended in double-strength powdered milk as a cryoprotectant, quick frozen in a dry ice bath, and exposure to a high vacuum removes the water. This process of freeze-drying is also known as lyophilization and is characterized as having water evaporate from a solid phase to a gas phase without becoming a liquid. Viable bacteria have been isolated from Siberian permafrost where cells were subjected to a natural freeze-dry procedure. In lyophilized cells free water is removed but bound water is retained to stabilize molecules and critical structures. In a similar situation, endospores have no free water but have about 20% as bound water with the cortex having $a_w$ of 0.83.

Xeritolerant cyanobacteria serve as a model for the study of bacterial survival in dry environments. Under water stress conditions *Nostoc commune* has been found to produce three acidic proteins with molecular masses of 32, 37, and 39 kDa. These water stress proteins (Wsp) are associated with extracellular glucan secreted by the cyanobacteria but their specific physiological activity remains to be established. As water becomes scarce in deserts, cyanobacteria, such as *Solentia stratosa*, grow on the surfaces and subsurfaces of rocks. By growing in the cracks or fissures of limestone, *Solentia stratosa* gains protection against extreme temperature shifts and the rate of desiccation is reduced.

# 8.10. Temperature Stress Response[28,29]

Changes in temperature of the environment will elicit a response in microorganisms even when the new temperature is not lethal to the organism. Temperature fluctuation may be relatively long term as observed with seasonal events or may be relatively short as is the case with heating of the Earth's surface during the day and cooling at night. Successful persistence of these organisms is, in part, a result of the capability of organisms to adapt to these temperature changes. Careful examination of bacterial response to heat shock and cold shock reveals the manner in which microorganisms not only adjust to temperature stress but, in fact, exploit these temperature shifts.

## 8.10.1. Heat Shock Response[30,31]

Throughout the biological world unique proteins and cell activities have been observed following a brief upshift in temperature. These responses are collectively referred to as heat shock response and are observed in autotrophs, heterotrophs, aerobes, anaerobes, mesophiles, thermophiles, and psychrophiles. In *E. coli*, a temperature shift from 30°C to 42°C induces production of 17 heat shock proteins within the first 5 min of exposure to the increased temperature. This induction is attributed, in part, to gene *rpoH*, which codes for a unique sigma factor designated $\sigma^{32}$. This sigma factor is required for the production of various heat shock proteins (HSP). There are a couple of heat shock systems in *E. coli* in addition to the $\sigma^{32}$ regulon and these are presented in Table 8.9. In *E. coli*, additional or minor sigma factors produced under conditions of heat shock include $\sigma^E$, also known as $\sigma^{24}$, produced from the *rpoE* gene and $\sigma^{54}$, which is a product of the *pspF* gene. Regulation of certain HSPs in *Bacillus subtilis* is controlled by the HrcA repressor which binds to a 9-bp segment in the operator called the CIRCE to influence expression of the vegetative $\sigma^A$ ($\sigma^{70}$) factor. A minor sigma factor ($\sigma^B$) controls more than 50 stress genes and it regulates numerous protein–protein interactions resulting from glucose starvation or oxygen stress.

HSPs are organized into five groups based on molecular size. The sHsp family of small HSPs is 12 to 43 kDa while Hsp60 is 60 kDa and is ubiquitous in prokaryotes. Hsp70 is 70 kDa and is found in some halophilic methanogens, halophiles, and moderate thermophiles. Hsp90 has a mass of 90 kDa and Hsp104 is 100 kDa. The most prevelant HSP in hyperthermophilic archaea is a member of the Hsp60 family and is

Table 8.9. Control of heat shock genes in bacteria.[a]

| Characteristic | Factor | Organism | DNA sequence | | Number of proteins produced |
| | | | −35 | −10 | |
| --- | --- | --- | --- | --- | --- |
| Promoter | $\sigma^{70}$ | *Escherichia coli* | TTGACA (16–18 bp) | TATAAT | — |
| | $\sigma^{32}$ | *Escherichia coli* | CTTGAA (13–17 bp) | CCCCAT-T | 34 (9-chaperone and 5-protease) |
| | $\sigma^{24}$ | *Escherichia coli* | GAACTT (16 bp) | TCTGA | 3 |
| | Others | *Escherichia coli* | — | — | 10 |
| | $\sigma^{B}$ | *Bacillus subtilis* | — | — | ~50 |
| Operator | CIRCE | *Bacillus subtilis* and various other bacteria | TTAGCACTC (N9) | GAGRGCTAA | — |

[a]Adapted from Yura et al., 2000.[30]

known as thermosome or archaeosome. These thermosomes consist of two dissimilar subunits and assembly requires $Mg^{2+}$ plus ATP. Under normal growth, considerable quantities of these HSPs are found in the archaea with *Sulfolobus shibatae* condensing HSP monomers into long filaments. When *E. coli* is grown at 46°C, 15% of the cell mass will consist of two proteins: Hsp60 and sHsp.

Heat shock has been extensively examined in the $\sigma^{32}$, where it appears that induction is attributed to increased temperature promotes melting of the *rpoH* mRNA structure. Under low temperatures, the initiation region of the mRNA does not have a stable conformation and virtually no formation of the $\sigma^{32}$ factor is observed; however, at elevated temperatures the secondary structure of the mRNA from *rpoH* enhances translation. Thus, the immediate heat stress response is attributed to the rapid translation event and not to activation of some precursor or factor. When the cell is introduced to a moderate temperature, translation is not favored and production of the HSPs does not occur.

It is considered that the functions of HSP include protection of proteins from denaturation, modification of inactivated proteins to make them functional, and promotion of events involved in degradation of denatured proteins that collect in stressed cells. The HSPs protect proteins by chaperone activities to promote correct assembly of proteins including protein folding. Specific activities associated with HSPs include production of enzymes for DNA replication, reactivation of RNA polymerase, proteolysis of denatured proteins, processing of proteins for export, enhanced flagellar synthesis, and repair of damage to DNA by mutagenic agents. The HSPs are not limited to induction by temperature stress but have been observed when cells are exposed to the following: agents that damage DNA, starvation for carbon source, denaturation of proteins in the outer membrane, misfolding of proteins in the periplasm, alkaline stress, exposure to alcohol, the stringent response, misfolded proteins resulting from the incorporation of amino acid analogs, and infection by λ phage. Clearly HSPs protect cells against numerous environmental stresses, and future studies will provide information to clarify the diversity of the genes for heat stress response.

## 8.10.2.  Cold Shock Response[32–35]

The cold stress response has been extensively studied in mesophilic bacteria because of the necessity of controlling food spoilage.  The general response observed with bacteria transferred to a near-lethal cold temperature results in the production of cold shock proteins (CSPs) which are distinct from the cold acclimation proteins (CAPs) produced in bacteria with continuous growth at a low temperature.  For example *Enterococcus faecalis* produces 11 CSPs and 10 CAPs; however, five of these CSPs are distinct from the CAPs.  *Bacillus psychrophilus* produces nine CSPs when shifted from 20°C to 0°C.

It has been suggested that membranes, nucleic acids, or ribosomes may have a role in sensing the cold environment.  With a temperature downshift in *E. coli*, the plasma membrane changes to a higher degree of unsaturated fatty acids.  There is an activation of the enzyme β-ketoacyl-acyl carrier protein synthase II, which functions as a desaturase for palmitoleic acid.  In *B. subtilis*, the acyl-lipid desaturase is inducible whereas in *Listeria monocytogenes*, there is a shift of the *iso* fatty acids to *anteiso* fatty acids.  It has been proposed that with a temperature downshift the site on the ribosome is blocked so that the charged tRNA cannot be added and the concentration of (p)ppGpp decreases.  Following cold shock, the ribosomes will not recognize mesophilic mRNA but will interact only with mRNA in CSPs.  One of the CSPs produced is a ribosomal factor that enables ribosomes to translate mRNA from mesophilic cultures as well as from cultures grown at low temperatures.  Following exposure of cells to cold, there is an increase in negative supercoiling of DNA which may influence the orientation of a twist-sensitive promoter and expose the DNA to interactions that lead to production of CSPs.  It has been noted that mesophilic bacteria subjected to freezing will die at a slower rate than if they were initially pretreated to cold exposure, and this is termed cryotolerance.

It has been estimated that 28 to 30 CSPs are produced by organisms such as the mesophilic *E. coli* and the psychrophilic *Arthrobacter globiformis*.  Examples of the diversity of activities for the CSPs are given in Table 8.10.  The major CSP produced by *E. coli* is CspA, which has also been found in more than 60 species of Gram-negative and Gram-positive bacteria, while more than 30 CSPs are produced in *E. coli* and nine proteins are associated with the CspA family.  The CspA protein functions as an RNA chaperone where the binding of CspA to RNA is of low affinity and without specificity.  This prevents the mRNA from condensing into a stable form at low temperatures but does not interfere with ribosomal movement.  The *cspA* gene is expressed with a shift from 37°C to 30°C but greater levels of CspA are produced with cold downshifts to 10°C.  The *cspA* mRNA is produced at 37°C; however, the half-life is only about 12 sec.  Immediately after a downshift in temperature, the half-life of the *cspA* mRNA is >20 min.  After some time at the cold temperature the instability of the *cspA* mRNA returns with little CspA protein produced and other genes in the CspA family are expressed.  The *cspA* gene has an unusually long 5' untranslated region which enhances the transcription.  Essential for induction are the following: (1) an AT-rich segment that is upstream of the −35 region, (2) a TG motif immediately upstream of the −10 segment, and (3) an 11-base sequence at +2 region

Table 8.10. Cold shock responses reported for different bacteria and archaea.

| Organism | Features | Characteristic of response | References (see Additional Reading) |
|---|---|---|---|
| *Anabaena* sp. | B, P | Induction of RNA helicase | Charmot and Owttrim, 2000 |
| *Bacillus subtilis* | B, M | Change in fatty acids in membrane | Klein et al., 1999 |
| *Escherichia coli* | B, M | Induction of polynucleotide phosphorylase | Zangrossi et al., 2000 |
| | | Induction of acyltransferase for palmitoleoyl-acyl carrier protein | Carty et al., 1999 |
| *Listeria monocytogenes* | B, M | Ferritin-like protein produced | Hebraud and Guzzo, 2000[16] |
| *Methanococcoides burtonii* | A | DEAD-box RNA helicase | Lim et al., 2000 |
| *Mycobacterium smegmatis* | B, M | Induction of histone-like protein | Shires and Steyn, 2001 |
| *Rhizobium* sp. | B, P | Increase in the total number of proteins produced | Cloutier et al., 1992 |
| *Shewanella violacea* | B, P | Two cold shock genes cloned | Fujii et al., 1999 |
| *Sulfolobus* sp. | A, T | DNA topoisomerase level increased | LopezGarcia and Forterre, 1999 |
| *Thermotoga maritima* | B, T | Induction of cold shock protein homolog | Welker et al., 1999 Wassenberg et al., 1999[16] |

A, Archaea; B, bacteria; M, mesophilic growth; P, psychrophilic growth; T, thermophilic growth.

that is called the "cold box." The cold box is a pause region for transcription and contributes to the repression of *cspA* resulting in the attenuation of transcription.

## 8.11. Transient to Prolonged Exposures

The majority of the prokaryotes do not produce spores, and therefore their persistence in the environment is dependent on continuous metabolism and cell division. However, nutrients are not continuously available but may be considered to be only transiently present. Successful organisms in the biopopulation requires microorganisms to adjust to this "feast or famine" phenomena. The only storage nutrients for prokaryotes are their internal reserves and only the species that can adjust to environmental stress will survive. This stress response is a relatively new area in physiology and this physiological activity is examined here from the perspective of nutrient limitation, oxidative stress, and anaerobic response.

### 8.11.1. Starvation–Stress Response[37,38]

Bacteria are commonly exposed to a limitation of carbon material as well as suboptimal concentrations of phosphorus and nitrogen. These limitations are collectively referred to as starvation–stress response (SSR). *Salmonella typhimurium, Vibrio* sp., and *E. coli* are organisms that have been extensively examined with respect to SSR. Nutrient limitation for these organisms occurs in nature but in laboratory cultures it will be observed only when growth is arrested due to a lack of carbon or some other nutrient. With a limitation of nutrients SSR is observed in the stationary phase; however, in rich medium with adequate nutrients the culture enters the stationary phase without a starvation response. Under conditions of limited nutrients, SSR is observed already in the latter portion of the log growth phase as well as in the start of the stationary phase. In general, the SSR follows a series of sequential gene expressions that start just prior to completion of log phase and continues for several days in the stationary phase.

One of the more recognizable features of SSR is the reduction in size of the starved cell. While growing cells may be 1.5 to 2.5 μm, starved cells are 0.2 to 1.0 μm. These cells that are transiently of smaller size are called ultramicrobacteria (UMB). In natural aquatic and soil environments, bacteria may exist as UMBs because direct measurements have revealed that bacteria are 0.2 μm to 0.8 μm and on cultivation the cells display a length of <1 μm. Even though the parent cells are of rod or vibrio shape, UMBs are spherical with a significantly enlarged periplasm.

Numerous phenotypic changes are observed in cells undergoing SSR, and marine vibrio have been extensively examined. With starvation, cells utilize internal reserves and polymers of β-hydroxybutyrate in the cytoplasm will gradually disappear so that UMBs will not have any of this stored energy. Within 30 min of starvation, *Vibrio* sp. will accumulate ppGpp and intracellular proteolysis increases but there is a reduction in the production of RNA, proteins, and peptidoglycan. Also, in this phase of multiple nutrient starvation 30 to 40 Sti (starvation-induced) proteins are induced. Changes in the periplasm occur, with some Sti proteins increased 100-fold and contributes to nutrient transport including chemotaxis toward nutrients. From 1 to 6 hr of starvation several changes are seen in the vibrio: a reduction of the intracellular proteolytic activity, a change in lipid content to maintain membrane fluidity that results in decreasing the length of the fatty acids and increasing unsaturated fatty acids, and an overall decrease in protein synthesis but an induction of about 20 Sti-specific proteins.

Following 6 hr of starvation, several cellular changes occur in addition to the production of several Sti-specific proteins and these include increased release of exoproteases, release of flagella with a loss of motility, and marked reduction of cell size. There is a marked change in the stability of certain mRNAs throughout the SSR in that the half-life is lengthened from the 2 to 10 min at the onset of starvation to >70 min once starvation has been in effect for several hours. Several hours into starvation the cell surface becomes more hydrophobic, and adhesion between the cell and the environment is increased owing to fimbrae production.

There is an efficient means of controlling these genes and it is through the use of two master regulators: an alternative sigma transcription factor known as $\sigma^s$ or $\sigma^{38}$ from the *rpoS* gene and regulation by cAMP:CRP. A large number of genes are influenced by either cAMP:CRP in a positive or a negative fashion and many new genes are expressed with by direct control of $\sigma^s$. Some of these genes induced in *E.*

*coli* contribute to increased resistance to heat, osmotic pressures, or acidic conditions and include production of the following: lipoprotein, periplasmic trehalase, trehalose synthesis, cytochrome oxidase, catalase, acid phosphatase, and exonuclease III.

## 8.11.2. Oxidative Stress Response[39–41]

Dioxygen ($O_2$) is the most oxidized form of oxygen while $H_2O$ represents the reduced form of oxygen. The intermediates in the reduction of $O_2$ to $H_2O$ result in the production of highly reactive compounds including superoxide ($O_2^{\cdot-}$), hydrogen peroxide ($H_2O_2$), and hydroxyl radical (HO•). The following scheme indicates the formation of these oxygen-containing intermediates and the reduction potentials of the reactions:

$$O_2 \rightarrow O_2^{\cdot-} \rightarrow H_2O_2 \rightarrow HO\bullet + H_2O \rightarrow H_2O$$
$$-330\ mV \quad +940\ mV \quad +380\ mV \quad +2{,}330\ mV$$

These reactive oxygen intermediates produce single-strand breaks in DNA, lipid peroxidation, as well as damage to proteins. While the DNA damage can prevent replication and ultimately cell division, cell death can result from permeability damage to the plasma membrane.

Oxidative stress in bacteria is attributed to the presence of either hydrogen peroxide, the superoxide anion ($O_2^{\cdot-}$), or the hydroxyl radical (HO•). Oxidative reactions and the response of prokaryotes to these radicals are listed in Table 8.11. Oxidative damage due to superoxide anions or hydroxyl radicals is highly reactive and will produce single-strand breaks with direct interaction or will interact with cellular materials to produce peroxyanion radicals that damage DNA. The formation of these free radicals is attributed to the reaction of molecular oxygen with reduced cellular constituents. When $O_2$ interacts with reduced cytochromes, flavins, or other electron carriers that deliver a single electron, the superoxide anion in the form of $O_2^{\cdot-}$ is produced. Aerobic, microaerophilic, and aerotolerant bacteria produce superoxide dismutase to cat-

Table 8.11. Selected examples of oxidative stress response in *E. coli*.[a]

| oxyR regulon | | soxRS regulon | |
|---|---|---|---|
| Gene | Activity | Gene | Activity |
| *katG* | Hydroperoxidase I | *sodA* | Manganese superoxide dismutase |
| *ahpC* | Alkyl hydroperoxidase | *fur* | Ferric uptake repressor |
| *gorA* | Glutathione reductase | *fldA* | Flavodoxin |
| *grxA* | Glutaredoxin | *fpr* | Ferredoxin/flavodoxin reductase |
| *trxC* | Thioredoxin | *nfsA* | Nitroreductase A |
| *fur* | Ferric uptake repressor | *arcAB* | Multidrug efflux repressor |
| *fhuF* | Ferric reductase | *fur* | Ferric uptake repressor |

[a]Data from Storz and Zheng, 2000.[41]

alyze the reaction between $O_2^{\cdot-}$ and 2 $H^+$ to produce $H_2O_2$. The degradation of hydrogen peroxide to water is catalyzed by catalase. Hydrogen peroxide generates HO• in the cells by the interaction with $Fe^{2+}$ and this hydroxyl radical is an extremely strong oxidizing agent with an $E_0' = $ of $+2,330$ mV for the HO•/$H_2O$ couple. The generation of a hydroxyl radical may result from the chemical interaction between hydrogen peroxide and $Fe^{2+}$ and this is known as the Fenton reaction. As a protective defense, about 40 genes are induced in bacteria to counteract the effect of the hydroxyl radical. If there is no mechanism to destroy the hydroxyl radical, the reactive nature of this radical can irreversibly destroy cellular material. Hydroxyl radicals cause cell death by the indiscriminate removal of hydrogen atoms from saturated carbons of purines, pyrimidines, or other biomolecules.

Superoxide dismutase (SOD) functions as a scavenger for the removal of the superoxide radical, and this enzyme is present in many aerobic prokaryotes. The production of SOD by anerobic organisms is considered to provide protection against superoxide radicals and 5% to 25% of anaerobes are induced to produce SOD when exposed to air. Some of the aerotolerant anaerobes reported to produce SOD include *Bacteroides*, *Chlorobium*, *Clostridium*, *Desulfovibrio*, and *Selenomonas*. A key feature of the activity of SOD is the presence of a metal ion with the apoprotein and a survey of the prokaryotes reveals that different organisms have a preference for a given metal. Only *Photobacterium leioyhathi*, *Micrococcus denitrificans*, and a few other prokaryotes have a Cu-SOD or a Zn-SOD, whereas these enzymes are found widely distributed in eukaryotes. Mn-SOD is found in many species of prokaryotes and eukaryotes whereas Fe-SOD has been isolated only from prokaryotes. Generally, each species produces only one type of SOD; however, *E. coli* produces a Mn-SOD that is located in the cytoplasm and a Fe-SOD that is in the periplasm. Presumably, the cytoplasmic SOD would protect the cell from internally generated oxygen radicals while the periplasmic SOD would protect against externally generated radicals.

To generate oxidative stress in bacteria under laboratory conditions, low concentrations (10 to 50 $\mu M$) of hydrogen peroxide, plumbagen, menadione, or paraquat are added to the bacterial cultures. Frequently the level of resistance to oxidative stress that is induced will protect cells from subsequent exposure to these agents at high (5 to 10 m$M$) concentrations. Plumbagen, menadione, and paraquat generate $O_2^{\cdot-}$ with the induction of numerous genes although some of these genes are the same as induced with hydrogen peroxide.

The two principal mechanisms for bacteria to control oxidative stress involve the SoxR regulon and the OxyR regulon systems. The antioxidant activities of SoxR, OxyR, and other systems in *E. coli* are given in Table 8.12. The OxyR protein is a tetramer composed of 34 kDa monomers. There are two cysteine residues on each monomer and these react with hydrogen peroxide to produce disulfide bonds. The oxidized form of the OxyR protein activates transcription. The SoxR protein is a dimer consisting of 17 kDa monomers and two [2Fe-2S] centers per dimer. The superoxide anion reacts with the iron-sulfur centers to produce the reduced form of SoxR and it is this form that induces transcription. The role of $H_2O_2$ in oxidation of reduced OxyR and oxidation of reduced SoxR by $O_2^{\cdot-}$ is presented in Figure 8.9.

The exclusion of Fe(II) from the cells during this oxidative stress is accomplished

Table 8.12. Reactions associated with free radicals in prokaryotes.

| Enzyme or reaction name | Reaction |
|---|---|
| Chemical reaction | Reduced flavin + $O_2$ → Oxidized flavin + $O_2^{\cdot-}$ + $H^+$ |
| Superoxide dismutase (with dismutation) | $2\ O_2^{\cdot-}$ + $2H^+$ → $H_2O_2$ + $O_2$ |
| Superoxide dismutase (without dismutation) | $1e^-$ + $O_2^{\cdot-}$ + $2H^+$ → $H_2O_2$ |
| NADH:rubrerythrin oxidoreductase (peroxidase) | NADH + $H^+$ + $H_2O_2$ → $NAD^+$ + $2H_2O$ |
| Fenton reaction | $Fe^{2+}$ + $H_2O_2$ → $Fe^{3+}$ + $OH^-$ + $HO^{\cdot}$ |
| Catalase | $H_2O_2$ → $H_2O$ + ½ $O_2$ |
| Alkyl hydroperoxide reductase | NADH + alkyl hydroperoxide → $NAD^+$ + alkyl + $H_2O$ |

by OxyR induction of *fur*, which represses the uptake of Fe(III). The importance of excluding iron from the cell is underscored by the observation that both SoxR and OxyR induce *fur*. In addition to conferring resistance to HO• and $O_2^{\cdot-}$, SoxR and OxyR also provide resistance to HOCl, organic solvents, reactive nitrogen species, numerous antibiotics, and some xenobiotics. Many of the bacteria exposed to starvation stress also induce antioxidant genes encoding for catalase and superoxide dismutase.

Another component in the antioxidant defense system used is the alkyl hydrope-

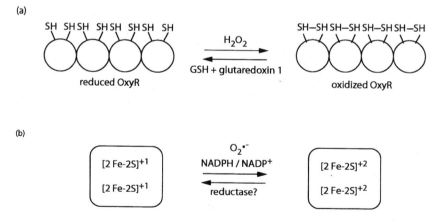

Figure 8.9. Reactions involving OxyR and SoxR. (**a**) Reduction of hydrogen peroxide by reduced OxyR. Reduced OxyR is regenerated from oxidized OxyR by reduced glutathione and glutaredoxin 1. (**b**) Reduction of oxygen radical by reduced SoxR and NADPH. The system for regeneration of reduced SoxR from oxidized SoxR has not been resolved. (From Storz and Zheng, 2000.[41] Used with permission.)

roxidase system, which reacts with organic hydroperoxides and hydrogen peroxide. The alkyl hydroperoxidase (Ahp) system has been reported in only a few organisms and future studies will be needed to determine if it is broadly distributed throughout the bacteria. In *Salmonella typhimurium*, the AhpC is a 22-kDa colorless protein that uses NADH for the reduction of ethyl hydroperoxide, cumene hydroperoxide, and hydrogen peroxide. Also part of the alkyl hydroperoxidase system in *Salmonella typhimurium* is AhpF, a 57-kDa FAD-containing protein, which is similar to the flavoprotein thioredoxin reductase produced by *E. coli*. The AhpF catalyzes the reduction of alkyl hydroperoxidase and hydrogen peroxide by transferring electrons from NADH to AhpC and the reduced AhpC interacts with the peroxides. Some organisms, such as *Mycobacterium tuberculosis*, produce two distinct proteins, AhpC and AhpD, which interact with alkyl hydroperoxides and hydrogen peroxide. Other bacteria that produce AhpC include *Helicobacter pylori, Amphibacillus xylanus, Xanthomonas campestris, Campylobacter jejuni*, and *Bacteroides fragilis*. In *B. subtilis*, genes for alkyl hydroperoxide reductase are controlled by a repressor designated as PerR. The PerR is a 21-kDa metalloprotein that is a Fur-like protein because it binds DNA at its N-terminal region. Other *fur* genes that contribute to oxidative stress resistance have been reported for *Desulfovibrio vulgaris, Mycobacterium tuberculosis*, and *Streptomyces venezuelae*.

Sulfate-respiring bacteria have a complex system that functions under conditions of oxidative stress. Using *Desulfovibrio vulgaris* is the model organism, several different proteins have been characterized that interact with oxygen molecules. A *Desulfovibrio* chemoreceptor (dcr) system contains at least 12 genes and one is a chemoreceptor, DcrH. The DcrH senses oxygen and by a mechanism involving methyl-accepting proteins influences the movement of *D. vulgaris*. Rubredoxin oxidoreductase (Rbo) is a dimer with each monomer containing a nonheme iron center, and this protein is proposed to catalyze a superoxide reductase reaction which results in only the production of hydrogen peroxide without the dismutation of oxygen found in other SOD reactions (see reactions in Table 8.12). Rubrerythrin (Rbr) is a nonheme iron protein that may have peroxidase related activity when coupled to a NADH-Rbr oxidoreductase (see the reaction in Table 8.12). While the metabolism of oxygen by sulfate-reducing bacteria is not fully understood, a model that links many of these interacting activities is given in Figure 8.10.

## *8.11.3. Anaerobic Response*[42,43]

Oxygen, or its absence, is often critical for the metabolism of many bacteria, and special systems have evolved for bacteria to respond to oxygen deprivation. In facultative anaerobes, the reduction of oxygen in the environment elicits new responses in the cells for several processes including anaerobic respiration, fermentation, and photosynthesis. In addition, some bacteria display considerable oxygen sensitivity to nitrogen fixation and production of virulence factors important for pathogenicity. Within the past few years it has been demonstrated that the control of genes of these diverse pathways is attributed to a few events that are associated with global regulation

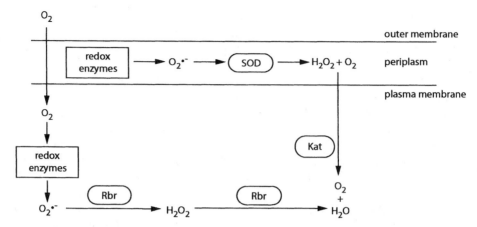

Figure 8.10. Oxidative stress response in a sulfate-reducing bacterium, *Desulfovibrio vulgaris*. SOD, Superoxide dismutase; Kat, catalase; Rbo, rubredoxin oxidoreductase, Rbr, rubrerythrin. (From Lumppio et al., 2001.[39] Used with permission.)

of this anaerobic response. It appears that the various pathways associated with this oxygen deprivation can be regulated by a few global regulatory pathways and $O_2$ sensing may be provided by specific cofactors.

The anaerobic response in *E. coli* is attributed to two global regulatory systems: ArcAB, where "arc" stands for *a*erobic *r*espiration *c*ontrol and FNR, which stands for *f*umarate and *n*itrate *r*eduction. The metabolic activities regulated include enzymes associated with the tricarboxylic acid (TCA) cycle or the membrane-associated electron transport chain. The ArcAB and FNR can either function independently or both can exhibit an effect on the genes. Examples of this type of regulation are presented in Table 8.13.

In general, the regulation of ArcAB by the anaerobic response is through a two-component system. For this two-component system to function, there must be a sensor for molecular oxygen and a modification of DNA expression. As in most two-component sensors, one specific activity is in the plasma membrane and the other in the cytoplasm. The one in the plasma membrane is attributed to a kinase that autophosphorylates using ATP to produce a histidine phosphate as a terminal segment of the protein. The cytoplasm contains a response protein that receives the phosphate from histidine and transfers it onto an aspartate moiety. The phosphorylated cytoplasmic protein functions as a regulator to alter the activity of gene expression. The ArcB is the kinase protein that is in the plasma membrane and senses $O_2$ by an unknown mechanism. ArcA is the cytoplasmic protein that regulates specific genes after binding to DNA through the following proposed binding site: 5'-[A/T]GTTAATTA[A/T]-3'.

With a transition from an aerobic environment to one that is anaerobic, there is a change in genetic expression as directed by the FNR regulatory protein. While FNR protein is produced in cells under both aerobic and anaerobic conditions, a stable

Table 8.13. Selected examples of enzymes under ArcA and FNR as regulators.[a]

| Gene | Enzymes | ArcA regulated | FNR regulated |
|------|---------|----------------|---------------|
| *fumB* | Fumarase B | Activation | Activation |
| *gltA* | Citrate synthase | Repression | ? |
| *sdhCDAB* | Succinate dehydrogenase | Repression | Repression |
| *sucAB* | α-Ketoglutarate dehydrogenase | Repression | Repression |
| *sucCD* | Succinyl coenzyme A synthetase | Repression | Repression |
| *fdnGHI* | Formate dehydrogenase-N | Repression | Activation |
| *hyaA-F* | Hydrogenase I | Activation | ? |
| *nuo* | NADH dehydrogenase | Repression | Repression |
| *cyd* | Cytochrome *d* oxidase | Activation | Repression |
| *cyoABCDE* | Cytochrome *o* oxidase | Repression | Repression |
| *frdABCD* | Fumarate reductase | ? | Activation |
| *narGHJI* | Nitrate reductase | ? | Activation |
| *focA-pfl* | Formate transport and pyruvate-formate lyase | Activation | Activation |

[a]Data from Patschkowski et al., 2000.[42]

interaction between DNA and FNR interaction occurs only in an anaerobic environment. The FNR protein is a dimer and contains two [4Fe-4S]$^{2+}$ clusters. In an aerobic environment, the FNR protein dissociates to a monomer containing a single [2Fe-2S]$^{2+}$ cluster with a corresponding decrease in binding to DNA. The amount of oxygen required to structurally modify the FNR protein and to reduce the interaction between DNA and the FNR protein is 1 to 10 μ*M* $O_2$. With return to anaerobic condition, the structure of FNR changes and the interaction between DNA and the structure of FNR protein increases.

There is a second level for control of anaerobic activity and this is by the Nar (*N*itrate *r*eductase) system, which is also a regulatory system consisting of two components. The membrane-bound sensory proteins are NarX and NarQ while the cytoplasmic responding proteins that bind DNA are NarL and NarP. The Nar regulon is not associated with assimilation of nitrate for biosynthesis but is responsible for regulation of nitrate or nitrite respiration, which is also known as dissimilatory nitrate reduction. Some of the functions associated with this regulation are given in Table 8.14. The NarX and NarQ are each about 65 kDa with the N-terminus extending into the periplasm and the C-terminal segment is in the cytoplasm. The C-end of the NarX and NarQ proteins has considerable similarity to that of the histidine protein kinase sensors. Both the NarL and NarP are each 24 kDa with the carboxyl end of the proteins binding to the DNA while the amino end is associated with the autophosphorylation reactions. The DNA binding sites for NarL protein have TACYNMT as a consensus sequence with M = A or C, N = is any nucleotide, and Y = C or T. Regulation by NarL or NarP can be either inductive or repressive.

## 8.11.4. Solvent Stress Response[44]

It has long been assumed that organic solvents such as toluene and xylene will disrupt membranes and prevent bacterial growth; however, several strains of Gram-positive

Table 8.14. Regulatory response attributed to NarL and NarP.[a]

| Activity | Operon | NarL regulation | NarP regulation |
|----------|--------|-----------------|-----------------|
| Nitrate reductase | narGHJI | Activation | |
| Nitrite export | narK | Activation | |
| Formate dehydrogenase H | fdnGHI | Activation | Activation |
| NADH-nitrite reductase | nirBDC | Activation | Activation |
| Molybdate uptake | modABCD | Activation | |
| Fumarate reductase | frdABCD | Repression | |
| Pyruvate-formate lyase | pfl | Repression | Repression |
| Alcohol dehydrogenase | adhE | Repression | |

[a]Data from Stewart and Rabin, 1995.[43]

and Gram-negative bacteria are capable of growing on various aromatic solvents. When bacteria are exposed to toluene and related aromatic compounds, the molecules of solvent accumulate between the lipid bilayer of the plasma membrane, causing it to swell and the fluidity to increase. The outer membrane of bacteria is less sensitive than the plasma membrane to organic solvents and alcohols. In 1989, Inoue and Horikoshi isolated a strain of *Pseudomonas putida* that grew in a two-phase water–toluene system. Previously, bacteria were grown on toluene vapors and not in a toluene-saturated water environment. In recent years, *Pseudomonas* isolates were found to grow on cyclohexane, hexane, or dimethyl phthalate and strains of *Bacillus* will grow on benzene, toluene, or *p*-xylene.

The solvent-stress response is not fully understood but it involves unique genes that encode for enzymes to change lipids and proteins for an increased efflux (export) of the solvent molecules. Solvent tolerance in *Pseudomonas putida* is attributed, in part, to the *cit* gene system that produces a *cis–trans* isomerase. This enzyme converts *cis* fatty acids into the *trans* isomer, thereby decreasing membrane fluidity. In addition, with this response to solvent stress is the production of a transport system that exports several different solvent molecules including toluene, *p*-xylene, octanol, ethylbenzene, cyclohexane, and hexane. For some unknown reason, the efflux system also exports several different types of antibiotics including tetracycline, chloramphenicol, erythromycin, novobiocin, rifampin, and penicillins. Based on sequence analysis of genes for the energy-dependent efflux of solvents in *P. putida*, considerable similarity has been found between the solvent efflux system and the resistance-nodulation cell division (RND) family. The RND efflux pumps extend across both the plasma and the outer membrane involving three protein units: a RND protein in the plasma membrane, a membrane fusion protein (MFP) that spans the periplasm, and an outer membrane protein (OMP). The RND is a transmembrane protein with a putative 12 loops through the membrane and it may be energized by proton-motive force. The attachment of the MFP to the RND protein and the OMP provides a direct conduit for export of these toxic solvent molecules. To date, four different efflux systems have been reported for different strains of *P. putida*, three different systems for *P. aeruginosa*, and one system for *E. coli*. Research is continuing in solvent resistant growth because bacteria grown in solvents would have proteins resistant to solvents.

# References

1. Kristjansson, J.K. (ed.). 1992. *Thermophilic Bacteria*. CRC Press, Boca Raton, FL.
2. Sundaram, T.K. 1986. Physiology and growth of thermophilic bacteria. In *Thermophiles* (T.D. Brock, ed.). John Wiley, New York, pp. 76–106.
3. Adams, M.W.W. 1993. Enzymes and proteins from organisms that grow near and above 100°C. *Annual Review of Microbiology* 47:627–658.
4. Duffield, M.L. and D. Cossar. 1995. Enzymes of *Thermus* and their properties. In *Thermus Species* (R. Sharp and R. Williams, eds.). Plenum Press, New York, pp. 93–141.
5. Vieille, C. and G.J. Zeikus. 2001. Hyperthermophilic enzymes: Sources, uses, and molecular mechanisms for thermostability. *Microbiology and Molecular Biology Reviews* 65:1–43.
6. Adams, M.W.W. and R.M. Kelly (eds.). 1992. *Biocatalysts at Extreme Temperatures*. American Chemical Society, Washington, D.C.
7. Ingraham, J. 1987. Effect of temperature, water activity, and pressure on growth. In *Escherichia coli and Salmonella typhimurium: Cellular and Molecular Biology* (F.C. Neidhardt, J.L. Ingraham, K. Brooks Low, B. Magasanik, M. Schaechter, and H.E. Umbarger, eds.). American Society for Microbiology Press, Washington, D.C., pp. 1543–1554.
8. Harold, F.M. and P.C. Maloney. 1996. Energy transduction by ion currents. In *Escherichia coli and Salmonella typhimurium: Cellular and Molecular Biology* (F.C. Neidhardt, J.L. Ingraham, K. Brooks Low, B. Magasanik, M. Schaechter, and H.E. Umbarger, eds.). American Society for Microbiology Press, Washington, D.C., pp. 283–306.
9. Krulwich, T.A. and D.M. Ivey. 1990. Bioenergetics in extreme environments. In *The Bacteria*, Vol. XII: *Bacterial Energetics* (T.A. Krulwich, ed.). Academic Press, New York, pp. 417–448.
10. Russell, J.B. and F. Diez-Gonzalez. 1998. The effects of fermentation acids on bacterial growth. *Advances in Microbial Physiology* 39:206–235.
11. Booth, I.R. 1985. Regulation of cytoplasmic pH in bacteria. *Microbiology Reviews* 49: 359–378.
12. Kashket, E.R. 1987. Bioenergetics of lactic acid bacteria: Cytoplasmic pH and osmotolerance *FEMS Microbiology Reviews* 46:233–244.
13. Foster, J.W. 2000. Microbial responses to acid stress. In *Bacterial Stress Response* (G. Storz and R. Hengge-Aronis, eds.). American Society for Microbiology Press, Washington, D.C., pp. 99–115.
14. Russell, J.B. and F. Diez-Gonzalez. 1998. The effects of fermentation acids on bacterial growth. *Advances in Microbial Physiology* 30:206–235.
15. Horikoshi, K. and T. Akiba. 1982. *Alkalophilic Microorganisms*. Springer-Verlag, Berlin.
16. Lanyi, J.K. 1990. Light-driven primary ionic pumps. In *The Bacteria*, Vol. XII: *Bacterial Energetics* (T.A. Krulwich, ed.). Academic Press, New York, pp. 55–86.
17. Ventosa, A., J.J. Nieto, and A. Oren. 1998. Biology of moderately halophilic aerobic bacteria. *Microbiology and Molecular Biology Reviews* 62:504–544.
18. Nicholls, D.G. and T.J. Ferguson. 1992. *Bioenergetics 2*. Academic Press, New York.
19. Russell, N.J. 1992. Physiology and molecular biology of psychrophilic micro-organisms.

In *Molecular Biology and Biotechnology of Extremophiles* (R.A. Herbert and R.J. Sharp, eds.). Chapman and Hall, New York, pp. 203–224.

20. Makarova, K.S., L. Aravind, Y.I. Wolf, R.L. Tatusov, K.W. Minton, E.V. Koonin, and M.J. Daly. 2001. Genome of the extremely radiation-resistant bacterium *Deinococcus radiodurans* viewed from the perspective of comparative genomics. *Microbiology and Molecular Biology Reviews* 65:44–79.

21. Smith, M.D., C.I. Masters, and B.E.B. Mosley. 1992. Molecular biology of radiation-resistant bacteria. In *Molecular Biology and Biotechnology of Extremophiles* (R.A. Herbert and R.J. Sharp, eds.). Chapman and Hall, New York, pp. 258–280.

22. Bremer, E. and R. Krämer. 2000. Coping with osmotic challenges: Osmoregulation through accumulation and release of compatible solutes in bacteria. In *Bacterial Stress Response* (G. Storz and R. Hengge-Aronis, eds.). American Society for Microbiology Press, Washington, D.C., pp. 79–97.

23. Czonka, L.N. 1989. Physiological and genetic responses of bacteria to osmotic stress. *Microbiological Reviews* 53:121–147.

24. Galinski, E.A. and H.G. Trüper. 1994. Microbial behaviour in salt-stressed ecosystems. *FEMS Microbiology Reviews* 15:95–108.

25. Prieur, D. 1992. Physiology and biotechnology potential of deep-sea bacteria. In *Molecular Biology and Biotechnology of Extremophiles* (R.A. Herbert and R.J. Sharp, eds.). Chapman and Hall, New York, pp. 163–202.

26. Brown, A.D. 1976. Microbial water stress. *Bacteriology Reviews* 40:803–846.

27. Potts, M. 1994. Desiccation tolerance of prokaryotes. *Microbiology Reviews* 58:755–805.

28. Neidhardt, F.C. and R.A. Van Bogelen. 1987. Heat shock response. In Escherichia coli *and* Salmonella typhimurium: *Cellular and Molecular Biology.* American Society for Microbiology Press, Washington, D.C., pp. 1334–1345.

29. Jones, P. and M. Inouye. 1994. The cold shock response. A hot topic. *Molecular Microbiology* 11:811–818.

30. Yura, T., M. Kanemori, and M.T. Morita. 2000. Heat shock response: Regulation and function. In *Bacterial Stress Response* (G. Storz and R. Hengge-Aronis, eds.). American Society for Microbiology Press, Washington, D.C., pp. 3–18.

31. Yura, T., N. Nagai, and H. Mori. 1993. Regulation of the heat-shock response in bacteria. *Annual Review of Microbiology* 47:321–350.

32. Hebraud, M. and J. Guzzo. 2000. The main cold shock protein of *Listeria monocytogenes* belongs to the family of ferritin-like proteins. *FEMS Microbiology Letters* 190:29–34.

33. Klein W., M.H.W. Weber, and M.A. Marahiel. 1999. Cold shock response of *Bacillus subtilis*: Isoleucine-dependent switch in the fatty acid branching pattern for membrane adaption to low temperatures. *Journal of Bacteriology* 181:5341–5349.

34. Phadtare, S., K. Yamanaka, and M. Inouye. 2000. The cold shock response. In *Bacterial Stress Response* (G. Storz and R. Hengge-Aronis, eds.). American Society for Microbiology Press, Washington, D.C., pp. 33–46.

35. Wassenberg, D., C. Welker, and R. Jaenicke. 1999. Thermodynamics of the unfolding of the cold-shock protein from *Thermotoga maritima. Journal of Molecular Biology* 289: 187–193.

36. Kolter, R., D.A. Siegele, and A. Tormo. 1993. The stationary phase of the bacterial life cycle. *Annual Review of Microbiology* 47:855–874.

37. Östling, J., L. Holmquist, K. Flärdh, B. Svenblad, Ä. Jouper-Jaan, and S. Kjellenber. 1993. Starvation and recovery of *Vibrio*. In *Starvation in Bacteria* (S. Kjellenberg, ed.). Plenum Press, New York, pp. 103–127.

38. Spector, M.P. 1998. The starvation-stress response (SSR) of *Salmonella*. *Advances in Microbial Physiology* 40:235–279.
39. Lumppio, H.L., N.V. Shenvi, A.O. Summers, G. Voordoow, and D.M. Kurtz, Jr. 2001. Rubrerythrin and rubredoxin oxidoreductase in *Desulfovibrio vulgaris*. A novel oxidative stress protection system. *Journal of Bacteriology* 183:101–108.
40. Poole, L.B., A. Godzik, A. Nayeem, and J.D. Schmitt. 2000. AhpF can be dissected into two functional units: Tandem repeats of two thioredoxin-like folds in the N-terminus mediate electron transfer from the thioredoxin reductase-like C-terminus to AhpC. *Biochemistry* 39:6602–6615.
41. Storz, G. and M. Zheng. 2000. Oxidative stress. In *Bacterial Stress Response* (G. Storz and R. Hengge-Aronis, eds.). American Society for Microbiology Press, Washington, D.C., pp. 47–59.
42. Patschkowski, T., D.M. Bates, and P.J. Kiley. 2000. Mechanism for sensing and responding to oxygen deprivation. In *Bacterial Stress Response* (G. Storz and R. Hengge-Aronis, eds.). American Society for Microbiology Press, Washington, D.C., pp. 61–78.
43. Stewart, V. and R.S. Rabin. 1995. Dual sensors and dual response regulators interact to control nitrate- and nitrite-responsive gene expression in *Escherichia coli*. In *Two-Component Signal Transduction* (J.A. Hoch and T.J. Silhavy, eds.). American Society for Microbiology Press, Washington, D.C., pp. 233–252.
44. Kieboom, J. and J.A.M. deBont. 2000. Mechanisms of organic solvent tolerance in bacteria. In *Bacterial Stress Response* (G. Storz and R. Hengge-Aronis, eds.). American Society for Microbiology Press, Washington, D.C., pp. 393–402.

# Additional Reading

Stephenson, M. 1949. *Bacterial Metabolism*. Longmans, Green and Co., New York.

## *Thermophiles and Hyperthermophiles*

Brock, T.D. (ed.). 1986. *Thermophiles*. John Wiley, New York.
Friedman, S.M. (ed.). 1978. *Biochemistry of Thermophily*. Academic Press, New York.
Grant, W.D. and K. Horikoshi, 1992. Alkaliphiles: Ecology and biotechnology applications. In *Molecular Biology and Biotechnology of Extremophiles* (R.A. Herbert and R.J. Sharp, eds.). Chapman and Hall, New York, pp. 143–162.
LopezGarcia, P. and P. Forterre. 1999. Control of DNA topology during thermal stress in hyperthermophilic archaea: DNA topoisomerase levels, activities and induced thermotolerance during heat and cold shock in *Sulfolobus*. *Molecular Microbiology* 33:766–777.
Sharp, R. and R. Williams (eds.). 1995. *Thermus Species*. Plenum Press, New York.
Sterner, R. and W. Liebl 2001. Thermophilic adaption of proteins. *Critical Reviews in Biochemistry and Molecular Biology* 36:39–106.

## *Osmotic Tolerance*

Dong, Z., M.J. Canny, M.E. McCully, M.R. Roboredo, C.F. Cabadilla, E. Ortega, and R. Rodés. 1994. A nitrogen-fixing endophyte of sugarcane stems. *Plant Physiology* 105:1139–1147.
Galinski, E.A. 1995. Osmoadaption in bacteria. *Advances in Microbial Physiology* 37:273–328.

## Halophiles

Vreeland, R.H. and L.I. Hochstein (eds.). 1993. *The Biology of Halophilic Bacteria.* CRC Press, Boca Raton, FL.


## pH Stress

Hall, H.K., K.L. Karem, and J.W. Foster. 1995. Molecular responses of microbes to environmental pH stress. *Advances in Microbial Physiology* 37:229–272.


## Heat Shock

Craig, E.A., B.D. Gambill, and R.J. Nelson. 1993. Heat shock proteins: Molecular chaperones of protein biogenesis. *Microbial Reviews* 57:402–414.


## Psychrophiles and Cold Shock

Beran, R.K., and R.W. Simons. 2001. Cold-temperature induction of *Escherichia coli* polynucleotide phosphorylase occurs by reversal of its autoregulation. *Molecular Microbiology* 39:112–125.

Carty, S.M., K.R. Sreekumar, and C.R.H. Raetz. 1999. Effect of cold shock on lipid A biosynthesis in *Escherichia coli*: Induction at 12 degrees C of an acyltransferase specific for palmitoleoyl-acyl carrier protein. *Journal of Biological Chemistry* 274:9677–9685.

Chamot, D. and G.W. Owttrim. 2000. Regulation of cold shock-induced RNA helicase gene expression in the cyanobacterium *Anabaena* sp. Strain PCC 7120. *Journal of Bacteriology* 182:1251–1256.

Cloutier, J., D. Prevost, P. Nadeau, and H. Antoun. 1992. Heat and cold shock protein synthesis in arctic and temperate strains of rhizobia. *Applied and Environmental Microbiology* 58: 2846–2853.

Fujii, S., K. Nakasone, and K. Horikoshi. 1999. Cloning of two cold shock genes, *cspA* and *cspG*, from the deep-sea psychrophilic bacterium *Shewanella violacea* strain DSS12. *FEMS Microbiology Letters* 178:123–128.

Huston, A.L., B.B. KriegerBrockett, and J.W. Deming. 2000. Remarkably low temperature optima for enzyme activity from arctic bacteria and sea ice. *Environmental Microbiology* 2: 383–388.

Lim, J., T. Thomas, and R. Cavicchioli. 2000. Low temperature regulated DEAD-box RNA helicase from the Antarctic archaeon, *Methanococcoides burtonii. Journal of Molecular Biology* 297:553–567.

Shires, K. and L. Steyn. 2001. The cold-shock stress response in *Mycobacterium smegmatis* induces the expression of a histone-like protein. *Molecular Microbiology* 39:994–1009.

Panoff, J.-M., D. Corroler, B. Thammavongs, and P. Boutibonnes. 1997. Differentiation between cold shock proteins and cold acclimation proteins in a mesophilic Gram-positive bacterium, *Enterococcus faecalis* JH2-2. *Journal of Bacteriology* 179:4451–4454.

Welker, C., G. Bohm, H. Schurig, and R. Jaenicke. 1999. Cloning, overexpression, purification, and physicochemical characterization of a cold shock protein homolog from the hyperthermophilic bacterium *Thermotoga maritima. Protein Science* 8:394–403.

Zangrossi, S., F. Briani, D. Ghisotti, M.E. Regonesi, P. Tortora, and G. Deho. 2000. Transcriptional and post-transcriptional control of polynucleotide phosphorylase during cold acclimation in *Escherichia coli*. *Molecular Microbiology* 36:1470–1480.

## Barophiles

Kato, C. and D.H. Bartlett. 1997. The molecular biology of barophilic bacteria. *Extremophiles* 1:111–116.

Kato, C., L. Li, Y. Nogi, Y. Nakamura, J. Tamaoka, and K. Horikoshi. 1998. Extremely barophilic bacteria isolated from the Mariana Tranch, Challenger Deep, at a depth of 11,000 meters. *Applied and Environmental Microbiology* 64:1510–1513.

Welch, T.J. and D.H. Bartlett. 1997. Cloning, sequencing and overexpression of the gene encoding malate dehydrogenase from the deep-sea bacterium *Photobacterium* species strain SS9. *Biochemica et Biophysica Acta* 1350:41–46.

## Oxygen Stress Response

Imlay, J.A. 2002. How oxygen damages microbes: Oxygen tolerance and obligate anaerobiosis. *Advances in Microbial Physiology* 46:112–155.

Kurtz, D.M., Jr. 2003. Oxygen and anaerobes. In *Biochemistry and Physiology of Anaerobic Bacteria* (L.G. Ljungdahl, M.W. Adams, L.L. Barton, J.G. Ferry, and M.J. Johnson, eds.). Springer-Verlag, New York, pp. 128–142.

Lefebre, M.D. and M.A. Valvano. 2001. In vitro resistance of *Burkholderia cepacia* complex isolates to reactive oxygen species in relation to catalase and superoxide dismutase production. *Microbiology* 147:97–109.

Michan, C., M. Manchado, G. Dorado, and C. Pueyo. 1999. In vivo transcription of the *Escherichia coli oxyR* regulon as a function of growth phase and in response to oxidative stress. *Journal of Bacteriology* 181:2759–2764.

Pomposiello, P.J. and B. Demple. 2002. Global adjustment of microbial physiology during free radical stress. *Advances in Microbial Physiology* 46:320–341.

Svensater, G., B. Sjogreen, and I.R. Hamilton. 2000. Multiple stress responses in *Streptococcus mutans* and the induction of general and stress-specific proteins. *Microbiology* 146:107–117.

Wood., D.R. and D.T. Jones. 1986. Physiological responses of *Bacteroides* and *Clostridium* strains to environmental stress factors *Advances in Microbial Physiology* 28:1–64.

# IV
## Bioenergetics

# 9
# Electron Transport and Coupled Phosphorylation

The obscure we see eventually, the completely apparent takes longer. P. Mitchell, *Science*, 1979

## 9.1. Introduction[1-4]

The unidirectional transport of electrons is a hallmark of living systems, and the concept of an electron transport chain can be traced to the cytochrome research conducted by D. Keilin in the 1920s. Just as in eukaryotic organisms, the electron transport systems for various bacteria have been found to contain three basic segments: (1) enzymes involved in the membrane-mediated electron donor process catalyzed by dehydrogenases or oxidoreductases; (2) cytochromes, lipids, and nonheme iron proteins that move electrons along within the membrane; and (3) terminal oxidases or reductases that move electrons from cytochromes to the final electron acceptor. While the theme of electron movement is essential for cellular viability, the specific molecules involved may vary. Some of the differences between the eukaryotic and bacterial electron transport chains are presented in Table 9.1. Mitochondrial systems have an inflexible electron transport chain in which each component is fixed; however, bacteria have considerable molecular variability that enables them to use many different types of electron donors and electron acceptors. This chapter addresses the function of the electron transport chain as well as the diversity of respiratory chains in the prokaryotic world.

Because there appears to be considerable latitude in the specific electron carriers involved in bacterial respiration, it should be noted that the direction of electron flow is not random but is regulated by thermodynamic principles. The tendency of molecules to act as electron acceptors or as electron donors is based on the properties of the chemical involved and is defined as the oxidation–reduction potential, which is measured in millivolts. The standard conditions used for the calculation of these reduction potentials are: The reaction occurs at 25°C, concentrations are 1 $M$, pressure is 1 atm, and the system is in equilibrium. The potentials of biological interest are expressed as $E_0'$ because these values are adjusted to physiological conditions of pH 7.0. When a system constitutes two components, the electron donor and the elec-

Table 9.1. Comparison of electron transport systems in nonphotosynthetic bacteria with mitochondrial systems.

| Characteristics | Bacteria | Mitochondria |
|---|---|---|
| Cytochrome types present | *a*, *b*, *c*, *d*, and *o* | *a*, *b*, and *c* |
| Heme type present | A, B, C, D, and O | A, B, and C |
| Cytochrome production | Type and quantity varies | Fixed ratio of *a:b:c* |
| Respiratory activity | Aerobic and anaerobic | Aerobic only |
| Redox-active lipid involved | Ubiquinones and menaquinones | Ubiquinone |
| Flow of electrons in electron transport chain | Linear, and branched | Linear |
| Coupled activity (ATPases) | Proton translocation (F-type and V-type) | Proton translocation (F-type) |
| Interfacing dehydrogenases | Type and number varies with bacterial species | Fixed and does not vary |

tron acceptor, the reference to each segment of the reaction is called the half-reaction and tables will list the millivolts for the half-reaction. The review by Thauer et al. (1977) contains many of the half-reactions encountered in anaerobic microbiology and serves as an outstanding contribution to microbial energetics.

When constructing an electron transfer chain for bacteria, it has been found in general that electrons flow from the most negative to the most positive. Examples of some of the reduction potentials for some molecules are given in Table 9.2. It should be recalled that the reduction potential values were established at equilibrium at 1 $M$ concentrations; neither of these conditions are found in biology. Thus, it may be necessary to make corrections from published information. Certainly, it is important to know the voltage potential for the different half-reactions involved so that the components can be appropriately organized.

Respiration in bacteria is the movement of electrons from an electron donor to an acceptor through quinones, cytochromes, and specific proteins in the plasma membrane. Because there are some fundamental differences between photosynthetic and nonphotosynthetic systems, each is presented separately. The approach in this chapter is to first examine the molecules involved in the electron transport before detailing specific electron transfer schemes of aerobes and anaerobes. Finally, oxidative phosphorylation and the role of proton-motive force in prokaryotic cells is discussed.

## 9.2. Systems for Movement of Electrons

The variability of electron transfer systems in bacteria is seen when different heterotrophic bacteria are examined. Several different schemes for electron transfer to molecular oxygen are used by bacteria, and in almost all cases bacteria have multiple electron donors that interface with the electron transport system. Some bacterial elec-

Table 9.2. Reduction potential for selected biological compounds.

| Half-reaction | $E_0'$ (mvolts, at pH 7.0) |
|---|---|
| Acetate + $CO_{2(G)}$ + 2 $H^+$ + 2 $e^-$ = Pyruvate + $H_2O$ | −699 |
| Succinate + $CO_2$ + 2 $H^+$ + 2 $e^-$ = α-Ketoglutarate + $H_2O$ | −670 |
| Acetate + 2 $H^+$ + 2 $e^-$ = Acetaldehyde + $H_2O$ | −600 |
| Ferredoxin$_{(ox)}$ + $e^-$ = Ferredoxin$_{(red)}$ | −430 |
| 2 $H^+$ + 2 $e^-$ = $H_2$ (1 atm) | −420 |
| Acetoacetate + 2 $H^+$ + 2 $e^-$ = β-Hydroxybutyrate + $H_2O$ | −350 |
| Pyruvate + $CO_2$ + 2 $H^+$ + 2 $e^-$ = Malate + $H_2O$ | −330 |
| $NADP^+$ + 2 $H^+$ + 2 $e^-$ = NADPH + $H^+$ | −327 |
| $NAD^+$ + 2 $H^+$ + 2 $e^-$ = NADH + $H^+$ | −320 |
| FMN + 2 $H^+$ + 2 $e^-$ = $FMNH_2$ | −219 |
| FAD + 2 $H^+$ + 2 $e^-$ = $FADH_2$ | −219 |
| Acetaldehyde + 2 $H^+$ + 2 $e^-$ = Ethanol | −200 |
| Pyruvate + 2 $H^+$ + 2 $e^-$ = Lactate | −197 |
| Dihydroxyacetone-phosphate + 2 $H^+$ + 2 $e^-$ = Glycerol-1-phosphate | −190 |
| Oxalacetate + 2 $H^+$ + 2 $e^-$ = Malate | −170 |
| α-Ketoglutarate + $NH_3$ + 2 $H^+$ + 2 $e^-$ = Glutamate | −140 |
| Menaquinone$_{(ox)}$ + 2 $H^+$ + 2 $e^-$ = Menaquinone$_{(red)}$ | −74 |
| Cytochrome $b_{(ox)}$ + $e^-$ = Cytochrome $b_{(red)}$[a] | −140 (−40) |
| Fumarate + 2 $H^+$ + 2 $e^-$ = Succinate | +30 |
| Ubiquinones$_{(ox)}$ + 2 $H^+$ + 2 $e^-$ = Ubiquinone$_{(red)}$ | +040 |
| Cytochrome $c_{(ox)}$ + $e^-$ = Cytochrome $c_{(red)}$[a] | +230 (+220) |
| Cytochrome $a_{(ox)}$ + $e^-$ = Cytochrome $a_{(red)}$[a] | +200 (+290) |
| Cytochrome $a_{3\,(ox)}$ + $e^-$ = Cytochrome $a_{3\,(red)}$[a] | +375 (+550) |
| $O_2$ (1 atm) + 2 $H^+$ + 2 $e^-$ = $H_2O$ + ½ $O_2$ | +820 |

[a]Values are for *Rhodopseudomonas sphaeroides*. Mitochondrial values are in parentheses.

tron transport systems are linear, such as the system found in *Escherichia coli* under high aerobic conditions. Some electron transport chains are branched, with different types of cytochromes interacting with molecular oxygen and this would include *E. coli* under low aeration and *Azotobacter vinelandii*. Some have a shortened electron transport chain, as is the case with *Paracoccus denitrificans*, which has a methanol dehydrogenase interfacing directly with *c*-type cytochromes.

Respiration under anaerobic conditions also provides additional variation. In some instances a bacterial species can respire under both under aerobic and anaerobic states while in other cases growth occurs only with anaerobic respiration. Fumarate interfaces with the electron transport system through a *b*-type cytochrome and is the most common organic compound to serve as an electron acceptor in anaerobic respiration. A large percentage of bacteria that grow aerobically can also use nitrate or nitrite as an electron acceptor. Elemental sulfur is used as a final electron acceptor, and in the *Desulfovibrio* the enzyme responsible for this reduction is a *c*-type cytochrome. A sizeable group of archaea require elemental sulfur for growth with the resulting formation of hydrogen sulfide (see Table 1.5). Many other compounds serve as electron acceptors for bacteria under anaerobic conditions with either the demonstrated or presumed use of cytochromes. In the methanogens, carbon dioxide is an electron acceptor; however, few of the methanogens have unique organic substances that carry electrons and few methanogens have cytochromes.

## 9.2.1. Electron Donors for Chemoorganotrophs[5,6]

Bacteria have many different electron donors and each interacts with specific enzymes found in the plasma membrane. Heterotrophic bacteria use organic compounds as electron donors with oxidation catalyzed by dehydrogenases. Electrons enter the plasma membrane through a primary dehydrogenase and exit the membrane by terminal reductases. Prosthetic groups that carry electrons in the dehydrogenases include iron in iron-sulfur proteins, flavins in FAD- or FNM-containing proteins, and molybdenum in molybdoproteins. The dehydrogenases are secured into the plasma membrane either by transmembrane helices or by nonspecific binding into the phospholipid bilayer. Solubilization of dehydrogenases is either by sonication procedures or by the use of surfactants as discussed in Chapter 2. Bacteria using organic compounds as electron sources commonly use molecular oxygen as the electron acceptor in the respiratory process. A partial list of dehydrogenases found in aerobic bacteria is given in Table 9.3. Frequently the ability of bacteria to grow on a specific electron donor is attributed to the presence of an appropriate enzyme for the dehydrogenation step. Although most dehydrogenases couple oxidation to reduction of $NAD^+$, there are several dehydrogenases that couple substrate oxidation to reduction of cytochrome or other redox active components in the plasma membrane. Generally, dehydrogenases are flavoproteins or metalloflavoproteins, and because of induction the concentration in bacteria may vary by a factor of 1,500. Most heterotrophic bacteria are capable of using various electron donors and the characteristics of some of the enzymes associated with electron donors for bacteria are briefly reviewed in this section.

### 9.2.1.1. Succinate Dehydrogenase

Succinate dehydrogenase (SDH) is essential for functioning of the tricarboxylic acid (TCA) cycle in that it oxidizes succinate to fumarate. In *E. coli*, SDH is produced only under aerobic conditions by a regulatory mechanism that is not yet established. SDH is a transmembrane protein that has been isolated from *E. coli*, *Bacillus subtilis*, and *Rhodospirillum rubrum* but purification of the enzyme has been difficult to achieve. Based on genomic analysis, the SDH from *E. coli* is known to consist of four polypeptides: 64.3 kDa, 26 kDa, 14.2 kDa, and 12.8 kDa. FAD is covalently bonded to the largest subunit and the 26-kDa subunit contains a 2Fe-2S center. The active site of the SDH is associated with these two large subunits, and the two smaller subunits are attached into the plasma membrane because of their hydrophobic character.

### 9.2.1.2. Glycerol-3-P Dehydrogenase

In *E. coli* oxidation of *sn*-glycerol-3-phosphate is under both aerobic and anaerobic cultivation by glycerol-3-P dehydrogenase (G-3-PDH). However, there are two dis-

Table 9.3. Dehydrogenases and electron donor reactions associated with respiratory systems of bacteria.

| I. Heterotrophic bacteria | Dehydrogenase | Reaction catalyzed |
|---|---|---|
| Many species | NADH dehydrogenase | $NADH + H^+ \rightarrow NAD^+ + 2\ e^- + 2\ H^+$ |
| | Succinic dehydrogenase | $Succinate \rightarrow Fumarate + 2\ e^- + 2\ H^+$ |
| *Mycobacterium phlei* | Malate: vitamin K | $Malate \rightarrow Oxalacetate + 2\ e^- + 2\ H^+$ |
| *Micrococcus lysodeikticus* | dehydrogenase | |
| *Escherichia coli* | Formate dehydrogenase | $Formate \rightarrow CO_2 + 2\ e^- + 2\ H^+$ |
| *Acetobacter* sp. | D-Lactate dehydrogenase | $Lactate \rightarrow Pyruvate + 2\ e^- + 2\ H^+$ |
| *Paracoccus denitrificans* | Methanol dehydrogenase | $Methanol \rightarrow Formaldehyde + 2\ e^- + 2\ H^+$ |
| *Pseudomonas* AM1 | | |

| II. Chemolithotrophic bacteria | Enzyme for electron donor | Reaction catalyzed |
|---|---|---|
| Sulfate-reducing bacteria | Hydrogenase | $H_2 \rightarrow 2\ e^- + 2\ H^+$ |
| Iron-reducing bacteria | | |
| *Nitrosomonas europaea* | Ammonia monooxygenase | $NH_3 + H_2O \rightarrow NH_2OH + 2\ e^- + 2\ H^+$ |
| | Hydroxylamine oxidoreductase | $NH_2OH^+ + H_2O \rightarrow HNO_2 + 4\ e^- + 4\ H^+$ |
| *Acidithiobacillus ferroxidans* | Ferrous oxidoreductase | $Fe^{2+} \rightarrow Fe^{3+} + 2\ e^- + 2\ H^+$ |
| *Nitrobacter hamburgensis* | Nitrite oxidoreductase | $NO_2^- + H_2O \rightarrow NO_3^- + 2\ e^- + 2\ H^+$ |
| *Hydrogenomonas carboxyovorans* | CO oxidoreductase | $CO + H_2O \rightarrow CO_2 + 2\ e^- + 2\ H^+$ |
| *Acidithiobacillus thiooxidans* | Thiosulfate oxidoreductase | $S_2O_3^{2-} + H_2O \rightarrow SO_4^{2-} + 2\ e^- + 2\ H^+$ |
| *Sulfolobus acidocaldarius* | | |

tinct enzymes for this activity, with one produced under aerobic conditions and the other produced under the anaerobic state. The G-3-PDH produced under aerobic conditions is a flavoprotein dimer with monomers of 58 kDa with FAD as the prosthetic group. Under anaerobic conditions, the G-3-PDH is composed of two nonidentical subunits. The larger subunit is 62 kDa and contains covalently linked FAD while the smaller subunit of 43 kDa contains a nonheme iron cluster. The necessity for distinct proteins under different electron acceptors is unknown but it may reflect the requirement of G-3-PDH forming a lipoprotein association in the membrane under anaerobic conditions to produce a proton pump complex.

## 9.2.1.3. NADH Dehydrogenase

NADH dehydrogenase (NDH) is a flavoprotein consisting of a single polypeptide and FAD. In *E. coli*, the NDH is 47.3 kDa while enzymes of different mass have been isolated from other bacteria. The NDH from *Bacillus subtilis* is 63 kDa while in an alkaliophilic *Bacillus* YN-1 it is 65 kDa. In the thermophilic *Bacillus caldotenax* it is 44 kDa, the enzyme in *Pseudomonas denitrificans* has two subunits (48 kDa and 25 kDa), in *Bacillus stearothermophilus* it is 43 kDa, and in *Thermus thermophilus* it is 53 kDa. The enzyme that catalyzes the reaction between NADH and a dye (artificial electron acceptor) is termed "diaphorase."

Examination of the NADH dehydrogenase proteins reveals that there are no transmembrane loops but the proteins are loosely attached to the cytoplasmic side of the plasma membrane bilayer. In the mitochondrial membrane system, NADH dehydrogenase is an integral segment of complex I which forms a proton pump. Thus far in bacteria, the presence of complex I has been demonstrated only in *E. coli*, *Pseudomonas denitrificans*, and *Thermus thermophilus*. It is unclear if the NADH dehydrogenase isolated from bacteria functions only in electron transport or if it has been dislodged from complex I.

### 9.2.1.4. D-Lactate Dehydrogenase

Lactic acid is an end product of fermentation of a group of bacteria that are collectively referred to as the lactic acid bacteria. Two distinct enzymes are involved in formation of lactic acid, with one producing L-lactic acid and the other producing D-lactic acid. In *E. coli*, L-lactate dehydrogenase (LDH) is a cytoplasmic FMN-containing flavoprotein of 43 kDa that uses $NAD^+$ as the electron acceptor. The D-LDH is a membrane-bound flavoprotein containing FAD and has a mass of 64.6 kDa. Although the D-LDH is constitutive in *E. coli* it is most important in anaerobic environments. The D-LDH oxidizes lactate to pyruvate and releases reducing equivalents (2 $e^-$ and 2 $H^+$) into the respiratory system.

### 9.2.1.5. Formate Dehydrogenase

Formate can be used by several organisms as an important energy source by providing electrons to energize membranes and drive anaerobic oxidative phosphorylation. The general reaction associated with formate dehydrogenase enzymes is as follows:

$$HCOOH \rightarrow CO_2 + 2 H^+ + 2 e^-$$

Many anaerobes use formate as the electron donor with fumarate, nitrate, or sulfate as final electron acceptors. Several aerobic bacteria couple the oxidation of formate to the reduction of $O_2$ and also use formate as the sole carbon source for biosynthesis. The use of formate as the carbon source for autotrophic bacteria is discussed in Chapter 11.

Organisms such as *E. coli* have two types of formate dehydrogenase activities and these are designated as $Fdh_H$ and $Fdh_N$. Both of these enzymes catalyze the following reaction where the redox couple is highly electronegative ($E_0' = -432$ mV). In *E. coli*, $Fdh_N$ is located on the periplasmic side of the plasma membrane. With respect to anaerobic respiration, $Fdh_N$ transfers electrons to a quinone with nitrate or fumarate as the final electron acceptor. The formate dehydrogenase ($Fdh_N$) purified from *E. coli* consists of three components: $\alpha$-subunit, 110 kDa; $\beta$-subunit, 32 kDa; and $\gamma$-subunit, 20 kDa. Also, the formate dehydrogenase from *E. coli* contains a molybdenum cofactor, cytochrome $b_{555}$, an iron-sulfur center, and selenium which is covalently bound in the enzyme. The role of selenium in formate dehydrogenase is discussed in Chapter 14. Both $\alpha$- and $\beta$-subunits are transmembrane and may partic-

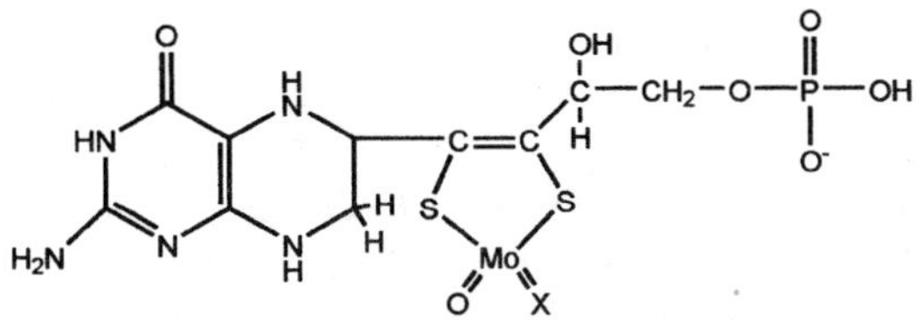

Figure 9.1. Structure of molybdenum co-factor. X = oxygen or sulfide group. Bacteriopterin contains an additional phosphate group and an aromatic moiety.

ipate in proton pumping. In the oxidation of formate, electrons are released to cytochrome $b_{555}$, which has a midpoint potential of $-105$ mV. The molybdenum cofactor appears to function as an intermediate in the reaction and the structure of this cofactor is given in Figure 9.1. The role of the nonheme iron is unresolved at this time. $Fdh_H$ is associated with formate-hydrogen lyase because it interfaces with a hydrogenase to unite electrons and protons from formate dehydrogenase with the production of $H_2$. The activity of formate-hydrogen lyase in enteric bacteria is discussed in Chapter 11. The reactions of these two enzymes in *E. coli* are given as follows:

$$Formate \rightarrow CO_2 + 2\ H^+ + 2\ e^-\ (Fdh_N, formate\ dehydrogenase)$$

$$Formate \rightarrow CO_2 + H_2\ (Fdh_H, formate\ dehydrogenase\ plus\ hydrogenase$$
$$to\ produce\ formate\text{-}hydrogen\ lyase)$$

These two formate dehydrogenases are highly regulated, with $Fdh_H$ repressed by nitrate but induced by formate and $Fdh_N$ repressed by fermentation but induced by nitrate. Unlike $Fdh_N$, $Fdh_H$ is not associated with the production of electrons for cell energetics.

### 9.2.1.6. PQQ-Containing Dehydrogenases[7]

Several bacteria contain dehydrogenases that use an unusual quinone to transfer the electrons from the donor to the electron transport system. The apoprotein for these dehydrogenases has one or two subunits and the prosthetic group is pyrroloquinoline quinole (PQQ). The structure of PQQ is given in Figure 9.2. PQQ carries two electrons plus two protons and the free $PQQ/PQQH_2$ has a redox potential of $+90$ mV but the potential may be more electronegative when linked to protein. Although the PQQ moiety is not commonly discussed in general texts, PQQ as an electron

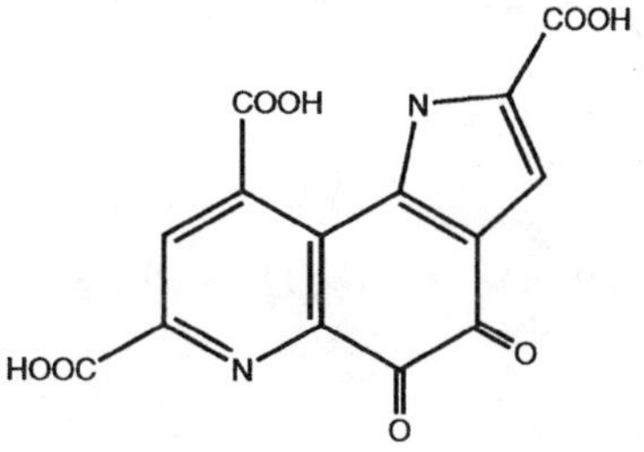

Figure 9.2. Structure of pyrroloquinoline quinole (PQQ).

carrier is essential for the bacterial dehydrogenases that catalyze reactions with a midpoint potential that is too electropositive to donate electrons to NAD(P)$^+$. Enzyme systems in bacteria that use the PQQ group include the following: methanol dehydrogenase and methylamine dehydrogenase in Gram-negative methylotrophic bacteria; alcohol dehydrogenase in *Pseudomonas aeruginosa* and *Gluconobacter suboxydans*; glucose dehydrogenase in *Acinetobacter*, *Pseudomonas*, and *E. coli*; aldehyde dehydrogenase in acetic acid bacteria and glycerol dehydrogenase in *Gluconobacter industruis*; and polyethylene glycol dehydrogenase in *Flavobacterium* and *Pseudomonas*. A rather curious feature is that some bacteria produce the dehydrogenase that requires the PQQ cofactor; however, the bacterium does not synthesize the PQQ cofactor but acquires this cofactor from the environment. Presumably, this is an example of partial gene transfer: The gene for glucose dehydrogenase is moved laterally in the environment independent of the gene for PQQ synthesis.

### 9.2.1.7. Pyruvate Oxidase

Pyruvate oxidase is distinct from pyruvate dehydrogenase in the phosphoroclastic complex and is induced by the presence of pyruvate in the growth medium. Pyruvate oxidase is a tetramer that contains one FAD in each polypeptide monomer. The action of the enzyme is to convert pyruvate to acetate plus carbon dioxide with 2 e$^-$ and 2 H$^+$ transferred into the aerobic electron transport system. Because of the reaction catalyzed, pyruvate oxidase has dehydrogenase character but retains the oxidase name initially used to describe reactions with a dye (i.e., methylene blue) as the electron acceptor.

## 9.2.2. Electron Donors for Chemolithotrophs[8]

Many prokaryotes grow by obtaining energy from the oxidation of inorganic compounds. These organisms generally are also autotrophs in that they use carbon dioxide as the carbon source for cellular materials. A list of the physiological types of chemolithotrophs and the energy released from the reaction with electron donor and acceptor is presented for the various physiological types in Table 9.4. The use of H$_2$ as an electron donor has broad implications because several strains of bacteria can grow under anaerobic conditions using H$_2$. Because the amount of energy required to produce ATP is equivalent to 7.3 kcal/mol, each of the reactions listed in Table 9.4 with the various inorganic electron donors provides enough energy for production of one to several molecules of ATP.

### 9.2.2.1. Hydrogenase[9]

The reaction catalyzed by hydrogenase results in either the production of molecular hydrogen or the uptake of molecular hydrogen and is characterized by the following activity:

$$H_2 = 2 H^+ + 2 e^-$$

Table 9.4. Physiological types of aerobic prokaryotes based on electron source for growth.

| | Reactions | kJ/reaction (kcal/reaction) |
|---|---|---|
| Hydrogen oxidizers | $H_2 + \frac{1}{2} O_2 \rightarrow H_2O$ | $-233.2$ $(-56.7)$ |
| Carboxydobacteria | $CO + \frac{1}{2} O_2 \rightarrow CO_2$ | $-253.1$ $(-61.5)$ |
| Sulfur oxidizers | $S^{2-} + 2 O_2 \rightarrow SO_4^{2-}$ | $-790.5$ $(-189.9)$ |
| | $S^0 + 1\frac{1}{2} O_2 + H_2O \rightarrow SO_4^{2-} + 2 H^+$ | $-580.9$ $(-139.8)$ |
| | $S_2O_3^{2-} + 2 O_2 + H_2O \rightarrow 2 SO_4^{2-} + 2 H^+$ | $-739.0$ $(-176.8)$ |
| Iron oxidizers | $Fe^{2+} + \frac{1}{4} O_2 + H^+ \rightarrow Fe^{3+} + \frac{1}{4} H_2O$ | $-44.4$ $(-10.6)$ |
| Ammonia oxidizers | $NH_3 + 2 O_2 \rightarrow NO_2^- + 2 H^+ + H_2O$ | $-266.7$ $(-64.7)$ |
| Nitrite oxidizers | $NO_2^- + \frac{1}{2} O_2 \rightarrow NO_3^-$ | $-73.4$ $(-18.4)$ |
| Methane oxidizers | $CH_4 + O_2 \rightarrow 2 H_2O + CO_2$ | $-782.6$ $(-186.3)$ |
| Heterotrophic oxidizers | $Glucose + 6 O_2 \rightarrow 6 H_2O + 6 CO_2$ | $-2870$ $(-686)$ |

The kJ/ reaction is calculated at pH 7.0 except for ferrous oxidation, which is at pH 0. The amount of energy released from each reaction is listed with kcal/mol in parentheses.

Although the reaction is reversible, the hydrogenase in anaerobic bacteria generally has a specific function for either hydrogen uptake or hydrogen evolution. Hydrogen evolution functions in fermentative reactions while hydrogen uptake is associated with electron donor activities of anaerobic respiration. Because of the low redox potential of the $H^+/H_2$ couple, uptake hydrogenases are important physiological electron donors that can reduce $NAD^+$ or a quinone. Hydrogenases in bacteria are of various sizes and may be found either in the cytoplasm, periplasm, or bound to plasma membranes. Generally, membrane-bound hydrogenases form a proton pump and translocate protons across the plasma membrane as electrons are moved within the membrane. The discussion here focuses on the hydrogenases that are associated with respiration and not fermentation.

There are three general classes of hydrogenases based on their metal content and active site chemistry: the [Fe] hydrogenase, the [NiFe] hydrogenase, and the [NiFeSe] hydrogenases. The [Fe] hydrogenase is found in several bacteria including sulfate-reducing bacteria where the electrons are carried by iron-sulfur centers consisting of 4Fe-4S or 2Fe-2S. In *Desulfovibrio vulgaris*, the [Fe] hydrogenase has a large subunit of 46 kDa and a small subunit of 46 kDa. The [NiFe] hydrogenases are widely distributed in sulfate-reducing bacteria, photosynthetic bacteria, hydrogen oxidizing bacteria, and methanogenic archaea. Generally the [NiFe] hydrogenase has two subunits, 60 kDa and 30 kDa, with one Ni atom and two 4Fe-4S centers. Nickel may shuttle electrons because it may occur in multiple oxidation states: $Ni^+$, $Ni^{2+}$, or $Ni^{3+}$. The [NiFe] hydrogenase produced by *Wolinella succonogenes* in unique in that it carries heme B as a cofactor and at this time it is the only hydrogenase known to be a cytochrome. The [NiFeSe] hydrogenases are found in some methanogens and in *Desulfovibrio*. The [NiFeSe] hydrogenase consists of two subunits, approximately 50 and 30 kDa, one atom of Se, one atom of Ni, and two 4Fe-4S centers. The role of Se is interesting and it appears to influence the activity of the hydride form in the active site of the enzyme. Selenium is positioned close to the active site in that selenocysteine forms the ligand for binding Ni to the enzyme. The rationale for three different classes of hydrogenase is unclear. Some suggest that [Fe] hydrogenase function is especially important for $H_2$ evolution reactions. There is a difference in inhibition of the three classes in that the [Fe] hydrogenase is most sensitive to CO and $NO_2^-$ with the [NiFe] hydrogenase being the most sensitive and the [NiFeSe] hydrogenase is intermediate in sensitivity. One sulfate-reducing bacterium, *Desulfovibrio vulgaris*, has all three classes of hydrogenase and it is difficult to reconcile the presence of these three forms. It has been suggested that the [Fe] hydrogenase takes up hydrogen because it is in the periplasm with electrons going to the reduction of adenosine phosphosulfate (APS). The membrane-bound [NiFe] hydrogenase may be important for proton pumping as well for the reduction of sulfite and thiosulfate while the [NiFeSe] hydrogenase is for $H_2$ production. These three hydrogenases are considered important for hydrogen cycling and this feature is discussed later in the chapter.

### 9.2.2.2. Carbon Monoxide Oxidoreductase[10]

The enzyme that catalyzes the oxidation of carbon monoxide with the formation of carbon dioxide is carbon monoxide oxidoreductase. This enzyme is distinct from CO dehydrogenase, which is involved in the anaerobic acetyl-CoA pathway. CO oxidoreductase generates electrons from the oxidation of carbon monoxide according to the following reaction:

$$CO + H_2O \rightarrow CO_2 + 2\ H^+ + 2\ e^-$$

*Pseudomonas carboxydovorans* is an example of a bacterium that grows with CO as the electron source and $O_2$ as the final electron acceptor. The CO oxidoreductase from *Pseudomonas carboxydovorans* is bound to the plasma membrane on the cytoplasmic side and contains a molybdenum cofactor and two (2Fe-2S) clusters. A flavin is also associated with the enzyme; however, it is not covalently bonded to the protein. The redox potential of the above reaction is $-540$ mV and electrons enter the electron transport system at the region of cytochrome *b*/ubiquinone. For the generation of NADH, the reverse of electron transfer from ubiquinone to $NAD^+$ occurs, and this is an energy-dependent process. Although most cytochrome *o* molecules are inhibited by CO, the cytochrome *o* of *Pseudomonas carboxydovorans* is unique in that it is not inhibited by CO.

Other bacteria that produce carbon monoxide oxidoreductase include *Alcaligenes carboxydus*, *Bacillus schlegeli*, *Moorella thermoacetica* (formerly *Clostridium thermoaceticum*), and several methanogenic archaea. The enzyme from *Moorella thermoacetica* is 440 kDa with an $\alpha_3\beta_3$ configuration, with the $\alpha$-subunit of 78 kDa and the $\beta$-subunit of 71 kDa. Electrons from oxidation of CO are diverted to a flavoprotein in the plasma membrane of *Moorella thermoacetica* which may be involved in proton pumping (see Figure 9.3). In comparison, the carbon monoxide oxidoreductases from *Methanosarcina barkeri* and *Methanococcus vannieli* have an $\alpha_2\beta_2$ configuration with the $\alpha$-subunit being about 20 kDa and the $\beta$-subunit being about 87 kDa.

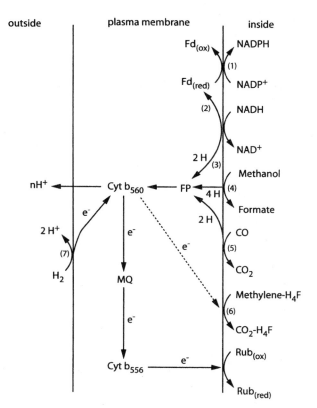

Figure 9.3. Proposed electron flow in *Moorella thermoacetica* (formerly *Clostridium thermoaceticum*) and *Moorella thermoautotrophica* (formerly *Clostridium thermoautotrophicum*). Enzymes are as follows: (1) NADPH-ferredoxin oxidoreductase, (2) NADH-ferredoxin oxidoreductase, (3) NADH dehydrogenase, (4) methanol dehydrogenase, (5) CO dehydrogenase, (6) methylene-$H_4$ folate reductase, and (7) hydrogenase. Fd, ferredoxin; Fp, flavoprotein; $H_4$, tetrahydrofolate, 2H, 2 $e^-$ + 2 $H^+$; Rub, rubredoxin; MQ, menaquinone. (From Wood and Ljungdahl, 1991.[10] Used with permission.)

### 9.2.2.3. Ferrous as an Electron Donor[11–13]

$Fe^{2+}$ is oxidized by several different types of aerobic bacteria according to the following reaction:

$$2\ Fe^{2+} + \tfrac{1}{2}\ O_2 + 2\ H^+ \rightarrow 2\ Fe^{3+} + H_2O$$

The midpoint redox potential for $Fe^{2+}/Fe^{3+}$ is +770 mV and under aerobic conditions, electrons from $Fe^{2+}$ proceed to $O_2$ because the midpoint potential of $O_2/H_2O$ is +820 mV. Microorganisms that use $Fe^{2+}$ as the electron donor for growth are acidophiles because $Fe^{2+}$ is not stable under neutral or alkaline conditions. The iron-oxidizing prokaryotes are placed into three categories: mesophiles, moderate thermophiles, and thermophiles that consist of Gram-negative bacteria, Gram-positive bacteria and archaea, respectively.

The best characterized iron oxidizer is *Acidithiobacillus* (formerly *Thiobacillus*) *ferroxidans*, which has several unique proteins for iron oxidation. In the outer membrane of *Acidithiobacillus ferroxidans* is a 92-kDa glycoprotein that interacts with $Fe^{2+}$ to produce an iron complex. Electrons from $Fe^{2+}$ are directed to cytochrome $c$

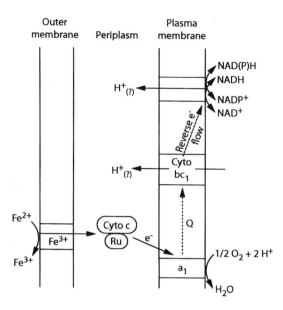

Figure 9.4. Proposed electron transfer in *Acidithiobacillus ferroxidans* for the oxidation of $Fe^{2+}$. Cyto c, Cytochrome $c_{552}$; Ru, rusicyanin (+680 mV). *Dashed line* indicates reverse electron flow. The proton pumps have not been established but are proposed. (Modified from Ehrlich et al., 1991.[13])

or rusticyanin, both of which are in the periplasm. Rustocyanin is a 16.3-kDa copper protein that contains one atom of copper for each molecule, and 5% of the cell protein may be rusticyanin when cells are grown autotrophically. The proposed electron transport of *Acidithiobacillus ferroxidans* is given in Figure 9.4. Rusticyanin has a mid-point redox potential of +680 mV and interfaces with $Fe^{2+}$ which is either as soluble $Fe^{2+}$ or $Fe^{2+}$ present in the iron complex in the plasma membrane. Electrons from rusticyanin are directed to cytochrome c in the cytoplasm. A cytochrome $c_4$ has been isolated from *Acidithiobacillus ferroxidans* and is a 20-kDa protein with two heme C moieties per molecule. Ultimately, electrons are passed to $O_2$ by way of a cytochrome $a_1$ in the plasma membrane. The functioning of the cytochrome $a_1$ oxidase results in the reduction of $O_2$ as well as the pumping of protons to the outer side of the plasma membrane. The production of NADH results from reverse electron flow by an energy-dependent process that is unresolved at this time.

While rusticyanin is associated with $Fe^{2+}$ oxidation in *Acidithiobacillus thiooxidans*, other bacteria use different electron carriers. In a bacterium tentatively identified as *Leptospirillum ferroxidans*, a unique electron carrier, cytochrome-579, is involved in $Fe^{2+}$ oxidation. This $cyto_{579}$ is 17.9 kDa and the reduced molecule adsorbs light at 579 nm and 442 nm. *Sulfobacillus thermosulfidooxidans* has a unique flavoprotein that is proposed to account for $Fe^{2+}$ oxidation. In other organisms such as *Metallosphaera sedula* and *Acidianus brierleyi*, both thermophilic archaea, the electron carrier is unlike any of those described for other $Fe^{2+}$ oxidizers. Clearly, more research is required to define the various processes for $Fe^{2+}$ oxidation in prokaryotes.

### 9.2.2.4. Sulfur Compounds[11-14]

Inorganic sulfur compounds are oxidized by chemolithotrophic organisms as well as by certain photosynthetic bacteria. Sulfur oxidizers generally involve the oxidation of sulfide, elemental sulfur, sulfite, or thiosulfate (see Table 9.5). Sulfur oxidizing chemolithotrophic bacteria include the various species of thiobacilli, *Beggiatoa*, and *Thiothrix* as well as the marine bacterium *Thiomicrospira pelophila* and the archeon *Sulfolobus*.

There are two principal reactions involving sulfite oxidation and only one of these is associated with respiratory activity. Sulfite oxidation by *Thiobacillus* is by a periplasmic sulfite: cytochrome *c* oxidoreductase. In the oxidation of sulfite, six electrons and seven protons are released for each sulfite produced. These electrons are moved from the periplasm to a cytochrome oxidase with the associated activity of proton pumping. In the anaerobic *Thiobacillus denitrificans* and *Chromatium vinosum*, sulfite is oxidized by a sulfite reductase in the periplasm.

The protein for elemental sulfur oxidation has been isolated from *Acidithiobacillus* (formerly *Thiobacillus*) *thiooxidans* and *Thiobacillus thioparus* and found to require reduced glutathione. On the other hand, the enzyme obtained from the thermophilic archaeons *Sulfolobus brierleyi* and *Desulfurolobus ambivalens* does not require reduced glutathione for activity. The sulfur-oxidizing enzymes in archaea are large; in *Sulfolobus brierleyi* they are 560 kDa with 35-kDa subunits and in *Desulfolobus ambivalens* they are 550 kDa with 40-kDa subunits. In contrast, the enzyme from *Acidithiobacillus thiooxidans* is only 40 kDa. While the location of these enzymes remains to be established, the role that sulfur oxidation plays in chemolithotrophic growth is unclear because no electrons are released from the reaction, as the enzyme is an oxygenase with the direct interaction with $O_2$. The reaction involving microbial oxidation of elemental sulfur is given in Table 9.5. It could be suggested that protons generated in the reaction may contribute to charging of the plasma membrane, especially if the enzyme is located on the periplasmic side of the plasma membrane.

Chemolithotrophic bacteria and sulfur phototrophic bacteria metabolize thiosulfate by rhodanese with the production of sulfite and elemental sulfur. Because the rhodanese is in the periplasm of Gram-negative bacteria, elemental sulfur accumulates in the periplasm as sulfur globules. The oxidation of sulfite to sulfate occurs in the periplasm and electrons released are diverted to the membrane for electron transport activity. The thiosulfate oxidizing species of phototrophic bacteria (*Chromatium*, *Thiocystis*, *Thiocapsa*, and *Amoebacter*) grow under microaerophilic conditions by

Table 9.5. Midpoint potentials of reactions associated with sulfur oxidizers.

| Reaction | mV |
|---|---|
| $H_2S \rightarrow S^0 + 2\ H^+ + 2\ e^-$ | −250 |
| $S^0 + 3\ H_2O \rightarrow SO_3^{2-} + 6\ H^+ + 2\ e^-$ | + 50 |
| $SO_3^{2-} + H_2O \rightarrow SO_4^{2-} + 2\ H^+ + 2\ e^-$ | −280 |
| $S_2O_3^{2-} + 3\ H_2O \rightarrow 2\ SO_3^{2-} + 6\ H^+ + 6\ e^-$ | + 20 |

conserving energy from the electron flow from thiosulfate to $O_2$. The system for this chemolithotrophic growth of anaerobic photosynthetic bacteria remains to be established but the presence of an oxygen-reducing cytochrome oxidase in aerobic growth would be consistent with the concepts of bacterial respiration.

Some heterotrophic bacteria oxidize thiosulfate in the presence of $O_2$ with the production of sulfate. These heterotrophic bacteria use an enzyme similar to thiosulfate: cytochrome $c$ oxidoreductase with electrons apparently diverted into the plasma membrane. In a few but not all growth is stimulated by thiosulfate oxidation. Presumably, the bacteria that lack appropriate proton pumps along the electron flow from thiosulfate to $O_2$ cannot receive an energetic benefit from this electron flow.

Sulfide oxidation occurs in several anaerobic bacteria and the resulting electron flow contributes to cell growth. Sulfide oxidation in *Wolinella succinogenes* is coupled to fumarate reduction with the end products being elemental sulfur and succinate. Sulfide oxidation in *Beggiatoa* and *Thiobacillus novellus* is accompanied with production of elemental sulfur and growth-promoting electron transport to $O_2$. In photosynthetic bacteria, sulfide oxidation occurs via a unique flavocytochrome. The flavocytochrome $c_{553}$ from *Chlorobium limicola f. thiosulfatophilum* and the flavocytochrome $c_{552}$ from *Chromatium vinosum* contain one heme C and one FAD moiety covalently bound per molecule. The $c_{553}$ flavocytochrome is 58 kDa and consists of a flavoprotein subunit that is 47 kDa and a cytochrome subunit that is 11 kDa. The $c_{552}$ flavocytochrome is 67 kDa with a 46-kDa flavoprotein subunit and a 21-kDa cytochrome subunit.

### 9.2.2.5. Nitrogen Compounds[15]

Some chemolithotrophic bacteria use $CO_2$ as the sole carbon source with electrons and energy obtained from the oxidation of inorganic compounds. There are two groups of highly specialized nitrogen oxidizers: bacteria that oxidize ammonia to nitrite which are members of genera that have the prefix "*nitroso-*" and bacteria that oxidize nitrite to nitrate and belong to the genera with the prefix of "*nitro-*". The biochemistry of these reactions is addressed in Chapter 13, and here the overall electron transport process is detailed.

Our knowledge of oxidation of ammonia is based primarily on the results obtained from activities of *Nitrosomonas europaea*. The substrate for metabolism is not ammonium ($NH_4^+$) but ammonia ($NH_3$) and the oxidation of ammonia to nitrite ($NO_2^-$) is a two-step process. The oxidation of ammonia with the production of nitrite is by the following two reactions:

$$NH_3 + 2\ H^+ + 2\ e^- + O_2 \rightarrow NH_2OH + H_2O$$
$$H_2O + NH_2OH \rightarrow NO_2^- + 5\ H^+ + 4\ e^-$$

The first reaction is the conversion of ammonia to hydroxylamine ($NH_2OH$) and the second step is the formation of hydroxylamine to nitrite. A unique feature is that the first step involves a 2 $e^-$ reduction while the second step is the release of 4 $e^-$ with the oxidation of hydroxylamine. Thus the 2 $e^-$ released from hydroxylamine are used for ammonia reduction and the remaining 2 $e^-$ go into the electron transport chain

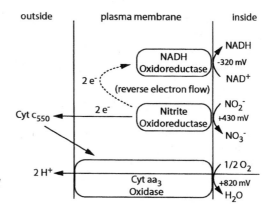

Figure 9.5. Electron transport in *Nitrobacter* with oxidation of nitrite. The *dashed line* reflects reverse electron flow.

for either the reverse electron flow to generate NADH or to $O_2$ with the coupled proton pumping to charge the membrane. This splitting of electrons from oxidation of hydroxylamine results in the low cell yields for chemolithotrophic bacteria growing on ammonia.

*Nitrobacter agilis* can grow by using energy released from the oxidation of nitrite with the formation of nitrate, and this organism has been used extensively to examine the electron transport processes. The oxidation of nitrite is given in the following reaction with the oxygen from water used as one of the oxygens in nitrate:

$$NO_2^- + H_2O \rightarrow NO^{3-} + 2\ H^+ + 2\ e^-$$

The midpoint potential of the $NO_3^-/NO_2^-$ reaction is $+430$ mV and electrons from this reaction would flow through a periplasmic cytochrome with midpoint potential of $+250$ mV as it proceeds to cytochrome $aa_3$, the cytochrome oxidase. A model for electron transport in *Nitrobacter* is given in Figure 9.5. Details of the electron transport in *Nitrobacter* remain to be established. It is apparent that for generation of NADH there must be a reverse electron flow but it is estimated that only 1/50 of all electrons from $NO_3^-$ are used for NADH generation.

### 9.2.2.6. Other Metal Oxidases[11–13]

The growth of several chemoautotrophs has been attributed to energy obtained from the oxidation of various metals. Considerable interest has been given to growth of *Pseudomonas* and other marine organisms at the expense of $Mn^{2+}$ oxidation because of the abundance of ferromanganese nodules on the ocean floor. The product of manganese oxidation is the insoluble $MnO_2$, and with a midpoint potential of $+1,230$ mV for $MnO_2/Mn^{2+}$ reverse electron flow would be required for the generation of NADH. *Acidithiobacillus ferroxidans* has been demonstrated to grow by coupling the oxidation of $Sn^{2+}$, $Cu^+$, $Se^{2-}$, or $Se^o$ to aerobic respiration. *Pseudomonas arsenitoxidans* grows with the oxidation of arsenite to arsenate while *Stibiobacter senarmontii* fixes $CO_2$ with energy obtained from antimony ($Sb^{3+}$) oxidation. Several bacterial isolates that remain unclassified have been reported to grow with $MoO_2$ to $MoO_4^{2-}$. In most of these cases, the metals used or produced are toxic so conditions must be employed that control the accumulation of toxic chemicals. The details of these various metal oxidase reactions need to be developed, although it is apparent that prokaryotes can couple the oxidation of redox active elements to biological energy production.

## 9.2.3. Quinones as Electron Carriers[16]

The plasma membranes contain a lipid that is capable of carrying both electrons and protons. Bacteria will produce either ubiquinone (U) or menaquinone (MK) and in some instances will produce both ubiquinone and menaquinone. Menaquinones have been called naphthaquinones by some because of the naphthalene-like structure (see Figure 9.6). Both ubiquinone and menaquinone function by carrying either one electron and one proton, at the semiquinone state, or two electrons and two protons when fully reduced. An important side chain for both the ubiquinone and the menaquinone structure is the isoprenoid unit which is a highly lipophilic unit. The different mo-

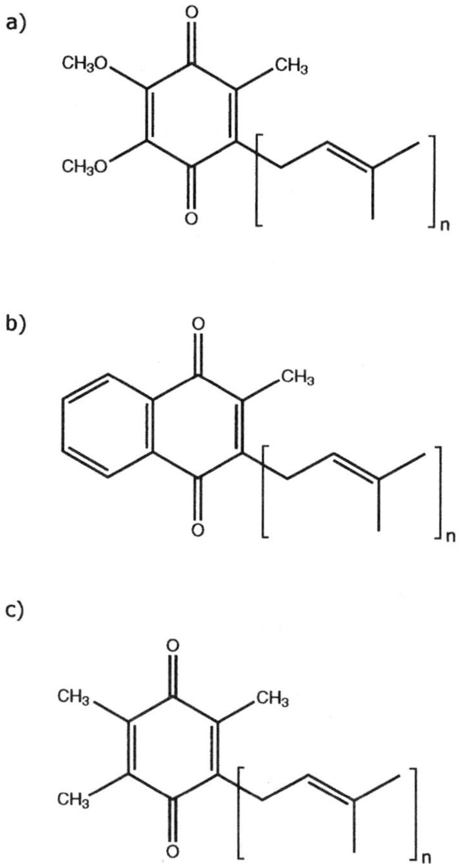

a)

b)

c)

Figure 9.6. Structures of important quinones in bacteria. (a) Ubiquinone, (b) menaquinone and (c) plastoquinone.

Table 9.6. Distribution of quinones in bacteria.

| Bacteria | Type of quinone present |
| --- | --- |
| *Acetobacter xylinum* | Q-10 |
| *Acidithiobacillus thiooxidans* | Q-8 |
| *Azotobacter chroococcum* | Q-8 |
| *Bacillus* sp. | MK-9 |
| *Bacteroides fragilis* | MK-10 and MK-11 |
| *Enterococcus faecalis* | DMK-9 |
| *Escherichia coli* | Q-9, MK-8 and DMK-8 |
| *Haemophilus parainfluenza* | DMK-6 |
| *Halobacterium halobium* | MK-8 |
| *Lactobacillus plantarum* | Lacks quinones (?) |
| *Lactobacillus lactus* | MK-9 |
| *Listeria monocytogenes* | MK-7 |
| *Myxococcus flavus* | MK-8 |
| *Mycobacterium tuberculosis* | MK-9(H$_2$) |
| *Pseudomonas aeruginosa* | Q-9 |
| *Pseudomonas maltophilia* | Q-8 |
| *Rhodopseudomonas acidophila* | Q-10 and MK-10 |
| *Vibrio costicola* | MK-8 and Q-8 |

DMK, Demethylmenaquinone; MK-9 (H$_2$), dihydromenaquinone.
The number following Q or MK reflects the number of isoprenoid subunits present. Only the quinone in greatest abundance in the membrane is given and most bacteria will also produce several additional MK or Q but in relatively low concentrations.

lecular types of ubiquinones and menaquinone is a reflection of the number of repeating isoprenoid units. A common designation of quinones is to indicate the number of isoprenoid side chains by a number following the quinone, as is the case for MK-6, MK-7, and MK-8. The side chains would function to retain the MK in the membrane owing to the lipophilic "tail" consisting of the isoprenoid side chain. The distribution of quinones in bacteria is given in Table 9.6. Menaquinones are commonly present in Gram-positive bacteria. In bacteria containing both ubiquinones and menaquinones, each quinone may have a specific and distinct electron-carrying role. For example, *E. coli* produces ubiquinone in aerobic culture but menaquinone under anaerobic conditions. While ubiquinone and menaquinone are interchangeable in some reactions, menaquinones are less reactive in aerobic systems and are irreplaceable in formate reduction and with trimethylamine-*N*-oxide reductase. Thus, determination of whether an organism has ubiquinone or menaquinone is based on functional requirements. Bacteria that do not have cytochromes also lack quinones, so it would appear that quinones are essential for cell respiration.

Quinones have also been demonstrated in archaea. *Halococcus* and *Halobacterium* contain MK-8 and dihydrogenated MK-8, which are sufficiently abundant to constitute 10% of the acid-soluble lipids in these extreme halophiles. *Thermoplasma acidophilum* contains MK-7 and small amounts of MK-6 and MK-5 along with the unique thermoplasmaquinone. All species of *Sulfolobus* produce a sulfur-containing heterocyclic quinone identified as benzo(*b*)thiopheno-4,7-quinone. The role of these quinones in electron transport systems remains to be established.

## 9.2.4. Proteins as Electron Carriers

The molecules that are important in carrying electrons in the electron transport system of microorganisms are either proteins or lipids. The lipids include the ubiquinones and meniquinones discussed in the previous section. Amino acids in proteins do not carry electrons but the metal centers or organic substances bound into the proteins are responsible for carrying electrons. A survey of electron-carrying proteins is given in Figure 9.7. Organic substances that carry electrons include FAD and FMN, which are found in flavodoxin and flavoprotein dehydrogenases and the quinols such as PQQ which are characteristic of quinoprotein dehydrogenases.

Several different transition metals are used to make up the redox centers of dehydrogenases, oxidases, and electron transfer proteins. Iron is essential for the Fe-S centers, and proteins with these centers are known as nonheme iron proteins. In 1964, Rieske discovered that the quinol *b/c* complex contained a Fe-S protein of about 20 kDa that mediated the electron transfer between quinone and cytochrome *b/c*. This nonheme iron protein is present in complex III of organisms with aerobic respiration and is commonly described as the Rieske protein. Other Fe-S centers are found in respiratory proteins such as ferredoxins and dehydrogenases. Cytochromes carry electrons because of the presence of iron in the heme complex as either the reduced form, $Fe^{2+}$, or the oxidized form, $Fe^{3+}$. Several dehydrogenases, especially those of dinitrogen fixation and nitrate reduction, use molybdenum as the electron mediator and these molybdoproteins are discussed in Chapter 13. Copper is an important redox center in proteins for specific roles in electron transport by the blue copper proteins and interface with molecular oxygen through the cytochrome oxidases. The diversity of cytochrome types underscores the different roles of cytochromes in microbial elec-

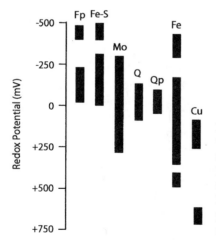

Figure 9.7. Redox potential of cofactors associated with dehydrogenases, oxidases, or reductases with inorganic and organic substrates. Fp, Flavoproteins; Fe-S, iron sulfur proteins; Mo, molybdoproteins; Q, ubiquinones or menaquinones; QP, quinoproteins; Fe, cytochromes; Cu, blue copper proteins or cuproproteins. (From Jones, 1988.[21] Used with permission.)

Table 9.7. Fe-S. Selected examples of iron proteins in prokaryotes.[a]

| Class of Fe-protein | Bacteria | kDa | mV |
|---|---|---|---|
| **Mononuclear-Fe protein** | | | |
| [1Fe] Rubredoxin | *Clostridium pasteuranium* | 6 | −570 |
| [2Fe] Desulforedoxin | *Desulfovibrio gigas* | 7.9 | −35 |
| **[2Fe-2S] protein** | | | |
| Rieske protein | *Thermus thermophilus* | 12.5 | −240 |
| Putidaredoxin-like | *Spirulina platensis* | 10.6 | −380 |
| | *Halobacterium* sp. | 14 | −345 |
| **[3Fe-4S] protein** | | | |
| Ferredoxin II | *Desulfovibrio gigas* | 24 | −130 |
| **[4Fe-4S] protein** | | | |
| Ferredoxin | *Rhodospirillum rubrum* | 14.5 | −430 |
| | *Bacillus polymyxa* | 9 | −420 |
| HiPiP | *Chromatium vinosum* | 10 | +350 |
| **2[4Fe-4S] protein** | | | |
| Ferredoxin | *Clostridium pasteuranium* | 6 | −390 |
| | *Chromatium vinosum* | 13 | −490 |
| **7[Fe-nS] proteins** | | | |
| [4Fe-4S], [3Fe-3S] protein | *Azotobacter vinelandii* | 14.5 | +340, −420 |
| [4Fe-4S], [3Fe-nS] protein | *Thermus thermophilus* | — | −530, −250 |

[a]Information for this table is from Hughes and Poole, 1989.[18]

tron transport. The use of various electron transport proteins provides the organisms with a versatility to utilize a wide range of redox active components such as electron donor/electron acceptor combinations to conserve energy released in the redox reactions.

## 9.2.4.1. Ferredoxin[17–19]

Ferredoxins and flavodoxins are electron carrier proteins that commonly participate in enzyme-driven low-potential reactions. Ferredoxin molecules are highly acidic non-heme iron proteins generally of 6 to 10 kDa and are found in various physiological types of microorganisms. A hallmark feature of the ferredoxins is the iron-sulfur (Fe-S) clusters which are found mostly as 2Fe-2S or 4Fe-4S clusters, and representative examples of these ferredoxins are listed in Table 9.7. Ferredoxin has been reported in about 100 prokaryotic species and is widely distributed throughout the microorganisms. Although an exhaustive listing is not included here, the breadth of ferredoxin distribution is apparent in that it has been isolated from aerobic bacteria (i.e., *Agrobacterium tumefaciens*, *Azotobacter vinelandii*, *Thermus thermophilus*, *Escherichia coli*, *Bacillus stearothermophilus*, and mycobacteria); anaerobic bacteria (i.e., *Moorella thermoaceticum*, *Clostridium kluyveri*, and *Clostridium thermosaccharolyticum*); archaea (i.e., *Methanosarcina barkeri*); and photosynthetic bacteria (i.e., *Chlorobium limicola* and *Spirulina platensis*). In ferredoxin, the sulfur atoms are bonded to iron

a)

b)

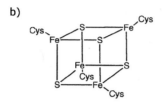

Figure 9.8. Iron sulfur clusters associated with proteins. (a) 2Fe-2S cluster and (b) 4Fe-4S cluster. Cys represents cysteine in the protein and iron is liganded to the cysteinyl-sulfhydral groups of protein.

and the iron-sulfur cluster is held in the protein by four cysteine residues from the apoprotein. A model indicating the iron-sulfur cluster arrangements in ferredoxin is given in Figure 9.8. Based on the binding of cysteine to iron, four different types of peptide sequences are associated with the 4Fe-4S clusters. The proposed consensus sequences of amino acids for the 4Fe-4S clusters are as follows: (1) type I— XCXGCXTCXXXCK; (2) type II—XCXXCXXPACXXXCP; (3) type III— XCXXCXXCXXXCP; and (4) type IV—KCXXCY-9-XCXEXCX. In comparison, the 2Fe-2S cluster has a consensus sequence of SCRXXXCGSCG-8-ACX. The presence of proline following the fourth cysteine in type II and type III is considered to produce a sharp fold in the peptide chain after the 4Fe-4S cluster. In both type IV of the 4Fe-4S cluster and the 2Fe-2S cluster there are numerous (10 to 11) amino acids between the third cysteine and the fourth cysteine. Currently, research is being directed to elucidate the pathway for synthesis of the Fe-S clusters.

Most ferredoxins carry 1 e$^-$ with the redox center changed from (2+) to (1+) and participate in redox reactions instead of NAD$^+$. Ferredoxins are soluble proteins that participate in reactions in the cytoplasm or with the photosynthetic apparatus. At physiological pH levels, all ferredoxins carry a pronounced acidic charge that is most distinctive for these molecules; however, the biological significance of this charge has not been established. There is a considerable range in examples of enzyme reactions requiring ferredoxin (see Table 9.4), and most of these reactions involve low potential electron donors. Neither oxidized nor reduced ferredoxins absorb light at a specific wavelength, so spectroscopy using visible light cannot be used to follow reactions. However, the (1+) state of the iron-sulfur cluster is paramagnetic and appearance or disappearance of the paramagnetic state can be followed using electron paramagnetic resonance (EPR) spectroscopy.

## 9.2.4.2. Flavodoxin[20]

Under conditions of iron limitation, flavodoxin will be produced by many bacteria instead of ferredoxin. Flavodoxins are proteins with a mass of about 15 kDa that do not contain any Fe-S clusters but instead have FMN bound to the protein. The structure of FMN is given in Figure 9.9. There is a covalent bond that attaches FMN to

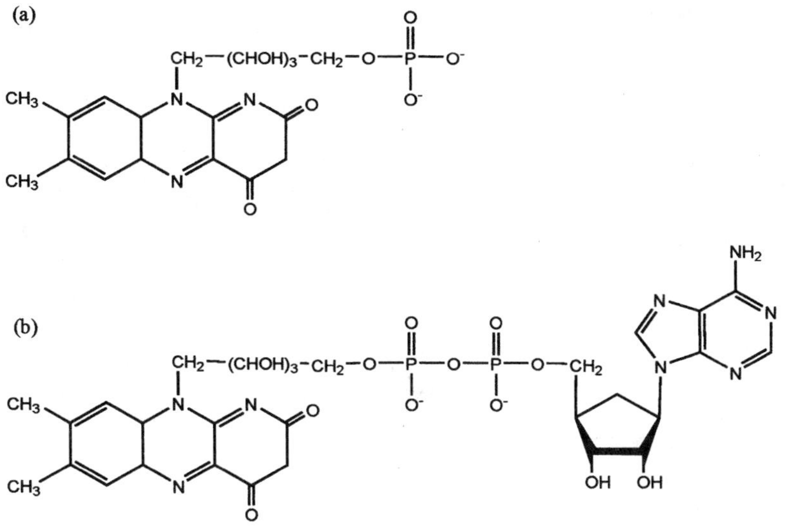

Figure 9.9. Structures of two electron carriers. (a) Flavin mononucleotide (FMN); and (b) flavin adenine dinucleotide (FAD).

the cysteine or to the tyrosine moieties of the apoprotein. FMN may be the semi-quinone, which is the partially reduced state as a result of accepting one electron and one proton with a redox potential of about $-150$ mV. In addition, FMN may be in the fully reduced form as a result of accepting two electrons and two protons. The equilibrium between the fully reduced FMN and the semiquinone is $-440$ mV, which is about the same as for ferredoxin$_{(oxidized)}$ = ferredoxin$_{(reduced)}$. Thus, flavodoxin can substitute for ferredoxin in the phosphoroclastic, NADP reductase, and other reactions. However, ferredoxin and flavodoxin are not identical in physiological activity because flavodoxin carries both electrons and protons while ferredoxin carries only protons. Examples of bacteria producing flavodoxin and some of their characteristics are given in Table 9.8. The reduction of flavodoxin can be measured by spectrophotometric analysis. Flavodoxin in the oxidized form strongly absorbs light at 445 nm while with the reduction of FMN in flavodoxin there is a loss of absorption. Flavodoxin is found only in the cytoplasm but in some organisms there may be some association with membranes because treatment of disrupted cells with a mild detergent enhances the amount of flavodoxin obtained.

### 9.2.4.3. Copper Blue Protein[21]

In some bacteria, the electron transport system contains copper blue proteins which are also known as cuproteins. These copper blue proteins are distinct from some of the cytochrome oxidases which contain copper in addition to heme moieties. A copper atom is coordinately attached through a nitrogen of histidine and sulfur of cysteine and methionine (see Figure 9.10). The blue copper proteins have a high

Table 9.8. Examples of flavodoxins produced by prokaryotes.[a]

| Source | $E_0'$ (mV) of | | Reaction |
| | Half reduced | Fully reduced | |
|---|---|---|---|
| *Anabaena nidulans* | −221 | −447 | Nitrogenase |
| *Anabaena variabilis* | −195 | −390 | Nitrogenase |
| *Azotobacter vinelandii* | −165 | −460 | Nitrogenase |
| *Clostridium beijerinckii* | −92 | −399 | Nitrogenase, phosphoro-clastic reaction |
| *Desulfovibrio gigas* | −150 | −440 | Phosphoroclastic reac-tion, sulfite reaction |
| *Desulfovibrio vulgaris* | −143 | −435 | Phosphoroclastic reaction |
| *Escherichia coli* | −240 | −410 | Phosphoroclastic reaction |
| *Klebsiella pneumoniae* | −158 | −412 | Nitrogenase, phosphoro-clastic reaction |

[a]From Vervoot et al., 1994.[20]

positive redox potential where the $Cu^{2+}/Cu^+$ couple is $\geq 230$ mV. Under aerobic conditions, the cuproprotein is a distinct blue color; however, the acquisition of one electron by the $Cu^{2+}$ atom results in the formation of $Cu^+$ and the reduced molecule is colorless. These copper blue proteins are localized in the periplasm, where they function in electron transfer. There are several examples of blue copper proteins in bacterial respiration, including azurin in *Pseudomonas aeruginosa*, amicyanin in *Pseudomonas denitrificans*, rusticyninin in *Acidithiobacillus ferroxidans*, auracyanin in *Chloroflexus aurantiacus*, and plastocyanin in cyanobacteria.

## 9.2.5. Cytochromes[22-28]

Cytochromes are found in almost all of the aerobic microorganisms and many of the anaerobic ones that use anaerobic respiration. In 1925, Keilin initially observed cytochromes in *Bacillus subtilis* and since then cytochromes have been found in a vast number of prokaryotes. Chemolithotrophs that rely on inorganic metabolism for an electron source or for inorganic compounds for electron acceptors rely heavily on cytochromes for electron transfer. Cytochromes have been found in bacterial endo-

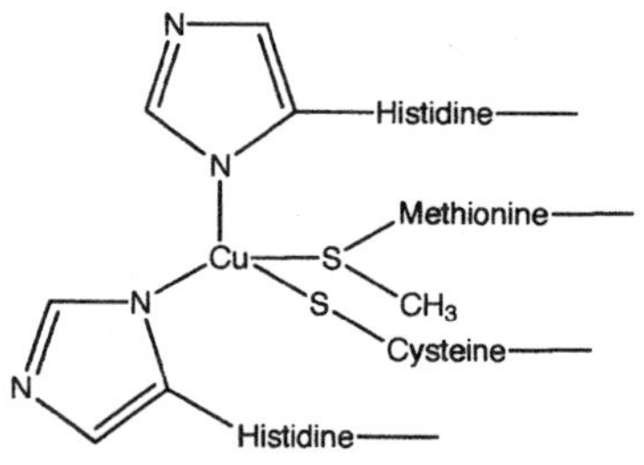

Figure 9.10. The Cu center of blue copper protein where copper occurs as $Cu^+$ or $Cu^{2+}$. The copper atom is coordinately bonded by amino acid moieties (histidine, cysteine, and methionine) of the protein.

spores and in many, but not all, anaerobic bacteria. Through independent efforts of Ishimoto and Postgate in 1954, the $c$-type cytochrome was demonstrated in the anaerobic bacterium, *Desulfovibrio*, and this provided the precedence to demonstrate cytochromes in other anaerobes including *Bacteroides*, *Campylobacter*, *Shewanella*, and *Geobacter*. Cytochromes of the $b$-type have been found in obligate anaerobes such as the clostridia-like bacteria associated with acetogenesis and in the archaeal strains of *Thermoplasma* and *Methanosarcina*. The archaeon *Halobacterium* has a complex cytochrome system that is comparable to that found in bacteria.

Cytochromes are unique structures in that they have a protein moiety and a heme structure. In terms of electron-carrying activity, the heme is important because it contains an iron atom that is coordinately bonded to the tetrapyrrole. Iron exists in heme as either the reduced form, $Fe^{2+}$, or the oxidized form, $Fe^{3+}$. In bacteria four different heme types (A, B, C, and D) are present and each has a specific tetrapyrrole structure (see Figure 9.11). Oxidized cytochrome molecules have a distinct absorption at about 390 to 420 nm and this portion of the visible spectrum is referred to as the Soret region. When reduced, cytochromes typically have three bands of absorption which are designated as the $\alpha$, $\beta$, and $\gamma$ (or Soret) absorption peaks, with the $\alpha$-peak at the highest wavelength and the $\gamma$-peak at the lowest wavelength. Whereas the reduced cytochromes of the $c$-type and $b$-type have three absorption bands ($\alpha$, $\beta$, and $\gamma$), the reduced cytochrome $a$ has only two absorption bands: $\alpha$ and $\gamma$. The absorption spectra for each of the reduced cytochromes is characteristic for a given cytochrome in that each has a different wavelength at which absorption occurs. The wavelengths at which absorption occurs are given in Table 9.9. The wavelength of maximum absorption at the $\alpha$ absorption band is distinctive for each cytochrome with cytochrome $a$ absorbing at 590 to 635 nm, cytochrome $b$ at 555 to 558 nm, and cytochrome $c$ at 550 to 555 nm. The spectra in Figure 9.12 are used to identify the cytochromes present in two different strains of bacteria.

Some bacteria have cytochrome $o$ which contains heme B and when cytochrome $o$ is reduced, it will react with cyanide or with carbon monoxide to produce a characteristic absorption spectra. The spectrum of the reduced cytochrome $o$ is shifted several nanometers to a higher wavelength as a result of CO binding to the reduced heme. The carbon monoxide adjunct with reduced heme is disrupted by exposure to high-intensity light at 400 to 450 nm. Cytochrome $d$ is found in some bacteria and it contains heme D, which has adsorption at 628 nm and a soret band at 442 nm. In comparison to other hemes, heme D has a reduced ring number 3 in the tetrapyrrole while in $D_1$ ring number 2 is reduced. Cytochrome $d$ will also form a shift in the spectrum of the reduced hemes; however, removal of carbon monoxide from heme D cannot be accomplished with bright light. Chlorins are hemes with one of the four pyrrole rings reduced and with this perspective, both heme D and heme $D_1$ are examples of chlorins. Over the years, there have been various classifications for the bacterial cytochromes. Perhaps one of the best ways of designating the bacterial cytochrome is to indicate the wavelength of absorption of the $\alpha$-band. For example, the bacterial $c$-type cytochrome which has absorption at the $\alpha$-band of 553 nm is designated as cytochrome $c_{553}$.

In addition, there is a difference in the attachment of the protein to the heme with

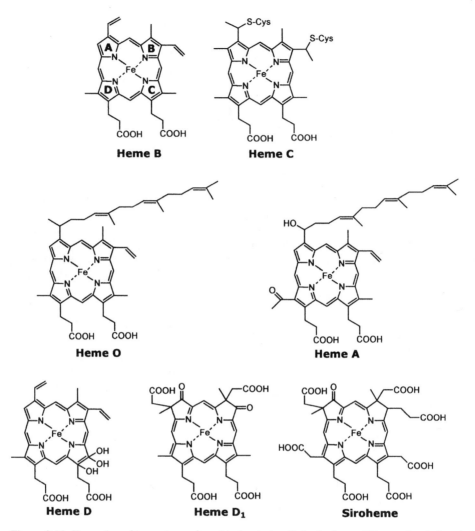

Figure 9.11. Examples of heme types found in bacteria. Only the heme C is covalently bonded to a protein and this linkage is through the two cysteine moieties on rings A and B.

the different cytochrome types. With cytochrome $c$, the protein is attached by covalent bonds through cysteine residues of the protein to rings 1 and 2 of the tetrapyrrole. The amino acid motif of Cys-X-X-Cys-His or Cys-X-X-X-X-Cys-His in the polypeptide chain is responsible for binding to heme $c$ and this sequence can be used to predict from DNA sequence the number of c-type cytochromes in bacteria. The absence of covalent bonds between the protein and the heme in cytochrome $b$ and cytochrome $a$ accounts for the extreme difficulty in isolating these two cytochromes. Coordination from the four pyrrole rings and from two amino acids of the protein

Table 9.9. Absorption characteristics of bacterial cytochromes.

| Cytochrome type | Potential (mV) | α-Peak[a] | β-Peak[a] | γ-Peak[a] |
|---|---|---|---|---|
| a (Heme A) | +225 | 590–635 | — | 439–445 |
| b (Heme B) | +80 | 555–558 | 523–530 | 422–430 |
| c (Heme C) | Varies | 550–555 | 521–526 | 416–420 |
| o (Heme B)[b] | +130 | 565–575 | 531–543 | 412–421 |
| d (Heme A)[b] | +270 | 632–642 | 537–539 | 440–442 |

[a]Wavelengths given are from experiments conducted at room temperature. When experiments are conducted at 77 K the absorption is generally shifted 2 or 3 nm to a lower wavelength.
[b]Results given are with the CO spectra which produces an absorbancy that is about 5 nm longer in wavelength than the reduced cytochrome without CO.

stabilizes the iron atom at either $Fe^{2+}$ or $Fe^{3+}$. The axial ligands for coordinating iron in $c$-type cytochromes are commonly histidine–heme–histidine or methionine–heme–histidine.

### 9.2.5.1.  $b$-Type and $c$-Type Cytochromes

The activities of the $b$-type cytochromes are associated with the plasma membrane, where the cytochrome $b$ moves electrons in the vicinity of the quinone in the electron transport chain. One of the most widely distributed systems throughout the prokaryotes is the cytochrome $bc_1$ unit which is part of complex II involved in the transfer of electrons from reduced quinone to cytochrome $c$. In *Paracoccus denitrificans* and *Rhodobacter capsulata*, there are three subunits in the complex; however, in the beef mitochondrial system there are 11 subunits. Contained in each subunit is cytochrome $b$ with two heme B, one cytochrome $c_1$, and the Fe-S (Rieske) protein. In *Rhodobacter capsulata*, the cytochrome $b$ is 42 kDa with midpoint potentials for the heme moieties of $-60$ mV and $+50$ mV. It is interesting to note that *E. coli* lacks the cytochrome $bc_1$ complex, and therefore is absent a critical proton pumping system. When the cytochrome $bc_1$ complex is isolated from *Anabaena variabilis* or *Rhodopseudomonas sphaeroides* and reconstituted in proteoliposomes, the complex is capable of pumping 2 $H^+$ across the membrane for each 2 $e^-$ moved through the system. In conjunction with the Q-cycle, it is considered that 4 $H^+$ are pumped out for each 2 $e^-$. Cytochrome $b$ has eight transmembrane helices with the two heme B moieties held between two adjacent transmembrane helices. Cytochrome $b$ interfaces with the quinone and the Fe-S protein. Cytochrome $c_1$ has a mass of 34 kDa with a single transmembrane helix and the cytochrome has a midpoint potential of $+285$ mV.

In addition, there are several examples of a situation in which microorganisms have a cytochrome $b$ that transfers electrons from an electron donor to the electron transport chain or from the electron transport chain to terminal electron acceptors. Examples of respiratory enzymes that have cytochrome $b$ associated with electron donor activity include formate dehydrogenase and fumarate reductase. Several reductases use cytochrome $b$ for transfer of electrons to the final electron acceptor and some of these are indicated in Figure 9.13.

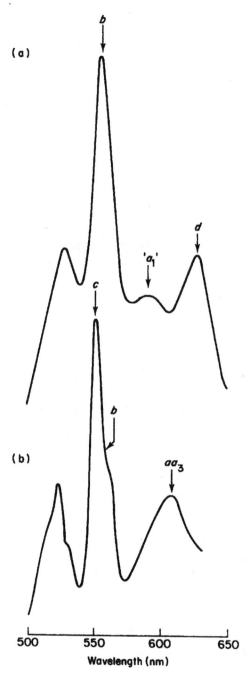

Figure 9.12. Spectra of reduced cytochromes present in (**a**) *Escherichia coli* and (**b**) *Methylophilus methylotrophus*. Reduced minus oxidized difference spectrum in (a) indicates the absorption peaks of cytochromes $b$, $a_1$, and $d$. Reduced minus spectrum in (b) indicates the α-absorption peak of cytochromes $c$, $b$, and $aa_3$. (From Jones, 1988.[21] Used with permission.)

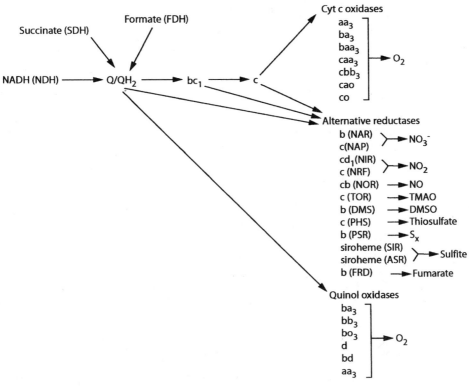

Figure 9.13. A composite of some of the cytochromes used by bacteria with a focus on electron acceptors. Under low concentrations of $O_2$, cytochrome *d* replaces cytochrome *o* or cytochrome *co* and cytochrome *o* or cytochrome *co* replaces cytochrome $aa_3$. (From Thony-Meyer, 1997.[28] Used with permission.)

Cytochrome *c* is a relatively stable molecule in that heme C moiety is covalently bound by the protein. Depending on the metabolism of the organism, cytochrome *c* may be found either in the cytoplasm, plasma membrane, or the periplasm. In addition, there is considerable variation in the cytochromes of the *c*-type in that some may be monoheme while others are multiheme with several heme groups per cytochrome molecule. Generally the *c*-type cytochrome has a role in transfer of electrons from complex III to a terminal oxidase and has a midpoint redox potential of about +200 mV. However, some cytochrome *c* have an extremely low redox potential of −200 mV and can interface with electron donor reactions that also have low redox potentials. Multiheme cytochrome *c* are present in some of the lithotrophic bacteria in which each heme has a distinct midpoint potential. Several examples of multiheme cytochromes present in a few bacteria are given in Table 9.10, and this is used to illustrate some of the cytochrome variability found in prokaryotes.

Table 9.10. Examples of polyheme cytochromes in selected bacteria.[a]

| Cytochrome | Wavelength of absorption | Bacteria | Number of heme groups | Molecular weight (kDa) | Midpoint potential of heme (mV) |
|---|---|---|---|---|---|
| $c$ | 553 nm | *Desulfovibrio vulgaris* | 1 | 9 | +20 |
| $c_3$ | 553 nm | *Desulfovibrio desulfuricans* | 4 | 13 | −375, −335, −305, −225 |
| $c_3$ | 553 nm | *Desulfovibrio vulgaris* | 8 | 26 | (−180, −270, −320, −370)₂ |
| $c$ | 553 nm | *Desulfovibrio vulgaris* | 16 | 75 | +60, +15, (−120 to −135)₇ (−190 to −205)₅, (−260)₂ |
| $c_{544}$ | 544 nm | *Nitrosomonas europaea* | 4 | 20 | +47, +47, −147, −276 |
| $c_{552}$ | 552 nm | *Nitrosomonas europaea* | 1 | 11 | +250 |
| Hydroxylamine oxidoreductase | 553 nm | *Nitrosomonas europaea* | 8 | 63 | +11, +288, −10, −162 −192, −260, −265, −412 |
| Flavocytochrome $c$ | 552 nm | *Chromatium vinosum* | 2 | 67 | +16, −33 |

[a]From Peck and LeGall, 1994[25]; Hooper et al.[23]

### 9.2.5.2. Oxidases

The oxidases can be divided into two groups of cytochromes based on natural substrates. One group includes those that accept electrons from reduced quinols and are also referred to as quinol oxidases. The second group of cytochromes accepts electrons from cytochrome $c$ and are referred to as cytochrome $c$ oxidases. As shown in Figure 9.13, prokaryotes have numerous cytochromes involved in each group of oxidases. Multiple heme moieties of varying types appears to be the rule. Cytochrome $aa_3$ is present in both groups as a quinol oxidase and as a cytochrome $c$ oxidase. However, it is significant to point out that only the cytochrome $aa_3$ associated with cytochrome $c$ oxidation has been demonstrated to be capable of serving as a proton pump. Other oxidases include multiple heme moieties per molecule as well as the unique cytochrome $o$ and cytochrome $d$.

Cytochrome $aa_3$ contains four redox centers that consist of cytochrome $a$ ($Fe_A$), cytochrome $a_3$ ($Fe_B$), $Cu_A$, and $Cu_B$. This cytochrome $aa_3$ is widely distributed throughout the prokaryotes in that it is present in cyanobacteria, sulfur and nitrogen bacteria, spore-forming bacillus, and archaea including *Sulfolobus* and *Halobacterium*. The structure of the cytochrome $aa_3$ found in prokaryotes is relatively simpler than that of eukaryotes. For example, in *Paracoccus denitrificans*, cytochrome $aa_3$ consists of three subunits: 62 kDa, 33 kDa, and 31 kDa. In contrast, the cytochrome $aa_3$ in yeast has nine subunits while in beef heart mitochondria there are 11 subunits.

In reduced $O_2$ concentrations, many bacteria produce and use cytochrome $o$ which contains heme B. This cytochrome has been purified from numerous bacteria and the molecular characteristics found in *E. coli* are typical of most $o$-type cytochromes. In

*E. coli*, cytochrome *o* has four subunits of 66 kDa, 36 kDa, 18 kDa, and 13 kDa. The prosthetic groups include a chromophore of cytochrome $b_{562}$, a cytochrome $b_{555}$ which may be the heme that binds carbon monoxide, and two Cu atoms. Cytochromes of the *o*-type are broadly distributed in the bacteria and appear to function as both quinol oxidases and as cytochrome *c* oxidases.

Cytochrome *d* is an example of another oxidase in bacteria and it is produced when the oxygen concentrations are very low. Cytochrome *d* is resistant to cyanide and azide while cytochromes *o* and $aa_3$ are sensitive to these respiratory inhibitors. There is a greater affinity for molecular oxygen by cytochrome *d* than for other oxidases and a $K_m$ of 0.37 μ*M* for $O_2$ has been calculated for this cytochrome. In *Photobacterium phosphoreum*, cytochrome *d* consists of two dissimilar subunits of 58 kDa and 43 kDa. The cytochrome *d* from *Photobacterium phosphoreum* has four heme groups: two *b*-type cytochromes (i.e., cytochrome $b_{558}$ and cytochrome $b_{594}$) and two heme D moieties. The cytochromes of the *d*-type are considered to be principally quinol oxidases.

### 9.2.5.3. Distribution

The distribution of cytochromes in bacteria is highly variable. Bacteria may contain cytochromes of *a*, *b*, and *c* types or may contain only one type or two types of cytochromes. Some bacteria may have several cytochromes of the *b*-type while others have several cytochromes of the *c*-type. Some examples of cytochromes in bacteria are given in Table 9.11. A few heterotrophic bacteria have special cytochromes that interact directly with oxygen and are designated as type-*o* and type-*d*. As shown in Figure 9.11, cytochrome *o* contains the heme B while cytochrome *d* contains heme D, which is a modified heme A.

A characteristic of cytochrome content in bacteria is the absence of fixed stoichiometry for cytochromes. In the mitochondrial systems, the ratio of cytochrome type-*b*: type-*c*: type-*a* is constant; however, in bacteria the concentration of cytochrome in

Table 9.11. Examples of cytochromes found in bacteria.

| Cytochrome | Prosthetic group | Inhibited by | | Molecular mass (kDa) |
|---|---|---|---|---|
| | | CO | CN⁻ | |
| *b* | 1 Heme B | No | No | 12–17.5 |
| *c* | 1 Heme C | No | No | 12–100 |
| $c_{co}$ | 1 Heme C | Yes | Yes | 12.5 |
| *ca* | 1 Heme C | Yes | Yes | 38–118 |
| | 1 Heme A | | | |
| | 1 Cu | | | |
| $aa_3$ | 2 Heme A | Yes | ≤20 μ*M* | 73 |
| | 2 Cu | | | |
| *o* | 2 Heme B | Yes | ≤0.3 m*M* | 28 |
| *d* | 2 Heme B | Yes | 10 m*M* | 350 |
| | 2 Heme D | | | |

a cell is not fixed. Classic experiments conducted by David C. White in the early 1960s demonstrated that the abundance of individual cytochromes in *Haemophilus parainfluenzae* may vary by a factor of 70 as the production of cytochromes is influenced by changes in aeration, types of nutrients present, and growth phase of culture. Cytochromes that interface with $O_2$ are produced by bacteria under aerobic cultivation but under anaerobic conditions, these cytochromes are not produced. The publications of D.C. White were highly instrumental in influencing the direction of cytochrome-related research for several decades.

## 9.2.6. Electron Acceptors[22–28]

A novel feature about electron transport in bacteria is anaerobic respiration in which various oxidized compounds can serve as the terminal electron acceptor. Various electron acceptors used by bacteria are listed in Table 9.12.

### 9.2.6.1. Molecular Oxygen: Oxidases

Molecular oxygen as the terminal electron acceptor functions to oxidize the cytochrome involved in transferring electrons to oxygen. In an aerobic state, several cytochromes function as oxidases and are inhibited by carbon monoxide and cyanide (see Table 10.4). Copper atoms, found in cytochrome $aa_3$ and cytochrome $ca$, function to assist in the transfer of electrons from the iron atom in heme A to molecular oxygen. Bacterial cytochromes $o$, $d$, $a_3$, and $c_{co}$ interact directly with molecular oxygen and also are oxidases. The reason for multiple oxidases in bacteria is not understood at this time. If the oxygen level in the growth solution is high, cytochrome $o$ is frequently used and if the oxygen level is low, cytochrome $d$ functions to transfer electrons from the respiratory chain to molecular oxygen. The apparent affinity $K_m$ for oxygen by bacterial oxidases such as cytochrome $aa_3$ and cytochrome $o$ is about 1 $\mu M$ for $O_2$ but cytochrome $d$ has a slightly higher affinity (0.37 $\mu M$) for molecular oxygen.

Table 9.12. Electron acceptors frequently involved with anaerobic respiration of bacteria.

| Reaction | Electron acceptor | Reaction product | $E_0'$ (mV) |
|---|---|---|---|
| $S° + 2(H) \rightarrow HS^+ + H^+$ | $S°$ | $H_2S$ | $-270$ |
| Fumarate $+ 2(H) \rightarrow$ Succinate | Fumarate | Succinate | $+30$ |
| $(CH_3)_3NO + 2(H) \rightarrow (CH_3)_3N + H_2O$ | Trimethylamine-$N$-oxide (TMAO) | Trimethylamine (TMA) | $+130$ |
| $AsO_4^{3-} + 2(H) \rightarrow AsO_3^{2-} + H_2O$ | Arsenate | Arsenite | $+139$ |
| $(CH_3)_2SO + 2(H) \rightarrow (CH_3)_2S + H_2O$ | Dimethyl sulfoxide (DMSO) | Dimethyl sulfide (DMS) | $+160$ |
| $2Fe^{3+} + 2(H) \rightarrow 2\ Fe^{2+} + 2\ H^+$ | $Fe^{3+}$ | $Fe^{2+}$ | $+200$ |
| $NO_3^- + 2(H) \rightarrow NO_2^- + H_2O$ | $NO_3^-$ | $NO_2^-$ | $+430$ |
| $SeO_4^{2-} + 2(H) \rightarrow SeO_3^{2-} + H_2O$ | Selenate | Selenite | $+475$ |
| $MnO_2 + 2(H) \rightarrow Mn^{2+} + 2\ H^+$ | $Mn^{4+}$ | $Mn^{2+}$ | $+798$ |
| $ClO_3^- + 6(H) \rightarrow Cl^- + 3\ H_2O$ | Chlorate | Chloride | $+1.030$ |

### 9.2.6.2. Fumarate Reductase

Fumarate reductase functions in anaerobic systems as a terminal electron acceptor with the production of succinate from fumarate. The reaction catalyzed by fumarate reductase is as follows:

$$\text{fumarate} + 2\,H^+ + 2\,e^- = \text{succinate}$$

Fumarate reductase consists of four dissimilar subunits. Subunit A is the largest, at 69 kDa, and it contains FAD covalently bound to the protein and the site for substrate binding. Subunit B is 25 kDa and contains three iron-sulfur centers. Subunits C and D are 15 and 13 kDa, respectively, and function to anchor fumarate reductase into the cytoplasmic side of the membrane. Fumarate reductase has a midpoint redox potential of +30 mV which is sufficiently electropositive that it can be used by various bacteria as a terminal electron acceptor. Some of the most commonly used electron donors for fumarate reductase are $H_2$, formate, glycerol-3-phosphate, and lactate. In terms of enzyme structure and activity, considerable similarity occurs between fumarate reductase and succinate reductase. However, these two enzymes can be distinguished by kinetic parameters with succinate dehydrogenase having a higher affinity for succinate than for fumarate while fumarate reductase has a higher affinity for fumarate than for succinate. An additional distinction between these two enzymes is attributed to genetics, with fumarate reductase being encoded on *frd* genes while succinate dehydrogenase is found on the *sdh* genes. It is considered that the ancestral gene system was for fumarate reductase.

### 9.2.6.3. TMANO Reductase

Widely distributed throughout bacteria is the capability of using trimethylamine *N*-oxide (TMANO) as an electron acceptor. TMANO is produced from the decomposition of fish and shellfish. In many organisms, TMANO reductase is constitutive; however, in some bacteria it may be inducible. TMANO reductase is not highly specific in that it may also reduce hydroxylamine and certain *N*-oxides of purines and pyrimidines. Reduction of TMANO is as follows with the formation of trimethylamine:

$$(CH_3)_3NO + 2\,H^+ + 2\,e^- \rightarrow (CH_3)_3N + H_2O$$

With a midpoint redox potential of +130 mV, TMANO is appropriate as a terminal electron acceptor. Commonly, NADH and formate serve as electron donors for TMANO reductase with cytochromes *b* and *c* as electron carriers.

### 9.2.6.4. DMSO Reductase

The reduction of dimethyl sulfoxide (DMSO) by several bacteria is associated with the respiratory system. It is unclear if reduction of DMSO is mediated by a distinct enzyme because TMANO reductase has been reported to also display DMSO reductase activity. The conversion of DMSO to dimethyl sulfide occurs according to the following reaction:

$$(CH_3)_2SO + 2 H^+ + 2 e^- \rightarrow (CH_3)_2S + H_2O$$

With a midpoint potential of $+160$ mV, DMSO reductase is well positioned to accept electrons from the oxidation of $H_2$, NADH, formate, and glycerol-3-phosphate. Although the mechanism is not established, it is assumed that 3 $H^+$ are translocated across the plasma membrane for each 2 $e^-$ directed to DMSO.

### 9.2.6.5. Respiratory Nitrate Reductase

Respiratory nitrate reductase is bound to the plasma membrane and displays a subunit structure of $A_2B_2C_4$. Subunit A is 145 kDa and contains a Mo-cofactor and a nonheme iron cluster. Subunit B is 60 kDa while subunit C is 20 kDa and contains a heme B moiety. In *E. coli*, the nitrate reductase consists of two heme groups that have a redox potential of $+20$ and $+120$ mV but both absorb at 556 nm. The nitrate reductase requires a quinone, either ubiquinone or menaquinone, to accept electrons from the oxidation of $H_2$, NADH, formate, lactate, or glycerol-3-phosphate. With electrons from $H_2$ or NADH, 4 $H^+$ are pumped out for each molecule of $NO_3^-$ reduced to $NO_2^-$. However, with lactate or glycerol-3-P coupled to nitrate reduction, only 2 $H^+$ are expelled with the reduction of one molecule of nitrate.

### 9.2.6.6. Respiratory Nitrite Reductase

Some anaerobic bacteria contain a membrane-bound nitrite reductase and nitrite is the final electron acceptor for growth with formate of molecular hydrogen as the electron donor. The reduction reaction is as follows:

$$NO_2^- + 7 H^+ + 6 e^- \rightarrow NH_3 + 2 H_2O$$

The respiratory nitrite reductase has been purified from *Desulfovibrio desulfuricans*, *Vibrio fischeri*, *E. coli*, and *Wolinella succinogenes*. The nitrite reductase is a hexaheme protein that has a molecular weight of about 66 kDa. Considerable specificity is shown with the enzyme in that nitrite reductase will not react with nitrate, selenite, or sulfite. Examples of respiratory systems for *D. desulfuricans* and *Wolinella succinogenes* with nitrite as the electron acceptor are given in Figure 9.14 and Figure 9.15, respectively.

### 9.2.6.7. Respiratory Sulfate and Sulfur Reduction

Sulfate is used as the terminal electron acceptor for the physiological group known as sulfate-reducing or sulfate-respiring organisms. The coupling of sulfate reduction to respiratory electron transport is called dissimilatory sulfate reduction and is in contrast to the assimilatory sulfate reduction found in biosynthesis of sulfur amino acids from inorganic sulfate. This property of dissimilatory sulfate reduction is found in the archaeon *Archaeoglobus* and numerous genera of bacteria. The sulfate-reducing bacteria carry the prefix "*Desulfo-*" in their generic names and the most studied organisms are members of the genus *Desulfovibrio*. The dissimilatory pathway for sulfate reduction is unique in that ATP is required for activation of sulfate

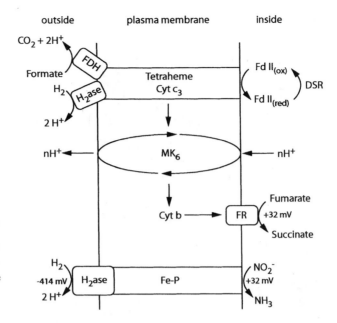

outside          plasma membrane          inside

Figure 9.14. Electron transport of *Desulfovibrio gigas* with a focus on the plasma membrane. FDH, Formate dehydrogenase; $H_2$ase, hydrogenase; Fe-P, nonheme iron containing protein; FR, fumarate reductase; Fd, ferredoxin; DSR, dissimilatory sulfate reductase; $MK_6$, menaquinone-6. (From Fauque et al., 1991.[9] Used with permission.)

before reduction can occur. These two steps of sulfate activation and reduction are shown in the following reaction:

$$SO_4^{2-} + ATP \rightarrow APS + PP_i$$
$$APS + 2\ H^+ + 2\ e^- \rightarrow HS^- + H^+ + AMP$$

APS and $PP_i$ refer to adenosine phosphosulfate and inorganic pyrophosphate, respectively. The enzymology of the dissimilatory sulfate reduction is discussed fully in Chapter 13. The enzymes of dissimilatory sulfate reduction are cytoplasmic and APS reductase interfaces with the plasma membrane to receive electrons from the oxidation of formate or $H_2$. A proposed scheme for the distribution of the electron transport components in the membrane of *Desulfovibrio gigas* is given in Figure 9.14. The sulfate-reducing bacteria have a highly complex electron transport system with thiosulfate and sulfite also acting as electron acceptors. While the schemes for the reduction of the sulfur oxy-anions are not established it is known that electron flow through the membrane will charge the membrane with nonuniform distribution of protons and ATP can be obtained by oxidative phosphorylation processes.

Elemental sulfur can serve as the terminal electron acceptor for various sulfate-reducing bacteria and anaerobic bacteria with the production of sulfide. The reaction of sulfur reduction is as follows:

$$S^0 + 2\ H^+ + 2\ e^- \rightarrow HS^- + H^+$$

In sulfate-reducing bacteria, the protein with sulfur reducing activity is a tetraheme *c*-type cytochrome that is 15 to 16 kDa. Sulfur-reducing cytochromes have been isolated

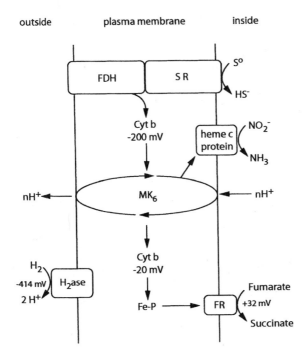

Figure 9.15. Electron transport in the plasma membrane of *Wolinella succinogenes*. FR, fumarate reductase; $MK_6$, menaquinone-6; SR, sulfur reductase; FDH, formate dehydrogenase. (From Fauque et al., 1991.[9] Used with permission.)

from *Desulfovibrio gigas*, *Desulfovibrio africanus*, *Desulfovibrio salexigens*, and *Desulfomicrobium baculatum*. Many cytochromes in the sulfate reducers do not function in transferring of electrons to elemental sulfur and these would include the monoheme $c_{553}$ from *Desulfovibrio vulgaris* and from *Desulfomicrobium baculatum*, the triheme $c_3$ from *Desulfomicrobium acetoxidans*, the tetraheme from $c_3$ from *Desulfovibrio vulgaris*, and the octaheme $c_3$ from *Desulfovibrio gigas* or from *Desulfovibrio vulgaris*. Sulfur is reduced in *Campylobacter* sp. and in several archaea (*Thermoproteus*, *Thermodiscus*, *Pyrodictium*, *Desulfurococcus*, *Thermofilum*, and *Thermococcus*) by $H_2$; however, the role of cytochrome in these reactions remains unresolved.

Other anaerobes use polysulfide as the electron acceptor and not elemental sulfur. Polysulfide is produced from elemental sulfur reacting with sulfide according to the following reaction:

$$nS^0 + HS^- \rightarrow S_{n+1}^{2-} + H^+$$

The polysulfide, $S_{n+1}^{2-}$ is the substrate for the sulfur:oxidoreductase which has been isolated from *Wolinella succinogenes*. The positioning of this electron receptor with respect to other activities in *Wolinella succinogenes* is given in Figure 9.15. In addition, this sulfur:oxidoreductase has been demonstrated in *Sulfurospirillum deleyianum*, a bacterium previously described as a free-living *Campylobacter*.

### 9.2.6.8. Other Electron Acceptors

Based on growth of microorganisms, several different compounds can serve as terminal electron acceptors and cells can obtain energy from the respiratory process.

Table 9.13. Physiological groups of prokaryotes based on energy yielding reactions with different electron acceptors for growth.

| Physiological group | Reactions | Energy yield kJ/ reaction |
|---|---|---|
| Denitrification | glucose $+ 8$ $NO_2^- + 8$ $H^+ \rightarrow 6$ $CO_2 + 4$ $N_2 + 10$ $H_2O$ | $-3144.0$ |
| Ammonia producers | $3$ $H_2 + NO_2^- + 2$ $H^+ \rightarrow NH_4^+ + 2$ $H_2O$ | $-436.8$ |
| Acetogens | $4$ $H_2 + 2$ $CO_2 \rightarrow$ Acetic acid $+ 2$ $H_2O$ | $-104.8$ |
| Fumarate reducers | $H_2 +$ Fumarate $\rightarrow$ Succinate | $-86.2$ |
| Methanogens | $4$ $H_2 + CO_2 \rightarrow CH_4 + 2$ $H_2O$ | $-136.0$ |
| Iron reducers | Acetate $+ 8$ $Fe^{3+} + 4$ $H_2O \rightarrow 2$ $HCO_3^- + 8$ $Fe^{2+} + 9$ $H^+$ | $-233.0$ |
| Sulfate reducers | $4$ $H_2 + SO_4^{2-} \rightarrow HS^- + 3$ $H_2O + OH^-$ | $-152.0$ |
| Sulfur reducers | $H_2 + S^0 \rightarrow HS^- + H^+$ | $-29.0$ |
| Arsenate reducers | Lactate $+ 2$ $HAsO_4^{2-} \rightarrow$ Acetate $+ H_2AsO_3^- + HCO_3^-$ | $-40.3$ |

Although many different compounds can be reduced by prokaryotes, not all of them support microbial growth. To be confident that bacteria and archaea can indeed grow with a specific electron acceptor it is necessary to measure growth after the organism has been transferred for three consecutive passages in that medium. A list of electron acceptors is given in Table 9.13, and it has been established that bacteria can obtain energy from these dissimilatory reduction processes. In many cases the enzymology of the specific reactions has not been established and coupling sites for proton pumping have not been established. Additional information concerning reduction associated with nitrogen and sulfur compounds is discussed in Chapter 13 and metal reduction is presented in Chapter 14.

## 9.3. Organization of the Respiratory System

Metabolic capabilities of bacteria are readily displayed by examination of their respiratory systems. Bacteria generally have several electron donors with numerous electron acceptors including oxygen at different concentrations. Inorganic compounds are used as electron donors by chemolithotrophic bacteria and these reactions are commonly catalyzed by oxidoreductases. Examples of oxidoreductase reactions that interface with the electron transport system are given in Table 9.2. Although many different compounds serve as electron donors, the best characterized are associated with the carbon, nitrogen, and sulfur cycles. Not only is the electron movement optimized to accommodate the different electron acceptors but it is also directed in a pattern through proton pumps. Electron transport may be linear but frequently it is branched to accommodate the different electron acceptors.

The electron transport chain of *E. coli* is a good example of interfacing with numerous electron donors and electron acceptors. As indicated in Figure 9.16, distinct dehydrogenases are used to oxidize different substrates under aerobic or anaerobic conditions and it is interesting to note that α-glycerol phosphate requires a different dehydrogenase based on the level of oxygen present. A distinction is made between lipid carriers with ubiquinone used under aerobic culture and menaquinone, which is

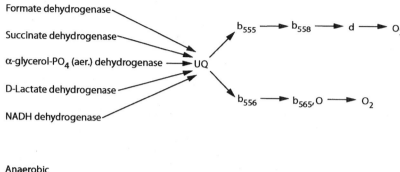

**Aerobic**

**Anaerobic**

Figure 9.16. Electron transport of *Escherichia coli* under aerobic or anaerobic growth. Menaquinone (MQ) is found under anaerobic culture while ubiquinone is under aerobic culture. Membrane-bound cytochrome *c* is absent in aerobic culture and in anaerobic electron transport, *b*-type cytochromes may link MQ to fumarate reductase and MQ to nitrate reductase. (From Lin and Kuritzkes, 1987[6]; Poole and Ingledew, 1987.[26])

used under anaerobic conditions. Under aerobic culture, the low-affinity cytochrome *o* is employed; however, under microphilic culture, cytochrome *d* is produced because it has a high affinity for molecular oxygen.

Electron transport chains have been established for many bacteria and a few of the schemes are presented in Figure 9.17. Linear and abbreviated electron transport schemes are found in *Acidithiobacillus ferroxidans*, *Methylophilus methylotrophus*, and *Nitrobacter agilis* with these examples of electron donors. *Paracoccus denitrificans* uses nitrate in one chain but a second electron transport branch uses nitrite and a further branching process uses nitric oxide and nitrous oxide as alternate electron acceptors branching from the cytochrome *c* chain. *Azotobacter vinelandii* has an efficient means of utilizing oxygen as the electron acceptor with an oxygen reduced by cytochrome *o* and in microaerophilic conditions uses cytochrome *d*. *Desulfovibrio desulfuricans* uses sulfate as the electron sink with distinct enzymology for the reduction of adenosine phosphosulfate (APS) and sulfite. *Mycobacterium phlei* has a commitment to using molecular oxygen with the numerous organic compounds and displays a branching at the electron donor region.

**a) Abbreviated respiratory chains**

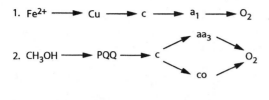

1. $Fe^{2+} \longrightarrow Cu \longrightarrow c \longrightarrow a_1 \longrightarrow O_2$

2. $CH_3OH \longrightarrow PQQ \longrightarrow c$ → $aa_3$ → $O_2$, → co →

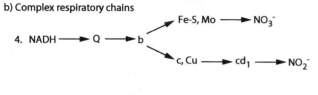

3. $NO_2^- \longrightarrow a_1 \, Mo \longrightarrow c \longrightarrow aa_3 \longrightarrow O_2$

**b) Complex respiratory chains**

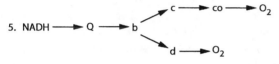

4. $NADH \longrightarrow Q \longrightarrow b$ → Fe-S, Mo → $NO_3^-$ ; → c, Cu → $cd_1$ → $NO_2^-$

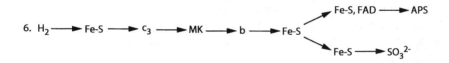

5. $NADH \longrightarrow Q \longrightarrow b$ → c → co → $O_2$ ; → d → $O_2$

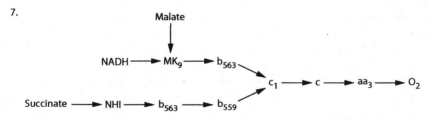

6. $H_2 \longrightarrow Fe\text{-}S \longrightarrow c_3 \longrightarrow MK \longrightarrow b \longrightarrow Fe\text{-}S$ → Fe-S, FAD → APS ; → Fe-S → $SO_3^{2-}$

7.

Malate

$NADH \longrightarrow MK_9 \longrightarrow b_{563}$

$Succinate \longrightarrow NHI \longrightarrow b_{563} \longrightarrow b_{559}$

$\longrightarrow c_1 \longrightarrow c \longrightarrow aa_3 \longrightarrow O_2$

Figure 9.17. Examples of electron transport in various bacteria. The electron donor is on the left of the reaction and the electron acceptor is on the right. (1) *Acidithiobacillus ferroxidans*, an iron oxidizer; (2) *Methylophilus methylotrophus*, a methylotroph; (3) *Nitrobacter agilis*, a nitrifying bacterium; (4) *Paracoccus denitrificans*, a denitrifying bacterium; (5) *Azotobacter vinelandii*, a soil bacterium; (6) *Desulfovibrio desulfuricans*, a sulfate-reducing bacterium; (7) *Mycobacterium phlei*, a heterotrophic bacterium. (From Jones, 1988.[21])

## 9.3.1. Generation of Proton-Motive Force[1-4]

The term *respiration* is applied to the transport of electrons from a donor through the cytochromes in the membrane to an electron acceptor. In bacterial respiration, as in mitochondrial respiration, production of ATP is coupled to the movement of electrons in the membrane which is referred to as oxidative phosphorylation. Peter Mitchell, the 1979 Nobel Prize laureate, provided insight into the chemiosmotic mechanism for oxidative phosphorylation. According to Mitchell, the driving force for the chemiosmotic mechanism was proton-motive force (pmf) which is defined using the Nernst equation where $\Delta\bar{\mu}^+_H$ is the electrochemical potential (cal mol$^{-1}$ or J mol$^{-1}$) resulting from the disproportional distribution of $H^+$ across the plasma membrane:

$$\text{pmf} = \Delta p = \Delta\bar{\mu}^+_H = \Delta\psi - Z\Delta pH$$

The components are defined as follows: $\Delta\psi$ is the difference in charge on the membrane expressed in mV and $\Delta pH$ is the difference in pH units across the plasma membrane. $Z = 2.3\ RT/nF$ where $R$ is the gas constant (1.9869 cal mol$^{-1}$°C$^{-1}$ or 8.314 J mol$^{-1}$°C$^{-1}$), $T$ is the absolute temperature (Kelvin), $n$ is the charge on the transported species which is the proton $= 1$, and $F$ is Faraday's constant (23.06 cal mV$^{-1}$ eq$^{-1}$ or 96.519 J mV$^{-1}$ eq$^{-1}$). At 25°C, $Z$ is 59. The basic premise is that protons are pumped outward across the plasma membrane and become localized at the outer face of the plasma membrane. Because the establishment of the chemiosmotic force is attributed to the spatial separation of hydroxyl ions from protons, both electric and chemical activities are involved. The coupling of ATP synthesis to electron flow provides the cell with energy for growth, and this chapter provides information about the contribution of chemiosmotic activities to bioenergetics of prokaryotes.

To envision how respiration can account for making the external region of the plasma membrane positive and acidic as a result of protons pumped out of the cell, it is best to examine the spatial organization of the electron transport components in the plasma membrane. There are certain features of enzymes in membranes that do not vary. For example, NADH dehydrogenase is always on the inner side of the plasma membrane and if aerobic respiration is employed, the passage of electrons from the cytochrome to molecular oxygen also occurs on the inner side of the plasma membrane. The distribution of electron transport components in membranes for aerobic bacteria is given in Figure 9.17.

For anaerobic respiration, the electron transport components must also be arranged in the membrane to promote the separation of protons from electrons. As with aerobic respiration the enzymes for electron donor and for electron acceptor may be located on the inner side of the plasma membrane. However, in anaerobes electron donor enzymes often may be located on the outer side of the membrane. With the oxidation of $H_2$ and formate at the outer face of the plasma membrane, the electron is shunted into the membrane but the protons may remain on the outside of the membrane. In this case, a proton gradient across the membrane is produced as a result of oxidizing the electron donor without requiring protons to be pumped out acros the membrane.

Some examples giving the spatial distribution of anaerobic electron transport are given in Figure 9.3 and Figures 9.13 to 9.16.

## 9.3.2. Proton Transduction Sites[21,29]

The role of respiration is to spatially separate electrons from protons. Electrons move within the membrane by flavins, nonheme iron proteins, quinones, and cytochromes but protons are transported across the plasma membrane. The number of sites for $H^+$ translocation may vary with the organism and the number of protons translocated may vary between the sites. For this reason, when electron transport schemes are presented, the quantity of protons per translocation site are generally listed as $nH^+$. Protons translocated outward across the membrane are from two sources. In one instance, protons are from the electron donor and the other possibility is that protons come from the ionization of water in the cytoplasm of the cell. Whereas the proton is expelled across the plasma membrane, the hydroxyl ion from ionization of water is not capable of translocating across the membrane but remains in the cytoplasm and generates a slightly alkaline environment inside the cell.

The $H^+/2\ e^-$ values reflect the number of protons pumped out from the cell for each pair of electrons moving through the membrane-bound electron transport system. The $H^+/2\ e^-$ value ranges from 4 to 8 depending on the bacterial species, the electron donor and electron acceptor used, and the nature of the test system. There are four $H^+$ pumping sites in the bacterial respiratory chain and each has unique characteristics. Proton pumping is at the transhydrogenase site, the NADH dehydrogenase site, the region of the quinone, and with specific cytochrome oxidases. As summarized in Table 9.14, the number of protons expelled across the membrane for each pair of electrons directed through the coupling site is frequently 2 but there are some exceptions. The specific $H^+$ pumping activities are as follows: $NADPH \rightarrow NAD^+$, site 0; $NADH \rightarrow$ quinone, site 1; quinone $\rightarrow$ cytochrome $c$, site 2A; quinone $\rightarrow O_2$, site 2B; cytochrome $c \rightarrow O_2$ by an oxidase not pumping protons, site 3A; and cytochrome $c \rightarrow O_2$ by an oxidase that pumps protons, site 3B. Because $2\ e^-$ are required for each atom of oxygen reduced, the number of protons translocated can be expressed as $H^+/O$ as well as $H^+/2\ e^-$. Obviously, the $H^+/O$ value is useful only when aerobic res-

Table 9.14. Sites in prokaryotes for proton pumping coupled to respiration.[a]

| Sites | Reactions | Number of $H^+$ pumped/$2\ e^-$ |
|-------|-----------|--------------------------------|
| 0 | Transhydrogenase | 2 |
| 1 | NADH $\rightarrow$ quinone | 2 |
| 2A | Quinone $\rightarrow$ cytochrome $c$ | 4 |
| 2B | Quinone $\rightarrow O_2$ | 2 |
| 3A | Cytochrome $c \rightarrow O_2$ | 0 |
| 3B | Cytochrome $c \rightarrow O_2$ (via $aa_3$ or $caa_3$) | 2 |

[a]From Jones, 1988.[21]

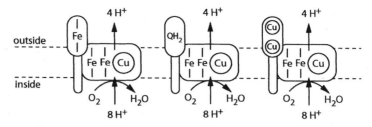

Figure 9.18. Model of heme-copper oxidase systems that serve as proton pumps. The model on the left is the $cbb_3$-type cytochrome $c$ oxidase, the model in the center is the $bo_3$-type quinol oxidase, and the model at the right is the $aa_3$-type cytochrome $c$ oxidase. Heme groups are as –Fe-, the quinone moiety is $QH_2$, and each copper atom is Cu. (From Richardson, 2000.[27] Used with permission.)

piration is discussed, and Figure 9.18 summarizes some of the variations in $H^+$ pumping sites.

Additional research is needed to provide a basis for generalizations concerning the $H^+/2\ e^-$ values for anaerobic bacterial respiration. Coupling of site 0, site 1, and site 2A would be expected to be present in some anaerobic respiratory chains; however, it may also be that some electron acceptor activities may also pump protons by a system analogous to site 3B. Because of the variations of electron donors and acceptors, generalizations of proton pumping sites are expected to vary with each physiological type of organism.

### 9.3.2.1.  Site 0

The transhydrogenase catalyzes the exchange of electrons and protons between nicotinamide nucleotides according to the following reaction:

$$NADPH + NAD^+ \leftrightarrow NADP^+ + NADH$$

The individual reactions are similar in that the redox potential of $NADP^+/NADPH$ is $-327$ mV and for $NAD^+/NADH$ is $-320$ mV. Some bacteria have a flavin-bound soluble transhydrogenase that through allosteric regulation transfers reducing potential for biosynthetic activities. However, the membrane-bound transhydrogenase is responsible for producing a proton-motive force by pumping protons with the reduction of $NADP^+$ by NADH. The transhydrogenase transfer of electrons and protons from NADPH to $NAD^+$ occurs only when the plasma membrane has a high proton motive force. This energy coupling activity of nucleotide transhydrogenase is referred to as site 0 and has been demonstrated in many chemoorganotrophs including *E. coli* and *Paracoccus denitrificans* as well as in the phototrophic *Rhodospirillum rubrum*.

### 9.3.2.2.  Site 1

A major component in cell respiration is the NADH:quinone oxidoreductase system which can be isolated as a unit and has been designated as complex I (see Chapter

2). NADH dehydrogenase is a small enzyme, 47.3 kDa in *E. coli*, and it is generally believed that this polypeptide is associated with Fe-S containing proteins which are used for proton pumping. Site 1 is not as important in bacteria as in mitochondrial systems because bacteria have numerous dehydrogenases that enter the electron transport chain at points other than at NADH while alternate dehydrogenase systems are not prevalent in mitochondria. It is not just the movement of protons and electrons to the quinone that is important because proton pumping is not observed with lactate dehydrogenase, glycerol-3-phosphate dehydrogenase, or with succinate dehydrogenase. It is to be recalled that succinate dehydrogenase isolates as complex II in mitochondrial systems. It should not be assumed that a proton pumping action occurs because a bacterium has an NADH dehydrogenase because other components are needed for the pumping activity. For example, Fe-S proteins are required for site 1 pumping of protons and the production of iron-sulfur centers is restricted if cells are grown under sulfur limitation.

### 9.3.2.3. Site 2

At site 2, electrons are moved from the quinone to an acceptor that is either cytochrome *c* (site 2A) or molecular oxygen (site 2B). The 2A site employs the Q-cycle in addition to the cytochrome $bc_1$ complex for proton pumping. In the Q-cycle, the quinone carries two protons plus two electrons with a proton released from the outer side of the plasma membrane in a sequential manner with the involvement of the semiquinone, which is a half-reduced quinone. With the final release of the second proton and the second electron from the quinone, the quinone is completely reduced. A model for the Q-cycle pumping of electrons/protons is given in Figure 9.19. Two additional protons are pumped from the cell with concomitant electron passage through the cytochrome $bc_1$ complex. These components accounting for $H^+$ pumping

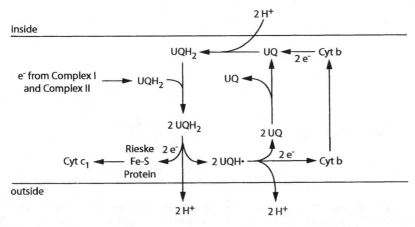

Figure 9.19. The proposed Q-cycle in bacteria with proton pumping activity. UQ, Oxidized ubiquinone; UQH, half-reduced ubiquinone; $UQH_2$, fully reduced ubiquinone.

at site 2A can be isolated as a unit and referred to as complex II. With two protons pumped out by the Q-cycle and two protons pumped by the cytochrome $bc_1$ complex, site 2A pumps out four protons per two electrons. The second pump associated with site 2 is designated as site 2B in which proton pumping is associated with electron transfer directly from reduced quinone to $O_2$ through cytochrome $o$, cytochrome $d$, or cytochrome $aa_3$. The configurations of the quinone–cytochrome $b$–cytochrome $o$ (or cytochrome $d$ or cytochrome $aa_3$) are such that two electrons are pumped for each pair of electrons moved.

### 9.3.2.4. Site 3

Cytochrome oxidase systems provide the potential for proton pumping; however, not all reduced cytochromes that react with $O_2$ will charge the membrane. Site 3A characterizes the electron transfer from cytochrome $c$ to $O_2$ by cytochrome $co$ or by cytochrome $aa_3$ without proton pumping. Site 3B refers to the pumping of protons with electrons moving through cytochrome $aa_3$ to $O_2$. Whereas cytochrome $aa_3$ transfers electrons to $O_2$ in the mitochondrial system, referred to as complex IV, some of these oxidases will not pump protons in prokaryotes. There have been many attempts to isolate the bacterial cytochrome oxidase (cytochrome $aa_3$) and place it in proteoliposomes to assess proton pumping. Thus far, the only systems that have been demonstrated to pump protons at site 3B are *Bacillus caldolyticus* and *Bacillus stearothermophilus*. A representation of the proton pumping by a composite aerobic electron transport chain is shown in Figure 9.20.

### 9.3.2.5. Sites in Anaerobic Respiration

The presence of proton pumping in anaerobic chemolithotrophic/chemoorganotrophic organisms is based primarily on the nature of the electron donor and the electron

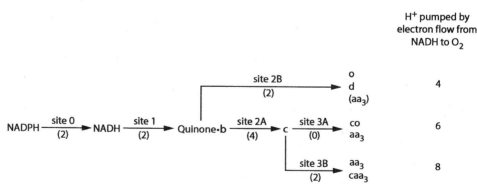

Figure 9.20. Sites for proton pumping coupled to electron flow in bacteria. The various oxidases are diagrammatically presented. The numbers in parentheses indicate $H^+$ pumped for each pair of electrons. Site "0" has not been examined in chemolithotrophic bacteria. (From Jones, 1988.[21])

acceptor used to support growth. In bacteria that grow under both aerobic and anaerobic conditions by respiratory processes, site 0, site 1, and site 2 may function in anaerobic pumping. In addition, it may be that there are certain sites for proton pumping that are unique to anaerobes. Achem Kröger and colleagues have provided one of the best characterized electron transport systems of anaerobes through their studies of fumarate and $S^0$ (polysulfide) reduction by *Wolinella succinogenes*. It has been found that *W. succinogenes* will grow with the following electron donor $\rightarrow$ electron acceptor systems: $H_2 \rightarrow$ fumarate, formate $\rightarrow$ fumarate, $HS^- \rightarrow$ fumarate, formate $\rightarrow S^0$, and $H_2 \rightarrow S^0$. In studies with artificial proteoliposomes containing hydrogenase, ATP synthase, and fumarate reductase from *W. succinogenes*, it was determined that menaquinone was needed for electron transport and that ATP synthesis was concomitant with hydrogen oxidation, thereby suggesting that protons were transported across the membrane. In proteoliposomes constructed with hydrogenase and $S^0$ reductase from *W. succinogenes*, it was determined that menaquinones were not required for electrons to move from $H_2$ to $S^0$ and some charging of the membrane followed electron flow. Many chemolithotrophic bacteria grow with molecular hydrogen as the electron donor and membrane-bound uptake hydrogenases may serve as proton pumps. It is apparent that more research is required before generalizations can be made about proton pumping sites in anaerobes.

## 9.4. Demonstration of Phosphorylation Coupled to Electron Flow[1-4]

Aerobic oxidative phosphorylation was demonstrated in cell-free extracts from several different strains of bacteria prior to the development of the chemiosmotic process of ATP synthesis. Using *Mycobacterium phlei*, Brodie and Gray in 1955 provided biochemical evidence for oxidative phosphorylation and soon after oxidative phosphorylation was demonstrated in numerous Gram-positive and Gram-negative bacteria. In 1966, H.D. Peck, Jr. demonstrated oxidative phosphorylation in the anaerobic bacterium *Desulfovibrio gigas* by following the production of ATP in cell-free extracts with $H_2$ as the electron donor and sulfite as the electron acceptor. In 1972, L. Barton, J. LeGall, and H. Peck, Jr. established the utility of fumarate reduction in anaerobic bacteria by demonstrating anaerobic oxidative phosphorylation coupled to electron flow from $H_2$ to fumarate in membrane fractions of *D. gigas*. Subsequently, several anaerobic bacterial species have been demonstrated to couple phosphorylation to anaerobic electron flow. Even though the electron flow may vary with the various bacterial species, it is apparent that the mechanism of ATP synthesis coupled to electron flow is the same and the unity of biochemistry/physiology is again observed.

The production of ATP coupled to electron transport is oxidative phosphorylation. The mechanism of this ATP synthesis from ADP and $P_i$ is the movement of protons inward across the plasma membrane through the $F_1F_0$ unit that functions as an ATP synthase. The number of protons needed for this response is of the magnitude of 2 or 3 for the formation of one ATP. Because of our lack of knowledge for this step, it is customary to indicate that for the synthesis of one ATP there is the requirement for "*n*" protons. A proton-motive charge of about $-210$ mV is required to enable proton entry to drive ATP synthesis.

## 9.4.1. F-Type ATPase[30,31]

The plasma membrane of bacteria contains the $H^+$-transporting system consisting of two structurally and functionally distinct components commonly referred to as $BF_1$ and $BF_0$, which are structurally and functionally similar to the $F_1$ and the $F_0$ of mitochondria. The $BF_1$ $BF_0$ is a homolog of the $F_1F_0$ unit, where "B" designates bacteria. In many published reports and throughout this book, the "B" designation is not provided; however, the origin of the ATPase is implied. The $F_1F_0$ unit accounts for ATP synthase coupled to $H^+$ translocation inward across the plasma membrane and as an ATPase, ATP is hydrolyzed to ADP plus inorganic phosphate with $H^+$ expelled through the plasma membrane. The $F_1$ moiety resides on the surface of the inner side of the plasma membrane and when the $F_1$ is removed it displays ATPase activity. The $F_0$ structure traverses the plasma membrane, providing the channel for $H^+$ translocation. Inhibitors such as dicyclohexyl-carbodiimide (DCCD) bind to one of the subunits and prevent $H^+$ pumping.

The F-type ATPase structure of *E. coli* is representative of bacteria and is presented in Figure 9.21. The $F_1$ is approximately 380 kDa and consists of five subunits with a stoichiometry of $\alpha_3\beta_3\gamma\delta\epsilon$. The $F_0$ structure is approximately 150 kDa and has three subunits in an $a_1b_2c_{10-12}$ ratio. In *E. coli*, the production of $\alpha$, $\beta$, $\gamma$, $\delta$, and $\epsilon$ is from genes A, D, G, H, and C, respectively, on the *unc* operon and the three subunits of

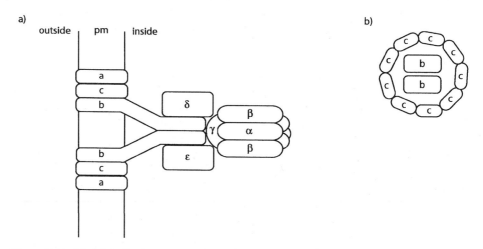

Figure 9.21. Model indicating structure of ATP synthase/ATPase of *Escherichia coli*. (**a**) The arrangement of the $\alpha$, $\beta$, $\gamma$, $\delta$, and $\epsilon$ proteins that make up the $F_1$ subunit and the a, b, and c proteins that make up the $F_0$ subunit. (**b**) The clockwise rotation of the "b" protein in the $F_0$ subunit results in the synthesis of ATP. With one rotation of the b subunit one ATP is produced and 3 $H^+$ enter the $F_0$. (From Jones, 1988.[21] Used with permission.)

$F_0$ are from *uncB*, *uncF*, and *uncE*. A comparison of F-type and V-type ATPases is given in Table 9.15. The analysis of the $V_1$-ATPase in archaea is not at the level of $F_1$-ATPase in bacteria and the values of the $V_1$ subunits in Table 9.15 are based on mammalian vacuolar ATPase. There is considerable similarity in the $V_1V_0$-ATPase subunits found in mammals, plants, *Neurospora*, and yeast. Because the $V_1$ unit is 500 to 600 kDa and the $V_0$ unit is 250 to 300 kDa, the V-type ATPase unit is significantly larger than the F-type ATPase.

There are two consensus sequences for ATP binding that are found in the $\alpha$- and $\beta$-subunits of the $F_1$ portion of the F-type ATPase. One motif is either [K/R HyGXGXXGK] or [K/R HyGXXG XK], where [Hy] is a hydrophobic residue and the number of [Hy] residues is either 4 or 6. The second binding site for ATP is [K/R XGXHyHyHyHyHyD] where X is either two or three residues. Other transport components containing these ATP-binding motifs, also known as Walker A and B motifs, include the V-type ATPase and the subunits of the ABC type transporters.

Table 9.15. Characteristics of F-type and V-type ATPases.[a]

| | F-type (*E. coli*) Subunits: name/number/size | | | V-type[b] Subunits: name/number/size | | |
|---|---|---|---|---|---|---|
| Peripheral complex re- | $F_1$: α | 2 | 55 kDa | $V_1$: A | 3 | 72 kDa |
| leased as water- | β | 3 | 50 kDa | B | 3 | 57 kDa |
| soluble ATPase | γ | 1 | 31 kDa | C | 1 | 44 kDa |
| | δ | 1 | 19 kDa[e] | D | 1 | 34 kDa |
| | ε | 1 | 15 kDa | E | 1 | 33 kDa |
| | | | | F | 1 | 13 kDa |
| | | | | G | 1 | 14 kDa |
| Hydrophilic segment in | $F_0$: a | 1 | 30 kDa | $V_0$: a | 1 | 107 kDa |
| the plasma membrane | b | 2 | 17 kDa | d | 1 | 38 kDa |
| | c | 10–12[c] | 8 kDa[d] | c | 6 | 16 kDa |
| | | | | cc" | 1 | 19 kDa |
| Inhibitors of proton | DCCD | | | DCCD | | |
| transport | Oligomycin | | | Nitrate | | |
| | Azide | | | 7-Chloro-4-nitrobenzo-2-oxa-1,3- | | |
| | 7-Chloro-4-nitrobenzo-2-oxa-1,3- | | | diazole | | |
| | diazole | | | *N*-ethylmaleimide | | |
| | Tributylin chloride | | | | | |

[a]From Nelson, 1992[17]; Stevens and Forgac, 1997.[17]

[b]Values are for mammalian vacuolar ATPase an similar size proteins are found for plants, *Neurospora*, and yeast vacuolar ATPase.

[c]*Escherichia coli* = 10 to 12 subunits but 4 subunits in mitochondria.

[d]Subunit binds DCCD (*N,N*'-dicyclohexylcarbiimide).

[e]Subunit for binding oligomycin in mitochondria but not in bacteria. Bacteria are insensitive to oligomycin.

## 9.4.2. V-type ATPase[30,31]

The principal type of $H^+$-ATP synthase in prokaryotes is either the F-type ATPase or the "V-type"-ATPase. The V-type ATPase is found in archaea and in the vesicles of fungal, plant, and animal cells. In eukaryotic systems, the V-type ATPase is primarily a proton pump that is energized by ATP hydrolysis. Like the F-type ATPase, the V-type ATPase consists of a $V_1/V_0$ subunit composed of eight proteins that are contained on a single operon. As indicated in Table 9.15, F-type ATPase is inhibited by both DCCD and a substituted nitrobenzene but are distinguished by sensitivities to azide, oligomycin, and nitrate. While the V-type ATPase is commonly found in archaea, it is also found in some bacteria including *Deinococcus*, *Thermus*, *Enterococcus*, spirochetes, and chlamydiae. A review of the genome analysis of bacteria has revealed that while the F-type ATPase is the predominant type some bacteria also contain genes for the V-type. Whether or not these bacteria can use both the V-type and the F-type in physiological activities in a single cell remains to be established. Many scientists consider that the distribution of the V-type ATPase in bacteria may have resulted from natural horizontal gene transfer. The physiological activity of V-type ATPases in plant vacuoles is the transport of protons across the vacuole membrane with energy derived from ATP hydrolysis. Some investigators suggest that the ATPases from archaea should be identified as a type distinct from that found in vacuoles, and future studies should provide useful information to clarify differences between vacuolar and archaeal ATPases.

## 9.4.3. ATP Synthesis and Hydrolysis[21]

The function of the F-ATPase/ATP synthase complex has been examined in *E. coli* and *Bacillus stearothermophilus*. The complex consists of two segments in which the $F_1$ component consists of five different subunits as $\alpha_3\beta_3\gamma\delta\epsilon$ while the $F_0$ unit has three different subunits described as $ab_2c_{10-12}$. A detailed model of the ATPase/ATP synthase is given in Figure 9.21. Dicyclohexylcarbodiimide binds into the $F_0$ to prevent entry of $H^+$. While oligomycin inhibits the entry of protons through the $F_0$ of mitochondrial systems it has no effect on inhibiting the proton passage through the $F_0$ of bacteria. ADP and inorganic phosphate bind to the $F_1$ and as protons reenter the cell through $F_0$, ATP synthesis occurs.

The mechanism of ATP synthesis is attributed to the clockwise rotation of the subunits in the membrane (see Figure 9.21). This rotation driven by reentry of 3 $H^+$ produces conformational changes of subunits in the $BF_0$ and accomplishes ATP formation. In the reactions in which three protons are pumped out as a result of one ATP hydrolyzed, segments of the $F_0$ rotate in a counterclockwise direction. Biochemists are studying the interactions between subunit *b* and subunit *c* of $F_0$ to establish mechanisms for this biorotational activity.

Table 9.16. Inhibitors of the bacterial electron transport system.

| Compound | Site of inhibition |
|---|---|
| Atebrin | Flavins in electron transport |
| Amytol | Flavins in electron transport |
| Antimycin A | Between cytochrome $b \to$ cytochrome $c$ |
| Bright light at 390 nm | Ubiquinone and Menaquinone |
| HOQNO | Between cytochrome $b \to$ cytochrome $c$ (aerobes); MK/UQ $\to$ cytochrome $b$ (anaerobes) |
| Cyanide | Cytochrome oxidases |
| Azide | Cytochrome oxidases |
| Carbon monoxide | Cytochrome oxidases |

# 9.5. Assessments and Measurements[32]

## 9.5.1. Inhibitors of Electron Transport

The proton-motive force of the membrane is dependent on electron flow, and inhibition of the respiratory system results in the inability to establish a charge on the membrane or to generate a pH differential across the membrane. The proton pump is tightly coupled to the flow of electrons and not dependent on the site of inhibition. Inhibitors of electron transport in bacteria are given in Table 9.16 and, in many instances, the site of inhibition in bacteria is the same for mitochondrial systems.

## 9.5.2. Artificial Electron Acceptors and Carriers

In some instances, water-soluble redox active chemicals can be used to pick up electrons from the electron transport system. In high concentrations, the molecules added can serve as electron acceptors for enzymes or cell suspensions while in low concentrations, these molecules serve as artificial electron carriers. A list of artificial electron acceptors is given in Table 9.17. The interaction of these electron acceptors with the intact electron transport chain is nonspecific but is at a region that is more electronegative than the half-cell reaction of the artificial electron acceptor. It is to be recalled that in a chemical reaction, the electron flow is directional in that it proceeds from an electronegative system to a more electropositive reaction. Because these artificial electron carriers can divert electrons around proton pumping reactions, the artificial electron carriers at catalytic levels often can serve as uncouplers of oxidative phosphorylation.

Historically, methylene blue has been used to test for respiratory activity of cells and the time required to reduce methylene blue to the colorless form was measured. For the stoichiometric reduction of methylene blue with an electron donor, the natural electron acceptors must be excluded. Most of the reduced electron acceptors such as viologens and methylene blue will be oxidized with oxygen, making respiratory measurements difficult. Tetrazolium salts produce a water-insoluble crystal when reduced and are not oxidized by exposure to air. Because the reduced tetrazolium compounds

Table 9.17. Artificial electron carriers suitable for use with microorganisms.

| Name | $E_0'$ | Molar extinction coefficient[a] $(cm^{-1}$ liter $mole^{-1})$ |
|---|---|---|
| Methyl viologen | −446 mV | $\varepsilon_{604}{}^b = 1.0 \times 10^4$ |
| Benzyl viologen | −359 mV | $\varepsilon_{604}{}^b = 1.0 \times 10^4$ |
| Safranine T | −290 mV | — |
| Phenosafranine | −252 mV | — |
| Indigodesulfonate | −125 mV | — |
| Triphenyltetrazolium | −80 mV | $\varepsilon_{485} = 1.2 \times 10^4$ |
| Phenazine methosulfate | + 80 mV | $\varepsilon_{387} = 2.38 \times 10^4$ |
| Methylene blue | +11 mV | $\varepsilon_{668} = 6.3 \times 10^4$ |
| 2,6-Dichlorophenolindophenol | +217 mV | $\varepsilon_{600} = 2.1 \times 10^4$ |
| Potassium ferricyanide | +430 mV | $\varepsilon_{420} = 1.01 \times 10^3$ |

[a]$\Sigma\varepsilon$ = the molar extinction coefficient; the specified wavelength of adsorption is given in nm.
[b]Reduced form is colored.

are soluble in toluene, the color intensity can be determined by standard spectrometric methods. Microbial reduction can be measured in soil and other heterogeneous systems by following the amount of tetrazolium reduced under controlled conditions. In cell-free experiments, artificial electron acceptors are used to measure enzyme activity. Because not all electron acceptors have the same efficiency in accepting electrons it is necessary to survey various molecules to determine the molecule for optimum measurements. Also, reduced electron acceptors can also serve as electron donors for certain reactions. For example, hydrogenase activity can be measured by following the reduction of methyl viologen or by measuring the amount of $H_2$ evolved in the presence of reduced methyl viologen. Artificial electron acceptors are widely used in purification and characterization of dehydrogenases where the natural electron acceptor is not known.

## 9.5.3. Protonophores as Uncoupling Agents

The electron transport system in bacteria is loosely coupled to phosphorylation because with the addition of uncoupling agents to cells, ATP synthesis stops but electron movement continues. Uncoupling agents may either prevent establishing of a proton charge on the membrane or may provide an electron circuit that circumvents the proton pumps. Certain chemicals, called protonophores, function to destroy the proton localization on the exterior of the plasma membrane that was established as a result of either aerobic or anaerobic microbial respiration. Artificial electron carriers or electron acceptors can pick up the electrons from the bacterial membrane and also serve as uncouplers of oxidative phosphorylation.

Ionophores promote the transmembrane movement of specific ions. These molecules have a nonpolar periphery that enables them to traverse the membrane and readily equilibrate across it. A specific inorganic cation is carried in the center of the molecule and these caged ions are referred to as clathrates. Carriers that promote the equilibration of protons across the plasma membrane are referred to as protonophores and include 2,4-dinitrophenol (DNP), pentachlorophenol (PCP), carbonyl cyanide *m*-chlorophenyl hydrozone (CCCP), carbonyl cyanide-*p*-trifluoromethyl hydrazone (FCCP), and gramicidin. These protonophores are also effective in diminishing the proton charge on membranes in mitochondrial systems. In addition, oligomycin disrupts proton segregation across the mitochondrial membrane but it has no effect on bacterial membrane systems. Chemicals that do not allow bacterial cells to produce ATP by oxidative phosphorylation are referred to as uncoupling agents. The inhibition of bacterial growth can be accomplished with uncouplers of oxidative phosphorylation and CCCP had been used rather effectively to prevent decomposition of wooden posts and piers. However, the lack of specificity for CCCP and its slow rate of decomposition in the environment have restricted its commercial use.

## 9.5.4. Measuring Membrane Charge or ΔpH[33]

The measurement of intracellular pH is important to assess the extent of chemiosmotic activity in cells. Methods used to estimate the intracellular pH have included the following: measuring pH of lysed cells, evaluation of cell color following the uptake of bromothymol blue or other pH indicators, following the distribution of weak acids or weak bases across the plasma membrane, measuring the amount of 9-aminoacridine or acridine orange taken up by cells, or the use of high-resolution nuclear magnetic resonance to follow $^{31}P$ of phosphate ionization or $^{1}H$ of imidazole ionization. Because nonionized acids or bases can diffuse across membranes, the partitioning of weak acids or weak bases across the plasma membrane has attracted numerous investigators and the sensitivity of the assays with the weak acids or weak bases is enhanced using $^{14}C$-labeled molecules. Weak acids frequently employed for pH measurements include 5,5-dimethyl-2,4-oxazolidinedione (DMO), acetic acid, acetylsalicylic acid, salicylic acid, and benzoic acid. If the internal pH of cells or vesicles is more acid than the surrounding medium, weak bases are used and the greatest success has been with methylamine, hexylamine, and benzylamine. To calculate ΔpH, the partitioning of the weak acid ($A^-$) or weak base ($B^+$) across the membrane is calculated by using the following equation:

$$\Delta pH = \log_{10} \{[A^-] \text{ inside the cells}/[A^-] \text{ outside the cells}\}$$
$$\Delta pH = pH_{in} - pH_{out}$$

The pH values inside the cells have been measured for various prokaryotes. For neutrophilic bacteria or archaea that grow in environments between pH 6 and pH 7, the internal pH is approximately 7.5 to 8.5. For acidophilic organisms growing in environments from pH 2 to 5, the internal pH is generally within the range of 5.5 to 6.5. Alkalophilic organisms growing in environments of pH 9 to 10 generally have an internal pH of 7.5 to 8.5. Specific examples of internal pH values for various bacteria and archaea are given in Table 8.4.

Membrane charge across the membranes of microbial cells, here has been calculated using permeant ions with a positive charge. One measurement involves following of

$K^+$ accumulation inside cells. Although $K^+$ does not readily diffuse across the plasma membrane, it will quickly move through the membrane when carried by the ionophore valinomycin. To increase the sensitivity of the measurement, it would be useful to employ an isotope of $K^+$; however, as there none is available, $^{86}Rb^+$ (an analog of $K^+$) is substituted for $K^+$. Valinomycin will carry $Rb^+$ equally as well as $K^+$ but valinomycin will not complex with $Na^+$. Another permeant ion system used to calculate membrane charge involves the use of lipophilic cation such as tetraphenyl phosphonium ($TPP^+$) or triphenylmethylphosphonium ($TPMP^+$). Once the amount of charged species (i.e., $Rb^+$, $TPP^+$) is calculated inside and outside of the cell, the charge on the membrane in mV is calculated using $TPP^+$ with the following equation:

$$\text{Membrane charge} = \Delta\psi = -60 \log_{10}\left\{[TPP^+]_{\text{inside the cells}}/[TPP^+]_{\text{outside the cells}}\right\}$$

Cationic cyanine dyes may also be used to measure membrane charge. The cyanine dyes have the unique property of being fluorescent and the fluoresence is quenched when they diffuse into a cell interior that is negatively charged. Of the different cyanine dyes available, 3,3'-dipropylthiodicarbocyanine and 3,3'-dihexyloxacarbocyanine are the most widely used. With appropriate calibration, the cyanine dyes give $\Delta\psi$ values that are similar to other measurements.

The transmembrane electrochemical gradient of protons has been examined with membrane vesicles of *E. coli* at pH 7 and found to account for a maximum $\Delta pH$ of 2 pH units with a maximum $\Delta\psi$ of 75 mV. Thus, the electrochemical force across the plasma membrane of *E. coli* may be as high as 230 mV. The proton-motive force of various prokaryotes is given in Table 9.18. Using thermodynamics, it is possible to calculate that it takes a charge of about 210 mV across the membrane to maintain a ATP/ADP ratio of 1. The electrochemical activity at pH 7.0 to 8.0 is attributed to $\Delta\psi$ while at pH 5.5 to 6.0 it is attributed to $\Delta pH$.

Table 9.18. Examples of proton-motive force measurements in prokaryotes.[a]

| Microorganism | pH$_{\text{outside}}$ | Z$\Delta$pH (mV) | $\Delta\psi$ (mV) | pmf (mV) |
|---|---|---|---|---|
| *Thermoplasma acidophilum* | 2.0 | 266 | +10 | −256 |
| *Acidithiobacillus ferrooxidans* | 1.0 | 340 | +70 | −270 |
| *Salmonella typhimurium* | 7.5 | 0 | −162 | −162 |
| *Vibrio alginolyticus* | 8.0 | 0 | −125 | −125 |
| *Halobacterium halobium* | 6.3 | 65 | −100 | −165 |
| *Mycoplasma gallisepticum* | 6.5 | 65 | −65 | −130 |
| *Streptococcus lactis* | 6.0 | 40 | −96 | −136 |
| *Moorella thermoacetica* | 6.0 | 71 | −95 | −166 |
| *Methanosarcina barkeri* | 6.7 | 10 | −130 | −140 |

[a]From Kashket, 1985.[33]

# 9.6. Unique Events in Bacterial Respiration[1–4]

With bacterial electron transport being highly varied, bacteria have pursued various parameters and tested the limits of bioenergetic possibilities. In some physiological types of bacteria, electron flow is reversed and in some systems proton cycling occurs. These variants in electron flow are briefly discussed.

## 9.6.1. Reverse Electron Flow

Characteristic of most electron transport systems is the movement of electrons from an oxidative reaction to an electron acceptor that is markedly more electropositive. There are some notable exceptions in which microbial growth occurs only because electron flow is reversed and the electron acceptor is more electronegative than the electron donor. In chemolithotrophic bacteria, reversed electron flow is associated with the reduction of $NAD^+$ by electrons from the oxidation of $Fe^{2+}$, nitrite, ammonia, and carbon monoxide. In anaerobic respiration, a reverse electron flow is also envisioned to explain $NAD^+$ reduction coupled to the oxidation of succinate. Mechanistic details have not been developed to explain these various systems of reverse electron flow; however, these would appear to be energy-dependent processes.

The reversal of electron flow is apparent in the aerobic oxidation of ammonia by *Nitrosomonas*. Ammonia oxidation with the formation of nitrite occurs by the following reaction:

$$NH_3 + 2\,H^+ + 2\,e^- \rightarrow NO_2^- + H_2$$

Hydroxylamine is a stable intermediate in the oxidation of ammonia and subsequent oxidation of hydroxalamine provides energy for the overall reduction of $NAD^+$ (see Figure 9.22). Two electrons are released in ammonia oxidation and based on redox potentials they enter the electron transport system at the region of cytochrome *b*. Hydroxylamine released from ammonia oxidation is oxidized to nitrite with the release of 4 $e^-$ and these electrons proceed through the electron transport system with some going for $NAD^+$ reduction and some to $O_2$. If the proton pump is associated with cytochrome *a* activity, the amount of electrons diverted to $NAD^+$ influences the amount of oxidative phosphorylation.

A similar situation occurs in the oxidation of nitrite by *Nitrobacter* with the formation of nitrate. The high midpoint potential of the $NO_3^-/NO_2^-$ couple (+430 mV) indicates that the electron flow would be to the cytochrome oxidase. Because $NAD(P)H$ is required for $CO_2$ fixation, reverse electron flow from nitrite oxidation would occur in that electrons would need to be diverted to a more electronegative acceptor. Details of this reverse electron flow with nitrite oxidation remain to be established.

Another interesting situation is presented by oxidation of $Fe^{2+}$ by *Thiobacillus*. As presented in Figure 9.4, the redox potential of $Fe^{2+}/Fe^{3+}$ is +770 mV which is near +820 mV, the midpoint potential of the $O_2/O^{2-}$ couple. The generation of NADH requires special reactions for diversion of electrons from $Fe^{2+}$ oxidation and specifics

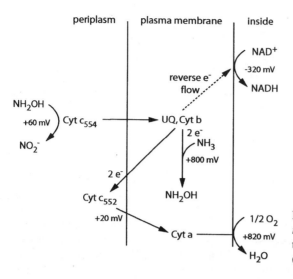

Figure 9.22. Electron transport in *Nitrosomonas* systems. Reverse electron flow is indicated by broken line. (From Bock et al., 1991.[34])

remain to be developed. Reverse electron flow is considered to be an energy-dependent process. Although there may be no direct input of ATP into the reaction, there is the use of electrons to reduce $NAD^+$ rather than use of electrons to pump protons across the membrane and establish a membrane charge. As about 80% of the energy of an autotroph is used to fix $CO_2$ with the production of carbohydrates, the amount of NADH needed is highly significant. The thermodynamic efficiencies of chemolithotrophic bacteria with reverse electron flow are markedly less than with other organisms. While chemoorganotrophs have an efficiency of about 66% with oxygen respiration or 33% with fermentation, bacteria dependent on $Fe^{2+}$, ammonia, and nitrite oxidation have energy efficiencies of 21%, 5.9%, and 7.9%, respectively. Generally bacteria using reverse electron flow have lower cell yields and grow significantly slower than heterotrophic bacteria.

## 9.6.2. Proton Cycling[34,35]

In 1981, Odum and Peck proposed a model of $H_2$ cycling that was proposed for sulfate-reducing bacteria. This energetic concern was proposed in part to explain the huge production of $H_2$ when *Desulfovibrio vulgaris* was growing on lactate in the absence of sulfate. *D. vulgaris* has multiple hydrogenases with one in the periplasm and one in the cytoplasm. It was proposed that the role of the cytoplasmic hydrogenase was to produce $H_2$ from $2 H^+$ and $2 e^-$ released from pyruvate and lactate oxidation. A scheme depicting proton cycling in *D. vulgaris* is given in Figure 9.23. It is proposed that $H_2$ is produced from $2 H^+$ and $2 e^-$ at the plasma membrane and in the periplasm another hydrogenase would direct electrons through the cytochrome $c_3$ back to the cytoplasm where sulfate is reduced to sulfide. This is a most intriguing concept and future studies are needed to determine how broad proton cycling may be used by microorganisms.

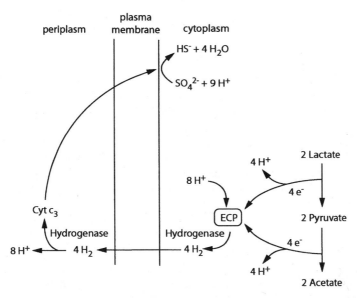

Figure 9.23. $H_2$ cycling in *Desulfovibrio vulgaris*. The lactate and sulfate are transported into the cell while sulfide, carbon dioxide, and acetate are exported. Electrons released from oxidation of lactate and pyruvate are recycled to reduce sulfate. ECP, Electron carrier protein. Used with permission (From Hamilton, 1988.[35])

## 9.6.3. Interspecies Hydrogen Transfer

$H_2$ is present in anaerobic environments as a result of bacterial fermentation. Many organisms have hydrogenase and can use $H_2$ in electron donor reactions. This phenomena of interspecies $H_2$ transfer has been demonstrated for several anaerobes and even occurs between bacteria and archaea. This syntrophic association can be readily demonstrated by using a coculture of two different physiological types of prokaryotes. One species produces $H_2$ and the other uses $H_2$ as the electron source. For example, a coculture of *Methanosarcina barkeri* and *D. vulgaris* can grow with methanol or acetate with sulfate as the final electron acceptor. *M. barkeri* cannot use sulfate as an electron acceptor and *D. vulgaris* cannot use methanol or acetate as an electron donor. However, both cultures grow because *M. barkeri* will produce $H_2$ when oxidizing acetate or methanol and the $H_2$ is ultimately used by *D. vulgaris* as an electron donor coupled to sulfate reduction. In both species, growth is coupled to electron flow. Another example is seen with proton reducers (i.e., *Syntrophobacter* and *Syntrophomonas*) which grow with different $H_2$ oxidizing organisms including sulfate reducers, methanogens, and acetogens. A broader implication is that one organism's end products may be another organism's energy source.

## 9.7. Electron Transport and Chlorophyll-Containing Photosynthesis[36,37]

Based on structural composition and functional activities, there are two fundamental groups of bacteria that use chlorophyll for photosynthetic reactions and chlorophyll-mediated reactions have not been reported for any of the members of the archaea. One of the bacterial groups is the oxygenic cyanobacteria that uses photosystem (PS) I and PS II and the other is the anoxygenic or anaerobic photosynthetic bacteria. Both of these systems are similar in that they both (1) use chlorophyll in the light-sensitive reaction, (2) couple ATP synthesis to the transport of electrons through photosynthetic membranes, and (3) use $NAD(P)^+$ as the electron acceptor. Both groups of chlorophyll-containing bacteria spatially separate protons and electrons to generate the $\Delta\psi$ and $\Delta pH$ across photosynthetic membranes. The generation of energy is attributed to proton-driven ATP synthase activity of the bacterial equivalent to $CF_0/CF_1$ ATPase found in chloroplasts. The two systems are distinguished by the different types of electron donors used and the specific molecules collecting light.

The principal molecule in photosynthesis is chlorophyll, which has a tetrapyrrole structure with $Mg^{2+}$ as the coordinated ion. Although several structural variations of chlorophyll are found in photosynthetic bacteria (see Figures 9.24 and 9.25), all chlorophylls have the same function. Each chlorophyll type absorbs light at a specific wavelength and this enables bacteria to grow by using light from different parts of the visible spectrum. This enables bacteria to grow deep in the water column as well as at the water surface because light is filtered by water with light at longer

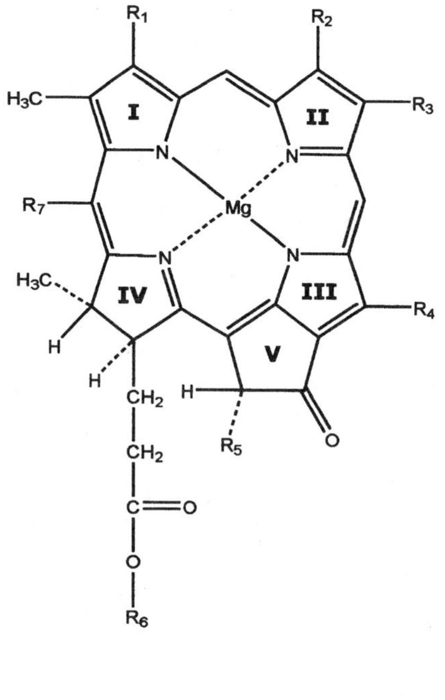

Figure 9.24. Structure of chlorophyll. The R-groups are listed in Figure 9.25.

| | $R_1$ | $R_2$ | $R_3$ | $R_4$ | $R_5$ | $R_6$ | $R_7$ |
|---|---|---|---|---|---|---|---|
| Chl a (680) | $CH=CH_2$ | $CH_3$ | $CH_2CH_3$ | $CH_3$ | $COOCH_3$ | Phytyl | H |
| Bchl a (650-910, 800-810) | $COCH_3$ | $CH_3$ | $CH_2CH_3$ | $CH_3$ | $COOCH_3$ | Phytyl | H |
| Bchl b (1020-1035, 835-850) | $COCH_3$ | $CH_3$ | $=CHCH_3$ | $CH_3$ | $COOCH_3$ | Phytyl | H |
| Bchl c (745-760) | $CHOHCH_3$ | $CH_3$ | $CH_2CH_3$ | $CH_2CH_3$ | H | Farnesyl | $CH_3$ |
| Bchl d (725-745) | $CHOHCH_3$ | $CH_3$ | $CH_2CH_3$ | $CH_2CH_3$ | H | Farnesyl | H |
| Bchl e (715-725) | $CHOHCH_3$ | CHO | $CH_2CH_3$ | $CH_2CH_3$ | H | Farnesyl | $CH_3$ |
| Bchl g (788, 670) | $CH=CH_2$ | $CH_3$ | $=CHCH_3$ | $CH_3$ | $COOCH_3$ | Farnesyl | H |

Figure 9.25. Side groups added to the basic chrolophyll structure given in Figure 9.24. Numbers in parentheses indicate the wavelength for maximum absorption for the corresponding to the specific bacteriochlorophyll.

wavelengths penetrating deeper into the water column than light at shorter wavelengths.

The chlorophyll molecules form a complex of 40 to 300 molecules that consists of reaction center (RC) and antennae chlorophyll. There are only a few RC chlorophyll to convert light energy to chemical energy; however, there are hundreds of antenna molecules that collect light and transfer energy to the RC chlorophyll. The chlorophyll antenna are important in harvesting light in sufficient quantities to drive photosynthesis, especially in nature where the illumination may be of low intensity.

## 9.7.1. Oxygenic Photosynthesis[38]

Cyanobacteria have a photosynthetic system similar to that found in green plants in that both generate $O_2$, use chlorophyll a, and employ PS I along with PS II. These characteristics, as well as the collection of cells into filamentous aggregates, contributed to their initial classification as blue-green algae. In the 1960s, the prokaryotic cell structure of the blue-greens became accepted and with the efforts of R. Stanier they became recognized as blue-green bacteria or cyanobacteria.

In addition to containing chlorophyll *a*, cyanobacteria also contain carotenoids and phycobiliproteins which serve as ancillary light receptors and transfer light energy to the chlorophyll molecule. The photosystems in freshwater and marine cyanobacteria have been extensively examined and have been found to be similar to chloroplast activity in green plants. In cyanobacteria, the membranes that contain the photosystem are not localized in chloroplasts but are in special membranes referred to as thylakoids.

The thylakoid membranes are layered in the cytoplasm with phycobilisomes located on the outer surface of the thylakoid membrane. Although there is some variation in size of phycobilisomes, the phycobilisomes in cyanobacteria are approximately 60 nm long, 30 nm high, and 10 nm wide. See Figure 9.26 for relevant structures.

The phycobilisome is a proteanous unit that consists of core proteins that form a cylinder and protein rods that radiate from the core. About 85% of the proteins in the phycobilisome are associated with light capture and transfer while the remaining protein functions to link the antennae proteins to the thylakoid membrane. The core protein consists of allophycocyanin and smaller linker proteins. In the cyanobacteria, there are commonly two or three core or cylindrical structures. Each core is composed of four separate protein complexes that vary in the number of subunits of allophycocyanin present and the linker proteins. The arrangement of these protein complexes appears important with little variation between species, thereby suggesting that there is a specific function in core protein configuration (Table 9.19). The rod segments of the phycobilisomes contain a series of protein discs that contain either phycoerythryin or phycocyanin along with linker proteins. Each half of the disc is a trimer $[(\alpha\beta)_3]$ of the phycobiliprotein that faces the other to form the completed disc. Either 3 or 4 discs make up each rod and six rods make up the phycobilisome. If different phycobiliproteins are present, the rings are arranged according to light absorbing maximum with phycocyanin closest to the core and phycoerythrin at the distal end of the rod. Linker proteins of 27 to 35 kDa attach the disc proteins to stabilize the structure

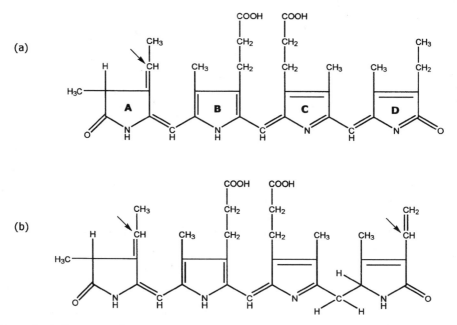

Figure 9.26. Structures of (**a**) phycocyanobilin and (**b**) phycoerythrobilin. Protein is attached to the pigments through a thioether linkage to the carbon at the point of the *arrow*. The letters refer to pyrroles in the heme structure.

Table 9.19. Light-sensitive proteins associated with phycobilisomes.

| Phycobiliproteins | Chromophore | Wavelength for absorption in cells | Location in phycobilisome |
|---|---|---|---|
| Allophycocyanin | Phycocyanobilin | 680 nm | Core |
| Phycocyanin | Phycocyanobilin | 648 nm | Disc on rod |
| Phycoerythrin | Phycoerythrobilin | 576 nm | Disc on rod |

and separate linker proteins attach the phycobilisome to the PS II site on the photo-synthetic membrane. Light is absorbed by the phycobiliproteins and transferred to chlorophyll in PS II, and the quantity of light transferred by the phycobilisomes is comparable to the amount of light collected by PS I.

The photo-electron transfer scheme for cyanobacteria is given in Figure 9.27. In PS II electrons are released from water by a mechanism requiring $Mn^{2+}$ and are directed to the RC of PS II. Relatively little is known about the process resulting in cleavage of water to yield electrons and protons but it is considered that it may involve sequential release of 4 $e^-$ with the oxidation of water because there are four Mn ions bound to the protein cluster. Energy is directed through the phycobilisomes to the RC chlorophyll $a$ and P680 is electronically excited. Electrons from the energized P680 go to pheophytin and on to two different plastiquinones bound to proteins before being donated to the cytochrome $bf$ complex. Cytochrome $f$ is a $c$-type cytochrome and the cytochrome $bf$ complex can pump protons in a manner similar to the cyto-chrome $bc$ complex in bacteria and mitochondria. Electrons from the cytochrome $bc$ complex are carried to the RC of PS I by plasticyanin, which is a blue copper protein. In PS I, electrons flow from chlorophyll $a$ through a series of $Fe_4$-$S_4$ centers to fer-redoxin where the reduced ferredoxin participated in the ferredoxin: $NADP^+$ oxido-reductase to generate NADPH. This system of electron flow from $H_2O \rightarrow$ PS II $\rightarrow$ PS I $\rightarrow NADP^+$ is known as linear or noncyclic electron transfer. There is also the possibility of cyclic electron transfer in cyanobacterial photosynthesis. Under appro-priate conditions such as accumulation of NADPH, electrons from the energized chlo-rophyll are returned to the plastiquinone and recycled through PS I. The net result is that the electrons are continuously cycled through PS I with the formation of ATP coupled to electron flow but no reducing equivalents (NADPH) are formed.

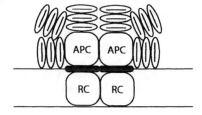

Figure 9.27. Model of the structure of a phycobilisome as found in cyanobacteria. RC, Reactive center; APC, allophycocyanin, rods consist of discs of phycocyanin or phycoerythrin.

## 9.7.2. Anaerobic Photosynthesis[39-42]

Anaerobic photosynthetic bacteria are generally divided into two groups based on color: purple bacteria and green bacteria. However, the color is not a valid characteristic because some purple bacteria are green, red, or even brown. Some organisms in the group of green phototrophs may also be brown. Color of the culture is dependent on both the type of chlorophyll present and the type of carotenoid present. Traditionally, purple and green photosynthetic bacteria are divided into sulfur and nonsulfur groups based on the characteristic that elemental sulfur is produced as an end product from $HS^-$ oxidation or that elemental sulfur is used as an electron donor. Generally, all purple and green bacteria can use $H_2$ or $HS^-$ as an electron donor in the photosynthetic process. Important characteristics to distinguish the phototrophic bacteria are given in Table 9.20. Anoxygenic photosynthesis by bacteria differs from oxygenic photosynthesis by several important features. Unlike cyanobacterial photosystems which use only chlorophyll $a$, the anaerobic photosynthetic bacteria do not have chlorophyll $a$ but have several other types of chlorophylls that function in photosynthesis. In the purple nonsulfur bacteria (*Rhodospirillum*) and purple sulfur bac-

Table 9.20. Characteristics of major photosynthetic prokaryotes.[a]

| Group | Cyanobacteria | Chlorobiaceae | Chloroflexaceae | Purple bacteria | Heliobacteria |
|---|---|---|---|---|---|
| Selected examples of genera | *Anabaena* *Oscillatoria* *Anacystis* | *Chlorobium* *Chloroherpeton* | *Chloroflexus* | *Rhodobacter* *Chromatium* *Rhodopseudomonas* *Ectothiorhodospira* | *Heliobacterium* *Heliospirillum* *Heliobacillus* |
| Photosynthesis | Oxygenic | Anaerobic | Anaerobic | Anaerobic | Anaerobic |
| Photosystems | I and II | I | II | II | I |
| Autotrophic $CO_2$ fixation | Calvin cycle | Reduced TCC[b] | Hydroxy-propionate | Calvin cycle | No |
| Electron donor | $H_2O$ | $H_2S$, $H_2$, $S^o$, thiosulfate | $H_2S$, organic compounds | $H_2S$, $H_2$, organic compounds | Organic compounds |
| Aerobic, dark growth | Some species | No | Yes | Some species | No |
| Localization of photopigments | Thylakoid Phycobilins | Chlorosome | Chlorosome | Internal membranes | Plasma membrane |
| Chlorophylls | Chl $a$ | Bchl $a,c,d,e$ | Bchl $a,c,d$ | Bchl $a,b$[c] | Bchl $g$ |
| Carotenoids[d] | | Chlorobactene | β-Carotene γ-Carotene | Neurospoene Lycopene Okenone | Diaponeurosporene |
| Sulfur accumulation | | | | Intracellular[e] Extracellular[f] | |

[a]Information for this table is from Imhoff, 1999.[42]
[b]TCC, Tricarboxylic acid cycle.
[c]*Thiocapsa pfennigii, Ectothiorhodospira halochloris,* and *Rhodopseudomonase virdis* contain only Bch *b*.
[d]This is not a comprehensive list but provides information on types of carotenoids present in prokaryotes.
[e]*Chromatium.*
[f]*Ectothiorhodospira.*

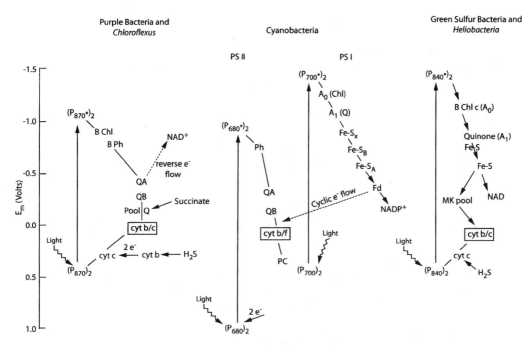

Figure 9.28. A comparison of three types of electron transfer in photosynthetic bacteria. The cytochrome $bc$ or $bf$ complex is assumed to be the proton pump for ATP photophosphorylation. For *Heliobacterium*, the photocenter (PC) would absorb light at 798 nm. A quinone-type of reaction center is found in the purple bacteria, *Chloroflexaceae*, and photosystem (PS) II of cyanobacteria. An iron-sulfur (Fe-S) protein system is associated with the reaction center of PS I of the cyanobacteria, green sulfur bacteria, and the *Heliobacterium chlorum*. Both cyclic and noncyclic photorespiration is given for each of the three systems. (From Blankenship et al., 1987.[39] Used with permission.)

teria (*Chromatium* and *Ectorhodospira*), Bchl *a* is located in internal membranes. Although both *Chlorobium* and *Chloroflexus* (green sulfur bacteria) have chlorosomes with various BChls, BChl *a* in these organisms is present in the plasma membrane and not in the chlorosome membranes. In addition, it should be emphasized that anaerobic photosynthetic organisms do not require accessory pigments such as phycobiliproteins which are present in cyanobacteria. The heliobacteria are taxonomically distinct from the Gram-negative purple and green bacteria in that they are Gram-positive and have Bchl *g* as the only chlorophyll type.

Although chloroplasts are the region for photosynthesis in plant systems, considerable variation in photosynthetic structure exists for bacteria. The photosynthetic apparatus in bacteria appears as an extension of the plasma membrane and the form varies with the bacterial species. Some of the forms of these photosynthetic arrangements are given in Figure 9.28. In purple phototrophs, the photosynthetic apparatus is present in membranes that are extensions from the plasma membrane as intracytoplasmic membranes. The arrangement of the photosyhthetic membranes is not random but species specific. Intracytoplasmic membranes that are organized as vesicles are characteristic of *Rhodospirillum rubrum*, *Chromatium vinosum*, and *Rhodobacter capsulatus*. The photosynthetic membrane extending from the plasma membrane assumes a cylinder-like form in *Thiocapsa pfennigii*. Organized stacks of thylakoid membranes are found in *Ectothiorhodospira shaposhnikovii* while irregular arrangements of thylakoid membranes are found in *Rhodopseudomonas palustrus* and *Rhodopseudomonas viridis*. Differentiation of the intracellular membranes from the plasma membrane is induced by low $O_2$ tension and light intensity.

In green photosynthetic bacteria, the light detecting photo-apparatus is in small vesicles known as chlorosomes and in *Heliobacteria* the photosensitive system is present in the plasma membrane and there are no intracytoplasmic membranes. Common for all of the species is that the ATP synthase is oriented into the cytoplasm so the newly produced ATP does not have to cross any membranes before it is used in the cytoplasmic reactions.

While the reaction center of cyanobacteria can be characterized as having both PS I and PS II, anoxygenic bacteria will have only one system in the reaction center. Members of the Chlorobiaceae and heliobacteria have activity in the reaction center that is similar to PS I while the purple bacteria and Chloroflexaceae have activities more similar to PS II. The pathway for electron movement in the membranes of purple bacteria and green bacteria is given in Figure 9.28. In both bacterial types, proton pumping across the photosynthetic membranes is coupled to electron flow and ATP synthesis results. Not only are there different electron donors for the purple nonsulfur bacteria and the green sulfur bacteria but also these bacteria have different *c*-type cytochromes. A great deal of information is yet to be gained concerning the spatial arrangement of electron carriers in the membranes and formation of energy.

### 9.7.2.1. Purple Phototrophs

The purple bacteria can be divided into the purple sulfur bacteria such as *Chromatium* and *Thiospirillum*, which are in the gamma subdivision of proteobacteria, and the nonsulfur purple bacteria, which use organic acids such as succinate as the electron donor. Bacteria in the latter group include *Rhodospirillum*, *Rhodopseudomonas*, *Rhodobacter*, *Rhodomicrobium*, and *Rhodopida* in the alpha subgroup of proteobacteria and *Rhodocyclus* in the beta subdivision.

Several of the purple photosynthetic bacteria have been extensively examined and they include *Rhodobacter sphaeroides* (formerly *Rhodopseudomonas sphaeroides*) and *Rhodobacter capsulatus* (formerly *Rhodopseudomonas capsulata*). *Rhodobacter capsulatus* is a highly versatile prokaryote in that it can (1) grow photosynthetically using light to energize $CO_2$ fixation, (2) grow photoheterotrophically using light to energize the use of organic compounds, (3) grow chemoautotrophically in the dark with $H_2$ as the electron donor with $O_2$ as the electron acceptor, and (4) grow heterotrophically in the dark using organic carbon compounds as electron donors with $O_2$ or alternate compounds as the electron acceptor. To account for these respiratory activities, the following electron transport scheme has been proposed for *Rhodobacter capsulatus* (Figure 9.29).

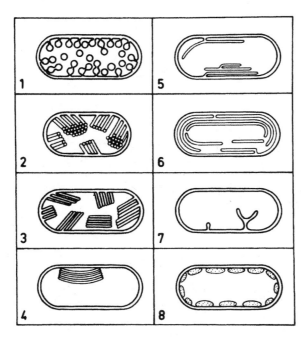

Figure 9.29. Structures in bacteria that contain the photosynthetic apparatus. In the purple phototrophic bacteria the intracellular membranes may be arranged in various forms with a few systems shown here: (1) vesicle type as found in *Rhodobacter capsulatus, Rhodospirillum rubrum*, and *Chromatium vinosum*; (2) tubular type as found in *Thiocapsa pfennigii*; (3) short membranes stacked at an angle to the plasma membrane as found in *Rhodospirillum molischianum*; (4) flat membrane-type where membranes are arranged in regular stacks as found in *Ectothiorhodospira* sp.; and (5) irregular membranes underlying the plasma membrane and occasionally branched as in *Rhodopseudomonas palustris*; (6) concentric layers of membrane invaginations as in *Rhodomicrobium vanielii*; (7) no or only a few irregular invaginations from the plasma membrane as in *Rhodospirillum gelatinosa, Heliobacterium chlorum*, and *Rhodospirillum tenue*; and (8) plasma membrane with attached chlorosomes as in *Chlorobium* sp., *Chloroflexux* sp. or *Prosthecochloris* sp. (From Oelze and Drews, 1981.[40] Used with permission.)

Purple phototrophs such as *Rhodobacter sphaeroides, Rhodobacter capsulatus*, and *Rhodospirillum rubrum* have a photosynthetic apparatus that is highly complex. These purple bacteria contain the following: (1) the reaction center (RC) which contains chlorophyll, (2) a light harvesting complex 1 (LH 1), (3) a light harvesting complex 2 (LH 2), and (4) the cytochrome $bc_1$ complex which is the same as in heterotrophic growth. In *Rhodobacter sphaeroides*, the RC has three polypeptides which are known as H, L, and M. Bound to the L and M polypeptide structures are four molecules of Bchl $a$, two molecules of bacteriopheophytin (bacteriochlorophyll minus the $Mg^{2+}$ ion), two molecules of ubiquinone, and one atom of Fe. The LH 1 is located adjacent to the RC and consists of the following: two polypeptide units, 24 molecules of Bchl, and 24 carotenoid molecules. LH 2 has two polypeptide units plus 18 mol-

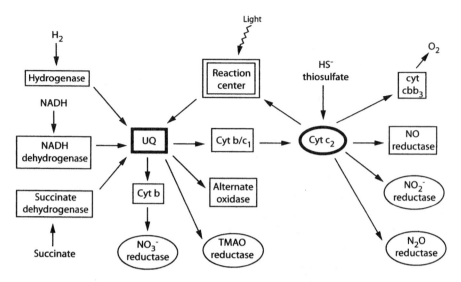

Figure 9.30. Generalized model for electron transport in *Rhodobacter sphaeroides*. Ovals are soluble enzymes in the periplasm and boxes are found in the membrane. (From Ferguson, 1988.[15] Used with permission.)

ecules of Bchl *a* and 9 carotenoid molecules. The LH 2 is located adjacent to LH 1 but somewhat removed from the RC. Energy flows from LH 2 to LH 1 and finally to the RC. The carotenoid function is not entirely clear because it assists light collection in the photo antenna but it can also protect against oxygen radicals. The $Fe^{2+}$ ion is positioned between the quinone molecules and appears to limit the quinones to single electron chemistry. As a result, one electron passes at a time through each of the two quinone molecules before the electrons collect in the quinone pool in the membrane (Figure 9.30).

In purple bacteria and *Chloroflexis*, there is no direct route for the generation of NADH. The reason is that the electrons from quinones ($Q_A$ and $Q_B$) go to cytochrome *bc* and on to cytochrome *c*. To produce NADH, there must be reverse electron transfer from the quinone pool to $NAD^+$. While this reverse flow does not employ ATP directly, energetics is important because the potential of the Q/QH reaction is $-88$ mV and the midpoint potential of $NAD^+$/NADH reaction is $-320$ mV. It is possible that through initial velocity reactions, kinteics would enable this reverse electron flow. An example of an electron donor for the system is succinate $\rightarrow$ fumarate $+ 2 H^+ + 2 e^-$ where the electron enters the system near the quinones. For systems employing $HS^-$ and thiosulfate as the electron donor, electrons would enter the system through cytochrome *c*.

### 9.7.2.2. Green Phototrophs

The green sulfur bacteria include *Chlorobium* and *Chloroherpeton* which are members of the Chlorobiaceae family. Green phototrophs are included in the Chloroflexaceae with *Chloroflexus* being the species most commonly examined. The photo antenna is in the chlorosome, which is a small structure attached to the inner face of the plasma membrane. A model of the chlorosome is given in Figure 9.31. While each green phototroph has a chlorosome of a specific size, the chlorosomes are within the size of 70 to 260 nm long and 30 to 100 nm wide. The covering or envelope of the chlorosome on the cytoplasmic side is 2 to 3 nm with a single electron-dense layer. Where the chlorosome and plasma membrane interface, there is a base plate about 5 to 6 nm thick. The core of the chlorosome contains rod-like structures that span the length of the chlorosome. In *Chlorobium*, the rods are highly organized with a diameter of 10 nm and there are 10 to 30 rods in each chlorosome.

In addition to Bchl, the chlorosomes contain specific lipids, carotenoids, and proteins. Glycolypids are present in higher levels in the chlorosome envelope than in the plasma membrane. It should be noted that glycolipids are not present in the photosensing membranes of the purple bacteria. Chlorosomes of both families of green bacteria contain monogalactosyl diglycerol (MGDG) which may have a role in organization of pigments in the chlorosome. Carotenoids are present in considerable quantities in the chlorosomes of green bacteria and may be concentrated in the envelope. In *Chlorobium limicola*, the principal carotenoids are as follows: chlorobactene, 69%; neurosporene, 21% and lycopene, 8%. In *Chlorobium phaeobacteroides*, isorenierate is the only carotenoid present. In *Chloroflexus aurantiacus*, the distribution of carotenoids in the chlorosomes is as follows: β-carotene, 36%; γ-carotene, 43% and hydroxy-γ-carotene, 12%. Carotenoids may function in light absorption and participate as the initial receptor in the photo antennae. However, not all carotenoids are involved with photo reception because close contact is required for this energy relay and this is not always possible. In *Chloroflexus aurantiacus*, the carotenoids are collected in structures known as "oleosomes" and although the function of these structures is unknown, it is not likely that oleosomes are used as photo antennae. Proteins

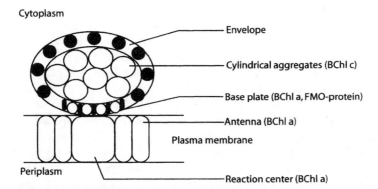

Figure 9.31. Chlorosome structure in *Chloroflexus aurantiacus*. Chlorosome structure and composition are highly conserved throughout the green photosynthetic bacteria while the reaction centers for the bacteria differ significantly.

of 5.8 kDa or 6.2 kDa have been isolated from chlorosomes and these proteins are proposed to interact with chlorophylls in some manner.

A base plate attaches the chlorosome to the plasma membrane and present in the base plate is Bchl $a$. In Chlorobiaceae, but not in chloroflexaceae, there is a protein that binds Bchl $a$ that is known as the Fenna–Mathews–Olson (FMO) protein. In the green phototroph, *Prosthecochloris aestrarii*, the FMO protein consists of 366 amino acids and 6 histidine residues are considered to be involved in complexing with Bchl $a$ molecules. Based on X-ray analysis, the FMO protein in *Chlorobium tepidum* is projected to be a trimer that is folded much like a taco shell. Seven Bchl $a$ molecules are surrounded by the FMO protein which contains 15 strands of β-sheet segments. The Bchl $a$ molecules are bonded into the FMO protein by hydrogen bonding, hydrophobic interactions with the phytyl tail of the Bchl $a$, and liganding of $Mg^{2+}$. The Bchl $a$ molecules are closely packed and approximately 24 Å separates individual molecules. The trimer proteins are closely packed with about 30 Å separating Bchl $a$ in adjacent FMO proteins. The Bchl $a$–FMO protein forms a crystal plate between the chlorosome and plasma membrane through which energy flows from the chlorosome to the RC in the plasma membrane.

The chlorosome vesicle is filled with hundreds of Bchl molecules of $c$, $d$, or $e$ type and through tight packing of Bchl-Bchl stacking a tube-like structure is formed that functions as an antenna in collecting light. Energy is transmitted from chlorophyll in the chlorosome, through Bchl $a$ in the base plate, through light harvesting (LH) Bchl $a$ in the plasma membrane and finally to Bchl $a$ in the reaction center (RC). Activity in the RC leads to production of reducing potential in the form of NADH and in proton pumping action through the cytochrome $bc$ complex.

## 9.8. Chlorophyll-Containing Aerobes[43,44]

Several aerobic bacteria other than cyanobacteria have bacteriochlorophyll (Bchl) $a$. In the past, these organisms were referred to as "paraphotosynthetic bacteria," "quasiphotosynthetic bacteria," "aerobic anoxygenic bacteria," and here the term aerobic chlorophyll-containing bacteria is used. There are at least 8 genera that include 20 different species of aerobic bacteria that contain chlorophyll. These chlorophyll-containing bacteria are distinguished from true phototrophic bacteria in several ways: (1) Unlike anaerobic phototrophs, they are obligate aerobes that cannot grow anaerobically in the light. (2) Synthesis of Bchl in aerobic chlorophyll bacteria occurs only under aerobic conditions. (3) The aerobic chlorophyll-containing bacteria do not contain accessory pigments for light harvesting nor do they have special structures for photo activity such as phycobilisomes or chlorosomes. The chlorophyll-containing bacteria can be placed in three physiological groups: nitrogen-fixing rhizobia, methylotrophic bacteria, and a highly diverse group containing various species of bacteria. Photosynthetic rhizobia have been found in nodules on the stems of various tropical legumes, and under special conditions, nodules will form on the stems of field beans, soybeans, and peanuts. Some of the chlorophyll-containing methanol utilizing bacteria are phylotrophic and enhance plant resistance to environmental stress.

The quantity of Bchl in various phototrophic bacteria is markedly less than phototrophic bacteria and this distribution of chlorophyll is given in Table 9.21. The total quantity of BChl in the aerobic anoxygenic bacteria is significantly less than that found in typical anaerobic photosynthetic bacteria. While Bchl $a$ contains $Mg^{2+}$ in the center of chlorin macrocyclic ring in typical photosynthetic bacteria, species of the genera *Acidiphilium* are known to produce a Zn-containing Bchl. *Acidiphilium rubrum* produces 810 and 120 nmol/g cell dry weight of a Zn-BChl and Mg-BChl, respectively. This level of Mg-BChl produced by *Acidiphilium rubrum* is about 100 times less than found in *Rhodopseudomonas acidophila*, a typical photosynthetic bacterium. The aerobic chlorophyll-containing bacteria also contain carotenoid molecules which produce brightly pigmented colonies of yellow, orange, and pink. Although most of these aerobic bacteria containing Bchl have only one or two different carotenoids, *Erythromicrobium ramosuma* has more than 10 different carotenoid molecules. The ratio of chlorophyll to carotenoid molecules ranges from 1.4 to 1.7 for most of the aerobic phototrophic bacteria. These carotenoids do not contribute to light harvesting because some contain highly polar groups and some carotenoids are present in the outer membrane. It is quite possible that these carotenoids participate in protecting the cells from stress due to oxygen radicals generated from light-catalyzed reactions.

The photosynthetic pigments in aerobic anoxygenic phototrophs aggregate in photosynthetic membranes located in the plasma membrane or intracytoplasmic membranes. The aerobic anoxygenic phototrophs have a simplified light harvesting system and reactive center containing cytochrome $c$ and carotenoids. The reactive center of *Roseobacter denitrificans* contains four molecules of Bchl $a$, two molecules of bacteriopheophytin, four molecules of cytochrome $c_{554}$, two molecules of ubiquinone-10, and carotenoids. In comparison, the reactive center of purple anoxygenic phototrophic bacteria contain four molecules of Bchl $a$, two molecules of bacteriopheophytin, two molecules of quinones, and one molecule of nonheme iron protein. Cyclic electron transfer between cytochrome $bc_1$ and the reactive center is mediated by cytochrome

Table 9.21. Bacteriochlorophyll content in select strains of bacteria.[a]

| Bacteria | Bchl (nmol/mg dry weight of cells) |
| --- | --- |
| **Anaerobic purple photosynthetic** | |
| *Rhodobacter sphaeroides* | 20 |
| **Aerobic anoxogenic phototrophic** | |
| *Acidiphilium rubrum* | 0.7 |
| *Erythrobacter longus* | 2 |
| *Erythromicrobium hydrolyticum* | 1.5–6 |
| *Roseococcus thiosulfatophilus* | 0.14–1.4 |
| *Sandaracinobacter sibiricus* | 1.5–6 |
| **Rhizobia** | |
| *Bradyrhizobium* strain BT Ai 1 | 0.4 |

[a]From Fleischmann and Kramer, 1998[44]; Yurkov and Beatty, 1998.[43]

$c_{551}$. The electron transport system works under aerobic conditions to charge the membrane; however, these aerobic anoxygenic phototrophs cannot grow by the photosynthetic process under anaerobic conditions. Photosynthetic growth has not been demonstrated in these chlorophyll-containing bacteria but there is some evidence that the photo process can enhance growth yields when cultures grow in the presence of light.

## 9.9. Paradigm of Prokaryotic Bioenergetics

Bacteria display considerable versatility with their electron transport systems. A large number of redox active molecules are used as electron donors or electron acceptors and a limiting requirement appears to be a reduction/oxidation couple within the biological range of about $-400$ mV to about $+1000$ mV. Unique dehydrogenases and reductases characterize bacterial respiration and cytochromes facilitate the unidirectional flow from electron donors to electron acceptors. In a few instances, bacteria may use light to energize electron flow through chlorophyll-mediated processes. A central theme in respiration is the coupling of ATP synthesis to membrane-mediated electron flow. As summarized in Figure 9.32, prokaryotes have several ATP synthase systems coupled to $H^+$ pumps. Proton-driven ATP synthesis occurs in bacteria and archaea using F-type ATPase and V-type ATPase, respectively. In a few systems and especially under alkaline conditions, $Na^+$ replaces $H^+$ for ATP synthesis using related but distinct ATPase structures. The ATPase structures appear to be highly versatile

Figure 9.32. Coupled transmembrane reactions of bioenergetic importance in prokaryotes. The numbers in the figure reflect activity in the following reactions: 1. F-type ATP synthase driven by $H^+$ in bacteria, 2. F-type ATP synthase driven by $Na^+$ as occurs in bacteria growing in alkaline environments, 3. V-type ATP synthase driven by $H^+$ in archaea, 4. V-type and F-type ATPase for proton export, 5. $H^+$ pump at site 0 of respiratory chain, 6. $H^+$ pump at site I of respiratory chain, 7. $H^+$ pump at site III of respiratory chain, 8. $H^+$ pump at site IV of respiratory chain, 9. $H^+$ pump by anaerobic bacteria including methanogens, acetogens, and sulfate-reducing bacteria, 10. $H^+$ pump by light-sensitive bacteriorhodopsin.

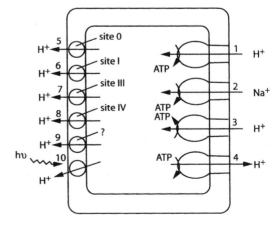

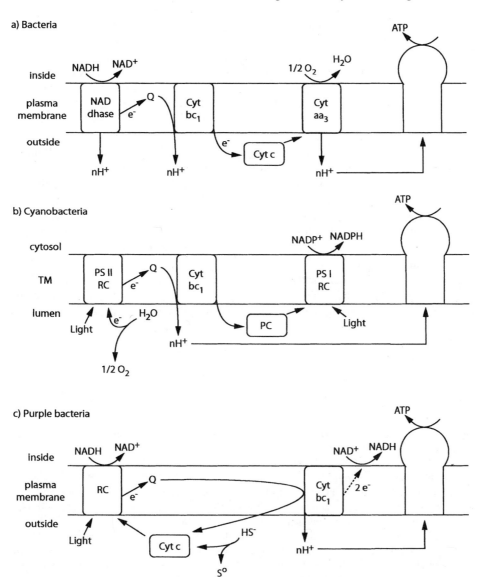

Figure 9.33. A distribution across the membranes of respiratory systems as found in certain heterotrophic bacteria and photosynthetic bacteria. Examples of electron transport schemes are given for (**a**) bacteria, (**b**) cyanobacteria, and (**c**) photosynthetic purple bacteria. PS I, Photosystem I; PS II, Photosystem II; RC, reactive center in photoresponse; PC, plastocyanin. See text for discussion. (From Baker et al., 1998[46]; Fromwald et al., 1999[47]; and Babic et al., 1999.[45])

and under nonrespiratory conditions can pump $H^+$ or $Na^+$ out of the cytoplasm to charge the plasma membrane.

Under conditions of respiration, bacteria export $H^+$ across the plasma through specific protein aggregates referred to as proton pumps. A common theme is that bacteria use one of the four common electron transport-driven proton pumps and while bacteria rarely have all four proton pumps, they commonly have at least one of the proton-driven pumps in respiration. Unique physiological groups of prokaryotes such as methanogens and acetogens may have an additional type of proton pump that has not yet been characterized. A unique proton pump is the rhodopsin-containing transmembrane protein that exports protons from the cytoplasm by a light-energized process.

The structural organizations of bacterial membranes that participate in respiration have several common features. The models in Figure 9.33 emphasize a theme found in heterotrophic bacterial respiration, photorespiration involving PS I and PS II, and anaerobic bacterial photorespiration. Heterotrophs and numerous chemolithotrophs have electron transport energizing the proton pumps in the plasma membrane and with the Gram-negative groups, the cytochrome $c$ in the periplasm is important in coupling the electron flow. In cyanobacterial photorespiration, electron flow is coupled either by a plastiquinone bound to the membrane or by a cytochrome. Despite the specific differences in electron transport or energizing the ATP synthases, the light driven and dark respiratory processes are highly conserved and strengthen the concept of unity in physiology.

# References

1. Jones, C.W. 1982. *Bacterial Respiration and Photosynthesis*. American Society for Microbiology Press, Washington, D.C.
2. Mitchell, P. 1979. Keilin's respiratory chain concept and its chemiosmotic consequences. *Science* 206:1148–1158.
3. Nicholls, D.G. and S.J. Ferguson. 1992. *Bioenergetics 2*. Academic Press, New York.
4. Thauer, R.K., K. Jungermann, and K. Decker. 1977. Energy conservation in chemotrophic anaerobic bacteria. *Bacteriological Reviews* 41:100–180.
5. Cronan, J.E., Jr., R.B. Gennis, and S.R. Maloy. 1987. Cytoplasmic membrane. In Escherichia coli *and* Salmonella typhimurium: *Cellular and Molecular Biology* (F.C. Neidhardt, J.L. Ingraham, K. Brooks Low, B. Magasanik, M. Schaechter, and H.E. Umbarger, eds.). American Society for Microbiology Press, Washington, D.C., pp. 202–221.
6. Lin, E.C.C. and D.R. Kuritzkes. 1987. Pathway for anaerobic electron transport. In Escherichia coli *and* Salmonella typhimurium: *Cellular and Molecular Biology* (F.C. Neidhardt, J.L. Ingraham, K. Brooks Low, B. Magasanik, M. Schaechter, and H.E. Umbarger, eds.). American Society for Microbiology Press, Washington, D.C., pp. 202–221.
7. Anthony, C. 1988. Quinoproteins and energy transduction. In *Bacterial Energy Transduction* (C. Anthony, ed.). Academic Press, New York, pp. 293–316.
8. Gottschalk, G. 1986. *Bacterial Metabolism*. Springer-Verlag, New York.
9. Fauque, G., J. LeGall, and L.L. Barton. 1991. Sulfate-reducing and sulfur-reducing bacteria. In *Variations in Autotrophic Life* (J.M. Shively and L.L. Barton, eds.). Academic Press, New York, pp. 271–339.

10. Wood, H.G. and L.G. Ljungdahl. 1991. Autotrophic character of the acetogenic bacteria. In *Variations in Autotrophic Life* (J.M. Shively and L.L. Barton, eds.). Academic Press, New York, pp. 201–250.

11. Wood, P.M. 1988. Chemolithotrophy. In *Bacterial Energy Transduction* (C. Anthony, ed.). Academic Press, New York, pp. 183–230.

12. Blake, R. II and D.B. Johnson. 2000. Phylogenetic and biochemical diversity among acidophilic bacteria that respire on iron. In *Environmental Microbe-Metal Interactions* (D.R. Lovley, ed.). American Society for Microbiology Press, Washington, D.C., pp. 53–79.

13. Ehrlich, H.L., W.J. Ingledew, and J.C. Salerno. 1991. Iron- and manganese-oxidizing bacteria. In *Variations in Autotrophic Life* (J.M. Shively and L.L. Barton, eds.). Academic Press, New York, pp. 147–170.

14. Kelly, D.P. 1988. Oxidation of sulfur compounds. In *The Nitrogen and Sulfur Cycles* (J.A. Cole and S.J. Ferguson, eds.). Forty-Second Symposium of the Society for General Microbiology. Cambridge University Press, Cambridge, pp. 65–99.

15. Ferguson, S.J. 1988. The redox reactions of the nitrogen and sulphur cycles. In *The Nitrogen and Sulfur Cycles* (J.A. Cole and S.J. Ferguson, eds.). Forty-Second Symposium of the Society for General Microbiology. Cambridge University Press, Cambridge, pp. 1–30.

16. Ratledge, C. and S.G. Wilkinson. 1988. *Microbial Lipids*, Vol. 1. Academic Press, New York.

17. Gennis, R.B. and V. Stewart. 1996. Respiration. In Escherichia coli *and* Salmonella typhimurium: *Cellular and Molecular Biology*, 2nd edit. (F.C. Neidhardt, R. Curtiss III, J.L. Ingraham, E.C.C. Lin, K. Brooks Low, B. Magasakie, W.S. Rezniroff, M. Riley, M. Schaechter, and H.E. Umbarger, eds.). American Society for Microbiology Press, Washington, D.C., pp. 217–261.

18. Hughes, M.N. and R.K. Poole. 1989. *Metals and Micro-organisms*. Chapman and Hall, New York.

19. Moura, J.J.G., A.L. Macedo, and P.N. Palma. 1994. Ferredoxins. In *Inorganic Microbial Sulfur Metabolism. Methods in Enzymology*, Vol. 243 (H.D. Peck, Jr. and J. LeGall, eds.). Academic Press, New York, pp. 165–187.

20. Vervoort, J., D. Heering, S. Peelen, and W. van Berkel. 1994. Flavodoxins. In *Inorganic Microbial Sulfur Metabolism. Methods In Enzymology*, Vol. 243 (H.D. Peck, Jr. and J. LeGall, eds.). Academic Press, New York, pp. 188–202.

21. Jones, C.W. 1988. Membrane-associated energy conservation in bacteria: A general introduction. In *Bacterial Energy Transduction* (C. Anthony, ed.). Academic Press, New York, pp. 1–82.

22. Cronan, J.E. Jr., R.B. Gennis, and S.R. Maloy. 1987. Cytoplasmic membrane. In Escherichia coli *and* Salmonella typhimurium: *Cellular and Molecular Biology* (F.C. Neidhardt, J.L. Ingraham, K. Brooks Low, B. Magasanik, M. Schaechter, and H.E. Umbarger, eds.). American Society for Microbiology Press, Washington, D.C., pp. 31–55.

23. Hooper, A.B., T. Vannelli, D.J. Bergmann, and D.M. Arciero. 1997. Enzymology of the oxidation of ammonia to nitrite by bacteria. *Antoine van Leeuwenhoek* 71:59–67.

24. Lin, E.C.C. and D.R. Kuritzkes. 1987. Pathways for anaerobic electron transport. In Escherichia coli *and* Salmonella typhimurium: *Cellular and Molecular Biology* (F.C. Neidhardt, J.L. Ingraham, K. Brooks Low, B. Magasanik, M. Schaechter, and H.E. Umbarger, eds.). American Society for Microbiology Press, Washington, D.C., pp. 201–221.

25. Peck, H.D., Jr. and J. LeGall (eds.). 1994. *Inorganic Microbial Sulfur Metabolism. Methods in Enzymology*, Vol. 243. Academic Press, San Diego.

26. Poole, R.K. and W.J. Ingledew. 1987. Pathways to oxygen. In Escherichia coli *and* Salmonella typhimurium: *Cellular and Molecular Biology* (F.C. Neidhardt, J.L. Ingraham, K.

Brooks Low, B. Magasanik, M. Schaechter, and H.E. Umbarger, eds.). American Society for Microbiology Press, Washington, D.C., pp. 170–200.

27. Richardson, D.J. 2000. Bacterial respiration: A flexible process for a changing environment. *Microbiology* 146:551–571.

28. Thöny-Meyer, L. 1997. Biogenesis of respiratory cytochromes in bacteria. *Microbiology and Molecular Biology Reviews* 61:337–376.

29. Sone, N. 1990. Respiration-driven proton pumps. In *Bacterial Energetics* (T.A. Krulwich, ed.). Academic Press, New York, pp. 1–32.

30. Fillingame, R.H. 1990. Molecular mechanisms of ATP synthesis by $F_1F_0$-type $H^+$-transporting ATP synthases. In *Bacterial Energetics* (T.A. Krulwich, ed.). Academic Press, New York, pp. 345–392.

31. Nelson, N. 1992. Evolution of organellar proton-ATPases. *Biochimia et Biophysica Acta* 1100:109–124.

32. Kaback, H.R. 1990. Active transport: Membrane vesicles, bioenergetics, molecules, and mechanisms. In *Bacterial Energetics* (T.A. Krulwich, ed.). Academic Press, New York, pp. 151–202.

33. Kashket, E.R. 1985. The proton motive force in bacteria: A critical assessment of methods. *Annual Review of Microbiology* 39:219–242.

34. Bock, E., H.-P. Koops, H. Harms, and B. Ahlers. 1991. The biochemistry of nitrifying bacteria. In *Variations in Autotrophic Life* (J.M. Shively and L.L. Barton, eds.). Academic Press, New York, pp. 171– 201.

35. Hamilton, W.A. 1988. Energy transduction in anaerobic bacteria. In *Bacterial Energy Transduction* (C. Anthony, ed.). Academic Press, New York, pp. 83–150.

36. Jackson, J.B. 1988. Bacterial photosynthesis. In *Bacterial Energy Transduction* (C. Anthony, ed.). Academic Press, New York, pp. 318–375.

37. D. White. 2000. *The Physiology and Biochemistry of Prokaryotes*. Oxford University Press, New York.

38. Sherman, L., T. Bricker, J. Guikema, and H. Pakrasi. 1987. The protein composition of the photosynthetic complexes form the cyanobacterial thylakoid membranes. In *The Cyanobacteria* (P. Fay and C. Van Baalen, eds.). Elsevier, New York, pp. 1–34.

39. Blankenship, R.E., D.C. Brune, J.M. Freeman, G.H. King, J.D. McManus, T. Nozawa, T. Trost, and B.P. Wittmershaus. 1987. Energy trapping and electron transfer in *Chloroflexus aurantiacus*. In *Green Photosynthetic Bacteria* (J.M. Olson, J.G. Ormerod, J. Amesz, E. Stackelbrandt, and H.G. Trüper, eds.). Plenum Press, New York, pp. 57–68.

40. Oelze, J. and G. Drews. 1981. Membranes of phototrophic bacteria. In *Organization of Prokaryotic Cell Membranes*, Vol. II (B.K. Ghosh, ed.). CRC Press, Boca Raton, FL, pp. 113–196.

41. Prince, R.C. 1990. Bacterial photosynthesis: From photons to Δp. In *Bacterial Energetics: The Bacteria*, Vol. XII (T.A. Krulwich, ed.). Academic Press, New York, pp. 111–149.

42. Immhoff, J.F. 1999. A phylogenetically orientated taxonomy of anoxygenic phototrophic bacteria. In *The Phototrophic Prokaryotes* (G.A. Peschek, W. Löffelhardt, and G. Schmetterer, eds.). Kluwer Academic/Plenum, New York, pp. 763–774.

43. Yurkov, V.V. and J.T. Beatty. 1998. Aerobic anoxygenic phototrophic bacteria. *Microbiology and Molecular Biology Reviews* 62:695–724.

44. Fleischman, D. and D. Kramer. 1998. Photosynthetic rhizobia. *Biochimica et Biophysica Acta* 1364:17–36.

45. Babic, S., C.N. Hunter, N. Rakhlin, R.W. Simmons, and M.K. Phillips-Jones. 1999. A putative translational regulator of photosynthesis gene expression. In *The Phototrophic*

*Prokaryotes* (G.A. Peschek, W. Löffelhardt, and G. Schmetterer, eds.). Kluwer Academic/Plenum, New York, pp. 139–147.

46. Baker, S.C., S.J. Ferguson, B. Ludwig, M.D. Page, O.-M.H. Richter, and R.J.M. Van Spanning. 1998. Molecular genetics of the genus *Paracoccus*: Metabolically versatile bacteria with bioenergetic flexibility. *Microbiology and Molecular Biology Reviews* 62:1046–1078.

47. Fromwald, S., M. Wastyn, M. Lübben, and G.A. Peschek. 1999. Extended heme promiscuity in the cyanobacterial cytochrome *c* oxidase. In *The Phototrophic Prokaryotes* (G.A. Peschek, W. Löffelhardt, and G. Schmetterer, eds.). Kluwer Academic/Plenum, New York, pp. 357–366.

# Additional Reading

## *Microbial Energetics*

Deckers-Hebestreit, G. and K.A. Hendorf. 1996. The $F_0F_1$-type of ATP synthase of bacteria: Structure and function of the $F_0$ complex. *Annual Review of Microbiology* 50:791–824.

Mitchell, P. 1979. Keilin's respiratory chain concept and its chemiosmotic consequences. *Science* 206: 1148–1158.

Stevens, T.H. and M. Forgac. 1997. Structure, function and regulation of the vacular ($H^+$)-ATPase. *Annual Review of Cell Developmental Biology* 13:779–808.

Trumpower, B.L. and R.B. Gennis. 1994. Energy transduction by cytochrome complexes in mitochondrial and bacterial respiration. *Annual Review of Biochemistry* 63:675–716.

## *Bacterial Respiration*

Knowles, C.J. (ed.). 1980. *Diversity of Bacterial Respiratory Systems*, Vols. 1 and 2. CRC Press, Boca Raton, FL.

## *Photosynthesis*

Blankenship, R.E., M.T. Madagin, and C.E. Bauer (eds.). 1995. *Anoxygenic Photosynthetic Bacteria*. Kluwer, Boston.

Fay, P. and C. Van Baalen (eds.). 1987. *The Cyanobacteria*. Elsevier, New York.

Olson, J.M., J.G. Ormerod, J. Amesz, E. Stackelbrandt, and H.G. Trüper (eds.). 1987. *Green Photosynthetic Bacteria*. Plenum Press, New York.

Peschek, G.A., W. Löffelhardt, and G. Schmetterer (eds.). 1999. *The Phototrophic Prokaryotes*. Kluwer Academic/Plenum, New York.

# 10
# Transmembrane Movement: Mechanisms and Examples

For insofar as bioenergetics is the study of the vital activities of living things, it is an essential element in any integrated view of life—whether we seek to understand it at the molecular, cellular, organismic, or even ecological level.
F.M. Harold, *The Vital Force*, 1986

## 10.1. Introduction

Many chemicals are required for microbial growth, and their acquisition by cells is essential for metabolism. The cell wall limits permeability of large molecules from interfacing with the plasma membrane, and only soluble organic or inorganic nutrients are acquired by the cells. In some instances, chemicals accumulate inside the cell and their release is essential for cell viability. These compounds entering or exiting the cell account for only a fraction of the aqueous environment and are referred to as solutes. The movement of solutes across the plasma membrane is frequently attributed to a "permease," a term that was proposed by G. Cohen, S. Rittenberg, and J. Monod in the mid-1950s to describe a protein in the plasma membrane that functions to move a solute across the plasma membrane. Not all transport systems are permeases because some have multiple components with binding proteins in the periplasm along with proteins in the membrane; these are frequently described as "transporter" systems. On occasion, solute uptake has been described as occurring by a "carrier" system; however, this should not be used as a substitute for a more descriptive designation. Transport is discussed in terms of dilute solutions where the concentration of solute is $<0.1$ $M$ because at low solute concentrations molecular interferences are at a minimum. There are two basic types of solute transport in bacteria, designated as active transport or passive transport according to energy requirements. A series of models are shown in Figure 10.1 to indicate some of the various types of transmembrane movement of solutes in prokaryotes. The active transport systems require energy while diffusion systems are not energy dependent. The movement of water as the solvent for the solute across a semipermeable membrane is commonly considered osmosis and is not discussed here.

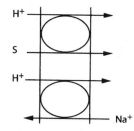

(a) **Diffusion**

Passive

Facilitated

(b) **Chemiosmotic**

Proton symport

Proton antiport

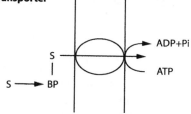

Figure 10.1. Activities associated with the major transport systems in prokaryotes. (**a**) Diffusion does not require energy from the cell. Passive diffusion is nonspecific for a solute while facilitated diffusion does have solute specificity. (**b**)–(**e**) Active transport systems that require some type of cellular energy. (b) Many solutes are transported in response to proton-motive force. (c) ATP-driven uptake requires solute-specific binding proteins (BP). (d) Sugars are phosphorylated as they are moved across the plasma membrane. A chain of proteins carry the phosphate moiety to the sugar molecule with phosphoenolpyruvate as the energy source. (e) Solute transport is coupled with uptake of a specific compound linked to export of another. See the text for specific details on these transport systems.

(c) **ABC uptake transporter**

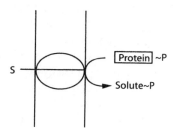

(d) **Group translocation (phosphotransferase system)**

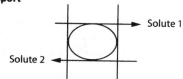

(e) **Exchange transport**

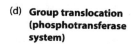

## 10.2. Diffusion Systems[1]

An empirical relationship for diffusion was established by Adolph Fick in 1855 that is known as Fick's first law and is expressed as follows:

$$J = dn/dt = -DA \, (dc/dx)$$

The flux, $J$, refers to the number of moles of solute that have diffused with time measured in seconds and is expressed as $dn/dt$. $D$ is the diffusion coefficient for that solute molecule which has units of cm$^2$/sec. $A$ is the area through which diffusion occurs while $dc/dx$ is the concentration gradient with the concentration, $c$, in mol/cm$^3$ and $x$ is the thickness through which diffusion has occurred. The negative sign indicates that the solute movement occurs in a direction from a higher to a lower concentration. Diffusion stops when there is no concentration gradient. The diffusion coefficients have been calculated for a large number of important biological molecules including salts, gases, peptides, and sugars. The shape and size of the molecule has considerable influence on the solute because long large molecules have higher frictional interactions than smaller spherical ones. For asymmetric molecules, the relationship of frictional coefficients to diffusion coefficient can be calculated using equations developed by Einstein and others.

### 10.2.1. Passive Diffusion

Diffusion through membranes is difficult to express mathematically because of membrane complexity. Because diffusion through the membrane is slower than in an aqueous solution, the concentration gradients outside the membrane are not rate limiting for diffusion. The transmembrane transport or permeability is dependent on the properties of the membrane and the concentration across the membrane. For mathematical consideration, it is assumed that the concentration across the membrane of approximately 10 nm is uniform and that the concentrations of solute at the outer ($c_o$) and inner ($c_i$) sides of the membrane are constant. The difference in the external and internal concentrations of solute is the concentration gradient across the membrane. The net flux of solute in mol/cm$^2$/sec across the membrane, $J$, can be expressed as:

$$J = D/x_m \, (c_o - c_i)$$

Because the functional thickness of the plasma membrane and the diffusion coefficient inside the plasma membrane are not known, the $D/x_m$ expression is replaced by $P$, which is the permeability coefficient with the units of cm/sec. The net solute flux across the membrane is expressed in the following equation.

$$J = P \, (c_o - c_i)$$

Diffusion is the movement of solute across a semipermeable membrane by energy-independent processes and is found in all cellular systems. The two fundamental types of diffusion found in microorganisms are passive or simple diffusion and facilitated diffusion. Passive diffusion is relatively slow and there is no solute specificity. Because membranes have a lipid interior and charged surfaces, molecules that have both lipophilic and hydrophilic character will traverse the membrane faster than molecules without both of these characteristics. Nonionized molecules with a molecular mass of 100 to 250 daltons will readily traverse the membrane. However, one or more ionized groups on the molecule will present difficulties unless the charge is removed. Weak acids will penetrate the cell when the external pH is sufficiently low to depress ionization and promote the formation of the free acid. Conversely, if the solute is a weak base penetration occurs only if the external pH is increased. The direction of solute flow is dependent on concentration with flow in the direction from a higher to a lower concentration. A hallmark characteristic of passive diffusion is that solute will not accumulate against a gradient but solute movement will stop when the concentration across a membrane is equal.

## 10.2.2. Facilitated Diffusion

Certain molecules are considered to pass through the membrane in a manner slightly different than by simple diffusion. The principal difference is that limited specificity is shown by the facilitated diffusion system for a given solute. That is, it would be expected that the facilitated diffusion system could distinguish between various molecular species but could not discriminate between enantiomorphs (D- or L-forms) of optically active compounds. The movement of glycerol in *Escherichia coli*, *Lactococcus lactis*, *Streptomyces coelicolor*, and *Bacillus subtilis* is considered to be an example of facilitated passive diffusion. As in passive diffusion, facilitated diffusion would not require energy for entry and no concentration of solute will occur against a gradient. Thus far the transmembrane pores for glycerol movement and the plasma membrane protein accounting for this transport have not been identified.

## 10.2.3. Accumulation of Solute

There are three factors to consider with respect to transmembrane diffusion: the concentration of solute outside the membrane, the movement through the membrane, and the concentration of solute on the cytoplasmic side of the membrane. If the solute is not allowed to accumulate in the cytoplasm, equilibrium will never be reached even though solute continues to cross the plasma membrane. Removal of solute may be the result of metabolism, as a large number of catabolic or biosynthetic enzymes are found soluble in the cytoplasm. In the early stages of the lag phase of growth, diffusion of substrates into the cell is important, as the induction of appropriate pathways for enzyme synthesis is crucial. Because the permeability coefficient is specific for a molecule, this difference in rate of diffusion by substrates may help explain the varying length of lag seen with different substrates.

Solute accumulation may be due to charges resulting from the interior of the bacterial cell being pH 7.5 to 8.0, which may be about 2 pH units higher than outside of the plasma membrane. Although molecules always display transport flux which is the bidirectional exchange across the membrane, the significant change in charge on the molecule once inside the cell may produce a molecular species that diffuses out at a reduced rate. As the alkaline pH environment inside the cell is a reflection of metabolic activity, accumulation of a solute inside the cell could result.

Diffusion appears to be an extremely important mechanism for the entry of many antibiotics into bacterial cells. Charges on the antibiotic are a consideration, with greatest transport occurring with noncharged species. For example, the minimal inhibitory concentration of erythromycin for bacteria decreases as the medium becomes more alkaline because erythromycin has a $pK_a$ of 8.8. Even though the rate of diffusion for antibiotics may be slow, the antibiotics will continue to enter the cell for a considerable time because the antibiotics bind to an enzyme, ribosome, or macromolecular structure and do not accumulate as the free molecule in the cytoplasm. Thus, the concentration of antibiotic outside the plasma membrane ($c_o$) never equals that inside ($c_i$) and equilibrium is not reached. The shifting of the equilibrium by antibiotic binding to target molecules in the bacterial cytoplasm may explain why antibiotics are effective in treating bacteria that grow inside animal cells.

In this consideration of diffusion transport, only the plasma membrane has been discussed; however, movement of the solute must also be through the outer membrane of Gram-negative cells or through the lipid cell walls of acid-fast bacteria. The different chemical properties of the cell walls of bacteria may also influence the rate of solute traversing the cell wall region because in many instances the solute in the extracellular environment may not be readily accessible to the plasma membrane. Models have not yet been developed to take into consideration the impact of cell wall properties on solute transport.

## 10.3. Chemiosmotic Activities[2]

Of paramount importance to the microbial cell is the coupling of energy produced by respiratory systems to solute transport. Based on Mitchell's concept of vectorial movement of protons in the plasma membrane, solute transport systems are organized into two groups: primary transport and secondary transport. These two categories of solute transport are summarized in Table 10.1. Basically, the primary transport process involves metabolism as the solute is transported into the cytoplasm. The nature of this metabolism may range from phosphorylation reactions attributed to high-energy phosphate-mediated reactions of the ATP binding cassette (ABC) transporter systems and ATPase-coupled transport to the phosphenol phosphate-driven reactions of the phosphotransferase system. The driving force for secondary transport is from ion currents that are established across the plasma membrane and produce the proton-motive force. The currents generally are associated with coupling solute movement to $H^+$ or $Na^+$ cycling across the membrane. Specific features of solute transport coupled to chemiosmotic processes are discussed in the following segments of this chapter.

### 10.3.1. Proton-Motive Force

The chemiosmotic transport of solutes couples transmembrane movement of solute to metabolism and this concept is attributed to Peter Mitchell, who received the Nobel

Table 10.1. Organization of the transport systems based on mechanisms of transport.[a]

**Primary transport systems**: Vectorial transport of a solute is concomitant with metabolism.
1. Primary solute transport: Transport of a solute is coupled to ATP hydrolysis Examples of this include the various P-type ATPase and ABC systems.
2. Group translocation: Chemical modification of solute as it is transported across the plasma membrane. An example of this transport system is the phosphotransferase reaction in which a sugar is phosphorylated as it traverses the plasma membrane.
3. Proton, sodium, and electron translocation: Vectoral movement of protons and sodium ions by multiprotein configurations. Examples of this include $H^+$ or $Na^+$ transport by F-type and V-type ATPases as well as $H^+$ pumps associated with electron transport complexes.

**Secondary transport systems**: Ionic gradients are used to drive transport without metabolism of solute. Generally the system uses protons and sodium ions which are generated independent of the solute to be transported.
1. Symport (also known as cotransport): Two solutes are moved in the same direction by a single carrier or transporter. An example of this is the proton-coupled transport of sugars in bacteria.
2. Anteport (also known as exchange reaction): Two solutes are moving in the opposite direction by a single carrier or transporter. An example of this in bacteria is the export of $Na^+$ driven by $H^+$ uptake and both use the same carrier.
3. Uniport: A solute is transported through a channel or carrier without an association of other ions. An example of this is the carrying of ions by valinomycin or gramicidin.

[a]Adapted from Harold, 1986.[2]

Prize for this work in 1979. In the chemiosmotic solute transport system, osmotic energy is created from exergonic reactions and a transmembrane gradient of protons and electrons is generated. As electrons and protons are removed from substrates by oxidative reactions, electrons move within the membrane to an electron acceptor while $H^+$ are spatially separated from the electrons and expelled outward across the plasma membrane. The result is the localizing of protons on the exterior surface of the plasma membrane and hydroxyl ions on the inner surface of the plasma membrane. A model detailing the charging of the bacterial plasma membrane is shown in Figure 10.2. Hydroxyl ions and protons come from the ionization of water and while protons are expelled the hydroxyl ions remain inside and contribute to the charge across the membrane. On the outside of the membrane, the electrical potential is positive and acidic owing to the presence of protons while the inner side of the membrane is negative and alkaline owing to the accumulation of hydroxyl ions. These membrane-related activities are summarized by the following reaction:

$$\text{proton-motive force} = \Delta\bar{\mu}_{H+} = \Delta\psi - Z\Delta pH$$

The proton-motive force is attributed to the membrane potential, $\Delta\psi$, and the transmembrane pH gradient, $\Delta pH$. In the equation, $Z$ is the factor used to convert pH into millivolts and at 25°C $Z = 59$. In bacteria of high metabolic activity, $\Delta\psi$ has been determined to be 150 to 200 mV which is approximately equivalent to the energy from the hydrolysis of a molecule of ATP. In growing bacteria, the $\Delta pH$ has been found to be about 2 pH units. Thus, there are two distinct but related features ($\Delta\psi$ and $\Delta pH$) contributing to chemiosmotic driving force.

It should be noted that $\Delta\bar{\mu}_{H+}$ is generated in bacterial membranes during respiration

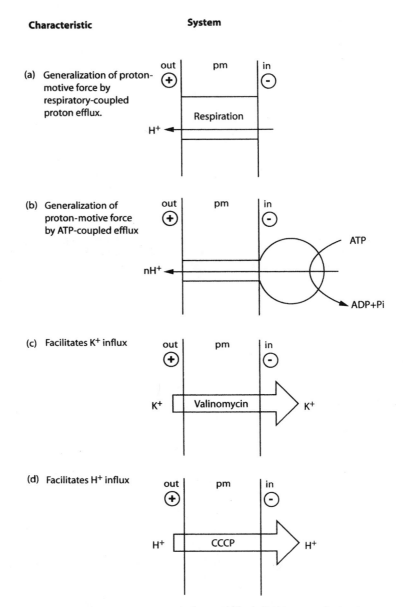

Figure 10.2. Chemiosmotic activities associated with solute transport in prokaryotes. (**a, b**) Examples of proton extrusion across the plasma membrane. (**c, d**) Examples of dissipation of proton-motive force by the use of ionophores.

and in fermenting bacteria an alternate method for generating $\Delta\tilde{\mu}_{H+}$ is used. Fermenting bacteria utilize ATP from substrate phosphorylation to reverse the $F_1/F_0$ ATPase and expel protons in a reaction that is driven by ATP hydrolysis. The generation of $\Delta\tilde{\mu}_{H+}$ by the ATPase is shown in Figure 10.2.

Using everted (right side out) vesicles from plasma membranes, $\Delta\psi$ and $\Delta pH$ have been measured when electron transport was active. For these measurements, the Nernst equation is used:

$$\Delta\psi = -59 \log ([K^+]_{in} / [K^+]_{out})$$

and

$$\Delta pH = -59 \log ([\text{weak acid}]_{in} / [\text{weak acid}]_{out})$$

## 10.3.2. Proton-Coupled Transport[3]

In bacteria, chemiosmotic driven solute movement is coupled to $H^+$ translocation and energized by $\Delta\tilde{\mu}_{H+}$ (see Figure 10.2). The entry of a negatively charged or uncharged solute along with $H^+$ is termed proton symport, the entry of $H^+$ with the exit of charged solute is termed proton anteport, and the entry of positively charged solute without any couple to proton flow is uniport. In general, sugars and other noncharged solutes are translocated on a proton symport involving both protons and membrane charge. Anions carrying a negative charge are translocated on a proton symport and are influenced only by the pH gradient; the transport is then electroneutral. Cations with a positive charge are translocated by a uniport where solute translocation is driven by membrane potential without any movement of protons. Table 10.2 provides a summary for chemiosmotic driven solute transport.

It is possible to determine if the driving force for transmembrane movement of a solute is attributable to $\Delta\psi$ or to $\Delta pH$ through the use of specific inorganic ion carriers referred to as ionophores. Most commonly, the organic molecule carries a specific inorganic cation in the center of the molecule and these caged ions are referred to as clathrates. One of these ionophores is valinomycin, which is a cyclic molecule consisting of α-hydroxyl acids and α-amino acids that has the following composition: (D-hydroxyvalerate–D-valine–L-lactate–L-valine)$_3$. The aliphatic side chains on the exterior of the molecule contribute to the lipid-soluble character and carboxyl groups projected toward the center of the molecule form a specific binding site for $K^+$.

Table 10.2. Transport characteristics associated with charged solutes.

| Solutes | Charge on solute at pH 7 | Mechanism of transport | Characteristic of flux | Driving force |
|---|---|---|---|---|
| Lysine | + | Uniport | Electrogenic | $\Delta\psi$ |
| Glucose | 0 | $H^+$-Symport | Electrogenic | $\Delta\psi - Z\Delta pH$ |
| Phosphate | − | $H^+$-Symport | Electroneutral | $-Z\Delta pH$ |
| $Na^+$ | + | $H^+$-Anteport | Electroneutral | $-Z\Delta pH$ |

Table 10.3. Influence of ionophores on chemiosmotic transport of solutes.

| Ionophore | Specificity |
|---|---|
| **Cationophore** | |
| Valinomycin | $K^+$, $Rb^+$ >> $Na^+$, $Li^+$, $H^+$ |
| Monensin | $Na^+$, $Li^+$, > $K^+$, $Rb^+$ |
| Nigericin | $K^+$, $Rb^+$, > $Na^+$, $Li^+$ |
| **Protonophores** | |
| 2,4-DNP (2,4-dinitrophenol) | $H^+$ >> $K^+$, $Rb^+$, $Na^+$ |
| PCP (Pentachlorophenol) | $H^+$ >> $K^+$, $Rb^+$, $Na^+$ |
| CCCP (*m*-Chlorophenyl carbonyl cyanide hydrozone) | $H^+$ >> $K^+$, $Rb^+$, $Na^+$ |
| FCCP (*p*-Trifluoromethoxyphenyl carbonyl cyanide hydrazone) | $H^+$ >> $K^+$, $Rb^+$, $Na^+$ |
| Gramicidin | $H^+$ > $K^+$, $Rb^+$, $Na^+$ |
| **Anionophore** | |
| Reduced benzyl viologen | $NO_3^-$ and $NO_2^-$ |

Other antibiotic ionophores commonly used in bacteria energetics include nigericin, which transports $K^+$ and other monovalent cations by a $H^+$ anteport and monensin, which moves $Na^+$ by a $H^+$ anteport system. Proton ionophores such as gramicidin, 2,4-dinitrophenol, and m-chlorophenyl carbonyl cyanide hydrozone (CCCP) are commonly used as uncouplers of oxidative phosphorylation but can also be used to disrupt the proton gradient across the membrane in solute transport studies. The specificities of some commonly used ionophores and protonophores are given in Table 10.3.

As shown in Figure 10.2, the nature of the driving force for chemiosmotic transport can be determined with the addition of ionophores that will disrupt the ion gradients. These ionophores are uncharged molecules that specifically complex with cations such as $H^+$, $K^+$, or $Na^+$. The result is a lipophilic complex that enables $H^+$ or the cation to freely traverse the membrane, disrupting either $\Delta\psi$ or $\Delta$pH. Examples of ionophores and their activities are given in Table 10.3. An ionophore that carries $H^+$ is termed a protonophore and its presence will disrupt the $H^+$ gradient across the membrane. In addition to disrupting chemiosmotic driven solute transport, protonophores will also stop bacterial flagellar rotation and oxidative phosphorylation.

In *Escherichia coli*, the proteins associated with proton-driven sugar transport have been studied extensively. The lactose: proton symporter has been isolated by Kaback and co-workers and the amino acid sequence is known. The lactose protein has 12 transmembrane helices just as is found with the proton symporters that have been isolated for arabinose, xylose, and galactose. Ongoing experiments with site-directed mutagenesis are being conducted to identify conclusively the proton-carrying feature of the transporter molecule.

## 10.3.3. Na$^+$ Ion Gradients[4,5]

There are many examples of $Na^+$ cotransport in bacteria (see Table 10.4). There are three mechanisms whereby the $Na^+$ electrochemical gradient, $\Delta\bar{\mu}_{Na^+}$, is generated:

Table 10.4. Examples of Na$^+$ cotransport systems in bacteria.

| Environment | Organism | Cosubstrate transported |
|---|---|---|
| Extreme salt (2.5 to 5 $M$ Na$^+$) | *Halobacterium halobium* | 19 different amino acids |
| Marine (0.46 $M$ Na$^+$) | *Altermonas haloplanktis* | Galactose, succinate, malate, citrate, alanine, leucine, glutamate, lysine, asparagine, arginine, phosphate |
| | *Methanococcus voltae* | Glycine, isoleucine, leucine, phenylalanine, proline, serine, valine |
| | *Vibrio fischeri* | Succinate, alanine, arginine, glycine |
| Rumen | *Streptococcus bovis* | Alanine, serine, threonine |
| | *Methanobrevibacter ruminantium* | Coenzyme M |
| Alkaline | *Bacillus alcalophilus* | Aspartate, leucine, serine, malate, α-aminobutyrate |
| Intestine | *Escherichia coli* | Glucose, melibiose, proline |
| | *Salmonella typhimurium* | Melibiose, proline |

(1) a Na$^+$/H$^+$ anteport system, (2) a primary Na$^+$ pump, and (3) a respiratory Na$^+$ translocating ATPase/synthase. The Na$^+$/H$^+$ anteporter is broadly distributed in alkalophiles, halophiles, marine bacteria, rumen bacteria, and heterotrophic bacteria. A primary Na$^+$ pump has been demonstrated in or suggested from inhibitor studies to occur in *Streptococcus faecalis*, *Acholeplasma laidlawii*, and *Methanococcus voltae*. Another example of a primary Na$^+$ pump is attributed to a Na$^+$ translocating decarboxylase. Decarboxylation of citrate by *Klebsiella aerogenes* and decarboxylation of oxaloacetate by *Salmonella typhimurium* are specific examples of Na$^+$ pumps coupled to decarboxylases broadly distributed in anaerobes. The third example is observed with *Propionigenium modestum* which ferments succinate to propionate and pumps two Na$^+$ ions out of the cell with each succinate molecule decarboxylated. In addition to using the resulting Na$^+$ electrochemical gradient formed to function in Na$^+$ symport systems, *Propionigenium modestum* is known to generate ATP by the Na$^+$ translocating ATP synthase, as the F$_o$ of the F$_1$F$_o$-type ATPase transports Na$^+$.

It is considered that the Na$^+$ cotransport systems evolved after the H$^+$ cotransport systems. Few mutations would be needed to change the amino acids to account for recognition of Na$^+$ in place of H$_3$O$^+$ and produce a low-affinity Na$^+$ cotransport system. The next mutation could have produced a high-affinity Na$^+$ cotransport system that lost affinity for H$_3$O$^+$. The use of Na$^+$ for cotransport in bacteria growing in high salt environments can be readily accepted from a physiological ecology perspective. The use of Na$^+$ transport by rumen bacteria may reflect the presence of approximately 0.1 $M$ NaCl in animal circulatory systems.

Unique bacteria that grow in alkaline environments rarely experience a proton gradient across the plasma membrane. These organisms establish a Na$^+$ gradient across the plasma membrane and use Na$^+$ to drive solute transport. It appears that the use of $Na^+$ driven transport in bacteria is limited to a few species while the proton-driven reactions are broadly distributed throughout the various genera of bacteria.

## 10.4. Transporter ATPases

With ATP serving as the energy commodity in cells, it is to be expected that there is a system of solute transport in cells that is driven by ATP. In fact, there are several different ATP-dependent transport systems that are distinguished by the type of solutes moved and the molecular architecture of the transporter. Previously, the V-type ATPase and the F-type ATPases were discussed with respect to their function in $H^+$ or $Na^+$ transport and the reader is referred to Chapter 9. Important features that distinguish V- and F-type ATPase from P-type ATPase are the sequence motifs for ATP binding. These sequences are also known as Walker motif A, $(K/R)(Hy)_{4-6}$ G[XGXX or XXGX]GK, and Walker motif B, $(K/R)X_{2-3}GX_{2-3}(Hy)_5D$. In these sequences, Hy is any hydrophobic residue, X is any amino acid, alternative sequences are enclosed in brackets, and nonspecific residues are in parentheses. These Walker A and B motifs are found only in $\alpha$- and $\beta$-subunits of $F_1$ subunit of F-type ATPase, in the $V_1$ subunit of V-type ATPase, and in ABC transporters.

### 10.4.1. P-Type ATPases[6–8]

In addition to the F-type and V-type proton translocating ATPase, bacteria also have a P-type ATPase. Generally, the P-type ATPase has only one large protein subunit of approximately 100 kDa but exists as two conformational forms, each with different reactivity to substrates and proteases. The P-type ATPases are so called because there is a phosphoryl intermediate involving an aspartyl phosphorylation site. Vanadate, a phosphate analog, inhibits the P-type ATPase by competing with the phosphate for the aspartic residue which resides in one of the extramembraneous regions. All P-type ATPases contain the conserved [TGDGVNDHyPAL] sequence for ATP binding and the aspartyl phosphorylation consensus is [HyCSDKTGTHyT]. Fully conserved residues are in bold and [Hy] refers to any hydrophobic residue. In addition, P-type ATPases have a conserved proline in the transmembrane segment that is considered to be important in ion translocation. At this time, there are three classes of P-type ATPases in bacteria and generalized models of these transport systems are given in Figure 10.3. In all classes, the P-type ATPases consist of three protein units; however, there is a distinction in the number of transmembrane helical peptide segments and the types of solutes transported. The number of transmembrane loops in P-3 class, P-2 class, and P-1 class are 6, 10, and 8, respectively. Although the specific site for metal binding is not resolved for the P-type ATPases, it is known that metal binding in the P-1 class is near the N-terminus of the molecule.

The P-type ATPases are broadly distributed in the biological world. Repesentative examples found in the eukaryotic organisms include the $Na^+,K^+$-ATPase of higher animals, $Ca^{2+}$-ATPase in sarcoplasmic reticulum, $K^+,H^+$-ATPase in gastric cells, and

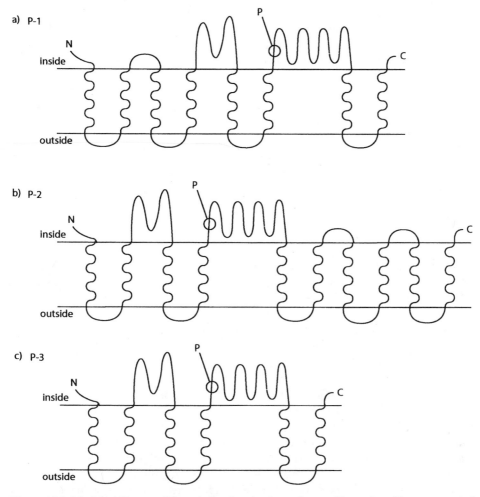

Figure 10.3. Models of P-type ATPase indicating the three classes. Examples of ions transported by class (**a**) P-1 are $Cd^{2+}$, $Cu^{2+}$, $Zn^{2+}$; (**b**) P-2 are $Na^+$, $K^+$, $K^+/H^+$, $Ca^{2+}$, $Na^+/K^+$; and (**c**) P-3 is $K^+$. The region encircled at "P" indicates the location of an aspartate residue that becomes phosphorylated by ATP. This is the site for inhibition by vanadate.

$H^+$-ATPase of fungi. The Kdp ATPase of *Escherichia coli* is an example of the P-type specific for $K^+$ transport and it has a structure consisting of three integral protein subunits with masses of 59 kDa, 72 kDa, and 20.5 kDa. The Kdp has a $K_m$ of 2 $\mu M$ and with respect to energetic considerations, it is suggested that one or two $K^+$ ions are transported for each ATP hydrolyzed. Using probes constructed from the Kdp genes of *Escherichia coli*, Southern blot analysis of DNA from Gram-negative bacteria revealed that Kdp is present in *Enterobacter, Pseudomonas, Salmonella, Shigella*, and *Rhizobium*. In *Salmonella typhimurium*, there is a 102-kDa protein that was derepressed in low Mg²⁺ and it is known as the MgtB system. In addition to Mg²⁺, the MgtB system also transports Ni²⁺, Co²⁺, and Mn²⁺. In *Enterococcus hirae* and in *Streptococcus lactis*, Ca²⁺ is exported by a P-type ATPase. Resistance to cadmium has been attributed to a plasmid-encoded Cd²⁺-ATPase, a P-type ATPase of 78 kDa that is responsible for the efflux of Cd²⁺ from the cytoplasm.

## 10.4.2. A-Type ATPases[9,10]

The A-type ATPase is a relatively small family that accounts for microbial resistance to arsenate ($AsO_4^{3-}$), arsenite ($AsO_3^{2-}$), antimonite ($SbO^+$), and tellurite ($TeO_3^{2-}$). This A-type ATPase is structurally distinct from F-type, V-type, or P-type ATPases. The efflux of these toxic oxy-anions is by an anion pump that is driven by ATP hydrolysis and a model of the efflux system is discussed in Chapter 14. Bacterial plasmids carry the genes for arsenate resistance on an *ars* operon of 4.3 kb that is induced by arsenate, arsenite, or antimonite. In plasmid R773 of *Escherichia coli*, there are three important genes in the *ars* operon: *arsA*, *arsB*, and *arsC*. The *arsA* encodes for a 63-kDa soluble protein, the *arsB* encodes for a 45.5-kDa membrane-bound protein, and the *arsC* designates a 15.8-kDa soluble protein. For arsenite export, the ArsA and ArsB proteins are required. ArsA is the protein that binds arsenite while ArsB is the transmembrane subunit that associates with ArsA and exports arsenite. ArsC is the arsenate reductase bound on the cytoplasmic side of the ATPase. The sequence of physiological activity is as follows: arsenate in the cytoplasm is reduced to arsenite which is rapidly exported from the cell.

The ArsA is a homodimer on the cytoplasmic side of the A-type ATPase and it binds arsenite ($As^{3+}$) or antimony ($Sb^{3+}$) through three specific cysteine residues. The hydrolysis of ATP provides the energy for arsenite export through ArsB. If arsenate is present, it is reduced by ArsC to arsenite. There appear to be differences in the supply of reducing power to the ArsC proteins. The arsenate reductase produced by plasmid pI258 (*Staphylococcus aureus*) interfaces with thioredoxin but not glutaredoxin while the product from plasmid R773 (*E. coli*) uses glutaredoxin but not thioredoxin. In both cases, the reducing power is transferred through cysteine residues to the arsenate molecule.

## 10.5. ABC Transporters

A major transport system for many prokaryotes involves proteins exterior to the plasma membrane and proteins in the plasma membrane with an ATP binding site called the ATP-binding cassette (ABC). Energy for solute transport comes from the hydrolysis of ATP. These ATP transporters are found in both Gram-negative and Gram-positive bacteria including archaea. One type of ABC transporter contains a binding protein exterior to the plasma membrane and functions as an importer while the other ABC transporter lacks a binding protein exterior to the membrane and has a role in export of cytoplasmic materials. The following section will first focus on the import transport system with binding proteins before addressing the ABC export system.

## 10.5.1. Binding Proteins and ABC Importers[11–13]

Solutes may be imported by a system consisting of binding proteins and ABC proteins. Binding proteins are induced when the solute is in low concentration and these proteins have been found to have a high binding value for a specific solute. More than 100 binding proteins have been isolated from bacteria and characteristics of some of these are given in Table 10.5. Binding proteins are loosely associated with the outer face of the plasma membrane and can be readily released from the periplasm. A cold osmotic shock procedure has been developed by Neu and Hepple to release binding proteins from the periplasm. Cells are washed with 30 m$M$ NaCl and 10 m$M$ Tris, pH 7.3, before they are resuspended in 20% sucrose containing 1 m$M$ EDTA in 33 m$M$ Tris buffer. Cells are collected by centrifugation and rapidly diluted in cold distilled water. In the procedure, sucrose collected in the periplasm generates high osmotic pressure against the outer membrane with ultimate rupture and release of periplasmic materials. The cells are collected by centrifugation and the centrifugate contains the shock fluid which contains about 25 different proteins or about 15% of the cellular proteins. Transport proteins not released by this shock procedure are usually of the solute: proton transporter class.

The ABC system consists of transmembrane proteins flanked on the cytoplasmic side of the membrane with two proteins with a specific binding site for ATP. The model used to explain ABC importer is either the histidine transport in *Salmonella typhimurium* or maltose in *E. coli*. In each case there are five proteins involved: a periplasmic binding protein, two transmembrane proteins, and two ATP-binding proteins.

Table 10.5. Examples of binding proteins of the ABC system isolated from enteric bacteria.[a]

| Solute | Organism | Molecular weight | $K_D$ |
|---|---|---|---|
| Sulfate | *Salmonella typhimurium* | 34.7 kDa | $3.6 \times 10^{-5}$ |
| Phosphate | *Escherichia coli* | 34.0 kDa | $2.0 \times 10^{-7}$ |
| L-Leucine | *E. coli* | 37.0 kDa | $3.0 \times 10^{-7}$ |
| L-Histidine | *S. typhimurium* | 25.0 kDa | $4.0 \times 10^{-7}$ |
| L-Lysine | *E. coli* | 28.0 kDa | $1.5 \times 10^{-7}$ |
| L-Glutamine | *E. coli* | 26.0 kDa | $6.0 \times 10^{-7}$ |
| D-Glucose | *E. coli* | 35.0 kDa | $4.0 \times 10^{-7}$ |
| L-Arabinose | *E. coli* | 33.0 kDa | $2.0 \times 10^{-6}$ |
| D-Ribose | *S. typhimurium* | 29.5 kDa | $3.0 \times 10^{-6}$ |

[a]From Furlong, 1987.[13]

$K_D$ = dissociation constant.

## 10.5.2. ABC Systems for Export[14]

The export or secretion of several enzymes, capsular polysaccharides, and peptide antibiotics may occur via the ABC exit transporter. In Gram-negative bacteria this ATP-dependent transport of protein outward across the plasma membrane and the outer membrane while in Gram-positive bacteria the protein is moved across the plasma membrane and the cell wall. This transporter is also known as the type I secretory system. Some prefer to designate the export across the plasma membrane as "transport" while export across both the outer and plasma membranes as "secretion." Proteins secreted by the ABC transport system have a specific C-terminal glycine-rich sequence that interacts with the ABC proteins in the plasma membrane. This glycine-rich sequence contains GGXGXD, and depending on the protein, the sequence is repeated 4 to 36 times. The protein molecules transported do not have a complex series of subunits and include (1) the pore-forming toxins of *E. coli, Bordatella pertussis*, and *Pasteurella hemolytica*; (2) the metalloproteases of *Pseudomonas aeruginosa, Serratia marcescens*, and *Erwinia chrysanthenium*; and (3) colicin V of *E. coli*. The genes for secretion of exoproteins in Gram-negative bacteria have at least five functional units: (1) a gene that encodes for the hydrophobic ABC protein that contains six transmembrane segments; (2) a gene for the protein that becomes localized in the outer membrane (OMP) and functions to secrete the exoprotein through the outer membrane; (3) a gene that produces the membrane fusion protein (MFP) that forms a channel between the ABC protein and the OMP; (4) a gene for the production of the protein to be secreted that contains the glycine-rich peptide for secretion recognition; and (5) a gene that encodes for either an inhibitor peptide to prevent the exoprotease or an activator for the toxins, thereby preventing the proteases or toxins from harming the cell that produces it. It is considered that a specific ABC transporter functions for only one exoprotein. Additional information on protein secretion in bacteria is found in Chapter 12.

## 10.6. Group Translocation System

Group translocation is a solute transport process that couples metabolism with transmembrane movement. In this case, the solute is changed as it crosses the plasma membrane and in a strict sense, there is never an accumulation of nonmetabolized solute inside the cell. The phosphorylation of sugars coupled to transport has been extensively characterized and is the model for group translocation. The metabolism of other solutes as transport occurs is extremely important but is poorly characterized. The following section addresses both types of group translocation.

## 10.6.1. Phosphotransferase System[15,16]

An active transport system unique to bacteria is the group translocation process, and phosphotransferase system (PTS) is a specific example of this activity. The PTS was initially reported by S. Roseman and colleagues in 1964 and it is important for the transport of several hexose and hexitol sugars in various species of bacteria. In this metabolic process, the sugar is phosphorylated as it is transported into the cell with phosphoenol pyruvate (PEP) serving as the source of the high-energy phosphate. As depicted in Figure 10.4, there are a series of proteins that serve to shuttle phosphate from PEP to the sugar. The first protein to interact with PEP is enzyme I, which is 80 kDa and carries the phosphate through a histidine residue. The second protein is Hpr, which is 9.6 kDa and it also becomes phosphorylated through histidine. Enzyme I and Hpr are constitutive and can interact with enzyme II, which displays considerable specificity for solutes. There are 10,000 to 30,000 molecules of enzyme I in a cell; however, there are 100,000 to 300,000 molecules of Hpr in each cell. Enzyme II-C transport proteins are in the plasma membrane and it appears that the membrane proteins require about 350 amino acid residues to produce a hydrophobic central channel in the plasma membrane. In the case of mannitol, enzyme II-C$^{Mtl}$ in the membrane and is initially phosphorylated to a histidine moiety before the phosphate group is transferred to mannitol. Enzyme II-C$^{Mtl}$ is 60 kDa, enzyme II-C$^{Glu}$ is 43 kDa, and Enzyme II-C$^{Man}$ is 27 kDa.

The EII proteins consist of multiple components of either three domains of IIA, IIB, and IIC or four domains where an additional domain, IID, is present. The protein

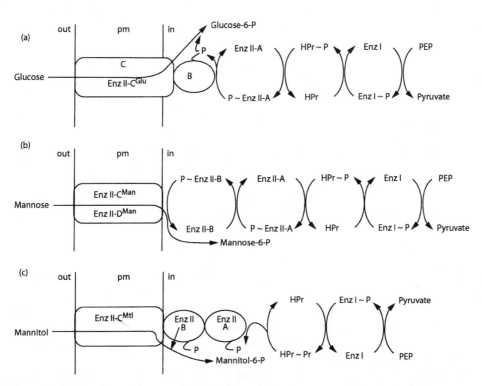

Figure 10.4. Phosphotransferase activities in bacteria for the transport of hexoses. (a) Transport of glucose. (b) Transport of mannose. (c) Transport of mannitol. (From Postma et al., 1996.[15])

domains of IIA, IIB, and IIC are either free or coupled together by covalent bonds. Examples of these variations of EII are given in Figure 10.4. The transport of glucose is by a free EIIA protein and EIIB which is bonded to EIIC. With mannose transport, both EIIA and EIIB are soluble or free while with mannitol transport, the subunits of EIIA, EIIB, and EIIC are attached together as as a single unit. Even though some of the EII domains are coupled together, each domain functions as if it were independent. In those instances where EII subunits are coupled together, it is attributed to translation as a single polypeptide chain. The amino acid chains attaching the EII domains are composed of either alanine-proline (A) rich linkers or glutamine (Q) rich linkers. These AP-rich peptides produce flexible segments that hold the EII domains. In several Gram-positive bacteria, Q-rich pepetides produce the flexible link between EII subunits.

Regulation of PTS over other sugars is proposed to be by either the control of entry of non-PTS sugars and by enhanced production of cAMP from ATP by adenylase cyclase. In both cases it appears that enzyme II $A^{Glu}$, but not the phosphorylated enzyme $A^{Glu}$ interacts with the lactose carrier and excludes lactose. With glycerol, the enzyme $A^{Glu}$ does not bind to the glycerol facilitator but instead it binds to the glycerol kinase. The phosphorylated component (P~enzyme II-$A^{Glu}$) stimulates adenylate cyclase to produce cAMP from ATP. The cAMP along with CAP (catabolite activating protein) stimulate the transcription of various genes for carbohydrate catabolism. However, owing to the rapid transfer of ~P from enzyme II-$A^{Glu}$ to glucose, the P~Enzyme II-$A^{Glu}$ inducer is rarely present.

The uptake of PTS sugars is subject to several types of regulation. One important feature is the ratio of phosphoenol pyruvate to pyruvate. With two or more PTS sugars available to a growing culture of bacteria, only one is metabolized with preference based on competition for the phospho-HPr. The components of the PTS regulate non-PTS sugar metabolism in Gram-positive and Gram-negative bacteria; however, the non-PTS sugars appear to have no effect on the PTS. The PTS may have evolved under anaerobic conditions when phosphoenol pyruvate was a product of glycolysis. Support for the concept that PTS is an ancient process is from the observation that expression of genes for PTS proteins are produced by bacteria under anaerobic and microaerophilic conditions, which are conditions in which a proton-motive force is a result of secondary events and not from respiratory coupled activity.

## 10.6.2. Transport of Non-Sugar Solutes[17,18]

The movement of carbohydrates by the phosphotransferase system involves metabolism linked with transport and a similar linking of catabolism with transport has been noted for fatty acid uptake by bacteria. The uptake of oleate and related long-chain fatty acids by *E. coli* is by a single saturable system which is characteristic of a specific membrane carrier. In mutants, the transport of oleate, dodecanate, decanoate, and octanate occurs only in bacteria with an active system for degradation of fatty acids. The enzymes in catabolism of fatty acids include acyl:CoA synthetase, thiolase, β-hydroxyacyl-CoA dehydrogenase, and enoyl-CoA hydrase. No free fatty acid is found in the cytoplasm at the time of fatty acid uptake and it was observed that acyl:CoA synthetase is involved in mediating translocation. The uptake of long-chain fatty acids is by vectoral acylation which is mechanistically similar to phosphotransferase.

Even though the transport of short-chain fatty acids is distinct from uptake of long-chain fatty acids, a metabolism-linked transport is suggested for several other solutes. The transport of butyrate in cells of *E. coli* requires the presence of acetyl CoA: butyrate transferase in a manner suggestive of vectorial acylation. Although the results are not absolutely conclusive, the transport of uracyl, adenine, guanine, xanthine, and hypoxanthine is suggested to occur by a vectoral phosphorylation system. In this case, free bases are not found inside cells but the saturable transport process results in accumulation of base-ribosyl phosphate in the cytoplasm.

The uptake of nucleosides by *E. coli* is characterized by concomitant deaminations and other steps of degradation, thereby suggesting metabolism-linked transport. Additional research is required to determine if metabolic enzymes are indeed comparable to enzyme II in the plasma membrane of sugar phosphotransferase systems. At this time, the transport of nucleosides, purines, and pyrimidines is suggested to occur by transport processes similar to a vectoral-metabolism system.

## 10.7. Transport in Response to Osmotic Pressure

While yeast and fungi can grow in a wide range of osmotic pressures, prokaryotes prefer a dilute osmotic environment for growth. The one group of organisms that prefer high osmotic conditions are the halophilic and halotolerant archaea and bacteria. However, these prokaryotes that grow in 10% to 20% salt are not known for their adaptation to rapid changes in osmotic pressures, and dilution of halophiles in distilled water results in cell rupture. Bacteria that grow in low to moderate levels of osmotic pressure can withstand rapid shifts in osmotic pressures and this rapid adjustment to the new osmotic environment is attributed to aquaporins and to gated channels. The demonstration of aquaporins and gated channels in bacteria is a recent event and future studies will provide a new dimension to prokaryotic survival in a changing environment.

### 10.7.1. Aquaporins[19]

Large quantities of water can be rapidly moved across the plasma membrane by unique proteins that create a water channel referred to as aquaporins. It has long been known that water can form molecular configurations that tranverse the bimolecular membranes found in biology and this process of moving single water molecules across a membrane is termed osmosis. However, osmosis is relatively slow and it does not explain the rapid transport of water across membranes as a result of an osmotic upshift or an osmotic downshift. While the presence of aquaporins has been recognized in different eukaryotic cells, the presence of aquaporins in prokaryotes has been established only recently. The aquaporin protein (AqpZ) is an integral protein in the plasma membrane that manages water flow either into or out of the cell by forming a water-selective channel across the plasma membrane. The movement of water through the aquaporin channels may be influenced by the degree to which the proteins are phosphorylated, with phosphorylated amino moieties of the aquaporins enabling more water to be transferred than if the proteins are nonphosphorylated. This movement of water is passive and direction of water flow is dependent on a water potential gradient. Water potential is the expression of the free energy associated with water which is dependent on the solute, pressure, and gravity. It is becoming important for bacterial and archaeal physiologists to understand osmotic potential in various environments. It is interesting to note that plant physiologists have been concerned with water potential since the late 1800s, when Wilhelm Pfeffer made his discoveries on osmotic potentials in plants. At this time, the distribution of aquaporins throughout prakaryotes is unknown. Thus far, aquaporins in *E. coli* have been characterized and have been demonstrated in *Brucella abortus*.

## 10.7.2. Gated Channels[20]

Exciting recently acquired information has provided evidence for the presence of a gated channel system in the plasma membrane of prokaryotes. Gated channels function to release cytoplasmic solutes from cells in extremely hypoionic environments. Through electrophysiological studies, it has been learned that gated channels occur in both bacteria (*E. coli* and *Mycobacterium tuberculosis*) and in archaea (*Halofax volcanni*). The channel is open for 0.1 to 3.0 nsec and there is no apparent specificity for ions or solutes in the channel. Not only do small molecules exit through these gated channels but proteins as large as 12 kDa have been observed to exit through these channels.

Three types of efflux proteins (MscL, MscM, and MscS) have been found to produce gated channels and some have been partially characterized. Studies on the crystal structure of MscL isolated from *Mycobacterium tuberculosis* reveals that it is a pentamer with each subunit containing two transmembrane helices. Both the carboxyl and the amino ends of the MscL channel proteins are located on the cytoplasmic side of the membrane and the protein extending into the cytoplasm is at an angle to produce a funnel shape. The cytoplasmic proteins form a gate where hydrophobic interactions hold adjacent proteins at the throat of the funnel. These gated channels respond to mechanosensitive activities. As pressure is extended on the funnel region of the transmembrane protein, the hydrophobic bonds between the monomers of MscL are disrupted and the channel opens. When a pressure reduction in the cytoplasm occurs, the channel collapses and the gate is secured by hydrophobic bonds.

## 10.8. Characteristics of Selected Transport Systems

Solutes pass across the plasma membrane by various passive and active processes and a composite model indicating solute transport is given in Figure 10.5. The following section provides a background for cellular transport activities and this is intended to provide insight into a few of the charged and noncharged solute molecules. To satisfy

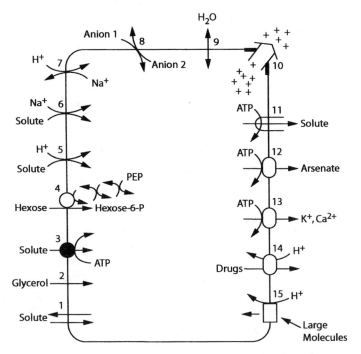

Figure 10.5. Model representing different systems of solute transport in prokaryotes. (1) Passive diffusion; (2) facilitative diffusion; (3) ABC uptake transport; (4) group translocation—shown here phosphotransferase of sugars; (5) solute uptake by proton symport; (6) solute transport driven by $Na^+$ symport; (7) cation-exchange system—shown here is $Na^+$ efflux driven by $H^+$ anteport; (8) anion-exchange system; (9) aquaporin; (10) gated channel; (11) export of solutes; capsules, and drugs by ABC transporter; (12) A type ATPase for arsenate export; (13) P-type ATPase for $Cd^{2+}$ or $K^+$; (14) proton-driven export of antibiotics, disinfectants, and other chemicals; and (15) proton-driven uptake of DNA, $B_{12}$, and sideophores which require special transport across the outer membrane.

the requirement of a diversity of solute molecules being transported into and out of the cell, unique features are generally added to a fundamentally basic transport process. As will be seen, some solute molecules have multiple systems for transport and this reflects a capability for that organism to grow and persist in several different environments. While transport always involves the plasma membrane, there are several instances in which large solute molecules require special systems to enable them to traverse the outer membrane.

## 10.8.1. Cations

The plasma membrane serves as an effective wall in preventing the diffusion of positively charged minerals required for microbial nutrition. Prokaryotic cells have acquired a specific process for importing the cations and it appears that every metal

## Box 10.1
## A Bug's-Eye View of the Periodic Table
A Perspective by Simon Silver

When I was a student, back in the salad days of molecular biology, everyone wanted to discover the secrets (genes) of life. I discovered instead by accident that basically all cells—of bacteria, plants, and animals—have highly specific protein membrane pumps, enzymes, and genes to govern interactions with the cations and anions formed by the elements of the Periodic Table. From a bacterial perspective it is not so much cations and anions, as nutrient good ions (such as $Mg^{2+}$, $K^+$, $Fe^{3+}$, $SO_4^{2-}$, and $PO_4^{3-}$) and toxic bad ions (such as $Hg^{2+}$, $Cd^{2+}$, and $AsO_4^{3-}$); there are genes and proteins for each. $Mg^{2+}$ and $Mn^{2+}$ uptake transporters were the first we found. The $Mn^{2+}$ NRAMP system of bacteria, plants, and animals is a major pathogenicity determinant as organisms fight over needed $Mn^{2+}$, just as they fight over $Fe^{3+}$. These transport systems can either respond to the cell membrane potential or be ATPases. Surprisingly, there are a few major classes of ATPases, firstly the 8-polypeptide ATP synthase of bacteria, mitochondria, and chloroplasts (it is all the same) that rotates one third of a circle for each ATP hydrolyzed (clockwise for synthesis and counterclockwise for hydrolysis, or the reverse depending on whether you are taking movies from the topside of bottom). I spent my first happy year in Australia studying the $\beta$ subunit of $F_1$ ATPase of *E. coli*. The $F_1F_0$ ATP synthase is basically a $H^+$ pump with proton movement secondarily driving ATP synthesis. The P-type ATPases that are basically all cation pumps are quite different: (1) they are formed with single large polypeptides with both membrane and cytoplasmic domains and (2) they have phosphorylated protein intermediates (hence the name) which can be made radioactive at a specific aspartate site in each. The first known P-type ATPase is the animal muscle $Ca^{2+}$ pump; we have been lucky enough to find and work with ones for $Cd^{2+}$, $Cu^{2+}$, and $Ag^+$. Finally, in this brief overview, we have found enzymes, mostly reductases and oxidases, with inorganic ions as substrates, converting $Hg^{2+} \leftrightarrow Hg^0$ and $AsO_4^{3-} \leftrightarrow AsO_3^-$. A bug's-eye view of the Periodic Table is that for each and every inorganic cation or anion needed for growth or found in the environments, there are separate genes and proteins. In fact and in total, these can represent about 10% of the genes and 10% of the metabolism of bacterial cells—a lot. And the mechanisms are amazingly the same or similar in bacteria, plant, and animal cells. There are 92 elements in the Periodic Table that are found naturally, plus some more only known from man-made isotopes. It takes us about 10 years to find out most of what we wish to know about the microbiology of the ions of any element. I have been at this topic less than a lifetime; and tomorrow is likely to be more exciting than yesterday. That is good to learn.

Table 10.6. Cationic transport systems commonly found in bacteria.

| System | Principal ion transported | Cation moved secondarily |
|---|---|---|
| Amt system | $NH_4^+$ | Methylammonium |
| CadA system | $Cd^{2+}$ | $Zn^{2+}$ |
| Cop operon | $Cu^+$ | $Ag^+$ |
| CorA system | $Mg^{2+}$ | — |
| CzcD system | $Zn^{2+}$ | $Co^{2+}$ |
| MgtA system | $Mg^{2+}$ | $Zn^{2+}$ |
| MgtB system | $Mg^{2+}$ | $Co^{2+}$, $Ni^{2+}$ |
| MntA system | $Mn^{2+}$ | $Cd^{2+}$ |
| Nik operon | $Ni^{2+}$ | — |
| ZntA system | $Zn^{2+}$ | $Cd^{2+}$ |
| Zur operon | $Zn^{2+}$ | — |

cation has its own unique system. Examples of some of the major cation transport systems in bacteria are listed in Table 10.6. However, the specificity for these cation transport systems is not absolute but alternate metal ions will be transported when the primary compound is not present. The direction of transport and the use of one transport system for more than one solute contributes to the difficulty in understanding metal transport in unicellular systems. This discussion of cation transport also includes the antibiotics and other ion carriers that function as ionophores.

### 10.8.1.1. Iron Uptake[21-25]

Even though iron is the fourth most abundant element in the Earth's crust and the second most abundant metal, bacteria are frequently limited by the level of available iron. In aerobic environments that are at pH 7 or higher, $Fe^{3+}$ is present as insoluble minerals, carbonates, hydroxides, or phosphates. At pH 3 or lower, $Fe^{3+}$ is highly soluble and organisms that grow in extremely acidic conditions rarely have difficulty in obtaining sufficient iron. Bacteria require a few million atoms per cell for the production of cytochromes, iron-proteins, and Fe-S clusters present in redox proteins.

*Siderophore Systems*

In aerobic conditions, $Fe^{3+}$ is insoluble in the environment and bacteria produce chelating agents that specifically sequester and mobilize $Fe^{3+}$. These iron-chelating agents have a molecular mass of 0.5 to 1.5 kDa and are collectivley referred to as siderophores. The word *siderophore* is derived from Greek and means "iron carrier." As chelators of ferric ion, siderophores generally have a six-coordinate octahedral complex with $Fe^{3+}$ held in the center of the molecular arrangement. Except for the lactic acid bacteria which have no requirement for iron, aerobic bacteria have siderophores that are characteristic for a given species. Although more than 100 different siderophores have been characterized for bacteria and fungi, a listing of a few siderophores produced by bacteria is given in Table 10.7. Bacterial siderophores are grouped according to the type of ligand that holds $Fe^{3+}$ and the most common are

Table 10.7. Examples of siderophores and their characteristics.[a]

| Compounds | H/C/M | Log binding constant | $E_0'$ | Organism producing siderophore |
|---|---|---|---|---|
| Aerobactin | H | 22.5 | −336 | *Escherichia coli*, *Salmonella* spp., *Shigella* spp. |
| Anguibactin | M | — | — | *Vibrio anguillarum* |
| Azotobactin | M | — | — | *Azotobacter vinelandii* |
| Chrysobactin | C | — | −350 | *Erwinia chrysanthemi* |
| Enterobactin | C | 52 | −759 | Enteric bacteria |
| Ferroxamine B | H | 30.6 | −468 | *Streptomyces* spp. |
| Parabactin | C | — | −673 | *Paracoccus denitrificans* |
| Pseudobactin | M | — | — | *Pseudomonas aeruginosa* |
| Staphyloferrin | M | — | — | *Staphylococcus aureus* |
| Vibriobactin | C | — | — | *Vibrio cholerae* |

[a]Modified from Crumbliss, 1991.[23]

H = hydroxamyate siderophore, C = catecholate siderophore, M = mixed chelating groups for $Fe^{3+}$ that consist of dihydroquinoline, hydroxamate, hydroxycarboxylate, or carboxylate moieties.

Chrysobactin = 2,3'-dihydroxybenzoylserine.

catecholates and hydroxylates. The binding groups for the catecholate and hydroxylate structures are given in Figure 10.6. The binding of $Fe^{3+}$ to the siderophores is great with dissociation constants of 22 to 52 (see Table 10.7). Thus, the siderophore acquires $Fe^{3+}$ because the affinity of the siderophore for $Fe^{3+}$ is greater than the solubility constant for the insoluble iron compounds.

When the concentration of soluble iron is less than 1 $\mu M$, *E. coli*, *Salmonella typhimurium*, *Shigella* spp., *Klebsiella* sp., and *Entrobacter* sp. produce enterobactin, which is also known as entrochelin. The entrobactin gene cluster resides on the chromosome of *E. coli* and it encodes for the following activities: (1) *ent* genes encode for biosysnthesis of enterobactin, (2) *fep* genes encode for transport of Fe-enterobactin into the cell, and (3) the *fes* gene produces an esterase that degrades Fe-enterobactin with the release of $Fe^{3+}$ into the cytoplasm. The activities of these gene products are given in Figure 10.7. Dihydroxybenzoate (DHB) is produced from chrismate by enzymes encoded on *ent* A,B,C and enterobactin is constructed from DHB with enzymes produced from *ent* C,D,E. Under iron stress conditions, *E. coli* will produce about 10 to 15 mg of enterobactin per liter. Owing to the high binding affinity for $Fe^{3+}$, enterobactin acquires iron from the environment. The ferri-enterobactin moves through the outer membrane by way of a special porin that is produced by the *fepA* gene. The FepA protein is a barrel-shaped protein with 22 antiparallel β-sheets and a series of specific protein loops that fold into the channel to the "barrel" to regulate molecular inflow.

Energy for opening of the channel to permit ferri-enterobactin entry comes from the TomB protein, which originates in the plasma membrane and extends across the plasma membrane to FepA in the outer membrane. The TonB protein is unique in that it contains two proline-rich segments of EPEPEPEPIPEP and KPKPKPKPKPKP. These proline sequences are considered to assist in formation of a rod-like structure

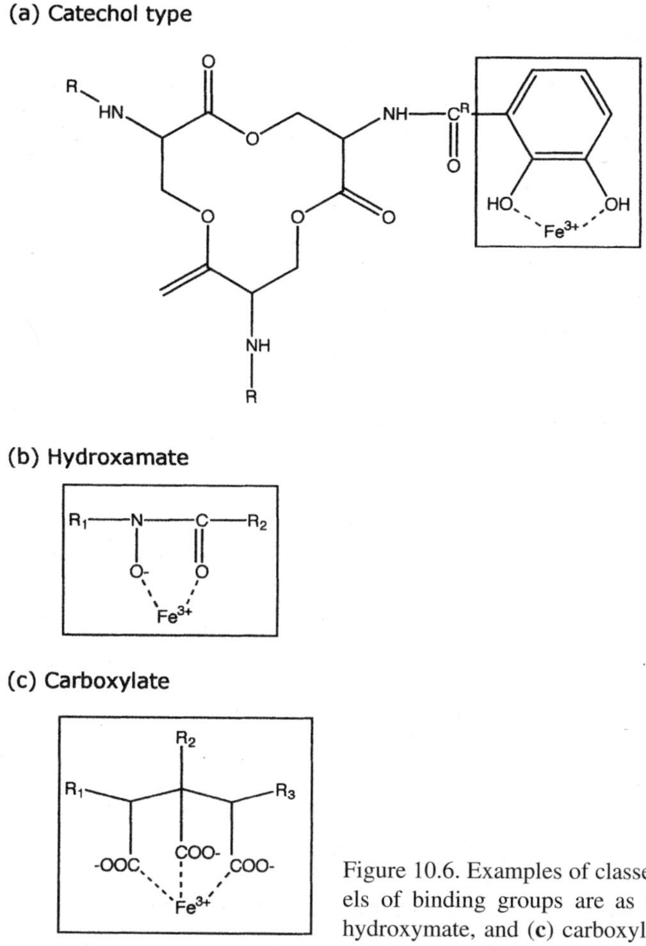

**(a) Catechol type**

**(b) Hydroxamate**

**(c) Carboxylate**

Figure 10.6. Examples of classes of chelators for $Fe^{3+}$. Models of binding groups are as follows: (**a**) catecholate, (**b**) hydroxymate, and (**c**) carboxylate.

in the TonB that extends across the periplasm. Owing to extensive proteolytic activity in the periplasm, the TonB protein is subject to degradation and for continued activity new synthesis of TonB is required. This mechanism of energy transfer across the periplasmic region is unresolved but may be due to a change in structural conformation that is initiated by energy from proton-motive force. Thus, the degradation of the TonB protein could be an important conservation of proton-motive force and membrane energetics.

Once the ferri-enterobactin is in the periplasm, it interacts with FepB which is a specific binding protein for Ferri-enterobactin. The FepB is part of an ABC transporter and it delivers the ferri-enterobactin to the cytoplasm through the use of dual transmembrane proteins of FepC. Proteins from the *fepD* gene are located on the cytoplasmic side of the ABC transporter and energize the transporter by energy from ATP hydrolysis. Once in the cytoplasm, the ferri-enterobactin is degraded by a spe-

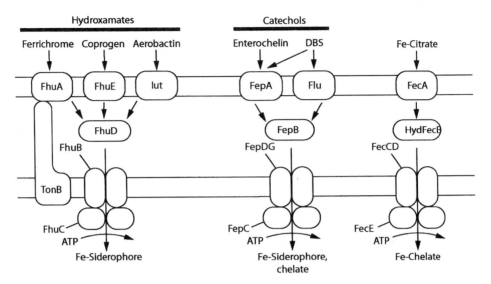

Figure 10.7. Examples of siderophores that can function to provide $Fe^{3+}$ to *Escherichia coli*. (From Braun, 1995.[21])

cific esterase (Fes) to release $Fe^{3+}$. Enterochelin is degraded to dihydroxybenzoic acid (DBS) which is released from the cell as an end product. In *E. coli*, the release of DBS does not constitute an energy drain as fewer than a million atoms of iron are required for the growth of a single cell. $Fe^{3+}$ is rapidly reduced to $Fe^{2+}$ by a NAD(P)H-driven ferric reductase and this reduced iron form is used for synthesis of cytochromes and proteins with Fe-S centers.

Bacteria have the capability of using several different types of siderophores including those produced by other microorganisms. Perhaps one of the more promiscuous consumers of chelated iron is *E. coli*, which can use several different ferri-hydroxymates, ferri-catechols, as well as ferri-citrate. A summary of these $Fe^{3+}$ transport systems for *E. coli* is given in Figure 10.7. Ton B is required for activation of porins in the outer membrane for transport of ferrichrome, aerobactin, $B_{12}$, and enterochelin but not for ferri-citrate. Apparently recognition between Ton B protein and the various porin proteins is between amino acid sequences near the N-terminal of the outer membrane receptor. This sequence is referred to as the "Ton B box" and has a consensus sequence of XTXXV. Also, it appears that each ABC transporter may be capable of recognizing several different ferri-siderophores.

There is not a clear picture of how $Fe^{3+}$ is released from siderophores other than enterobactin. Reduction of $Fe^{3+}$ to $Fe^{2+}$ is a possibility for those ferri-siderophore complexes that have a half reaction potential that is relatively close to that of NAD(P)H, which is $-320$ mV. The redox potential for a few of the siderophores is given in Table 10.7. Because the midpoint potentials are calculated at equilibrium, it is possible that NAD(P)H could slowly reduce siderophores that may be $-400$ mV. Certainly the redox potential of $-759$ mV for enterobactin would be too electronegative for NAD(P)H reduction, and therefore the degradation of the siderophore releases $Fe^{3+}$ and permits reduction of nonchelated $Fe^{3+}$ to be achieved by reduced pyridine nucleotides. It should be noted that the binding of $Fe^{3+}$ to a siderophore is dependent on several chemical characteristics. $Fe^{3+}$ has a smaller ionic radius than $Fe^{2+}$ and $Fe^{3+}$ is a strong Lewis acid with a binding affinity for $O^-$, which is a good electron donor for $Fe^{3+}$. $Fe^{2+}$ is not a hard acid and has low affinity for catecholates or hydroxymates.

*Heme as a Source of Iron*

An important source of iron in human and animal pathogens is heme. Free heme originates in serum from hemoglobin which may be released from erythrocytes as a response to bacterial-derived hemolytic virulence factors. Because heme is only slightly soluble, it rapidly binds to serum proteins and relatively little is free in the fluid of the blood. The general theme of iron acquisition from hemoglobin is as follows:

1. On the surface of bacterial cells are proteins that bind hemoglobin and function also as proteases to release heme from hemoglobin. In Gram-negative bacteria, the binding of hemoglobin is by proteins in the outer membrane while in Gram-positive bacteria, binding of hemoglobin is by proteins anchored into the outer side of the plasma membrane and extending through the cell wall.
2. Heme is transferred across the plasma membrane into the cytoplasm by binding-protein ABC transporters. Heme binding proteins are unique porins in the outer membrane of Gram-negative bacteria which is referred to as a hemophore. Once heme is transferred into the periplasm, it is bound to a binding protein that is specifically associated with the ABC transporter.
3. $Fe^{3+}$ is released from heme in the cytoplasm as a result of enzymatic decomposition of heme. In *Corynebacterium diphtheriae* and *Neisseria meningitis*, it has been determined that heme is degraded by specific oxygenases. Finally, $Fe^{3+}$ is reduced to $Fe^{2+}$ by a NAD(P)H specific reductase. Reductases can reduce $Fe^{3+}$ to $Fe^{2+}$; however, unlike siderophores, the deferri-heme has considerable binding affinity for $Fe^{2+}$ and iron is not released.

The best characterized systems of heme uptake is with *Yersinia enterocolitica*, in which the genes for uptake have all been characterized. Variations of the *Yersinia enterocolitica* genetic system for heme uptake have been demonstrated in about 20 different species and some of these bacteria include the following: *Pseudomonas aeruginosa, Haemophilus influenzae, Neisseria meningitidis,* and *Porphyromonas gingivalis*.

*Transferrin and Lactoferrin as an Iron Source*

The human body has transferrin in the serum and lactoferrin in leukocytes. These eukaryotic proteins bind $Fe^{3+}$ and reduce the level of free $Fe^{3+}$ to less than $10^{-12}$ $\mu M$. When bacteria invade the body, bacterial cells are confronted with a low concentration of free iron and bacteria that are able to get $Fe^{3+}$ are capable of growth. Transferrin is a glycoprotein of 76.5 kDa that binds two ions of $Fe^{3+}$ per molecule of transferrin. Members of *Pasteurellaceae* and *Neisseriaceae* produce outer membrane proteins that bind iron-loaded transferrin and lactoferrin. TbpA and TbpB proteins are specific for transferrin while LbpA and LbpB proteins recognize lactoferrin. As a result of this binding of transferrin and lactoferrin to their respective proteins in the outer membranes, $Fe^{3+}$ is released and $Fe^{3+}$ enters the periplasm where it is bound to a $Fe^{3+}$-binding protein. Transport across the plasma membrane is by an ABC transporter system. Transferritin utilization has been shown in Gram-positive bacteria with *Staphylococcus aureus* and *Staphylococcus epidermidis* having a transferrin-binding protein in the cell wall. An ABC transporter specific for $Fe^{3+}$ has also been proposed for iron acqusition form transferrin in staphylococci.

Ovotransferrin is found in egg whites and it keeps the iron level low to prevent growth of many bacteria that gain entry into the egg. Lactoferrin is also found in milk and dairy products and to some extent controls the growth of iron-requiring bacteria. Many of the bacteria that grow in environments containing ovotransferrin or lactoferrin acquire $Fe^{3+}$ from these proteins through the use of siderophores. Siderophore-mediated acquisition of $Fe^{3+}$ from lactoferrin or ovotransferrin is less efficient than the systems that bind transferrin to the surface of the bacterial cell. On release, siderophores must locate and mobilize $Fe^{3+}$ before the ferri-siderophore returns to the cell. When iron-sequestering proteins are bound onto the surface of the cell, iron is released into that cell.

*Iron Uptake in Anaerobes*

Under anaerobic conditions, bacteria use $Fe^{2+}$ and not $Fe^{3+}$. Unlike $Fe^{3+}$, $Fe^{2+}$ is highly soluble and iron uptake by anaerobes proceeds without siderophores. A high-affinity $Fe^{2+}$ transporter has been found in *E. coli*, *Salmonella typhimurium*, and *Helicobacteri pylori*. This $Fe^{2+}$ uptake system is designated as encoded on the *feo* gene and contains a large transmembrane protein of 84 kDa that has binding regions appropriate for ATPase activity. Regulation and biochemistry of this *feo* system in anaerobic bacteria require further study.

### 10.8.1.2. Magnesium Transport[26]

Cells require $Mg^{2+}$ for ribosome function, to maintain membrane integrity, and as a cofactor for numerous enzymes. In enteric organisms such as *E. coli* and *Salmonella typhimurium*, there are multiple systems for $Mg^{2+}$. *S. typhimurium* has three transport systems for $Mg^{2+}$ and they include the CorA, the MgtA, and the MgtB systems. Characteristics of these three transport systems are given in Table 10.8. The major uptake system for $Mg^{2+}$ transport is CorA, which is is constitutively produced while MgtA and MgtB are produced under conditions of low extracellular $Mg^{2+}$ concentration. Only CorA is associated with homoeostasis because it also serves as a system for $Mg^{2+}$ efflux. Based on kinetic values, CorA system has a greater velocity for $Mg^{2+}$ transport than MgtA and MgtB. As all three $Mg^{2+}$ transport systems have similar affinity for $Mg^{2+}$ with similar $K_m$ values, no one system provides greater benefit in sequestering $Mg^{2+}$. While CorA is highly specific for $Mg^{2+}$, the MgtA and B systems have broader cation recognition. In conditions of reduced $Mg^{2+}$ concentra-

Table 10.8. Characteristics of magnesium transport systems in *Salmonella typhimurium*.[a]

| Characteristic | Transport systems | | |
| --- | --- | --- | --- |
| | CorA | MgtA | MgtB |
| $K_m$[b] | 15 | 12 | 6 |
| $V_{max}$[b] | 250 | 24 | 75 |
| Cation specificity[c] | $Mg^{2+}>Mn^{2+}>Co^{2+}$ $>Ni^{2+}>Ca^{2+}$ | $Zn^{2+}>Mg^{2+}>Ni^{2+}$ $>=Co^{2+}>Ca^{2+}$ | $Mg^{2+}=Co^{2+}=Ni^{2+}$ $=Mn^{2+}>>Ca$ |
| Characteristics | Constitutive, influx and efflux | Regulated by repression | Regulated by repression |

[a]From Hmiel et al., 1989.[26]
[b]Measurements were at 20° C for CorA and MgtA but at 37° C for MgtB.
[c]Based on inhibition of $Mg^{2+}$ transport.

tions, it is likely that the MgtA may function to transport $Zn^{2+}$ while the MgtB system may also function to transport $Co^{2+}$ or $Ni^{2+}$.

### 10.8.1.3. Manganese Transport[27]

Manganese is required by microorganisms to support a variety of physiological roles. Cyanobacteria require four atoms of Mn to oxidize water with the liberation of electrons in photosystem II. In many bacteria, $Mn^{2+}$ is an essential ion for activity of enzymes including superoxide dismutase, DNA and RNA polymerase, peptidases, GTPase, phosphatases, esterases, and ligases. To acquire $Mn^{2+}$ for these enzymes, different types of permeases have evolved: (1) a P-type-ATPase, (2) an ABC transporter, and (3) a proton-dependent uptake system. These three systems are outlined in Table 10.9 and are briefly discussed in the following segment.

The P-type ATPase for import of $Mn^{2+}$ has been extensively characterized in *Lactobacillus plantarum*, which is a bacterium that relies on substrate-level phosphorylation without cytochromes or proteins that contain Fe-S clusters. Lactic acid bacteria (i.e., members of the *Lactobacillaceae*, *Streptococcaceae*, and *Enterococcaceae*) have a great dependency on $Mn^{2+}$; an intracellular accumulation of 30 m*M* $Mn^{2+}$ is not uncommon. In *Lactobacillus plantarum*, the MntA system is involved in importing $Mn^{2+}$ and $Cd^{2+}$. Proteins of the MntA system contain a phosphorylation domain, an ATP-binding domain, and an ion transduction region. On comparison of the gene sequences of the MntA system to genomic sequences, this system is suggested to be present in archaea (*Thermoplasma acidophilum* and *Methanococcus jannachii*) and bacteria including *Acidithiobacillus ferroxidans*, *Chlorobium tepidum*, and *Geobacter sulfurreducens*.

The import of $Mn^{2+}$ into bacterial cells by a process involving an ABC transporter has been reported for several bacteria. Characteristic of this process is the presence of a metal-binding protein in the periplasm of Gram-negative bacteria or a metal-binding protein anchored into the outer side of the plasma membrane of Gram-positive bacteria such as the *Streptococcaceae* and *Enterococcaceae*. The transmembrane structure consists of two identical proteins, and two proteins that bind ATP are attached on the inner side of the plasma membrane. Transport is attributed to ATP hydrolysis. As shown in Table 10.9, several other cations may be transported by this ABC transporter.

Table 10.9. Selected examples of $Mn^{2+}$ transport systems in bacteria.[a]

| Transporter system | Bacteria | Metals recognized by transporter |
|---|---|---|
| **P-type ATPase** | | |
| MntA | *Lactobacillus plantarum* | $Mn^{2+}$, $Cd^{2+}$ |
| **Protein-binding ABC system** | | |
| PsaA | *Streptococcaceae* | $Mn^{2+}$, $Zn^{2+}$, $Fe^{3+}$ |
| Mnt ABC | *Sycecchocystis* 6803 | $Mn^{2+}$, $Cd^{2+}$, $Co^{2+}$, $Zn^{2+}$ |
| Sca ABC | *Streptococcus gordonii* | $Mn^{2+}$, $Zn^{2+}$, $Cu^{2+}$ |
| Yfe ABCD | *Yersinia pestis* | $Mn^{2+}$, $Fe^{3+}$ |
| **Proton-driven transporter** | | |
| MntH | *Escherichia coli* | [b]$Mn^{2+}$, $Cd^{2+}$, $Zn^{2+}$, $Co^{2+}$, $Fe^{2+}$, $Ni^{2+}$, $Cu^{2+}$ |
| | *Bacillus subtilis* | $Mn^{2+}$, $Cd^{2+}$ |

[a]From Cellier, 2001.[27]
[b]Listed in preference for uptake with $Mn^{2+}$ highest and $Cu^{2+}$ lowest.

Metal ion uptake, including $Mn^{2+}$ transport, has been demonstrated in several bacteria by a system that is driven by proton-motive force. Based on sequence analysis, there is considerable variation in this chemiosmotic driven transport of $Mn^{2+}$, and the transporters are grouped into three categories designated as group A, group B, and group C. Group A is broadly distributed in bacteria including both Gram-positive and Gram-negative bacteria. The ancestral organism of bacteria probably contained MntH group A as the $Mn^{2+}$ chemiosomotic transporter and groups B and C may have emerged as time progressed. *Chloribium tepidium* and *Clostridium acetobutylicum* are members of group B and appear to have 11 transmembrane helices for $Mn^{2+}$ uptake. Group C consists of bacteria that have a system most closely related to eukaryotic organisms and this includes *Enterococcus faecalis* and *Pseudomonas aeruginosa*, which have the *MntH* gene.

### 10.8.1.4. Calcium Transport[28]

While $Ca^{2+}$ channels in membranes of eukaryotic cells has been well accepted, there is little information to indicate the presence of $Ca^{2+}$ voltage-gated channels in prokaryotes. The best evidence for the presence of $Ca^{2+}$ channels in bacteria is from the use of inhibitors that affect $Ca^{2+}$ channels in eukaryotic organisms. These inhibitors include verapamil, ω-conotoxin, dimethylbenzamil, trifluoperazine, and nitrendipine which have various effects on bacteria, including growth, chemotaxis, membrane synthesis, and $N_2$ fixation, to name a few. Sporulating bacterial cells have a unique requirement of $Ca^{2+}$ and it would be interesting to determine if inhibitors that block $Ca^{2+}$ channels in eukaryotes would also inhibit heat resistance in sporulation. The requirement for $Ca^{2+}$ by sporulating bacteria requires the presence of a $Ca^{2+}$ transporter in the plasma membrane of the mother cell as well as a $Ca^{2+}$ transporter in the spore membrane. Although $Ca^{2+}$ influx is not well characterized, it may be driven by proton anteport and uptake of $Ca^{2+}$ across the membranes of the forespore may be a symport with diaminopimelic acid.

Efflux of $Ca^{2+}$ in prokaryotes is apparent because typical bacterial cells when subjected to chemical analysis have only 0.025 m$M$ $Ca^{2+}$ while $Mg^{2+}$ is 25 m$M$. This value includes the amount of calcium in the free form plus that which is bound to cell walls, membranes, and other cellular constituents. Gangola and Rosen have used fluorescent probes to determine that the amount of free $Ca^{2+}$ in *E. coli* is approximately $10^{-7}$ $M$ and that chelated calcium is readily released into the cytoplasm as the soluble form. The presence of high concentrations of free $Ca^{2+}$ can present considerable inhibition to cellular metabolism and bacteria have developed several different systems for $Ca^{2+}$ export. One system involves a P-type ATPase of the P-2 class and, thus far, it has been reported in *Streptococcus sanguis*, *Streptococcus lactis*, *Streptococcus faecalis*, *Flavobacterium ororatum*, and *Anabaena varabilis*. A second calcium efflux system is the $Ca^{2+}/H^+$ anteporter which is driven by proton-motive force. This $Ca^{2+}/H^+$ transport system has been demonstrated in *Azotobacter vinelandii*, *Bacillus megaterium*, *Bacillus subtilis*, *Mycobacterium phlei*, *Chromatium vinosum*, and *Rhodobacter capsulata*. A third efflux system has been demonstrated in alkalophilic organisms such as *Halobacterium halobium* and *Bacillus* sp. A-007. This system for organisms in an alkaline environment is the $Ca^{2+}/Na^+$ anteporter system. The presence of numerous different systems for $Ca^{2+}$ efflux indicates the importance to the microorganism of reducing the amount of intracellular $Ca^{2+}$.

### 10.8.1.5. Nickel Transport[29]

Various bacteria and archaea have nickel-containing hydrogenase, carbon monoxide dehydrogenase, urease, and in methanogens the nickel-containing coenzyme M reductase and factor $F_{430}$ are present. This dependency on nickel for activity required that cells be capable of acquiring this unusual metal cation. There is a nonspecific $Mg^{2+}$ transport system that will transport $Ni^{2+}$, $Co^{2+}$, and $Mn^{2+}$ in bacteria; however, high concentrations of $Mg^{2+}$ will exclude the transport of other divalent cations. There would be relatively few environmental settings in which the concentration of $Ni^{2+}$ would be expected to equal or exceed the concentration of $Mg^{2+}$. There is a high-affinity transport system in prokaryotes that is specific for $Ni^{2+}$ and is a member of the ABC transport systems. In *E. coli*, the nickel operon encodes for five proteins: NikA is a periplasmic binding protein, NikB and NikC are membrane proteins, and NakD and NakE bind ATP on the cytoplasmic side of the nickel membrane proteins. Other high-affinity transport systems with eight transmembrane helices are being studied in *Bradyrhizobium japonicum*, *Helicobacter pylori*, and *Ralstonia metallidurans*.

### 10.8.1.6. Copper Transport[30]

Copper plays an important role in the redox activity of several microbial enzymes including cytochrome *c* oxidase, superoxide dismutase, rusticyanin, nitrite reductase, nitrous oxide reductase, ammonia monooxygenase, and dopamine β-hydrolase. Generally less than 10 μ$M$ copper in the culture medium is required to satisfy the metabolic requirements. For copper transport, *Enterococcus hirae* has a *cop* operon consisting of four genes and a model of these gene products are given in Figure 10.8. *CopY* produces a hydrophilic protein that functions as a copper-induced repressor

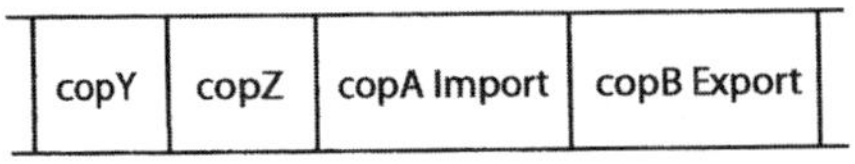

| copY | copZ | copA Import | copB Export |
|------|------|-------------|-------------|

Figure 10.8. Operon for copper (cop) transport in *Enterococcus hirae*. (From Kakinuma, 1998.[30])

while *copZ* produces an activator protein for the *cop* operon. It is considered that *copA* codes for copper uptake while *copB* codes for copper extrusion.

Both the CopA and CopB proteins have the Asp-Lys-Thr-Gly-Thr sequence considered to be the phosphorylation site in P-type ATPases. In the N-terminus of the CopA protein, there is the sequence of Gly-X-Cys-X-X-Cys which is similar to that found in the mercury reductase of *Acidothiobacillus thiooxidans*. The CopA protein also appears to transport $Ag^+$ into the cell and it is not known if $Cu^{2+}$ is reduced in the periplasm or inside the cell. The CopB transport system recognizes $Cu^+$ with an apparent $K_m$ of 1 $\mu M$ and functions to prevent toxic levels of copper to accumulate in the cytoplasm. CopB transport protein will also transport $Ag^+$.

### 10.8.1.7. Zinc Transport[31]

Zinc is a essential micronutrient that stimulates cell growth by stabilizing membranes, interacting with nucleic acids, and serving as a cofactor for numerous metalloenzymes. $Zn^{2+}$ uptake has been demonstrated in numerous bacteria and it is under control of the Zur (*z*inc *u*ptake *r*egulation) system. The system is regulated by a Zur protein reported to have nine cysteine moieties which are reported to account for the binding of two or three atoms of $Zn^{2+}$ for each protein molecule. Although the Zur protein is not well characterized, there are numerous similarities between it and the Fur protein that regulates iron uptake.

One example of a high-affinity $Zn^{2+}$ uptake system is an ABC transporter with high specificity for $Zn^{2+}$. This system consists of a binding protein (ZnuA) that is present in the periplasm and recognizes $Zn^{2+}$, a transmembrane segment (ZnuB) that facilitates the movement of $Zn^{2+}$ from the periplasm to the cytoplasm, and a protein in the cytoplasm attached to ZnuB that has ATP binding sites and energizes the transport system. A model of this $Zn^{2+}$ transporter is given in Figure 10.9. This system functions in dilute $Zn^{2+}$ environments to provide the cell with $Zn^{2+}$ and it is widely distributed in Gram-positive and Gram-negative bacteria.

In environments with moderate levels of $Zn^{2+}$, $Zn^{2+}$ may enter the cell by way of the Pit system which functions in phosphate transport. In this case, $HPO_4^{2-}$ moves with a divalent cation, like $Zn^{2+}$, into or out of the cell with the phosphate ion. The Pit transport system could also function in $Zn^{2+}$ export; however, it is difficult to envision a condition in which a high level of phosphate would be expelled from bacteria.

Prokaryotes growing in environments of high levels of $Zn^{2+}$ have several mechanisms to achieve homeostasis and prevent accumulation of $Zn^{2+}$ to a toxic level. The amount of $Zn^{2+}$ associated with cells is ~$6 \times 10^{-4}$ $M$ if all bound zinc would be free; however, the amount of free $Zn^{2+}$ is closer to $10^{-8}$ $M$. Some systems used to maintain a low intracellular concentration of $Zn^{2+}$ in Gram-negative bacteria are shown in Figure 10.9. In *E. coli*, there is a ZntA system that is a class-1 of the P-type

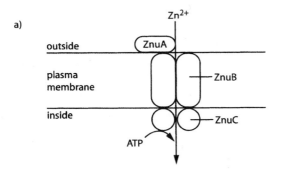

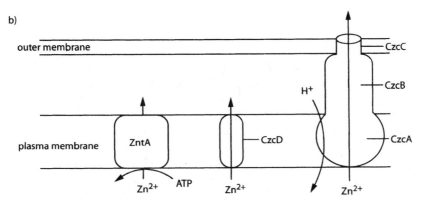

Figure 10.9. Transport systems for $Zn^{2+}$ in bacteria. (**a**) $Zn^{2+}$ uptake system by and ABC transporter in a dilute $Zn^{2+}$ environment. (**b**) Efflux systems for $Zn^{2+}$ in environments of high $Zn^{2+}$ levels. The systems are as follows: ZntA, CzcD, and CzcABC. (From Hantke, 2001.[31])

ATPase. In *Staphylococcus aureus*, the export of $Zn^{2+}$ is attributed to a P-type ATPase which also exports $Cd^{2+}$ and has been referred to as the CadA system. This CadA system with both $Zn^{2+}$ and $Cd^{2+}$ efflux is found also in *Listeria monocytogenes*, *Stenothophomonas maltophila*, and *Bacillus subtilis*. Intracellular binding of $Zn^{2+}$ as thiolates of cysteine or glutathionine are preferred substrates for the CadA and the ZntA systems. The principal function of the CadA and the ZntA systems is efflux of $Cd^{2+}$ and $Zn^{2+}$, with $Zn^{2+}$ efflux being a secondary activity.

A second efflux mechanism for bacteria growing in high concentrations of $Zn^{2+}$ is the CzcD system. The CzcD system is a member of the diffusion facilitator family that exports $Zn^{2+}$, $Co^{2+}$, and $Cd^{2+}$. From homology studies, it is evident that the CzcD protein is present in various bacteria including *E. coli* and *Ralstonia metallidurans*.

A third export system for $Zn^{2+}$ is the CzcD system, which is highly efficient in expelling cations across both the outer membrane and the plasma membrane. The CzcC protein is in the plasma membrane while the CzcC is a channel in the outer membrane. Connecting the channel in the outer membrane and the inner membrane is the CzcB protein, which extends through the periplasm. A model of the Czc system is given in Figure 10.9. This unique system accounts for the efficient efflux of the $Zn^{2+}$ ion without accumulation in the periplasm. Energy for the export of $Zn^{2+}$ through CzcA is attributed to proton entry. This $Zn^{2+}$/proton anteporter consists of CzcA, B, C and is primarily a $Zn^{2+}$ exporter but has secondary function in export of $Co^{2+}$ and $Cd^{2+}$.

### 10.8.1.8. Cation Anteporters[32]

There are many examples where prokaryotes display a process of cation exchange, and several are presented in Table 10.10. Bacterial systems shown to expel cations in exchange for protons include the $K^+/H^+$ anteporter, the $Na^+/H^+$ anteporter, and the $Ca^{2+}/H^+$ anteporter. Each of these anteport systems is proposed to have a specific function. In case of the $Na^+/H^+$ anteporter, one function is to generate a $Na^+$ gradient for $Na^+$/solute symport and another activity is to regulate intracellular pH alkalophilic bacteria. The $Na^+/H^+$ anteporter has been found in *Deinococcus radiodurans*, *Bacillus subtilis*, *Rickettsia prowazekii*, and thermophiles. The $K^+/H^+$ anteporter is proposed to regulate cytosolic pH when bacteria are grown in high ionic strength medium and with an alkaline pH. High levels of calcium are detrimental to many metabolic activities of bacteria and an appropriate low intracellular concentration of calcium is maintained by a $Ca^{2+}/H^+$ anteporter.

Another example of cation movement functions with a $Na^+$ or $K^+$ gradient as is the case with the $K^+/NH_4^+$ anteporter and $Ca^{2+}/Na^+$ anteporter. A $K^+/NH_4^+$ anteporter has been demonstrated in *E. coli* and more research is required to establish the function of this activity in bacteria. The $Ca^{2+}/Na^+$ anteporter has been demonstrated in halophilic *Halobacterium halobium* and *Bacillus* sp. A-007. It appears that an important function of the $Ca^{2+}/Na^+$ anteporter is to reduce the level of $Na^+$ in these cells.

Table 10.10. Solute exchange reactions associated with transport.

| Anteporter systems (uptake/export) | Function |
|---|---|
| **Cations** | |
| $K^+/H^+$ | Regulate pH in cytoplasm |
| $Na^+/H^+$ | Generate $Na^+$ gradient for solute transport |
| $Ca^{2+}/H^+$ | Expel $Ca^{2+}$ from cytoplasm |
| $K^+/NH_4^+$ | Regulate intracellular pH |
| $Ca^{2+}/Na^+$ | Reduce level of $Na^+$ in cell |
| Arginine/ornithine | Supply of arginine for metabolism |
| **Anions** | |
| Oxylate/formate | $H^+$ pump to generate proton-motive force |
| ATP/ADP | ATP uptake supports energetics of symbionts |
| Organophosphate/phosphate | Uptake of sugar-phosphate |
| Nitrate/nitrite | Export of $NO_2^-$ with $NO_3^-$ respiration |

Clearly, there is an interesting interconnection of different ions in these secondary transport systems.

### 10.8.1.9. Transport of Ammonium[33]

Ammonia ($NH_3$) rapidly diffuses through the plasma membrane of prokaryotes; however, the conditions for free ammonia in the environment are relatively limited. Ammonia in slightly alkaline conditions will form ammonium ($NH_4^+$) and this is the nitrogen form used by many prokaryotes as a nitrogen source. At low concentrations of $NH_4^+$, many bacteria and archaea display a specific transport process that has been designated the "Amt system." In *E. coli* and several bacteria, the ammonium transporter has 10 to 12 transmembrane helices and it is proposed to contain an essential aspartic acid moiety that binds ammonium in the transport process. Because of the lack of a good isotope for nitrogen, methylammonium labeled with $^{14}C$ has been used as a structural analog of ammonium.

The Amt of *E. coli* and of *Corynebacterium glutamium* is specific for the cationic molecule and transports either $NH_4^+$ or $CH_3NH_4^+$ but not $NH_3$ or $CH_3NH_3$. These systems with Amt transporter have the bioenergetic status of a uniport system in which $NH_4^+$ is driven across the plasma membrane by charge potential across the membrane. However, not all prokaryotes have an Amt-type transporter because genomic analysis has revealed the absence of appropriate Amt homologies in *Chlamydia trachomatis*, *Haemophilus influenzae*, *Helicobacter pylori*, *Mycoplasma pneumoniae*, *Rickettsia prowaskii*, *Treponema pallidum*, and *Pyrococcus horikoshii*. This suggests that prokaryotes have systems alternate to the Amt for transport of $NH_4^+$. Several of these $NH_4^+$ transport systems are exchange processes and examples of $NH_4^+$ transport are given in Figure 10.10. Ammonium uptake in acidic conditions may occur by a proton symport and in neutral alkaline water transport by a $Na^+$ symport is favored.

A possible ramification of $NH_4^+$ accumulation in cells is the loss of nitrogen as $NH_3$.

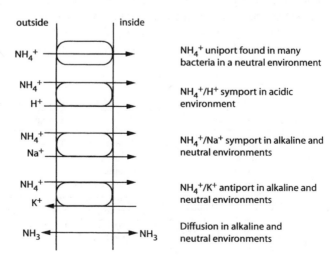

Figure 10.10. Ammonium transport systems in bacteria. (From Van Dommelen et al., 2001.[32])

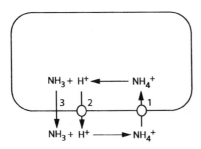

Figure 10.11. Model of ammonium movement in bacteria. Transported at "1" is ammonium by the Amt system which is driven by membrane charge. Proton export at transporter "2" is ATPase which exports $nH^+$ with hydrolysis of ATP. Ammonia exits across the membrane at "3" without a transporter with movement driven by diffusion. (From Van Dommelen et al., 2001.[32])

A model indicating this activity is given in Figure 10.11. As $NH_3$ is formed from $NH_4^+$ in the alkaline cytoplasm, $H^+$ is exported from the cell by an ATPase. The proton extruded will combine with $NH_3$ that has diffused across the membrane and the $NH_4^+$ that is formed will be transported into the cell by the Amt transporter. Such recycling of $NH_3$ conserves nitrogen but at the expense of consuming ATP.

## *10.8.2. Anions*

Prokaryotes require minerals and carbon structures that carry a negative charge. Acquisition of anions is by systems as diverse as the anions to be transported and examples of some anionic transport systems are given in Table 10.10 and Table 10.11. Generally, anion transport follows the bioenergetic requirements of the proton-motive force or ATP hydrolysis; however, in a few cases, the anion may traverse the plasma membrane as a weak acid. The diversity of anion transport systems includes exchange systems and herbicides (viologen dyes) as a ionophore for anions.

### 10.8.2.1. Nitrate and Nitrite Transport[34]

Nitrate is used by bacteria as a nitrogen source for synthesis of organic nitrogen compounds and in some microorganisms as an electron acceptor. The dissimilatory activities and enzymology of nitrate reduction are covered in Chapter 13. For nitrate to be used for either assimilatory or dissimilatory processes, it must be reduced to

Table 10.11. Examples of anionic transport systems in bacteria.

| Ion imported | System |
|---|---|
| Nitrate | *Nar* operon |
| Nitrite | *Nir* operon |
| Sulfate | Cys system |
| Phosphate | Pst system for inorganic phosphate—inducible |
|  | Pit system for inorganic phosphate—constitutive |
|  | Glpt system for α-glycerol phosphate |
|  | Uhp system for hexose phosphate |
| Molybdate | *Mod* operon |

nitrite by a nitrate reductase. Several different systems for nitrate transport are present in the prokaryotes. A model of nitrate and nitrite transport systems in bacteria is given in Figure 10.12. An important feature is that nitrite can diffuse across membranes as the lipid-soluble nitrous acid. It has been determined that reduced benzyl viologen functions as an ionophore for both nitrate and nitrite. This is the only demonstrated ionophore for anions and it is anticipated that in the future other chemicals will be demonstrated to have ionophore activity for other anions.

Based on purification of proteins and analysis of gene information, *E. coli* has been determined to have at least three different systems for nitrate reduction. One system of nitrate reduction is controlled by the *narGHJI* operon and the enzyme is localized in the cytoplasm. Induction occurs in anaerobic environments where the nitrate level is appreciable and no enzyme is produced under aerobic conditions. A second system is controlled by the *narZYWV* operon and nitrate reductase is produced as *E. coli* enters the stationary phase. Although the enzyme is produced under both aerobic and anaerobic conditions, the induction of nitrate reductase is relatively weak and little reduction of nitrate is attributed to this *narZYWV* operon system. The third nitrate reductase is under control of the *nap* operon and the enzyme is produced in the periplasm under anaerobic conditions. This periplasmic location of nitrate reductase would not require nitrate transport. Nitrate transport would not be required for nitrate to reach the nitrate reductases produced by the *narGHJI* and *narZYWV* operons.

In *E. coli*, there are three separate systems for the transport of nitrate and nitrite. One is the NarK system which is involved in the uptake of nitrate, uptake of nitrite, and export of nitrite. The second nitrate transport system is NarU, which deals with uptake of nitrate, export of nitrite, but no uptake of nitrite. The *narU* gene is physically associated with the *narZWYV* operon. Both the NarK and the NarU systems are appropriate for the nitrate dissimilatory reduction system with nitrate transported

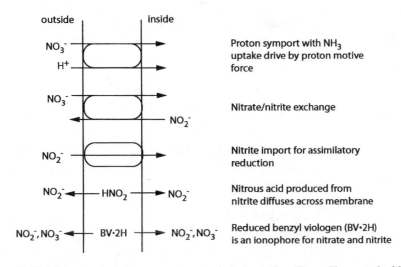

Figure 10.12. Nitrate and nitrite transport systems in bacteria. (From Clegg et al., 2002.[34])

into the cell and nitrite exported without accumulation in the cytoplasm. The third system is NirC, which is a membrane protein with six transmembrane helices. The NirC is considered to be important in uptake of nitrite at a rate sufficient to form ammonium needed for biosynthesis. NirC may also function for efflux of nitrite.

In bacteria other than enterics, the transport of nitrate is separated into three systems and each is under separate genetic control. NRT1 is a nitrate transporter found in many bacteria and the gene sequence analysis reveals a close relationship between the bacterial and plant nitrate transport systems. NRT2 is a nitrate transporter that is driven by proton-motive force and sequence analysis indicates that it is unrelated to NRT1. A third nitrate transport system is a product of the *nas* locus and the mechanism of nitrate entry into the bacterial cell is by the ABC import system. Using *Klebsiella oxytoca* (formerly *pneumoniae*) as an example, the ABC nitrate transport system consists of a periplasmic binding protein, NasF, that interfaces with a transmembrane protein, NasE, and the cytoplasmic ATPases proteins, NasD. A model of the ABC transport system of *Klebsiella oxytoca* is given in Chapter 13.

### 10.8.2.2. Sulfate Transport[35]

Two systems for sulfate transport are used in prokaryotes, with one used when sulfate is limiting for growth and the other when an adequate level of sulfate is present. Under sulfate-limiting conditions, the transport of sulfate into enteric bacteria is mediated by a periplasmic binding protein and transmembrane uptake is assumed to employ the ABC transport mechanism. With *Salmonella typhimurium*, the apparent $K_m$ for sulfate is 36 $\mu M$ and transport is inhibited by thiosulfate, sulfite, selenate, chromate, and molybdate. It is believed that in the absence of sulfate or other potential sulfate transport inhibitors (thiosulfate, sulfite, selenate, chromate, and molybdate) may be substrates for this transport system. In *E. coli*, the *cysA* region codes for a 32-kDa protein that binds a single sulfate molecule with a $K_d$ of $2 \times 10^{-8} M$. There are about 10,000 molecules of sulfate-binding per cell and the binding protein system of sulfate functions when sulfate is limiting in the environment. The *cysZ* locus, which is physically near the *cysA* gene, is part of the cysteine regulon, and is also required for sulfate transport in *E. coli*. At 0.1 m$M$ sulfate in the medium, the ratio of sulfate inside the cell to sulfate outside the cell is about 9:1.

Under levels of sulfate that provide sufficient anion for growth, sulfate entry into the bacterial cells is not driven by an ABC system but by a chemiosmotic driven mechanism. The unique physiological group of bacteria known as the sulfate-reducing bacteria grow with sulfate at m$M$ levels and produce extremely high quantities of sulfide. In bacteria that grow in freshwater environments, sulfate is transported by a proton symport system while in marine organisms and moderate halophiles sulfate is transported by a $Na^+$ symport system.

### 10.8.2.3. Phosphate Transport Systems[36]

There are at least four distinct systems for the uptake of phosphate in *E. coli* and other enteric organisms. Of these, two of these sulfate transport systems are inducible and two are constitutive. The inducible systems are for the uptake of hexosephosphate

(Uhp) and the α-glycerol phosphate transport (Glpt) system. One constitutive system is the Pit ($P_i$ transport) system and the other is the Pst (phosphate-specific system). The mechanism for phosphate transport in prokaryotes is dependent on the concentration. At high phosphate concentrations, transmembrane movement is attributed to a $P_i/H^+$ symport system designated as "Pit." At low concentrations of phosphate, a high-affinity ABC transporter (Pst) is used. The Pst is only partially constitutive with 20% of activity constitutive and 80% repressible. The Pst system has characteristics of an ABC system with the *pstS* gene producing about 25,000 periplasmic binding proteins per cell and each molecule has a mass of 38 kDa. From structural analysis, it has been determined that the phosphate ion is held in the center of the binding protein by 12 hydrogen bonds. The PstS protein binds to PstA and PstC which are membrane proteins and each have six transmembrane helices. As presented in Figure 10.13, the tandem PstB proteins are attached to PstA and PstC proteins and through the energy from ATP hydrolysis, phosphate is transported across the plasma membrane into the cytoplasm.

The Pit system does not have a periplasmic binding protein and phosphate transport is driven by proton-motive force. As indicated in Figure 10.13, phosphate transport is by a proton symport and the $HPO_4^{2-}$ ion is complexed with a divalent cation. Suitable metal cations for binding to phosphate ion include $Mg^{2+}$, $Ca^{2+}$, $Mn^{2+}$, and $Co^{2+}$. The $K_m$ of the Pit system is 25 $\mu M$ and once the concentration of inorganic phosphate is <0.16 $\mu M$, the Pit system can no longer effectively transport phosphate. With an absence of phosphate, alkaline phosphatase production is derepressed and phosphate is released from organophosphates in the periplasm. The uptake of this phosphate released by phosphatase action is dependent on the concentration of phosphate.

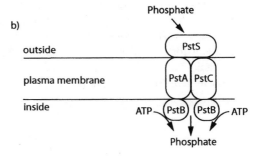

Figure 10.13. Systems for orthophosphate transport in bacteria. The Pit system functions at elevated inorganic phosphate levels. (a) Proton symport with phosphate anion complexed with divalent cation (Me). Suitable divalent cations would include: $Mg^{2+}$, $Co^{2+}$, $Mn^{2+}$, $Ca^{2+}$, and $Zn^{2+}$. (b) The PstS system is an example of an ABC transporter and functions with limiting phosphate. The PstS binds phosphate which is present without counterions, and inorganic phosphate is transported through the channel created by PstA and PstC.

### 10.8.2.4. Molybdenum Transport[37]

Molybdenum is an important element used by biological systems as a cofactor in a variety of enzymes. Select prokaryotes have a nitrogenase that has a molybdenum–iron–sulfur cofactor, referred to as FeMoco, which consists of $MoFe_3S_3$ and $Fe_4S_3$. Microorganisms have oxidoreductases containing molybdopterin as a cofactor and the structure of this cofactor is given in Chapter 9. Unlike molybdopterin found in eukaryotes, the molybdo-cofactor in prokaryotes contains a nucleotide.

Acquisition of molybdenum by prokaryotes involves transporting molybdate into the cell from the surrounding medium. It is considered that a low-affinity transport system with a $K_m$ of 300 $\mu M$ functions at molybdate concentrations greater than 10 $\mu M$; however, this system is poorly understood. In several bacteria, it has been established that there is a high-affinity transport system for molybdate with a $K_m$ of 20 $\mu M$. This high-affinity system functions when molybdate is about 10 n$M$ and, depending on the bacterial system, is saturated at 1 to 10 $\mu M$. The genes associated with high-affinity molybdate transport in *E. coli*, *Azotobacter vinelandii*, and *Rhodobacter capsulatus* are given in Chapter 9. The periplasmic binding protein, ModA, is 26 to 27 kDa, has a $K_d$ of 3 $\mu M$ for molybdate, and does not bind arsenate, sulfate, phosphate, nitrate, selenate, or chlorate. The 24-kDa hydrophobic membrane protein, ModB, is composed of five transmembrane segments and there is similarity in protein sequence for sulfate, thiosulfate, phosphate, and molybdate transporters. On the inner side of the plasma membrane is a peripheral membrane protein of 38 kDa that binds ATP. With ATP hydrolysis, molybdate is transported into the cytoplasm where, in excess, it binds to repressor proteins that are soluble in the cytoplasm.

### 10.8.2.5. Anion Antiporters[38–40]

Some of these reactions involving anion exchange resemble the transport found in the inner membrane of the mitochondria. The reactions thus far reported are concerned primarily with intermediary metabolism. In most cases, the exchange reactions (also known as anteporters) involve phosphates or carboxylic acids.

#### *Oxalate: Formate Transporter*

*Oxalobacter formigenes* is found in the rumen of herbivores and is important in degradation of oxalate. P.C. Maloney and co-workers demonstrated that oxalate is taken up as the dianion and decarboxylated to produce formate plus $CO_2$. Formate exits the cell by exchanging with another oxalate and $CO_2$ exits as a gas and not as $HCO_3^-$. This reaction would constitute a proton pump and generate proton-motive force. See the model for this exchange reaction (Figure 10.14). Similarly, a reaction associated with malate/lactate exchange occurs in *Lactococcus lactis* with the generation of a proton-motive force. Malate is converted to lactate by a cytoplasmic decarboxylation reaction.

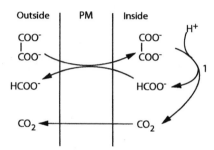

Figure 10.14. Oxalate-formate anteporter system that results in export of a proton. Reaction (1) is decarboxylation. (Redrawn with permission from Nicholls and Ferguson, 1992.[40])

### ATP: ADP Transporter

For many years, it was known that rickettsia had limited ability to produce ATP and the means whereby this intracellular bacterium produced ATP was unresolved. When H.H. Winkler demonstrated that *Rickettsia prowazekii* has an ATP/ADP exchange system with ATP from the host cell being exchanged for ADP from the rickettsia cell, the energetics of this organism became understandable. Anteporter activity is stimulated by the presence of inorganic phosphate. If the nucleotide exchange is an electroneutral reaction, inorganic phosphate or another anion would be required to be cotransported with ADP to exchange for ATP. While mitochondria also contain a nucleotide exchange, the gene for ATP: ADP antiporter is markedly distinct from that in rickettsia. An ATP: ADP transporter system has been reported to be present in *Chlamydia psittaci*, an obligate intracellular pathogen of animal cells, and *Frankia* sp., a plant symbiont.

### Phosphate: Sugar 6-Phosphate Transporter

It has been observed by several investigators that internal inorganic phosphate, $P_i$, exchanges for organic phosphates to transport phosphorylated substrates into *E. coli*, *Salmonella typhimurium*, *Staphylococcus aureus*, and *Streptococcus lactis*. An exchange of phosphate was reported for glucose 6-phosphate, glyceraldehyde 3-phosphate, phosphoenol pyruvate, 3-phosphoglyceric acid, and mannose 6-phosphate. In several cases, arsenate is the only analog of phosphate that functions in this exchange reaction and contributes to arsenate detoxification. These exchanges were observed using nonmetabolizable substrates, such as 2-deoxyglucose 6-phosphate and reconstituted systems in which the isolated transporter was inserted into proteoliposomes.

### Amino Acid-Linked Anteporter

In *Streptococcus lactis* and *Pseudomonas aeruginosa*, there is an electroneutral exchange of arginine for ornithine. The transporter catalyzes exchanges of ornithine/ ornithine and arginine/ornithine with the concentration gradient of ornithine serving as the driving force for the reaction. A model of this activity is given in Figure 10.15. In addition, an electroneutral exchange of agmatine and putrescine has been suggested for *Enterococcus faecalis*.

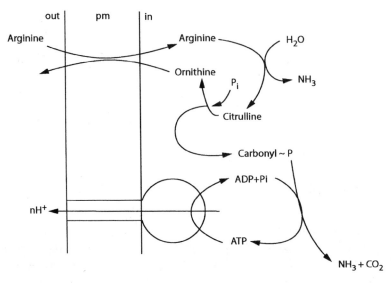

Figure 10.15. Transport of arginine into bacterial cells. (Redrawn with permission from Dressen and Konings, 1990.[40])

### 10.8.3. Sugars[41]

The uptake of sugars into the bacterial cell occurs either by systems that transport the unaltered molecule or by the PTS system that phosphorylates sugars as they are transported across the plasma membrane. In the phosphotransferase system for carbohydrate transport, the specific sugars transported may vary considerably. In *S. aureus* and *E. coli*, PTS-sugars include glucose, fructose, mannitol, sorbitol, and *N*-acetylglucosamine. PTS-sugars found in *S. aureus* but not *E. coli* include lactose, sucrose, galactose, trehalose, melibiose, maltose, and glycerol. In *E. coli*, lactose, maltose, glycerol, and galactose are non-PTS-sugars. Lactose transport by a permease encoded by the *lac* operon was one of the first sugar transport systems extensively examined and it was established by Kaback and colleagues to be a specific transmembrane protein. A commonly employed system for acquisition by sugars in dilute concentration is to employ binding proteins with sugar transport by the ABC system, and examples of a few of the carbohydrate binding proteins are given in Table 10.5. Another system of sugar uptake without phosphorylation involves the chemiosmotic system with sugar transport energized by $H^+$ or $Na^{2+}$ cotransport. In *E. coli*, sugars energized by the chemiosmotic system include galactose, arabinose, and melibiose. Activities of these transport systems have been reported earlier in this chapter.

### 10.8.4. Amino Acids and Peptides[42]

Prokaryotic cells generally have highly active systems for acquisition of amino acids and peptides even though they may not be the nitrogen or carbon source for cell growth. The function for accumulation of amino acids and oligopeptides is a recycling of material hydrolyzed in the periplasm as a result of proteolysis of leader protein sequences of transported material or hydrolysis of proteins in the cell wall region damaged by various factors. Some of the major features of a selected number of amino acid transport systems are covered in the following section to provide examples of this important cellular process.

### 10.8.4.1. Amino Acid Transport

Based on chemical characterization of isolated proteins and studies with mutants, it has been established that there are several different amino acid transport systems in bacteria. These transport systems are distinguished by substrate specificity and mechanism for transport. With respect to specificity, there are two classes: (1) systems with low affinity for substrate that can transport several different amino acids and (2) systems of high specificity that recognize only one amino acid. A list of both general and specific amino acid transport systems is given in Table 10.12. The transport systems of lower specificity function when the cell requires numerous different amino acids for growth and the amino acids are readily available in the environment. When

Table 10.12. Transport systems for amino acids in bacteria.[a]

| System | Amino acids transported | Bacteria with this system | Protein for recognition | Energy driving transport |
|---|---|---|---|---|
| **A. Low-affinity systems** | | | | |
| 1. Aromatic amino acid system | Tyrosine, phenylalanine, tryptophan | *Escherichia coli* | — | — |
| 2. Arginine-ornithine system | Arginine, ornithine | *E. coli* | Periplasmic | ATP |
| 3. LAO | Lysine, arginine, ornithine | *Salmonella typhimurium* | Periplasmic | ATP |
| 4. LIV-II | Leucine, isoleucine, valine | *E. coli, S. typhimurium* | Membrane | pmf (?) |
| 5. PP-II | Proline | *E. coli, S. typhimurium* | Membrane | pmf |
| **B. High-affinity systems** | | | | |
| 6. Alanine | Alanine | Thermophile strain PS3 | Membrane | Na$^+$ or H$^+$ |
| 7. Glutamine specific | Glutamine | *E. coli* | Periplasmic | ATP (?) |
| 8. Histidine specific | Histidine | *S. typhimurium* | Periplasmic | ATP |
| 9. LS | Leucine | *E. coli* | Periplasmic | ATP (?) |
| 10. PP-I | Proline | *E. coli* | Membrane | Na$^+$ |
| 11. Phenylalanine specific | Phenylalanine | *E. coli* | — | — |
| 12. Tyrosine specific | Tyrosine | *E. coli* | Membrane | pmf |

[a]Source is Furlong, 1987.[13]
pmf, Proton-motive force.

a specific amino acid is limiting for intracellular reactions, a specific amino transport system is induced.

*Aromatic Amino Acids*

Multiple transport systems for a given amino acid are relatively common in bacteria. The transport of aromatic amino acids in bacteria occurs by overlapping systems: one of low affinity and others of high affinity. For high-velocity uptake of L-tyrosine, L-phenylalanine, and L-tryptophan, there is the low-affinity system. However, when one of these amino acids becomes limiting a specific transport system of high affinity is produced for only one amino acid. Similarly, the transport of proline is by two different systems: PP-I and PP-II. The high-affinity system (PP-I) displays considerable regulation with repression by L-proline. Proline transport by PP-I is moved by a $Na^+$ or $Li^+$ symport. The PP-II system has a low affinity for proline and it is on a region of the chromosome distinct from PP-I system.

The transport of histidine has been extensively examined in *Salmonella typhimurium*. This transport system consists of three membrane-bound proteins (Q, M, and P) and one periplasmic binding protein (J) (see Figure 10.16). The first protein to

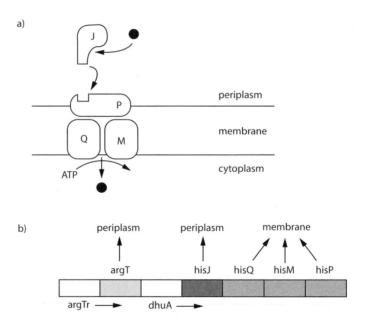

Figure 10.16. Histidine transport system for *Salmonella typhimurium*. (**a**) Model for histidine transport. The small dark oval represents histidine and histidine binds onto protein J in the periplasm. The binding protein (J) attaches to the membrane localized P and histidine movement across the membrane is guided by proteins M and Q. Energy for the transport of histidine is attributed to ATP hydrolysis. (**b**) Arrangement of histidine transport genes. *dhuA* is the promoter for histidine transport and proteins for transport are products from *hisJ*, *hisQ*, *hisM*, and *hisP*. (Redrawn with permission from Antonucci and Oxelender, 1986.[42])

interact with histidine is J and as a result of the binding a conformational change occurs in J to expose a recognition site with P in the membrane. The passage of histidine from J through the membrane is facilitated by the proteins P, Q, and M. Proteins M and P are considered to be involved in ATP binding and ATP hydrolysis energizes the uptake of histidine. The genes for histidine transport are organized in a linear sequence with *dhuA* serving as the promoter (Figure 10.16). These genes for the histidine transport system are located adjacent to the binding protein gene (*argT*) for the LAO transport system. However, each of these systems has a distinct promoter.

### Branched-Chain Amino Acids

There are three separate pathways for branched chain amino acid transport in *E. coli* and *Pseudomonas aeruginosa*. The LIV-I system transports L-leucine, L-isoleucine, L-valine, L-threonine, L-alanine, and L-serine by a periplasmic binding protein that has a single binding site per molecule. A second system, LIV-II, transports L-leucine, L-isoleucine, and L-valine but it has a lower affinity for these amino acids than the LIV-I system. The lower affinity system found in LIV-II provides for high-velocity uptake of amino acids and enables the bacteria to achieve the highest rates of growth. A third transport system is LS, which is specific for the transport of L-leucine or D-leucine but will not recognize isoleucine, valine, threonine, or alanine. The LIV-I and LS systems are highly regulated and are induced only when amino acid pools in the cytoplasm are diminished.

### Basic Amino Acids

There are three major systems for the transport of lysine, arginine, and ornithine in *E. coli*. The LAO system involves a periplasmic binding protein with greatest affinity for arginine and lesser affinity for lysine and for ornithine. As measured by $K_d$, the affinity of the periplasmic binding protein for arginine is 0.15 $\mu M$ while for lysine it is 3.0 $\mu M$ and for ornithine it is 5.0 $\mu M$. Even though there is this difference in affinity of the binding protein for the three different basic amino acids, all three are transported by this process. If lysine is available in limiting amounts, a specific system for lysine is produced and this involves only a membrane-bound protein without involvement of a periplasmic binding protein. When either arginine or ornithine is limiting, a transport system specific for these two amino acids is produced and this system involves a protein binding system.

### Glutamate-Aspartate Transport

Because glutamate and aspartate are important in reactions involving nitrogen assimilation, there is considerable redundancy in transport of these amino acids, with five different systems present in *E. coli*. One system involves a periplasmic binding protein with equal affinity for glutamate and for aspartate. In the absence of adequate levels of aspartate, a specific periplasmic binding protein is produced for aspartate and if the level of glutamate is limiting, a specific periplasmic binding system for glutamate is induced. Glutamate can also be moved on a chemiosmotic system by being coupled to a Na$^+$ symport. Finally, there is a dicarboxylic acid transport system that moves

4-C dicarboxylic acids and aspartate is one of these acids transported. Although glutamine is also linked to nitrogen metabolism, it is transported by only one system and it involves a periplasmic binding protein.

### Cystine Transport

In *E. coli*, there are two transport systems for cystine and they differ in substrate specificity and in mechanism of transport. One is a general system that is osmotic sensitive and with a $K_m$ of $3 \times 10^{-7}M$ for cystine; however, it also recognizes diaminopimelic acid and related diaminodicarboxylic acids. *E. coli* also has a specific system that transports only cystine and is not sensitive to osmotic shock. The specific transport system has a $K_m$ of $2 \times 10^{-8}M$ for cystine and does not recognize diaminodicarboxylic acids. Enteric bacteria also have a solute transport system for cysteine, however, it remains poorly characterized.

### 10.8.4.2.  Peptide Transport[43]

In *E. coli* and *Salmonella typhimurium*, peptides traverse the outer membrane through the general class of porins. Peptide movement across the plasma membrane appears to be by several different systems: (1) an oligopeptide system that includes peptide length up to six amino acid residues, (2) the tripeptide system, and (3) the dipeptide system. Based on sequence analysis, the oligopeptide system in *Salmonella typhimurium* is the product of five genes (*oppA*, *oppB*, *oppC*, *oppD*, and *oppF*). The oligopeptide transport system involves a periplasmic binding protein of about 60 kDa that is a product of *oppA*. The oppA protein binds a wide range of peptides that may originate in the periplasm due to degradation of signal peptides. Alanine–alanine–alanine and alanine–phenylalanine–glycine exhibit strong binding to the oppA protein. In general, peptides involved in binding to oppA protein have polar or hydrophobic side chain residues while amino acids that have ionic side chains are either not bound or are very weakly bound to the periplasmic binding protein. Proteins oppB and oppC are integral membrane proteins with a mass of 33.4 kDa and 33.1 kDa, respectively. Proteins oppD and oppF are 36.9 kDa and 37.1 kDa, respectively, with ATP binding sites on the cytoplasmic side of the proteins.

A transport system for tripeptides is present in many bacteria including *E. coli* and *Salmonella typhimurium*. This transport system is produced from the *tpp* (tripeptide protein) gene cluster and it appears to be a member of the ABC transport system. In terms of specificity, the tripeptide transport system is distinct from the oligopeptide system and from the dipeptide system. The tripeptide transport system has a preference for tripeptides with reduced activity with dipeptides and tetrapeptides or oligopeptides. It has a distinct preference for hydrophobic amino acids and especially peptides with valine or methionine at the N-terminus. Transport requires an N-terminus with a positive charge but the C-terminal carboxyl group may be esterified or aminated. Expression of the *tpp* genes appears to be controlled by anaerobiosis and thus may be important in the rumen, intestinal tract, as well as in anaerobic sediments.

A dipeptide transport system has been demonstrated in various bacteria and this system has low specificity for amino acid moieties and will transport dipeptides containing D-amino acids. The dipeptide protein (Dpp) in *E. coli* is encoded on the *dppA* gene and is about 50 kDa. Under certain conditions, it is the most abundant protein in the periplasm and it has been recorded to account for 90% of the periplasmic proteins. A major role of this peptide transport may be to recycle the peptides of signal proteins released in the periplasm.

Gram-positive bacteria such as *Lactococcus, Bacillus, Staphylococcus, Enterococcus*, and *Streptococcus* also have peptide transport systems. The most common substrates for transport are dipeptides and oligopeptides. Transport of peptides in these bacteria is similar to that in Gram-negative bacteria in that peptide binding proteins are present; however, these binding protiens are firmly bound into the outer side of the plasma membrane. The mechanism of transport has not yet been resolved but there is some evidence for ATP-mediated transport.

A curious feature of peptide transport in bacteria is the potential for uptake of antibacterial compounds attached to one or more amino acids. Although the absence of a specific transport system for the antibacterial compound limits the quantity of material that can traverse the plasma membrane, the attachment of this antibacterial compounds to peptides enables the chemical to be taken up by dipeptide or oligopeptide transport systems. Once the complex is internalized, the peptide segment is removed from the antibacterial molecule by cytoplasmic aminopeptidases and the target enzyme is inhibited. A model for this transport system is given in Figure 10.17. A few examples of these antibacterial–peptide complexes are given in Table 10.13. In these situations, the antibacterial molecule is an amino acid analog that is recognized by an enzyme owing to structural similarity with the normal substrate. The analog compound remains bound in the active site of the enzyme, and thereby inhibits the enzyme. The antibacterial compounds that are normally excluded from the cell enter by a carrier complex and this peptide-mediated accumulation of antibacterial compounds has been dubbed "smugglin" by J.W. Payne. There had been some speculation

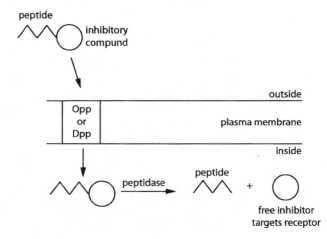

Figure 10.17. The smugglin model for bacteria with the antibiotic attached to a peptide. Opp, Oligopeptide permease; Dpp, dipepetide permease. (From Payne and Smith, 1994.[43])

Table 10.13. Example of amino acid analogs in peptides transported by bacteria.

| Chemical (source) | Transported peptide | Analog of active chemical | Action |
|---|---|---|---|
| Bacilysin: a dipeptide (*Bacillus subtilis*) | ECH-alanine | Glutamine | Glucosamine synthetase in peptidoglycan synthesis |
| Bialaphos: a tripeptide (*Streptomyces hygroscopicus*) | PPT-alanyl-alanine | Glutamic acid | Glutamine synthetase |
| Phaseolotoxin: a tripeptide (*Pseudomonas syringae*) | Phosphotriamide-ornithyl-alanyl-homoarginine | Ornithine | Ornithine carbonyltransferase |
| Alafosfalin: a dipeptide (synthetic) | L-1-aminoethylphosphonic acid-alanine | Alanine | Alanine racemase in peptidoglycan synthesis |
| A dipeptide | β-Chloroalanyl-alanine | Alanine | Alanine racemase |
| A dipeptide | L-Propargylglycine-alanine | | Cystathionine β-synthase |

[a]Data from Payne and Smith, 1994.[43]
ECH, L-β-(2,3-Aepoxycyclohexanono-4); PPT, L-2-amino-4-methylphosphinobutyric acid.

about increasing uptake of pharmaceutically important antibacterial compounds through attachment to peptides; however, it has been determined that the attachment of amino acids to chloramphenicol, nalidixic acid, or erythromycin does not enhance the activity of these compounds. In some cases, it is likely that steric problems occur at the membrane surfaces because the peptide-bound antibiotic is too large for the porin or transport channels.

## 10.8.5. Uptake Systems for Drugs and Antibiotics in Bacteria[44]

Antibiotics and related chemicals that are involved in the inhibition of bacterial growth are an excellent example of diverse transport processes. Although these molecules range from biological products to unique chemicals synthesized in the laboratory, all have an inhibitory action on some aspect of bacterial growth or metabolism. In some instances, the transport processes are attributed to a similarity of structures while in other instances there is transport by a single system of structurally unrelated molecules. Uptake systems for selected antibiotics are given in Table 10.14 and it is evident that many different systems are employed. In some instances, there is biomimicry because the antibiotic transported is a structural analog of a nutrient and in other instances diffusion is the means of transfer across the plasma membrane.

The process of acquisition of drugs and antibiotics ranges from diffusion to active transport. In Gram-negative bacteria, the initial barrier for the cell is the outer membrane and for antibiotics to traverse this structure it is considered that the molecule is (1) soluble in water, (2) has a molecular weight less than 600 Da to traverse the general porin or protein *ompF*, and (3) at neutral pH the molecule is either a cation or an unchanged species. Highly lipophilic antibiotics such as rifamycin and fusidic

Table 10.14. Systems for the uptake of antibiotics in bacteria.[a]

| Antibiotic | Action of antibiotic | Characteristic of transport |
| --- | --- | --- |
| Chloramphenicol | Binds to 50S ribosome | Diffuses across the outer and plasma membranes |
| D-Cycloserine | Inhibits D-alanine formation | Is structurally similar to D-alanine. Uses the proton symport activity of the D-alanine transport system. |
| 3,4-Dihydroxybutyl | Inhibits phosphoglyceride-1-phosphonate metabolism | An analog of glycerol-3-phosphate and enters on the glycerol-3-phosphate transport as a proton symport. |
| Fosfomycin | Inhibits formation of N-acetylmuramic acid | Enters the bacterial cell on the glycerol-3-phosphate transport system as well as on the hexose phosphate transport process. |
| Norjirimycin | Inhibits polysaccharide use where α- and β-glucosidic bonds are hydrolyzed | Since it is structurally similar to N-acetylglucosamine it is assumed to be transported by the phosphotransferase system. |
| Showomycin | Inhibits nucleotide synthesis by inhibiting uridine monophosphate kinase | Passes through outer membrane through the *ompF* pores. Uses uridine or nucleotide transport system to pass through the plasma membrane. |
| Sideromycins (e.g., albomycin, ferrimycin A) | Enzyme digest of sideromycins inhibits bacterial growth | Transported on ferrichrome or siderophore systems |
| Tetracyclines | Binds to 50S ribosome and inhibits protein synthesis | Transfer across the outer membrane is by passive diffusion.<br>Transfer across the plasma membrane is by an energy-dependent process similar to streptomycin uptake. |
| β-Lactams (e.g., penicillins and cephalosporins) | Inhibits cross-wall formation in peptidoglycan | Rate of diffusion across the outer membrane is size of antibiotic, electric charge, and hydrophobicity of molecule. Binds into cell wall and no transfer across plasma membrane is needed. |

[a]Data from Chopra and Ball, 1982.[44]

acid are excluded by the outer membrane. In those instances where the antibiotic gains access to the cell, this is proposed to be attributed to the special structures that form a junction between the outer membrane and the plasma membrane.

Drugs or antibiotics that are greater than 100 daltons cannot effectively diffuse across the plasma membrane. Antibiotics or drugs that inhibit peptidoglycan synthesis are required to traverse only the outer membrane protein and not the plasma membrane while the inhibitors of protein synthesis or metabolism must traverse both the plasma membrane and the outer membrane. Bulky peptidoglycan layers are unable to limit molecules unless the molecular mass exceeds 100 kDa; however, the negatively charged teichoic acids found in the cell walls of Gram-positive bacteria may bind cationic antibiotics or drugs.

At neutral pH the aminoglycosides (streptomycin, gentamicin, kanamycin, and neomycin) are positively charged and diffuse across the outer membrane. The accumulation of streptomycin, and structurally related antibiotics, is by three phases: (1) an energy-independent phase in which the antibiotic is bound to the cell surface, (2) an energy-dependent activity that appears driven by $\Delta\psi$ and involves cytochromes or other components in the electron transport chain, and (3) a second energy-dependent process attributed to binding of the antibiotic to the bacterial ribosomes. The result is that streptomycin may accumulate inside the bacterial cell to more than 400 times the external concentration.

## 10.8.6. DNA Uptake[45]

The uptake of DNA in bacteria follows a series of steps; however, owing to structural differences of the cell wall in Gram-positive and Gram-negative bacteria, each group employs different specific components. A model for uptake of DNA by Gram-positive bacteria is given in Figure 10.18. In the uptake of DNA by Gram-positive bacteria, three steps are recognized. Step one is the binding of DNA to the cell surface. In the case of *Bacillus subtilis* or *Streptococcus pneumoniae* there are about 50 binding sites per cell. The second step is fragmentation of DNA by a specific protein as the DNA is attached to the cell surface. The single strands are nicked sequentially to produce a segment of DNA estimated to be 23 kilobases (kb) in *Bacillus subtilis* and 6 kb in *Streptococcus pneumoniae*. There is no base preference in binding or fragmentation. However, it has been observed that upstream from the DNA-binding region of ComEA is a conserved sequence of QQGGGG that may make the ComEA molecule flexible and capable of bending to engage DNA into the transport channel. The third step is transport in which one strand is degraded while the other single-stranded (ss) DNA is transported across the plasma membrane. With *Bacillus subtilis*, uptake of ss-DNA is linear at a rate of approximately 180 nucleotides/sec with some evidence to indicate that strands of either polarity (5' $\rightarrow$ 3' or 3' $\rightarrow$ 5') are transported. With *Streptococcus pneumoniae* the uptake of ss-DNA has a specific polarity at 3' $\rightarrow$ 5' and proceeds at a rate of 90 to 100 nucleotides/sec.

In Gram-negative bacteria, the transformation steps would be expected to include: (1) binding of DNA to cell surface, (2) uptake of DNA across the outer membrane and localized in the periplasm, and (3) the transport of ss-DNA across the plasma membrane with degradation of the remaining ds-DNA in the periplasm. Unfortunately, binding of DNA to the cell surface has not been demonstrated. Uptake of double-stranded (ds) DNA into the periplasm occurs and one strand is degraded while the other strand is transported into the cell. Transport of DNA across the plasma membrane is polarity dependent with the 3' $\rightarrow$ 5' end transported. In many Gram-negative bacteria the uptake sequences are present as inverted repeats between genes. Some variation in DNA uptake occurs with *Haemophilus influenzae* using a 29-basepair consensus sequence to promote recognition whereas *Acinetobacter calcoace* displays no preference for base sequence in DNA uptake. The transport of DNA into *Haemophilus influenzae* has been reported to proceed at the rate of 500 to 1000 nucleotides/sec and there are only 4 to 8 sites/competent cell.

The mechanism of DNA uptake is inconclusive and the energetics is not resolved

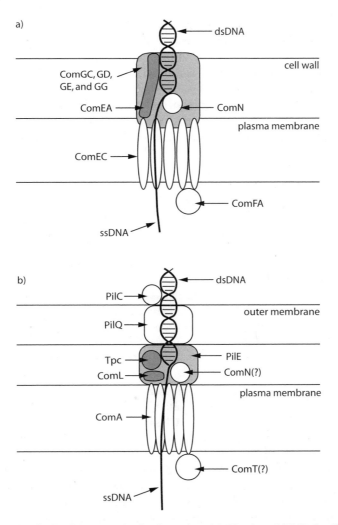

Figure 10.18. Models for DNA uptake by bacteria. (**a**) Uptake of DNA by Gram-positive bacteria. A structure that permeates the cell wall is formed by ComGC, ComGD, ComGE, and ComGG. The ComEA protein binds to ds-DNA and one strand of DNA is hydrolyzed by ComN. ComEC forms a pore that enables ss-DNA to traverse the plasma membrane. (**b**) Uptake of DNA by the Gram-negative *Haemophilus influenzae*. PilC is required to assist in formation of a pore consisting of PilQ and this pore enables ds-DNA to pass through the outer membrane. A structure in the periplasm is formed by PilE with ComL and Tpc needed for this formation which permeates the peptidoglycan. Ss-DNA passes into the cytoplasm through a pore formed by ComA. The presence of proteins comparable to ComN, ComT, and ComEA has not been reported but they may be present. (From Dubnau, 1999.[45])

but all signs indicate that DNA uptake is an active process. It would appear that as DNA moves from the exterior of the cell to the outer side of the plasma membrane and finally to the cytoplasm there would be a different charge on the molecule. Outside the cell the DNA would be ionized with an alkaline charge, at the plasma membrane surface the addition of protons would occur to produce a weak acid, and finally as the DNA enters the cytoplasm the DNA would become ionized. Studies concerning the role of ATP in transport of DNA are inconclusive at this time. It has been shown that in both *Bacillus subtilis* and *Haemophilus influenzae* the proton-motive force influences DNA uptake with both pH gradient and membrane potential driving DNA uptake.

In both Gram-negative and Gram-positive bacteria, a relatively complex system occurs that involves many different components to provide competence and DNA uptake. One group of proteins employed in DNA transfer is produced from the ComG operon. These ComG proteins have similarity with proteins from Gram-negative bacteria that are associated with type 2 secretion, assembly of type 4 pili, and for twitching motility. These are known as PSTC proteins which refer to molecules involved in *p*ilus/*s*ecretion/*t*witching/*c*ompetence. Based on known activities these PSTC proteins can be divided into five discrete classes. Class 1 includes membrane-bound proteins that have a consensus site for nucleotide binding. ComGA protein in *Bacillus subtilis* is a peripheral membrane protein on the inner side of the plasma membrane and has similarities to those Gram-negative activities of twitching motility, secretion of type 2 proteins, and assembly of pilus. (See Chapter 12 for discussion of some of these activities.) Class 2 includes those proteins that are predicted from DNA sequence to have three membrane-spanning helices. Included in this class are competence proteins found in both Gram-negative and Gram-positive systems as well as pilus assembly and type-2 secretions. Class 3 includes those proteins involved in secretion and pilus assembly as well as the protein from the ComG operon in *Bacillus subtilis*. These ComGC, ComGD, ComGE, and ComGG (see Figure 10.18) are produced by a protease processing from larger proteins. Initially, all of the ComG proteins are anchored in the plasma membrane by a specific N-terminal sequence. On processing into the functional proteins of ComGC, D, E, and G these proteins are released from the plasma membrane and translocated to the cell wall to provide a pore or channel for DNA passage. Class 4 consists of those proteins that function as proteases to release class 3 proteins from the initial protein. Members of class 4 have peptidase activity that recognizes a conserved sequence in the N-terminal region of the class 3 proteins. Following cleavage, the peptidases will frequently transfer a methyl group from *S*-adenosylmethionine to the terminal amino group produced from peptidase activity. Class 5 includes those proteins that have similarities at the C-terminus of the molecule and are associated with secretory activities and are frequently known as secretins. Class 5 proteins function to move proteins across the outer membranes and therefore are found in Gram-negative bacteria but not Gram-positive bacteria. Secretins have an essential role in outer membrane passage by bacterial virus, proteins, and pili as well as in DNA. Thus, the transformation of Gram-negative and Gram-positive bacteria is attributed to protein activities used for other cellular processes as well as for DNA transfer.

## 10.8.7. Vitamin $B_{12}$ Transport[46]

In transport into Gram-negative cells, cobalamine must pass through the outer membrane as well as through the plasma membrane. Based on information from investigations with *E. coli* and *Salmonella typhimurium*, the transport of cobalamin is by proteins encoded on the *btu* gene complex. A model of vitamin $B_{12}$ transport is given in Figure 10.19. The BtuB protein is present in the outer membrane and in addition to binding vitamin $B_{12}$, it also binds phage BF 32 and colicin E. Passage of $B_{12}$ through the outer membrane is a result of action by the protein known as TonB which is located in the plasma membrane but extends across the plasma membrane to the BtuB protein in the outer membrane. The interaction of TonB with BtuB produces a conformational change in the BtuB structure and vitamin $B_{12}$ is moved into the periplasm. TonB is energized by the chemiosmotic driven proton reentry across the plasma membrane. The periplasmic binding protein is presumed to be the BtuF protein and it releases vitamin $B_{12}$ to the BtuCD proteins in the plasma membrane. Movement across the plasma membrane is attributed to energy from ATP hydrolysis, as the protein BtuCD has an ATP binding site.

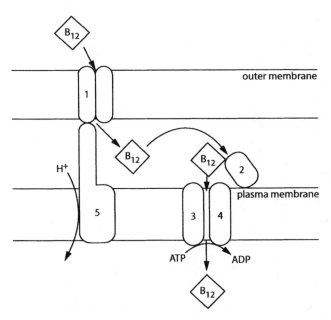

Figure 10.19. Model to depict transport of vitamin $B_{12}$ in *Salmonella typhimurium*. Proteins are identified as follows: OM, Outer membrane; PM, plasma membrane. Numbered proteins are as follows: 1, BtuB binding to protein in outer membrane; 2, BtuF, which is a periplasmic binding protein; 3 and 4, BruCD proteins in the plasma membrane; and 5, TonB protein, which is activated by proton influx to activate the transport of vitamin $B_{12}$ by BtuB protein through the outer membrane. (From Roth et al., 1996.[46])

## 10.9. Grouping of Transport Activities[7,47]

With all the various types of transport systems in microorganisms, it became important to organize the transport systems into groups or families. Saier has used computer-assisted clustering of the various membrane-bound proteins of the transport systems and has produced a significant outline for transport activities in prokaryotes. The various groups of transport systems are given in Table 10.15. These integral membrane proteins are often referred to as permeases or transporters and the direction of transport can be out of the cell as well into it. Peripheral membrane proteins may be involved in the transport process but these are not used for the various cluster arrangements. Many of the assessments were made from analysis of genes and not from characterization of solubilized permeases or transporters.

When comparing the projected molecular structures of the integral proteins of the various transport systems, the number of proposed transmembrane spanning proteins

Table 10.15. Characteristics of transport families in prokaryotes.[a]

Transport group (abbreviation/system)—selected activities

I.  Channel for diffusion
   1. Outer membrane pores or channels (porin)—movement of complex sugars
   2. Voltage sensitive ion channels (VIC)—specific for $Na^+$, $K^+$, or $Ca^{2+}$
   3. Major intrinsic protein family (MIP)—includes the glycerol facilitator

II.  ATP-driven transport
   1. P-type ATPase—Cation transport driven by ATP hydrolysis
      Group 1—Transport of $Mg^{2+}$ coupled to ATP hydrolysis
      Group 2—Transport of $Ca^{2+}$, $Na^+$, $H^+$ coupled to ATP hydrolysis
      Group 3—Uptake of $K^+$, or uptake and export of $Cu^{2+}$, or export of $Cd^{2+}$
   2. F-type ATPase—Reversible, respiratory coupled $H^+$ translocating
   3. V-type ATPase—Reversible, respiratory coupled $H^+$ translocating
   4. A-type ATPase—Export of arsenate, arsenite, tellurite, and antimonite
   5. ATP binding cassette (ABC type)
      A. Periplasmic binding proteins required for uptake:
         Group 1—Maltose as well as $\alpha$-glycerol phosphate, multiple sugars, iron hydroxymate
         Group 2—Galactose as well as arabinose, ribose, multiple sugars
         Group 3—Histidine as well as lysine, arginine, ornithine, glutamine, basic amino acids, octopine, and nopaline
         Group 4—Oligopeptides, as well as dipeptides and nickel.
      B. Lack periplasmic binding proteins and function for export
         Group 5—Peptides and proteins
         Group 6—Capsular polysaccharides, oligosaccharides, and drugs

III. Electron flow
   1. $H^+$ proton translocating electron carriers for prokaryotes with aerobic respiration
      Complex I—the NADH dehydrogenase complex
      Complex III—the quinone-cytochrome $c$ reductase complex
      Complex IV—the cytochrome oxidase complex
   2. $H^+$ Translocating transhydrogenase

Table 10.15. *Continued*

IV. Carrier mediated

    1. Major facilitator superfamily (MFS)—a superfamily containing the following:

        Family 1—Efflux of toxic substances including quinolone, antiseptics, tetracycline, methymen-omycin A, aminotriazole, and various drugs

        Family 2—Exporters of tetracycline, chloramphenicol, and various drugs

        Family 3—Importers of simple sugars and organic ions including glucose, galactose, arabinose, xylose, lactose, maltose, quinate, and other drugs

        Family 4—Transporters of citric acid, $\alpha$-ketoglutarate, proline/betaine, bialaphos

        Family 5—Antiporters with phosphorylated organic esters including hexose phosphates, glycerol-3-phosphate, and phosphoglycerates

        Family 6—Importers of oligosaccharides including raffinose, lactose, and sucrose

    2. Amino acid/polyamine/choline (APC)—transports amino acids, polyamines, and choline

    3. $Na^+$ solute symport family (SSF)

    4. Resistance/nodulation/cell division (RND)

        Group 1—The $Co^{2+}$, $Ni^{2+}$, $Cd^{2+}$, and $Zn^{2+}$ resistance pump of *Alcaligenes eutrophus*

        Group 2—Nodulation of alfalfa by *Rhizobium meliloti*

        Group 3—Cell division and resistance to antibiotics, detergents, and crystal violet

V. Group translocation also known as the phosphotransferase system (PTS)

        Group 1—Enzyme II is specific for sucrose or trehalose

        Group 2—Enzyme II is specific for glucose or *N*-acetylglucosamine

        Group 3—Enzyme II is specific for fructose or mannitol

        Group 4—Enzyme II is specific for lactose or cellibiose

VI. Light-driven transport

        Group 1—Reaction center (RC) in photosynthesis

        Group 2—Bacteriorhodpsin (BR)

[a]Data from Saier, 1994[48]; Paulsen et al., 1996.[47]

was found to vary with the system. Proteins in the membrane that have six $\alpha$-helix transmembrane units would include the voltage-sensitive $K^+$ channels, voltage-sensitive $Na^+$ channels, voltage-sensitive $Ca^{2+}$, symporters and antiporters of the amino acid/polyamine/cholate (APC) and $Na^+$ solute symport (SSF) families, the ABC-type systems, and the ABC-type transport systems. The PTS permease has six to eight transmembrane spanners per molecule, the RND has 12 spanning segments per unit, and the bacteriorhodopsin has seven transmembrane spanners per molecule. The major facilitator superfamily (MFS) of *E. coli* has 12 spanner segments per unit, as does the arsB proteins, the iron-hydroxamate permease, and the lactose permease. It is unclear if the 12 spanners is the result of gene duplication of a 6-spanner segment. The multitransmembrane spanners, such as the 12 or 14 segments, may assume a spherical arrangement in the membrane to provide a channel for transport of solutes. Through the application of molecular biology techniques, some interesting export (efflux) systems have been established and some of these examples are given in Table 10.16. It is not readily apparent why a single system allows molecules of different structure (e.g., antibiotics, detergents, and dyes) to be transported by a common process. Many of the systems listed in Table 10.16 will export penicillin or other antibiotics active against cell wall synthesis and this is indeed perplexing because resistance to these antibiotics is not achieved by excluding these molecules from the cytoplasm.

Using gene sequence information as a basis for evaluation, predictions on evolution have been made by Saier. The various transport segments may have had an independent evolution with evolution of the ABC superfamily about 3.5 billion years ago. The MIP family may have emerged about 2.5 billion years ago, and gene duplication to provide the 12 spanners may have evolved about 2 billion years ago. The RND family arose about 1.5 billion years ago after the emergence of eukaryotes, with the voltage-sensitive channels of $Na^+$ and $Ca^{2+}$ having evolved about 1 billion years ago. As gene analysis continues, the evolution of transport proteins is sure to be continuously developed.

Table 10.16. Selected examples of proton-dependent drug efflux systems in bacteria.[a]

| Representative group | Organism | Substrates | Number of TMS | Protein |
|---|---|---|---|---|
| MFS, Family 1 | *Bacillus subtilis* | TET | 14 | BsTet |
| | *Escherichia coli* | CCCP, TCS, nalidixic acid, and organic mercury compounds | 14 | EmrB |
| | *Mycobacterium smegmatis* | AC, BC, EC, and fluoroquinolones | 14 | LfrA |
| | *Streptomyces pristinaespirali* | Pristinamycin I and II, rifampin | 14 | Ptr |
| | *Staphylococcus aureus* | BC, CH, CT, EB, and PE | | |
| | *S. aureus* | TET | 14 | TetK |
| MFS, Family 2 | *B. subtilis* | AC, CML, CT, EB, fluoroquinolones, Rhodamine 6G and TPP | 12 | Blt |
| | *E. coli* | CCCP | 12 | EmrD |
| | *Pseudomonas aeruginosa* | TET | 12 | TetC |
| | *S. aureus* | AC, CML, CT, EB, and TPP | 12 | NorA |
| | *Vibrio anguillarum* | TET | 12 | TetG |
| RND Family | *Alcaligenes eutrophus* | Cobalt, cadmium, and zinc | 12 | CzcABC |
| | *Alcaligenes xylosoxidans* | Cadmium, cobalt, and nickel | 12 | NccA |
| | *E. coli* | AC, CV, detergents, EB, erythromycin | 12 | AcrB |
| | *Neiserria gonorrhea* | detergents, hydrophobic antibiotics and dyes | 12 | MtrD |
| | *Pseudomonas aeruginosa* | Penicillins, fluoroquinolones and TET | 12 | MexB |
| SMR Family | *E. coli* | CT, CV, EB, MV, TET, and TPP | 4 | EmrE |
| | *S. aureus* | CT, CV, EB | 4 | Smr |

[a]Data from Paulsen et al., 1996.[47]

AC, Acriflavin; BC, benzalkonium chloride; CH, chlorhexidine; CCCP, carbonyl cyanide *m*-chlorophenylhydrazone; CML, chloramphenicol; CT, cetyltrimethylammonium bromide; CV, crystal violet; EB, ethidium bromide; MV, methyl viologen; PE, pentamidine isethionate; TET, tetracycline; TCS, tetrachlorosalicylanilide; TPP, tetraphenylphosphonium; TMS, transmembrane segments.

# References

1. Driessen, A.J.M. and W.N. Konings. 1990. Energetic problems of bacterial fermentations: Extrusion of metabolic end products. In *The Bacteria*, Vol. XII: *Bacterial Energetics* (T.A. Krulwich, ed.). Academic Press, New York, pp. 449–478.

2. Harold, F.M. 1986. *The Vital Force: A Study of Bioenergetics*. W.H. Freeman, New York.

3. Kaback, H.R. 1990. Active transport: Membrane vesicles, bioenergetics, molecules, and mechanisms. In *The Bacteria*, Vol. XII: *Bacterial Energetics* (T.A. Krulwich, ed.). Academic Press, New York, pp. 151–224.

4. Maloy, S.R. 1990. Sodium-coupled cotransport. In *The Bacteria*, Vol. XII: *Bacterial Energetics* (T.A. Krulwich, ed.). Academic Press, New York, pp. 203–224.

5. Unemoto, T., H. Tokuda, and M. Hayashi. 1990. Primary sodium pumps and their significance in bacterial energetics. In *The Bacteria*, Vol. XII: *Bacterial Energetics* (T.A. Krulwich, ed.). Academic Press, New York, pp. 33–54.

6. Epstein, W. 1990. Bacterial transport ATPases. In *The Bacteria*, Vol. XII: *Bacterial Energetics* (T.A. Krulwich, ed.). Academic Press, New York, pp. 87–110.

7. Saier, M.H., Jr. 1994. Computer-aided analysis of transport protein sequences. *Microbiological Reviews* 58:75–93.

8. Silver, S. 1996. Transport of inorganic cations. In Escherichia coli *and* Salmonella: *Cellular and Molecular Biology*, 2nd edit. (F.C. Neidhardt, R. Curtiss III, J.L. Ingraham, E.C.C. Lin, K.B. Low, B. Magasanik, W.S. Reznikoff, M. Riley, M. Schaechter, and H.E. Umbarger, eds.). American Society for Microbiology Press, Washington, D.C., pp. 1091–1102.

9. Kaur, P. 2001. Microbial arsenite and antimonite transporters. In *Microbial Transport Systems* (G. Winkelmann, ed.). Wiley-VCH, Weinheim, pp. 377–395.

10. Silver, S. and Le T. Phung. 1996. Bacterial heavy metal resistances. *Annual Review of Microbiology* 50:753–789.

11. Ames. G.F.-L. 1990. Energetics of periplasmic transport systems. In *The Bacteria*, Vol. XII: *Bacterial Energetics* (T.A. Krulwich, ed.). Academic Press, New York, pp. 225–246.

12. Boos, W. and J.M. Lucht. 1996. Periplasmic binding protein-dependent ABC transporters. In Escherichia coli *and* Salmonella: *Cellular and Molecular Biology*, 2nd edit. (F.C. Neidhardt, R. Curtiss III, J.L. Ingraham, E.C.C. Lin, K.B. Low, B. Magasanik, W.S. Reznikoff, M. Riley, M. Schaechter, and H.E. Umbarger, eds.). American Society for Microbiology Press, Washington, D.C., pp. 1175–1209.

13. Furlong, C. 1987. Osmotic-shock-sensitive transport systems. In Escherichia coli *and* Salmonella typhimurium: *Cellular and Molecular Biology* (F.C. Neidhardt, J.L. Ingraham, K.B. Low, B. Magasanik, M. Schaechter, and H.E. Umbarger, eds.). American Society for Microbiology Press, Washington, D.C., pp. 768–796.

14. Wandersman, C. 1996. Secretion across the bacterial outer membrane. In Escherichia coli *and* Salmonella: *Cellular and Molecular Biology*, 2nd edit. (F.C. Neidhardt, R. Curtiss III, J.L. Ingraham, E.C.C. Lin K.B. Low, B. Magasanik, W.S. Reznikoff, M. Riley, M. Schaechter, and H.E. Umbarger, eds.). American Society for Microbiology Press, Washington, D.C., pp. 953–966.

15. Postma, P.W., J.W. Lengeler, and G.R. Jacobson. 1996. Phosphoenolpyruvate:carbohydrate phosphotransferase systems. In Escherichia coli *and* Salmonella: *Cellular and Molecular Biology*, 2nd edit. (F.C. Neidhardt, R. Curtiss III, J.L. Ingraham, E.C.C. Lin K.B. Low, B. Magasanik, W.S. Reznikoff, M. Riley, M. Schaechter, and H.E. Umbarger, eds.). American Society for Microbiology Press, Washington, D.C., pp. 1149–1174.

16. Saier, M.H., Jr. and A.M. Chin. 1990. Energetics of the bacterial phosphotransferase system in sugar transport and the regulation of carbon metabolism. In *The Bacteria*, Vol. XII: *Bacterial Energetics* (T.A. Krulwich, ed.). Academic Press, New York, pp. 273–300.

17. Cronan, J.E., Jr., R.B. Gennis, and S.R. Maloy. 1987. Cytoplasmic membrane. In *Escherichia coli and Salmonella typhimurium: Cellular and Molecular Biology* (F.C. Neidhardt, J.L. Ingraham, K.B. Low, B. Magasanik, M. Schaechter, and H.E. Umbarger, eds.). American Society for Microbiology Press, Washington, D.C., pp. 31–55.

18. Hays, J.B. 1978. Group translocation transport systems. In *Bacterial Transport* (B.P. Rosen, ed.). Marcel Dekker, New York, pp. 43–103.

19. Booth, I.R. and P. Louis. 1999. Managing hypoosmotic stress: Aquaporins and mechanosensitive channels in *Escherichia coli*. *Current Opinion in Microbiology* 2:166–169.

20. Bremer, E. and R. Krämer. 2000. Coping with osmotic challenges: Osmoregulation through accumulation and release of compatible solutes in bacteria. In *Bacterial Stress Response* (G. Storz and R. Hengge-Aronis, eds.). American Society for Microbiology Press, Washington, D.C., pp. 79–97.

21. Braun, V. 1995. Energy-coupled transport and signal transduction through the Gram-negative outer membrane via TonB-ExbB-ExbD-dependent receptor proteins. *FEMS Microbiology Reviews* 16:295–307.

22. Braun, B. and K. Hantke. 2001. Mechanisms of bacterial iron transport. In *Microbial Transport Systems* (G. Winkelmann, ed.). Wiley-VCH, Weinheim, pp. 289–313.

23. Crumbliss, A.L. 1991. Aqueous solution equilibrium and kinetic studies of iron siderophore and model siderophore complexes. In *Handbook of Microbial Iron Chelates* (G. Winkelmann, ed.). VCH, New York, pp. 177–233.

24. Earhart, C.F. 1987. Ferrienterobactin transport in *Escherichia coli*. In *Iron Transport in Microbes, Plants and Animals* (G. Winkelmann, D. van der Helm, and J.B. Neilands, eds.). VCH, Weinheim, pp. 67–84.

25. Neilands, J.B., K. Konopka, B. Schwyn, M. Coy, R.T. Francis, B.H. Paw, and A. Bagg. 1987. Comparative biochemistry of microbial iron assimilation. In *Iron Transport in Microbes, Plants and Animals* (G. Winkelmann, D. van der Helm, and J.B. Neilands, eds.). VCH, Weinheim, pp. 3–34.

26. Hmiel, S.P., M.S. Snavely, J.B. Florer, M.E. Maguire, and C.G. Miller. 1989. Magnesium transport in *Salmonella typhimurium*: Genetic characterization and cloning of three magnesium transport loci. *Journal of Bacteriology* 171:4742–4751.

27. Cellier, M. 2001. Bacterial genes controlling manganese accumulation. In *Microbial Transport Systems* (G. Winkelmann, ed.). Wiley-VCH, Weinheim, pp. 325–346.

28. Smith, R.J. 1995. Calcium and bacteria. *Advances in Microbiol Physiology* 37:83–133.

29. Eitinger, T. and B. Fredrich. 1997. Microbial nickel transport and incorporation into hydrogenases. In *Transition Metals in Microbial Metabolism* (G. Winkelmann and C.J. Carrano, eds.). Harwood Academic, Amsterdam, pp. 235–256.

30. Kakinuma, Y. 1998. Inorganic cation transport and energy transduction in *Enterococcus hirae* and other Streptococci. *Microbiology and Molecular Biology Reviews* 62:1021–1045.

31. Hantke, K. 2001. Bacterial Zn transporters and regulators. *Biometals* 14:239–249.

32. Van Dommelen, A., R. DeMot, and J. Vanderleyden. 2001. Ammonium transport: Unifying concepts and unique aspects. *Australian Journal of Plant Physiology* 28:959–967.

33. Kleiner, D. 1985. Bacterial ammonium transport. *FEMS Microbiology Reviews* 32:87–100.

34. Clegg, S., F. Yu, L. Griffiths, and J.A. Cole. 2002. The roles of the polytopic membrane proteins NarK, NarU and NirC in *Escherichia coli* K-12: Two nitrate and three nitrite transporters. *Molecular Microbiolgy* 44:143–155.

35. Webb, D.C. and G.B. Cox. 1994. Proposed mechanism for phosphate translocation by the phosphate-specific transport (Pst) system and role of the Pst system in phosphate regulon. In *Phosphate in Microorganisms* (A. Torriani-Gorini, E. Yagil, and S. Silver, eds.). American Society for Microbiology Press, Washington, D.C., pp. 37–43.

36. Van Veen, H.K., T. Abee, G.J.J. Kortstee, W.N. Konings, and A.J.B. Zehnder. 1994. Phosphate inorganic transport (Pit) system in *Escherichia coli* and *Acinetobacter johnsonii*. In *Phosphate in Microorganisms* (A. Torriani-Gorini, E. Yagil, and S. Silver, eds.). American Society for Microbiology Press, Washington, D.C., pp. 43–62.

37. Pau, R.N., W. Klipp, and S. Leimkhler. 1997. Molybdenum transport, processing and gene regulation. In *Transition Metals in Microbial Metabolism* (G. Winkelmann and C.J. Carrano, eds.). Harwood Academic, Amsterdam, pp. 217–234.

38. Ambudkar, S.V. and B.P. Rosen. 1990. Ion-exchange systems in prokaryotes. In *The Bacteria*, Vol. XII: *Bacterial Energetics* (T.A. Krulwich, ed.). Academic Press, New York, pp. 247–272.

39. Driessen, A.J.M. and W.N. Konings. 1990. Extrusion of metabolic products. In *The Bacteria*, Vol. XII: *Bacterial Energetics* (T.A. Krulwich, ed.). Academic Press, New York, pp. 449–478.

40. Nicholls, D.G. and S.J. Ferguson. 1992. *Bioenergetics 2*. Academic Press, New York.

41. Maloney, P.C. and T.H. Wilson. 1996. Ion-coupled transport and transporters. In Escherichia coli *and* Salmonella: *Cellular and Molecular Biology*, 2nd edit. (F.C. Neidhardt, R. Curtiss III, J.L. Ingraham, E.C.C. Lin K.B. Low, B. Magasanik, W.S. Reznikoff, M. Riley, M. Schaechter, and H.E. Umbarger, eds.). American Society for Microbiology Press, Washington, D.C., pp. 1130–1148.

42. Antonucci, T.K. and D.I. Oxender. 1986. The molecular biology of amino acid transport in bacteria. *Advances in Microbial Physiology* 28:146–180.

43. Payne, J.W. and M.W. Smith. 1994. Peptide transport by micro-organisms. *Advances in Microbiological Physiology*. 36:2–80.

44. Chopra, I. and P. Ball. 1982. Transport of antibiotics into bacteria. *Advances in Microbiological Physiology* 23:184–240.

45. Dubnau, D. 1999. DNA uptake in bacteria. *Annual Review of Microbiology* 53:217–244.

46. Roth, J.R., J.G. Lawrence, and T.A. Bobik. 1996. Cobalamin (coenzyme $B_{12}$): Synthesis and biological significance. *Annual Review of Microbiology* 50:137–181.

47. Paulsen, I.T., M.H. Brown, and R.A. Skurray. 1996. Proton-dependent multidrug efflux systems. *Microbiology Review* 60:575–608.

# Additional Reading

## *Transport*

Aquero, J., C. Delamanche, and J.M. Garcia-Lobo. 2000. A functional water channel protein in the pathogenic bacterium *Brucella abortus*. *Microbiology* 146:3251–3257.

Braibnt, M., P. Gilot, and J. Content. 2000. The ATP binding cassette (ABC) transport systems of *Mycobacterium tuberculosis*. *FEMS Microbiology Reviews* 24:449–467.

Driessen, A.J.M., J.L.C.M. van de Vossenberg, and W.N. Konings. 1996. Membrane composition and ion-permeability in extremophiles. *FEMS Microbiology Review* 18:139–149.

Kawano, M., R. Abuki, and K. Igarashi. 2000. Potassium uptake with low affinity and high rate in *Enterococcus hirae* at alkaline pH. *Archives for Microbiology* 175:41–45.

Rabus, R., D.L. Jack, D.J. Kelly, and M.H. Saier, Jr. 1999. TRAP transporters: An ancient family of extracytoplasmic solute-receptor-dependent secondary active transporters. *Microbiology* 145:3431–3445.

Takano, J., K. Kaidoh, and N. Kamo. 1995. Fructose transport by *Haloferax volcanii*. *Canadian Journal of Microbiology* 41:241–246.

# V
# Metabolic Processes

# 11
# Pathways of Carbon Flow

Let us not forget, in the conduct of research, that "The reward of the general is not a bigger tent, but command,"* and that the goal of our studies is not solely related fact, but understanding. W.W. Umbreit, *Bacterial Physiology*, 1951 (*O.W. Holmes)

## 11.1. Introduction

Metabolism refers to all of the chemical changes in a cell, and these changes are not random but are directed by about a thousand different enzymes. To achieve cell growth and reproduction, most of the enzymes function collectively for biosynthesis while others are involved with oxidation of organic compounds as energy sources. Some of the enzymes interface with either catabolic (degradative) or anabolic (synthetic) processes and are referred to as amphibolic (from the Greek, *amphi*, meaning "either"). Given the large number of possible organic compounds that can be converted to a few common end products by prokaryotes, catabolism is a convergent process. On the other hand, bacteria use relatively few precursor molecules to produce a wide variety of cellular structures and anabolism is a divergent process. This chapter examines some of the principal catabolic activities of bacteria as well as a few of the anabolic processes important in carbon flow.

Although enzymes function independently of each other, they are organized into pathways to effectively promote a sequence of reactions and genes. Research approaches that are invaluable for constituting metabolic pathways include the following: (1) the use of radiolabeled substrates to establish enzymatic reactions, (2) the isolation of enzymes to establish kinetic parameters, and (3) the use of metabolic mutants to determine intermediates in the pathways. Over the years catabolic enzymes found in prokaryotic cells were established as similar or identical to enzymes found in eukaryotic cells and acceptance of this biochemical uniformity of cells came with the construction of similar metabolic pathways for the various forms of life. With this demonstration of similar enzymatic reactions in microbial and eukaryotic cells, the concept of Unity in Biochemistry gained momentum.

In catabolism, there are generally two major objectives: generation of energy and conversion of the carbon material to building blocks for biosynthetic reactions. Substrate level phosphorylation accounts for the formation of ATP from energy-yielding reactions in the various steps of the metabolic pathway. The carbon compounds in the pathway are oxidized with the release of $CO_2$, 2 $e^-$, and 2 $H^+$. It is relatively obvious that for stoichiometric reactions an accounting of carbons is needed to evaluate the metabolic event. It is equally important to account for 2 $e^-$ and 2 $H^+$ that are released as the carbon compounds are oxidized. With respect to the protons, caution must be exercised because at a physiological pH of 7.0 organic acids are ionized and the free acid form does not exist. These concerns of balancing equations for carbons, electrons, and protons are addressed when microbiologists construct fermentation balances to evaluate anaerobic catabolism.

## 11.2. Pathways of Catabolism in Heterotrophs[1–3]

As one may expect, there are various different routes for molecules to take in microbial catabolism of sugars and other compounds used as carbon and energy sources. Traditionally, discussion of the pathways centers on glucose because it is the most abundant form of hexose on Earth largely because of the presence of cellulose in plants. Glucose is the preferred carbon source in many heterotrophs. The utilization of hexoses other than glucose and pentoses is included because sugar utilization is a principal means of identifying bacteria by fermentation tests. In addition, amino acids, nucleic acids, and organic acids are commonly used as carbon sources and it is important to understand the pathways that carbon skeletons utilize in energy production.

### 11.2.1. Glycolysis

Although heterotrophic bacteria will utilize many different carbon compounds to support growth, most bacteria can metabolize glucose by the glycolytic pathway which is also known as the Embden–Meyerhof–Parnas pathway, or the fructose bisphosphate aldolase pathway. This metabolic system is given in Figure 11.1 and the overall reaction for glycolysis is given as follows:

$$\text{Glucose} + 2 \text{ ATP} + 2 \text{ NAD}^+ \rightarrow 2 \text{ pyruvate} + 4 \text{ ATP} + 2 \text{ NADH}$$

All enzymes for glycolysis are soluble in the cytoplasm and none are bound to the plasma membrane. Glycolysis functions under both aerobic and anaerobic conditions with the only distinction being the fate of NADH generated in the pathway which is influenced by the condition of growth. Under aerobic conditions, electrons from NADH are diverted to $O_2$ while the electron acceptor in an anaerobic environment is commonly an organic compound.

The first step of glycolysis requires energy from ATP. Hexokinase, which catalyzes the phosphorylation of glucose to produce glucose-6-phosphate, is one of the few enzymes in glycolysis that is not reversible. Another enzyme that is not reversible is phosphofructokinase, which phosphorylates fructose-6-phosphate to produce fructose-1,6-bisphosphate. Both hexokinase and phosphofructokinase are regulatory enzymes for glycolysis. While hexokinase is inhibited by glucose-6-phosphate, phosphofruc-

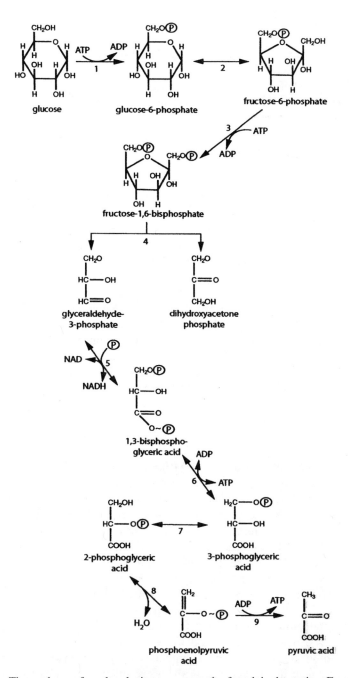

Figure 11.1. The pathway for glycolysis as commonly found in bacteria. Enzymes are as follows: (1) hexokinase, (2) isomerase, (3) phosphofructokinase, (4) fructose-1,6-bisphosphate aldolase, (5) glyceraldehyde-3-phosphate dehydrogenase, (6) phosphoglycerate kinase, (7) mutase, (8) enolase, and (9) pyruvate kinase. Reactions of 1, 3, and 9 are unidirectional. For simplicity, organic acids are indicated as the free acid but at physiological pH these acids would be ionized.

tokinase is inhibited by ATP or citrate, and phosphofructokinase is activated by AMP or ADP.

In the second part of glycolysis, fructose-1,6-bisphosphate is cleaved and with subsequent metabolism ATP is produced with electrons carried by NADH. The enzyme responsible for the cleavage of fructose-1,6-bisphosphate is an aldolase, and based on the requirement for metal cations, there are two different classes of bacterial fructoaldolase. Class I fructoaldolase is found in *Bacillus stearothermophilus* and it does not require metal cofactors for reaction. Class II fructoaldolase uses either $Fe^{2+}$, $Co^{2+}$, $Mn^{2+}$, or $Zn^{2+}$ and it has been isolated from *Brucella suis*, *Lactobacillus bifidus*, *Clostridium perfringens*, and *Anacystis nidulans*. An interconversion between glyceraldehyde-3-phosphate and dihydroxyacetone phosphate is catalyzed by triose phosphate isomerase. The first oxidation reaction in glycolysis is the conversion of glyceraldehyde-3-phosphate to 1,3-bisphosphoglycerate with the formation of NADH by glyceraldehyde-3-phosphate dehydrogenase. An ATP molecule is generated by the conversion of 1,3-bisphosphoglycerate to 3-phosphoglycerate by phosphoglycerate kinase. The phosphoglycerate mutase moves the phosphate from the 3 position to the 2 position on glycerate and the enolase forms phosphoenol pyruvate (PEP). Pyruvate is formed from PEP by pyruvate kinase with the formation of ATP. The metabolism of dihydroxyacetone phosphate occurs only after it is converted to glyceraldehyde-3-phosphate. Thus, in accounting for all the carbon atoms from the cleavage of fructose-1,6-bisphosphate, it is important to remember that there are two sets of glyceraldehyde-3-phosphate traversing the triose segment of glycolysis. An important inhibitor for glycolysis is fluoride, which inhibits fructoaldolase.

## 11.2.2. Hexose Monophosphate Pathway

The hexose monophosphate pathway (HMP), also called the phosphogluconate pathway, is a complex set of reactions using glucose-6-phosphate to produce pentose phosphate and generate NADPH for the cell. Enzymes for the HMP are found in the cytoplasm of both anaerobic and aerobic heterotrophic bacteria. The HMP is composed of a set of oxidative reactions and a series of reactions with the interconversion of sugars (see Figure 11.2). The initial reaction oxidizes glucose-6-phosphate to 6-phosphogluconate with 2 $e^-$ and 2 $H^+$ used to produce NADPH. The second step involves decarboxylation of 6-phosphogluconate with the release of $CO_2$, the production of ribulose-5-phosphate, and formation of another NADPH. These oxidative reactions cannot be reversed but subsequent reactions in this pathway are reversible. Reaction involving the condensation of ribose-5-phosphate and xylulose-5-phosphate yield sedoheptulose-7-phosphate and glyceraldehyde-3-phosphate. When these phosphorylated seven-carbon and three-carbon compounds interact, erythrose-4-phosphate and fructose-6-phosphate are produced. With the reaction of the phosphorylated four-carbon and six-carbon sugars, there is the formation of xylulose-5-phosphate and glyceraldehyde-3-phosphate. Keeping track of the carbon atoms in the HMP is essential.

The glycolytic and HMP pathways function simultaneously in bacteria; however, bacteria do not have an obligatory requirement for HMP. Mutants that have been blocked from using glycolysis can utilize HMP; however, the rate of growth is reduced

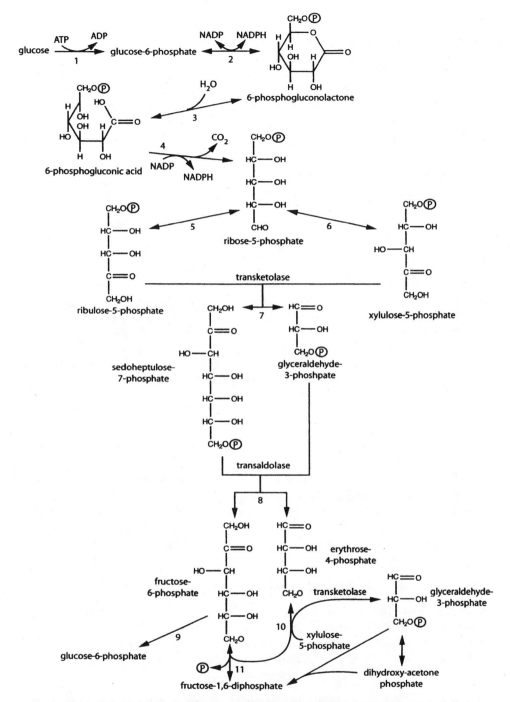

Figure 11.2. Carbon flow in the pentose phosphoketolase pathway, also known as the hexose monophosphate pathway. Enzymes are as follows: (1) hexokinase, (2) glucose-6-phosphate dehydrogenase, (3) gluconolactonase, (4) 6-phosphogluconate dehydrogenase, (5) ribose-5-phosphate isomerase, (6) ribulose monophosphate epimerase, (7) transketolase, (8) transaldolase, (9) isomerase, (10) transketolase, (11) phosphofructokinase.

substantially. Mutants blocked for HMP continue to grow on glucose using glycolysis at the same rate as when both glycolysis and HMP function. The HMP is a supplemental pathway for glycolysis in that the starting compound for HMP is from glycolysis and intermediate compounds for the glycolytic pathway are end products of HMP. Because ribose-5-phosphate from HMP is an important precursor for purine and pyrimidine biosynthesis, the shunting of carbon molecules from HMP to biosynthetic processes would effectively reduce the amount of glyceraldehyde-3-phosphate and fructose-6-phosphate formed that would enter the glycolytic pathway.

## 11.2.3. Entner–Doudoroff Pathway

Unlike the glycolytic pathway and the HMP, the Enterner–Doudoroff (ED) pathway is found only in bacteria. The ED pathway is an alternate pathway to metabolize glucose, by way of 6-phosphogluconate, to glyceraldehyde-3-phosphate and pyruvate (see Figure 11.3). Initially, glucose is phosphorylated by hexokinase to glucose-6-phosphate which is subsequently oxidized to gluconolactone-6-phosphate with formation of NADPH. An intermediate formed in this ED pathway is 2-keto-3-deoxy-6-phosphogluconate and it is cleaved by a specific aldolase to produce the triose phosphate and pyruvate. Glyceraldehyde-3-phosphate may be further metabolized by enzymes in the last portion of the glycolysis pathway to produce another pyruvic acid and in the process 2 mols of ATP and 1 mol of NADH will be formed with each mole of hexose. The overall reaction for the ED pathway is as follows:

$$\text{Glucose} \rightarrow \text{ATP} + 2\ \text{NADH} + \text{NADPH} + 2\ \text{pyruvate}$$

The presence of the three glucose metabolizing pathways in bacteria is not completely clear but represents a diversity in metabolism. The specific catabolic pathways used by bacteria are given in Table 11.1. Also, there is evidence that when two pathways for glucose catabolism are present in a cell, there is a preference with one pathway used primarily in log phase and another used in stationary phase. It would be reasonable to consider that there is not an abrupt end point for use of a specific pathway but that a transition period is part of the process. Thus, the energetics of glucose utilization may change considerably as cells go through the growth phases.

## 11.2.4. Pentose Phosphoketolase Pathway

A small group of bacteria use pentoses by the pentose phosphate pathway which is a variation of the HMP pathway. In this pathway, ribose, xylose, or arabinose are transformed to xylulose-5-phosphate and in the presence of inorganic phosphate, a phosphoketolase produces glyceraldehyde-3-phosphate and acetyl phosphate (see Figure 11.4). The overall reaction is as follows:

$$\text{Pentose} + \text{ATP} + P_i \rightarrow \text{glyceraldehyde-3-phosphate} + \text{acetyl}{\sim}\text{phosphate} + \text{ADP}$$

The pentose phosphoketolase pathway is unique in that it enables five-carbon sugars to be utilized and it ends with high-energy compounds. The generation of NADH, ATP, and pyruvate results when glyceraldehyde-3-phosphate is metabolized by en-

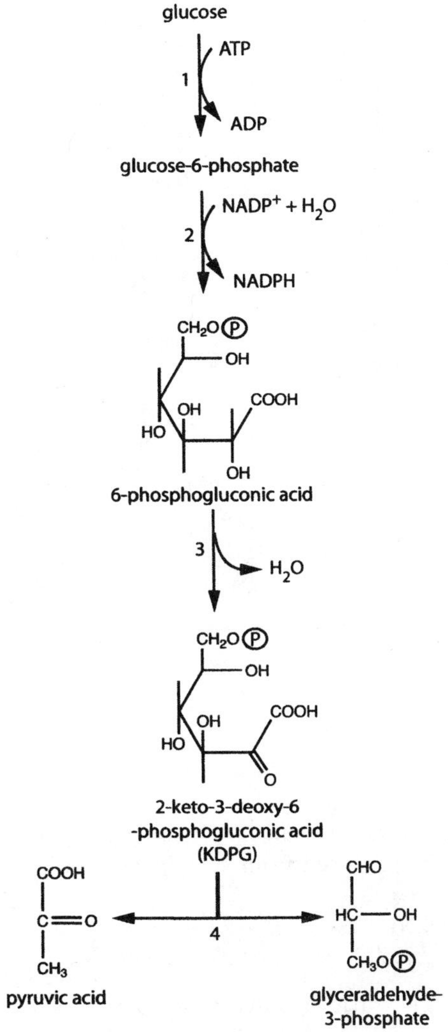

Figure 11.3. Entner–Douderoff pathway in bacteria. The enzymes are as follows: (1) hexokinase, (2) 6-phosphogluconate dehydratase, and (4) KDPG aldolase. Glyceraldehyde-3-phosphate can be converted to a second pyruvate by using enzymes commonly found in glycolysis.

Table 11.1. Catabolic pathways used by bacteria.

| Bacteria | EMP (%) | HMP (%) | ED (%) |
|---|---|---|---|
| *Streptomyces griseus* | 97 | 3 | |
| *Bacillus subtilis* | 74 | 26 | |
| *Escherichia coli* | 72 | 28 | |
| *Sarcina lutea* | 70 | 30 | |
| *Pseudomonas aeruginosa* | | 29 | 71 |
| *Acetomonas oxidans* | | 100 | |
| *Pseudomonas saccharophilus* | | | 100 |
| *Zymomonas mobilis* | | | 100 |

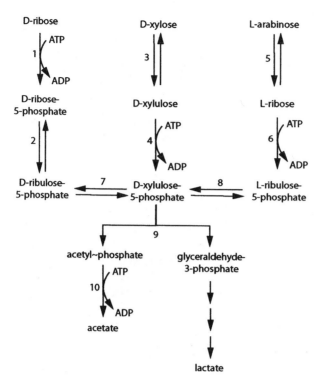

Figure 11.4. Utilization of sugars by heterofermentative lactobacilli and *Bifidobacteria* using the pentose phosphoketolase pathway. Enzymes are as follows: (1) ribulokinase, (2) ribose-5-phosphate isomerase, (3) xylose ketol isomerase, (4) xylulose kinase, (5) arabinose isomerase, (6) ribulokinase, (7) ribulose-5-phosphate 3-epimerase, (8) ribulose phosphate-4-epimerase, (9) phosphoketolase, and (10) acetokinase.

zymes associated with the glycolytic pathway. In addition, it is possible for ATP to be produced from acetylphosphate with the formation of acetate.

## 11.2.5. Citric Acid Cycle

The citric acid cycle (also know as the Krebs cycle, tricarboxylic acid cycle, or TCA cycle) is responsible for oxidation of acetyl~CoA to $CO_2$ and $H_2O$ and is thereby important for catabolism. The citric acid cycle functions only in aerobic bacteria and while the cycle does not function in anaerobic bacteria, anaerobes may use some of reactions associated with this cycle. The principal entry of the citric acid cycle is the acetyl group, from acetyl~CoA, which is generated by decarboxylation of pyruvate. The cycle is often called the Krebs cycle because in 1937 H.A. Krebs provided crucial information concerning the condensation of the two-carbon unit with oxalacetate to form citric acid. The citric acid cycle and the oxidative decarboxylation of pyruvate is given in Figure 11.5.

The production of acetyl~CoA from pyruvate is not a part of the citric acid cycle but is crucial because it provides the two-carbon unit needed for citric acid formation. The decarboxylation of pyruvate is by a multistep reaction involving three enzymes and five coenzymes making up the pyruvate dehydrogenase complex. The overall

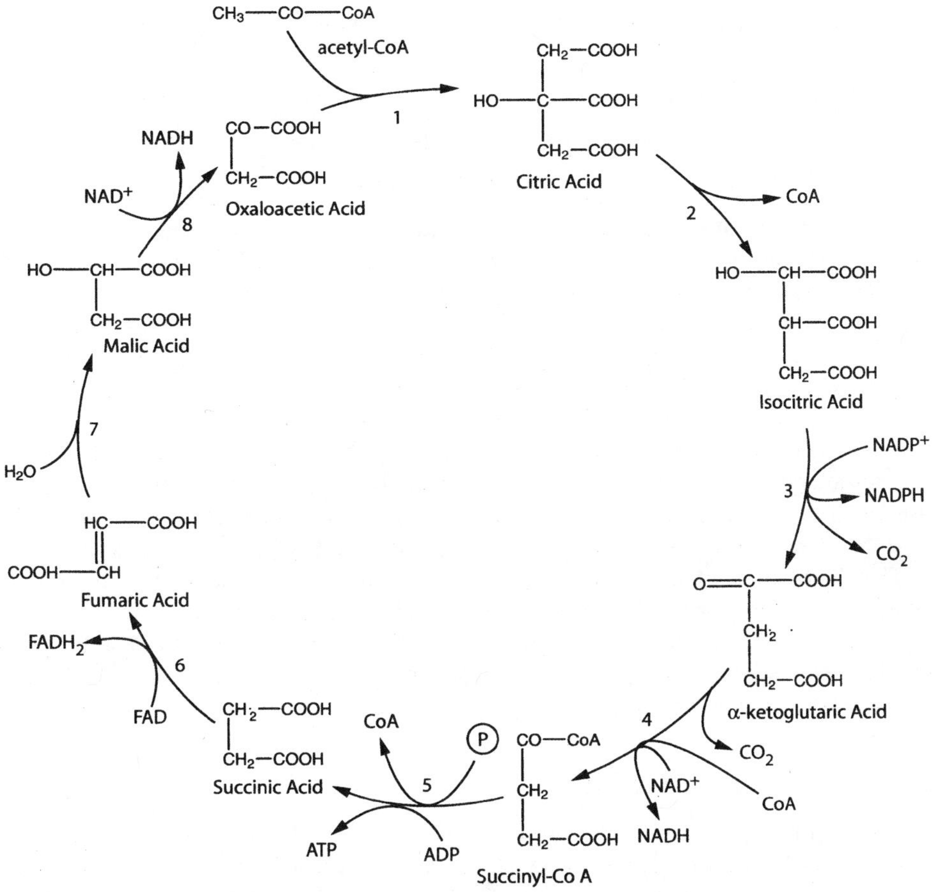

Figure 11.5. The citric acid cycle in aerobic bacteria. Enzymes are as follows: (1) citrate synthase; (2) aconitase; (3) isocitrate dehydrogenase; (4) α-ketoglutarate dehydrogenase; (5) succinate thiokinase; (6) succinate dehydrogenase; (7) fumarase; and (8) malate dehydrogenase. Not shown are the following intermediates: *cis*-aconitate, which is between citrate and isocitrate, and oxalosuccinate, which is between isocitrate and α-ketoglutarate. *Cis*-aconitate is an unstable intermediate that is found only associated with aconitase. Oxalosuccinate is produced from the oxidation of isocitrate with the generation of NADH. With the decarboxylation of oxalosuccinate, α-ketoglutarate is produced. Both the formation of oxalosuccinate from isocitrate and the production of α-ketoglutarate from oxalosuccinate are catalyzed by isocitrate dehydrogenase.

reactions of the pyruvate decarboxylation reaction and for the citric acid cycle are given in the following reactions:

$$\text{Pyruvate} + NAD^+ + CoA \rightarrow Acetyl{\sim}CoA + CO_2 + NADH + H^+$$

$$Acetyl{\sim}CoA + 3\ NAD^+ + FAD + GDP + P_i \rightarrow 2\ CO_2$$
$$+\ 3\ NADH + 3\ H^+ + FADH_2 + GTP + CoA$$

There are four points for regulating the activities of enzymes involved in the conversion of pyruvate to $CO_2$: (1) The pyruvate dehydrogenase complex is inhibited by acetyl$\sim$CoA and by elevated levels of NADH. (2) Citrate synthase, the enzyme that catalyzes the condensation of acetyl$\sim$CoA and oxalacetate, is inhibited by ATP, NADH, succinyl$\sim$CoA, and acyl$\sim$CoA derivitives of fatty acids. (3) Isocitrate dehydrogenase accounts for the first decarboxylation reaction in the citric acid cycle and is irreversible. This enzyme is one of the rate-limiting enzymes with activation by ADP and inhibition by ATP or NADH. (4) The oxidative decarboxylation of $\alpha$-ketoglutarate to succinyl$\sim$CoA is attributed to the $\alpha$-ketoglutarate dehydrogenase complex which uses a mechanism similar to pyruvate decarboxylase. The $\alpha$-ketoglutarate dehydrogenase is inhibited by ATP, GTP, NADH, and succinyl$\sim$CoA.

The distribution of the citric acid cycle is not universal in bacteria. Frequently, anaerobic bacteria lack a functioning citric acid cycle with the inability to convert $\alpha$-ketoglutarate to succinate. The citric acid cycle does not function in Mycoplasmas (Mollicutes) because they do not have citrate synthase; however, they do contain many of the other enzymes of the cycle.

## 11.2.6. Glyoxylic Acid Cycle

Fairly common in bacteria is the capability of growing with acetate as the sole carbon source. The glyoxylate cycle accounts for acetate utilization by inserting enzymes into the citric acid cycle (see Figure 11.6). The two additional enzymes are isocitrate

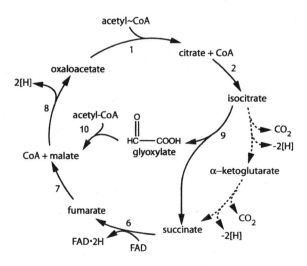

Figure 11.6. The glyoxylate cycle in bacteria. Enzymes are as follows: (1) citrate synthase, (2) aconitase, (6) succinate dehydrogenase, (7) fumarase, (8) malate dehydrogenase, (9) isocitrate lyase, and (10) malate synthase. Reactions 1, 2, 6, 7, and 8 are from the TCA cycle. Dashed lines are reactions in the TCA cycle that are bypassed in the glyoxylate cycle.

lyase, which cleaves isocitrate to glyoxylate plus succinate and malate synthase, which condenses glyoxylate with acetyl~CoA to produce malate. Acetyl~CoA will also condense with oxalacetate using enzymes of the citric acid cycle and it can be envisioned that for every three acetyl~CoA units entering the glyoxylate/citric acid cycle one is broken down to $CO_2$ with the concomitant formation of NADH. The overall reaction of the glyoxylate and citric acid cycle is as follows:

$$3 \text{ Acetyl~CoA} \rightarrow \text{malate} + 2 \text{ } CO_2 + 2 \text{ NADH} + 3 \text{ CoA}$$

## 11.2.7. Determination of Major Catabolic Pathways

An important characteristic in pathways that break down sugars to smaller units is that the process is specific and highly reproducible. There are three principal ways to evaluate which catabolic process is used by prokaryotes: (1) Follow the individual

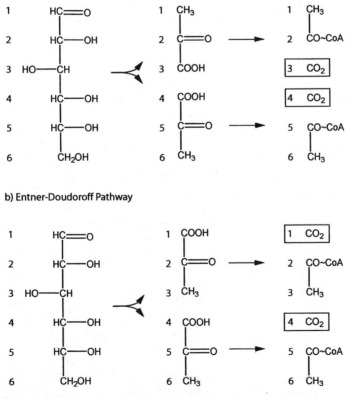

Figure 11.7. Distribution of labeled carbon atoms in catabolism of glucose by (**a**) glycolysis or (**b**) Entner–Doudoroff pathway. Numbers indicate the carbon in glucose. $CO_2$ are products of the pyruvate dehydrogenase. (From Gottschalk, 1986.[1])

carbon atoms of the sugars using radiolabeled molecules, (2) demonstrate the presence of key enzymes for the pathway, and (3) demonstrate the presence of intermediate molecules that are distinctive for that pathway.

The carbon label in the end product will reveal the enzymes involved and the pathways through which the catabolism proceeds. Consider that each of the carbon atoms in glucose can be identified as carbon 1 through 6 (see Figure 11.7). Catabolism by glycolysis results in cleavage of the six-carbon unit of fructose-1,6-bisphosphate into two carbon units each with three carbons, and oxidation occurs in the center of the molecule. With the conversion of dihydroxyacetone phosphate to glyceraldehyde-3-phosphate, and the subsequent conversion to pyruvate, it would be found that C-3 and C-4 atoms in the hexose would be found in the carboxyl of pyruvate. Complete oxidation of glucose occurs with the citric acid cycle following glycolysis. As pyruvate is decarboxylated to produce acetyl~CoA, there is the loss of $CO_2$ with the carbon coming from the carboxyl of pyruvate which represents C-3 and C-4 in the original glucose molecule. The remaining carbon atoms that would be from carbon number 1, 2, 5, and 6 of glucose are liberated as $CO_2$ in the citric acid cycle.

If the hexose monophosphate pathway is followed, C-1 of 6-phosphogluconic acid is lost in decarboxylation with the formation of ribulose-5-phosphate. In the Entner–Douderoff pathway with the formation of the carboxyl of pyruvate from C-1 of the hexose. As the glyceraldehyde-3-phosphate which is also formed in the Entner–

Table 11.2. Key enzymes and intermediates that can be used to indicate the presence of catabolic pathways or cycles.

| Metabolic pathway | Allosteric enzymes | Allosteric effectors[a] |
|---|---|---|
| Glycolytic pathway | Phosphofructokinase | (−)-Citrate |
|  |  | (−)-ATP |
|  |  | (−)-Phosphoenolpyruvate |
|  |  | (+)-ADP |
|  | Fructoaldolase | (−)-Dihydroxyacetone |
|  | Pyruvate kinase | (+)-AMP |
| Hexose monophosphate pathway | Glucose 6-phosphate deydrogenase | (−)-NADPH |
| Decarboxylation of pyruvate | Pyruvate dehydrogenase | (−)-NADH |
|  |  | (−)-Acetyl~CoA |
| Citric acid cycle | Citrate synthase | (−)-ATP |
|  |  | (−)-NADH |
|  |  | (−)-Succinyl-CoA |
|  | Isocitrate dehydrogenase | (−)-NADH |
|  |  | (−)-ATP |
|  |  | (+)-ADP |
|  | α-Ketoglutarate dehydrogenase | (−)-ATP |
|  |  | (−)-GTP |
|  |  | (−)-NADH |
|  |  | (−)-Succinyl-CoA |

[a] (+) indicates positive action in stimulation of reaction by allosteric effector and (−) indicates negative action on the enzyme involved.

Table 11.3. Key enzymes and intermediates that can be used to indicate the presence of catabolic pathways or cycles.

| Metabolic pathway | Unique enzymes | Unique compounds |
|---|---|---|
| Glycolytic pathway | Fructoaldolase | Fructose-1,6-bisphosphate |
| Hexose monophosphate pathway | Glucose 6-P dehydrogenase | Gluconolactone 6-P |
|  | 6-Phosphogluconate dehydrogenase | Sedoheptulose 7-P |
| Entner–Douderoff pathway | 2-Keto-3-deoxy-6-phosphogluconic acid aldolase | 2-Keto-3-deoxy-6-phosphogluconic acid |
| Citric acid cycle | Citrate synthase | Citrate |
|  | Isocitrate dehydrogenase | Isocitrate |
|  | α-Ketoglutarate dehydrogenase | Oxalacetate |
| Glycoxylate cycle | Isocitrate lyase | Glyoxylate |
|  | Malate synthase |  |
| Pentose phosphate pathway | Phosphoketolase | Pentoses |

Douderoff pathway is metabolized to pyruvate, the carboxyl of pyruvate comes from carbon number 4 of the hexose. Thus, as these two pyruvate molecules enter decarboxylation reactions, the liberation of two molecules of $CO_2$ would be C-1 and C-4 of the hexose.

To assist in determining which metabolic pathway is being used, the presence of key enzymes and carbon compounds becomes important. Using cell-free extracts, biochemical tests can be conducted to demonstrate the presence of all the enzymes present in a given pathway. Alternately, the demonstration of enzymes found exclusively with a pathway would be used as an indicator that a specific metabolic pathway is being used. Because many pathways generate glyceraldehyde-3-phosphate with subsequent metabolism to pyruvate, it would not be beneficial to look for these enzymes in identifying the pathway being used. Several carbon compounds are found in only one pathway and the demonstration of their presence would indicate that the pathway is indeed being used. Listed in Table 11.2 are some of the key enzymes and intermediate compounds associated with different metabolic pathways. As enumerated in Table 11.3, many of the key enzymes are highly regulated by allosteric effectors and they control the rate at which the metabolic pathway processes substrates.

## 11.3. Specialty Reactions

Bacteria in the environment are capable of catabolizing a vast array of carbon compounds. Anaerobic organisms have established many different pathways for decomposition of organic materials. It is beyond the scope of this book to detail all the various metabolic pathways employed by bacteria; however, two reactions merit some attention because of their uniqueness and because they are potentially present in many different bacteria. These two activities are the Stickland reaction and the phosphoryclastic reaction which are found in anaerobic bacteria.

## 11.3.1. Stickland Reaction[4]

A unique system for coupling deamination of amino acids to form NADH and ATP synthesis occurs in certain proteolytic *Clostridium* species. Although the process was described initially by Stickland in 1934, certain aspects of this reaction remain to be elucidated as one amino acid is the reducing agent and the other is the oxidizing agent. The overall activity in the Strickland reaction is shown in the following reactions:

$$\text{L-Alanine} + 2 \text{ glycine} + 3 \text{ ADP} + 3 \text{ P}_i \rightarrow 3 \text{ acetate} + 3 \text{ ATP} + 3 \text{ NH}_3 + \text{CO}_2$$

$$\text{L-Alanine} + 2 \text{ L-proline} \rightarrow \text{acetate} + 2 \text{ } \delta\text{-aminovalerate} + \text{NH}_3 + \text{CO}_2$$

L-Alanine is oxidatively deaminated to produce ammonia plus pyruvate which is subsequently decarboxylated to acetate plus $CO_2$. The oxidative decarboxylation of pyruvate to acetate requires the presence of coenzyme A, thiamine pyrophosphate, inorganic phosphate plus $NAD^+$ with the intermediate formation of acetyl~CoA and acetyl~phosphate. This reaction series is similar to the phosphoroclastic reaction found in numerous anaerobic bacteria. Glycine, the oxidizing agent, is deaminated to acetate plus ammonia in the presence of NADH. The Stickland reaction occurs with many amino acids with one amino acid serving as the donor and the other as the acceptor. Suitable donor amino acids include alanine, leucine, valine, phenylalanine, cysteine, serine, histidine, aspartic acid, and glutamic acid while acceptor amino acids are glycine, proline, hydroxyproline, ornithine, arginine, and tryptophan.

## 11.3.2. Phosphoroclastic Reaction[5]

In aerobic bacteria, the electrons from pyruvate oxidation are directed to the respiratory system where a proton-motive force may be generated, whereas in various anaerobes, substrate-level phosphorylation is coupled to the formation of acetate. Many anaerobes display the following reaction:

$$\text{Pyruvate} \rightarrow \text{acetate} + \text{CO}_2 + \text{H}_2$$

In clostridia and related organisms, the above reaction consists of a series of steps that are given in Figure 11.8. The formation of acetyl~P from pyruvate was proposed by Lipmann, who suggested the term "phosphoroclastic split" to describe the process. At the time the phosphoroclastic reaction was named, it was not known that acetyl~CoA was an intermediate preceding the formation of acetyl~phosphate. Characteristic of the set of reactions that includes the phosphoryoclastic reaction is the production of ATP. As seen from the reactions presented in Figure 11.8, electrons from the oxidation of pyruvate are carried by ferredoxin (Fd) which has a half-reaction ($Fd_{(red)}/Fd_{(ox)}$) of $-410$ mV. The midpoint reaction for ferredoxin is more electronegative than the NADH/NAD couple, which is $-320$ mV. Thus, $Fd_{(red)}$ is capable of reducing $H^+$ to $H_2$ which has a half-reaction ($H_2/2 \text{ H}^+$) potential of $-420$ mV while NADH would not readily reduce $H^+$ to $H_2$.

A distinction is made with respect to the one-carbon unit produced in the oxidation

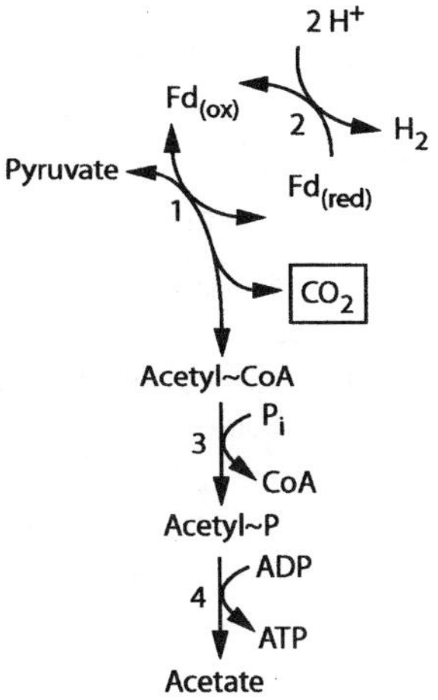

Figure 11.8. Scheme for phosphoroclastic split of pyruvate. Enzymes are as follows: (1) pyruvate:ferredoxin 2-oxidoreductase, (2) hydrogenase, (3) phosphotransacetylase, and (4) acetokinase.

of pyruvate. The pyruvate reaction resulting in the generation of $CO_2$ plus $H_2$ along with acetate is markedly different from the reaction in which pyruvate is cleaved to formate plus acetate. Enteric bacteria displaying mixed acid fermentations use pyruvate-formate lyase as a means of cleaving pyruvate to acetyl~CoA plus formate. Subsequently, formate is hydrolyzed by formate-hydrogen lyase to produce $CO_2$ and $H_2$. The reactions involving the pyruvate-formate lyase and formate-hydrogen lyase are discussed later in this chapter in Section 11.4.4. In general, clostridia are described as having pyruvate oxidation producing $CO_2$ plus $H_2$ without formate as an intermediate. Although acetyl~phosphate is produced from pyruvate by enteric bacteria, only the reversible transformation of pyruvate to acetyl~phosphate plus $H_2$ and $CO_2$ is termed the phosphoroclastic reaction.

# 11.4. Fermentation[1–3]

Fermentation occurs in anaerobic organisms where electrons from metabolism are diverted to a final electron acceptor without membrane-associated respiration. Anaerobes as well as aerobes have a unidirectional flow of electrons to a final electron acceptor with the regeneration of oxidized electron carriers. In anaerobic prokaryotes, organic carbon compounds are the final electron acceptor and the carriers of electrons from the glycolytic pathway are $NAD^+$. With the formation of NADH, reactions become important that will accept electrons from NADH and regenerate $NAD^+$. In fermentation, cytochromes are not involved nor are membranes involved in electron transport. Pathways for generation of carbon compounds as products of fermentation are given in Figure 11.9. The variety of end products of fermentation suggests that the regeneration of $NAD^+$ from NADH is a driving strategy in evolution and not the generation of a specific end product. While the end products of fermentation of bac-

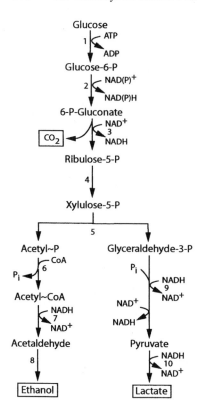

Figure 11.9. Homofermentation by lactic acid bacteria. Reactions are as follows: (1) hexokinase, glucose-6-phosphate dehydrogenase, (3) 6-phosphogluconate dehydrogenase, (4) ribulose-5-phosphate 3-epimerase, (5) phosphoketolase (reaction requires thiamine pyrophosphate), (6) phosphotransacetylase, (7) acetaldehyde dehydrogenase, (8) alcohol dehydrogenase, (9) enzymes from glycolysis, and (10) lactate dehydrogenase.

teria are extremely varied, several patterns in metabolism have emerged and some of the major groups of fermentation are summarized in Table 11.4.

## 11.4.1. Ethanol Production

Fermentation of glucose by yeast results in the production of ethanol and carbon dioxide. Pyruvate is a product of the Embden–Meyerhof–Parnas pathway and two enzyme reactions are required to produce ethanol from pyruvate. The reactions in yeast contributing to alcohol fermentation are as follows:

Glucose $\rightarrow$ 2 pyruvate + 2NADH + 2ATP (Embden–Meyerhof–Parnas pathway)
2 Pyruvate $\rightarrow$ 2 acetaldehyde + 2 $CO_2$ (pyruvate decarboxylase)

2 Acetaldehyde + 2 NADH $\rightarrow$ 2 ethanol + 2 $NAD^+$ (alcohol dehydrogenase)

Although pyruvate decarboxylase is present in various species of yeast, *Zymomonas mobilis* is one of the few anaerobic bacteria that have this enzyme. Unlike yeast, *Zymomonas mobilis* uses the pentose pathway to produce pyruvate from glucose. Common to both the yeast and *Zymomonas mobilis* systems is the use of acetaldehyde

Table 11.4. Bacterial fermentation groups.

| Type of metabolism | Organism | End products |
|---|---|---|
| **Ethanol fermentation** | *Zymomonas mobilis* | ethanol, $CO_2$ |
| **Lactic acid fermentation** | | |
| Homofermentation | *Streptococcus cremoris, Lactobacillus delbrueckii* | L-(+)-Lactic acid D-(−)-Lactic acid |
| Heterofermentation | *Lactobacillus plantarum* | DL-Lactic acid, ethanol, $CO_2$ |
| **Butyric acid and solvent fermentation** | *Clostridium acetobutylicum, Clostridium beijerinckii* | Acetone, butanol, isopropanol, butyrate, acetate, $CO_2$, $H_2$ |
| | *Thermoanaerobacter ethanolicus Clostridium thermosaccharolyticum* | Ethanol, acetate, $CO_2$, $H_2$ |
| | *Clostridium butyricum* | Acetate, butyrate, $CO_2$, $H_2$, 1,3-propanediol |
| **Mixed acid and butanediol fermentation** | *Escherichia coli* | Acetate, succinate, ethanol, formate, lactate, $CO_2$, $H_2$ |
| | *Enterobacter aerogenes* | 2,3-Butanediol, ethanol, formate, lactate, $CO_2$, $H_2$ |
| **Propionic acid fermentation** | *Megasphera elsdenii Clostridium propionicum* | Propionate, acetate, $CO_2$ |

From Mitchell, 1998[7] and Gottschalk, 1986.[1]

as the final electron acceptor with ethanol produced as a result of the action of alcohol dehydrogenase.

A second mode of ethanol production occurs in various bacteria including clostridia, enteric bacteria, and lactic acid bacteria. These bacteria do not have pyruvate decarboxylase to cleave pyruvate to acetaldehyde and carbon dioxide but produce acetyl~CoA and carbon dioxide from pyruvate. According to the reactions below, the formation of ethanol from acetyl~CoA requires the action of aldehyde dehydrogenase and alcohol dehydrogenase:

Acetyl~CoA + NADH → acetaldehyde + CoA + NAD$^+$
(aldehyde dehydrogenase)

Acetaldehyde + NADH → ethanol + NAD$^+$ (alcohol dehydrogenase)

Because acetyl~CoA and pyruvate may interface with other reductive pathways, many of the bacteria lacking pyruvate decarboxylase produce acetate and lactate in addition to ethanol. Reactions leading to the formation of acetyl~CoA from pyruvate are covered in another section of this chapter.

An inhibitor of ethanol production is bisulfite which complexes with acetaldehyde to produce an acetaldehyde–bisulfite complex that is not metabolized. The NADH what would have been used to reduce acetaldehyde to ethanol is available for reduction of dihydroxyacetone phosphate to glycerol phosphate. With action of a phosphatase on glycerol phosphate, glycerol is formed. This manipulation of microbial alcoholic glycolysis was the basis for industrial production of glycerol during the World Wars.

## 11.4.2. Lactic Acid Production[6]

Certain species of *Lactobacillus*, *Streptococcus*, and *Leuconostoc* produce lactate in appreciable quantities and are collectively termed lactic acid bacteria. One of the fermentation reactions with the fewest steps involves the production of lactate from pyruvate. The reaction catalyzed by lactate dehydrogenase is as follows:

$$Pyruvate + NADH \rightarrow lactate + NAD^+$$

Owing to specific mechanisms of reduction, a single type of lactic dehydrogenase is capable of producing only the L(+) or the D(−) configuration of lactic acid. This is the case when *Streptococcus lactis* produces L(+)-lactic acid and when *Lactobacillus lactis* produces D(−)-lactic acid. The production of DL-lactic acid by *Lactobacillus plantarum* is attributed to two different lactic acid dehydrogenases, each with a different stereospecificity. With *Lactobacillus curvatus*, the lactic acid dehydrogenase produces L(+)-lactic acid; however, the organism also produces a lactate racemase that changes the L(+) enantiomer to the D(−) form.

When only lactic acid is produced from the fermentation of glucose by the lactic acid bacteria, this is homofermentation and the reaction is:

$$Glucose \rightarrow 2 \text{ lactic acid}$$

In heterofermentation by lactic acid bacteria, ethanol and $CO_2$ are produced along with lactate as indicated in the following overall scheme:

$$Glucose \rightarrow lactate + ethanol + CO_2$$

The production of equal quantities of lactate and ethanol is a result of a metabolic process that is the combination of parts of several different pathways: pentose pathway, glycolysis, and alcoholic fermentation. Heterofermentation of glucose by lactic acid bacteria is shown in Figure 11.10. The overall for this heterofermentation scheme is as follows:

$$Glucose \rightarrow lactate + ethanol + CO_2 + ATP$$

The enzymes of glycolysis are important in producing ATP as well as NADH while alcoholic fermentation accounts for two sequential steps in use of NADH. The amount of ATP produced by the lactic acid bacteria is not great, and most of the organisms have considerable nutritional requirements and limited biosynthetic capabilities.

Some lactic acid bacteria produce minor end products in addition to lactate and ethanol. When *Leuconostoc mesentroides* grows on fructose, some fructose is reduced to mannitol and the end products of fermentation are lactate, ethanol, $CO_2$, and mannitol. *Streptococcus lactis* and *Leuconostoc cremoris* produce diacetyl and acetoin in addition to the usual end products of lactate fermentation. The production of diacetyl and acetoin results from the fermentation of citric acid that is present in milk. The pathway for production of diacetyl and acetoin by lactic acid bacteria is given in Figure 11.11. The synthesis of diacetyl in *Bacillus* sp. and in enterobacteria occurs by pathways distinct from that of the lactic acid bacteria.

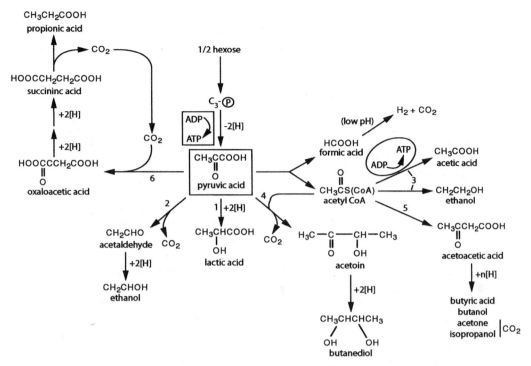

Figure 11.10. Generalized scheme of fermentation in bacteria. Pyruvate is the central metabolite for numerous types of fermentations. The abbreviation of "+ 2[H]" indicates the addition of reducing equivalents as 2 e⁻ and 2 H⁺. The fermentation types are as follows: (1) lactic acid, (2) ethanolic or alcoholic, (3) mixed acid, (4) butanediol, (5) butyric acid, and (6) propionic acid.

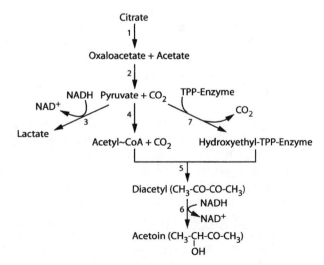

Figure 11.11. Fermentation of citrate to diacetyl and acetoin by lactic acid bacteria. TPP, Thiamine pyrophosphate. Enzymes are as follows: (1) citrate lyase, (2) oxalacetate decarboxylase, (3) lactate dehydrogenase, (4) pyruvate dehydrogenase complex, (5) acetoin dehydrogenase, and (7) enzyme not characterized.

## 11.4.3. Butyric Acid and Solvent Production[7]

Butyric acid fermentation was initially reported by Pasteur in 1861, making it one of the first bacterial metabolic processes discovered. A relatively small group of obligate anaerobic bacteria produce butyrate and they are generally from the following genera: *Clostridium*, *Butyrivibrio*, *Eubacterium*, and *Fusobacterium*. Based on studies with *Clostridium acetobutylicum*, *Clostridium beijerinckii*, *Clostridium tetanomorphum*, and *Closstridium aurantibutyricum*, the pathway for butyric acid production and solvent production has been established (see Figure 11.12). The solvent producing bacteria are capable of utilizing a variety of hexoses with initial metabolism using the Embden–Meyerhof–Parnas pathway. Pyruvate is the product of glycolysis and the conversion of pyruvate to acetyl~CoA with the release of $CO_2$ and reduction of ferredoxin. The reduced ferredoxin interacts with hydrogenase to produce $H_2$, thereby alleviating the requirement for additional organic compounds to serve as an electron donor. The use of ferredoxin here instead of $NAD^+$ is critical because reduced ferredoxin has a lower midpoint potential than NADH and this lower potential is needed for the production of $H_2$. As indicated in Figure 11.12, reduced ferredoxin, $Fd_{(red)}$, may also transfer electrons to $NAD^+$ or to $NADP^+$ by specific enzyme reactions. These reactions for reoxidation of reduced ferredoxin are needed, because there are environments in which $H_2$ will accumulate or where hydrogenase is inhibited by CO. The acetyl~CoA is a key intermediate in that it serves as the branch point in the production of three acids (butyric acid, lactic acid, and acetic acid) and four solvents (butanol, ethanol, isopropanol, and acetone). In the pathway acetyl~CoA and butyryl~CoA precede the production of acetate and butyrate, respectively. Because these compounds bonded to CoA contain a high level of energy, simple hydrolysis to produce the organic acids would result in an energy loss. Thus, the conversion of acetyl~CoA and butyryl~CoA to acetyl~P and butyryl~P conserves the energy and substrate level phosphorylation is used to trap the energy with the production of ATP. The formation of acetyl~P from pyruvate is known as the phosphoroclastic reaction and it is discussed in an earlier section of this chapter. The butyrate and solvent fermentation set of reactions is more than an electron sink because it can provide supplemental levels of ATP.

Considerable quantities of end products are formed by butyrate and solvent fermentation. The products from glucose utilization by *Clostridium acetobutylicum*, *Clostridium butryicum* and *Clostridium thermosaccharolyticum* are listed in Table 11.5. *Clostridium thermosaccharolyticum* produces high levels of butyrate but no solvents while *Clostridium acetobutylicum* produces high levels of butanol and acetone but low quantities of butyrate. As is seen in Table 11.5, each bacterial species has its own characteristic end products.

The butyrate/solvent pathway is highly regulated and to some extent is controlled by pH and the concentration of butyric acid. In organisms such as *Clostridium acetobutylicum*, acid production occurs at pH 6.5 or higher and below pH 4.4, solvent production occurs. The details of pathway regulation remain to be established; however, it is known that lactate dehydrogenase has a high pH optimum and will not produce lactate if the cellular pH is below 5.0. In addition, there is the requirement

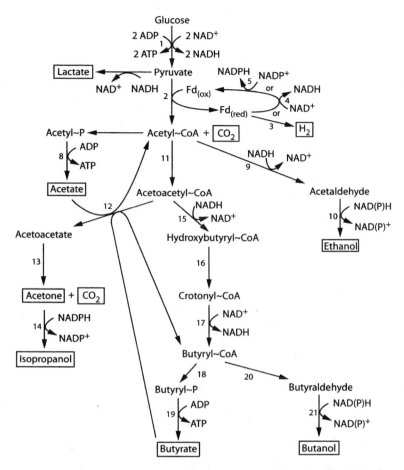

Figure 11.12. Scheme for butyrate and solvent fermentation in sacchrolytic clostridia. End products are boxed. Enzymes are as follows: (1) enzymes of the Embden–Meyerhof–Parnas pathway, (2) pyruvate:ferredoxin oxidoreductase, (3) hydrogenase, (4) ferredoxin:NAD oxidoreductase, (5) ferredoxin:NADP reductase, (6) lactic dehydrogenase, (7) phosphotransacetylase, (8) acetate kinase, (9) acetyladehyde dehydrogenase, (10) ethanol dehydrogenase, (11) thiolase, (12) CoA transferase, (13) acetoacetate dehydrogenase, (14) isopropanol dehydrogenase, (15) β-hydroxybutyryl~CoA dehydrogenase, (16) crotonase, (17) butyryl~CoA dehydrogenase, (18) phosphotransbutyrylase, (19) butyrate kinase, (20) butyraldehyde dehydrogenase, and (21) butanol dehydrogenase.

that butanol production will not proceed until the intracellular concentration of butyrate reaches 17 m$M$. Presumably, butyrate depresses the intracellular pH as well as serving as an exchange with acetoacetyl~CoA to produce higher concentrations of butryl~CoA which is a substrate for butanol formation. Solvent production follows acid synthesis and in batch cultures, solvent formation may be regulated by the same factors that control endospore formation.

Table 11.5. Examples of butyrate and solvent fermentation by *Clostridium*.[a]

| Substrate and products | Mol/100 mol of glucose fermented | | |
|---|---|---|---|
| | *C. acetobutylicum* (mol/100 mol of substrate) | *C. butylicum* (mol/100 mol of substrate) | *C. thermosaccharolyticum* (mol/100 mol of substrate) |
| **Substrate** | | | |
| Glucose | 100 | 100 | 100 |
| **Products** | | | |
| $CO_2$ | 220.0 | 207.0 | 174.0 |
| $H_2$ | 165.9 | 111.1 | 229.4 |
| Acetate | 24.8 | 20.3 | 48.5 |
| Butyrate | 7.1 | 14.5 | 59.5 |
| Lactate | | | 25.7 |
| Ethanol | | 4.9 | |
| Butanol | 47.4 | 50.2 | |
| Acetone | 22.3 | | |
| Isopropanol | | 18.0 | |
| Acetoin | 5.7 | | |
| Percent of carbon recovered | 98.0% | 93.5% | 98.0% |
| O/R total | 1.01 | 1.05 | 0.99 |

[a]From Moat and Foster, 1995.[3]

Several other bacteria produce butyrate without the formation of butanol or acetone. One organism with an interesting fermentation reaction is *Clostridium kluyveri*, which uses ethanol and acetate as the substrate according to the following reaction:

$$6 \text{ Ethanol} + 3 \text{ acetate} \rightarrow 3 \text{ butyrate} + 1 \text{ caproate} + 2 \text{ H}_2 + 4 \text{ H}_2\text{O} + \text{H}^+$$

Certain anaerobes use amino acids for butyrate fermentation. Butyrate and acetate are produced by *Clostridium tetanomorphum* and *Peptostreptococcus asaccharolyticus* from L-glutamate as the substrate while *Clostridium sticklandii* and *Clostridium subterminale* produce the same end products from L-lysine. The mechanism for production of energy for these bacteria is unclear but may involve the phosphoroclastic reaction because acetyl~CoA is an intermediate.

## 11.4.4. Mixed Acid and Butanediol Production

The fermentation products of several Gram-negative facultative anaerobes are relatively diverse and consist of four acids (formate, acetate, lactate, and succinate), two gases ($CO_2$ and $H_2$), ethanol, and 2,3-butanediol. The pathway for fermentation of glucose to mixed acids and butanediol is given in Figure 11.13. Glucose is metabolized by the Embden–Meyerhof–Parnas pathway and phosphoenol pyruvate (PEP) or pyruvate is used for the production of mixed acids or butanediol. For the generation of succinate, PEP provides the three-carbon unit and the remaining carbon comes from $CO_2$. Oxalacetate is produced from PEP plus $CO_2$ by PEP carboxylase and

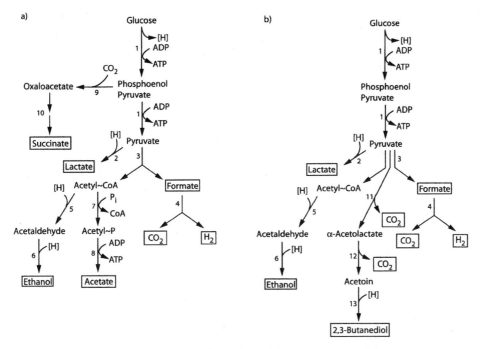

Figure 11.13. Metabolic pathway for mixed acid (**a**) and butanediol (**b**) fermentation. [H], Electron carrier. Enzymes are as follows: (1) enzymes of glycolysis; (2) lactate dehydrogenase; (3) pyruvate:formate lyase; (4) formate:hydrogen lyase; (5) acetaldehyde dehydrogenase; (6) alcohol dehydrogenase; (7) phosphotransferase; (8) acetate kinase; (9) phosphoenol pyruvate carboxylase; (10) malate dehydrogenase, fumarase, and fumarate reductase; (11) α-acetolactate synthase; (12) α-acetolactate decarboxylase; and (13) 2,3-butanediol dehydrogenase.

succinate is produced from oxalacetate by relevant enzymes from the tricarboxylic acid cycle. This production of succinate occurs only with butanediol fermentation. Other end products originate from pyruvate through several independent reactions. In one case, pyruvate is converted to lactate by lactate dehydrogenase and this reaction is greater in mixed acid fermentations than with butanediol fermentations. A second reaction involves the conversion of pyruvate to acetyl~CoA by pyruvate-formate lyase, which is produced only under anaerobic conditions. The pyruvate-formate lyase reaction system is used only in anaerobic cells while pyruvate dehydrogenase is present only in aerobic cells, where it provides acetyl~CoA for the TCA cycle. Formate is cleaved to $CO_2$ and $H_2$ by a formate-hydrogen lyase reaction. The formate-hydrogenase lyase system is distinct from the formate dehydrogenase enzyme which is used to serve as an electron donor for certain chemolithotrophic bacteria. With the formation of acetyl~CoA by pyruvate lyase, there is no generation of NADH, as there is with the pyruvate dehydrogenase complex. There is an advantage for fermenting cells to not overly produce NADH because extra NADH increases the burden of finding an electron sink. The acetyl~CoA that is produced is readily converted

Table 11.6. End products associated with mixed acid and butanediol fermentation. Carbon and fermentation balances are given for each type of metabolism.[a]

| Substrate and products | Escherichia coli (mixed acid) | | | | Enterobacter aerogenes (butanediol) | | | |
|---|---|---|---|---|---|---|---|---|
| | 100 mol of substrate | mol of carbon | O/R value | O/R value (mol/100 mol) | mol/100 mol of substrate | mol of carbon | O/R value | O/R value (mol/100mol) |
| **Substrate** | | | | | | | | |
| Glucose | 100 | 600 | — | — | 100 | 600 | — | — |
| **Products** | | | | | | | | |
| 2,3-Butanediol | 0.3 | 1.2 | −3 | −0.9 | 66.4 | 265.6 | −3 | −199.2 |
| Ethanol | 49.8 | 99.6 | −2 | −99.6 | 69.5 | 139.0 | −2 | −139.0 |
| Formate | 2.4 | 2.4 | +1 | +2.4 | 17.0 | 17.0 | +1 | +17.0 |
| Acetate | 36.5 | 73.0 | 0 | 0 | 0.5 | 1.0 | 0 | 0 |
| Lactate | 79.5 | 238.5 | 0 | 0 | 2.9 | 8.7 | 0 | 0 |
| Succinate | 10.7 | 42.8 | +1 | +10.7 | — | — | | |
| $CO_2$ | 88.0 | 88.0 | +2 | +176.0 | 172.0 | 172.0 | +2 | +344 |
| $H_2$ | 75.0 | — | −1 | −75 | 35.4 | — | −1 | −35.4 |
| Carbon products total | | 545.5 | | | | 603 | | |
| O/R total | | | | −175.6 +189.1 | | | | −373.4 +361 |

[a]From Gottschalk, 1986.[1]
Carbon recovered for mixed acid fermentation is 545.5/600 = 90.9% and for butanediol fermentation it is 603.3/600 = 100.5%. O/R balance for mixed acid fermentation is 189.1/175.6 = 1.076 and for butanediol fermentation it is 361/373.4 = 0.967.

to ethanol. A third pyruvate metabolizing system produces α-acetolactate by α-acetolactate synthase which causes the condensation of two pyruvate to produce acetolactate plus $CO_2$. Acetolactate is decarboxylated to produce acetoin and finally is reduced to produce 2,3-butanediol. The end products produced by *Escherichia coli*, which displays mixed fermentation, and by *Enterobacter aerogenes*, which has butanediol fermentation, are given in Table 11.6.

## *11.4.5. Propionic Acid Production*

An alternate to the previously mentioned fermentation reactions is the production of propionate by anaerobic bacteria. There are two principal pathways for the production of propionate and both prefer to use lactate as the substrate. One is the acrylate pathway, which derives its name from acrylyl~CoA which is an intermediate in production of propionate. Organisms that employ this pathway include *Megasphaera elsdenii* and *Clostridium propionium*. The overall reaction that produces propionate as the major end product is:

$$3 \text{ Lactate} \rightarrow 2 \text{ propionate} + \text{acetate} + CO_2 + H_2O$$

The second pathway for propionate production is the succinate-propionate pathway. In this scheme, propionate is produced from pyruvate with a C-1 unit originating from succinate. The scheme for this reaction is given in Figure 11.14. The propionibacteria use lactate as the substrate for fermentation and the overall reaction is:

$$\text{Lactate} + \text{NADH} + \text{ADP} + \text{P}_i \rightarrow \text{propionate} + \text{NAD} + \text{ATP} + 2\ \text{H}_2\text{O}$$

Because succinate is not present in the chemical medium for the bacteria, it must be produced by a specific reaction. The origin of succinate can be traced to the addition of $CO_2$ to pyruvate to produce oxalacetate. Through enzymes of the TCA pathway, succinate is produced from oxalacetate. With the decarboxylation of methylmalonyl~CoA which originates from succinyl~CoA, propionyl~CoA is produced and ultimately propionate is released from the cell.

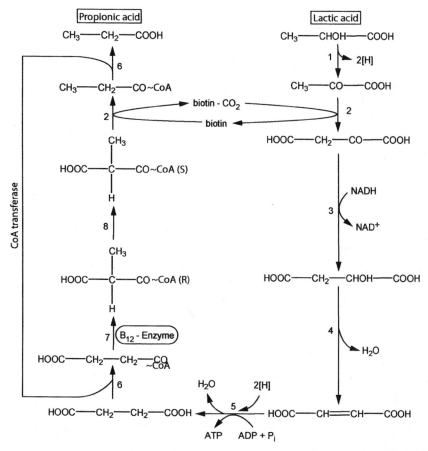

Figure 11.14. Production of propionic acid from lactate by propionibacteria. [H], Electron carrier. Enzymes are as follows: (1) lactate dehydrogenase, (2) (*S*)-methylmalonyl~CoA-pyruvate transcarboxylase, (3) malate dehydrogenase, (4) fumarase, (5) fumarate reduction, (6) CoA transferase, (7) (*R*)-methylmalonyl~CoA mutase, and (8) methylmalonyl~CoA racemase. For reaction 7, coenzyme $B_{12}$ is required.

## 11.4.6. Acetic Acid Production[8]

There is a group of anaerobes that have acetic acid as the predominant end product of fermentation and these organisms are collectively known as acetogenic bacteria or acetogens. Bacteria that account for production of acetate are found in several different genera and characteristics of these organisms are given in Table 11.7. Most of the bacteria in Table 11.7 are homoacetogens, except members of *Acetonema* and *Eubacterium*, which produce butyrate along with acetate as the end products.

### 11.4.6.1. The Ljungdahl–Wood Reaction

Hexoses are converted to acetate by most of the acetogens according to the overall fermentation reaction:

$$C_6H_{12}O_6 \rightarrow 3 \ CH_3COOH$$

The unique feature in this fermentation is that every carbon atom of the hexose is converted into acetate. The pathway for this conversion has been extensively studied and initially hexose is converted to two pyruvate molecules as a result of the Embden–Meyerhof–Parnas pathway. Each pyruvate moiety is oxidized to acetate and $CO_2$ by a series of reactions with the phosphoroclastic reaction identified as accounting for the formation of acetyl~P. The phosphoroclastic reaction is discussed in Section 11.3.2. of this chapter. The reactions associated with acetate synthesis from a hexose are given in Figure 11.15. One acetate is produced directly from each of the pyruvate molecules; however, the third acetate molecule is the result of a complex set of reactions utilizing $CO_2$ released from pyruvate by the action of pyruvate-ferredoxin oxidoreductase. The set of reactions leading to the production of acetate from two molecules of $CO_2$ is termed the Ljungdahl–Wood pathway in recognition of their efforts in establishing the process. The individual reactions for acetate synthesis from

Table 11.7. Characteristics of acetogenic bacteria.

| Genera | Endospores produced | Gram reaction | End products |
|---|---|---|---|
| *Acetobacterium* | No | Positive | Acetate |
| *Acetitomaculum* | No | Positive | Acetate |
| *Acetoanaerobium* | No | Positive | Acetate |
| *Acetohalobium* | No | Negative | Acetate |
| *Acetonema* | Yes | Negative | Acetate, butyrate |
| *Acetogenium* | No | Positive | Acetate |
| *Clostridium* | Yes | Positive | Acetate |
| *Eubacterium* | No | Positive | Acetate, butyrate |
| *Moorella* | Yes | Positive | Acetate |
| *Peptostreptococcus* | No | Positive | Acetate |
| *Sporomusa* | Yes | Positive | Acetate |
| *Syntrophococcus* | No | Positive | Acetate |

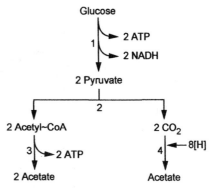

Figure 11.15. Acetogenesis with fermentation of glucose. Enzymes for reactions are as follows: (1) enzymes of glycolysis, (2) pyruvate:ferredoxin oxidoreductase, (3) sequential action of phosphotransacetylase and acetokinase, and (4) acetyl~CoA pathway.

$CO_2$ are given in Figure 11.16. Reactions that are important for the conversion of $CO_2$ use tetrahydrofolate as one of the C-1 carriers and through the use of a $B_{12}$–enzyme complex add the C-1 unit to CO with the production of acetate.

In addition to acetate being produced from hexoses and pentoses, acetogens produce acetate from autotrophic growth on $H_2$ and $CO_2$ according to the following reaction:

$$4 \, H_2 + 2 \, CO_2 \rightarrow CH_3COOH + 2 \, H_2O$$

This autotrophic growth is accomplished because ATP is produced as electrons from hydrogen oxidation are carried by a cytochrome system producing oxidative phosphorylation coupled to electron transport. The details of this respiratory driven ATP synthesis are given in Chapter 9.

The versatility of acetogens is displayed by their capability of using methanol as an electron donor of $H_2$. The acetogenic reaction with methanol occurs as follows:

$$4 \, CH_3OH + 2 \, CO_2 \rightarrow 3 \, CH_3COOH + 2 \, H_2O$$

In other reactions, acetogens will replace $CO_2$ with formate or CO. The reaction indicating acetate formation with CO is as follows:

$$CH_3OH + CO \rightarrow CH_3COOH$$

## 11.4.6.2. The Glycine Reductase System[9]

Thus far, the acetogenic reactions discussed have involved acetyl~CoA pathways and there is another acetate producing system that uses acetyl~P as the intermediate. This second system for the production of acetate employs the glycine reductase system. Glycine is converted by *Clostridium purinolyticum, Peptostreptococcus glycinophilus, Eubacterium acidaminophilum* and other anaerobes to acetyl~P by glycine reductase according to the following reaction:

$$Glycine + P_i \rightarrow acetyl{\sim}P + NH_3$$

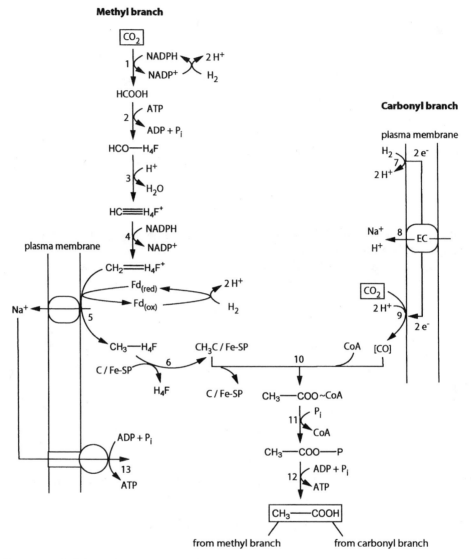

Figure 11.16. Scheme for acetogenesis in autotrophic growth of *Acetobacterium woodii*, a bacterium that does not contain cytochromes. Enzymes are as follows: (1) formate dehydrogenase, (2) formyl tetrahydrofolate synthase, (3) methylene tetrahydrofolate cyclohydrolase, (4) methylene tetrahydrofolate dehydrogenase, (5) methylene tetrahydrofolate reductase, (6) methyltransferase, (7) hydrogenase, (8) EC = electron carrier is driven by an electrochemical gradient of $\Delta\bar{\mu}_{Na^+}$ or $\Delta\bar{\mu}_{H^+}$, (9) carbon monoxide dehydrogenase, (10) acetyl~CoA synthase "CO dehydrogenase," (11) phosphotransacetylase, (12) acetate kinase, and (13) $Na^+$ driven ATP synthase.

Acetate is produced from acetyl~P by acetate kinase according to the following reaction:

$$Acetyl{\sim}P + ADP \rightarrow acetate + ATP$$

The production of acetate from glycine is potentially an important reaction in anaerobes because glycine is generated from the decomposition of several different compounds including purines, serine, threonine, and collagen.

The glycine reductase system has been extensively examined by T. Stadtman and colleagues. There are three proteins that make up glycine reductase in *Clostridium purinolyticum* and they are known as $P_A$, $P_B$, and $P_C$. $P_A$ is a glycoprotein of 15.7 kDa that readily aggregates owing to the presence of carbohydrate. A unique moiety in $P_A$ is selenocysteine, which is produced by a UGA codon. $P_A$ is a redox active protein and the redox center is similar to thioredoxin and thioredoxin reductase. $P_B$ of *Clostridium sticklandii* is 240 kDa and it interacts directly with the glycine and forms a Schiff base. The protein $P_C$ in *Clostridium sticklandii* is a dimer with subunits of 49 kDa and 58 kDa. The $P_C$ connects a carboxymethylselenoether from $P_A$ to a high-energy system with the formation of an acetyl~P ester. The electron donor for the glycine reductase is thioredoxin.

## 11.4.7. Fermentation Balance[1-3]

With the array of products that can be produced by fermentation, there have been several methods of evaluating end products produced from a given substrate. The oxidation/reduction (O/R) balance and carbon recovery are two evaluations commonly used to determine if all fermentation products are measured. The O/R balance is an assessment of the number of oxygen and hydrogen atoms present in each of the products. The O/R value is determined by assigning a value of $+1$ for each oxygen atom and $-1$ for every two hydrogen atoms and subtracting the value for hydrogen from the value for oxygen. For example, glucose ($C_6H_{12}O_6$) is a common substrate for bacterial fermentation and it has an O/R value of zero because of the following calculation:

$$
\begin{aligned}
\text{O/R value} &= (\text{number of O atoms } x + 1) + (\text{number of H atoms}/2x - 1) \\
&= (6 \text{ O atoms } x + 1) + (12 \text{ H atoms}/2x - 1) \\
&= (+6) + (-6) \\
&= 0
\end{aligned}
$$

To establish the O/R balance of a fermentation activity, the O/R value for each end product is determined and multiplied by the number of moles of product formed for each 100 moles of substrate used. An example of this calculation is shown in Table 11.6, in which the O/R balance has been calculated for mixed acid fermentation and for butanediol fermentation. Using the numbers in Table 11.6, the O/R balance for the mixed acid fermentation by *Escherichia coli* is 1.076 while the O/R balance for butanediol fermentation by *Enterobacter aerogenes* is 0.967. Theoretically, if all end products are determined and all of the substrate is used to produce end products, the O/R balance for the fermentation would be 1.00.

Another means of balancing fermentation reactions is to determine the amount of carbon present in the end products and compare that to the amount of carbon in the substrate. For comparison, these carbon values are on a per mole basis. The total number of moles of carbon produced is determined by multiplying the moles of product times the number of carbons in that molecule. For example, if 7 moles of ethanol are produced from 100 moles of substrate, the number of moles of carbon would be 14 because there are two carbon atoms in ethanol and $2 \times 7$ moles produced is 14. For a point of demonstration of determining carbon balance in reactions, the values have been established for the fermentation reactions in Table 11.6. Summing the moles of carbon in the products for the example of mixed acid fermentation, only 90.9% of the carbon is determined while with the example of butanediol fermentation, 100.5% of the substrate can be accounted for in the products. The amount of carbon recovered from butyrate and solvent fermentation is given in Table 11.5 and the same table provides a summary of O/R balance for these fermentation reactions. Perhaps one of the greatest problems in these types of balance procedures is to have complete conversion of substrate into end product without secondary metabolism assimilating some of the product.

## 11.5. Metabolism of the One-Carbon Compounds

Several physiological groups of bacteria and archaea that use one-carbon ($C_1$) compounds as a carbon source and some even use $C_1$ molecules for energy. Examples of prokaryotes that use $C_1$ molecules include autotrophs, methylotrophs, carboxydobacteria, and methanogens. The organisms that grow on $CO_2$ as the sole carbon source are termed autotrophs and by analogy those microorganisms that use reduced $C_1$ compounds or compounds containing no carbon-to-carbon bonds are termed methylotrophs. Methylotrophic organisms can metabolize formaldehyde, formate, formamide, methylamine, methanol, methane, and carbon monoxide. Methylotrophs also use compounds that have one or more carbon atoms and these include dimethyl ether, dimethylamine, trimethylamine, trimethylamine *N*-oxide, tetramethylammonium, and trimethylsulfonium. Autotrophic organisms are distinguished by the energy source, with chemolithotrophs obtaining energy from inorganic electron donors and phototrophs receiving energy from chlorophyll-based reactions. Autotrophs and some heterotrophs fix $CO_2$; however, the reaction sequences vary considerably from species to species. A relatively small group of bacteria known as carboxydobacteria are capable of using CO as both an electron source and a carbon source. A unique set of metabolic reactions is found only in some of the archaea and these reactions result in the production of methane. The next section of the chapter examines some of the specialized groups of organisms and the $C_1$ compounds that they consume.

### 11.5.1. CO₂ Fixation and Related Activities

Carbon dioxide is an important carbon source for phototrophs, sulfate reducers, methanogens, acetogens, and chemolithotrophic bacteria and archaea. There are four major systems used by autotrophs to fix $CO_2$ and they include: (1) the Calvin cycle, (2) the

Table 11.8. Characteristics of $CO_2$ fixation pathways.

| Pathway: autotrophic reactions | Organisms | ATP required for Synthesis of 1 triose-P | $CO_2$ fixing enzymes |
|---|---|---|---|
| Calvin–Benson–Bassham cycle | Chemolithotrophs, Phototrophs | 9 | Ribulose-1,5-bisphosphate carboxylase |
| Reductive citric acid cycle | *Chlorobium* *Desulfobacter* *Thermoproteus* *Hydrogenobacter* | 5 | (1) 2-Oxoglutarate synthase (2) Isocitrate dehydrogenase (3) Pyruvate synthase |
| Acetyl-CoA pathway | *Archaeoglobus* *Desulfobacterium* Acetogenic bacteria Methanogens | 4–5 | (1) Acetyl-CoA synthase/CO dehydrogenase (2) Formate dehydrogenase (3) Pyruvate synthase |
| 3-hydroxypropionate cycle | *Chloroflexus* *aurantiacus* | ? | (1) Acetyl-CoA carboxylase (2) Propionyl-CoA carboxylase |

reductive citric acid cycle, (3) the acetyl~CoA pathway, and (4) the 3-hydroxy-propionate cycle. The distribution of these $CO_2$ fixing pathways in prokaryotes is given in Table 11.8. Each of these systems has a unique mechanism for assimilation of $CO_2$ and this would imply that there were multiple periods for the evolution of carbon dioxide fixation. This section examines the characteristics of each of these four autotrophic systems and the utilization of carbon monoxide.

### 11.5.1.1. Carbonic Anhydrase[10,11]

While carbonic anhydrase has been isolated from only a few prokaryotes, sequence analysis of microbial genomes reveals that most bacteria and archaea contain this enzyme. The enzyme carbonic anhydrase promotes the formation of bicarbonate according to the following reaction:

$$CO_2 + H_2O = HCO_3^- + H^+$$

The prokaryotic enzyme is a zinc-containing metalloprotein of 25 to 30 kDa. Based on similarity of gene sequences, carbonic anhydrases in prokaryotes are grouped into three classes: $\alpha$, $\beta$, and $\gamma$. The distribution of carbonic anhydrase in bacterial and archaeal cells is given in Table 11.9. Class $\alpha$ is found primarily in animal cells and a few bacteria while class $\beta$ enzyme is found in plant cells and in various species of bacteria. Carbonic anhydrase of the $\gamma$ class is found in both archaeal and bacterial cells. Not only is there a structural difference in these three classes of carbonic anhydrases but there also appears to be different cellular distributions and physiological functions. The enzyme of the $\alpha$ class appears to be located outside of the bacterial cell and is associated with either the cell wall, periplasm, or the plasma membrane. Enzymes of the $\beta$ class are also external to the cell and it has a lipoprotein attachment, which suggests that it is associated with the periplasm or the plasma membrane. Carbonic anhydrases of the $\gamma$ class are involved in carbon concentration within the cells, and in cells with $CO_2$ fixation, the enzyme of the $\gamma$ class appears to be associated with the RuBisCo subunits. The specific roles of carbonic anhydrase in prokaryotes

Table 11.9. Carbonic anhydrase in prokaryotes located by gene sequence analysis.[a]

| Organism | α Class[b] | β Class | γ Class |
|---|---|---|---|
| **Bacteria** | | | |
| *Bacillus subtilis* | 0 | 2 | 1 |
| *Clostridium acetobutylicum* | 0 | 1 | 1 |
| *Escherichia coli* | 0 | 2 | 2 |
| *Helicobacter pylori* J99 | 1 | 1 | 0 |
| *Mycobacterium tuberculosis* | 0 | 3 | 1 |
| *Rickettsia prowazekii* | 0 | 0 | 1 |
| *Synechococcus* sp. strain PCC7942 | 1 | 1 | 1 |
| **Archaea** | | | |
| *Methanobacterium thermoautotrophicum* | 0 | 1 | 1 |
| *Methanococcus jannaschii* | 0 | 0 | 1 |

[a]Information reflects survey of completed genomes.
[b]Numbers of genes present for each class are as indicated.

are not altogether clear but it may be that the enzyme may shift the equilibrium of enzymes where $CO_2$ is an end product. In addition, the enzyme may adjust the intracellular pH of the cell and it may participate in cellular regulation. In at least one organism, a cyanobacterium, the role of the carbonic anhydrase has received considerable attention and a model of its role is given in Figure 11.17. Carbonic anhydrase is frequently associated with cell membranes but it would not be needed for $CO_2$ transport because a gas readily moves across membranes.

### 11.5.1.2. The Calvin–Benson–Bassham Cycle[12–14]

The Calvin–Benson–Bassham cycle is commonly referred to as the Calvin cycle and it uses a reductive pentose phosphate cycle to assimilate $CO_2$. This system can be divided into three processes: (1) carboxylation, (2) reduction, and (3) regeneration of pentose. The Calvin–Benson–Bassham cycle for $CO_2$ assimilation is given in Figure 11.18. The initial reaction is the irreversible carboxylation of ribulose-1,5-bisphosphate to produce 2 moles of 3-phosphoglycerate. The second sequence is the reduction of 3-P-glycerate to glyceraldehyde-3-phosphate. In bacteria, the reducing power for the cycle is NADH while in plants the reduction is by NADPH. Several reactions using enzymes of the pentose phosphate cycle and gluconeogenesis are involved in the metabolism.

Unlike the pentose phosphate cycle in heterotrophic bacteria which has sedoheptulose-7-phosphate in carbon rearrangement, the autotrophic Calvin cycle produces ribulose-1,7-bisphosphate. A phosphatase is used to produce ribulose-7-phosphate from ribulose-1,7-bisphosphate and this phosphatase reaction is not reversible accounts for the irreversible character of the Calvin cycle. The two key enzymes are ribulose-1,5-bisphosphate carboxylase (RuBisCo) and phosphororibulokinase. RuBisCo is widely distributed in chemolithotrophic and phototrophic bacteria. In *Rhodospirillum rubrum*, RuBisCo may account for as much as 50% of total cell protein while in most other bacteria the RuBisCo level may be 3% to 10% of cell protein. The RuBisCo

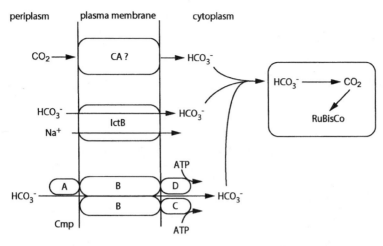

Figure 11.17. The role of carbonic anhydrase in the concentration of $CO_2$ in cyanobacteria under conditions of low concentration of $CO_2$. The Cmp system consists of proteins A, B, C, and D of the ABC transporter found in cell membranes under conditions of low concentrations of $CO_2$. Under conditions of high concentrations of $CO_2$, the ABC transporter for carbonate is absent. Present in both low and high concentrations of $CO_2$ is the protein IctB which appears to be a $Na^+$ driven uptake transporter for bicarbonate. Also, present in both high and low concentrations of $CO_2$ is a carbonic anhydrase (CA) that converts $CO_2$ to bicarbonate. The enzyme ribulose bisphosphate carboxylase (RuBisCo) in the carboxysome recognizes $CO_2$ and not bicarbonate. (From Smith and Ferry, 2000.[10])

isolated from bacteria has a structure similar to that of eukaryotic RuBisCo in that it has a mass of approximately 500 kDa and consists of eight large (8L) subunits and eight small (8S) subunits. This RuBisCo complex of 8L8S is referred to as type I and several species of *Rhodospirillum* have a second RuBisCo which has a 6L structure and is designated as type II. The 8L and 6L subunits are immunologically distinct and are products of different genes. Both form I and form II enzymes are found in carboxysomes. However, not all bacteria have RuBisCo in carboxysomes because *Ralstonia eutrophus,* which has form I, lacks carboxysomes. The function of these carboxysomes remain somewhat unclear but it appears that the carboxysomes protect the the $CO_2$ fixing enzyme from various types of potential inhibitors. The RuBisCo is a bifunctional enzyme which reacts with either $CO_2$ or $O_2$ according to the following reaction:

$$\text{Ribulose-1,5-bisphosphate} + CO_2 \xrightarrow{\text{Carboxylase}} 2 \text{ (3-phosphoglycerate)}$$

$$\text{Ribulose-1,5-bisphosphate} + O_2 \xrightarrow{\text{Oxygenase}} \text{3-phosphoglycerate} + \text{2-phosphoglycolate}$$

The oxygenase reaction depletes the cell from carbon material because the 2-phosphoglycolate is converted to glycolate which is not used by the cell and is exported

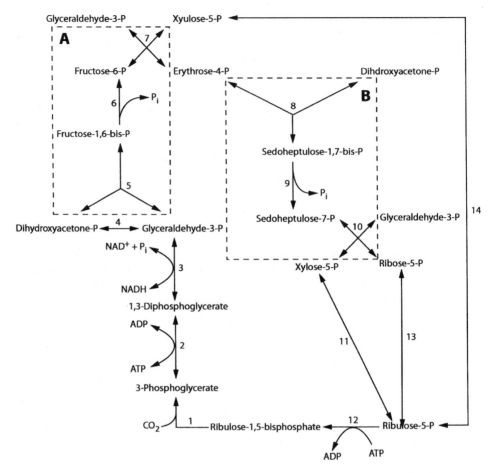

Figure 11.18. $CO_2$ fixation in prokaryotes using the Calvin–Benson–Bassham cycle. Rearrangement reactions are highlighted and contain an aldolase, a phosphatase, and a ketolase (APT) reaction. The APT reactions are outlined as A and B. Enzymes are as follows: (1) ribulose bisphosphate carboxylase/oxygenase (RuBisCo); (2) phosphoglycerate kinase; (3) glyceraldehyde-3-phosphate dehydrogenase; (4) triosephosphate isomerase; (5) fructose-1,6-bisphosphate aldolase; (6) fructose-1,6-bisphosphatase; (7) transketolase; (8) aldolase; (9) sedoheptulose bisphosphatase; (10) transketolase; (11) pentose-5-phosphate epimerase; (12) phosphoribulokinase; and (13) pentose-5-phosphate isomerase.

from the cell. Some have proposed that RuBisCo may function in detoxification of $O_2$; however, definite evidence of this protection has not been presented.

### 11.5.1.3. Reductive Tricarboxylic Acid Cycle[15,16]

The reductive citric acid cycle functions in a few bacteria and archaea for autotrophic $CO_2$ fixation. The metabolic scheme for this $CO_2$-fixing cycle is given in Figure 11.19.

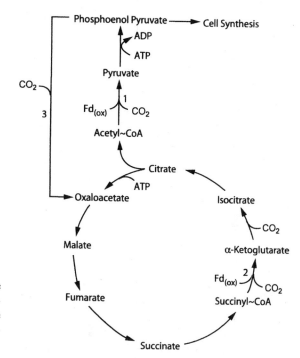

Figure 11.19. Reductive TCA cycle for assimilation of $CO_2$ by *Chlorobium*. Key enzymes: (1) pyruvate synthase, (2) α-ketoglutarate synthase, and (3) PEP carboxylase.

There are three enzymes in the reductive tricarboxylic acid cycle that are distinct from the standard citric acid cycle and these enzymes enable the citric acid cycle to be reversed because they replace nonreversible enzymes. These enzymes that enable the citric acid cycle to be reversed are fumarase, α-ketoglutarate:ferredoxin oxidoreductase, and citrate lyase.

## 11.5.1.4. 3-Hydroxypropionate Cycle[15,16]

The 3-hydroxypropionate cycle has been demonstrated to function in assimilation of $CO_2$ by a relatively few different bacteria. This metabolic scheme for this cycle is given in Figure 11.20. Some details remain to be established before the energy requirements for this cycle can be determined.

## 11.5.1.5. The Acetyl~CoA Pathway[17-19]

The acetyl~CoA synthase is a complex structure that displays several different activities essential for autotrophic $CO_2$ fixation. The synthesis of acetyl~CoA from $CO_2$ by *Acetobacterium woodii* is given in Figure 11.16. In the formation of $CH_3CO~CoA$, $CO_2$ provides carbon for the methyl (-$CH_3$) and the carbonyl (-CO ) groups. In the formation of the methyl group, $CO_2$ is reduced to formate before the $C_1$ unit is attached to tetrahydrofolate, where it is reduced to given methyl-tetrahydrofolate. The transfer of the methyl group from methyl-tetrahydrofolate to

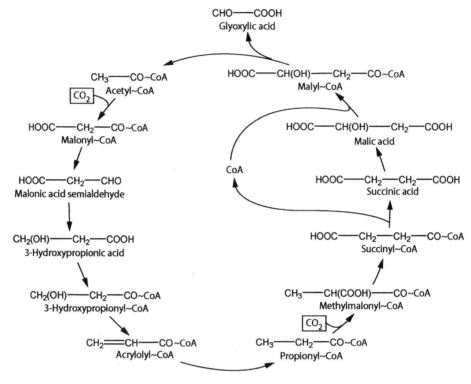

Figure 11.20. Assimilation of $CO_2$ in *Chloroflexus aurantiacus* by the 3-hydroxypropionate cycle.

the corroinoid enzyme (also described as the corroinoid protein) is by a methyltransferase. This corrinoid protein is 58 kDa that has an $\alpha\beta$ structure with subunits of 34 and 55 kDa, respectively. Present in the corrinoid enzyme is vitamin $B_{12}$ (a corrinoid) and the Co in the corrinoid carries the methyl group. This corrinoid protein also has several [4Fe-4S] clusters that participate in the redox reaction. As indicated in Figure 11.21. The $Co^{2+}$ ion of vitamin $B_{12}$ is reduced to $Co^+$ by electrons from the oxidation of $H_2$, CO, or pyruvate. The $Co^+$ of the corrinoid accepts the methyl group from methyl tetrahydrofolate and in the process, $Co^+$ is oxidized to $Co^{3+}$. The methyl group is transferred to a Ni atom on the CO dehydrogenase. A second $CO_2$ molecule is combined with the methyl from $CH_3$-corrinoid by carbonylcyclase activity. The cysteine residues on the carbon monoxide dehydrogenase are considered important in carrying the methyl group and coenzyme A for the condensing reaction. The functional regions have been designated by Ljungdahl and colleagues as X, Y, and Z. It has been proposed that the methyl group binds to the cysteine at X, the carbonyl group binds to Ni at Y, and coenzyme A binds at region Z. Through channeling activity, the units are assembled into acetyl~CoA which serves as the source for cell carbon. Cysteine residues are considered to be an active moiety at X and Z while a nickel

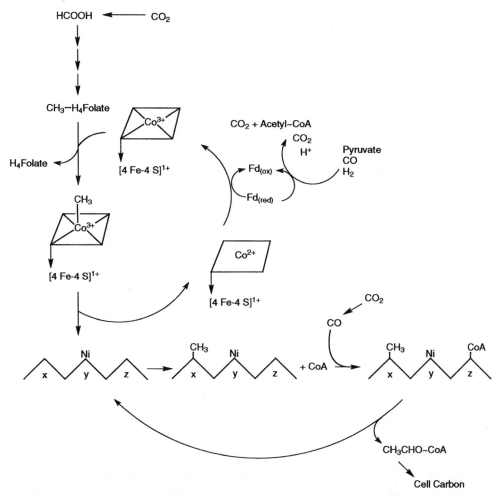

Figure 11.21. Activities of acetyl-CoA pathway in $CO_2$ assimilation. Activities: Reaction (1) methyltransferase, (2) carbon monoxide reductase, and (3) acetyltransferase.

atom is present at Y. The CO reductase complex isolated from *Methanosarcina thermophila* has five subunits: 89, 70, 60, 58, and 19 kDa along with Ni, Zn, Co, Fe, and acid labile sulfur. This complex structure would support the proposed multifunctional feature required for conversion of 2 $CO_2$ to acetyl~CoA.

### 11.5.1.6. CO Utilization[20-22]

The utilization of CO by prokaryotes is an important feature for the carbon cycle and prokaryotes of various physiological types oxidize CO to $CO_2$. Some examples of organisms with this capability are listed in Table 11.10. There are several different

Table 11.10. Examples of organisms that oxidize carbon monoxide.[a]

Physiological groups

| **Bacteria** | Acetogens | *Acetobacterium woodii* |
| | | *Clostridium pasteurianum* |
| | Carboxydo-trophs | *Alcaligenes carboxydus* |
| | | *Bacillus schlegelii* |
| | | *Pseudomonas carboxydoflava* |
| | | *Pseudomonas compransoris* |
| | Methanotrophs | *Pseudomonas methanica* |
| | | *Methylosinus methanica* |
| | | *Methylococcus capsulatus* |
| | Nitrogen fixers | *Azomonas* B1 |
| | | *Azospirillum lipoferum* |
| | | *Bradyrhizobium japonicum* |
| | Phototrophs | *Rhodocyclus gelatinosa* |
| | | *Rhodospirillum rubrum* |
| | | *Spirulina platensis* |
| | Sulfate reducers | *Desulfobacterium autotrophicum* |
| | | *Desulfotomaculum acetoxidans* |
| | | *Desulfovibrio desulfuricans* |
| | | *Desulfovibrio vulgaris* |
| **Archaea** | Methanogens | *Methanobacterium thermoautotrophicum* |
| | | *Methanosarcina barkeri* |
| | | *Methanothrix soehngenii* |

[a]From Barton et al., 1991.[11]

molecular forms that display CO dehydrogenase activity and properties of some of these enzymes are given in Table 11.11. Numerous metals are associated with these CO dehydrogenases and based on catalytic activity, there is a Mo-containing group and a Ni-containing group. The Mo-containing CO dehydrogenases are present in aerobic bacteria while the Ni-containing enzyme are found in anaerobic organisms. The discussion in this section will examine the CO dehydrogenase of carboxydobacteria, acetogens, and methanogens.

*Carboxydobacteria*

CO is used an the energy and carbon source by several different species of bacteria including *Pseudomonas carboxydovorans*, *Alcaligenes carboxydus*, and *Bacillus schlegelii*. Oxidation of CO by the carboxydobacteria proceeds as follows:

$$CO + H_2O \rightarrow CO_2 + 2\ H^+ + 2\ e^-\ (E_0' = -540\ mV)$$

The enzyme for the above reaction has been referred to as CO oxidase; however, it is more appropriately designated as CO dehydrogenase. The carboxydobacteria are

Table 11.11. Characteristics of CO dehydrogenases from selected physiological groups of prokaryotes.[a]

| Physiological group | Organism | Molecular weight | Subunit structure | Subunit size | Metals present | Type of Fe-S centers |
|---|---|---|---|---|---|---|
| **Carboxydobacteria** | | | | | | |
| | *P. carboxydovorans* | 300 kDa | $(\alpha\beta\gamma)_2$ | 86 kDa | Mo, Fe | [2Fe-2S] I |
| | | | | 34 kDa | | [2Fe-2S] II |
| | | | | 17 kDa | | |
| | *P. carbocydohydrogena* | 400 kDa | $(\alpha\beta\gamma)_3$ | 85 kDa | Mo, Fe | [2Fe-2S] I |
| | | | | 28 kDa | | [2Fe-2S] II |
| | | | | 14 kDa | | |
| **Acetogens** | | | | | | |
| | *Mo. thermoacetica* | 440 kDa | $(\alpha\beta)_3$ | 78 kDa | Ni, Fe | [4Fe-4S] I |
| | | | | 71 kDa | | [4F3-4S] II |
| | *A. woodii* | 460 kDa | $(\alpha\beta)_3$ | 80 kDa | Ni, Fe | [4Fe-4S] I |
| | | | | 68 kDa | | |
| **Methanogen** | | | | | | |
| | *M. barkeri* | 232 kDa | $(\alpha\beta)_3$ | 92 kDa | Ni | |
| | | | | 18 kDa | | |
| **Sulfate reducer** | | | | | | |
| | *D. desulfuricans* | 180 kDa | | | Ni | [2Fe-2S] |

[a]Adapted from Meyer and Fiebig, 1985.[21]
*P.*, *Pseudomonas*; *Mo.*, *Moorella*; *A.*, *Acetobacterium*; *M.*, *Methanosarcina*; *D.*, *Desulfovibrio*.

capable of growing with CO as the electron donor, obtaining the oxygen for carbon dioxide from water and not from molecular oxygen. The CO oxidase is characteristically associated with the plasma membrane and electrons generated are directed to the electron transport system with $O_2$ as the final electron acceptor. A model proposed by O. Meyer for the membrane-related reactions of CO dehydrogenase in *Pseudomonas carboxydovorans* is presented in Figure 11.22. For carboxydobacteria, CO oxidation is coupled to three functions: (1) generation of $CO_2$ for fixation by the Calvin cycle, (2) generation of ATP, and (3) production of NADH for biosynthetic activities. Bacteria such as *Pseudomonas carboxydovorans* have a branched cytochrome chain and at least one of the cytochrome oxidase systems is not inhibited by CO.

*Acetogens*

In an earlier section concerning acetogens, it was indicated that CO is an intermediate in fermentation, resulting in the production of acetic acid. In that case, CO does not accumulate in the environment but is stoichiometrically consumed through acetyl~CoA synthase as soon as it is formed. As shown in Table 11.11, the CO dehydrogenase of *Moorella thermocetica* (formerly *Clostridium thermoaceticum*) and

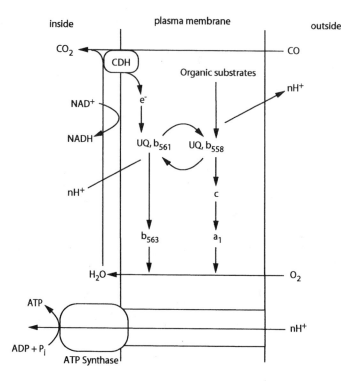

Figure 11.22. Diagrammatic representation of CO oxidation by carboxydotrophs. CDH, carbon monoxide dehydrogenase, UQ, ubiquinone, $a$, $b$, $c$ = cytochromes. (From Meyer and Fiebig, 1985.[21])

*Acetobacterium woodii* contains Ni but not Mo. The CO dehydrogenase from *Moorella thermocetica* is 440 kDa with an $\alpha_3\beta_3$ structure that also contains 2 Ni, 1 Zn, 11 Fe, and 14 acid-labile sulfides.

*Methanogens*

CO is an energy source for numerous methanogens with disproportionation according to the following reactions:

$$CO + H_2O \rightarrow CO_2 + H_2 \ (\Delta G^{0\prime} = -20 \text{ kJ/mol CO})$$

$$CO_2 + 4 H_2 \rightarrow CH_4 + 2 H_2O \ (\Delta G^{0\prime} = -131 \text{ kJ/mol CH}_4)$$

The oxidation of CO to $CO_2$ is catalyzed by an enzyme complex that is known as the CO dehydrogenase. In the case of *Methanosarcina*, the CO dehydrogenase complex contains multiple components with one segment consisting of a Ni/Fe-S protein and the other subunit has a corrinoid and a Fe-S protein. The protein accounting for CO utilization by methanogens is commonly referred to as a "CO dehydrogenase"; however, scientists working in methanogenesis prefer to refer to this as CO dehydrogenase activity because CO incorporation into acetyl~CoA is catalyzed by the enzyme acetyl~CoA synthase. In general, the carbon monoxide dehydrogenase found in methanogens is approximately 220 kDa with an $\alpha_2\beta_2$ structure containing Ni, Zn, and Fe-S clusters.

CO production in the environment may be attributed to burning of carbonaceous compounds and to diverse degradation activities of methanogens. Although the magnitude of CO production by methanogens remains to be determined, the following reaction has been reported for *Methanobacterium thermoautotrophicum* when it is actively producing methane:

$$CO_2 + H_2 \rightarrow CO + H_2O \qquad \Delta G^{0'} = +20 \text{ kJ/mol } CO_2$$

CO is often considered a potent inhibitor of microbial hydrogenase enzymes; however, this inhibition is selective for the Fe-hydrogenases. CO does not inhibit FeNi-hydrogenases nor does it inhibit the FeNiSe-hydrogenases.

*Methanotrophs*

Bacteria that produce a soluble monomethane oxidase, sMMO, have an enzyme of low substrate specificity and under appropriate conditions this sMMO will oxidize CO to $CO_2$. This activity is not conducive for bacterial growth because it does not promote the formation of NADH, nor does it produce energy. The oxidation of CO by sMMO is an example of a futile reaction in bacteria.

## 11.5.2. Methane, Methanol, and Formaldehyde Utilization[23,24]

Microorganisms that utilize methane are termed methanotrophs. Because methanol and formaldehyde are intermediates in the oxidation of methane to $CO_2$, methanotrophs would be a specialized group of bacteria that are within the broad group of methylotrophs. The methanotrophs are commonly found in freshwater, marine environments, and water-saturated terresterial regions at the interface between the aerobic and anaerobic zones. This is because the first enzyme in methane metabolism requires $O_2$ while methane and methanol are produced under anaerobic conditions. Thus far, all methanotrophs belong to the *Bacteria* domain and none are *Archaea*. Characteristics of the methanotrophs are given in Table 11.12. Although the first methylotroph was isolated by Söhngen in 1906, these organisms did not become popularized until the reports by Whittenbury and Wilkinson in the 1970s. Obligate methylotrophs utilize methane, methanol, and formaldehyde while facultative methylotrophs use methanol and formaldehyde but not methane. Because not all methylotrophs use methane as a carbon source, the term *methanotroph* is frequently used to describe those organisms that grow with methane oxidation.

A series of reactions dealing with methane oxidation with methanol, formaldehyde, and formic acid as intermediates is given below:

$CH_4 + NADH + H^+ + O_2 \rightarrow CH_3OH + NAD^+ + H_2O$ (Methane monooxygenase)

$CH_3OH + PQQ_{oxidized} \rightarrow HCHO + PQQ \cdot 2H$ (Methanol dehydrogenase)

$HCHO + NAD^+ + H_2O \rightarrow HCOOH + NADH + H^+$ (Formaldehyde dehydrogenase)

$HCOOH + NAD^+ \rightarrow CO_2 + NADH + H^+$ (Formate dehydrogenase)

Table 11.12. Examples of autotrophs growing with methanol or formate as a substrate.[a]

| Physiological groups | Genera | Formate used | Methanol used | CO2 pathway |
|---|---|---|---|---|
| **Bacteria** | | | | |
| Methylotrophs | *Acetobacter methanolicus* | | + | ? |
| | *Ralstonia eutrophus* | + | − | Calvin |
| | *Blastobacter viscosus* | + | + | Calvin |
| | *Mycrocyclus aquaticus* | + | + | Calvin |
| | *Pseudomonas oxalaticus* | + | + | Calvin |
| Nitrogen-cycle bacteria | *Azospirillum brasilense* | | + | ? |
| | *Nitrobacter agilis* | + | + | Calvin |
| | *Paracoccus denitrificans* | + | + | Calvin |
| Phototrophs | *Rhodopseudomonas acidophila* | + | + | Calvin |
| | *Rhodopseudomonas palustria* | + | + | Calvin |
| Sulfur-cycle bacteria | *Desulfobacterium autotrophicum* | + | | Acyl~CoA |
| | *Desulfococcus multivorans* | + | | ? |
| | *Desulfotomaculum orientis* | + | + | ? |
| | *Thiobacillus novellus* | + | + | Calvin |
| | *Archaeoglobus* | | | Acyl~CoA |
| Acetogens | *Moorella thermoacetica* | + | | Acyl~CoA |
| | *Moorella thermoautotrophica* | + | | Acyl~CoA |
| Methanogens | *Methanobacterium formicicum* | + | | Acyl~CoA |
| | *Methanococcus vannielii* | + | | Acyl~CoA |
| | *Methanohalophilus mahii* | | + | Acyl~CoA |
| | *Methanosarcina barkeri* | | + | Acyl~CoA |

[a]From Barton et al., 1991.[11]

These reactions are important for methanotrophs as well as for the methylotrophs. The enzymology of the above reactions is well established (see Figure 11.23); however, the mechanism of the reactions requires additional study. Regulation of the above $C_1$ reactions is a key factor for maintaining a balance in biosynthesis and energetics of the bacteria.

### 11.5.2.1. Methane Monooxygenase (MMO)[25,26]

Methane is oxidized to methanol by methanotrophs according to the following reaction:

$$CH_4 + 0.5\ O_2 \rightarrow CH_3OH \qquad \Delta G^{0\prime}\ (pH\ 7) = -109.7\ kJ/mol$$

This is an energy releasing reaction that can support the growth of bacteria. There are two forms of MMO produced in bacteria with a soluble form, sMMO, present in the cytoplasm while the other is immobilized in the plasma membrane and is designated as pMMO. The presence of copper in the growth medium stimulates the production of pMMO while the synthesis of sMMO is inhibited. The best characterized enzyme of the methane oxidation process is the soluble MMO which has been purified

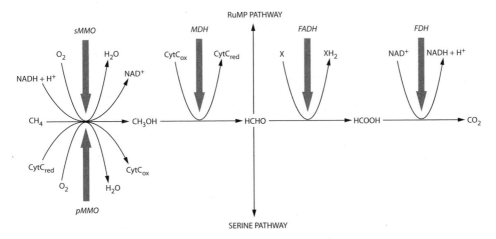

Figure 11.23. Pathway for oxidation of $C_1$ compounds by methanotrophs. CytC, Cytochrome c; FADH, formaldehyde dehydrogenase; FDH, formate dehydrogenase; MDH, methanol dehydrogenase; pMMO, particulate (membrane) methane monooxygenase; sMMO, soluble methane monooxygenase; RuMP, ribulose-5-phosphate. (From Hanson and Hanson, 1999,[24] with permission.)

and the crystal structure at 1.7 Å is available. A model of the soluble MMO complex from *Methylococcus capsulatus* strain Bath is given in Figure 11.24. The soluble MMO consists of three components: a hydroxylase, a reductase, and a regulatory protein. The hydroxylase is 251 kDa and consists of three different polypeptides as $\alpha_2\beta_2\gamma_2$ where the $\alpha$ subunit is 60.6 kDa, the $\beta$ subunit is 45 kDa, and the $\gamma$ subunit is 19.8 kDa. The hydroxylase also contains two atoms of Fe, called di-iron, which is involved in the insertion of an $-OH$ moiety into methane to produce methanol. The reductase is a 39-kDa protein that contains FAD plus a [2Fe-2S] cluster. As shown in the model in Figure 11.24, there is intimate interaction between the hydroxylase and a regulator protein known as protein B. The regulator protein is 16 kDa and its activity is influenced by proteolysis from the N-terminus. At high concentrations of protein B, the rate of methanol formation is increased but at low concentrations the hydroxylase action is favored over the oxidase action. The reducing power comes from NADH, which interacts with the reductase segment of the sMMO. Characteristic of sMMO, but not pMMO, is the broad range of substrates including alkanes, alkenes, and aromatic compounds. As evident from the large number of substrates used by MMO (Table 11.13), this enzyme lacks substrate specificity.

The pMMO is encoded on an operon distinct from the sMMO operon. The pMMO consists of three proteins: 46 to 47 kDa, 23 to 27 kDa, and 20 to 25 kDa. The difficulty of maintaining an active preparation following purification has hampered our understanding of the pMMO system. The genes for pMMO have been isolated and are designated as *pmoB, pmoA,* and *pmoC*. Based on the evaluation of proteins expected to be produced from these gene sequences, the pmoA protein would be ex-

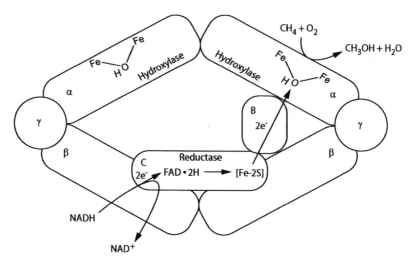

Figure 11.24. Model of the sMMO complex from *Methylococcus capsulatus*. Electrons are proposed to flow from NADH to the di-iron center of the hydroxylase. The complex contains three subunits ($\alpha$, $\beta$, and $\gamma$) of the MMO plus a protein B and a protein C. Protein B converts the $\alpha$-subunit from an oxidase to a hydroxylase. Protein C is a reductase that transfers electrons from NADH to the hydroxylase. (From Murrell et al., 2000.[25])

Table 11.13. Substrates for methane monooxygenase from *Methylococcus capsulatus* (Bath).[a]

| One-carbon compounds | *n*-Alkanes |
|---|---|
| Bromomethane | Butane |
| Carbon monoxide | But-1-ene |
| Chloromethane | *cis*-But-2-ene |
| Cyanomethane | *trans*-But-2-ene |
| Dichloromethane | Ethane |
| Methanethiol | Ethene |
| Methanol | Hexane |
| Nitromethane | Heptane |
| Trichloromethane | Propene |
| | |
| **Alicyclic, aromatic, and** | |
| **heterocyclic compounds** | **Ethers** |
| Benzene | Dimethyl ether |
| Chlorohexane | Diethyl ether |
| PyridineStyrene | |
| Toluene | |

[a]From Anthony, 1988.[27]

Table 11.14. Compounds that are known to be
oxidized by aminomonooxidase.[a]

| Small molecules | Aromatic compounds |
|---|---|
| Ammonium | Aniline |
| Methane | Benzene |
| Methanol | Iodobenzene |
| Carbon dioxide | Naphthalene |
| | Nitrobenzene |
| | Styrene |
| **Sulfur compounds** | **Alkanes and alkenes** |
| Methyl sulfide | N-Alkanes to C-8 |
| Ethyl sulfide | 1-Alkenes to C-5 |
| Thiophene | Cyclohexane |
| **Halogenated alkanes** | **Halogenated alkenes** |
| Fluoromethane | Chloroethylene |
| Chloroethane | 3-Iodopropene |
| Iodoethane | 2,3-Dichloropropene |
| Dibromomethane | Tribromoethylene |

[a]From Hooper et al., 1997.[30]

pected to have four membrane spanning segments while the pmoB protein is expected
to have two membrane spanning segments. The pMMO protein complex contains 2
Fe atoms and approximately 15 Cu atoms per molecule. The hydrolysis of methane
by pMMO is proposed to involve the Cu atoms.

Several comparisons have been made between the obligate methanogens and ammonia oxidizing bacteria such as *Nitrosomonas europeae*. Both groups of bacteria
are capable of oxidizing ammonia, carbon dioxide, hydroxylamine, ethylene, methane,
propylene, phenol, and trichloroethylene. A listing of substrates for AMO is given in
Table 11.14. The contribution of metabolism by ammonia oxidizers, which use ammonia moonoxidase (AMO), in the presence of methanotrophs can be determined with
the addition of 2-pyridine carboxylic acid at 0.5 m$M$ which selectively inhibits MMO
but not AMO. Also, methane at elevated concentrations will prevent the growth of
ammonia oxidizers presumably because of the competition by both species for oxygen.
There are some similarities between the pMMO and the AMO in that both enzyme
systems use cytochromes but not NADH as the electron donor in the oxidation process.
Both enzymes are inhibited by acetylene.

## 11.5.2.2. Methanol Dehydrogenase[23,24,27]

Several different microbial species are capable of using methanol as a carbon source
and some of these are listed in Table 11.12. Methylotrophs growing on methanol
obtain energy from aerobic oxidation of methanol according to the following reaction:

$$CH_3OH + \frac{1}{2} O_2 \rightarrow HCHO + H_2O \qquad \Delta G^{0'} \text{ (pH 7)} = -188.2 \text{ kJ/mol}$$

The enzyme that oxidizes methanol to formaldehyde in methylotrophs is the methanol
dehydrogenase and it is capable of catalyzing the oxidation of numerous primary

alcohols in addition to methanol. This enzyme is localized in the periplasm and accounts for up to 15% of total soluble protein. It is considered that cells can control the exposure of cytoplasmic constituents to toxic activities of formaldehyde by regulating formaldehyde transport. The prosthetic group for the methanol dehydrogenase is pyrrolo-quinoline quinone (PQQ) (see structure in Figure 9.2), and this cofactor is held into the enzyme by ionic bonding. The methanol dehydrogenase is a homodimer with a mass of 60 kDa and has the following activity:

$$\text{Methanol} + \text{PQQ} \rightarrow \text{formaldehyde} + \text{PQQ·2H}$$

The midpoint redox potential of formaldehyde/methanol is $-182$ mV, which is significantly more electronegative than the midpoint redox potential of PQQ/PQQH, which is $+90$ mV. Electrons are readily transferred from methanol to PQQ but are unable to reduce $NAD^+$, as the $NAD^+/NADH$ couple is $-320$ mV. The reduced PQQ donates electrons to a periplasmic c-type cytochrome $c_L$ which is found only in methylotrophs. The $c_L$ cytochrome is a monoheme protein of 17 to 21 kDa with histidine and methionine as ligands for the heme group. The midpoint potential of the $c_L$ cytochrome is $+250$ mV to $+310$ mV and the $c_L$ chtochrome is well positioned to transfer electrons to an o-type cytochrome oxidase. In *Methylophilus methylotrophus*, the methanol dehydrogenase and $c_L$ cytochrome are estimated to reach concentrations of 0.5 m*M*. In *Methylophilus methylotrophus*, the methanol dehydrogenase activity maintains a charge on the plasma membrane of about $-165$ mV.

### 11.5.2.3. Formaldehyde Oxidation and Assimilation[23,24,27]

As seen in the following reaction, the coupling of formaldehyde oxidation to an aerobic electron transport system produces a considerable quantity of energy.

$$\text{HCHO} + \tfrac{1}{2}\,O_2 \rightarrow \text{CHOO}^- + H^+ \quad \Delta G^{0'} \text{ at pH 7} = -240 \text{ kJ/mol}$$

However, in methanotrophs, the reaction would not be functioning because formaldehyde is used as a carbon source. It is possible that under certain conditions, methylotrophs would utilize energy from the oxidation of formaldehyde to formate to provide additional levels of ATP by oxidative phosphorylation or to assist the cell in charging of the plasma membrane. The different systems for oxidation of formaldehyde are first discussed and then the utilization of formaldehyde as a carbon substrate is addressed.

Two different types of formaldehyde dehydrogenases have been isolated from methylotrophs. One is a $NAD^+$-dependent formaldehyde dehydrogenase that has been obtained from *Methylococcus capsulatus* with a molecular weight of 115 kDa and consists of two subunits. Another type of enzyme is considered to be a cytochrome-linked dehydrogenase and it has been found in *Methylomonas methylovora*. Neither of these dehydrogenases is highly specific for formaldehyde and will oxidize several other aldehydes.

There is a unique activity of methanol dehydrogenase that is important because it will also oxidize formaldehyde to formic acid. Because formaldehyde is used for the synthesis of carbon compounds, the reaction of methanol dehydrogenase oxidizing

Table 11.15. Major groups of methanotrophs based on metabolic pathways.

| Characteristics | Designated types | | |
|---|---|---|---|
| Methanogen type | I | II | X |
| Location of MMO | Membrane only | One membrane enzyme plus | One membrane enzyme plus |
| | | One cytoplasmic enzyme | One cytoplasmic enzyme |
| RuMP pathway present | Yes | No | Yes |
| Serine pathway present | No | Yes | Sometimes |
| RuBisCO present | No | No | Yes |
| Subdivision of proteobacter | Gamma | Alpha | Gamma |
| Examples of isolates | *Methylomonas methanica* | *Methylocystis parvus* | *Methylococcus capsulatus* (Bath) |
| | *Methylobacter albus* | *Methylosinus sporium* | |
| | *Methylococcus thermophilus* | | |

formaldehyde must be highly controlled. In all methylotrophs tested, an M-protein is the regulatory protein used to prevent formaldehyde oxidation by methanol dehydrogenase and the "M" is used to designate a modifier protein. This M-protein alters the affinity of the methanol dehydrogenase for formaldehyde, and thereby decreases the loss of formaldehyde by this reaction.

Because methanotrophs can use methane as the sole carbon source, the carbon from methane oxidation must provide carbon compounds of metabolic use and formaldehyde is the only carbon intermediate that is diverted into cellular material. The ribulose monophosphate (RuMP) pathway and the serine pathway are two distinct metabolic pathways used by bacteria for formaldehyde utilization. Organisms that use the RuMP pathway are referred to as type I methylotrophs while those that use the serine pathway for metabolism are designated as type II methylotrophs. The characteristics of methanotrophs with multiple assimilatory pathways are given in Table 11.15. Organisms in the type I category are members of the gamma subdivision of the 16S rRNA Proteobacteria taxonomic group while those found in type II are of the alpha subdivision of Proteobacteria. In at least one methanotroph (type-X), multiple pathways for $C_1$ are present and these include the RuMP pathway, the serine pathway, and the Calvin cycle. Both type I and type II methanotrophs have extensive layering of membranes throughout the cytoplasm of the cell which, no doubt, is to assist the cells in the oxidative process. Internal membranes of the type I methanotrophs have a disc-shaped form distributed throughout the cytoplasm while those of the type II have paired membranes that are along the periphery of the cell. An additional difference in these methanotrophs is that the type I organisms lack α-ketoglutarate dehydrogenase, and therefore have an incomplete citric acid cycle. Type II methanotrophs have a functional citric acid cycle. The two different systems for the fixation of formaldehyde were determined by the use of $^{14}C$-labeling experiments.

*The Ribulose Monophosphate (RuMP) Pathway*

The RuMP pathway for formaldehyde utilization is found in type I methanotrophs such as *Methylomonas*, *Methylobacter*, *Methylomicrobium*, *Methylocaldum*, and *Meth-*

*ylococcus*.  The action of RuMP is to convert three molecules of formaldehyde to a three-carbon unit that interfaces with the biosynthetic system of the bacteria.  There are three parts to the RuMP pathway: fixation, cleavage, and rearrangement.  There are several variants of the RuMP pathway or cycle and Figure 11.25 provides the formaldehyde assimilation scheme used by many obligate methylotrophs.

In the fixation phase of the RuMP cycle, formaldehyde is condensed with ribulose-5-phosphate to produce hexulose-6-phosphate.  The enzyme for this condensation is 3-hexulose phosphate synthase and it has been purified from the cytoplasm of several bacteria including *Methylomonas aminofaciens* and *Methylophilus methylotrophus*.  The 3-hexulose phosphate synthase requires $Mg^{2+}$ or $Mn^{2+}$ for activation and consists of dimer with each subunit being about 22 kDa.  The second step in the RuMP cycle

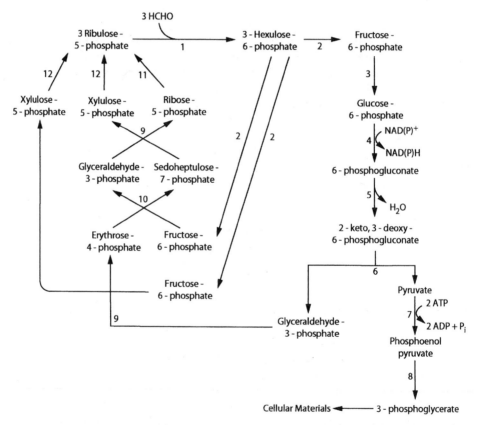

Figure 11.25.  Formaldehyde assimilation by the ribulose monophosphate (RuMP) pathway.  Enzymes are as follows: (1) hexulose phosphate synthase, (2) hexulose phosphate isomerase, (3) glucose phosphate isomerase, (4) glucose phosphate dehydrogenase, (5) phosphogluconate dehydratase, (6) 2-keto,3-deoxy,6-phosphogluconate aldolase, (7) phosphoenolpyruvate synthetase, (8) enolase and phosphoglyceromutase, (9) transketolase, (10) transaldolase, (11) pentose phosphate isomerase, and (12) pentose phosphate epimerase.

is the conversion of hexulose-6-phosphate to fructose-6-phosphate by a 6-hexulose phosphate isomerase. Fructose-6-phosphate is converted to glucose-6-phosphate which is oxidized to 2-keto-3-deoxy-6-phosphogluconate (KDPG). The cleavage of the KDPG by a specific aldolase produces pyruvate and glyceraldehyde-3-phosphate. In the rearrangement phase, enzymes of the pentose phosphate pathway are employed (see Figure 11.25); however, unlike the pentose phosphate pathway these reactions are not reversible. Finally, ribulose-5-phosphate is produced and the cycle in initiated to fix another formaldehyde molecule. The overall reaction of this pathway is presented below:

$$3 \text{ HCHO} + \text{NAD(P)}^+ \rightarrow \text{pyruvate} + \text{NAD(P)H} + \text{H}^+$$

Pyruvate would be used in the biosynthetic processes and the reduced pyridine nucleotide could be used either for biosynthesis or be directed toward the electron transport system.

A second variant of the RuMP cycle is found in facultative methylotrophs. The fixation of formaldehyde is the same but there are different reactions for the cleavage phase and the rearrangement phase. This scheme is given in Figure 11.26. One formaldehyde molecule is fixed by ribulose-5-phosphate with the ultimate formation of fructose-6-phosphate. Through the action of phosphofructokinase, fructose-6-phosphate is converted to fructose-1,6-bisphosphate. Fructose disphosphate aldolase is the key enzyme in the cleavage process and produces dihydroxyacetone phosphate plus glyceraldehyde-3-phosphate. Dihydroxyacetone phosphate is metabolized further by enzymes commonly associated with glycolysis while the condensation of glyceraldehyde-3-phosphate with fructose-6-phosphate is attributed to an enzyme associated with the pentose phosphate pathway. In one of the sugar rearrangements, sedoheptulose 1,7-bisphosphate is produced and to prevent it from being a final metabolic product, it is converted back to sedoheptulose-7-phosphate by a phosphatase. The overall reaction for this RuMP pathway using fructose bisphosphate aldolase and sedoheptulose bisphosphatase is as follows:

$$3 \text{ HCHO} + \text{ATP} \rightarrow \text{glyceraldehyde-3-phosphate} + \text{ADP}$$

The metabolism of glyceraldehyde-3-phosphate by enzymes of glycolysis results in biosynthesis as well as generation of reducing potential and production of ATP. There is considerable interest in the RuMP cycle because some consider it to be an evolutionary precursor of the ribulose-1,5-bisphosphate pathway for $CO_2$ fixation.

*The Serine Pathway*

The serine pathway for formaldehyde utilization is found in type II methanotrophs which include *Methylosinus*, *Methylocystis*, and *Methylocella*. The key enzyme in the serine pathway is serine transhydroxymethylase, which combines formaldehyde with glycine to produce serine (see Figure 11.27). Formaldehyde does not react directly with glycine but unites with tetrahydrofolate and the product of methylene tetrahydrofolate is the substrate for the condensation reaction. As shown in Figure 11.27, serine is transformed through several steps to phosphoenol pyruvate and at this time

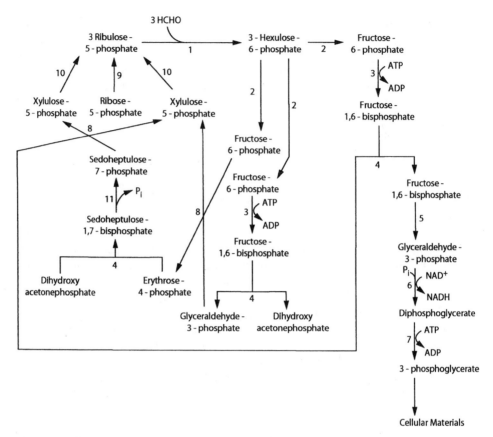

Figure 11.26. RuMP pathway 2. Enzymes are as follows: (1) hexulose phosphate synthase, (2) hexulose phosphate isomerase, (3) phosphofructokinase, (4) aldolase, (5) triosephosphate isomerase, (6) glyceraldehyde phosphate dehydrogenase, (7) phosphoglycerate kinase, (8) transketolase, (9) pentose phosphate isomerase, and (10) pentose phosphate epimerase.

the fixation of $CO_2$ occurs to produce oxalacetate. This four-carbon unit is transformed further to produce glyoxylate for another formaldehyde reaction and acetyl~CoA which is used for biosynthesis. The overall reaction for the serine pathway for fixation of formalehyde is as follows:

$$HCHO + CO_2 + 2\ NADH + 2\ H^+ + 2\ ATP + CoA \rightarrow$$
$$Acetyl{\sim}CoA + 2\ NAD^+ + 2\ ADP + 2\ P_i + 2\ H_2O$$

### 11.5.2.4. Formate Dehydrogenase[28]

Formate is an important substrate for aerobic as well as anaerobic prokaryotes and a partial listing of organisms with this activity is given in Table 11.12. The oxidation

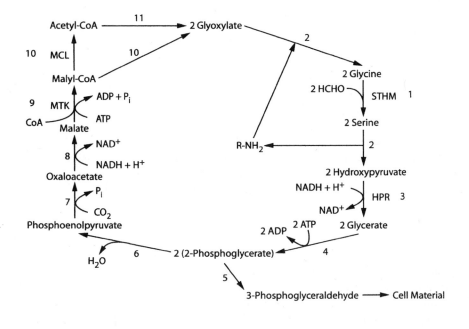

$$(2\ HCHO + 1\ CO_2 + 3\ ATP + 2\ NADH \longrightarrow 2\text{-Phosphoglycerate} + 3\ ADP + P_i + 2\ NAD^+)$$

Figure 11.27. Formaldehyde assimilation by the serine pathway. The enzymes are as follows: (1) serine transhydroxymethylase, (2) serine-glyoxylate aminotransferase, (3) glycerate kinase, (5) phosphoglycerate mutase, (6) enolase, (7) phosphoenolpyruvate carboxylase, (8) malate dehydrogenase, (9) malate thiokinase, (10) malonyl~CoA lyase, and (11) several enzymes of the citric acid cycle.

of formate is catalyzed by formate dehydrogenase and can be expressed in the following reaction:

$$HCOO^- \rightarrow CO_2 + 2H^+ + 2\ e^-$$

The midpoint potential of the $CO_2/HCOO^- + H^+$ couple is $-460$ mV, which is sufficiently electronegative for the transfer of electrons to $NAD^+$ or other electron carriers that have a negative midpoint redox potential. Microorganisms have two fundamental uses for formate and they are as follows: (1) $CO_2$, the product of formate dehydrogenase, is used by the organism for the biosynthesis of cellular material; and (2) electrons produced from the low potential reaction are used to drive energy-yielding reactions through coupling to cytochrome $b$ or coupled to the reduction of $NAD(P)^+$. The contribution of formate dehydrogenase to cell energetics is presented in Chapter 9 and is not reviewed here. The fate of $CO_2$ released by the formate dehydrogenase depends, to some extent, on the physiological type of organisms involved and three groups are addressed here: methylotrophs, acetogens, and methanogens.

*Methylotrophs*

As listed in Table 11.12, there are several different physiological types of autotrophic organisms that grow on formate that require the use of both $CO_2$ and reducing equivalents. For example, aerobic methylotrophs use formate according to the following reaction where sufficient energy is released to support cell metabolism:

$$HCOO^- + H^+ + \tfrac{1}{2}\, O_2 \rightarrow CO_2 + H_2O \quad \Delta G^{0'} = -244.7 \text{ kJ/mol}$$

The oxidation of formate to formaldehyde in methylotrophs is catalyzed by a soluble formate dehydrogenase that is specific for formate and $NAD^+$. It has been reported that *Pseudomonas oxalaticus*, a methylotroph, has a formate dehydrogenase that is 315 kDa with two FMN prosthetic groups and about 20 nonheme iron and sulfide groups. Many of the methylotrophs use the Calvin cycle for $CO_2$ fixation.

*Acetogens*[12]

Anaerobic bacteria (see Table 11.12), which produce acetic acid as a product of fermentation, have a unique soluble formate dehydrogenase that contains selenium, Fe-S centers, and molybdenum or tungsten. From the research activities of Ljungdahl and colleagues, a great deal is known about the formate dehydrogenase enzyme produced by *Moorella thermocetica*. This enzyme is 340 kDa and consists of two dissimilar subunits with a composition of $\alpha_2\beta_2$. The $\alpha$ subunit is 96 kDa and it also contains selenocysteine. The $\beta$-subunit is 76 kDa and does not have any selenium. The formate dehydrogenase contains 2 mol of tungsten, 2 mol of selenium, 36 mol of iron, and about 50 mol of sulfur. Tungsten is present as a cofactor and the structure is proposed to be similar to molybdopterin. The function of the formate dehydrogenase in acetogens is to synthesize formate from $CO_2$ with acetyl~CoA produced from formate. This formate reduction for acetogens is discussed in another section of this chapter.

*Methanogens*

Several archaea grow autotrophically with formate as the substrate for methane production, and a partial list of these organisms is given in Table 11.12. The use of formate by methanogens requires activity of a dehydrogenase to produce $CO_2$. The studies of T. Stadtman and her colleagues provide an important insight into the formate dehydrogenase activity in *Methanococcus vannielii*. There are two forms of formate dehydrogenase in *Methanococcus vannielii* and they can be distinguished by the presence of selenium. One enzyme is selenium dependent and it contains iron and molybdenum in addition to a residue of selenosysteine. The selenium-independent enzyme is 105 kDa and it also contains iron and molybdate. Electron acceptors for the selenium-independent enzyme include FMN and FAD but not $NAD^+$ nor $NADP^+$. The formate dehydrogenase of *Methanobacterium formicicum* is 288 kDa and does not contain selenium although it does possess iron and molybdenum. These enzymes from the methanogens are highly sensitive to oxygen with more than 90% of activity lost following a 2-min exposure to air. The assimilation of $CO_2$ by methanogens is discussed later in this chapter.

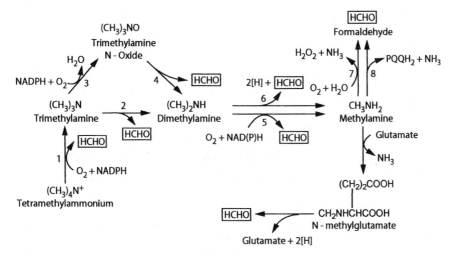

Figure 11.28. Oxidation of methylamines with the production of formaldehyde. [H], $e^-$ + $H^+$; PQQ, pyrroloquinoline quinone. Enzymes are as follows: (1) tetramethylammonium monooxygenase, (2) trimethylamine dehydrogenase, (3) trimethylamine monooxygenase, (4) trimethylamine N-oxide demethylase, (5) dimethylamine monooxygenase, (6) dimethylamine dehydrogenase, (7) amine oxidase, (8) methylamine dehydrogenase, (9) N-methylglutamate synthase, (10) N-methylglutamate dehydrogenase. (From Anthony, 1982,[23] with permission.)

### 11.5.2.5. Oxidation of Methylamines[27]

The oxidation of methylated amines by methylotrophs is carried out by a diverse group of enzymes that have been studied in several genera of bacteria. A composite metabolic scheme for oxidation of methylamine compounds is given in Figure 11.28. The specific enzymes associated with the oxidation of the different methylamines and the distribution of these enzymes are given in Table 11.16.

*Oxidation of Tetramethylammonium*

Tetramethylammonium oxidation is by a tetramethylammonium monooxygenase according to the following reaction:

$$(CH_3)_4N^+ + O_2 + NAD(P)H \rightarrow (CH_3)_3N + HCHO + NAD(P)^+$$

This enzyme is a monooxygenase and it appears to be distinct from the trimethylamine monooxygenase.

*Oxidation of Trimethylamine*

Trimethylamine metabolism with the production of dimethylamine can occur by either of two possible reactions. One system involves the direct conversion of trimethylamine to dimethylamine and the other is indirect in the production of dimethylamine with the formation of trimethylamine N-oxide as an intermediate. The trimethylamine dehydrogenase is 147 kDa with two nonidentical subunits of 70 kDa and 80 kDa.

Table 11.16. Oxidation of methylamine compounds by bacteria.

| Substrate | Organisms | Enzyme system used in oxidation |
| --- | --- | --- |
| Tetramethylammonium | Various methylotrophs | Tetramethylammonium monooxygenase |
| Trimethylamine dehydrogenase | (a) Obligate mehylotrophs<br>(b) Facultative methylotrophs | (a) NAD-independent trimethylamine<br>(b) Trimethylamine monooxygenase |
| Trimethylamine N-oxide | *Bacillus* PM6 and *Pseudomonas aminovorans* | Trimethylamine N-oxide demethylase |
| Dimethylamine | (a) Aerobic methylotrophs<br>(b) Anaerobic methylotrophs | (a) Dimethylamine monooxygenase<br>(b) Dimethylamine dehydrogenase |
| Methylamine | (a) Obligate methylotrophs<br>(b) *Arthrobacter* sp.<br>(c) *Pseudomonas* MA | (a) methylamine dehydrogenase<br>(b) amine oxidase<br>(c) N-Methylglutamate synthase |
| N-Methylglutamate | *Pseudomonas* sp. and *Hyphomicrobiaum* sp. | N-Methylglutamate dehydrogenase |

This unusual dehydrogenase has a prosthetic group of 6-$S$-cysteinyl-FMN, a [4Fe-4S] center, and has been reported to donate electrons to a FAD-containing protein as it does not use pyridine nucleotides (e.g., $NAD^+$ or $NADP^+$). This binding of FMN to the protein is unusual because FMN is either not covalently bonded or it is is bonded by an 8-methylene bridge. The trimethylamine dehydrogenase reaction is given below:

$$(CH_3)_3N + H_2O \rightarrow (CH_3)_2NH + HCHO + 2\ e^- + 2\ H^+$$

Various substrates are suitable for the reaction catalyzed by trimethylamine dehydrogenase and include N-methyl or N-ethyl secondary or tertiary amines. Compounds not appropriate as substrates for the trimethylamine dehydrogenase include diamines, quaternary ammonium salts, and primary amines.

An alternate route for oxidation of trimethylamine to dimethylamine is by the sequential reactions catalyzed by trimethylamine monooxygenase followed by trimethylamine N-oxide demethylase. The first enzyme in this indirect route of oxidation is catalyzed by the aerobic reaction involving trimethylamine monooxygenase as indicated below:

$$(CH_3)_3N + NADPH + H^+ + O_2 \rightarrow (CH_3)_3NO + H_2O + NADP^+$$

An additional anaerobic reaction involving trimethylamine N-oxidase demethylase is utilized to produce formaldehyde as given below:

$$(CH_3)_3NO \rightarrow (CH_3)_2NH + HCHO$$

This enzyme is neither oxidative nor is it hydrolytic but can be characterized as an aldolase. Trimethylamine N-oxidase demethylase has been characterized from *Bacillus* PM6 as a single polypeptide of about 45 kDa while a larger enzyme of 240 kDa to 280 kDa was isolated from *Pseudomonas aminovorans*.

*Oxidation of Dimethylamine*

The formation of methylamine from dimethylamine is by either of two enzymes: a dimethylamine monooxygenase or a dimethylamine dehydrogenase. The dimethylamine monooxygenase catalyzes the following reaction and is the only enzyme employed by aerobic organisms such as *Pseudomonas aminovorans:*

$$(CH_3)_2NH + NAD(P)H + H^+ + O_2 \rightarrow CH_3NH_2 + HCHO + NAD(P)^+ + H_2O$$

Organisms such as *Hyphomicrobium* X can grow by nitrate respiration with dimethylamine as the electron donor and the dimethylamine dehydrogenase is used in the anaerobic growth response. The dimethylamine dehydrogenase from *Hyphomicrobium* X is 176 kDa and consists of two identical subunits of 91 kDa. The reaction with dimethylamine dehydrogenase is given below:

$$(CH_3)_2NH + H_2O \rightarrow CH_3NH_2 + HCHO + 2\ e^- + 2\ H^+$$

Prosthetic groups for this enzyme are an [4Fe-4S] center and the unusual flavin of 6-*S*-cysteinyl-FMN.

*Oxidation of Methylamine*

The oxidation of methylamine to formaldehyde is by either of three enzyme systems: methylamine dehydrogenase, methylamine oxidase, and *N*-methylglutamate synthase. Methylamine dehydrogenase is unable to use pyridine nucleotides as the electron acceptor but instead uses PQQ acceptor according to the following reaction:

$$CH_3NH_2 + PQQ + H_2O \rightarrow HCHO + NH_3 + PQQH_2$$

The methylamine dehydrogenase from *Pseudomonas* AM1 and *Pseudomonas* J is 105 kDa with an $\alpha_2\beta_2$ configuration where the $\alpha$-subunit is 13 kDa and the $\beta$-subunit is 40 kDa. As with the methanol dehydrogenase the PQQ shuttles the electrons to the quinone in the membrane.

In the aerobic culture of *Arthrobacter* P1, there is a reaction catalyzed by methylamine oxidase which is also known as the amine oxidase. This reaction is as follows:

$$CH_3NH_2 + O_2 + H_2O \rightarrow HCHO + NH_3 + H_2O_2$$

The purified enzyme is considered to contain copper, which may be important for utilization of $O_2$.

A final pathway for the oxidation of methylamine involves methylated glutamate as an intermediate leading to the formation of formaldehyde. The *N*-methylglutamate synthase has been purified and is 350 kDa which is composed of 12 subunits consisting of 30 to 35 kDa. The reaction is as follows:

$$\text{methylamine} + \text{glutamate} \rightarrow N\text{-acetylglutamate} + \text{ammonia}$$

In the formation of *N*-acetylglutamate, there are two mechanisms for the conversion of the methyl group to glutaraldehyde involving a dehydrogenase reaction of two distinct types. In one case, electrons pass from the *N*-methylglutamate to $NAD^+$ and some bacteria produce an enzyme that passes electrons to a cytochrome or a flavin system. The general reaction is as follows:

$$\text{N-methylglutamate} + H_2O \rightarrow \text{glutamate} + HCHO + 2\ e^- + 2\ H^+$$

Since the end product of methylamine metabolism is formaldehyde, it is apparent that bacteria must be capable of using formaldehyde as the carbon source. Thus, the RuMP pathway and serine pathway become indispensable for methylamine utilization.

## 11.5.3. Methane Production

The formation of methane by prokaryotes is a highly complex activity not found in members of the *Bacteria* domain but is limited to a physiological group of the *Archaea* generally known as methanogens. Examples of methanogens and their substrates are given in Table 11.17. As expressed in the genus name, the methanogens have considerable diversity in cellular forms and can be divided into the chemolithotrophic group and the methylotrophic group. The chemolithotrophic methanogenic group includes those organisms growing on $H_2$ and $CO_2$ as well as those that use formate because the initial step in formate utilization is the conversion to $H_2$ and $CO_2$. The methylotrophic methanogens produce methane from the methyl groups of the substrate and do not utilize the carbon from $CO_2$ for $CH_4$ production. The energy yields with the methanogenic reactions are given in Table 11.18, where it is seen that these reactions release enough energy to account for cell growth.

### 11.5.3.1. Unique Coenzymes from Methanogens[29]

In the 1970s, R. Wolfe and colleagues provided considerable insight into the biochemistry of methanogens through the isolation and characterization of novel coenzymes. These coenzymes are either electron carriers or carriers of $C_1$ moieties for methane production and thermodynamics of reactions involving these cofactors are

Table 11.17. Methanogens and nutritional requirements for methane production.

| Archaea | Chemolithotrophs | | | Methylotrophs | | |
|---|---|---|---|---|---|---|
| | $H_2$ | $CO_2$ | CO | Methanol | Acetate | Methylamines |
| *Methanobacterium thermoautotrophicum* | + | + | + | − | − | − |
| *Methanobrevibacter arboriphilus* | + | + | − | − | − | − |
| *Methanococcus vannielli* | + | + | − | − | + | − |
| *Methanospirillum hungatei* | + | + | − | − | + | − |
| *Methanosarcina barkeri* | + | + | − | − | + | + |
| *Methanosarcina mazei* | − | − | − | + | + | + |
| *Methanothrix soehngenii* | − | − | − | − | + | − |
| *Methanolobus tindarius* | − | − | − | + | − | + |

Table 11.18. Overall reactions resulting in methane production.[a]

| Reactions | $\Delta G^{0''}$ |
|---|---|
| $CO_2 + 4\ H_2 \rightarrow CH_4 + 2\ H_2O$ | $-136$ kJ |
| $4\ HCOOH \rightarrow CH_4 + 3\ CO_2 + 2\ H_2O$ | $-144$ kJ |
| $CH_3COOH \rightarrow CH_4 + CO_2$ | $-37$ kJ |
| $CH_3OH + H_2 \rightarrow CH_4 + H_2O$ | $-112$ kJ/mol |
| $2\ CH_3CH_2OH + CO_2 \rightarrow CH_4 + 2\ CH_3COOH$ | $-111$ kJ/mol |
| $4\ CO + 2\ H_2O \rightarrow CH_4 + 3\ CO_2$ | $-448$ kJ/mol |
| $4\ CH_3OH \rightarrow 3\ CH_4 + CO_2 + 2H_2O$ | $-319$ kJ/mol $CH_4$ |
| $4\ CH_3NH_3 + 2\ H_2O \rightarrow 3\ CH_4 + CO_2 + 4\ NH_3$ | $-235$ kJ/mol $CH_4$ |

[a]From Fuchs, 1999.[16]

---

## Box 11.1
## From Anaerobes to Vitamin B$_{12}$ and Selenium Enzymes
A Perspective by Thressa C. Stadtman

I was born February 12, 1920, and my youngest brother and I enjoyed the benefits of being raised on a large family dairy and fruit farm near Oswego, New York. Based on my "A" average in high school I received a New York State Regents Scholarship, but the depression and my father's sudden death changed our finances. My mother learned that in the College of Agriculture at Cornell University I could have free tuition as a bacteriology major. My scholarship, together with work for room and board, allowed me to finish in 1940 with a B.S. degree. Studies on the vitamin B requirements of lactic acid bacteria at the Cornell-affiliated Agriculture Experiment Station in Geneva, New York led to an M.S. degree in bacteriology and nutrition in 1942. I transferred to the University of California in Berkeley and started work in the Department of Food Technology early in 1943. During World War II many courses at Berkeley were suspended, and work on stability of foods destined for consumption by the military in the South Pacific theater was of primary concern. At the end of the war, I started studies for the Ph.D. degree under the direction of H.A. Barker. My research on methane fermentations used [14]C, rare and only recently available. I isolated a new methane-producing species that grew on formate as the sole organic nutrient form San Francisco Bay mud. We named this organism *Methanococcus vannielii* in honor of C.B. van Niel. Another organism co-enriched in the formate cultures was isolated and later identified as *Clostridium sticklandii*.

Another of Barker's students, Earl R. Stadtman, and I were married in 1943, and we both were granted our Ph.D. degrees in June 1949. After completing postdoctoral studies at Massachusetts General Hospital under Fritz Lipmann (ERS) and at Harvard Medical School (TCS) with C.B. Anfinsen, we both took positions in the newly formed National Heart Institute at NIH in Bethesda. The aim of my research on bacterial degradation of cholesterol was production of precursors for chemical synthesis of steroid hormones, but this approach did not prove feasible. One enzyme that was identified in these studies is used chemically for measurement of cholesterol levels in blood.

In later research using extracts of *M. vannieli* and *C. sticklandii*, we discovered three new vitamin $B_{12}$–dependent enzymes involved in methane production and the anaerobic degradation of lysine. We showed that the glycine reductase reaction is coupled to energy conservation in the form of ATP. During purification of the enzyme components of glycine reductase, a serendipitous observation that selenium was required for formation of the small protein A of the complex suddenly changed my direction of research and led me into the unfamiliar realm of selenium chemistry. Requirements of selenium in the protein A component of the complex for catalytic activity and identification of the selenium moiety as a selenocysteine residue provided the first identification of this selenoamino acid in a selenoenzyme. Since then, we have shown that replacement of selenocysteine with cysteine in *E. coli* formate dehydrogenase leads to dramatic loss of catalytic activity and the cysteine mutant of glycine reductase protein A is inactive. Certain bacterial molybdopterin-containing enzymes have been shown to contain an essential dissociable selenium cofactor. Our recent discovery of the presence of selenium in mammalian thioredoxin reductase isolated from human tumor cells is another example of serendipity in research. The selenocysteine residue identified in the protein was shown to be essential for catalytic activity. This human selenoenzyme has many important biochemical roles.

given in Table 11.19. The structures of these unique coenzymes are given in Figure 11.29 and the following section provides some additional information about them.

## Coenzyme $F_{420}$

This soluble factor is an analog of FMN and in methanogens it functions like $NAD^+$ as an electron carrier. The oxidized structure has maximum absorption at 420 nm which accounts for the name of this metabolic factor. There is a similarity in structure between flavins and $F_{420}$; however, while flavins carry 2 $e^-$ and 2 $H^+$, $F_{420}$ carries 2 $e^-$ plus 1 $H^+$ which is due to the absence of a nitrogen atom at position 5. For this reason, $F_{420}$ is referred to as having a deazoflavin structure. In the oxidized form, $F_{420}$ is fluorescent but when reduced, the structure has a loss of fluorescence. In methanogens, $F_{420}$ participates as an electron carrier for hydrogenase and as a cofactor for methyl-coenzyme M reductase. The midpoint potential of the redox reaction ($F_{420(ox)}$ / $F_{420(red)}$) is $-360$ mV. Although $F_{420}$ is characteristic of methanogens, it is also found in various bacteria and archaea. In *Streptomyces griseus*, $F_{420}$ functions in DNA-related photoreactions.

Table 11.19. Reactions involving cofactors for reduction of $CO_2$ to $CH_4$.

**Methanofuran (MF)**

(1)[a] $CO_2 + MF + H_2 \rightarrow HCO\text{-}MF + H_2O$     $+16$ kJ/mol

**Tetrahydromethanopterine ($H_4MPT$)**

(2) $HCO\text{-}MF + H_4MPT \rightarrow HCO\text{-}H_4MPT + MF$     $-5$ kJ/mol

(3) $HCO\text{-}H_4MPT + H^+ \rightarrow CH\equiv H_4MPT^+ + H_2O$     $-2$ kJ/mol

**Factor 420 ($F_{420}$)**

(8) $H_2 + F_{420} \rightarrow F_{420}\cdot H_2$     $-13.5$ kJ/mol

(4) $CH\equiv H_4MPT^+ + F_{420}\cdot H_2 \rightarrow CH=H_4MPT + F_{420} + H^+$     $+6.5$ kJ/mol

(5) $CH=H_4MPT + F_{420}\cdot H_2 \rightarrow CH\text{-}H_4MPT + F_{420}$     $-5$ kJ/mol

**Coenzyme M**

(6) $CH\text{-}H_4MPT + HS\text{-}CoM \rightarrow CH_3\text{-}S\text{-}CoM + H_4MPT$     $-29$ kJ/mol

**7-Mercaptoheptanoy threonine phosphate (HTP)**
  **(coenzyme B)**

(7) $CH_3\text{-}S\text{-}CoM + HS\text{-}HTP \rightarrow CH_4 + CoM\text{-}S\text{-}S\text{-}HTP$     $-43$ kJ/mol

(9) $CoM\text{-}S\text{-}S\text{-}HTP + H_2 \rightarrow HS\text{-}CoM + HS\text{-}HTP$     $-43$ kJ/mol

[a]Numbers refer to reactions in Figure 11.29.

### Tetrahydromethanopterin

The tetrahydromethanopterin ($H_4MPT$) is found in methanogens where it is a 2 $e^-$ and a 2 $H^+$ carrier plus a $C_1$ carrier. In terms of structure and function, $H_4MPT$ resembles tetrahydrofolate. The midpoint potential for $H_4MPT$ is $-450$ mV, which is considerably more electronegative than the pyridine nucleotide (NAD and NADP) couple. In methane production, the $C_1$ moiety is bound to the $H_4MPT$ structure where it is carried first as a formyl group and ultimately as a methyl group.

### Methanofuran

Methanofuran (MFR) occurs in methanogens, where it exhibits various structural forms according to the types of side chain presents. The parent structure of the methanofuran is given in Figure 11.29. The free–$NH_2$ group serves as the point for attachment of the $C_1$ group to produce formylmethanofuran. The methanofuran molecule is the first carrier of the $C_1$ group in the conversion of $CO_2$ to $CH_4$.

### Coenzyme M

Coenzyme M (Co M) has been determined to be 2-mercaptoethanesulfonic acid. It is the smallest of known cofactors and is found in all methanogens at a concentration of 0.2 m$M$ to 2.0 m$M$. Owing to the active –SH group, CoM is readily oxidized in air to CoM–S–S–CoM. Because CoM–SH is the active form, it can be regenerated from (CoM–S)$_2$ by an NADPH oxidoreductase. CoM interacts with the $C_1$ group and accounts for the release of the methyl group from $CH_3$-$H_4MPT$. It appears that CoM is an obligate requirement in $CH_4$ production from $CO_2$. CoM is found only in methanogens.

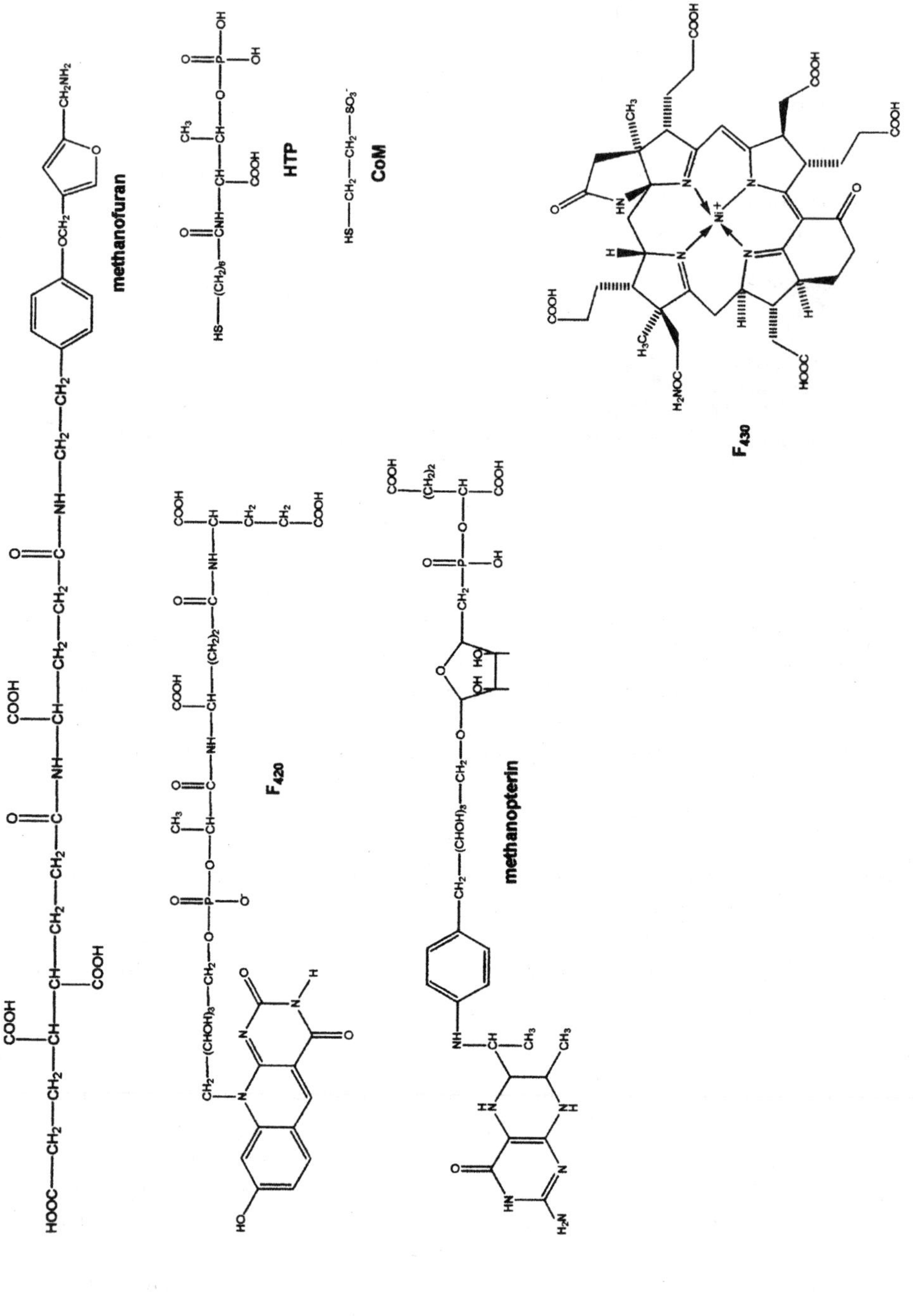

Figure 11.29. Cofactors for methanogens.

### 7-Mercaptohepanoyl threonine phosphate

An obligatory requirement in methane production is the presence of 7-mercaptohepanoyl threonine phosphate (HTP), which is also known as coenzyme B. The HTP has a free –SH group that in the final step of methanogenesis releases the methyl group from $CH_3$-S-CoM to produce $CH_4$. In this reaction, a heterodisulfide is produced and it consists of CoM–S–S–HTP. The reduced cofactors are regenerated by a membrane-bound heterodisulfide reductase. HTP is found only in methanogens.

### Coenzyme $F_{430}$

The coenzyme $F_{430}$ is a nickel protoporphyrin that is involved in methyl transfer activity leading to the production of methane. Two moles of the $F_{430}$ are bound tightly but without covalent linkages to the methyl-coenzyme M reductase. The nickel atom in the molecule ranges form $Ni^+$ to $Ni^{2+}$ and the midpoint potential of the $Ni^{2+}F_{430}$/$Ni^+F_{430}$ couple is $-500$ mV. Coenzyme $F_{430}$ is found only in methanogens.

### 11.5.3.2. Methanogenesis with $H_2$ and $CO_2$[20–22]

The pathway for $CO_2$ reduction to $CH_4$ requires the participation of several unique coenzymes and enzymes. The metabolic sequence of activities can be divided into three major events: (1) $CO_2$ fixation with the $C_1$ binding, (2) reduction of the $C_1$ moiety, and (3) release of the methyl group to produce methane. An outline of the pathway for $CO_2$ conversion to $CH_4$ is given in Figure 11.30.

### $CO_2/C_1$ Binding

The initial reaction involves the reduction of $CO_2$ with electrons from a membrane hydrogenase. This activity is influenced by the membrane potential on the membrane and leads to the uptake of $Na^+$ by a specific pump. In this case, electron flow is driven by ion gradients. The nature of the electron transfer in the membrane and to the enzyme for reduction of $CO_2$ is unresolved. Reduction of $CO_2$ by $H_2$ requires methanofuran (MFR) and three enzyme complexes: (1) a hydrogenase that is unable to react with $F_{420}$, (2) the formylmethanofuran dehydrogenase which unites $CO_2$ with MFR to produce CHO–MFR, and (3) an oxidoreductase complex that transfers reducing equivalents from the hydrogenase to an oxidized factor that is released as the CHO–MER complex is initiated. At the present time there is no information concerning the oxidoreductase complex.

The $F_{420}$ nonreducing hydrogenase contains both Ni and Fe. *Methanobacterium thermoautotrophicum* produces a Ni-Fe hydrogenase that consists of three subunits (α, β, and γ) which are 52 kDa, 41 kDa, and 17 kDa. One mole of Ni is bound into the α-subunit, which is the largest subunit, by binding motifs of Arg-X-Cys Gly-X-Cys-$X_3$-His at the N-terminus and Asp-Pro-Cys-$X_2$-His at the C-terminus. At the carboxyl end of the α-subunit of hydrogenase from *Methanococcus voltare*, there is a substitution of selenocysteine for cysteine.

Two types of formylmethanofuran dehydrogenases are involved in establishing the CHO–MFR structure. One contains molybdenum as a cofactor and the other uses

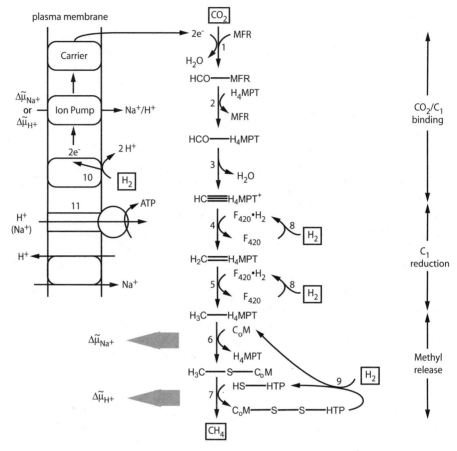

Figure 11.30. Model for methane production from $H_2$ and $CO_2$ by methanogens. The electron carrier for reaction 1 is unresolved. Reaction 6 and reactions 7 and 9 are considered to be responsible for generating membrane potentials. Enzymes are as follows: (1) formylmethanofuran dehydrogenase, (2) formyltransferase, (3) cyclohydrolase, (4) $F_{420}$-dependent $N^5$, $N^{10}$-methenyl methanopterin dehydrogenase, (5) $F_{420}$-dependent $N^5,N^{10}$-methylene methanopterin dehydrogenase, (6) methyltransferase, (7) methyl-S-CoM reductase, (8) hydrogenase, (9) hydrogenase, and (10) hydrogenase, (11) ATP synthase, and (12) $H^+/Na^+$ anteporter. PM, Plasma membrane.

tungsten as the cofactor. In *Methanobacterium thermoautotrophicum*, the enzyme is an $\alpha_1\beta_1$ dimer with subunits of 60 kDa and 45 kDa. Also present in the enzyme from *Methanobacterium thermoautotrophicum* is molybdenum, molybdopterin, nonheme iron, and sulfur. The enzyme from *Methanobacterium barkeri* consists of six different subunits: 65 kDa, 50 kDa, 37 kDa, 34 kDa, 29 kDa, and 17 kDa. Each enzyme from *Methanobacterium barkeri* consists of 1 mol molybdenum, 1 mol of molybdopterin, 28 iron atoms, and 28 sulfur atoms. *Methanobacterium wolferi* has two distinct for-

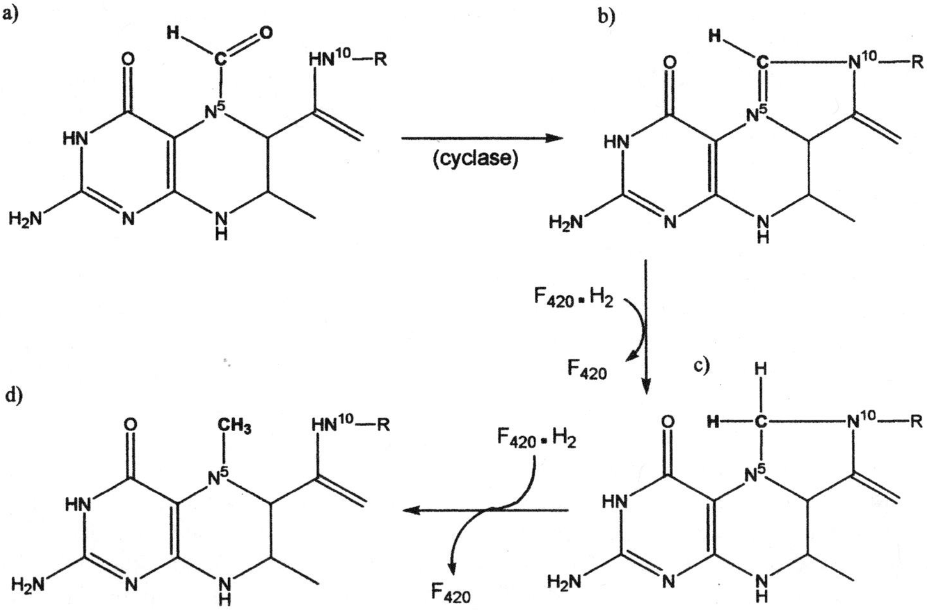

Figure 11.31. Reactions with $C_1$ unit attached to $H_4MPT$. Structures are as follows: (a) $N^5$-formyl-$H_4MPT$, (b) $N^5$, $N^{10}$-methenyl-$H_4MPT$, (c) $N^5$, $N^{10}$-methylene-$H_4MPT$, and (d) $N^5$, $N^{10}$-methyl-$H_4MPT$.

mylmethanofuran dehydrogenases: one contains molybdenum (enzyme I) and the other contains tungsten (enzyme II). Enzyme I has three subunits: 63 kDa, 51 kDa, and 31 kDa with 0.3 mol of molybdenum, 0.3 mol of molybdopterin, and four to six Fe and S atoms. Enzyme II is of content similar to Enzyme I except tungsten is substituted for molybdenum. *M. wolfi* produces enzyme I when Mo is readily abundant but when Mo becomes limiting, enzyme II is formed. The tungsten enzyme is highly specific for the substrate while the molybdenum enzyme will recognize several different molecular types as substrates.

The final step in $C_1$ binding before reduction involves the transfer of the CHO–group from CHO–MFR to $H_4MPT$. The formylmethanofuran: $H_4MPT$ formyl transferase is a 32- to 41-kDa protein that catalyzes this reaction. The CHO– group is added to the nitrogen at position number 5 on the ring. A second enzyme, $N^5$, $N^{10}$ methenyl-$H_4MPT$ cyclohydrolase, removes water and completes the bond between the $C_1$ unit and a second nitrogen at position number 10. These steps of $C_1$ associated with the formation of $N^5$, $N^{10}$ methenyl-$H_4MPT$ are shown in Figure 11.31.

*Reduction of the $C_1$ Compound*

The conversion of the methenyl moiety attached to $H_4MPT$ to a methylene group is accomplished by a $F_{420}$-dependent methylene-$H_4MPT$ dehydrogenase which uses reduced $F_{420}$ ($F_{420} \cdot H_2$) for the reduction process. $F_{420}$ is reduced by a specific hydrogenase. This methylene-$H_4$MPT dehydrogenase is a small molecule with a mass of 32 kDa and without prosthetic groups. A second reduction step results in the conversion of $N^5$, $N^{10}$-methylene-$H_4$MPT to $N^5$-methyl-$H_4$MPT. This enzyme has been isolated from several methanogens as it exists as either a pentamer or a hexamer with a polypeptide chain of 35 kDa.

*Methyl Release*

The methyl group from methyl-$H_4$MPT is transferred to coenzyme M by a two-step process. In the first reaction, the methyl group of $CH_3$-$H_4$MPT is transferred to a corrinoid associated with the enzyme. A corrinoid consists of a Co atom complexed in a tetrapyrrole in a manner similar to vitamin $B_{12}$. Cobalt, as $Co^+$, accepts the methyl group and in the process forms $CH_3$-$Co^{3+}$-corrinoid. The second part of the transfer is the interaction of $CH_3$-$Co^{3+}$-corrinoid with reduced CoM to produce methyl-CoM.

Methane is released from methyl-CoM by the addition of reduced 7-mercaptoheptanoyl threonine phosphate (HS-HTP) and the action of methyl-CoM reductase. A unique cofactor, $F_{430}$, is present in methyl-CoM reductase and the nickel atom of $F_{430}$. The product of methyl-CoM reductase is $CH_4$ and a heterodisulfide of CoM and HTP, which is CoM-S-S-HTP. The heterodisulfide is reduced to produce the individual coenzymes by action of heterodisulfide reductase and a membrane-bound hydrogenase. This phase of the methane production scheme is highly exergonic and is considered by many to account for establishing a charge on the plasma membrane.

Several other reactions that produce $CH_4$ and some of these are given in Table 11.18 and Table 11.19. In these reactions a membrane potential is also established as a result of either a $H^+$- or $Na^+$-motive force. It is assumed that ATP synthase provides energy coupled to $H^+$ or $Na^+$ entry. In some of the methanogens, no cytochromes are present and in organisms such as *Methanosarcina barkeri* both cytochrome *b* and cytochrome *c* are found in the plasma membrane.

### 11.5.3.3. Energy Production Coupled to Carbon Reduction

One of the unresolved issues in microbial physiology is explaining the energy production in methanogens. For autotrophic methanogens growing on $H_2$ and $CO_2$ the only form of ATP produced must be associated with electron transport and oxidative phosphorylation because there is no known system for substrate level phosphorylation. To have oxidative phosphorylation in methanogens, there must be appropriate energy yielding reactions, electron carriers, ion pumps, and ATP synthase reactions. Autotrophic methanogens growing on $H_2$ and $CO_2$ use $H_2$ as the electron source with hydrogenase-coupled reactions in the membrane providing an energy release sufficient to produce ATP. The electron carriers in the membrane are not resolved at this time; however, methanogens do not have quinones (ubiquinones or menaquinones) and only methanotrophic methanogens contain cytochrome. *Methanosarcina barkeri* contains two *b*-type cytochromes and two *c*-type cytochromes. The midpoint potentials of the two cytochrome *b* redox reactions are $-135$ mV and $-240$ mV while for the two cytochrome *c* they are $-140$ mV and $-230$ mV. Reactions that are proposed to involve *b*-type cytochromes are the membrane-bound hydrogenases. In methanogens where acetate and not $H_2$ is the electron donor, it has been proposed that ferredoxin serves as the electron carrier.

*Energetics*

There are two types of ion pumps in methanogens with one specific for $H^+$ and the other for $Na^+$. The formation of $\Delta\bar{\mu}_H^+$ is by a redox-driven proton pump associated with heterodisulfide reduction. There are two components in this $H_2$-dependent heterodisulfide reduction, with one being the hydrogenase and the other component is the heterodisulfide reductase. $Na^+$ is actively pumped across the plasma membrane in response to membrane-bound redox reactions. The two possible reactions that may function as primary sodium pumps are the methyl-$H_4MPT$: coenzyme M methyltransferase and formyl methanofuran dehydrogenase. Experiments with resting cells and inhibitors indicate that both of these reactions are coupled to $Na^+$ transport; however, studies with purified enzyme inserted into proteoliposomes are required for definitive proof that these are indeed the appropriate $Na^+$ pumping reactions.

ATP synthases are present in methanogens and they are V (archaeal)-type ATP synthases. For a considerable period, these archaeal ATPase synthase were considered as V-type structures based on size of subunits and binding to N,N'-Dicyclohexylcarbodiimide (DCCD). However, V-type ATPases do not synthesize ATP and are only proton pumps energized by ATP hydrolysis. The archaeal ATP synthase is driven by proton import which in methylotrophic methanogens is inhibited by DCCD, but not in marine methanogens such as *Methanococcus voltae*. $H^+$-driven ATP synthase in methanogens is inhibited by vanadate (such as P-type ATPases) but not DCCD. In *Methanococcus voltae*, both $H^+$ and $Na^+$ drive ATP synthase.

*Methanoreductosomes*[31]

When cells of *Methanosacrina barkeri* (strain Gö1) and *Methanococcus voltae* were examined by electron microscopy, small structures were observed adjacent to the plasma membrane. Negative stained membranes, as well as membranes subjected to freeze etching, revealed a structure of 19 to 36 nm in diameter with a stalk-like attachment to the inner side of the plasma membrane. By immunolocalization, this structure has been determined to contain methyl coenzyme M reductase and has been termed methanoreductosome. This methanoreductosome structure is on the same side as the ATP synthase; however, the methanoreductosome is much larger than the ATP synthase. The molecular weight of this proteinaceous complex is $5 \times 10^5$ to $1 \times 10^6$. There is some suggestion that protein translocation across the plasma membrane is coupled to heterodisulfide reduction. If this is substantiated, the methanoreductosome would have an important role in the chemiosmotic process.

### 11.5.3.4. Methanol and Acetate Fermentation[32,33]

Acetate accumulates in nature as a result of metabolic activity by fermentative bacteria and by acetogenic bacteria. The methanogens ferment acetate by reduction of the methyl group of acetate to methane with the carbonyl group oxidized to carbon dioxide. This metabolic activity is presented in Figure 11.32. The conversion of acetate to methane is the result of organisms that are members of the *Methanosarcina* and *Methanothrix* genera. As originally suggested by T. Stadtman in 1967, the activation of acetate is required for acetate fermentation to methane. Through the efforts of several research laboratories, it is now known that 1 mol of ATP is required for each mole of acetate fermented and the reactions for these activities are given in Table 11.20. While the presence of formate and $H_2$ appear be important inhibitors of the production of methane from acetate, the mechanism of inhibition remains to be established. It is considered that this production of methane from acetate produces ATP through the chemiosmotic mechanism of bioenergetics. As a result of acetate conversion to methane, *Methanosarcina barkeri* produces a proton-motive force on the plasma membrane of $-120$ mV. Several *b*-type cytochromes are present in the plasma membrane of *Methanosarcina* and these cytochromes have midpoint potentials of $-180$ mV to $-330$ mV. Some scientists have proposed that the cytochromes of *b*-

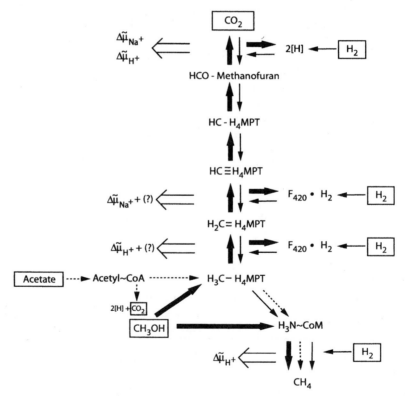

Figure 11.32. Abbreviated scheme for methanogenesis from different starting carbon sources. Specific methanofurans and tetrahydropterins may vary with the species. Pathways are as follows: acetate, *dotted line*; $CO_2$ and $H_2$, *solid line*; methanol, *heavy line*. (From Keltjens and Vogels, 1993.[33])

Table 11.20. Metabolic pathway for the production of methane from acetate by *Methanosarcina* and *Methanothrix*.[a]

| Reactions | $\Delta G^{0'}$ (kJ/mol) |
|---|---|
| Acetate + ATP → Acetyl-phosphate + ADP | +13 |
| Acetyl-phosphate + CoA → Acetyl-CoA + phosphate | −9 |
| Acetyl-CoA + $H_4$SPT → CO + Methyl-$H_4$SPT + CoA | +62 |
| Methyl-$H_4$SPT + HS-CoM → methyl-S-CoM + $H_4$SPT | −29 |
| Methyl-S-CoM + HS-HTP → methane + CoM-S-S-HTP | −43 |
| CO + $H_2$O → $CO_2$ + $H_2$ | −20 |
| $H_2$ + CoM-S-S-HTP → HS-CoM + HS-HTP | −42 |

[a]From Ferry, 1993.[32]

type are important for generation of proton-motive force as they are membrane bound and participate in the electron transport with methane synthesis from acetate.

Methanol is converted to methane by about 20 species of methanogens. As summarized in Figure 11.32, the metabolism of methanol includes a disproportionation of methanol with part of the molecule oxidized and another part of the molecule reduced. To generate reducing equivalents for methanol reduction to methane, one fourth of the methanol used must be oxidized to carbon dioxide. Although both *b*-type and *c*-type cytochromes have been detected in methanogens that utilize methanol, their role in cell energetics remains to be confirmed.

# 11.6. Biochemical Pathways: Redefinition and Summation[34,35]

The establishment of a specific catabolic or amphilic pathway has required considerable evaluation and this has frequently focused on a few organisms such as *Escherichia coli*, *Bacillus subtilis*, and *Pseudomonas aeruginosa*. The presence of a metabolic pathway in other organisms has frequently resulted from extrapolation of results obtained from biochemical demonstration of key enzymes and following the products of radiolabeled carbon compounds used as substrates. The laborious biochemical enzyme assays have been replaced with analysis of genomes using DNA sequences of a specific enzyme from a specific organism as the probe. Most commonly the DNA sequence for a specific enzyme is based on the genetic information in *Escherichia coli*. In a study conducted by S.J. Cordwell, he determined that relatively few prokaryotes have the enzymes for the TCA cycle, glycolysis, or pentose phosphate pathway (see Table 11.21). For example, aconitase, which is an important enzyme in the TCA cycle, is apparently not present in several of the prokaryotes that have several other enzymes of this pathway. Similar statements can be made for hexokinase in the glycolytic pathway and transaldolase in the pentose phosphate pathway. This raises the possibility that in many prokaryotes new reactions (or enzymes) have evolved to perform the needed reactions. For illustration, *Helicobacter pylori* uses a succinyl-

Table 11.21. Presence of metabolic enzymes in prokaryotes as determined by genomic annotation.[a]

| Enzymes | Aquifex aeolicus | Archaeoglobus fulgidus | Bacillus subtilis | Borrelia burgdorferi | Chlamydia trachomatis | Escherichia coli | Haemophilus influenza | Helicobacter pylori | Methanobacterium thermoautotrophicum | Methanococcus jannaschii | Mycobacterium tuberculosis | Mycoplasma genitalium | Mycoplasma pneumoniae | Pyrococcus horikoshii | Rickettsia prowazekii | Synechocystis sp. PCC6803 | Treponema pallidum |
|---|---|---|---|---|---|---|---|---|---|---|---|---|---|---|---|---|---|
| **Tricarboxylic acid cycle** | | | | | | | | | | | | | | | | | |
| Citrate synthase | + | + | + | − | − | + | − | + | + | − | + | − | − | − | + | + | − |
| Aconitase | + | + | + | − | − | + | − | + | − | − | + | − | − | − | + | − | − |
| Isocitrate dehydrogenasease | + | + | + | − | − | + | − | + | + | + | + | − | − | − | + | + | − |
| α-Ketoglutarate dehydrogenase | − | − | + | − | + | + | + | − | − | + | + | − | − | − | + | − | − |
| Succinyl-CoA synthetase | + | + | + | − | + | + | + | − | + | + | + | − | − | − | + | + | − |
| Succinate dehydrogenase | + | + | + | − | + | + | + | + | + | + | + | − | − | − | + | + | − |
| Fumarase | + | + | + | − | + | + | + | + | + | + | + | − | − | − | + | + | − |
| Malate dehydrogenase | + | + | + | − | + | + | + | − | + | + | + | − | − | + | + | + | − |
| **Glycolysis** | | | | | | | | | | | | | | | | | |
| Hexokinase | − | − | − | − | − | + | − | + | − | − | − | − | − | − | − | + | + |
| Phosphoglucose isomerase | + | − | + | + | + | + | + | + | − | + | + | + | + | − | − | + | + |
| Phosphoglucose kinase | + | − | + | + | + | + | + | − | − | − | + | + | + | + | − | + | + |
| Fructose-1,6 bis-PO$_4$ aldolase | + | − | + | + | − | + | + | + | − | − | + | + | + | − | − | + | + |
| Triose-PO$_4$ isomerase | + | + | + | + | + | + | + | + | + | + | + | + | + | + | − | + | + |
| Glyceraldehyde-3-PO$_4$ dehydrogenase | + | + | + | + | + | + | + | + | + | + | + | + | + | + | − | + | + |
| Phosphoglycerate kinase | + | + | + | + | + | + | + | + | + | + | + | + | + | + | − | + | + |
| Phosphoglyceromutase | + | − | + | + | + | + | + | + | − | − | + | + | + | − | − | + | + |
| Enolase | + | + | + | + | + | + | + | + | + | + | + | + | + | + | − | + | + |
| Pyruvate kinase | + | − | + | + | + | + | + | − | − | + | + | + | + | + | + | + | + |
| **Pentose phosphate pathway** | | | | | | | | | | | | | | | | | |
| Glucose-6-PO$_4$ dehydrogenase | + | − | + | + | + | + | + | + | − | − | + | − | − | − | − | + | + |
| 6-PO$_4$-gluconate dehydrogenase | + | − | + | + | + | + | + | + | − | − | + | + | − | − | − | + | + |
| Ribose-5-PO$_4$ isomerase | + | + | + | + | + | + | + | − | + | + | + | − | − | + | − | + | + |
| Ribulose-PO$_4$-3-epimerase | + | − | + | − | + | + | + | + | − | + | + | + | + | − | − | + | + |
| Transketolase | + | − | + | − | + | + | + | + | − | + | + | + | + | − | − | + | + |
| Transaldolase | + | − | + | − | + | + | + | + | − | + | + | − | − | − | − | + | + |

[a]From Cordwell, 1999.[35]

CoA:acetoacetate-CoA transferase to produce succinyl-CoA rather than succinyl-CoA synthetase which is the conventional TCA cycle enzyme.

In addition, some prokaryotes that lack "classical" phosphofructose kinase in the production of fructose-1,6 bisphosphate may rely on the action of 1-phosphofructose kinase to phosphorylate fructose-1-phosphate. There appears to be an inherent danger in relying on the generalization that the enzymes found in *Escherichia coli* are also found in all other prokaryotes that use the same pathway. Exciting times prevailing in prokaryotic proteome analysis and future observations should provide missing details of microbial evolution.

Thus, there is diversity in the major metabolic pathways and this should not be a surprise considering the amount of genotypic change that could have occurred over time in the prokaryotes. There are a few good illustrations of changes in metabolic intermediates and one of these is seen in the Entner–Douderoff pathway. Illustrated in Figure 11.33 is an abbreviated Entner–Douderoff pathway used by several bacteria and three alternate processes used by prokaryotes in the conversion of glucose to pyruvic acid plus glyceraldehyde-3-phosphate. In the hyperthermophile *Pyrococcus*, ferredoxin functions as the electron carrier instead of a pyridine nucleotide. M.W. Adams and colleagues have proposed calling this unique oxidation of glucose to pyruvate the "pyroglycolytic" pathway. Ferredoxin-dependent enzymes in this unique

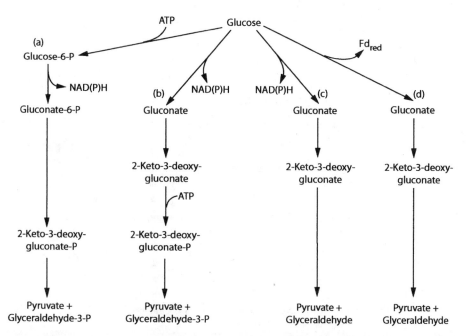

Figure 11.33. Variations of Entner–Douderoff pathway in bacteria and archaea. Pathways most commonly found in (**a**) *Pseudomonas*, (**b**) *Halobacterium* and *Clostridium aceticum*, (**c**) *Sulfolobus* and *Thermoplasma*, and (**d**) *Pyrococcus*.

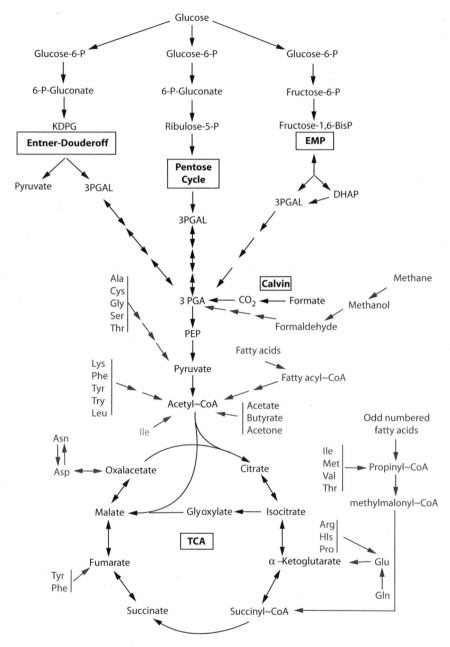

Figure 11.34. The convergence of carbon flow from various catabolic processes into compounds with three carbons and the TCA cycle.

pathway include the following: (1) glucose:ferredoxin oxidoreductase which oxidizes glucose to gluconate, (2) aldehyde:ferredoxin oxidoreductase that converts glyceraldehyde to glycerate, and (3) pyruvate:ferredoxin oxidoreductase for the conversion of pyruvate to acetyl~CoA. Some have speculated that the hyperthermophile employs ferredoxin because it is more stable than NAD(P) at elevated temperatures. Also, in the glycolytic pathway, *Pyrococcus* has ADP-dependent enzymes for the phosphorylation of glucose and fructose-6-phosphate in the formation of glucose-6-phosphate and fructose-1,6-bisphosphate, respectively. In *Escherichia coli* and other bacteria, these enzymes for phosphorylation of the hexoses use ATP but not ADP.

However, to return to the unitary concept of physiology that is the thrust of this book, the major pathways maintain the same function even though some diversity occurs in the intermediary processes. The fueling of the biosynthetic activities is through catabolism of various compounds and is illustrated in Figure 11.34. This convergence of carbon compounds from glycolysis, Entner–Douderoff pathway, or from the hexose monophosphate pathway/pentose pathway into three-carbon compounds is a most efficient means of providing energy and centralizing the carbon compounds for biosynthesis. Also, the catabolism of the carbon backbone of amino acids and fatty acids results in utilization of the TCA cycle or related metabolic systems to generate glyceraldehyde-3-phosphate or acetyl~CoA. Thus, prokaryotes have relatively few catabolic pathways but they may be a few unique enzymes to detail the intermediary metabolic events.

# References

1. Gottschalk, G. 1986. *Bacterial Metabolism.* Springer-Verlag, Berlin.
2. Lengeler, J.W., G. Drews, and H.G. Schlegel (eds.). 1999. *Biology of the Prokaryotes.* Blackwell, New York.
3. Moat, A.G. and J.W. Foster. 1995. *Microbial Physiology.* John Wiley & Sons, New York.
4. Seto, B. 1980. The Stickland reaction. In *Diversity of Bacterial Respiratory Systems*, Vol. II (C.J. Knowles, ed.). CRC Press, Boca Raton, FL, pp. 49–64.
5. Barton, L.L. 1994. The pyruvic acid phosphoclastic reaction. *Methods in Enzymology* 243: 94–103.
6. Litchfield, J.H. 1996. Microbial production of lactic acid. *Advances in Applied Microbiology* 42:45–96.
7. Mitchell, W.J. 1998. Physiology of carbohydrate to solvent conversion by clostridia. *Advances in Microbial Physiology* 39:33–131.
8. Drake, H.L. 1994. Acetogenesis, acetogenic bacteria, and the acetyl-CoA "Wood/Ljungdahl" pathway. In *Acetogenesis* (H.L. Drake, ed.). Chapman & Hall, New York, pp. 3–62.
9. Andreesen, J.R. 1994. Acetate via glycine: A different form of acetogenesis. In *Acetogenesis* (H.L. Drake, ed.). Chapman & Hall, New York, pp. 568–629.
10. Smith, K.S. and J.G. Ferry. 2000. Prokaryotic carbonic anhydrases. *FEMS Microbiology Reviews* 24:335–366.
11. Barton, L.L., J.M. Shivley, and J. Lascelles. 1991. Autotrophs: Variations and versatilities. In *Variations in Autotrophic Life* (J.M. Shivley and L.L. Barton, eds.). Academic Press, San Diego. pp. 1–23.

12. Codd, G.A. 1988. Carboxysomes and ribulose bisphosphate carboxylase/oxygenase. *Advances in Microbial Physiology* 29:115–164.

13. Shively, J.M., G. van Keulen, and W.G. Meijer. 1998. Something from almost nothing: Carbon dioxide fixation in chemoautotrophs. *Annual Review of Microbiology* 52:191–230.

14. Tabita, F.R. 1995. The biochemistry and metabolic regulation of carbon metabolism and $CO_2$ fixation in purple bacteria. In *Anoxygenic Photosynthetic Bacteria* (R.E. Blankenship, M.T. Madigan, and C.E. Bauer, eds.). Kluwer Academic, Dordrecht, The Netherlands, pp. 885–914.

15. Sirevåg, R. 1995. Carbon metabolism in green bacteria. In *Anoxygenic Photosynthetic Bacteria* (R.E. Blankenship, M.T. Madigan, and C.E. Bauer, eds.). Kluwer Academic, Dordrecht, The Netherlands, pp. 871–883.

16. Fuchs, G. 1999. Assimilation of macroelements and microelements. In *Biology of the Prokaryotes* (J.W. Lengeler, G. Drews, and H.G. Schlegel, eds.). Blackwell, New York, pp. 163–187.

17. Das, A. and L.G. Ljungdahl. 2003. Electron transport system in acetogens. In *Biochemistry and Physiology of Anaerobic Bacteria* (L.G. Ljungdahl, M.A. Adams, L.L. Barton, J.G. Ferry, and M.K. Johnson, eds.). Springer-Verlag New York, pp. 191–204.

18. Müller, V. and G. Gottschalk. 1994. The sodium ion cycle in acetogenic and methanogenic bacteria: Generation and utilization of a primary electrochemical sodium ion gradient. In *Acetogenesis* (H.L. Drake, ed.). Chapman & Hall, New York, pp. 127–156.

19. Wood, H.G. and L.G. Ljungdahl. 1991. Autotrophic character of the acetogenic bacteria. In *Variations in Autotrophic Life* (J.M. Shivley and L.L. Barton, eds.). Academic Press, San Diego, pp. 201–250.

20. Ferry, J.G. 1994. CO dehydrogenase of methanogens. In *Acetogenesis* (H.L. Drake, ed.). Chapman & Hall, New York, pp. 539–556.

21. Meyer, O. and K. Fiebig. 1985. Enzymes oxidizing carbon monoxide. In *Gas Enzymology* (H. Deng, R.P. Cox, and H. Toftlund, eds.). Reidel, Dordrecht, The Netherlands, pp. 147–168.

22. Whitman, W.B. 1994. Autotrophic acetyl coenzyme A biosynthesis in methanogens. In *Acetogenesis* (H.L. Drake, ed.). Chapman & Hall, New York, pp. 521–538.

23. Anthony, C. 1982. *The Biochemistry of Methylotrophs.* Academic Press, London.

24. Hanson, R.S. and T.E. Hanson. 1996. Methanotrophic bacteria. *Microbiology Reviews* 60: 439–471.

25. Murrell, J.C., B. Gilbert, and I.R. McDonald. 2000. Molecular biology and regulation of methane monooxygenase. *Archives for Microbiology* 173:325–332.

26. Rosenzweig, A.C. and S.J. Lippard. 1997. Structure and biochemistry of methane monooxygenase enzyme system. In *Transition Metals in Microbial Metabolism* (G. Winkelmann and C.J. Carrano, eds.). Harwood Academic, Amsterdam, pp. 257–280.

27. Anthony, C. 1988. Quinoproteins and energy transduction. In *Bacterial Energy Transduction* (C. Anthony, ed.). Academic Press, New York, pp. 293–316.

28. Grahame, D.A. and T.C. Stadtman. 1993. Redox enzymes of methanogens: Physiological properties of selected, purified enzymes. In *Methanogenesis* (J.G. Ferry, ed.). Chapman & Hall, New York, pp. 335–359.

29. Thauer, R.K., R. Hedderich, and R. Fisher. 1993. Reactions and enzymes involved in methanogenesis from $CO_2$ and $H_2$. In *Methanogenesis* (J.G. Ferry, ed.). Chapman & Hall, New York, pp. 209–252.

30. Hooper, A.B., T. Vannelli, D.J. Bergmann, and D.M. Arciero. 1997. Enzymology of the oxidation of ammonia to nitrite by bacteria. *Antonie van Leeuwenhoek* 71:59–67.

31. Sprott, G.D. and T.J. Beveridge. 1993. Microscopy. In *Methanogenesis* (J.G. Ferry, ed.). Chapman & Hall, New York, pp. 81–127.

32. Ferry, J.G. 1993. Fermentation of acetate. In *Methanogenesis* (J.G. Ferry, ed.). Chapman & Hall, New York, pp. 304–334.

33. Keltjens, J.T. and G.D. Vogels. 1993. Conversion of methanol and methylamines to methane and carbon dioxide. In *Methanogenesis* (J.G. Ferry, ed.). Chapman & Hall, New York, pp. 253–303.

34. Adams, M.W.W., J.-B. Park, S. Mukund, J. Blamey, and R.M. Kelly. 1992. Metabolic enzymes from sulfur-dependent extremely thermophilic organisms. In *Biocatalysis at Extreme Temperatures* (M.W.W. Adams and R.M. Kelly, eds.). American Chemical Society, Washington, D.C., pp. 4–22.

35. Cordwell, S.J. 1999. Microbial genomes and "missing" enzymes: Redefining biochemical pathways. *Archives for Microbiology* 172:269–279.

36. Fuchs, G. 1999. Oxidation of organic compounds. In *Biology of the Prokaryotes* (J.W. Lengeler, G. Drews, and H.G. Schlegel, eds.). Blackwell, New York, pp. 187–233.

37. Perry, J.J. and J.T. Staley. 1997. *Microbiology Dynamics and Diversity*. Saunders College Publishing, Philadelphia.

## Additional Reading

Umbreit, W.W. 1951. Significance of autotrophy for comparative physiology. In *Bacterial Physiology* (C.H. Werkman and P.W. Wilson, eds.). Academic Press, New York, pp. 566–575.

### *One-Carbon Metabolism*

Drake, H.L. 1994. *Acetogenesis*. Chapman & Hall, New York.

Ferry, J.G. 1993. *Methanogenesis*. Chapman & Hall, New York.

Lipscomb, J.D. 1994. Biochemistry of the soluble methane monooxygenase. *Annual Review of Microbiology* 48:371–399.

Stadtman, T.C. 2002. Discoveries of vitamin $B_{12}$ and selenium enzymes. *Annual Review of Biochemistry* 71:1–16.

Wolfe, R.S. and I.J. Higgins. 1979. Microbial biochemistry of methane—a study in contrasts. In *International Review of Biochemistry Microbial Biochemistry*, Vol. 21 (J.R. Quayle, ed.). University Park Press, Baltimore, pp. 293–353.

### *Organic Compounds*

Pollack, J.D., M.V. Williams, and R.N. McElhaney. 1997. The comparative metabolism of the Mollicutes (Mycoplasmas). *Critical Reviews in Microbiology* 23:269–354.

Xavier, K.B., M.S. da Costa, and H. Santos. 2000. Demonstration of a novel glycolytic pathway in hyperthermophilic Archaeon *Thermococcus zilligii* by 13C-labeling experiments and nuclear magnetic resonance analysis. *Journal of Bacteriology* 182:4632–4636.

# 12
# Organization and Cellular Processing

A living cell is not an undifferentiated mass of protoplasm, but is rather an integrated multi-component whole; many biological functions of the living cell depend on the maintenance of the integrity of this intracellular organization. It is this integration of structure that may be considered as a basis for the differentiation of "living" from "non-living" systems. J.S. Fruton and S. Simmonds, *General Biochemistry*, 1958

## 12.1. Introduction

Carbon flux in bacteria results from sugar or organic acids moving through metabolic pathways or cycles. This intermediary metabolism provides the cell with energy in the form of ATP, reducing power as NADH or NADPH, and small organic molecules to be used as building blocks for biosynthesis. At times of rapid cellular growth under conditions in which synthesis of amino acids, nucleic acids, or other carbon compounds occurs at a high rate, there is a demand for structures from the tricarboxylic acid (TCA) cycle and the pentose phosphate pathway. Figure 12.1 provides an overview of the metabolic activities of heterotrophs where there is a balance of catabolism, amphibolic (from the Greek, *amphi*, meaning "either") activities, and anabolism. While glycolysis, pyruvate oxidation, and TCA cycle are important in catabolism, they are equally important in biosynthesis. Metabolism must be well balanced with adequate catabolism to provide sufficient carbon for biosynthesis of proteins, nucleic acids, carbohydrates, and lipids as well as to provide ATP and reducing power to drive these biosynthetic reactions. The integration of these metabolic stages into a single organized process accounts for cellular growth.

It would be too idealistic for this chapter to provide a comprehensive study of all the individual systems that are integrated into a single process. Instead, a few cellular processes are addressed to provide an insight into the various types of coordination present in prokaryotes.

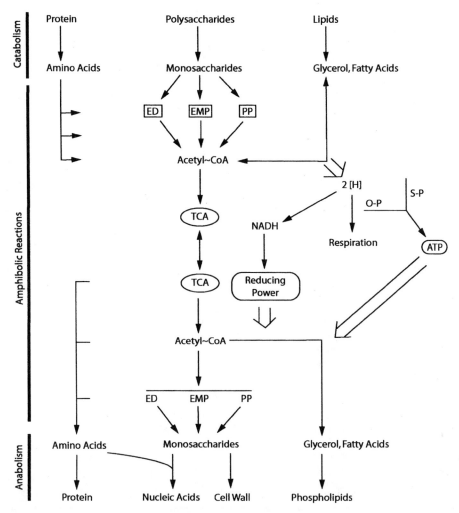

Figure 12.1. Stages in carbon metabolism in heterotrophs: catabolism, amphibolic activity, and anabolism. ED, Entner–Douderoff pathway; EMP, Embden-Meyerhof-Parnas pathway; PP, pentose phosphate pathway; TCA, tricarboxylic acid cycle; 2[H], electrons plus protons; O-P, oxidative phosphorylation; S-P, substrate-level phosphorylation.

## 12.2. Intracellular "Pools" of Nutrients[1]

The molecules that must be formed in a heterotrophic bacterium such as *Escherichia coli* have been estimated and their abundance is given in Table 12.1. These processes are considered to be representative of all prokaryotic organisms. When considering the energy requirements for these various molecules, it is apparent that the single greatest level is attributed to protein synthesis. The several thousand genes in bacteria

Table 12.1. Intracellular concentrations of constituents important for biosynthetic reactions.[a]

| Chemical | μmol/g | μ$M$ |
|---|---|---|
| ATP | 7.00 | 3,000 |
| ADP | 0.58 | 250 |
| AMP | 0.24 | 105 |
| dAMP | 0.41 | 175 |
| cAMP | 0.01 | 6 |
| GTP | 2.15 | 923 |
| GDP | 0.30 | 128 |
| GMP | 0.05 | 20 |
| dGTP | 0.28 | 122 |
| IMP | 0.38 | 162 |
| CTP | 1.20 | 515 |
| CDP | 0.19 | 81 |
| dCTP | 0.15 | 65 |
| UTP | 2.08 | 894 |
| UDP | 0.22 | 93 |
| UMP | 0.33 | 143 |
| dTTP | 0.18 | 77 |
| ADPGlc | 0.01 | 5 |
| GDPMan | 0.04 | 18 |
| UDPGlu | 1.33 | 570 |
| UDPGlcNAc | 0.38 | 164 |
| UDPAcMurP$_5$ | 0.55 | 234 |
| NAD$^+$ + NADH | 1.88 | 806 |
| NADP$^+$ + NADPH | 0.47 | 200 |
| FAD | 0.12 | 51 |
| FMN | 0.21 | 88 |
| Acetyl-CoA | 0.54 | 231 |
| Succinyl-CoA | 0.03 | 15 |
| Malonyl-CoA | 0.03 | 15 |
| ppGpp | 0.07 | 31 |
| pppGpp | 0.04 | 18 |
| PRPP | 1.24 | 472 |

[a]Data from Neuhard and Nygaard, 1987.[1]

produce regulatory, structural, or enzymatic proteins. For cells to regulate energy flow most efficiently, it is necessary to control protein synthesis because it consumes the greatest quantity of biosynthetic energy. Regulatory events used by bacteria to control synthesis of proteins are discussed in this chapter.

## 12.3. Families of Amino Acids and Their Synthesis

Except for a few fastidious organisms, bacteria have the genes for synthesis of amino acids. Information presented here relies principally on the biosynthetic activities found in *Escherichia coli* and *Salmonella* because these systems have been studied most extensively. Owing to a similarity of carbon structures of the various amino acids, synthesis of the common 20 amino acids can be grouped into six families. Characteristics of these six families are given in Table 12.2. The amphibolic pathways are the origin of the carbon structure for the amino acids and Figure 12.2 indicates the intermediates that are the source for each of the various amino acid families. Structural similarity is a means of assigning an amino acid to a given family. An example of structural similarity of amino acids is the presence of an aromatic nucleus in tryptophan, phenylalanine, and tryosine. Another structural similarity is observed with leucine and valine, both of which contain a branched carbon skeleton. Histidine has a unique structure and as a result is synthesized by a metabolic path distinct from other amino acids. Because several amino acid families are associated with the TCA cycle, the bidirectional movement of carbon intermediates provides the necessary intermediates for synthesis of glutamate and aspartate families under anaerobic conditions. The incomplete functioning of the TCA cycle by anaerobes is provided in Figure 12.3 and it should be noted that the precursors for amino acids are the same under aerobic and anaerobic growth.

The synthesis of amino acids is generally attributed to class II metabolic reactions while the reactions generating the small molecular weight precursors are considered as class I reactions. Regulation of class II reactions is important to balance the quantity of amino acids used in biosynthesis with the amount of amino acid synthesized. To accommodate biosynthesis, cells maintain a "pool" of amino acids and if the quantity of amino acids imported from the external environment increases or the rate of biosynthesis decreases, the rate of amino acid synthesis is appropriately diminished. The overall goal of regulation of synthesis of class II compounds is to conserve energy by not synthesizing materials that are not needed by the cells for growth. All amino acids discussed in this chapter are in the L-form unless designated otherwise.

Table 12.2. Amino acid families in bacteria.

| Family | Number of members | Amino acids as members | Starting substrate |
|---|---|---|---|
| 1. Aromatic | 3 | L-Phenylanine, L-typrosine, L-tryptophan | 3-Deoxy-7-phospho-D-arabinoheptulosonate |
| 2. Aspartate | 6 | L-Aspartate, L-asparagine, L-lysine, L-methionine, L-threonine, L-isoleucine | Oxaloacetate |
| 3. Glutamate | 4 | L-Glutamine, L-glutamate, L-proline, L-arginine | $\alpha$-Ketoglutarate |
| 4. Serine | 3 | L-Serine, glycine, L-cysteine | 3-Phosphoglycerate |
| 5. Branched chain | 3 | L-Isoleucine, L-leucine, L-valine | Pyruvate |
| 6. Histidine | 1 | L-Histidine | Ribose-5-phosphate |

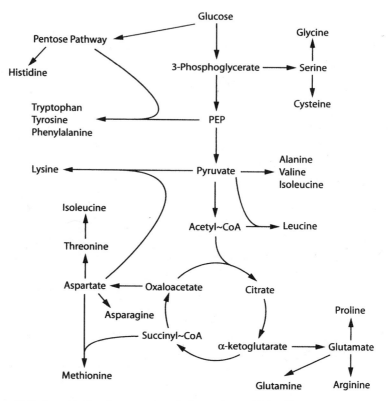

Figure 12.2. Source of carbon compounds for amino acid families in aerobic bacteria.

## 12.3.1.  Regulation of Biosynthesis[2]

Amino acid biosynthesis is regulated through both repression and end product inhibition.  The synthesis of amino acids is subject to regulation at the transcription level by induction and repression activities.  The time for expression of a functional enzyme following induction is only a few minutes; however, in cellular biosynthesis this would be a lengthy period of time and would represent a "coarse adjustment" in amino acid biosynthesis while other regulatory events that respond in seconds to fraction of seconds would be characterized as a "fine adjustment."  Once mRNA is produced, the process of attenuation will regulate the expression or nonexpression of a protein. Other "fine adjustments" in biosynthesis may include the activity of allosteric enzymes that are strategically positioned in the pathway.  In general, these regulatory enzymes in the amino acid biosynthetic pathway use the end product of the pathway, that is, the amino acid, as the effector to down-regulate the activity of an enzyme at a critical point of the metabolic pathway, and these critical points are either at the start or the branch of the pathway.  In general usage, induction/repression refers to production of mRNA while feedback inhibition refers to reduction of enzyme activity.

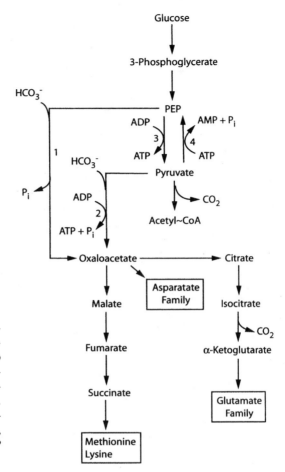

Figure 12.3. Under anaerobic conditions, activity of the TCA cycle supplies carbon structures for amino acids. Heterotrophic $CO_2$ fixation assists in generating oxaloacetate according to reactions 1 and 2. Enzymes are as follows: (1) PEP carboxylase, (2) pyruvate carboxylase, (3) pyruvate kinase, and (4) PEP synthetase.

There are three general mechanisms for regulation of biosynthetic pathways in bacteria through inhibition. The isofunctional enzymes, concerted feedback inhibition, or sequential feedback inhibition and schemes for these regulations are given in Figure 12.4. Bacteria have isofunctional enzymes that serve as critical steps in biosynthetic pathways and these are analogous to isoenzymes commonly found in eukaryotes. These isofunctional enzymes are allosteric molecules that have the same catalytic function but are distinguished by the effectors that they recognize. With one end product of biosynthesis serving to reduce the activity of the initial enzyme, the quantities of intermediary components in the pathway are correspondingly reduced. The pathway is shut off when both end products accumulate and both isofunctional enzymes are down-regulated.

Concerted feedback inhibition functions when the key enzyme in the branched metabolic pathway has two allosteric sites. One regulatory site recognizes one amino acid product and the other site recognizes the other amino acid. In the case of con-

(a)

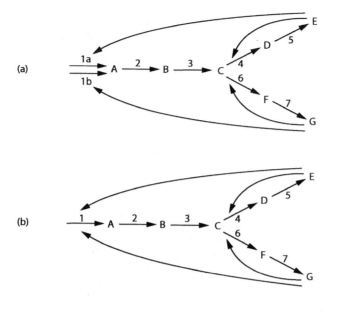

(b)

(c)

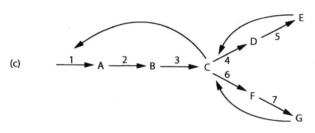

Figure 12.4. Models to indicate types of regulation found in biosynthetic pathways: (**a**) isofunctional enzymes, 1a and 1b subject to feedback inhibition; (**b**) concerted feedback inhibition; (**c**) sequential feedback inhibition.

certed feedback, one end product interacting with the allosteric enzyme will reduce the activity of the enzyme by approximately a half and when both end products, or effectors, are bound to the key allosteric enzyme the enzyme becomes inactivated. Another scheme for regulation of the enzyme is sequential feedback inhibition in which each of the end products inhibits the enzyme accountable for respectable formation and when both enzymes at a branch point are inhibited, the intermediate accumulates and functions as a negative allosteric effector to inhibit the first enzyme in the pathway.

## 12.3.2. The Aromatic Family[3]

The origin of the aromatic amino acids is 3-deoxy-7-phosphate-D-arabinoheptulosonate which is produced by the condensation of D-erythrose-4-phosphate from the pentose phosphate pathway and phosphoenol pyruvate (PEP) from the EMP pathway. The biosynthetic scheme for synthesis of phenylalanine, tyrosine, and tryptophan is

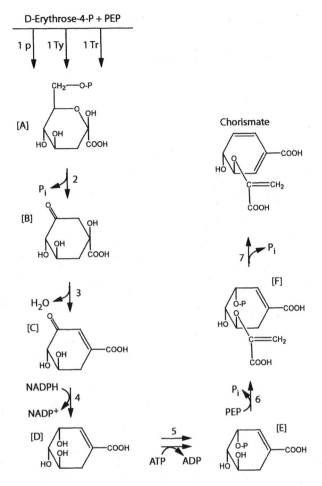

Figure 12.5. Central pathway for biosynthesis of aromatic amino acids with formation of chorismate. Structures are as follows: [**A**] 3-deoxy-7-phospho-arabinoheptulosonate, [**B**] 3-dehydroquinate, [**C**] 3-dehydroshikimate, [**D**] shikimate, [**E**] 3-phosphoshikimate, and [**F**] 5-enolpyruvyl-3phosphoshikimate. Enzymes are as follows: ($1_P$,$1_{Ty}$, and $1_{Tr}$) three isofunctional phospho-2-keto-3-deoxoheptonate aldolases; (2) 5-dehydroquinate synthase; (3) 5-dehydroquinate dehydratase; (4) shikimate dehydrogenase; (5) shikimate kinase; (6) 3-enolpyruvate-shikimate-5-phosphate aldolase; and (7) chlorismate synthase.

given in two parts; see Figures 12.5 and 12.6. Figure 12.5 provides the central pathway for these amino acids with the production of chorismic acid. This central metabolic path is used to synthesize chorismate which is the carbon structure for these three aromatic acids (Figure 12.6). The initial enzyme for this biosynthesis is phospho-2-keto-3-deoxyheptonate aldolase and the product of this reaction is 3-deoxy-7-phospho-arabinoheptulosonate. There are three isofunctional forms of this aldolase and each enzyme is regulated by one of the amino acids produced in the metabolic pathway.

In terms of regulation, the activities of a few of the enzymes of the aromatic family pathway are inhibited by phenylalanine, tyrosine, and tryptophan or repression of mRNA synthesis. The points of regulation are either at the origin of the metabolic pathway or where the formation of individual amino acids branches from the central metabolic scheme (see Figure 12.7). Each of the three forms of phospho-2-keto-3-

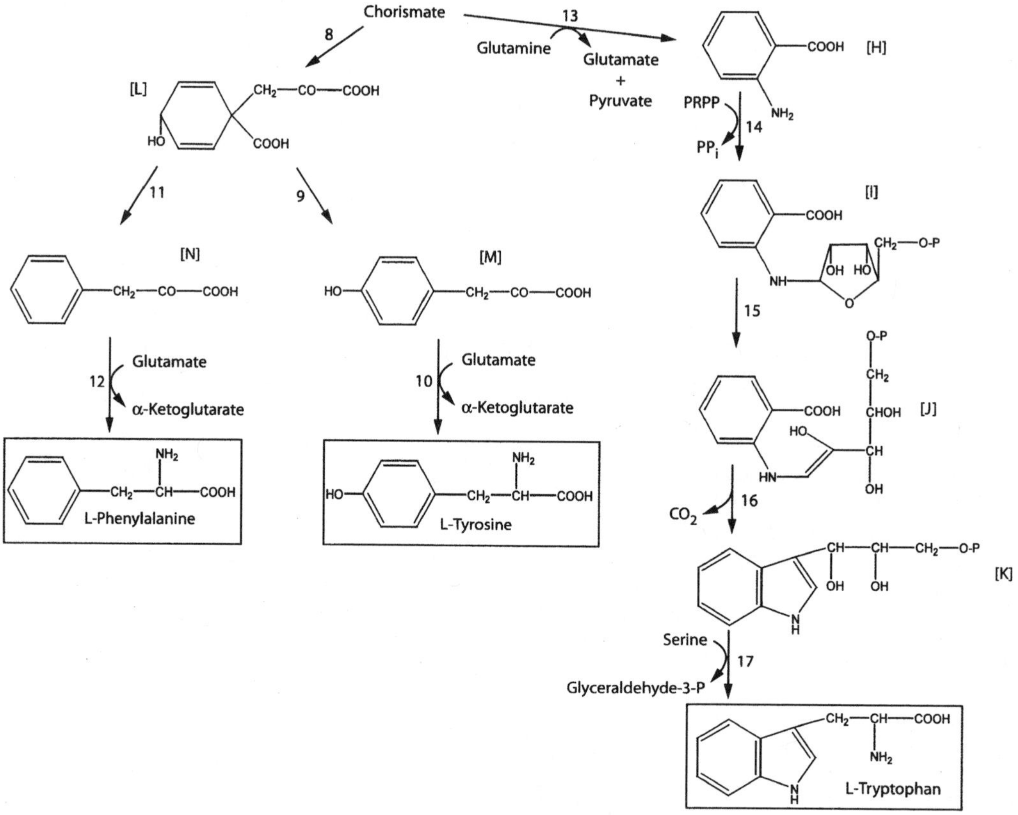

Figure 12.6. Biosynthesis of L-phenylalanine, L-tyrosine, and L-tryptophan from chorismate. PRPP, Phosphoribosylpyrophosphate. Structures are as follows: [H] anthraniate, [I] N-(5'-P-ribosyl) anthranilate, [J] 1-(O-carboxyphenylamino)-1'-deoxyribulose-5'-phosphate, [K] N(5'-P-ribosyl)-anthranalate, [L] prephenate, [M] 4-hydroxyphenyl pyruvate, and [N] phenylpyruvate. Enzymes are as follows: (8) chlorismate mutase; (9) prephenate dehydrogenase; (10) tyrosine aminotransferase; (11) prephenate dehydratase; (12) phenylalanine aminotransferase; (13) anthranilate synthase; (14) anthranilate phosphoribosyltransferase; (15) phosphoribosyl-anthranilate isomerase; (16) indole glycerol phosphate synthase; and (17) tryptophan synthase.

deoxyheptonate aldolase is subject to feedback inhibition or repression by the corresponding amino acid. Alternate points of inhibition and repression occur at the branch points for production of phenylalanine, tyrosine, or tryptophan. In the presence of tryptophan, chorismate is not converted to tryptophan because expression of the operon encoding the relevant enzymes is repressed. Transient regulation of the production of tryptophan is achieved as tryptophan inhibits the activity of anthranilate synthase, the enzyme that converts chorismate to anthranilate. The lack of an appropriate level of anthranilate diminishes the amount of tryptophan produced.

Both phenylalanine and tyrosine regulate subsequent metabolism of chorismate by

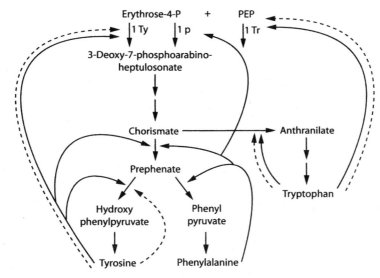

Figure 12.7. Regulation of aromatic amino acids. *Dashed lines* indicate feedback inhibition of specified enzymes and *solid lines* indicate repression of enzyme biosynthesis.

repressing the formation of chorismate mutase which produces prephenate. Prephenate dehydratase and prephenate dehydrogenase are required for the synthesis of phenylalanine and tyrosine, respectively. In the presence of phenylalanine and tyrosine, the synthesis of prephenate dehydratase and prephenate dehydrogenase is repressed. Feedback inhibition of chorismate mutase and prephenate dehydratase is by phenylalanine while tyrosine inhibits only prephenate dehydrogenase. This biosynthetic regulation prevents unnecessary production of aromatic amino acids.

## 12.3.3. The Aspartic Acid Family[4,5]

Oxalacetate is an essential carbon structure precursor for amino acids in the aspartate family and the metabolic scheme for this family is given in Figure 12.8. Asparagine is produced from aspartate by aspartate kinase which requires $NH_3$ and ATP. In *Lactobacillus arabinosus* and related bacteria, asparagine inhibits and represses synthesis of aspartate kinase. In *E. coli*, the formation of methionine, isoleucine, threonine, and lysine is dependent on three isofunctional aspartokinases with each enzyme regulated by a single amino acid. One aspartokinase ($1_{Meth}$, in Figure 12.9) is repressed by methionine and another aspartokinase ($1_{Lys}$) is repressed by lysine. Aspartoklinase $1_{Thre}$ is inhibited by threonine and is repressed by the presence of both threonine and isoleucine. In *Bacillus polymyxa* and *Rhodopseudomonas capsulatus*, aspartokinase occurs as a single enzyme which is inhibited by cooperative feedback by end products of the pathways.

There are several other regions for regulation of the common pathway used in the synthesis of aspartate family amino acids and reference should be made to Figure 12.9

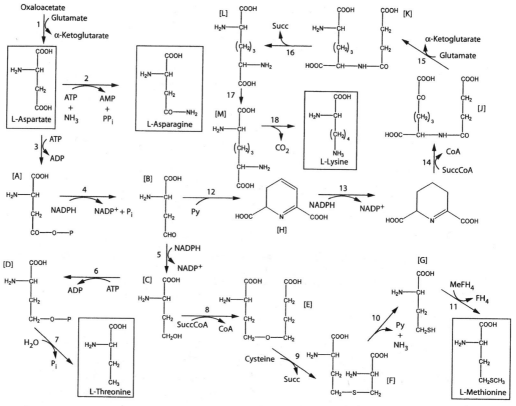

Figure 12.8. Synthesis of L-asparagine, L-threonine, L-methionine, and L-lysine in the aspartate family. Succ, Succinate; Py, pyruvate; FH$_4$, tetrahydrofolate. Structures are as follows: [A] 4-phosphoaspartate, [B] aspartate-4-semialdehyde, [C] homoserine, [D] homoserine $O$-phosphate, [E] $O$-succinylhomoserine, [F] cystathionine, [G] homocysteine, [H] dihydropicolinate, [I] piperideine-2,6-dicarboxykate, [J] $N$-succinyl-6-keto-L-2-aminopimelate, [K] $N$-succinyl-L,L-2,6-diaminopimelate, [L] L,L-2,6-diaminopimelate, and [M] $meso$-2,6-diaminopimelate. Enzymes are as follows: (1) transamidase, (2) asparagine synthetase, (3) aspartate kinase, (4) aspartate semialdehyde dehydrogenase, (5) homoserine dehydrogenase, (6) homoserine kinase, (7) threonine synthase, (8) homoserine $O$-succinyltransferase, (9) cystathionine γ-synthase, (10) cystathionine β-lyase, (11) homocysteine methyltransferase, (12) dihydropicolinate synthase, (13) dihydropicolinate reductase, (14) succinylketoaminopimelate synthase, (15) succinyl diaminopimelate aminotransferase, (16) succinyldiaminopimelate desuccinylase, (17) diaminopimelate epimerase, and (18) $meso$-diaminopimelate decarboxylase.

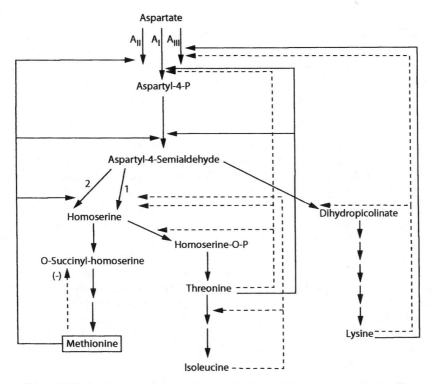

Figure 12.9. Regulation of the synthesis of amino acids in the aspartate family.

to evaluate the points of regulation. The combined presence of isoleucine plus thre-onine represses aspartate-semialdehyde dehydrogenase. Homoserine dehydrogenase has two isofunctional enzymes and one of these is subject to multivalent repression attributed to the presence of both threonine and isoleucine. The second form of hom-oserine dehydrogenase is repressed by methionine. Feedback inhibition by methionine targets homoserine O-succinyltransferase. Accumulation of threonine inhibits both homoserine dehydrogenase, which produces homoserine and homoserine kinase, which is involved in production of the precursor for threonine. Lysine inhibits the activity of dihydrodipicolinate synthase.

A specific example of biosynthetic pathway for biosynthesis of lysine, methionine, isoleucine, and threonine from aspartate is given in Figure 12.9. In this case there are three isofunctional enzymes that can catalyze the conversion of aspartate to aspartyl phosphate with equal ease. There is another isofunctional enzyme that changes as-partate semialdehyde to homoserine. With the formation of lysine, methionine, iso-leucine, or threonine, there is feedback to key enzymes to inhibit formation of that specific amino acid. Threonine and lysine each inhibit a specific isofunctional enzyme that initiates the pathway. Threonine also inhibits one of the allosteric enzymes involved in the formation of homoserine. The result is that one enzyme in the set of isofunctional enzymes will be uninhibited and precursors for methionine will be produced. Thus, through regulation of biosynthetic pathways by allosteric enzymes allows for a mechanism of fine control for synthesis of enzymes from the aspartate family.

## 12.3.4. The Glutamic Acid Family[6-8]

The key structure for entry into the glutamate family of amino acids is α-ketoglutarate from the TCA cycle. The various reactions in the glutamate family are indicated in Figure 12.10. Glutamate formation is produced by glutamate dehydrogenase and it is the principal reaction for fixation of ammonia in bacteria. Glutamine is produced by the ATP-requiring addition of ammonia to glutamate and this reaction functions as a major means of assimilating ammonia from the environment. Reactions of glutamine and glutamate with ammonia are discussed in considerable detail in Chapter 13. Proline is produced from glutamate by two reduction-based reactions that use NADH in the first reaction and NADPH in the second reduction. The formation of arginine from glutamate requires eight reactions with ornithine and citrulline formed as intermediates in the pathway. In a series of steps, the γ-carboxyl of glutamate is reduced by a NADH-coupled reaction and this is followed by the addition of an amino group onto the γ-carbon to produce ornithine. The subsequent addition of a carbamoyl group onto ornithine produces citrulline with an amino group from aspartate contributing to the formation of arginine.

The regulation of the glutamate family is controlled by both repression of mRNA synthesis and by feedback inhibition. Glutamine is the key enzyme in transferring the amide nitrogen for the synthesis of numerous compounds including histidine, tryptophan, carbamoyl phosphate, and glucosamine. In *E. coli*, formation of glutamine synthetase is controlled by numerous metabolites. Glutamine synthetase occurs as two enzyme structures: form I and form II. Form I is fairly insensitive to feedback inhibition while form II is readily inhibited by tryptophan, histidine, AMP, and CTP. However, form I may be converted to form II by an enzyme step involving adenylylation with ATP providing the needed carbon structure. Under appropriate conditions, form II can be deadenylylated and returned to form I.

Both arginine and proline control enzymes that are present at the major branch point of the biosynthetic pathway. Proline is a feedback inhibitor to limit the synthesis of glutamate-5-semialdehyde and to repress the enzymes important in converting glutamate to proline. Arginine inhibits glutamate acetyltransferase which is the first enzyme leading to the arginine synthesis starting from glutamate. All of the genes of arginine biosynthesis are under control of the regulatory gene, *arg R*, even though some of the genes are dispersed around the chromosome and not in a single operon. When arginine accumulates in the cell, there is the simultaneous repression of arginine synthesis.

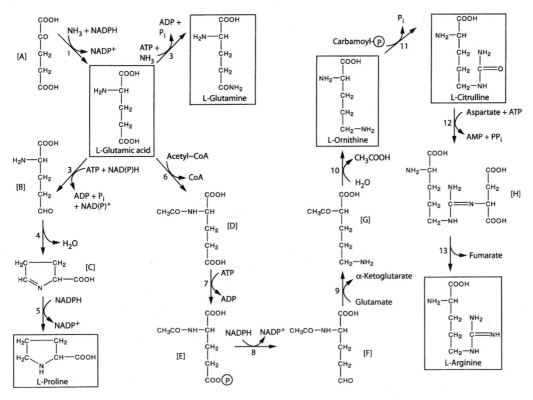

Figure 12.10. Synthesis of amino acids in the glutamate family producing L-glutamine, L-proline, L-ornithine, L-citrulline, and L-arginine. Compounds are as follows: [A] α-keto-glutarate, [B] glutamic-γ-semialdehyde, [C] Δ¹-pyrroline-5-carboxylate, [D] *N*-acetylglutamate, [E] *N*-acetylglutamyl phosphate, [F] *N*-acetylglutamate-γ-semialdehyde, [G] *N*-acetylornithine, and [H] L-argininosuccinate. Enzymes are as follows: (1) glutamate synthase, (2) glutamine synthetase, (3) γ-glutamyl kinase plus glutamate-γ-semialdehyde dehydrogenase, (4) spontane-ous, (5) Δ¹-pyrroline-5-carboxylate reductase, (6) *N*-acetylglutamate synthase, (7) *N*-acetyl-glutamate kinase, (8) *N*-acetylglutamate phosphate reductase, (9) *N*-acetylornithine aminotransferase, (10) *N*-acetylornithinase, (11) ornithine carbamoyl transferase, (12) argini-nosuccinate synthetase, and (13) argininosuccinase.

## 12.3.5. The Serine Family[9]

The precursor for serine synthesis is 3-phosphoglycerate which is an intermediate in glycolysis. In *Salmonella typhimurium* and *E. coli*, the biosynthetic pathway for serine production includes phosphohydroxypyruvate and phosphoserine as given in Figure 12.11. Some bacteria produce serine with glycerate and hydroxypyruvate as inter-mediates in the path of serine synthesis from 3-phosphoglycerate. A few bacteria produce serine from dehydrogenase action on threonine. From serine to glycine is a

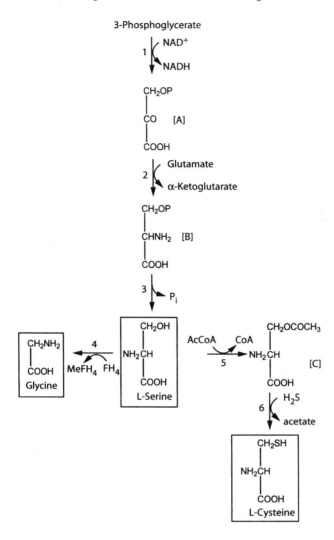

Figure 12.11. Synthesis of amino acids in the serine family. Ac, Acetate; MeHF$_4$, methylene tetrahydrofolate. Compounds are as follows: [A] phosphohydroxy pyruvate, [B] phosphoserine, and [C] O-acetylserine. Enzymes are as follows: (1) phosphoglycerate dehydrogenase, (2) phosphoserine aminotransferase, (3) phosphoserine phosphatase, (4) serine hydroxymethyltransferase, (5) serine O-acetyltransferase, and (6) cysteine synthase.

single enzyme step with one carbon atom removed from serine by a tetrahydrofolate reaction. Cysteine is formed from serine by the condensation of sulfide with O-acetylserine. Activation of sulfate to produce hydrogen sulfide is discussed in Chapter 13. The presence of serine inhibits 3-phosphoglycerate dehydrogenase by feedback inhibition. Cysteine inhibits activity of serine O-acetyltransferase to reduce the amount of O-acetylserine produced and thereby reduce the production of cysteine. The regulation of cysteine synthesis suggests that it may be a global regulator of metabolism.

## 12.3.6. The Branched-Chain Amino Acid Family[10]

Pyruvate is the precursor for the synthesis of alanine and the branched-chain amino acids of valine, leucine, and isoleucine. Alanine is produced from pyruvate by either of two reactions: (1) transamination with glutamate by alanine aminotransferase and (2) addition of $NH_3$ by the NADPH-dependent alanine dehydrogenase. Reactions accounting for the formation of branched amino acids are presented in Figure 12.12. The synthesis of L-valine and L-isoleucine is due to four reactions with three enzymes catalyzing comparable reactions in each of the two biosynthetic pathways. In the final reaction producing L-valine and L-isoleucine, there is a transamination step with each reaction having a preference for a specific amino donor. For L-valine production, a condensation of two pyruvate molecules produces acetolactate which is dehydrated before glutamate provides the amino group for L-valine. In the production of L-isoleucine, there is the condensation of pyruvate with α-ketobutyrate to produce the carbon skeleton for L-isoleucine. It is important to note that three enzymes are shared by both the L-isoleucine and the L-valine pathways. The rationale for this use of the same enzymes is that the substrate for enzymes in these pathways differs by only one carbon unit and the enzymes are unable to distinguish between these carbon skeletons. This is one of the few biosynthetic activities in which the same enzymes function in two different biosynthetic pathways. For L-leucine formation, there is a series of reactions that stem from the valine pathway.

While both L-valine and L-isoleucine require pyruvate for synthesis, synthesis of L-isoleucine requires α-ketobutyrate which is produced from the deamination of L-threonine. In *E. coli* and *S. typhimurium*, there are three isofunctional enzymes (e.g., I, II, and III) that catalyze the formation of L-valine, L-isoleucine, and L-leucine. These isofunctional enzymes are similar in structure in that all consist of two small subunits of approximately 10 kDa and two large subunits each of approximately 60 kDa. The difference in function of each enzyme is in recognition of amino acids that act as allosteric effectors and not in the recognition of precursor substrates. This difference in regulation is attributed to a specific sequence in the C-terminus of the large subunit and these sequences recognize L-valine, L-isoleucine, or L-leucine. Thus, these enzymes differ in recognition of amino acids in feedback inhibition and not in catalytic activity.

## 12.3.7. Histidine Synthesis[11]

Histidine has a carbon structure that is unique and not shared with other amino acids. The biosynthesis of histidine takes place in a series of 11 reactions involving 9 genes (see Figure 12.13). There are two genes, *hisB* and *hisI*, in *E. coli*, that produce bifunctional proteins with each protein having two enzyme functions. This is the reason that 9 genes can produce enzymes for 11 reactions. Histidine is produced from 5-phosphoribosyl-1-pyrophosphate (PRPP) with a nitrogen and a carbon atom obtained from adenine of ATP. The PRPP attaches to ATP at the N1 of the purine ring and by enzyme action the ring is opened to produce imidazoleglycerol phosphate (IGP) and aminoimidazolecarboxamide ribonucleotide (ALCRP). As IGP is processed

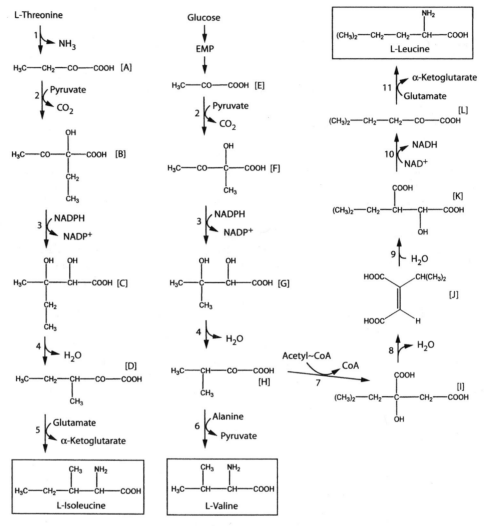

Figure 12.12. Synthesis of branched-chain amino acids. Compounds are as follows: **[A]** α-ketobutyrate, **[B]** α-aceto-α-hydroxybutyrate, **[C]** α,β-dihydroxy-β-methylvalerate, **[D]** α-keto-β-methylvalerate, **[E]** pyruvate, **[F]** α-acetolactate, **[G]** α,β-dihydroxyisovalerate, **[H]** α-ketoisovalerate, **[I]** α-isopropylmalate, **[J]** *cis*-dimethylcitraconate, **[K]** β-isopropylmalate, and **[L]** α-ketoisocaproate. Enzymes are as follows: (1) threonine deaminase, (2) acetohydroxy acid synthase, (3) acetohydroxy acid isomeroreductase, (4) dihydroxy acid dehydrase, (5) isoleucine aminotransferase, (6) valine aminotransferase, (7) isopropylmalate synthase, (8) isopropylmalate dehydrase, (9) isopropylmalate hydrase, (10) isopropylmalate dehydrogenase, and (11) leucine aminotransferase.

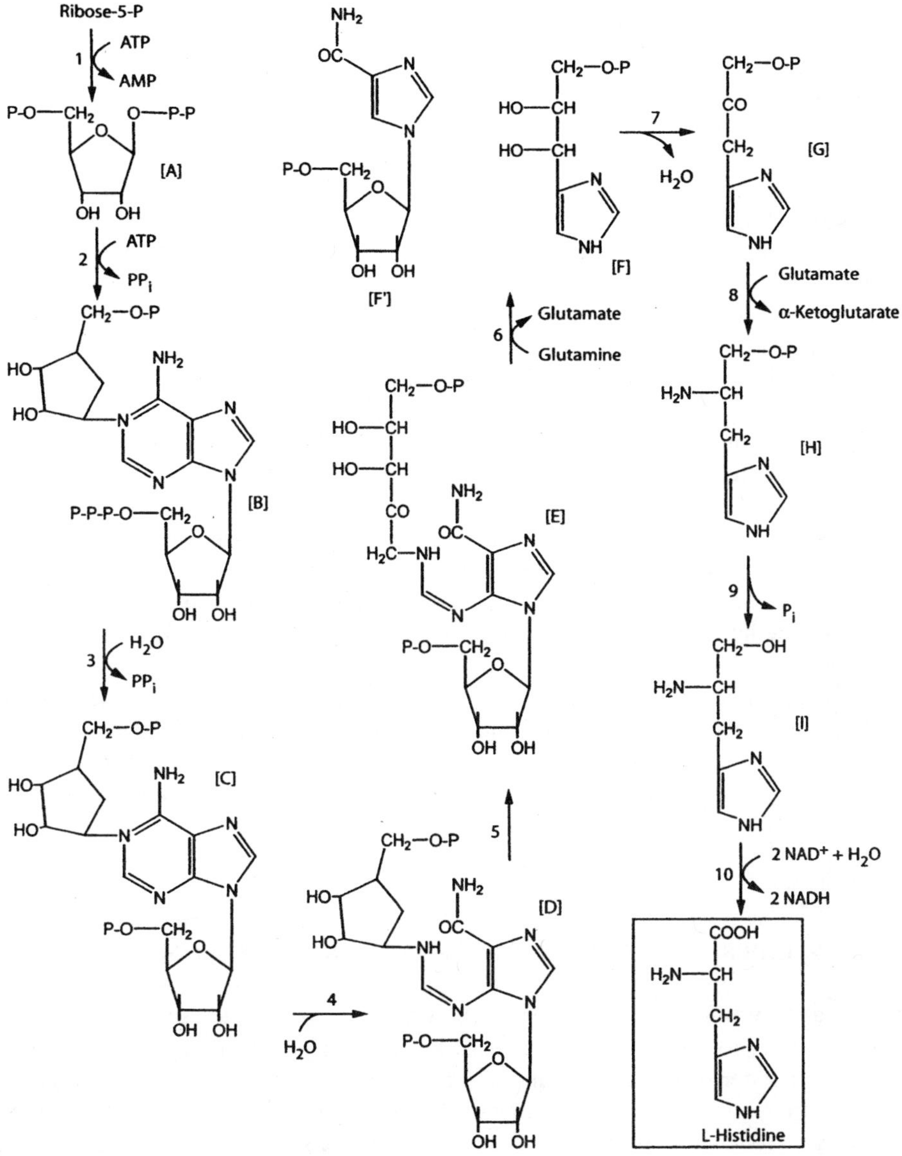

Figure 12.13. Pathway for the synthesis of histidine in bacteria. Structures are as follows: [A] phosphoribosyl pyrophosphate, [B] $N^1$-(5'-phosphoribosyl)-ATP, [C] $N^1$-(5'-phosphoribosyl)-AMP, [D] pro-phosphoribosyl formimino-5-aminoimidazole-4-carboxamide ribonucleotide (5'-pro FAR), [E] phosphoribosyl formimino-5-aminoimidazole-4-carboxamide ribonucleotide (PRFAR), [F] imidazole glycerol phosphate (IGP), [F'] 5-aminoimidazole-4-carboxamide ribonucleotide (ALCAR), [G] imidazoleacetol phosphate, [H] L-histidinol phosphate, and [I] L-histidinol. Enzymes are as follows: (1) phosphoribosyl pyrophosphate synthetase, (2) ATP phosphoribosyltransferase, (3) phosphoribosyl-ATP pyrophosphohydrolase, (4) phosphoribosyl-AMP cyclohydrolase, (5) 5'-ProFAR isomerase, (6) imidazoleglycerol phosphate synthase, (7) imidazoleglycerol phosphate dehydratase, (8) L-histidinol phosphate aminotransferase, (9) histidinol phosphate phosphatase, and (10) histidinol dehydrogenase.

further to produce histidine, a second nitrogen atom is added and this time it is from glutamine. ALCRP enters the pool of intermediate compounds and through reactions associated with purine biosynthesis, ATP is reformed.

It is paramount that bacteria closely regulate the histidine pathway because the amount of energy expended for the synthesis of a single histidine molecule is equivalent to about 41 ATP molecules and if histidine accumulates in the cell, this would constitute a considerable loss of energy. From studies with *E. coli* and *S. typhimurium*, feedback inhibition can account for about a 30-fold difference in pathway activity while repression can diminish histidine production by about 200-fold. Feedback inhibition as a form of regulation is primarily attributed to the action of histidine on the first enzyme, ATP phosphoribosyltransferase, in the biosynthetic pathway. Transcriptional control that coordinately represses all genes of the *his* operon are influenced by (1) attenuation which is a response to the availability of histidyl-tRNA[his] and (2) the accumulation of ppGpp, guanosine tetraphosphate, which serves as a general metabolic signal molecule. Attenuation is a transient but immediate control of the RNA polymerase and is attributed to base pairing in stem-loop structures which reflects the concentration of the histidine pool in the cell. The effect of ppGpp is that of a positive effector for *his* operon transcription but not translation. Under conditions of amino starvation, the intracellular pool of ppGpp increases and this results in stimulating the expression of the *his* operon. In *S. typhimurium* and *Staphylococcus aureus*, the nine structural genes constitute a unit with the *hisO* gene functioning as the operator that regulates the histidine operon. It is interesting to note that in *S. typhimurium*, the gene order does not follow the same sequence as in the enzyme reactions. In some bacteria the genes for histidine are not unified in a single unit but are distributed throughout the chromosome.

## 12.4. Synthesis of Nucleic Acids and Their Regulation

Purines and pyrimidines are important molecules in bacteria because they serve as building blocks for nucleic acid synthesis, serve as coenzymes, and function as activators for transfer of sugars for various types of synthesis. Purines and pyrimidines contain nitrogen atoms in a ring structure and the addition of ribose or deoxyribose to this base produces a nucleoside. A nucleotide is the structure produced after phosphate groups are added to the nucleoside. The de novo synthesis of purines and pyrimidines uses PRPP as an important precursor. PRPP is produced from ribose-5-phosphate with a pyrophosphate moiety added from ATP (see Figure 12.14). Not only is PRPP used in the synthesis of purines and pyrimidines but it is also required for the synthesis of histidine, tryptophan, and nicotinamide nucleotides which include $NAD^+$ and $NADP^+$. As can be seen in Table 12.1, there is a relatively high abundance of PRPP in growing cells. It has been estimated that of the PRPP produced 30% to 40% is for purines, 30% to 40% is for pyrimidines, 10% to 15% is for histidine and tryptophan, and 1% to 2% for $NAD(P)^+$ synthesis. Ribose phosphate pyrophosphate

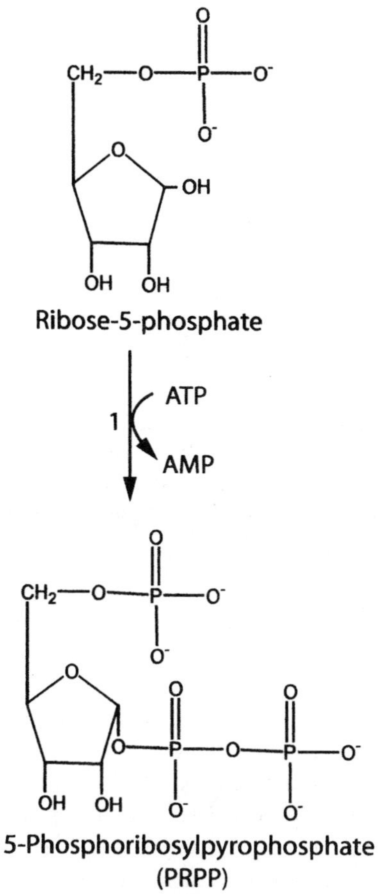

Ribose-5-phosphate

5-Phosphoribosylpyrophosphate
(PRPP)

Figure 12.14. Synthesis of PRPP from ribose-5-phosphate. Enzyme: (1) ribose phosphate pyro-phosphokinase.

kinase is the enzyme required for synthesis of PRPP and in *E. coli* the enzyme has a subunit mass of 35 kDa. The accumulation of either ADP, CTP, GTP, or UTP will reduce the rate of PRPP production and in some cases, the presence of tryptophan also influences rate synthesis of PRPP. The general role of PRPP in production of pyrimidines and purines is given in Figure 12.15. It should be noted that PRPP is used as the initiator structure for the synthesis of purines while PRPP is added relatively late in the pathway for the synthesis of pyrimidines. Uridine monophosphate (UMP) is the precursor for pyrimidine nucleotides of cytosine triphosphate (CTP) and uridine triphosphate (UTP) while inosine monophosphate (IMP) is the precursor for the purine nucleotides of adenosine triphosphate (ATP) and guanine triphosphate (GTP).

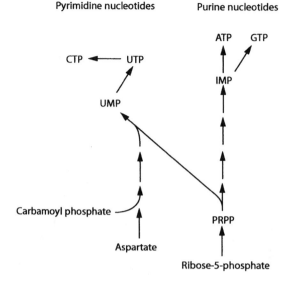

Figure 12.15. The general pathway for synthesis of PRPP used in production of purines and pyrimidines in bacteria. PRPP, 5'-Phosphoribosyl-pyrophosphate.

## 12.4.1. Biosynthesis of Pyrimidines

Owing to the vast amount of information known about pyrimidine synthesis, *E. coli* serves as an excellent model. From investigations with various bacteria, it appears that pyrimidine in many bacteria follows the general process employed by *E. coli*. The biosynthesis of pyrimidines is initiated by condensation of aspartate and carbamoylphosphate. The positions at which these precursors are incorporated into the pyrimidine base are given in Figure 12.16. The pathway for the synthesis of pyrimidines is given in Figure 12.17 and there are four allosteric enzymes in the pathway that modulate the biosynthetic activity. The synthesis of carbamoyl phosphate from glutamine, bicarbonate, and 2 mol of ATP is catalyzed by carbamoylphosphate synthase (CPSase). In *E. coli*, the CPSase is a heterodimer with subunits of 42 kDa and 118 kDa. Two distinct binding sites for the two atoms of ATP are required for the reaction. CSPase is an allosteric enzyme which uses IMP, PRPP, or ornithine as positive effectors and UMP as a negative effector. Aspartate carbamoyltransferase (ATCase) catalyzes the condensation of aspartate and carbamoyl phosphate with the formation of carbamoylaspartate. In *E. coli*, ATCase is a large allosteric enzyme with a mass of 310 kDa and a dodecamer that consists of two trimers (102 kDa) and three dimers (34 kDa). ATP is a positive effector for ATCase while CTP inhibits the enzyme by increasing the apparent $K_m$ of ATCase for aspartate. The other two allosteric enzymes in pyrimidine biosynthesis are CTP synthetase and UMP kinase. In *E. coli*, CTP synthase is a tetramer with identical subunits of 60 kDa. Positive effectors for CTPase are ATP, UTP, and GTP. The UMP kinase is a hex-

(a)

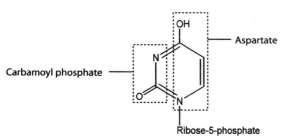

(b)

Figure 12.16. The origin of atoms in pyrimidine and purine structures. (a) IMP, a purine and (b) UMP, a purine.

amer with subunits having a mass of 26 kDa and it displays considerable specificity for UMP in that it will not recognize CMP, dCMP, or UMP as substrates.

## 12.4.2. Biosynthesis of Purines

The synthesis of purine nucleotides is complicated by the possibility that nucleotides are produced by "salvage pathways" as well as by de novo synthesis. In salvage pathways, there is an interconversion of purines in the cytoplasm of the cell or of purines that are acquired from the culture medium. There is considerable diversity of the type of purine interconversion with the various bacteria. However, the model used for de novo purine synthesis in bacteria is based on the observations of that found in *E. coli*. As shown in Figure 12.16, the atoms in the purine rings come from three different amino acids. The de novo pathway for purine biosynthesis is given in Figure 12.18. For purine production, the formation of the purine structure starts with PRPP serving as a precursor for 5-phosphoribosylamine (PRA). The enzyme for production of PRA is glutamine PRPP aminotransferase which is an allosteric enzyme with AMP and GMP as negative effectors. Another site for enzyme regulation of purine biosynthesis is the branch point at IMP leading to GMP and AMP (see Figure 12.19). Adenylo-succinate synthetase is the first enzyme converting IMP to AMP and inhibition is by AMP, GMP, and ppGpp. The inhibition by ppGpp, which is produced along with amino acid starvation, is an example of a mechanism of the stringent response. IMP dehydrogenase is important because it is the first enzyme in synthesis

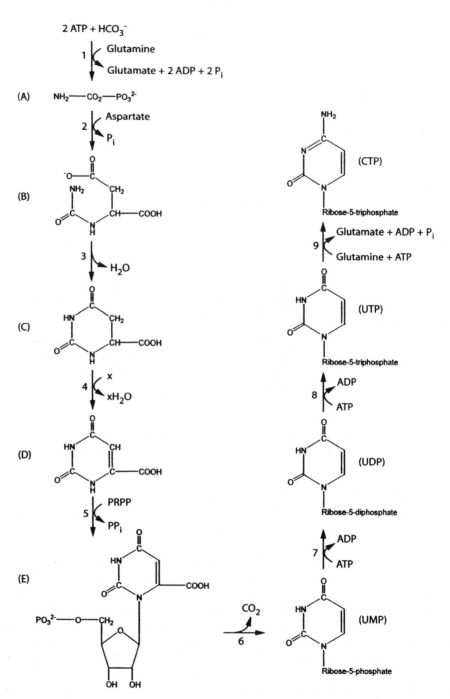

Figure 12.17. Pathway for synthesis of CTP and UTP from aspartic acid and carbamyl phosphate. Structures are as follows: (**A**) carbamoyl phosphate, (**B**) carbamoyl aspartate, (**C**) dihydroorotate, (**D**) orotate, and (**E**) orotidine-5-phosphate (OMP). Enzymes are as follows: (1) carbamoyl phosphate synthetase, (2) aspartate carbamoyltransferase, (3) dihydroorotase, (4) dihydroorotate dehydrogenase, (5) orotate phosphoribosyltransferase, (6) OMP decarboxylase, (7) UMP kinase, (8) UDP kinase, and (9) CTP synthetase. (From Neuhard and Kellin, 1996.)

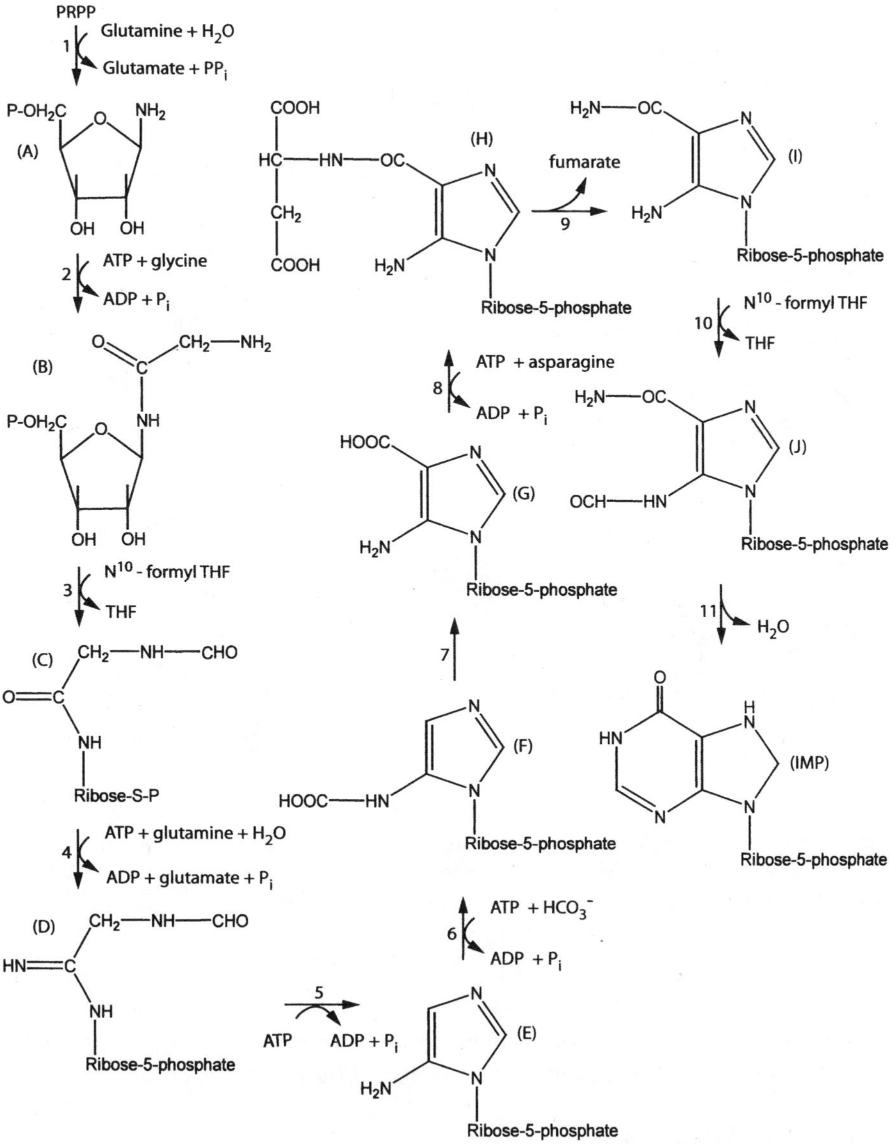

Figure 12.18. Pathway for the synthesis of IMP starting with PRPP. Structures are as follows: (**A**) 5-phosphoribosylamine, (**B**) 5-phosphoribosylglycinamide, (**C**) 5-phosphoribosyl-*N*-formylglycinamide (FGAR), (**D**) 5-phosphoribosyl-*N*-formylglycinamidine (FGAM), (**E**) 5-phosphoribosyl-5-aminoimidazole (AIR), (**F**) 5-phosphoribosyl-5-carboxyaminoimidazole (NCAIR), (**G**) 5-phosphoribosyl-5-aminoimidazole-4-carboxylate (CAIR), (**H**) 5-phosphoribosyl-4-(*N*-succinocarboxamide)-5-aminoimidazole (SAICAR), (**I**) 5-phosphoribosyl-4-carboxamide-5-aminoimidazole (ALCAR), (**J**) 5-phosphoribosyl-4-carboxamide-5-formamidoimidazole (FAICAR). Enzymes are as follows: (1) glutamine PRPP amidotransferase, (2) GAR synthetase, (3) GAR transformylase, (4) FGAM synthetase, (5) AIR synthetase, (6) NCAIR synthetase, (7) NCAIR mutase, (8) SAICAR synthase, (9) adenylosuccinate lyase, (10) AICAR transformylase, and (11) IMP cyclohydrolase.

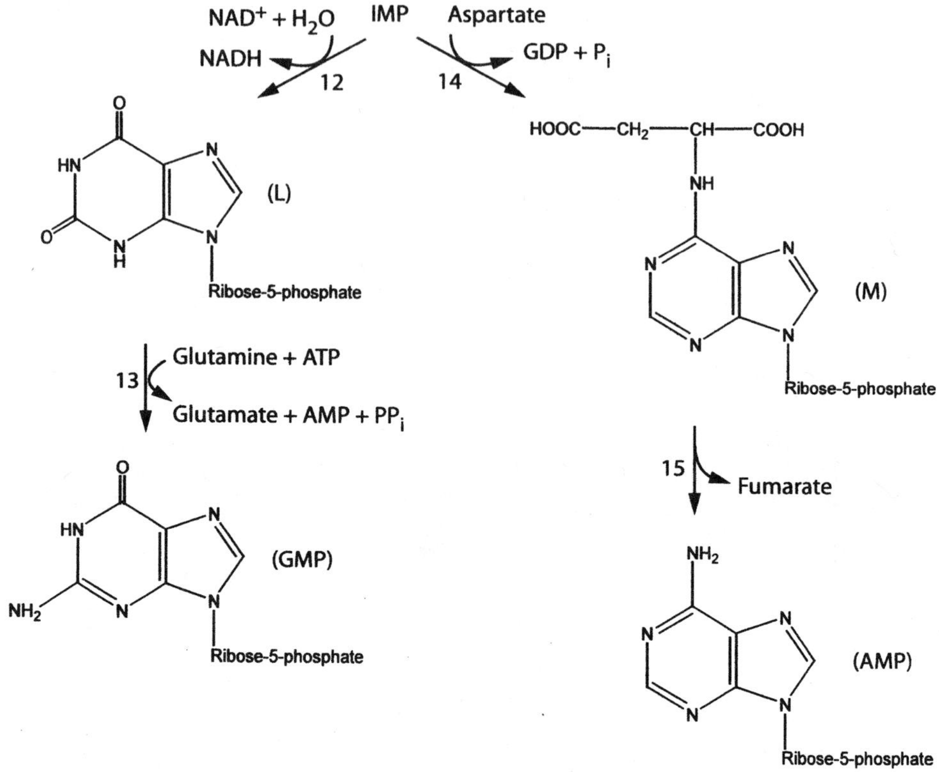

Figure 12.19. Synthesis of GMP and AMP from IMP. Structures are as follows: (**L**) xanthosine monophosphate (XMP) and (**M**) adenylosuccinate. Enzymes are as follows: (12) IMP dehydrogenase, (13) GMP synthetase, (14) adenylosuccinate synthetase, and (15) adenylosuccinate lyase.

of GMP from IMP and it is assumed to be negatively regulated by high concentrations of GMP; however, details remain to be established for this regulation.

## 12.4.3. Cross Talk Between Pathways

There are several instances in which intermediates or products of one pathway have an effect on another pathway. Clearly, this is seen in amphibolic systems where catabolism is balanced by biosynthesis. Another example is seen here where the synthesis of nucleotides is dependent on amino acid synthesis. As illustrated in Figure 12.20, there is a dependency on carbamoyl phosphate for both the synthesis of arginine and CTP. This regulation of carbamoyl phosphate synthetase is attributed to repression of enzyme formation as well as to changes in enzyme activity. If the UMP level in cells increases as a result of the acquisition of pyrimidines from the culture medium,

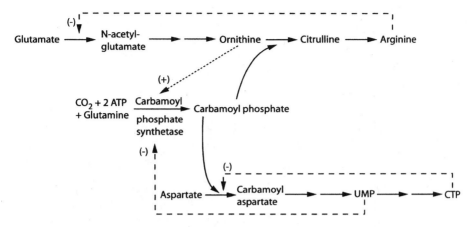

Figure 12.20. Interaction of pathways for biosynthesis of arginine and pyrimidines. *Dashed line* indicates negative (−) allosteric effector and *dotted line* indicates positive (+) allosteric effector.

carbamyl phosphate synthetase is inhibited. With a reduction of the intracellular pool of carbamoyl phosphate, ornithine will accumulate and activate carbamoyl phosphate synthetase. If arginine is acquired from the culture medium, the production of *N*-acetylglutamate is inhibited which results in a decline in the concentration of ornithine and in the activity of carbamoyl phosphate synthetase. Thus, this interfacing of activites of different metabolic pathways ensures the synthesis of CTP and arginine under different growth conditions.

## 12.4.4. Synthesis of Deoxyribonucleotides

The presence of deoxyribonucleotides in bacteria is paramount for the synthesis of DNA, and four different deoxynucleotides are required in relatively balanced quantities. In bacteria, the synthesis of deoxyribonucleotides is generally constructed from the corresponding diphosphate ribonucleotide and not the mono- or triphosphate nucleotide. The enzyme that replaces the 2' hydroxyl group on the ribose ring with a hydrogen atom is ribonucleotide reductase. The general scheme for synthesis of deoxyribonucleotides is given in Figure 12.21. Because of this preference for substrate by most bacterial and animal systems, the enzyme is referred to as ribonucleoside diphosphate reductase. In cyanobacteria and a few bacteria, the substrate for ribonucleotide reductase is the nucleoside triphosphate. To ensure a balance of DNA precursors, a highly complex regulatory process is present that relies on activation and inhibition of the ribonucleotide reductase. Because thymidine is present in DNA but not in RNA, thymidine serves as an allosteric effector in the synthesis of dATP, dGTP, and dCTP. The absence of thymidine nucleotides in RNA also is the reason that dTTP

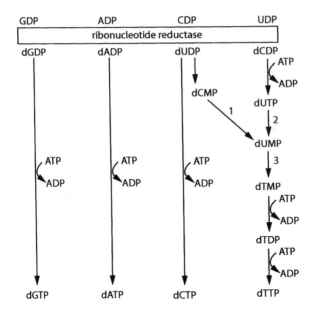

Figure 12.21. Synthesis of deoxyribonucleotides. Reaction 1 is a deamination system that converts cytosine to uracil. Reaction 2 is a phosphatase reaction that releases a pyrophosphate moiety. Reaction 3 is the reductive methylation reaction associated with the conversion of uracil to thiamine.

cannot be produced by a ribonucleotide reductase acting on a ribonucleotide precursor. The subsequent sections address the activities of ribonucleotide reductase and the synthesis of dTTP.

### 12.4.4.1. Nucleotide Reductase[12]

At this time, three different classes or molecular forms of ribonucleotide reductases are known to be produced by prokaryotes; each class is capable of using four different ribonucleotides to produce the needed four different deoxyribonucleotides. The type of metal cofactor involved in the reduction process is used to distinguish the different ribonucleotide reductases. Class I enzyme has two ferric ions as cofactor and this class is distributed throughout the eukaryotes. *Bacillus subtilis*, *S. typhimurium*, and aerobically grown *E. coli* produce class I ribonucleotide reductase. In the class II ribonucleotide reductase, the metal cofactor is the cobalt atom of adenosylcobalamin (coenzyme $B_{12}$). Class II ribonucleotide reductase is found in *Lactobacillus leichmannii*. The class III ribonucleotide reductase has an iron-sulfur cluster as a cofactor and is produced by *E. coli* under anaerobic culture.

The class I enzyme from *E. coli* is a $\alpha_2\beta_2$ tetramer in which the mass of the $\alpha$ polypeptide chain is 87 kDa and that of the $\beta$ polypeptide chain is 43 kDa. A model representing the two pairs of dissimilar proteins and the two ferrous ions is provided in Figure 12.22. Each half of this enzyme has two allosteric sites which are referred to as an "S" site and "A" site plus a catalytic site. Each catalytic site contains a tyrosine free radical moiety that extends from the $\beta$-subunit and two thiols that extend from the $\alpha$-subunit. To achieve a balance of deoxynucleotides, the enzyme uses two activity or A-sites and two specificity or S-sites. These four sites on the allosteric

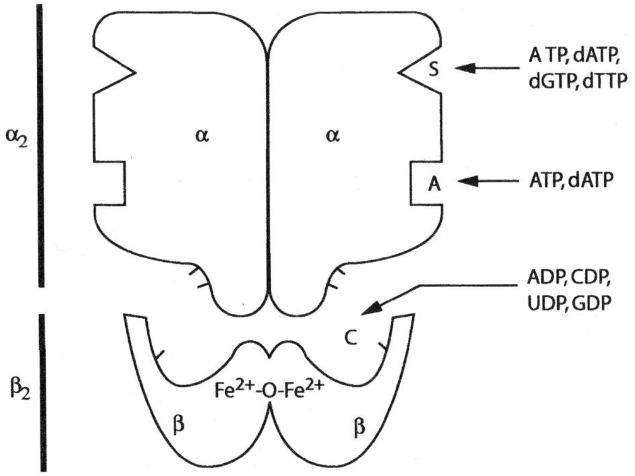

Figure 12.22. Model of ribonucleotide reductase from *E. coli* with distinct sites on the enzyme. The catalytic site is "C." Each β-subunit contains a tyrosine residue in the catalytic site and each α-subunit has a pair of thiols (-SH) in the catalytic site. "A" is the activity site and "S" is the specificity site.

protein influence the selection of substrate. The A-site weakly binds ATP and dATP while the S-site has a high affinity for dGTP, dTTP, dATP, or ATP. The reducing power to produce the 2' deoxy structure comes from NADPH. There are at least two mechanisms for transferring electrons from NADPH to ribonucleotide reductase in *E. coli*. The most commonly used system for electron transfer is accomplished by thioredoxin reductase, and a model of this electron transfer is given in Figure 12.23. Thioredoxin is a small protein of 12 kDa and it has a pair of redox active thiol groups in the following sequence: Cys-Gly-Pro-Cys. The reduced thiol groups interact with the oxidized thiol group in the catalytic site of the ribonucleotide reductase to participate in the reduction of the substrate. NADH is the source of the reducing power for thioredoxin and this action is mediated by thioredoxin reductase, a flavoprotein enzyme. The role of the two ferrous atoms in the ribonucleotide reductase is not resolved at this time.

In mutants of *E. coli* that do not produce thioredoxin, an alternate system for reduction of nucleotide reductase has been observed. A series of reactions involving glutaredoxin and glutathione relay reducing power from NADPH to ribonucleotide reductase. These reactions are as follows:

Glutathione$_{(ox)}$ + NADPH + H$^+$ → 2 glutathione$_{(red)}$ + NADP$^+$

Glutaredoxin$_{(ox)}$ + 2 glutathione$_{(red)}$ → glutaredoxin$_{(red)}$ + glutathione$_{(ox)}$

Ribonucleotide reductase$_{(ox)}$ + glutaredoxin$_{(red)}$ → ribonucleotide reductase$_{(red)}$ + glutaredoxin$_{(ox)}$

### 12.4.4.2. Biosynthesis of dTTP

Several enzyme systems can account for the production of dTTP. As illustrated in Figure 12.21, dTMP is produced from dUMP by thymidylate synthetase with reductive methylation of dUMP by $N^5,N^{10}$-methylenetetrahydrofolate. Subsequent production

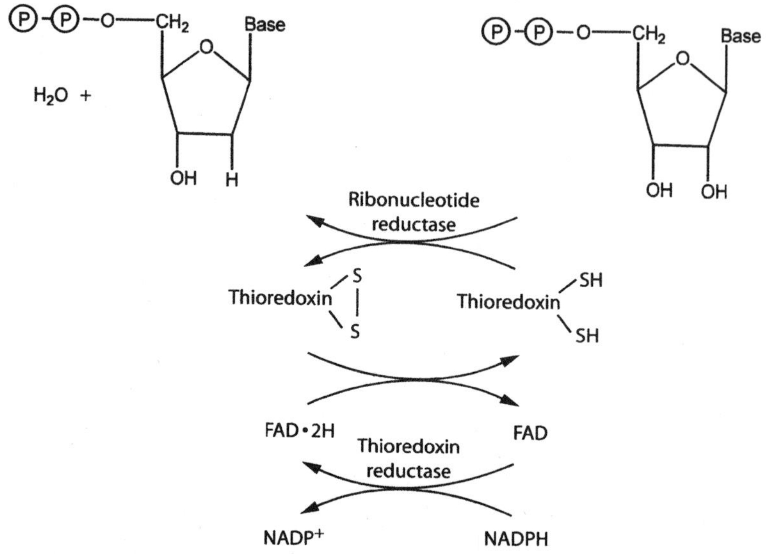

Figure 12.23. Source of electrons for ribonucleotide reductase.

of dTDP and dTTP from dTMP is accomplished by specific kinases and energized by ATP. In *E. coli*, the thymidylate synthase is regulated primarily by gene transcription and not allosteric effectors. This metabolic process ensures that dUTP does not accumulate in the cells, thereby preventing the accidental incorporation of dUTP into DNA.

## 12.5. Nucleotide-Activated Compounds

In several instances, nucleoside diphosphates (NDP) participate in biosynthetic activities through their association with sugars, lipids, or amino acids. These nucleotide compounds involve a variety of reactions that include activation of sugars, phospholipids, or glutamic acid; modification of hexoses, or transfer of a sugar moiety. Several examples of reactions in which NDP-sugars or lipids are employed by bacteria are given in Table 12.3. With substrates such as sugar phosphates, specific enzymes catalyze the reaction with NTP to produce a NDP-sugar and the release of an inorganic pyrophosphate ($PP_i$) moiety. This is the process that produces UDP-*N*-acetylglucosamine and UDP-*N*-acetylmuramic acid. Once these nucleotide sugars are formed, the sugar moieties are transferred to lipid moieties that move them through the membrane into the region of peptidoglycan synthesis. The activity of nucleotide diphosphate sugars in cell wall synthesis is covered in Chapter 3. In the synthesis of teichuronic acid, UDP-gluconic acid and GDP-aminomannuronic acid are required to provide the backbone structure of this polymer. A similar requirement for NDP-sugars is seen with the synthesis of lipopolysaccharide and several examples are given in Table 12.3.

Table 12.3. Biosynthetic systems using nucleotide-compounds.

| Nucleotide-sugar/lipid | Structure where sugar/lipid is deposited |
| --- | --- |
| UDP-*N*-acetylglucosamine | Peptidoglycan in the cell wall |
| UDP-*N*-acetylmuramic acid | Peptidoglycan in the cell wall |
| UDP-glucuronic acid | Teichuronic acid in the cell wall |
| GDP-aminomannuronic acid | Teichuronic acid in the cell wall |
| CMP-2-keto-3-deoxyoctonate | Lipopolysaccharide in the cell wall |
| TDP-rhamnose | Lipopolysaccharide in the cell wall and capsule |
| GDP-fucose | Lipopolysaccharide in the cell wall |
| CDP-abequose | Lipopolysaccharide in the cell wall |
| CDP-tyvelose | Lipopolysaccharide in the cell wall |
| CDP-diacylglycerol | Phospholipid in cell membranes |

Although UDP is the nucleotide carrier in peptidoglycan synthesis, the most commonly used nucleotides for biosynthesis of lipopolysaccharide are TDP, GDP, and CDP. The synthesis of phospholipids in most bacteria is attributed to CDP-diacylglycerol, which functions as an intermediate in the biosynthetic pathway for phospholipids.

The rationale for a specific type of nucleotide in construction of a specific sugar polymer is not clear at this time. A highly unusual situation is observed in archaea in which UDP is required for synthesis of a homopeptide of UDP-L-glutamate. After a polymer of glutamate is formed, this peptide of three or four amino acids is transferred into the cell wall region, where it is used to provide bonds to stabilize the wall structure. In several situations, NDP-hexoses are used for the transformation of sugars and reactions are catalyzed by epimerases, dehydrogenases, or reductases. Specific examples of situations in which modification of the sugar occurs only with the sugar-nucleotide and not with a free sugar are shown in Table 12.4. The reason that sugars or lipids are bonded to a nucleotide is not altogether understood but it may be that it is difficult to control bond linkage of a small molecule added onto a growing polymer. The presence of a nucleotide-sugar creates a bulky nucleotide that may be more readily bound into the enzyme. Furthermore, several enzymes have binding sites for nucleotides and through evolution systems have selected for this combination of NDP-compounds. Nucleotide-sugars are carriers for sugars in a manner roughly analogous to tRNA carrying amino acids. It should be emphasized that enzymes for catabolism of sugars or lipids do not require nucleotide carriers; therefore, a nucleotide added to a compound targets the molecule for biosynthesis.

Table 12.4. NDP-sugars for modification reactions.

| Enzyme-based activities | Reactions with substrates and products |
| --- | --- |
| Epimerase | UDP-glucose → UDP-galactose |
| Dehydrogenase | UDP-glucose → UDP glucuronic acid |
| Reductase | dTDP-glucose → dTDP-talose |
| Reductase | dTDP-glucose → dTDP-rhamnose |

## 12.6. Information Transfer

Replication proceeds in both direction along the circular chromosome with each parental strand of DNA serving as a template for synthesis of new DNA. The result is a semiconservative process containing one parental strand and one newly synthesized strand. If an error is made in the formation of a new strand of synthesized DNA, the physiological effect of the error is not observed until cell division has occurred and the chromosomes have been segregated. Because only one strand has the error, the corresponding gene in the complementary strand of DNA that is not damaged can provide the base sequence for appropriate protein synthesis. The requirement for cell division to occur before an error occurs on both strands of DNA and a mutation is produced is referred to as phenotypic lag.

Although it is commonly considered that bacteria have single circular chromosomes, some exceptions to this statement do occur and the extent of these exceptions can be determined from genome analysis. Linear chromosomes are present in *Streptomyces* (7.8 Mb) and in *Borrelia* (9.5 Mb). It will be interesting to learn if the replication of these linear chromosomes is similar to DNA replication in eukaryotic systems. Multiple chromosomes have been demonstrated in *Burkholder cepacia* which has three closed circle chromosomes of 3.6 Mb, 3.2 Mb, and 1.1 Mb. Preliminary analysis suggests that several other bacteria may have multiple chromosomes. The discussion here concerning DNA replication will center on bacteria that have a single chromosome.

### 12.6.1. DNA Replication[13]

The DNA macromolecule in bacteria is an extremely large complex structure that must be copied with great precision to produce an exact copy. The two fundamental types of DNA replication in bacteria are referred to as the "θ-type" and the "σ-type." Models of these two types of replication are given in Figure 12.24. The bidirectional replication of DNA from a single site on the chromosome is referred to as the θ-type because models of a replicating circle appear similar to the "θ" symbol. θ-type replication of chromosomal DNA consists of three major activities: initiation, formation of replication fork, and elongation. In contrast, rolling circle replication is designated as the σ-type and is characteristic of plasmids or bacteriophages, which are examples of small units of closed circle DNA.

#### 12.6.1.1. Initiation Site

With the circular bacterial chromosome, the first requirement in replication is to establish the site where DNA synthesis is initiated. Bacterial replication of the chromosome always starts at a single site which is called the origin of replication. In *E. coli* and *Bacillus*, the origin is the *oriC* site which consists of 260 bp. At this site several histone-like proteins shape the initiation complex and stabilize the DNA strands as relatively sharp bends are made in the molecule. DNA-associated proteins include HU, IHF (integration host factor), FIS (factor for inversion stimulation), and H-NS

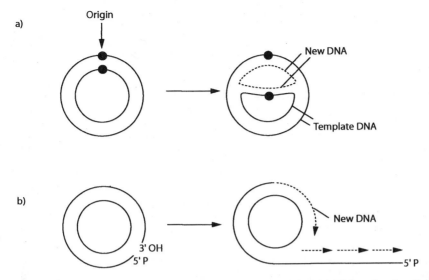

Figure 12.24. Models of two types of DNA replication in bacteria. (**a**) Bidirectional replication of circular bacterial chromosome indicating the apparent similarity of DNA replication to the Greek letter theta ($\theta$). Thus this is often referred to as $\theta$-replication. (**b**) Rolling circle replication of small circular plasmid DNA and this is often described as $\sigma$-replication. Diagrams are not to scale.

(histone-like nucleoid structuring protein). Addition of the DnaA protein to the *oriC* site, which contains an AT-rich region, leads to the opening of DNA and the histone-like proteins stabilize the DNA against possible breaks. These histone-like proteins are not unique to a given region but are distributed along the chromosome including the *oriC* site. The size and abundance of these histone-like proteins are given in Table 12.5.

## 12.6.1.2. Formation of the Replication Fork

For replication to occur, the double-stranded DNA helix must be unwound to expose a single strand of DNA. A model of replication is given in Figure 12.25. The formation of a replication fork is accomplished by a series of proteins that unwind the double helix as replication advances along the chromosome. The fork length in *E. coli* is about 4,200 kb and the movement of the fork is about 50,000 bp/min. Three separate proteins are required for the formation of the replication fork and they are as follows: (1) the DnaA protein, (2) the single stranded DNA-binding proteins, and (3) DNA helicase. Characteristics of these proteins for DNA replication are given in Table 12.5. The protein known as DnaA binds to double-stranded DNA with approximately 50 DnaA monomers binding to the replication origin and exposing a single strand of DNA. Single-stranded binding (SSB) proteins interact with DNA to stabilize the single strands of DNA and prevent reassociation of the helix. The SSB

Table 12.5. Proteins required for DNA replication in *Escherichia coli*.[a]

| Proteins | | Mass (kDa) | Native form | Molecules/cell (Division rate = 60 min) |
|---|---|---|---|---|
| **Initiation** | | | | |
| DnaA | | 52 | Monomer | 1,000 |
| HU | Histone-like proteins | 19 | α,β | 40,000 |
| IHF | Histone-like proteins | 21.8 | α,β | 3,000 |
| FIS | Histone-like proteins | 22.4 | Dimer | 30,000 |
| H-NS | Histone-like proteins | 15.5 | Monomer | 20,000 |
| SSB | | 75.6 | Quatramer | 800 |
| **Fork activities** | | | | |
| DnaA | | 52 | Monomer | 1,000 |
| SSB | | 75.6 | Quatramer | 800 |
| DNA gyrase | | 374 | $A_2 + B_2$ | 500 |
| RNA polymerase | | 450 | $\alpha_2\beta\beta'\sigma$ | ? |
| DnaA(θ)-type primosome | | | | |
| DnaB | $5' \rightarrow 3'$ helicase | 300 | Hexamer | 20 |
| DnaC | | 174 | Hexamer | 20 |
| DraG | primase for RNA synthesis | 60 | — | 50–100 |
| ΦX (σ)-type primosome | | | | |
| DnaB | $5' \rightarrow 3'$ helicase | 300 | Hexamer | 20 |
| DnaT | Primosome assembly | 58.5 | Trimer | — |
| PriA | $3' \rightarrow 5'$ helicase | 81.7 | Monomer | — |
| PriB | Primosome assembly | 22.8 | Dimer | — |
| PriC | Primosome assembly | 20.3 | Monomer | — |
| **Elongation of DNA** | | | | |
| DNA polymerase I | Replication and repair | 103 | Monomer | 300 |
| DNA polymerase II | Repair | 120 | — | 75 |
| DNA polymerase III | Replication | 918 | Multiunit | 10 |
| DNA ligase | Links DNA | 75 | Monomer | 3 |
| Topoisomerase I | Breaks ss-DNA | 100 | — | 1,000 |
| Topoisomerase II | Breaks ds-DNA | 374 | $A_2 + B_2$ | 500 |
| Topoisomerase III | Breaks ss-DNA | 74 | — | 10 |
| Topoisomerase IV | Breaks ds-DNA | 308 | $C_2 + E_2$ | 500 |

[a]From Linn, 1996,[13] Marians, 1996.[14]
See text for abbreviations.
Values are characteristic for *Escherichia coli* with cell division every 60 min.

protein is a tetramer of identical subunits that binds to a 32-base segment of DNA and there are about 200 SSB proteins at the replication fork. The SSB protein exhibits cooperative binding in that the addition of a binding protein to DNA becomes easier with the addition of each SSB protein. Not only do the binding proteins maintain a separation of the two DNA strands but also they protect single-stranded DNA from endonuclease attack. The DNA helicase binds to a single strand of DNA as it moves toward the replication fork with ATP providing the energy for the separation of the two DNA strands. The untwisting of DNA resulting from helicase action occurs at a rate of 50 turns/sec, which is equivalent to about 3,000 rpm.

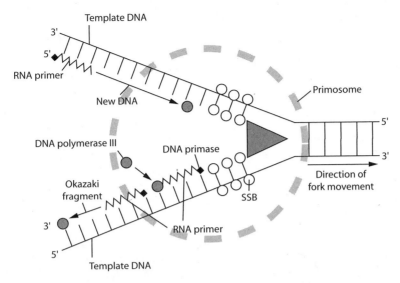

Figure 12.25. Model for DNA replication at the replication fork of bacteria. SSB, single-strand proteins that stabilize single strands of DNA through their binding activity.

The DNA polymerase makes copies along the single strand of parental DNA from the 3' → 5' direction. Because of the orientation of these strands, the DNA polymerase on one strand moves toward the replicating fork in a continuous manner and this is referred to as the leading strand. The other strand is oriented in the direction of 5' → 3' and because DNA replication is only from 3' → 5', only short intermittent replication of DNA occurs. This slow and discontinuous activity is associated with the lagging strand of DNA replication. To initiate DNA replication, a primosome structure is constructed on the single-stranded DNA. This promosome is frequently referred to as the DnaA-type because it employs the DnaA protein to displace the SSB proteins and insert the priming proteins of helicase (DnaB) and DnaC. These priming proteins enable the primase (DnaG) to add to the single-stranded DNA and synthesize RNA that is complementary to the 3' → 5' segment of the single-stranded DNA. The RNA segment synthesized in rapidly growing cells is relatively small in that it is only about two or three nucleotides in length and it is called the RNA primer. Only a single RNA sequence is needed to initiate DNA synthesis on the leading strand but numerous primer RNA sequences are established at regular intervals on the lagging strand.

## 12.6.1.3. Elongation of the Copied DNA

DNA polymerases contribute to elongation of strands by making additions to the RNA primer. DNA polymerase III uses a large enzyme of 918 kDa and is extremely complex in that it consists of 10 different subunits. The composition of DNA polymerase III is given in Table 12.6. As an enzyme with polymerase function, DNA polymerase III adds a deoxyribonucleotide to the 3'-hydroxyl moiety of the RNA primer. New strand growth is in the 5' → 3' direction which is antiparallel to template (parental)

Table 12.6. Subunit composition of DNA polymerase III.[a]

| Protein subunits | Mass (kDa) | Number of protein subunits/ DNA polymerase III |
|---|---|---|
| α | 129.9 | 2 |
| ε | 27.5 | 2 |
| θ | 8.6 | 2 |
| τ | 71.1 | 2 |
| γ | 47.5 | 2 |
| δ | 38.7 | 1 |
| δ' | 36.9 | 1 |
| χ | 16.6 | 1 |
| ψ | 15.2 | 1 |
| β | 40.4 | 4 |

[a]From Marians, 1996.[14]

strand which is read in the 3' → 5' direction. The substrates for DNA synthesis are 5'-deoxyribonucleoside triphosphates and pyrophosphate ($PP_i$) is released as the phosphodiesterase bond is formed. The lagging strand has short fragments of DNA that extend from the RNA primer until it is blocked by the next RNA primer. These short fragments are known as Okazaki fragments. Okazaki fragment synthesis starts every 1 to 2 sec and these Okazaki fragments are approximately 2 kb in length.

The RNA primer is removed by DNA polymerase I and DNA fills the gap. The DNA polymerase I not only polymerizes DNA in the 5' → 3' direction but it has exonuclease activity that is both 3' → 5' and 5' → 3'. Activity of the 5'→3' endonuclease is used to remove the RNA primer. The DNA polymerase I removes one RNA nucleotide and adds one DNA nucleotide until the entire RNA primer is replaced. The DNA polymerase I also has proofreading responsibilities in that it can remove from 1 to 10 nucleotides in the repair of damaged DNA. The frequency of errors in replicated DNA is only about $10^{-8}$, which is an indication of the highly efficient proofreading of DNA polymerase I and III. Finally, a phosphodiester bond is formed between the 5'-phosphate moiety produced by DNA polymerase III and the 3'-hydroxyl segment of DNA produced by DNA polymerase I. In the joining of DNA fragments by a DNA ligase energy for formation of a phosphodiesterase bond is derived from ATP, which is cleaved to produce AMP and $PP_i$. In bidirectional DNA replication, the two replicating forks encounter two termination (*ter*) sites halfway around the chromosome from the *oriC* site. The helicase (DnaB) stops and the newly synthesized DNA strands become joined to produce two covalently closed circular DNA molecules.

## 12.6.1.4. Supercoiling and Topoisomerases

In the bacterial cell, DNA is in the B-form where each turn of the right-handed double-stranded DNA is 10 base pairs and the helical pitch is 3.4 nm. In twisting and in unwinding of the chromosome, the DNA molecule is subjected to either negative or positive supercoiling. The native DNA helix turns in a clockwise direction which is opposite of the double-stranded turn, and this is an example of negative supercoiling. If the twisting occurs in the same direction of the clockwise turns of the double helix, this is an example of supercoiling. When about 5% of the DNA molecule is replicated, all negatively superhelicity is used and topological problems arise. The motion of the helicase promotes supertwisting of the double helix and bacteria use several enzymes to reduce the tension on the DNA molecule. These endonucleases are better known as topoisomerases and they function to produce temporary breaks in the DNA that remove the supercoiling of the helix structure. The topoisomerases are classified as type I if they hydrolyze and ligate only a single strand of the chromosome but they are type II if they produce transient breaks in both strands of DNA. Type I topoisomerases reduce the tension of negative supercoiling and are represented by enzymes of topoisomerase I and topoisomerase III. Type II topoisomerases relieve tension on both negative and positive supercoiling. Examples of type II enzymes are topoisomerase II and topoisomerase IV.

In bacteria, topoisomerase II is better known as the DNA gyrase. The prokaryotic topoisomerase II is the only enzyme that relaxes the supercoils on DNA to require ATP. Other topoisomerases use the energy released from hydrolysis of the phosphodiester bond to energize the reformation of the phosphodiester bond. The bacterial gyrase is the target of the quinolones, which are effective antibacterial agents. Naladixic acid and ciprofloxacin are two of the best known quinolones currently used to combat bacterial infections in humans.

### 12.6.1.5. Rolling Circle Replication

The mechanism for replicating of small closed double-stranded DNA is a variation of that observed for replicating bacterial chromosomes and is referred to as the "rolling circle" process. Rolling circle ($\sigma$-type) replication occurs in numerous double-stranded DNA systems in bacteria including the bacterial virus $\lambda$, the F-factor plasmid, the Ti-plasmid, and the Hfr-mediated conjugation. In $\sigma$-type replication, a primosome is used that is distinct from that found in chromosomal ($\theta$-type) replication. This primosome is often called the $\Phi$X-type because it is the type of replication that occurs in the bacterial virus $\Phi$X174 which has a single-stranded DNA viral chromosome. Proteins of the $\sigma$-type primosome are listed in Table 12.5. To initiate DNA replication, the double-stranded molecule from $\lambda$ is "nicked" by an endonuclease that breaks one strand of the DNA at the origin site. A replicating fork is created as the 5' end of the DNA is released and the 3' end of the single-stranded DNA remains in the circular DNA molecule. This single-stranded DNA segment in the circular viral genome functions like the leading strand template in chromosome replication. New DNA synthesized by the leading strand gives a continuous strand that restores the double-stranded character of the viral genome. As the 5' end of the single-stranded DNA is released, it is covered with SSB proteins and DNA synthesis proceeds in the direction of 3' $\rightarrow$ 5' from the replicating fork as is characteristic of a lagging strand template.

With replication of DNA from a small circular double-stranded DNA, a linear strand of DNA is produced without termination of replication. For the purpose of illustrating termination in the rolling circle process, the replication of DNA in $\lambda$ will be presented.

There is no termination of replication in the rolling circle process and the single strand of DNA continues to unroll from the circular double-stranded DNA. As the 5' end of the DNA lengthens, several copies of the λ chromosome are found in the linear array and this structure is called a concatamer. To form circular λ chromosomes, the concatamer is processed to release an appropriate DNA segment. An endonuclease produced from the *ter* (terminus-generating activity) gene recognizes a *cos* (cohesive ends) sequence that is at each end of the λ chromosome. The *cos* site is an 18-base sequence that produces a 12-base single-stranded end after endonuclease digestion. A circular λ chromosome is formed by ligation of the complementary ends of the 12-base single-strand ends.

## 12.6.2. RNA Synthesis

Bacteria produce several different types of RNA structures and these are classified as tRNA, rRNA, mRNA, and small RNAs. Some of characteristics of these RNA molecules are given in Table 12.7. While both RNA and DNA polymers consist of nucleotide monomers linked by phosphodiester bonds, RNA is distinguished from DNA in that it does not contain deoxyribose or thymine but instead has ribose and uracil. The presence of a hydroxyl at the 3-position of ribose accounts for considerable chemical instability of RNA as compared to DNA. Within the RNA group, mRNA persists in the bacterial cell for only 45 sec to 3 min while tRNA or rRNA are present in the cell throughout the cell lifetime. Numerous small RNA molecules are present in the cell and some of these are poorly characterized. As shown in Table 12.7, RNA synthesis does not require considerable energy but RNA plays a critical role in protein synthesis, which is the most costly energy requirement in the cell. This section briefly characterizes the RNA structures present in bacteria.

Table 12.7. Characteristics of RNA molecules in prokaryotes.

| Type | Sediment coefficient | Molecular weight ($\times 10^3$) | Number of nucleotides/molecule |
|------|---------------------|-----------------------------------|-------------------------------|
| **Large RNAs** | | | |
| mRNA | 6-50S | 25–1,000 | 100–10,000 |
| tRNA | 4S | 23–30 | 75–90 |
| rRNA | 5S | 48 | 120 |
|  | 16S | 616 | 1,540 |
|  | 23S | 1,200 | 3,000 |
| **Small RNAs** | | | |
| M1 | — | — | 377 |
| Primer RNA | — | — | 2–3 |
| 6S | 6S | — | 180 |
| 4.5S | 4.5S | — | 114[a] |

[a]Assumed to be important in cells because a cell contains 1,000 copies.

### 12.6.2.1. Synthesis of tRNA[15–17]

Although there are only 20 amino acids in proteins, there are 61 different codons for these 20 amino acids and each tRNA carries an anticodon for a specific amino acid. Each tRNA is produced by a specific gene and about 60 different tRNA molecules are found in a given bacterial species. Frequently, the tRNA genes are intermingled with rRNA genes. In *E. coli*, the locations of 79 genes for tRNA are found in more than 30 different operons. Some of these operons are multicistronic, consisting of genes in addition to tRNA while other operons contain only the tRNA genes. In *B. subtilis*, there are 62 tRNA genes with one group having 21 genes for tRNA. However, only 30 genes for tRNA production occur in *Mycoplasma capricolum*, and this economizing is a reflection of the diminished codon recognition requirement associated with a small genome size. Following transcription of precursor RNA, there is considerable processing to form the final tRNA molecule (see Figure 12.26). The final tRNA molecule is 75 to 90 nucleotides long and has a molecular weight of about 25,000. In processing, the short leader sequence of bases is removed at the 5' end of the precursor RNA by RNase P to produce a 5' monophosphate and not a 5' triphosphate end. Another endonuclease, RNase Q, removes unneeded bases at the 3' end and there are generally four unpaired nucleotides at this terminus where the amino acid is covalently bonded through the 3' -OH of the tRNA structure. A model of the typical tRNA in bacteria is given in Figure 12.27. A distinctive feature of the tRNA is the modified nucleosides present and these are summarized in Table 12.8 for all the tRNAs present in *E. coli* and *S. typhimurium*.

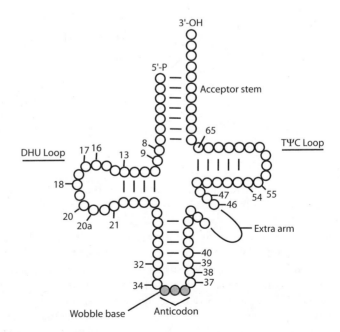

Figure 12.26. A model of the "L-form" of tRNA derived from X-ray analysis of tRNA structures.

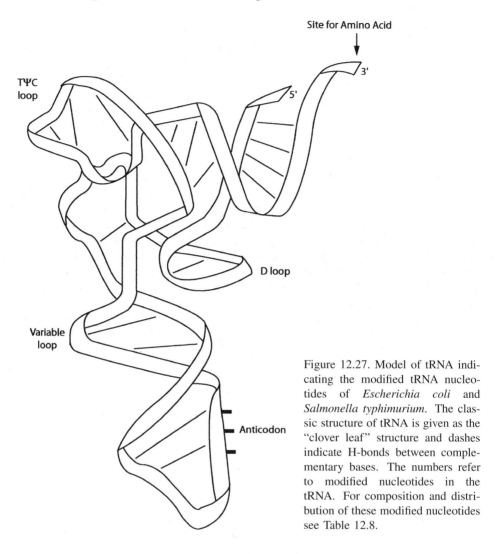

Figure 12.27. Model of tRNA indicating the modified tRNA nucleotides of *Escherichia coli* and *Salmonella typhimurium*. The classic structure of tRNA is given as the "clover leaf" structure and dashes indicate H-bonds between complementary bases. The numbers refer to modified nucleotides in the tRNA. For composition and distribution of these modified nucleotides see Table 12.8.

A common triplicate of nucleotides at the 3' end of the tRNA is CCA and these are part of the transcribed information and not added by posttranscription processes. In model presentations, the tRNA molecule is presented as a clover leaf structure; however, X-ray analysis has indicated that the molecule assumes a tertiary structure that is more of an L-form in which the loops of the tRNA molecule are less exposed. The typical bacterial tRNA structure has four regions where hydrogen bonding occurs between complementary bases and there are four loop structures: loop I is also known as the DHU arm, loop II is the anticodon region, loop III is a short segment that is present in several bacterial tRNA, and loop IV contains several unusual bases and is also known as the TψC arm. The portion of the tRNA structure that holds the amino

Table 12.8. Modified nucleosides in tRNA of *Escherichia coli*.[a]

| Nucleoside | Location in tRNA | Number of tRNAs with modified nucleosides | Modified nucleosides |
|---|---|---|---|
| 8 | Between acceptor stem | 26 | $S^4U$ |
| 9 | and DHU loop | 2 | $S^4U$ |
| 13 | DHU loop | 1 | $\Psi$ (Pseudouridine) |
| 16 | DHU loop | 14 | D |
| 17 | DHU loop | 16 | D |
| 18 | DHU loop | 12 | Gm |
| 20 | DHU loop | 26 | D |
| 20a | DHU loop | 14 | D |
| 21 | DHU loop | 1 | D |
| 32 | Anticodon loop | 16[b] | 4 Types[b] |
| 34 | Wobble base | 29[c] | 9 Types[c] |
| 37 | Anticodon loop | 41[d] | 8 Types[d] |
| 38 | Anticodon loop | 5 | 4 |
| 39 | Anticodon loop | 11 | $\Psi$ |
| 40 | Anticodon loop | 3 | $\Psi$ |
| 46 | Extra arm | 22 | $m^7G$ |
| 47 | Extra arm | 8 | $acp^3U$ |
| 54 | T$\Psi$C loop | 40 | $m^5U$ |
| 55 | T$\Psi$C loop | 40 | $\Psi$ |
| 65 | T$\Psi$C loop | 2 | $\Psi$ |

[a]From Björk, 1996.[15]

[b]Number and name as follows: 3 and $\Psi$; 4 and Um; 5 and Cm; 4 and $s^2C$.

[c]Number and name as follows: 1 and U; 4 and $cmo^5U$; 1 and $mnm^5U$; 1 and $mnm^5Um$; 3 and $mn^5s^2U$; 1 and $ac^4c$; 1 and I (inosine); 5 and Q (queuosine); 12 and G.

[d]Number and name as follows: 6 and $m^2A$; 1 and $m^6A$; 1 and $i^6A$; 8 and $ms^2i^6A$; 7 and $m^1G$; 7 and $t^6A$; 1 $m^6t^6A$; 10 and A.

acid at the 3' end is often referred to as the acceptor stem. Enzymatic processing of bases at loops I and IV, serve to produce unique nucleotides and these provide the molecule with distinctive physiological activities of unknown function. Most tRNA modifying enzymes are position specific. About 45 tRNA modifying enzymes are needed for tRNA in *E. coli* and this consumes about 0.25% of the genomic DNA. In a comparative sense, it can be stated that about four times the amount of genes is required to produce modifying enzymes than to synthesize the tRNAs. Some scientists have suggested that the unique bases may assist in stabilizing the tRNA into the ribosome while others have posed the possibility that they may serve to protect the single-stranded RNA loops against endonuclease attack. In addition, the role of these unusual bases in evolutionary development is extremely curious. The three nucleotides in the anticodon segment recognize the triplet in the mRNA. The two bases in the anticodon from the 3' end are essential for an appropriate match to the codon. There is some variation in recognition of the base that is on the 5' side of the anticodon triplicate and this variable recognition of the third base of the codon–anticodon is referred to as the wobble process.

### 12.6.2.2. Synthesis of rRNA[18]

The prokaryotic ribosome consists of a 30S subunit that consists of proteins plus a 16S rRNA molecule while the 50S subunit contains proteins plus a 23S rRNA and a 5S rRNA molecule. In *E. coli* and perhaps other bacteria, the genes for 5S rRNA, 16S rRNA, and 23S rRNA are on a single transcription unit which ensures proper stoichiometry for ribosome assembly. Bacteria generally have several of these gene sets for rRNA synthesis, and *E. coli* has seven such regions. A model indicating transcription of rRNA in bacteria is given in Figure 12.28. The long segment of transcribed RNA is referred to as the precursor rRNA or the p30S RNA and it is cleaved by RNase III into units of approximately 17S rRNA, approximately 27S rRNA, and 5⁺S rRNA. Through the action of specific enzymes, the precursor molecules are further processed into the individual (mature) 16S rRNA, 23S rRNA, and 5S rRNA molecules. Several modifications are made to the rRNA molecules to produce ribosomes that have full translation function. At the decoding region of the 16S rRNA, one pseudouridine nucleotide and 10 methylated nucleotides are formed. In the 23S rRNA, there are 20 modifications at the peptidyltransferase region and three additional modifications are present elsewhere in the molecule. To accomplish these

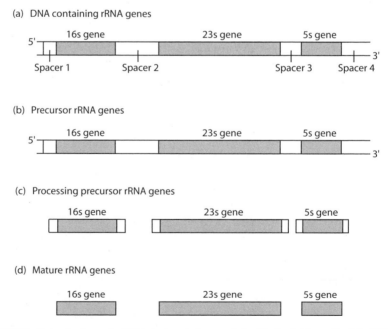

Figure 12.28. Organization of rRNA transcription unit in *Escherichia coli*. (**a**) tRNA genes may be present in the region of spacer 2 and spacer 4 but are not included in the model showing processing activity. (**b**) A single precursor rRNA molecule physically includes the three rRNAs (16S, 23S, and 5S). (**c**) Precursor rRNA is initially processed by RNase III. (**d**) Secondary processing results in production of native rRNAs.

modifications of rRNA, organisms such as *E. coli* have 36 rRNA modifying genes which account for 36 kb or about 0.8% of genetic information in *E. coli*. Methylation of 16S rRNA is an early transcription event while methylation of 23S rRNA is late and occurs at the time when the 50S subunit is assembled.

### 12.6.2.3. Synthesis of mRNA[19]

In the synthesis of proteins, mRNA functions as a transient template that is transcribed from DNA. There are three phases of mRNA synthesis: initiation, elongation, and termination. The production of mRNA requires that the DNA-dependent RNA polymerase (hereafter called RNA polymerase) recognize specific sites in the promoter section. Upstream of the site for initiation are two segments in the promoter that are at positions −30 and −10 bp. In vegetative cells, the −10 segment has the characteristic sequence of TATAAT and the −35 sequence is TTGACA. The −10 segment is frequently referred to as the Pribnow box, in honor of the scientist who made the discovery. These two sites with conserved nucleotides serve to bind the core RNA polymerase (see Figure 12.29). In some systems, an AT-rich segment at the −43 region serves to initially bind RNA polymerase and facilitate the matching of the enzyme. Promoters that have an AT-rich segment at −43 are considered to have special properties of initiating transcription and are described as having strong promoter activity. After the core RNA polymerase is attached to the promoter, a protein known as the sigma (σ) factor is added to produce a functional RNA polymerase. Transcription is initiated by reading on the 3' end of the single strand of DNA at the +1 nucleotide. Elongation of the mRNA proceeds along the single-stranded DNA template, and after 10 to 12 nucleotides, the σ-factor dissociates from the RNA polymerase. The core RNA polymerase continues transcription of DNA using a motion that is subject to frequent interruptions. Pauses are most apparent when hairpin-like structures are formed by linear RNA binding with itself.

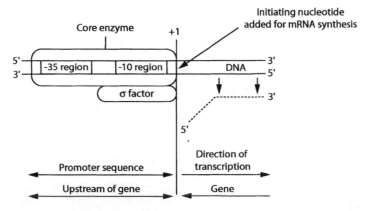

Figure 12.29. Initiation of transcription with assembly of RNA polymerase on bacterial DNA. RNA polymerase consists of the core enzyme and the σ-factor.

A number of special features contribute to termination of mRNA synthesis. One termination signal is attributed to an intrinsic termination sequence that produces an elongated pause in the movement of RNA polymerase. At the termination site there is a GC-rich segment that produces a stem-loop form of RNA that causes the RNA polymerase to stall and terminate the synthesis of mRNA. A second termination mechanism is attributed to the action of a specific protein known as Rho which binds to RNA and stops movement of RNA polymerase. Rho is a hexamer consisting of 46-kDa monomers and was discovered by Jeffery Roberts in 1969. Rho has a preference for C-rich regions of RNA and it is proposed that when it binds, it hydrolyzes ribonucleoside triphosphates to nucleoside diphosphates. Also, Rho has the capacity to unwind the RNA–DNA complex. When Rho has moved along the RNA by either of the two mechanisms, it promotes the release of RNA polymerase from DNA. Unlike tRNA and rRNA, mRNA is relatively unstable and is readily degraded. In *E. coli*, the principal enzyme involved in degradation of mRNA is RNase E; however, other endonucleases may also have a minor role in release of nucleotides from partially degraded mRNA.

### 12.6.2.4. Enzymes of RNA Synthesis[20,21]

*RNase P*

One of the highly unusual enzymes produced by bacteria is RNase P, which is often referred to as a ribozyme because it is a ribonucleoprotein. In *B. subtilis*, this enzyme contains 401 nucleotides while in *E. coli* the enzyme contains 377 nucleotides. In *E. coli*, the nucleotide segment is referred to as the M1 RNA and the protein segment is 18.5 kDa. Both the protein and M1 RNA of the nucleoprotein complex are involved in enzymatic removal of nucleotide sequences at the 5' end of precursor tRNA. If either the RNA or protein segment of RNase P is subjected to degradation, there is a loss of enzymatic activity. These enzymes are highly efficient because there are only about 20 to 50 copies in each cell.

*Aminoacyl-tRNA Synthetases*

Amino acid addition onto tRNA is a highly specific process that requires a separate aminoacyl-tRNA synthetase for each of the amino acids. In *E. coli*, there are 21 tRNA synthetase enzymes with one for each amino acid, except lysine which has two. There appears to be one enzyme for each 20 ribosomes and in growing cells there may be 800 molecules of each of the 21 tRNAs. The addition of an amino acid to a cognate tRNA requires energy according to the following reactions:

$$\text{Amino acid} + \text{ATP} \rightarrow \text{AMP-amino acid} + \text{PP}_i$$
$$\text{AMP-amino acid} + \text{tRNA} \rightarrow \text{tRNA-amino acid} + \text{AMP}$$

There are two domains for each enzyme and generally there is only one gene for each domain of tRNA synthetase. For addition of an amino acid onto the tRNA, one domain of the enzyme recognizes the anticodon loop of the tRNA molecule and the other domain, which contains the catalytic region, adds the amino acid onto the 3' end of the tRNA. An interesting variation of adding an amino acid to tRNA is found in several different physiological groups of bacteria in which the amino acid is modified after it is added to the tRNA. Addition of an amino acid onto the tRNA is referred to as "charging" of the tRNA molecule. In charging of the tRNA with glutamine, it has been determined that some archaea, cyanobacteria, *Bacillus*, and *Lactobacillus* do not add glutamine directly to tRNA but an amidotransferase is used to convert glutamyl-tRNA to glutaminyl-tRNA$^{Gln}$.

*RNA Polymerase*

Three classes of RNA are produced from a precursor-RNA molecule that is transcribed from DNA. Through chemical modification, precursor-RNA is processed to produce rRNA and tRNA structures. As illustrated in Figure 12.28, the operons in *E. coli* for precursor-RNA contains genes for 5S rRNA (*rrf*), 16S rRNA (*rrs*), 23S rRNA (*rrf*), and tRNA. Spacer segments are present between each RNA segment on the precursor structure, and RNase III hydrolyzes RNA to release individual rRNA segments and tRNA. Whereas eukaryotic organisms have multiple RNA polymerases, bacteria have a single enzyme that transcribes rRNA, tRNA, and mRNA.

The bacterial RNA polymerase is a large molecule with a mass of about 500 kDa and five polypeptide subunits. In the polymerase there is a single polypeptide chain, known as "β'," that has a basic charge, a 155-kDa mass, and appears to be important in binding to DNA. A second polypeptide is designated as "β," which is an acidic protein of 151 kDa mass and contains the active site of the enzyme. The RNA polymerase also contains two molecules of an α polypeptide with each molecule having a mass of 36.5 kDa. The $\alpha_2$ component appears to be the site for activator interaction involving *c*atabolite *a*ctivator *p*rotein (CAP). The fifth polypeptide is the sigma (σ) factor which functions to recognize the promoter and initiate RNA synthesis.

The fully assembled RNA polymerase with the five polypeptide subunits is referred to as the holoenzyme but the $\alpha_2\beta\beta'$ structure without the σ-factor is designated as the core polymerase. To initiate transcription, the holoenzyme associates with the double-stranded DNA at the −35 region upstream of the gene that is to be transcribed. Initially, this binding is relatively weak but is strengthened by secondary binding at the −10 segment. To expose the −10 region, the DNA must be untwisted and the AT-rich segment at the −10 region is important because AT base-pair binding has only two H-bonding interactions whereas GC has three H-bonding sets. With the RNA polymerase properly oriented on the DNA, the σ-factor is added and with appropriate recognition of the promoter sequence, RNA synthesis is initiated. After RNA synthesis proceeds for about 12 nucleotides, the sigma factor is dissociated and the holoenzyme proceeds along the DNA in synthesis of RNA.

The holoenzyme may recognize several different sigma factors and the sigma factor selected will determine which genes are to be transcribed. Each sigma factor has a specific −35 sequence and −10 sequence on the DNA that it recognizes. The sigma factor used by vegetative cells for growth (also referred to as housekeeping duties) is $\sigma^{70}$, which recognizes the −35 sequence of TTGACA and the −10 sequence of TA-TAAT. As summarized in Table 12.9, there are several other sigma factors in bacteria and each has a specific function. Presented in Chapter 7 is the information pertaining

Table 12.9. Consensus sequences for *Escherichia coli* sigma factors.

| Sigma factor | Genes transcribed | −35 Consensus sequence | −10 Consensus sequence |
|---|---|---|---|
| $\sigma^{70}$ | Major factor for housekeeping activities in growing cells | TTGACA | TATAAT |
| $\sigma^{54}$ | Nitrogen metabolism, formate metabolism | ——CTGGCAC | TTGCA |
| $\sigma^{32}$ | Used to express heat shock genes | CCCTTGAA | CCCCATTTA |
| $\sigma^{28}$ | Chemotaxis, motility, flagellin genes | ——TAAA | GCCGATAA |
| $\sigma^{24}$ | Cytoplasmic stress | GAACTT | TCTCA |
| $\sigma^{38}$ | Major factor in stationary phase, some important for oxidative and osmotic stress, and virulence gene expression | — | — |
| $\sigma^{19}$ | Iron citrate transport | — | — |

to spore-forming *Bacillus*, which have several different sigma factors to regulate enzymes associated with different stages of sporulation.

There is no standard terminology to describe the different sigma factors produced by bacteria, and scientists working on different organisms tend to have their own designations. In *E. coli*, the superscript used with the sigma symbol reflects the molecular mass of the protein. For example, sigma factors with a molecular mass of 70 kDa are indicated as $\sigma^{70}$ and those with a mass of 54 kDa are designated as $\sigma^{54}$. In the case of *Bacillus* and other bacteria, the sigma factors are identified by letters in the alphabet. Sigma factors associated with sporulation in fruiting body bacteria (*Myxococcus xanthus* and *Stigmatella aurantiaca*) are designated as $\sigma^{B}$ and $\sigma^{C}$. In *Ralstonia eutrophus*, the sigma factor $\sigma^{X}$ is related to the sequence $\sigma^{19}$ of *E. coli*, and $\sigma^{X}$ is important for cobalt and nickel resistance.

## 12.7. Ribosomal Mediated Process[22–24]

Rapidly growing cells may have 15,000 ribosomes per cell and this may account for as much as 25% of cell dry weight. The ribosome composition is remarkably constant throughout the bacteria and the ribosomal subunits from *E. coli* serve as useful models for all bacteria. The abundance of RNA and protein in the ribosome is given in Table 12.10. The 30S and 50S subunits are assembled in the cytoplasm and are condensed to produce a 70S ribosome on the mRNA. On completion of translation, the ribosomal subunits are released from the mRNA and return to the cytoplasm. The subunits are highly resistant to enzyme attack and as a result have a relatively long lifetime in the cell. This section addresses the structure and function of bacterial ribosomes.

Table 12.10. Characteristics of *Escherichia coli* ribosomes.

| Characteristic | Small subunit | Large subunit |
|---|---|---|
| Sedimentation coefficient | 30S | 50S |
| Mass (RNA + protein) | 930 kDa | 1,590 kDa |
| Mass (RNA) | 560 kDa | 1,104 kDa |
| Mass (protein) | 370 kDa | 487 kDa |
| rRNA | 16S = 1542 bases | 23S = 2,904 bases |
|  |  | 5S = 120 bases |
| Number of different proteins present | 21 proteins | 31 proteins |
| RNA (fraction of ribosome) | 60% | 70% |
| Protein (fraction of ribosome) | 40% | 30% |

## 12.7.1. Ribosome Architecture

Fifty-two different proteins are associated with bacterial ribosomes: 21 are with the 30S subunit and 31 are with the 50S subunit. The standard terminology is to designate proteins from the small (30S) subunit with the prefix "S" and proteins associated with the large (50S) with the prefix "L." The proteins with the ribosomes are generally small, 29.4 to 5.3 kDa in the large subunit and 61.1 to 8.3 kDa in the small subunit. It should be noted that in the small subunit there is only one protein larger than 26.6 kDa. The ribosomal proteins have various activities in assembling the ribosome structure and in contributing to ribosomal function. The structure and function of the ribosomal proteins have been described and some of these are given in Table 12.11. In *E. coli*, genes for the 52 proteins are in 20 discrete operons containing from 1 to 11 genes and 31 genes are in a single cluster at 73' on the chromosome.

The architecture of the *E. coli* ribosome has received considerable examination and each subunit contains some discrete areas for recognition. The compact structure of the ribosomal structure is a result of tertiary interactions between rRNA and specific proteins. The rRNA structure may contain helical domains but also has interior loops, hairpin loops, and tightly twisted forms that serve as a scaffold to compress the ribosome structure. Proteins interact with specific sites on the rRNA to produce the two subunits of 30S and 50S. The formation of the 30S and the 50S ribosomal subunits is entirely dependent on self-assembly. As indicated in Figure 12.30, there are relatively few proteins bound directly into the 16S RNA and the addition of the remaining proteins into the ribosome follows a precise sequence to obtain the correct binding. The interaction of proteins making up the 50S ribosomal subunit are also known and systems similar to that shown in Figure 12.30 have been constructed. An interesting feature in the 50S protein configuration is that extensive interaction between certain proteins of the ribosome. For example, L-15 interacts with 15 proteins, L-17 interacts with 13 proteins, L-5 interacts with 7 proteins, and L-22 interacts with 6 proteins.

Table 12.11. Structure and function of some of the ribosomal proteins in *Escherichia coli*.[a]

| Protein | Size (kDa) | Class of structure | Presumed function |
|---|---|---|---|
| S-1 | 61.1 | OB fold | mRNA binding |
| S-4[b] | 23.1 | Helical | Essential for 30S folding |
| S-5 | 17.1 | Left hand β-α-β | mRNA binding and tRNA binding |
| S-7[b] | 17.9 | Helical | Essential for 30S folding |
| S-8[b] | 13.9 | RRM | Essential for 30S folding |
| S-12 | 13.6 | ? | tRNA binding and streptomycin binding |
| S-15[b] | 10.0 | Helical | Essential for 30S folding |
| S-17[b] | 9.5 | OB fold | Essential for 30S folding |
| L-1[c] | 24.6 | RRM and Rossmann | ? |
| L-2[c] | 29.4 | OB fold and SH3-like barrel | Peptidyl transfer |
| L-5[d] | 20.1 | ? | ? |
| L-6[c] | 18.8 | RRM | Peptidyl transfer |
| L-9[c] | 15.4 | RRM | ? |
| L-12 | 12.1 | RRM and four helix bundle | Binding protein factors, translocation |
| L-14 | 13.5 | β-Barrel | ? |
| L-18[d] | 12.7 | ? | ? |
| L-22 | 12.2 | RRM | ? |
| L-25[d] | 10.7 | β-Barrel | Peptidyl transferase (?) |
| L-27 | 8.9 | ? | Part of P-site, peptidyl transfer |

[a]Noller and Nomura, 1996[24]; Mathews and van Holde, 1990[23]; Al-Karadaghi et al., 2000.[22]
[b]Binds to 16S rRNA.
[c]Binds to 23S rRNA.
[d]Binds to 5S rRNA.
OB, Oligonucleotide-oligosaccharide binding; RRM, hinge connects C-terminus domain to the N-terminus domain; SH3, short $3_{10}$ helix; Rossmann, nucleotide binding fold.

## 12.7.2. Protein Synthesis[25–27]

Translation of mRNA to protein is essential for cells to generate proteins required for catalytic processes, regulation of transcription, and stabilization of cellular structures. Although the general theme of ribosomal-mediated protein synthesis is similar for prokaryotes and eukaryotes, the two systems are not interchangeable but have significant structural differences. This discussion focuses on the process of protein synthesis in bacteria and not in the differences between bacteria and eukaryotes. The translation process can be divided into three phases: initiation, elongation, and termination.

### 12.7.2.1. Initiation Phase

The initial phase is concerned with the assembly of the numerous factors needed for transformation. The numerous factors needed for transformation are assembled on the 30S ribosomal subunit and this is followed by the binding of the ribosome onto the mRNA. These activities of the initiation phase of translation are depicted in Fig-

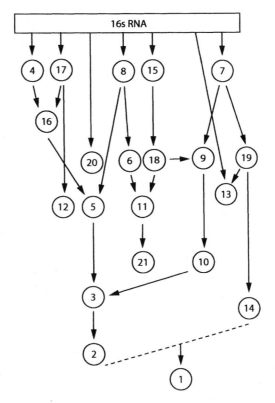

Figure 12.30. Assembly of 30S ribosome
of *Escherichia coli*.

ure 12.31. Initiation factors have been isolated from *E. coli* and have been extensively characterized. IF-1 is a small protein of 8.1 kDa and it stimulates the activity
of IF-2 and IF-3. There are two forms of the IF-2 forms, and they both promote the
binding of fMet-tRNA. In addition, the two IF-2 forms have GTPase activity. This
similarity in activity is noteworthy because IF-2$\alpha$ is 97.3 kDa and IF-2$\beta$ is 79.7 kDa.
IF-3 has a mass of 21.6 kDa and is important in binding of mRNA to the 30S ribosomal subunit.

*Step 1*

In the cytosol of bacteria are the components needed to initiate the assembly of the
ribosome. These components include the following: the 30S ribosomal subunit, the
50S ribosomal subunit, initiation factors (IF) IF-1, IF-2, IF-3, and GTPase. The three
IFs and the GTPase are bound onto the 30S ribosomal subunit before the 30S subunit
becomes associated with the mRNA.

*Step 2*

Key for ribosomal assembly is the Shine–Dalgarno sequence of nucleotide bases of
5'-UAAGGAGG-3' near the 5' end of the mRNA. This Shine–Dalgarno sequence is

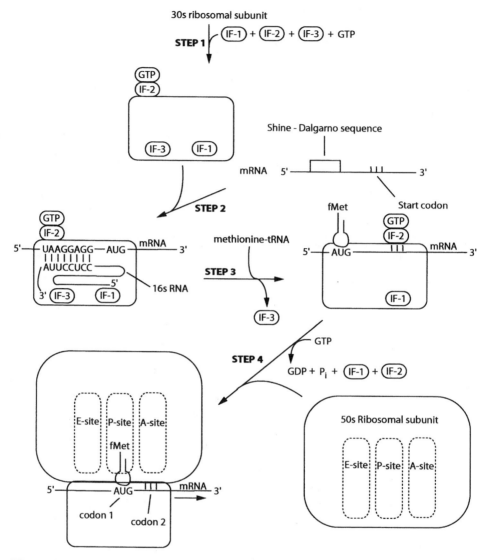

Figure 12.31. Model summarizing the initiation of protein synthesis in bacteria. See text for discussion.

6 to 10 bases upstream of the start codon of AUG. While AUG is the most common codon for initiation, GUG is used about 10% of the time while UUG and AUU are used with less frequency. Near the 3' end of the 16S rRNA of the 30S ribosomal subunit is a region that is complementary to the Shine–Dalgarno sequence. This matching of RNA bases between mRNA and rRNA provides appropriate positioning of the ribosome on the mRNA.

*Step 3*

The third step is the addition of methionyl-tRNA^Met onto the 30S ribosomal subunit and it is facilitated by IF-2. This methionyl-tRNA^Met carries the appropriate anticodon to AUG and interfaces with the mRNA. With the methionyl-tRNA^Met in position on the 30S–mRNA complex, a formyl group is added onto methionine. The enzyme transformylase transfers the C-1 unit from $N^{10}$-formyltetrahydrofolate to methionyl-tRNA^Met to produce formylmethionyl-tRNA^Met which is written as f-Met-tRNA^Met. A final activity in this third step is the release of IF-3 from the 30S ribosomal subunit to provide space for the addition of the 50S ribosomal subunit.

*Step 4*

In the final activity of ribosome assembly, the 50S ribosomal subunit is added onto the 30S subunit–mRNA complex. There are two sites on the 50S ribosomal subunit with f-Met-tRNA^Met at the P site and both the A site and E site are empty. To accommodate the assembly of the 70S ribosome, IF-1 and IF-2 are released. Energy for this assembly is from the hydrolysis of GTP.

### 12.7.2.2. Elongation Phase

The addition of amino acids onto the formylmethionine is accomplished by amino acids supplied from the binding of amino acyl-tRNAs into the A site of the 50S ribosomal subunit. This succession of amino acyl-tRNAs is dictated by codons on the mRNA. To facilitate the addition of amino acyl-tRNA, several elongation proteins participate in three discrete steps as outlined in Figure 12.32. One protein associated with elongation of the protein is unstable and therefore has been named EF-Tu. In rapidly growing bacteria, there are about seven times the number of EF-Tu molecules as there are ribosomes, and this accounts for about 5% of the soluble cell protein. The mass of the EF-Tu protein is about 43 kDa, it displays GTPase, and functions to bind the amino acyl-tRNA to the A site. The EF-Ts is an elongation factor of about 30 kDa and is relatively stable as compared to EF-Tu. The function of EF-Ts is to promote an exchange of GDP/GTP on the EF-Tu. With respect to abundance in cells, there appears to be about one EF-Ts molecule for every ribosome. EF-G is an elongation factor that is about 77 kDa, binds to the ribosome, and displays GTPase activity.

*Step 1*

The first event in the elongation of the peptide is the addition of the second amino acyl-tRNA. Specificity for this addition is the match of the anticodon on the tRNA with the codon on the mRNA. To facilitate the addition of the amino acyl-tRNA, several elongation factors (EF) are required. EF-Tu and GTP form a complex with the amino acyl-tRNA and the amino acyl-tRNA is added at the A site. EF-Tu is released and GTP is hydrolyzed to GDP + $P_i$. To regenerate the EF-Tu-GTP complex, another protein component, EF-Ts, is used to displace GDP and the result is the formation of EF-Tu–EF-Ts. The final activity in energizing of the elongation factors is replacement of EF-Ts by GTP and the regeneration of the EF–Tu–GTP complex.

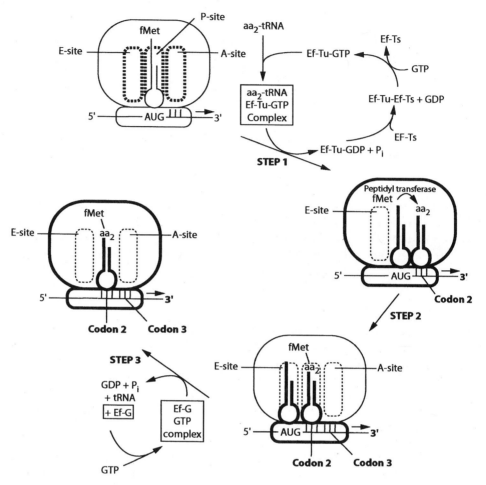

Figure 12.32. Model reflecting the elongation of peptide in bacterial translation. See text for description.

*Step 2*

A relatively short direct action occurs with the transfer of amino acid from the P-site to the A-site. Peptidyltransferase accomplishes the elongation of the peptide by one amino acid. The tRNA with two amino acids moves to the P site. The uncharged tRNA$^{\text{Met}}$ is moved to the exit (E) site.

*Step 3*

The final step in this elongation sequence is the release of the unloaded tRNA from the E site and the movement of the peptidyl-tRNA from the A site to the P site. For this step, it can be considered that the ribosome moves one codon toward the 3' end of the mRNA. This translocation of the ribosome requires energy in the form of GTP and an additional elongation factor, EF-G, is required to accommodate the interfacing of GTP with the 50S subunit. With the peptidyl-tRNA occupying the P site, the A site is vacant and a new amino acyl-tRNA can be accepted. Step 1 is reinitiated and subsequent activities of steps 2 and 3 occur with the addition of a second amino acid to formylmethionine. Thus, the elongation cycle continues, adding to the growing peptide chain one amino acid at a time until there is a termination of the message.

### 12.7.2.3. Termination Phase

The termination of the translation of the mRNA with the release of the protein from the ribosome proceeds by specific processes. Termination of polypeptide elongation is attributed to specific termination codons and the action of several release factors (RFs) as depicted in Figure 12.33. There are four release proteins with about 600 molecules of each type in a cell. RF-1 is about 36 kDa and RF-2 is 38 kDa. Both

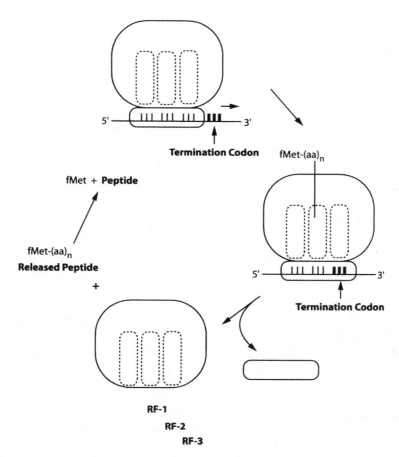

Figure 12.33. Termination of protein synthesis in bacteria. See text for discussion.

RF-1 and RF-2 participate in termination by recognizing the termination codons of UAA and UAG. These two termination codons do not have tRNAs with appropriate anticodons for their recognition and the A site on the 50S subunit remains open because there is no tRNA carrying an amino acid to the growing peptide chain. A third release factor is RF-3, which does not recognize codons but stimulates RF-1 and RF-2 in their recognition of termination codons. RF-3 is 46 kDa and may stimulate peptidyltransferase to hydrolyze the polypeptide from the tRNA at the P site. Finally, a ribosomal release factor (RRF), 23.5 kDa, is involved in releasing tRNA and mRNA from the ribosome. In the absence of stabilizing factors, the 70S ribosome dissociates into the 50S subunit and the 30S subunit. The 30S ribosomal subunit remains temporarily attached to the mRNA until it is also dissociated from the mRNA. In most instances, formylmethionine is cleaved from the polypeptide shortly after it is released from the ribosome.

In some cases, the mRNA is polycistronic with information carried from several genes on a single molecule. In this case, the ribosome does not dissociate from the mRNA when the end of a gene is reached but the polypeptide is released and the 70S ribosome continues toward the 3' end of mRNA. As the polycistronic message is translated, some of the 70S ribosomes will dissociate from the mRNA at the end of each structural gene so that the greatest amount of protein produced from each gene is greatest for the gene at the 5' end of the message on the mRNA. Ribosomes are assembled only at the 5' end of the mRNA and not along the internal regions of the polycistronic message. In the translation process, there are several ribosomes on each mRNA at any one time and these multiple ribosomes along the mRNA are known as polysomes. The presence of ribosomes on the unstable mRNA is important in protecting the mRNA against destruction by endonuclease action. The short half-life of mRNA is a direct reflection of the ease with which endonucleases can hydrolyze the ribonucleotide polymer.

### 12.7.2.4. Inhibition by Antibiotics

The inhibition of protein synthesis by antibiotics is an important means of controlling bacterial growth. Antibiotics may target the smaller ribosomal subunit, the larger ribosomal subunit, or one of the protein factors (e.g., initiation factors or elongation factors) of the ribosome. In addition, some antibiotics may also interfere with protein synthesis by binding to RNA that is associated with the ribosome subunit. Some of the important inhibitors for protein synthesis in bacteria are listed in Table 12.12 and Figure 12.34 provides a model that localizes the action of antibiotics on the ribosome. The action of these antibiotics is a result of specific recognition by a protein on the susceptible ribosome. For example, streptomycin is considered to bind to S-12 which is a protein of the 30S subunit. As indicated in Table 12.12, other antibiotic molecules bind to a protein elongation factor (EF) or to an initiation factor (IF) of the 70S ribosome. These antibiotics active against bacteria are not effective against eukaryotic protein synthesizing systems because the 80S ribosome or its subunits (40S and 60S) does not have the required target proteins to bind the antibiotics. Many of the key proteins involved with the structure of the bacterial 70S ribosome are not present in archaeal ribosomes.

Table 12.12. Examples of antibiotics and their specific action inhibiting protein synthesis in bacteria.

| Antibiotic | Inhibitory action |
|---|---|
| Aurintricarboxylic acid[a] | Prevents binding of mRNA to 30S ribosomal subunit. |
| Kasugamycin | Some bind to 16S RNA and block 30S addition to mRNA. |
| Streptomycin, neomycin, kanamycin | Binds to 30S ribosomal subunit and promotes improper codon/anticodon interaction at A site. |
| Tetracycline, aureomycin, terramycin, pactamycin, botlromycin | Blocks binding of amino acyl-tRNA at the A site. |
| Puromycin | Promotes premature termination of peptide from ribosome. |
| Chloramphenicol, sparsomycin, amicetin, gougerotin, lincomycin, spiramycin, pristinamycin IA and IIA | Inhibits peptidyltransferase and stops elongation of protein. |
| Erythromycin | Prevents movement of acyl-tRNA from A site to P site. |
| Spectinomycin | Prevents movement of acyl-tRNA by acting at the 30S/50S interface. |
| Fusidic acid | Prevents release of EF-G from ribosome. |
| Thiostrepton, siomycin, micrococcin | Blocks binding of IF-Tu, IF-G, and inhibits GTPase. |

[a]A triphenyl methane compound with a structure similar to crystal violet.

Resistance to antibiotics that inhibit protein synthesis can be attributed to several different mechanisms. In the streptococci, there is a multifunctional export system that actively exports the various antibiotics from the cytoplasm. Resistance to macrolides (e.g., erythromycin, clarithromycin, and HMR 3647) is attributed to methyltransferase activity which methylates 23S rRNA in the 50S subunit near the point of macrolide–rRNA interaction. The resistance of *Helicobacter pylori, Mycoplasma pneumoniae, Mycobacterium intracellulare,* and *Brachyspira hyodysenteriae* to clarithromycin appears to be due to methylation of the rRNA at the drug-binding pocket.

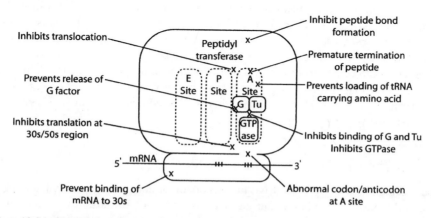

Figure 12.34. Sites for antibiotic inhibition of bacterial translation. See Table 12.10 for the specific antibiotics. Factors G and Tu are involved in elongation of mRNA in bacteria. (Modified from Mandelstam et al., 1986.[27])

In some instances the macrolide antibiotic molecule is modified through the action of specific enzymes that provide additions to the antibiotic molecule through phosphorylatation, glycosylatation, or esterification. Thus, enzymatic change of the antibiotic or modification of the target site for the antibiotic is responsible for bacterial resistance of many inhibitors of protein synthesis.

## 12.8. Systems for Secretion of Proteins and Other Biomolecules[28,29]

The synthesis of proteins and other biopolymers occurs within the cytoplasm of prokaryotes but several of these molecules are moved across the plasma membrane to the exterior of the cell. In general, the outward movement across the plasma membrane is designated as "export" and this term focuses on the mechanism of transmembrane movement without considering the ultimate location of the molecules. If one considers proteins outside of the plasma membrane, it is apparent that prokaryotes have several specific targets where proteins are located. Specific proteins are used for construction of S-layers which are found in all prokaryotes or for production of porins which are found in Gram-negative bacteria. Some proteins are released as extracellular enzymes or toxins, some enzymes are localized in the cell wall for peptidoglycan synthesis, some are hydrolytic enzymes that are localized in the periplasm, and others are involved in flagellar or pili assembly. In addition to proteins being moved across the plasma membrane, oligosaccharides are transported to the cell surface for capsule formation. In some bacteria, the mechanism for detoxification of the cytoplasm is the export of heavy metal cations, antibiotics, and antimicrobial molecules. There are several instances where proteins are moved from bacterial cells to eukaryotic cells. Not only are there different systems for movement of materials into and through the cell wall of bacteria but also there are considerable differences in the cell wall structure of Gram-positive and Gram-negative bacteria. The term "secretion" is used to describe the process of movement of biomolecules out of the cell without restriction to the molecular type or the destination of molecules transported. Secretion is inclusive of the various processes for transmembrane movement as well as movement to the extracellular regions where the biomolecules are deposited. The secretion of proteins and other biomolecules from bacteria involves interesting problems concerning management of cellular components. A summary of the various secretory processes used by bacteria is given in Table 12.13 and this section addresses the major mechanisms in bacteria for secretion of biomolecules.

### 12.8.1. Type I: ABC Secretion System[30,31]

Type I secretion uses ATP to drive the export of proteins, drugs, and metal cations through the plasma membrane and the outer membrane. Because this system uses a protein that is similar in activity to ABC uptake activity, it is frequently referred to as the ABC secretion system. In several cases, it has been determined that adjacent

Table 12.13. Characteristics of secretion systems in prokaryotes.[a]

| System | Selected examples of bacteria and secreted protein |
| --- | --- |
| **Type I: ABC system** | *Escherichia coli* produces α-hemolysin. |
| | *Bordatella pertussis* produces adenylate cyclase-hemolysin. |
| | *Pseudomonas aeruginosa* produces alkaline protease. |
| | *Caulobacter crescentus* produces S-layer proteins. |
| | *Azotobacter vinelandii* produces mannuronan epimerases. |
| **Type II: Signal sequences** | |
| (1) Autotransporter | *Neisseria gonorrhoeae* produces IgA1 protease. |
| | *Shigella* spp. produces SopA protease. |
| (2) Single helper protein | *Serratia marcescens* produces ShlA hemolysin. |
| (3) Multiple helper proteins | *Klebsiella oxytoca* produces pullulanase. |
| **Type III: Needle-like structure** | *Yersinia* spp. produces toxins that block phagocytosis and induce apoptosis. |
| | *Pseudomonas aeruginosa* produces a toxin that induces cytotoxicity. |
| **Type IV: Pilus-like structure** | *Rickettsia prowazekii* produces a toxin that regulates phagolysosome fusion. |
| | *Legionella pneumophila* produces a pore-forming toxin in animal membranes. |
| | *Helicobacter pylori* produces a toxin that induces IL-8 in animal cells. |
| | *Agrobacterium tumefaciens* exports tumor inducing genes to plant cells. |

[a]From Delepelaire and Wandersman, 2001.[35]

to the protein to be secreted are three genes that encode for the ABC secretion protein in the plasma membrane, the *m*embrane *f*usion *p*rotein (MFP) that is an auxiliary to the plasma membrane, and an *o*uter *m*embrane *p*rotein (OMP). An overview of the mechanisms involved in bacterial type I secretion is given in Figure 12.35. The three secretion proteins form a channel for the protein to be secreted with energy derived from ATP hydrolysis. On the cytoplasmic side of the ABC secretory protein is a nucleotide-binding site where ATP is presumed to be localized. One protein in the outer membrane has attracted considerable attention and this is the TolC, outer membrane protein of *E. coli*, used for secretion of cytotoxin. The TolC has been structurally defined as a trimer with each monomeric unit composed of four β-strands that are located in the outer membrane. In addition, the TolC protein has four α-helices with a length of 140 Å and these extend into the periplasm. At the tip of the α-helix of the TolC protein is a constricted pore that is about 30 to 35 Å wide, and this opening is too small for the transport of folded proteins. The mechanism for regulation of movement through the pore of the ABC secretory system is unresolved at this time.

The secretion of proteins has been shown to be specific for a certain type I unit and secretion through both membranes of Gram-negative bacteria is a single event. No binding specificity is known for proteins making up the channel in the periplasm or the outer membrane. The intact protein is secreted and it is assumed that chaperone

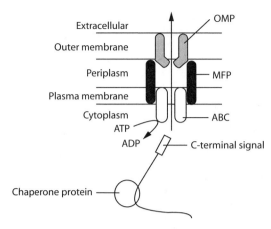

Figure 12.35. Model of type I secretory system in bacteria. OMP, Outer membrane proteins; MFP, membrane fusion protein; ABC, ATP binding cassette protein. *Arrow* indicates movement of protein. (From Delepelaire and Wandersman, 2001.[35])

proteins maintain the protein to be secreted in the unfolded form but with secretion the protein becomes folded. In Gram-negative bacteria, these ABC secretion systems proteins are aggregated into a single transmembrane unit and it is possible that this structure can form adhesion zones between the outer membrane and the plasma membrane. It has been proposed that these ABC secretion systems may account for the adhesion zones (Bayer patches) between the plasma membrane and outer membrane, discussed in Chapter 2.

Secreted proteins have a glycine-rich motif of GGXGXD that is located about 60 amino acid residues from the C-terminal end of the protein. This glycine-rich sequence is found to be repeated from 4 to 36 times and may be the signal that is recognized by the ABC secretory protein and may signal the opening of the pore of the outer membrane protein. Alternately, the glycine structure forms a β-parallel–roll structure that binds $Ca^{2+}$ to stabilize a specific conformation, and some have suggested that it forms a chaperone for the remainder of the protein to be secreted.

Type I secretion has been studied in various bacteria since the 1980s and its presence in archaea has been demonstrated only recently. In Gram-positive bacteria, type I is used for export of numerous structurally unrelated drug compounds (see Chapter 10), secretion of lantibiotics (nisin and subtilin), and release of cytolysin from *Enterococcus faecalis*. In Gram-negative bacteria, the ABC secretory system has been shown to be used by *Pseudomonas aeruginosa*, *Erwinia chrysanthemi*, and *Serratia marcescens* for metalloprotease secretion. *E. coli*, *Pasteurella hemolytica*, and *Bordatella pertussis* use type I secretion for the release of pore-forming toxins active against eukaryotic cells. The metalloproteases are prevented from acting on the cells that produces them because specific protease inhibitors are produced from genes adjacent to the metalloprotease genes. As the toxin is secreted, the pre-toxin protein is activated by a small cytoplasmic protein. Thus, with the bacterial production of highly reactive proteins a defense system is also produced for that cell.

## 12.8.2. Type II: Signal Peptides[32,33]

Leader peptides provide a signal sequence for the secretion of proteins in Gram-negative bacteria from the cytoplasm to the periplasm or the outer membrane. In Gram-positive bacteria secretion is from the cytoplasm to the extracellular medium. One of the best characterized systems for protein secretion is the *sec* (secretion) process, which is also known as the *general secretory pathway* (GSP). The structural features of this pathway are given in Figure 12.36. A unique sequence of amino acids at the N-terminus of the protein provides the signal for export across the plasma membrane. Although this signal may vary in the exact number and type of amino acids, there are several structural details common for signal peptides. A model for proteins secreted by the *sec* system is given in Figure 12.37. In *E. coli*, signal sequences are 18 to 30 amino acids in length and have the following structural features: (1) There are at least three to eight basic amino acids near the N-terminus. (2) There is a hydrophobic region in the middle of the signal peptide that is at least 15 to 18 amino acids in length. (3) The final amino acids of the leader sequence are commonly glycine or alanine, which attach to the protein, and these amino acids are the site of peptidase activity. It is considered that a role of the signal peptide is to maintain the protein in an unfolded state to enable protein export to be initiated.

In the *sec* system, the plasma membrane contains seven different proteins that collectively function in protein secretion and their activities are summarized in Table 12.14. The *sec* system for secretion of proteins generally adheres to the following events. As the leader sequence is released from the ribosome, the SecB protein in the cytoplasm moves along the protein to prevent protein folding. The basic amino acids of the leader segment have a positive charge and associate with the negatively charged phospholipids on the inner side of the plasma membrane. The hydrophobic region of the leader sequence extends through the membrane and the first few amino acid residues of the mature protein extend through a pore formed by SecY and SecE. SecA is energized by ATP hydrolysis and translocates about 20 amino acid residues through the membrane. An additional ATP enables the SecA to move another 20 amino acids residues and this is repeated until all the protein is exported. Also associated with secretion by the type II system is the requirement of a charge on the membrane; however, it is not understood how membrane charge contributes to protein

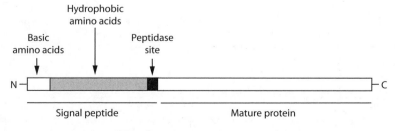

Figure 12.36. General structure of protein translocated by type II secretion using the *sec* system.

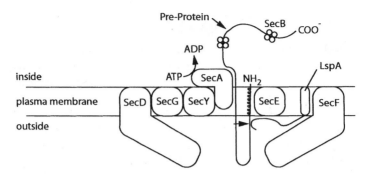

Figure 12.37. Model of protein secretion using the *sec* system. The protease (Lsp A) is bound to the membrane and hydrolyzes the pre-protein at the *arrow*. SecA pushes the protein through the pore in the membrane and is energized by ATP hydrolysis. The transmembrane pore is formed by SecE and SecY. SecD and SecF are proposed to assist in folding of the mature protein. SecB is a chaperone protein that permits premature folding of the pre-protein.

translocation. A signal peptidase is bound to the membrane and the active site of the peptidase is on the outer region of the membrane where it hydrolyzes the peptide bond between the leader sequence and the mature protein. The leader sequence is degraded by various peptidases and the amino acids are imported for future biosynthetic activities. The mature protein is stabilized in the periplasm by SecD and SecF segments and ultimately the protein assumes a folded structure. In *E. coli*, LspA is the leader

Table 12.14. Characteristics of proteins required for the *sec* secretion system.

| Protein | Characteristic |
|---------|----------------|
| SecA | The membrane-bound protein is a tetramer with 17-kDa monomers. It is actively involved in protein translocation and contains an ATP binding site. |
| SecB | A cytoplasmic protein that is a tetramer and functions to direct protein to the secretion site. It may function as a general chaperone for export of several different proteins. |
| SecY | This protein has 10 membrane spanning helices. It recognizes a good signal sequence and allows a protein to be exported. |
| SecE | This protein contains three transmembrane helices and is a part of the translocation reaction. Its function is unknown. |
| SecD | This protein has from one to three transmembrane segments depending on the particular secretion system. It extends from the membrane into the periplasm and it may assist in the release of the mature protein from the signal peptide protein. Some consider it may have a role in protein folding. |
| SecF | The same as Sec D |
| SecG | The same as Sec E |
| Lep | A serine protease attached to the plasma membrane by two membrane-spanning segments and its function is to hydolyze the leader peptide from the mature protein. It requires small amino acids at position $-1$ and $-3$. This proteinase may be replaced by Lsp. |

peptidase used for lipoproteins and Lep is the peptidase that removes the leader sequence of various proteins secreted by the *sec* system.

The mechanism for moving the protein from the periplasm to the outer membrane and ultimate insertion into the outer membrane is an area receiving considerable attention at this time. The protein to be secreted is moved across the plasma membrane as a result of the N-terminal signal sequence and in Gram-negative bacteria this secretion protein is transiently present in the periplasm. As the signal peptide is removed by a specific peptidase, the new N-terminus of the mature protein is methylated by the multifunctional protease. The methylated mature protein is secreted from the cell by one of three systems: (1) autotransporter, (2) assistance from a single protein, or (3) assistance from multiple proteins.

*The Autotransporter System*

A model for the autotransporter system is the IgA1 protease produced by *Neisseria gonorrhoeae*. The IgA1 protease contains a *sec*-leader signal sequence, a 106-kDa proteinase domain, and a 45-kDa domain at the C-terminus. The *sec*-signal sequence provides for the transport through the plasma membrane into the periplasm. The C-terminal domain binds into the outer membrane and assists in export of the protease from the periplasm through the outer membrane. The IgA1 protease self-hydrolyzes the secreted molecule to release the mature protein from the C-domain and the C-domain remains in the outer membrane. In those cases where the secreted protein does not undergo autodigestion, the mature protein remains attached to the outer membrane.

*The Single Protein Helper System*

A model for the single helper protein secretion is the hemolysin produced by *Serratia marcescens*. In addition to the hemolysin protein with a *sec*-leader sequence, a ShlB protein is produced that facilitates the passage of the hemolysin through the outer membrane. The ShlB is a monomer that has several transmembrane segments consisting of a β-sheet conformation that produces a small pore in the outer membrane.

*The Multiple Protein Helper System*

In this system, 11 to 12 types of proteins are involved in producing a pilus-like structure. One or two proteins are placed in the outer membrane of the bacterium, one is a peripheral plasma membrane protein, and nine proteins are associated with the plasma membrane but these nine proteins have major domains in the periplasm. Proteins to be secreted by this system are moved across the plasm membrane as a result of the *sec*-leader sequence, and export through the outer membrane continues through the pilus-like structure. Additional details for secretion by this system are not clear at this time.

Another secretory system using a signal peptide sequence is the Tat system, which has a consensus motif of (S/T)RRXFLK. Because of the presence of two arginine moieties (R) at the N-terminus in the signal peptide, this pathway is commonly referred to as the twin arginine transfer (Tat) system. A characteristic of this Tat system is that it translocates folded proteins with metal cofactors attached. In contrast, the *sec* system translocates only unfolded proteins. The secretion of trimethyl ammonium oxide (TMAO) reductase and dimethyl sulfoxide (DMSO) reductase are examples of enzymes that are translocated by this *sec*-independent pathway. The DMSO reductase is encoded on *dmsA, dmsB, dmsC*, and *dmsE* genes with protein expression occurring under anaerobic culture. The dmsA protein has a molybdocofactor, and when synthesized, it has a double arginine motif in the signal peptide. Export of the folded proteins across the plasma membrane takes about 5 min while only 30 sec are required for translocation of *sec*-proteins.

## 12.8.3. Type III: Needle-Like Structures[34-39]

The principal function of the type III secretion system is to enable bacteria to establish a symbiotic relationship with a eukaryotic cell or to enhance bacterial infection in a host cell. In this system there is the direct transfer of protein (referred to as effector protein or exoprotein) from the cytoplasm of bacteria to the cytosol of eukaryotic cells through the use of a specialized apparatus. A model representing the type III secretion system is given in Figure 12.38. Physical contact is required for transfer of effector protein and an essential step for transfer is the interfacing of bacteria with a host cell. Close contact between the host cell and bacterial cell is achieved by the outer membrane protein of bacteria docking with a protein on the surface of a eukaryotic cell. Specific docking of outer membrane protein with host cells is observed with *Yersinia* spp., *Salmonella* spp., *Shigella flexneri, Pseudomonas aeruginosa*, and enteropathogenic *Escherichia coli* (EPEC). In plant-related bacterial association, the pathogenicity of *Erwinia amylovora, Pseudomonas syringae, Ralstonia solanacearum*, and *Xanthomonas campestris* is enhanced by exoproteins introduced into the host cells by the hollow needle structures. In the formation of a plant–symbiotic association, *Rhizobium* spp. injects protein into cells of root hairs to induce nodule formation.

The exoprotein to be transferred either carries a specific signal sequence found in the first 100 amino acids or a specific signal sequence resides on the 5' end of the mRNA. Each species of bacteria can produce several different exoproteins. In cases

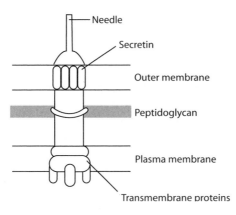

Needle

Secretin

Outer membrane

Peptidoglycan

Plasma membrane

Transmembrane proteins

Figure 12.38. Model of structures associated with type III secretion pathway based on electron microscopy of apparatus isolated from *Salmonella*.

in which signal sequences are carried on the protein, a specific chaperone protein protects the exoprotein from degradation by intracellular proteases and serves to pilot the exoprotein to the secretion region. The chaperone proteins have a negative charge and an α-helix. Genes for exoprotein production and chaperone protein appear to be linked so that the chaperone is produced when the exoprotein is released. The exoprotein–chaperone complex may accumulate in the cytoplasm of the bacteria before the exoprotein is secreted because inhibitors of protein synthesis do not prevent secretion of exoproteins. In the case of *Yersinia* spp., exoprotein is not detected in the cytoplasm of the bacteria but translation of message with production of exoprotein is coupled with secretion.

From observation of *Salmonella* spp. a needle-like structure can be observed and this structure has considerable resemblance to the flagellum structure of Gram-negative bacteria. Gene analysis is being used to assess the protein composition of the type III unit and it appears that bacteria have at least 12 to 20 genes contributing to this secretion unit. The apparatus in Gram-negative bacteria extends through the plasma membrane and the outer membrane and ring-like structures are used to hold the apparatus in the membranes. The needle-like structure is about 80 nm in length and 13 nm wide. The unit is hollow and it would need to be gated to prevent back flow of material into the cell and to account for selective exit. Although the energetics of this secretory system is unresolved, the presence of an ATP binding site in protein associated with the plasma membrane suggests that ATP may be involved in this translocation. It has been proposed that the type III secretory systems are activated by contact with eukaryotic host cells with a response attributed to a two-component signal system.

## 12.8.4. Type IV: Pilus-Like Secretion[40]

The type IV secretion pathway has been proposed for the transfer of DNA and certain proteins from bacterial to eukaryotic cells. The transfer of DNA between bacteria of different species by conjugation has been extensively characterized and provides a backdrop for this system of secretion. The transfer of DNA from bacteria to eukaryotic cells has received less attention than bacterial conjugation and the transfer of DNA between organisms of different Domains (or Kingdoms) is limited to the research on crown gall tumor production in plants. *Agrobacterium tumefaciens* carries a tumor-inducing (Ti) plasmid that is transferred to a plant cell. *Agrobacterium tumefaciens* is not host specific in that it can establish tumors in members of several different plant families. It is now recognized that infection follows a series of steps and these are briefly outlined as follows: (1) Plant cells that will become infected are adjacent to plant wounds and when wounded, plant cells release phenolic compounds. (2) *Agrobacterium tumefaciens* is attracted to the wound site by a two-component regulatory system that consists of VirA/VirG proteins. VirA senses the phenolic compounds and transfers the signal to VirG which interfaces with the Ti-plasmid. (3) The signal response activity results in stimulating transcription of *vir* genes present on the Ti megaplasmid. (4) As a result of action by endonuclease VirD1 and endonuclease VirD2, a single-stranded DNA molecule of 20 kb is produced from the Ti plasmid

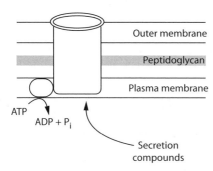

Figure 12.39. Model of apparatus for type IV secretion pathway in bacteria.

(T-DNA). (5) Endonuclease VirD2 remains attached at the 5' end of the single-stranded T-DNA and this serves to promote the attachment of a DNA-binding protein (VirE2) to the single stranded T-DNA. (6) The nucleoprotein (consisting of 20-kb T-DNA, plus VirD2 protein, plus VirE2 protein) is secreted from the bacterial cell into the plant cell. The VirE2 and VirD2 proteins serve to direct the nucleoprotein to the nucleus of the plant cell where the T-DNA becomes incorporated into the host DNA. (7) Compounds resulting from the integrated T-DNA include *trans*-zeatin, indole-3-acetic acid, an opine, and some proteins that interfere with normal signal transducing pathways in the plant cell. Zeatin (a cytokinin) and indole-3-acetic acid (an auxin) stimulate cell proliferation; plant tissues greatly enlarge and the plant displays considerable deformity. Opines are a class of nitrogen-containing compounds that are not produced in plants unless infected by *A. tumefaciens*. The invading pathogen, *A. tumefaciens*, uses opines as a nitrogen source while opines are inert to plant catabolism. The DNA transfer that occurs with *Agrobacterium tumefaciens* is similar to DNA transfer by conjugation in that a specific pilus structure is used and only single-stranded DNA is transferred. At least 11 different proteins are known to be involved in synthesis of the type IV pilus and some similarity is observed with proteins of the F-pilus. It is of interest to point out that the assembly of the type IV structure requires the structural proteins to be secreted by the *sec* signal sequence (type II) secretion system.

Other bacteria that use the type IV secretion system include some that are pathogenic for animal cells. These proteins are virulence-enhancing factors that have no signal sequences and are highly toxic to the host cell. Figure 12.39 summarizes some of the activities of the pathogenic bacteria with the type IV secretory system. Although it is well established that proteins are transferred from the pathogens by the type IV system, there is some suggestion that DNA from these bacteria may also be transferred. At this time, the transfer of virulence factors to animal cells is a highly active research area.

## 12.9. Posttranslational Modifications of Proteins

Until recently, it was widely believed that prokaryotes had proteins made exclusively of amino acids and that there was no posttranslational processing. As molecules were subjected to a higher degree of analysis, it has become apparent that prokaryotes in special instances will modify a limited number of proteins. Although only a few examples are discussed here, it is highly possible that future studies will reveal many additional systems in prokaryotes in which posttranslation modification of proteins occurs.

## 12.9.1.  Glycoproteins[41]

From 2% to 50% of the mass of glycoproteins may consist of carbohydrates which commonly include glucose and mannose.  The number of sugar chains may vary from 1 to more than 30 per molecule.  Evidence indicates that addition of sugars to the nascent polypeptide chain occurs after the protein leaves the ribosome.  Several different functions have been proposed for the carbohydrate moiety including facilitation of peptide chain folding, prevention of degradation by proteolytic enzymes, and promotion of thermostability of the protein structure.

Glycoproteins are widely distributed in eukaryotic organisms but their presence in prokaryotic cells has only recently received attention.  Glycine reduction in *Clostridium sticklandii*, *Clostridium purinolyticum*, and *Eubacterium acidaminophilum* is attributed to glycine reductase.

Of the three components making up glycine reductase, the $P_A$ component is considered to be a glycoprotein.  The thermophilic *Sulfolobus acidocaldarius* produces cytochrome $b_{558/566}$, which is a 64-kDa protein with more than 20% of the mass being sugar residues consisting of mannose, glucose, and *N*-acetylglucosamine.  Cytochrome $b_{558/566}$ is an integral membrane protein that resides on the outer side of the plasma membrane.  Based on the periodic acid–Schiff staining reactions employed for glycoproteins, flagellins from several archaeal strains have been demonstrated to be glycoproteins and these include flagellin from *Methanospirillum hungatei*, *Merthanothermus feridus*, *Methanococcus deltae*, and *Sulfolobus shitatae* (see Table 12.15).  The sugars are attached to the protein through Asn with a specific amino acid sequence, Asn-X-Thr (Ser), which is the same sequence found in eukaryotes with glycoproteins.  The S-layer of both bacteria and archaea contain glycoproteins (see Chapter 3) and these molecules contribute to the hydrophobic face of the cell wall structure.  Based on the presence of numerous *N*-glycosylation sites in thermopsin, it is highly likely that this acidic protease from *Sulfolobus acidocaldarius* is also a glycoprotein.  Although there is anecdotal discussion about bacteria amylases and

Table 12.15. Carbohydrates attached to flagella of *Halobacterium salinarum*.[a]

| Glycoconjugates[b] | Amino acid used for attachment |
| --- | --- |
| Glucose 1→4 hexuronic acid 1→4 hexuronic acid 1→4 glucose | Asparagine |
| Hexuronic acid 1→4 hexuronic acid 1→4 hexuronic acid 1→4 glucose | Asparagine |

[a]From Sumper, 1987.[41]
[b]Hexuronic acid is either glucuronic acid or iduronic acid.

phosphatases being glycoproteins, the presence of sugar moieties bound to enzymes of mesophilic bacteria has not been established.

## 12.9.2. Modification of Amino Acids[42]

Another posttranslational modification that has been observed in several prokaryotes is the methylation of lysine with the formation of $N$-ε-methyl-L-lysine. *Sulfolobus solfataricus* produces a glutamate dehydrogenase (46 kDa) that has $N$-ε-methyl-L-lysine in six different locations on the molecule and this has been proposed to promote thermostability. The ferredoxin in *Sulfolobus acidocaldarius* also contains $N$-ε-methyl-L-lysine residues. Members of the *Sulfolobus* genera produce the Sac7 proteins which are DNA-stabilizing proteins of about 7 kDa. In *Sulfolobus acidocaldarius*, the Sac7 protein contains 12 to 14 lysine residues and many of these lysines are methylated. Unfortunately, protein structure deduced from DNA sequences can predict only the number of lysines per molecule and not the number of posttranslational modifications resulting in $N$-ε-methyl-L-lysine formation. Future research is certain to provide interesting information concerning the the contribution of methylation to structure and function of proteins in prokaryotes.

## 12.9.3. Lantibiotics[43,44]

Peptide antibiotics may be produced by bacteria as a result of either ribosomal or nonribosomal processes. Antibiotics produced as a result of activities attributed to enzyme complexes and not of ribosomal activities include gramicidin, tyrocidine, and bacitracin. Antibiotics produced by ribosomal processes include the colicins and microcins of Gram-negative bacteria and the lantibiotics synthesized by Gram-positive bacteria. The lantibiotics have received considerable attention because they are em-

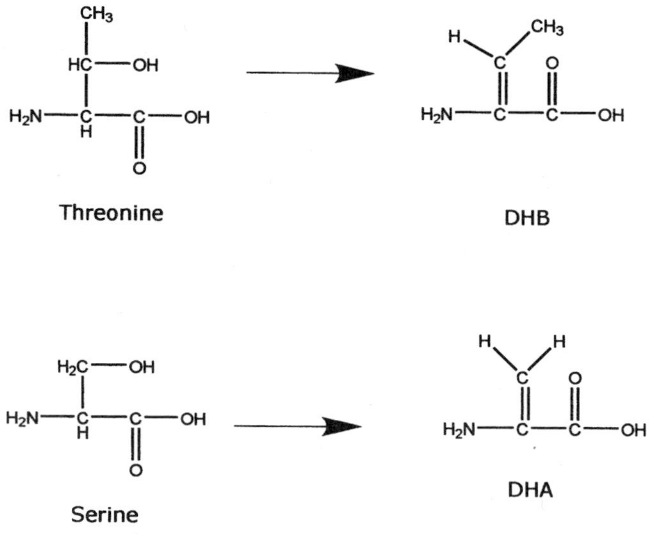

Figure 12.40. Source of dehydroalanine (DHA) and dehydrobutyrine (DHB) found in lantibiotics.

ployed as topical antibiotics and food additives. These antibiotics derive their name from the formation of lanthionine, which has the structure of Ala-s-Ala where "s" is the thioether crosslinkage. The thioethers in the lantibiotics produces a heterocyclic structure that is highly thermostable. Some peptide antibiotics contain methyl lanthionine, which is aminobutyric acid cross linked to alanine by a thioester bond. Some lantibiotics contain the unusual amino acids of dehydroalanine (DHA) and dehydrobutyrine (DHB) which are produced from serine and threonine, respectively. The structures and the synthesis of DHB and DHA are given in Figure 12.40. The formation of DHB, DHA, and thioether crosslinkages are examples of posttranscriptional processes in these bacteria.

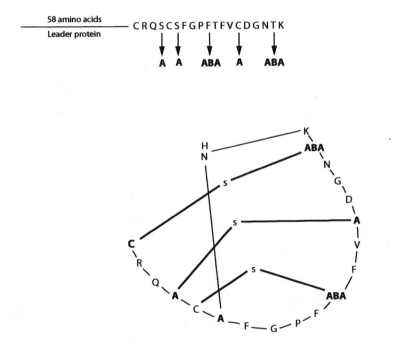

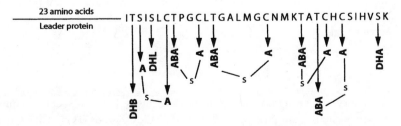

Figure 12.41. Lantibiotics produced by bacteria. The precursor peptide and the final peptide structures are given for nisin and cinnamycin. (Modified from Zuber et al., 1993.[44])

Table 12.16. Selected examples of lantibiotics produced by bacteria.[a]

| Antibiotic | Number of amino acids | Source | Activities |
|---|---|---|---|
| **Type A lantibiotics** | | | |
| Nisin | 34 | *Lactococcus lactis* | Antibacterial, effective against clostridia and mycobacteria. Used in food preservation. |
| Subtilin | 32 | *Bacillus subtilis* | Antibacterial, used in agriculture |
| Subtilosin | 32 | *B. subtilis* | Antibacterial |
| Pep5 | 34 | *Staphylococcus epidermidis* | Antibacterial |
| Epidermin | 22 | *S. epidermidis* | Antibacterial |
| Gallidermin | 22 | *Staphylococcus gallinarum* | Antibacterial |
| **Type B lantibiotics** | | | |
| Cinnamycin | 19 | *Streptoverticillium giseoverticilla-tum* | Phospholipase inhibitor |
| Ancovenin | 19 | *Streptomyces* spp. | Phospholipase inhibitor |
| Duramycin B | 19 | *Streptoverticillium* | Phospholipase inhibitor |

[a]From Zuber et al. 1993.[44]

Although the requirement for ribosomal activities had been proposed, direct evidence was obtained through experiment in molecular biology. Once the genes that encode for galliderman, subtilin, cinnamycin, nisin, and epidermin had been isolated, it was determined from analysis of open reading frames that the precursor peptides were longer than the final antibiotic. The peptide lantibiotics contain about 19 to 34 amino acid residues and their final structure is a result of processing of the larger precursor peptide. For example, cinnamycin contains only 19 amino acids with several thioester bonds and is produced from a precursor of 77 amino acid (see Figure 12.41). Similarly, nisin is a lantibiotic that has only 34 amino acids but has a precursor peptide of 57 amino acids. Although more than 100 different peptide antibiotics are produced by bacteria, a small number of these are lantibiotics. At this time there are two major types of lantibiotics. Type A antibiotics have a structure similar to nisin while type B are smaller and have a structure similar to cinnamycin. Several of these lantibiotics, and their activities, are listed in Table 12.16. Once the lantibiotics are produced, they are transported into the extracellular region by export systems associated with the ABC transport family.

## 12.9.4. Active Site of Sulfatases[45]

Like the eukaryotic systems, the bacterial sulfatases have a specific amino acid sequence at the active site which is C/S-X-P-X-R-X-X-X-X-TG. Through a post-translational modification system, the initial cysteine or serine is converted to formylglycine (F-Gly also known as 2-oxoalanine). If the cysteine or serine is not modified, there is no sulfatase activity.  While the modification of serine residue to produce F-Gly requires only a single oxidative step, the change of cysteine to F-Gly requires an inital step which is the oxidation to produce a thio-aldehyde and a second step which is hydrolytic.  To distinguish the oxidation of cysteine and serine further, the modification of cysteine occurs in the cytoplasm while the process for modification of serine is associated with the plasma membrane.  The enzymes and genes for these modification activities remain to be isolated.

### 12.9.5. Changes in Spore Coat Proteins[46]

A series of protein layers make up the inner coat and the outer coat of bacterial endospores.  By high resolution electron microscopy, it is seen that the protein coats are fibrous and the protein layers are often at right angles to each other.  It is now known that the strength of the spore coat is attributable to cross linkages between adjacent protein layers.  Disulfide bonds and dityrosine bonds are the most common cross linked arrangements and these could be produced by posttranslational modification of proteins.  Numerous tyrosine residues are present in the coat proteins and it is quite possible that the CotE, a coat protein that has peroxidase activity, may be responsible for the dityrosine cross linkages.  An additional cross linkage could be attributed to bond formation resulting from a $\varepsilon$-($\gamma$-glutamyl)lysine cross linkage.  This transglutaminase activity is encoded on the *tgl* gene found in *Bacillus subtilis*.  In addition, it is evident that the spore proteins have carbohydrates added to them.  It has been found that the *cge* genes which encode a glucosyltransferase may influence the terminal step of spore coat formation.  The presence of glycoproteins on the surface of the endospore would promote hydrophobicity and influence the adhesive property of the endospore.

### 12.9.6. Phosphorylation of Proteins

For a long time it was considered that phosphorylated proteins were present only in eukaryotic systems; however, it is now known that there several activities in the cell in which phosphate is attached to proteins.  In group translocation, phosphate is shuttled from phosphoenol pyruvate to the sugar to be transported through intermediary carriers of Hpr, enzyme I, and enzyme II.  Another system in which phosphorylated proteins are present is in the sensory system associated with chemotaxis.  In directed motility, phosphorylated proteins serve to regulate the motor for flagellar rotation.  Numerous systems require two-component signal transduction and these activities involve the presence of phosphorylated proteins that serve as intracellular switches.  These activities with phosphorylated proteins are all discussed earlier in this book.

## 12.10.  Histone-Like Proteins[47,48]

A distinguishing characteristic of eukaryotic and prokaryotic cells is organization of the nuclear material into chromosomes.  Eukaryotic cells have DNA packaged into chromosomes through the presence of histones while prokaryotes do not have distinguishable chromosomes. It is known that several members of the archaea do have small basic proteins that interact with the DNA but without the formal banding found in chromosomes. These DNA-interacting structural proteins are distinct from the regulatory proteins for DNA expression and are resolved into two separate groups: proteins with homology to histones and proteins without apparent homology to histones.

The histone-like proteins have been found in thermophilic archaea. The protein (HTa) isolated from *Thermoplasma acidophilum* and the two proteins (Hmf1 and Hmf2) from *Methanothermus fervidus* display considerable homology with eukaryotic histones. These proteins appear to stabilize the DNA structure in these cells to prevent separation of the DNA strands at the high temperatures of growth.

Several bacteria are known to produce DNA-binding proteins that function in a manner similar to histones and function in stabilizing DNA. The role of these chromosome-binding proteins in replication is covered in Section 12.6.1. The best characterized histone-like protein from bacteria is the HU protein produced by non-stressed cells of *Escherichia coli*. HU is 9.7 kDa with an amino acid composition and physical properties that are similar to those of eukaryotic histone. The HU pro-

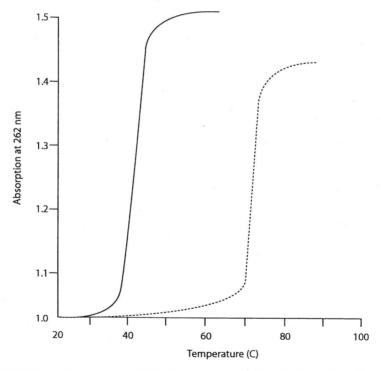

Figure 12.42. Thermal protection of DNA from denaturation by Sac7 proteins. The *solid line* is the melting curve of poly [dAdT] • poly [dAdT] and the *dashed line* is the melting curve of poly [dAdT] • poly [dAdT] after it was mixed with Sac7c and Sac7d proteins. (Redrawn with permission from McAfee et al., 1995.[47])

teins are distributed throughout the bacteria and have been best characterized from *E. coli*, *S. typhimurium*, *B. subtilis*, various cyanobacteria, and *Thermus thermophilus*. In the mesophilic bacteria the DNA-binding proteins would appear to function in packaging the DNA with the formation of supercoiled loop domains (see Chapter 1).

The second class of DNA-binding proteins do not have any structural holology to histones. Organisms such as *Sulfolobus acidocaldarius* and *Sulfolobus solfataricus* grow at 80° to 85°C and have a G + C base composition of DNA that is about 40%. It appears that these cells would require proteins to stabilize the DNA and avoid the consequences of DNA melting. The *S. acidocaldarius* produce large quantities of DNA-binding proteins which are about 7 kDa and are designated as Sac7a, b, c, d, and e. These Sac7 proteins are helix-stabilizing nucleoid proteins that bind to native or denatured DNA but not to double-stranded RNA. As presented in Figure 12.42, thermal denaturation experiments using synthesized DNA indicate that Sac7 substantially elevates the melting temperature of DNA.

## 12.11. Heme Synthesis[49,50]

The synthesis of hemes, chlorophylls, phycobilins, factor $F_{430}$, and cobalamins involves the use of tetrapyrroles with δ-aminolevulinic acid as the precursor. The general scheme for tetrapyrrole synthesis and heme products is given in Figure 12.43. δ-aminolevulinic acid is generated from either glutamate or from glycine plus succinyl~CoA. The production of δ-aminolevulinic acid from glutamate is the pathway used by plants, algae, and most bacteria including the enterics. A unique segment of this pathway is the role of tRNA$^{glu}$, which is the specific carrier for glutamate. The following reaction requires a ligase with energy for bond formation coming from ATP:

$$\text{Glutamate} + \text{tRNA}^{glu} + \text{ATP} \rightarrow \text{glutamyl-tRNA}^{glu} + \text{ADP} + \text{P}_i$$

The glutamyl-tRNA is reduced to glutamyl-1-semialdehyde by the following reaction which requires a reductase enzyme:

$$\text{Glutamyl-tRNA} + \text{NADPH} \rightarrow \text{glutamyl-1-semialdehyde} + \text{tRNA} + \text{NAD}^+$$

Finally, δ-aminolevulinic acid is produced from glutamate-1-semialdehyde by a specific amino transferase. Animals, yeast, fungi, and members of α-proteobacteria produce δ-aminolevulinic acid from glycine plus succinyl~CoA by δ-aminolevulinic acid synthetase by the following reaction:

$$\text{Glycine} + \text{succinyl}{\sim}\text{CoA} \rightarrow \text{δ-aminolevulinic acid} + \text{CoA}$$

A series of condensation reactions are required to produce tetrapyrroles that are the basis of the heme structures. Two δ-aminolevulinic acid molecules are combined to produce porphobilinogen which is a nitrogen-containing five-member heterocyclic structure containing numerous side chains. Four molecules of porphobilinogen are combined to produce a tetrapyrrole structure. The side chains of the tetrapyrrole and the ring structures are modified before $Fe^{2+}$ is added to protoporphyrinogen IX to

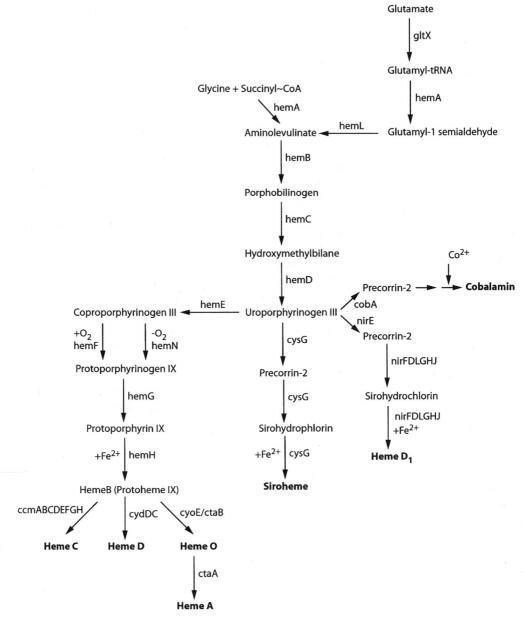

Figure 12.43. Generalized pathway for heme synthesis in prokaryotes. Genes for individual steps are indicated.

produce heme B, which is the precursor for heme C, heme D, heme O, and heme A. The structures for hemes and their incorporation into cytochromes are given in Chapter 9. Siroheme and heme $D_1$ do not originate from protoporphyrin IX but from sirohydrophlorin and sirohydrochlorin, respectively. Sirohemes are prosthetic groups in sulfite reductase and in nitrite reductase. Cobalamin and vitamin $B_{12}$ are produced from precorrin and $Co^{2+}$ is added to produce a functional cofactor. Bacteriochlorophyll and chlorophyll are synthesized from protoporphyrin IX with $Mg^{2+}$ coordinately held by the tetrapyrroles. The phycobilins of cyanobacteria are produced from protoheme and do not have a metal ion in the molecule.

It is important to note that bacteria have cytochromes in both aerobic and anaerobic cultivation. In aerobes, protoporphyrinogen IX is formed from coproporphyrinogen III (see Figure 12.43) is by an $O_2$-requiring reaction using coproporphyrinogen oxidase encoded on gene *hemF*. Under anaerobic culture, *E. coli* produces an anaerobic coproporphyrinogen oxidase that is a product of *hemN* and electrons from the reaction are directed to the anaerobic respiratory system.

## 12.12. Global Regulators[51,52]

The genes for metabolic processes, cellular differentiation, and adaption of bacterial populations are organized into three levels of coordinated expression. An operon is a cluster of genes under control of a single operator region and promoter region with a regulator protein interacting with the operator. A regulon constitutes several operons that encode for proteins with related activities and these operons are controlled by a common regulator protein. The coordination of a regulon and operons by a single regulator constitutes a modulon and modulons are required to achieve rapid adjustments to changes in the environment or in cellular processing.

*Global regulator* is a term used to characterize the protein that regulates a modulon. These global regulators are a type of control that function in addition to the standard regulators for operons and regulons. Several processes under global regulation in *E. coli* are listed in Table 12.17. The number of genes involved in a global control

Table 12.17. Examples of global regulation in *Escherichia coli*.[a]

| System | Regulatory protein | Number of genes regulated |
|---|---|---|
| Catabolite repression | CAP | >300 |
| Anaerobic respiration | FNR | >70 |
| Aerobic respiration | ArcA | >50 |
| Heat shock | $\sigma^{32}$ (alternative sigma factor) | 36 |
| Oxidative stress | OxyR | >30 |
| SOS response | LexA | >20 |
| Nitrogen utilization | $\sigma^{54}$ (alternative sigma factor) | >12 |

[a]From Madigan et al., 2003.[52]

Table 12.18. A selection of global control networks in prokaryotes.[a]

| System | Proteins as sensors or regulators | Activity or product or regulation |
|---|---|---|
| General starvation | CreC/CreB; $\sigma^{38}$ | Promotes survival in stationary phase |
| Stationary phase response | PilB/PilA; RcsC/RcsB; RpoS ($\sigma^{38}$) | Responsible for colony formation and surface adhesion |
| Cell division | Sfx (?) | DNA replication, cell cross wall and septa formation, cell division |
| Stringent response | RelA, SpoT | tRNA pool influences translation, common in anabolic systems |
| Phosphate utilization | PhoP,Q/PhoB,P | Operons of inorganic and organic phosphorus transport from the environment |
| Motility | CheA/CheY | Cell movement and chemotaxis |
| Sporulation | KinA/SpoOF; DivJ/DivK; PleC/SpoC | Endospore formation in *Bacillus* |
| Competence | ComP/ComA | Uptake of DNA |
| Photosynthesis | RegB/RegA PrrB/PrrA | Photopigments and structure of photosynthetic apparatus |
| Osmotic response | EnvZ/OmpR | Porin production |
| Cold shock | ? | With a shift to cold environment, cells are able to continue to grow. |

[a]From Lengeler and Postma, 1999.[51]

network can be extremely large, as seen with cAMP influencing expression of more than 300 genes. Another global regulator is CysB and it influences the expression of genes for cysteine synthesis and several other activities. CysB is a class I transcriptional activator that binds to the activation areas ($-35$ region) of the promoters for *cysJ*, *cysK*, and *cysP*. CysB also influences transcription by interacting with the C-terminus of the $\alpha$-subunit of RNA polymerase. In addition, by mechanisms that are not clear, CysB mediates acid resistance in *Escherichia coli*, synthesis of cellulolytic and pectinolytic enzymes in *Erwinia carotovora*, and alginate synthesis in *Pseudomonas aeruginosa*.

In many instances, the components of the global regulatory system are not fully resolved and the phenotypic activities are apparent but the number of genes involved are not known. As seen in Table 12.18, many of the activities observed in cellular growth, metabolism, and differentiation are under global regulation. Frequently, these global control networks are integrated with the two-component signal transduction sensory systems. Thus, global regulation functions as a master switch in the cell and produces a highly efficient control process with a quick response to environmental changes.

# References

1. Neuhard, J. and P. Nygaard. 1987. Purines and pyrimidines. In Escherichia coli *and* Salmonella typhimurium: *Cellular and Molecular Biology* (F.C. Neidhardt, J.L. Ingraham, B. Magasanik, K.B. Low, M. Schaechter, and H.E. Umbarger, eds.). American Society for Microbiology Press, Washington, D.C., pp. 445–473.

2. Stanier, R.Y., J.L. Ingraham, M.L. Wheelis, and R.R. Painter. 1986. *The Microbial World*, 5th edit. Prentice-Hall, Englewood Cliffs, N.J.

3. Pittard, A.J. 1996. Biosynthesis of the aromatic amino acids. In Escherichia coli *and* Salmonella: *Cellular and Molecular Biology* (F.C. Neidhardt, R. Curtiss III, J.L. Ingraham, E.C.C. Lin, K.B. Low, B. Magasanik, W.S. Reznikoff, M. Riley, M. Schaechter, and H.E. Umbarger, eds.). American Society for Microbiology Press, Washington, D.C., pp. 458–484.

4. Greene, R.C. 1996. Biosynthesis of methionine. In Escherichia coli *and* Salmonella: *Cellular and Molecular Biology* (F.C. Neidhardt, R. Curtiss III, J.L. Ingraham, E.C.C. Lin, K.B. Low, B. Magasanik, W.S. Reznikoff, M. Riley, M. Schaechter, and H.E. Umbarger, eds.). American Society for Microbiology Press, Washington, D.C., pp. 542–560.

5. Patte, J.-C. 1996. Biosynthesis of threonine and lysine. In Escherichia coli *and* Salmonella: *Cellular and Molecular Biology* (F.C. Neidhardt, R. Curtiss III, J.L. Ingraham, E.C.C. Lin, K.B. Low, B. Magasanik, W.S. Reznikoff, M. Riley, M. Schaechter, and H.E. Umbarger, eds.). American Society for Microbiology Press, Washington, D.C., pp. 528–541.

6. Glansdorff, N. 1996. Biosynthesis of arginine and polyamines. In Escherichia coli *and* Salmonella: *Cellular and Molecular Biology* (F.C. Neidhardt, R. Curtiss III, J.L. Ingraham, E.C.C. Lin, K.B. Low, B. Magasanik, W.S. Reznikoff, M. Riley, M. Schaechter, and H.E. Umbarger, eds.). American Society for Microbiology Press, Washington, D.C., pp. 408–433.

7. Leisinger, T. 1996. Biosynthesis of proline. In Escherichia coli *and* Salmonella: *Cellular and Molecular Biology* (F.C. Neidhardt, R. Curtiss III, J.L. Ingraham, E.C.C. Lin, K.B. Low, B. Magasanik, W.S. Reznikoff, M. Riley, M. Schaechter, and H.E. Umbarger, eds.). American Society for Microbiology Press, Washington, D.C., pp. 434–441.

8. Reitzer, L.J. 1996. Ammonia assimilation and the biosynthesis of glutamine, glutamate, aspartate, asparagine, L-alanine, and D-alanine. In Escherichia coli *and* Salmonella: *Cellular and Molecular Biology* (F.C. Neidhardt, R. Curtiss III, J.L. Ingraham, E.C.C. Lin, K.B. Low, B. Magasanik, W.S. Reznikoff, M. Riley, M. Schaechter, and H.E. Umbarger, eds.). American Society for Microbiology Press, Washington, D.C., pp. 391–407.

9. Stauffer, G.V. 1996. Biosynthesis of serine, glycine and one-carbon units. In Escherichia coli *and* Salmonella: *Cellular and Molecular Biology* (F.C. Neidhardt, R. Curtiss III, J.L. Ingraham, E.C.C. Lin, K.B. Low, B. Magasanik, W.S. Reznikoff, M. Riley, M. Schaechter, and H.E. Umbarger, eds.). American Society for Microbiology Press, Washington, D.C., pp. 506–513.

10. Umbarger, H.E. 1996. Biosynthesis of the branched-chain amino acids. In Escherichia coli *and* Salmonella: *Cellular and Molecular Biology* (F.C. Neidhardt, R. Curtiss III, J.L. Ingraham, E.C.C. Lin, K.B. Low, B. Magasanik, W.S. Reznikoff, M. Riley, M. Schaechter, and H.E. Umbarger, eds.). American Society for Microbiology Press, Washington, D.C., pp. 442–457.

11. Winkler, M.E. 1996. Biosynthesis of histidine. In Escherichia coli *and* Salmonella: *Cellular and Molecular Biology* (F.C. Neidhardt, R. Curtiss III, J.L. Ingraham, E.C.C. Lin,

K.B. Low, B. Magasanik, W.S. Reznikoff, M. Riley, M. Schaechter, and H.E. Umbarger, eds.). American Society for Microbiology Press, Washington, D.C., pp. 485–505.

12. Karlsoson, M. and M. Sahlin. 1997. Microbial ribonucleotide reductases—essential and diverse. In *Transition Metals in Microbial Metabolism* (G. Winkelmann and C.J. Carrano, eds.). Harwood Academic, Amsterdam, pp. 435–470.

13. Linn, S. 1966. The DNAses, topoisomerases, and helicases of *Escherichia coli*. In Escherichia coli *and* Salmonella: *Cellular and Molecular Biology* (F.C. Neidhardt, R. Curtiss III, J.L. Ingraham, E.C.C. Lin, K.B. Low, B. Magasanik, W.S. Reznikoff, M. Riley, M. Schaechter, and H.E. Umbarger, eds.). American Society for Microbiology Press, Washington, D.C., pp. 764–772.

14. Marians, K.J. 1996. Replication fork propagation. In Escherichia coli *and* Salmonella: *Cellular and Molecular Biology* (F.C. Neidhardt, R. Curtiss III, J.L. Ingraham, E.C.C. Lin, K.B. Low, B. Magasanik, W.S. Reznikoff, M. Riley, M. Schaechter, and H.E. Umbarger, eds.). American Society for Microbiology Press, Washington, D.C., pp. 749–763.

15. Björk, G.R. 1987. Stable RNA modification. In Escherichia coli *and* Salmonella typhimurium: *Cellular and Molecular Biology* (F.C. Neidhardt, J.L. Ingraham, B. Magasanik, K.B. Low, M. Schaechter, and H.E. Umbarger, eds.). American Society for Microbiology Press, Washington, D.C., pp. 861–886.

16. Green, C.J. and B.S. Vold. 1993. tRNA, tRNA processing, and aminoacyl-tRNA synthesases. In Bacillus subtilis *and Other Gram-Positive Bacteria* (A.L. Sonenshein, J.A. Hoch, and R. Losick, eds.). American Society for Microbiology Press, Washington, D.C., pp. 683–698.

17. King, T.C. and D. Schlessinger. 1987. Processing of RNA transcripts. In Escherichia coli *and* Salmonella typhimurium: *Cellular and Molecular Biology* (F.C. Neidhardt, J.L. Ingraham, B. Magasanik, K.B. Low, M. Schaechter, and H.E. Umbarger, eds.). American Society for Microbiology Press, Washington, D.C., pp. 703–718.

18. Russell, P.J. 1996. *Genetics*, 4th edit. HarperCollins, New York.

19. Richardson, J.P. and J. Greenblatt. 1996. Control of RNA chain elongation and termination. In Escherichia coli *and* Salmonella: *Cellular and Molecular Biology* (F.C. Neidhardt, R. Curtiss III, J.L. Ingraham, E.C.C. Lin, K.B. Low, B. Magasanik, W.S. Reznikoff, M. Riley, M. Schaechter, and H.E. Umbarger, eds.). American Society for Microbiology Press, Washington, D.C., pp. 822–848.

20. Lonetto, M.A. and C.A. Gross. 1996. Nomenclature of sigma factors from *Escherichia coli* and *Salmonella typhimurium* and relationships to sigma factors from other organisms. In Escherichia coli *and* Salmonella: *Cellular and Molecular Biology* (F.C. Neidhardt, R. Curtiss III, J.L. Ingraham, E.C.C. Lin, K.B. Low, B. Magasanik, W.S. Reznikoff, M. Riley, M. Schaechter, and H.E. Umbarger, eds.). American Society for Microbiology Press, Washington, D.C., p. 821.

21. Record, M.T., Jr., W.S. Reznikoff, M.L. Craig, K.L. McQuaide, and P.J. Schlax. 1996. *Escherichia coli* RNA polymerase ($E\sigma^{70}$), promoters, and the kinetics of the steps of transcription initiation. In Escherichia coli *and* Salmonella: *Cellular and Molecular Biology* (F.C. Neidhardt, R. Curtiss III, J.L. Ingraham, E.C.C. Lin, K.B. Low, B. Magasanik, W.S. Reznikoff, M. Riley, M. Schaechter, and H.E. Umbarger, eds.). American Society for Microbiology Press, Washington, D.C., pp. 792–820.

22. Al-Karadaghi, S., N. Davydova, I. Eliseikina, M. Garber, A. Liljas, N. Nevskaya, S. Nikonov, and S. Tishchenko, 2000. Ribosomal proteins and their structral transitions on and off the ribosome. In *The Ribosome Structure, Function, Antibiotics and Cellular Interactions* (R.A. Garrett, S.R. Douthwaite, A. Lilijas, A.T. Matheson, P.B. Moore, and H.F. Noller, eds.). American Society for Microbiology Press, Washington, D.C., pp. 65–72.

23. Mathews, C.K. and K.E. van Holde, 1990. *Biochemistry.* Benjamin/Cummings, Redwood City, Calif.

24. Noller, H.F. and M. Nomura. 1996. Ribosomes. In Escherichia coli *and* Salmonella: *Cellular and Molecular Biology* (F.C. Neidhardt, R. Curtiss III, J.L. Ingraham, E.C.C. Lin, K.B. Low, B. Magasanik, W.S. Reznikoff, M. Riley, M. Schaechter, and H.E. Umbarger, eds.). American Society for Microbiology Press, Washington, D.C., pp. 104–125.

25. Douthwaite, S. and B. Vester, 2000. Macrolide resistance conferred by alterations in the ribosome target site. In *The Ribosome Structure, Function, Antibiotics and Cellular Interactions* (R.A. Garrett, S.R. Douthwaite, A. Lilijas, A.T. Matheson, P.B. Moore, and H.F. Noller, eds.). American Society for Microbiology Press, Washington, D.C., pp. 431–439.

26. Hershey, J.W. 1987. Protein synthesis. In Escherichia coli *and* Salmonella typhimurium: *Cellular and Molecular Biology* (F.C. Neidhardt, J.L. Ingraham, B. Magasanik, K.B. Low, M. Schaechter, and H.E. Umbarger, eds.). American Society for Microbiology Press, Washington, D.C., pp. 614–647.

27. Mandelstam, J., K. McQuillen, and I. Dawes. 1986. *Biochemistry of Bacterial Growth.* Blackwell, Boston.

28. Baron, C., D. O'Callaghan, and E. Lanka. 2002. Bacterial secrets of secretion: Euro-Conference on the biology of type IV secretion process. *Molecular Microbiology* 43:1359–1365.

29. Fernández, L.A., and J. Berenguer. 2000. Secretion and assembly of regular surface structures in Gram-negative bacteria. *FEMS Microbiology Reviews* 23:21–44.

30. Saier, M.H. 2001. Families of transporters: A phylogenic overview. In *Microbial Transport Systems* (G. Winkelmann, ed.). Wiley-VCH, New York, pp. 1–22.

31. Saier, M.H., M.J. Fagan, C. Hoischen, and J. Reizer. 1993. Transport mechanisms. In Bacillus subtilis *and Other Gram-Positive Bacteria* (A.L. Sonenshein, J.A. Hoch, and R. Losick, eds.). American Society for Microbiology Press, Washington, D.C., pp. 133–156.

32. Binet, R., S. Létoffé, J.M. Ghigo, P. Delepelaire, and C. Wandersman. 1997. Protein secretion by gram-negative bacterial ABC exporters—a review. *Gene* 192:7–11.

33. Tjalsma, H., A. Bolhuis, J.D.H. Jongbloed, S. Bron, and J. van Dijl. 2000. Signal peptide-dependent protein transport in *Bacillus subtilis*: A genome-based survey of the secretome. *Microbiology and Molecular Biology Reviews* 64:515–547.

34. Aizawa, S.-I. 2001. Bacterial flagella and type III secretion systems. *FEMS Microbiology Letters* 202:157–164.

35. Delepelaire, P. and C. Wandersman. 2001. Protein export and secretion in Gram-negative bacteria. In *Microbial Transport Systems* (G. Winkelmann, ed.). Wiley-VCH, New York, pp. 165–208.

36. Christie, P.J. and A. Covacci. 2000. Bacterial type IV secretion systems: DNA conjugation machines adopted for export of virulence factors. In *Cellular Microbiology* (P. Cossart, P. Boquet, S. Normark, and R. Rappuoli, eds.). American Society for Microbiology Press, Washington, D.C., pp. 265–273.

37. Schesser, K., M.S. Francis, Å. Forsberg, and H. Wolf-Watz. 2000. Type III secretion systems in animal- and plant-interacting bacteria. In *Cellular Microbiology* (P. Cossart, P. Boquet, S. Normark, and R. Rappuoli, eds.). American Society for Microbiology Press, Washington, D.C., pp. 239–263.

38. Wandersman, C. 1996. Secretion across the bacterial outer membrane. In Escherichia coli *and* Salmonella: *Cellular and Molecular Biology* (F.C. Neidhardt, R. Curtiss III, J.L. Ingraham, E.C.C. Lin, K.B. Low, B. Magasanik, W.S. Reznikoff, M. Riley, M. Schaechter, and H.E. Umbarger, eds.). American Society for Microbiology Press, Washington, D.C., pp. 955–966.

39. Murphy, C.K. and J. Beckwith. 1996. Export of proteins to the cell envelope in *Escherichia coli*. In Escherichia coli *and* Salmonella: *Cellular and Molecular Biology* (F.C. Neidhardt, R. Curtiss III, J.L. Ingraham, E.C.C. Lin, K.B. Low, B. Magasanik, W.S. Reznikoff, M. Riley, M. Schaechter, and H.E. Umbarger, eds.). American Society for Microbiology Press, Washington, D.C., pp. 967–978.

40. Hettmann, T., C.L. Schmidt, S. Anemüller, U. Zähringer, H. Moll, A. Petersen, and G. Schäfer. 1998. Cytochrome b$_{558/566}$ from the archaeon *Sulfolobus acidocoldarus*. *Journal of Biological Chemistry* 273:12032–12040.

41. Sumper, M. 1987. Halobacterial glycoprotein synthesis. *Biochimie Biophysica Acta* 907: 69–79.

42. Maras, B., V. Consalvi, R. Chiaraluce, L. Politi, M. De Rosa, F. Bossa, R. Scandurra, and D. Barra. 1992. The protein sequence of glutamate dehydrogenase from *Sulfolobus solfataricus*, a thermoacidophilic archaebacterium. *European Journal of Biochemistry* 203: 81–87.

43. Sahl, H.-G. and G. Bierbaum. 1998. Lantibiotics: Biosynthesis and biological activity of uniquely modified peptides from Gram-positive bacteria. *Annual Review of Microbiology* 52:41–79.

44. Zuber, P., M.M. Nakano, and M.A. Marahiel. 1993. Peptide antibiotics. In Bacillus subtilis *and Other Gram-Positive Bacteria* (A.L. Sonenshein, J.A. Hoch, and R. Losick, eds.). American Society for Microbiology Press, Washington, D.C., pp. 897–916.

45. Kertesz, M.A. 1999. Riding the sulfur cycle—metabolism of sulfonates and sulfate esters in Gram-negative bacteria. *FEMS Microbiology Reviews* 24:135–175.

46. Diks, A. 1999. *Bacillus subtilis* spore coat. *Microbiology and Molecular Biology Reviews* 63:1–29.

47. McAfee, J.G., S.P. Edmondson, P.K. Datta, J.W. Shriver, and R. Gupta. 1995. Gene cloning, expression, and characterization of the Sac7 proteins from the hyperthermophile *Sulfolobus acidocaldarius*. *Biochemistry* 34:10063–10077.

48. Sandman, K. and J.N. Reeve. 2001. Chromosome packing by archaeal histones. *Advances in Applied Microbiology* 50:75–100.

49. Beale, S.I. 1996. Biosynthesis of Hemes. In Escherichia coli *and* Salmonella: *Cellular and Molecular Biology* (F.C. Neidhardt, R. Curtiss III, J.L. Ingraham, E.C.C. Lin, K.B. Low, B. Magasanik, W.S. Reznikoff, M. Riley, M. Schaechter, and H.E. Umbarger, eds.). American Society for Microbiology Press, Washington, D.C., pp. 731–748.

50. Thöny-Meyer, L. 1997. Biogenesis of respiratory chtochromes in bacteria. *Microbiology and Molecular Biology Reviews* 61:337–376.

51. Lengeler, J.W. and P.W. Postma. 1999. Global regulatory networks and signal transduction pathways. In *Biology of the Prokaryotes* (J.W. Lengeler, G. Drews, and H.G. Schlegel, eds.). Blackwell, New York, pp. 491–523.

52. Madigan, M.T., J.M. Martinko, and J. Parker. 2003. *Brock Biology of Microorganisms*. Prentice Hall, Upper Saddle River, N.J.

## Additional Reading

Fruton, J.S. and S. Simmonds. 1958. *General Biochemistry*. John Wiley, New York.

## Secretion Systems

Goosney, D.L., S. Gruenheid, and B.B Finlay. 2000. Gut feelings: Enteropathogenic *E. coli* (EPEC) interactions with the host. *Annual Review of Cell Developmental Biology* 16:173–189.

Hueck, C.J. 1998. Type III protein secretion systems in bacterial pathogens of animals and plants. *Microbiology and Molecular Biology Reviews* 62:379–433.

Putman, M. H.W. van Veen, and W.N. Konings. 2000. Molecular properties of bacterial multidrug transporters. *Microbiology and Molecular Biology Reviews* 64:672–693.

Salmond, G.P.C. and P.J. Reeves. 1993. Membrane traffic wardens and protein secretion in gram-negative bacteria. *Trends in Biochemical Sciences* 18:7–12.

## Posttranslational Processing

Faguy, D.M., S.F. Koval, and K.F. Jarrell. 1992. Correlation between glycosylation of flagellin proteins and sensitivity of flagellar filaments to Triton X-100 in methanogens. *FEMS Microbiology Letters* 90:129–134.

R.A. Garrett, S.R. Douthwaite, A. Lilijas, A.T. Matheson, P.B. Moore, and H.F. Noller (eds.). 2000. *The Ribosome Structure, Function, Antibiotics and Cellular Interactions.* American Society for Microbiology Press, Washington, D.C.

Moens, S. and J. Vanderleyden. 1997. Glycoproteins in prokaryotes. *Archives of Microbiology* 168:169–175.

## Histone-Like Proteins

Xue, H., R. Guo, Y. Wen, D. Liu, and L. Huang. 2000. An abundant DNA binding protein form the hyperthermophilic archaeon *Sulfolobus shibatae* affects DNA supercoiling in a temperature-dependent fashion. *Journal of Bacteriology* 182:3929–3933.

## Transcription

Soppa, J. 2001. Basal and regulated transcription in archaea. *Advances in Applied Microbiology* 50:171–219.

## Heme and Cobalamin Synthesis

Dailey, H.A. (ed.). 1990. *Biosynthesis of Hemes and Chlorophylls.* McGraw-Hill, New York.

O'Toole, G.A., M.R. Rondon, J.R. Trzebiatowski, S.-J. Suh, and J.C. Escalante-Semerena. 1996. Biosynthesis and utilization of adenosyl-cobalamin (coenzyme $B_{12}$). In Escherichia coli *and* Salmonella: *Cellular and Molecular Biology* (F.C. Neidhardt, R. Curtiss III, J.L. Ingraham, E.C.C. Lin, K.B. Low, B. Magasanik, W.S. Reznikoff, M. Riley, M. Schaechter, and H.E. Umbarger, eds.). American Society for Microbiology Press, Washington, D.C., pp. 710–720.

# 13

# Metabolism of Nitrogen and Sulfur

> While there appears to be a remarkable "unity of biochemistry" throughout life yet there are divergent pathways for both the assimilatory and the dissimilatory reactions in the same or different species. C.E. Clifton, *Bacterial Physiology*, 1951

## 13.1. Introduction

Two major themes are displayed by prokaryotes in the utilization of inorganic nitrogen and sulfur compounds: one feature is for biosynthetic reactions and the other is for energy-related activity. Most bacteria and archaea do not require preformed organic compounds containing sulfur or nitrogen but these microbial cells can synthesize amino acids, nucleic acids, and various organic compounds from simple inorganic compounds. A few specialized bacteria can use sulfate as the terminal electron acceptor with the production of $H_2S$, and several different species of bacteria can use nitrate as the final electron acceptor with the formation of either nitrite or $N_2$. A key feature of both the biosynthetic and the electron acceptor activities is the requirement of sulfate and nitrate reduction prior to microbial utilization. Thus, there are the distinct processes referred to as assimilatory reduction of nitrate, dissimilatory reduction of nitrate, assimilatory reduction of sulfate, and the dissimilatory reduction of sulfate. The dissimilatory processes are used in anaerobic respiration and the assimilatory processes are used for production of cellular materials.

## 13.2. The Nitrogen Cycle[1]

The cellular material of prokaryotes is approximately 60% protein and if one uses the conversion that protein is 16% N, then approximately 10% of the average bacterial cell is N. Environmental engineers have analyzed bacterial cells for different elements and have expressed bacterial composition by the following chemical formula: $C_6H_7O_2N$. Thus, for growth bacteria would require C:H:O:N at a ratio of 6:7:2:1. To support this requirement for cell nitrogen, prokaryotes have developed several different systems for utilization of various forms of inorganic nitrogen and these activities are part of the nitrogen cycle (Figure 13.1). Oceans, for example, contain nitrate:nitrite: ammonium in a ratio of 1,140:1:14. As the soluble forms of nitrogen are removed from land and ocean environments, prokaryotes fix nitrogen in terresterial and ocean environments at the level of $1.4 \times 10^{14}$ g $N_2$/year and $1.0 \times 10^{14}$ g $N_2$/year, respectively. However, not all of nitrogen metabolism is associated with synthesis of biomass because some bacteria exploit the energy released from the nitrogen reactions to support growth. For example, although $N_2$ is fixed into $NH_3$, considerable quantities of ammonia are oxidized by microorganisms to nitrate, and ammonia is the electron donor for growth. In addition, microbial denitrification accounts for microbial production of $N_2$ from nitrate with the release of $1.2 \times 10^{14}$ g $N_2$/year from land and $9.0 \times 10^{13}$ g $N_2$/year from the oceans. For many bacteria, nitrate serves as a terminal electron acceptor, and thereby enables continuous oxidation of substrate with coupled ATP production. The flux of nitrogen in the biosphere not only provides nitrogen for cellular biosynthesis but also contributes to the bioenergetics of many prokaryotes.

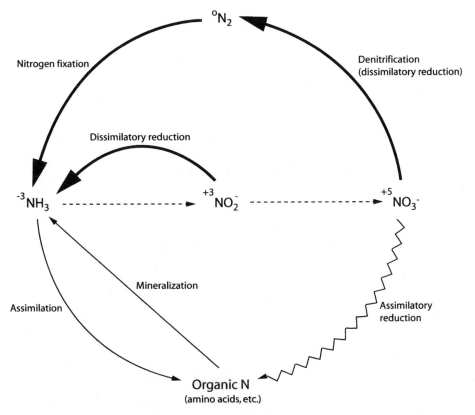

Figure 13.1. Major features of the nitrogen cycle that involve prokaryotes.

## 13.2.1. Dinitrogen Fixation[2]

Some of the earliest cultures of bacteria focused on the conversion of atmospheric $N_2$ to $NH_3$. In the 1890s, Winogradsky isolated the nitrogen-fixing anaerobe *Clostridium pasteurianum* and in 1901, Beijerinck isolated the aerobic nitrogen-fixing soil bacterium *Azotobacter chroococcum*. Gradually, the breadth of nitrogen fixation activity became known and it is now recognized that prokaryotes are extremely important for enrichment of the environment to support plant growth. Nitrogen fixation is found only in prokaryotes and is characteristic of some species of cyanobacteria (i.e., *Anabaena* and *Aphanazonemon*), heterotrophic aerobic soil bacteria (i.e., *Azotobacter*, *Azospirillum*, *Klebsiella*, *Beggiatoa*, and *Frankia*), anaerobic bacteria (i.e., *Clostridium* and *Desulfovibrio*), symbiotic bacteria growing on legume roots (i.e., *Sinorhizobium*, *Rhizobium*, and *Bradyrhizobium*), and archaeal species (i.e., *Methanosarcina* and *Methanococcus*). The general reaction for nitrogen fixation is as follows:

$$N_2 + 8\ e^- + 8\ H^+ + 16\ ATP \rightarrow 2\ NH_3 + H_2 + 16\ ADP + 16\ P_i$$

The reaction accounting for dinitrogen fixation contains two units: "component I" and "component II." These two components are designated by using the first letter of the genus and species. For example, components I and II of nitrogenase for *Azotobacter vinelandii* would be Av1 and Av2, respectively. Alternate terminology designates component I as the dinitrogenase reductase and component II as the dinitrogenase. As diagrammatically presented in Figure 13.2, molecular hydrogen is formed and this is attributed to reduction by component I of nitrogenase and $H_2$ is utilized by an uptake hydrogenase.

The source of electrons for the Fe-protein is derived from the conversion of pyruvate to acetyl-CoA by a pyruvate dehydrogenase and the half-reaction elicits about $-610$ mV. Electrons from this reaction may be carried by either reduced ferredoxin or reduced flavodoxin to a Fe-S cluster on the Fe-protein. These half-reactions for ferredoxin$_{ox}$ + 1 $e^-$/ferredoxin$_{red}$ and flavodoxin$_{ox}$ + 1 $e^-$/flavoxodin$_{red}$ are $-388$ mV and $-360$ mV, respectively.

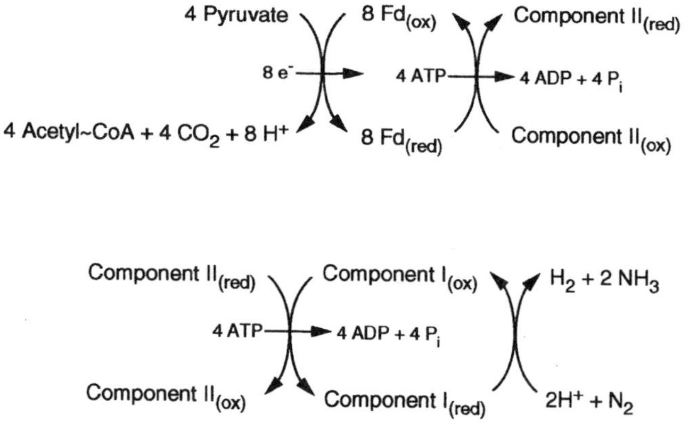

Figure 13.2. Role of components I and II in nitrogen fixation.

Component II is an Fe-protein of approximately 60 kDa that is a dimer composed of identical subunits. A 4Fe-4S cluster is attached through cysteinyl ligands that are located at the amino end of the dimer and this iron-sulfur center is attached at the subunit-subunit interface. The Fe-protein has two binding sites per dimer. One site contains the amino acid sequence GXXXXGKS (Walker's motif A; X is any amino acid) characteristic for the binding of nucleotides, MgATP, and MgADP, and a second motiff is DXXG which binds the Mg-phosphate protein interaction. The second functional site is the [4Fe–4S] cluster which involves a one-electron redox step with interconversion of the [2 $Fe^{2+}-2$ $Fe^{3+}$] and [3 $Fe^{2+}-Fe^{3+}$] states. The Fe protein subunit binds two molecules of MgATP and the result is a drop in the redox potential of about 100 mV. The mechanism proposed for the Fe-protein is a coupling of ATP hydrolysis with electron transfer. Because the redox potential of this component is $-380$ mV and only the reduction of the triple bond in $N_2$ to the double bond state has a slightly lower redox potential, there is no apparent use of ATP for thermodynamic considerations but rather ATP is suggested to be used as a "gated avenue" that regulates the flow of electrons through the Fe-protein.

Component I is a MoFe-protein of approximately 240 kDa consisting of an $\alpha_2\beta_2$ tetramer with minimal amino acid sequence homology between α- and β-subunits. The MoFe cofactor for each of the two MoFe-protein units appears to be formed from clusters composed of a FeMo-cofactor, 4Fe–3S and Mo–3Fe–3S (see Figure 13.3). Also present in component I is a "P cluster" that consists of two linked 4Fe-4S centers with cysteine ligands from each subunit involved in holding both 4Fe-4S centers. Although the electron flow in nitrogenase is not yet fully resolved, there is sufficient evidence to suggest that electrons would be passed from the iron-sulfur center of the Fe-protein to the P cluster and finally to the FeMo-cofactor which binds the substrate, dinitrogen. Not all prokaryotes have the same component I of nitrogenase but there

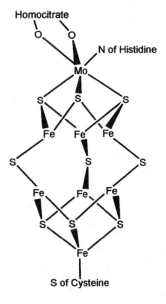

Figure 13.3. The FeMo-cofactor in nitrogenase.

are some variations. For example, the Av1 (component I of *Azotobacter vinelandii*) and the Cp1 (component I of *Clostridium pasteuranium*) differ in that Av1 has a deletion of approximately 50 amino acids in the β-subunit while there is an insert of approximately 50 amino acids in the α-subunit.

*Azotobacter vinelandii* and *Azotobacter chroococcum* are remarkable organisms because they produce a vanadium nitrogenase that contains a VFe-protein as an alternate to the MoFe-protein. In addition, if both Mo and V are limiting nutrients, a third nitrogenase is produced by these bacteria, the FeFe-protein. Under normal conditions it will produce a molybdenum-iron cofactor in component I but if molybdenum is limiting a vanadium-iron containing component I will be produced. The VFe-protein and the FeFe-protein are about 220 kDa and are proposed to have $\alpha_2\beta_2\gamma_2$ composition. Although the role of the γ-subunit is unknown, this 13-kDa protein is essential for nitrogenase activity. The VFe-protein contains 2 V atoms, 23 Fe atoms, and 20 acid-labile sulfides. V is substituted for Mo in the formation of the MoFe cofactor.

The great amount of interest in the nitrogenase reaction has stimulated scientists to examine the genetics of this system in *A. vinelandii*. It appears there may be as many as 20 genes in the bacterial *nif* (nitrogen fixation) regulon with a gene for each of the following: the Fe-protein (*nifH*), the α-subunit of the MoFe-protein (*nifD*), the β-subunit of the MoFe-protein (*nifK*), flavodoxin (*nifF*), and pyruvate dehydrogenase (*nifJ*). The *nifA* and *nifL* genes are involved in regulation and prevent expression of DNA to produce nitrogenase if ammonia is available. No doubt several other genes code for regulatory proteins that participate in processing FeMo cofactor and maturation of components I and II.

The utilization of $H_2$ by nitrogenase systems of bacteria is attributed to a $H_2$ uptake hydrogenase that is encoded on the *hup* gene. Most of these uptake hydrogenases are membrane-bound enzymes composed of a 30-kDa and a 60-kDa subunit. These hydrogenases contain FeS clusters and require $Ni^{2+}$ for activity. Excess $H_2$ is produced because the $2\ e^- + 2\ H^+ \rightarrow H_2$ reaction proceeds at a rate 100 times greater than the $H_2 \rightarrow 2\ e^- + 2\ H^+$ reaction of the uptake hydrogenases. The physiological role of the uptake hydrogenase has been suggested to be protecting the complex against damage by molecular oxygen and producing ATP by recycling $H_2$ through the electron transport chain. The nitrogen fixation system is rapidly inactivated by more than 1 ppm oxygen; therefore, aerobic bacteria have evolved systems to exclude air from the cytoplasm where nitrogenase resides.

The assay of nitrogenase using $^{13}N$, a β-decay isotope, is not routinely conducted because the isotope has an extremely short half-life of only about 10 min. With the use of $^{15}N_2$, a stable nonradioactive isotope as the dinitrogen source, the nitrogen-fixing enzyme produces $^{15}NH_3$. Nuclear magnetic resonance (NMR) analysis can be used to measure the amount of $^{15}N$ produced; however, these experiments are relatively tedious and not suited for a large number of samples. It was discovered that nitrogenase will react with acetylene, azide, HCN, alkyl cyanides, isocyanides, and other molecules that have a triple bond. The use of acetylene, $C_2H_2$, has proven to be a most effective means of assaying nitrogenase because the product of the reaction is ethylene, $C_2H_2$, which can be easily measured by gas chromatographic methods. Not only is the acetylene reduction assay highly sensitive, but it is also relatively inexpensive to test environmental samples.

## 13.2.2. *Assimilatory Nitrate Reduction*[3]

Nitrate assimilation has been extensively studied in *Klebsiella oxytoca* (previously known as *Klebsiella pneumoniae*), *Bacillus subtilis, A. vinelandii*, and various strains of cyanobacteria. The general theme found in these bacteria is that nitrate assimilation proceeds in several phases: (1) nitrate is transported into the cell by a specific binding protein, (2) nitrate is reduced to nitrite, and (3) nitrite is reduced to ammonia. Of the bacterial systems thus far examined, the genes for nitrate assimilation are designated *nrt* for nitrate and nitrite transport, *nar* for nitrate reduction, and *nir* for nitrite reduction. In the *Klebsiella* bacterial system, activities of nitrate and nitrite assimilation are encoded on *nas* genes (see Figure 13.4). The genes and enzymes for assimilatory nitrate reduction are distinct from those for dissimilatory reduction of nitrate.

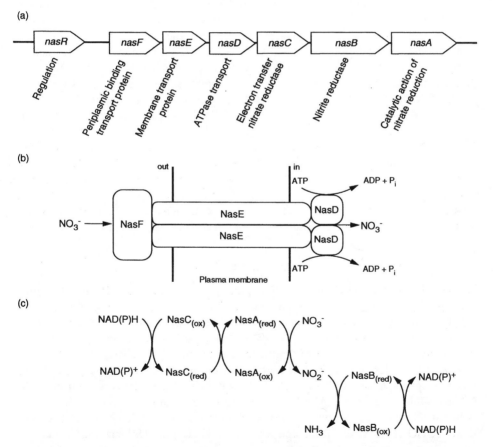

Figure 13.4. Assimilatory nitrate reduction systems in *Klebsiella*. (**a**) Gene organization and gene products. (**b**) Transport system for nitrate uptake by an ABC transporter system in bacteria. NasF is a periplasmic nitrate binding protein while NasE is a transmembrane protein and NasD is the ATPase of the ATP driven uptake system. (**c**) Sequence of reactions.

Table 13.1. Cofactors associated with the assimilatory nitrate reduction system in *Klebsiella oxytoca*.

| Enzyme | Protein | Cofactor |
|--------|---------|----------|
| Nitrate reductase | NasC | FAD |
| Nitrate reductase | NasA | 2Fe-2S, 4Fe-4S, and Mo-MGD |
| Nitrite reductase | NasB | FAD, 4Fe-4S, and a siroheme |

The first enzyme for nitrate assimilatory reduction is nitrate reductase, and three different locations of this enzyme have been found in enterobacteria. One form of the nitrate reductase is soluble and is in the cytoplasm, another is found in the periplasm, and the third is bound to the plasma membrane. The soluble cytoplasmic enzyme, approximately 140 kDa, accounts for assimilatory reduction of nitrate and uses a unique molybdopterin as a cofactor. The electron source for the assimilatory nitrate reductase varies with the organism and may include reduced ferredoxin, reduced flavodoxin, or NAD(P)H. The periplasmic enzyme, approximately 90 kDa, may be important in maintaining a proper redox state with anaerobic grown cells, and the enzyme in the plasma membrane is used for dissimilatory nitrate reduction. The periplasmic nitrate reductase may also have a role in signal activity for chemotaxis.

The second step in the bacterial nitrate assimilatory pathway is the cytoplasmic nitrite reductase with the production of $NH_3$. Whereas the plant and cyanobacterial nitrite reductase are ferredoxin dependent for electron reduction, the bacterial and fungal systems use NADH or NADPH as the source of electrons. In *Klebsiella oxytoca*, the movement of electrons in this six-electron step through the nitrite reductase is not a simple mechanism of reduction because it involves numerous cofactors. The cofactors for the various assimilatory reduction processes are given in Table 13.1. For nitrite reductase the general flow of electrons is considered to be as follows: NADH → FAD → a 4Fe-4S cluster → seroheme → nitrite with the formation of ammonium.

## 13.2.3. Assimilation of Ammonia[4,5]

Nitrogen is an essential element for microbial growth, with about 10 mmol required for 1 g of prokaryotic cells. Assimilation of ammonia is the central avenue for the supply of nitrogen to carbon chains produced from intermediary metabolism. The major pathways for assimilation of ammonia involves the production of glutamate and glutamine. As listed in Table 13.2, glutamate dehydrogenase (GDH) will catalyze the formation of L-glutamate from α-ketoglutarate and ammonia. Reducing power for the reaction comes from pyridine nucleotides with GDH from some organisms using NADH while others use NADPH. A second major system for assimilation of ammonia involves the combined action of glutamine synthetase (GS) and glutamate synthase (GOGAT, derived from the common name for this enzyme: *glutamine amide-2-oxoglutarate amino transferase*). The synthesis of glutamine by GS requires ATP for the addition of ammonia to glutamate. The production of 2 mol of glutamate is a reductive process of combining 1 mol of α-ketoglutarate with 1 mol of L-glutamine.  These reactions of GS and GOGAT are given in Table 13.2.  Both GDH and GS-GOGAT enzymes produce L-glutamate.  Genes for these enzymes are broadly distributed throughout the bacteria and the specific enzymes employed are dependent on the amount of ammonia present.  The GDH has a low affinity for ammonia and if the concentration is too low, GDH does not function.  The GS-GOGAT system is induced if the concentration of ammonia is less than 1 m$M$, making this system important at low ammonia concentrations.  A few other enzymes may be used by prokaryotes for utilization of ammonia and these constitute an alternative pathway for ammonia assimilation.  Some of these alternative pathways include asparagine synthetase, pyruvate dehydrogenase, and fumarase.

To produce the amino acids other than those produced by the initial reactions of ammonia assimilation, there are many reactions involving the transfer of $-NH_2$ moieties.  These activities have been termed aminotransferase reactions, with glutamate and glutamine as the primary molecules donating the amine groups.  The generalized activity involving transaminase reactions is expressed as follows, where the amine moiety is transferred from one amino acid to a keto acid which results in the generation of a second amino acid:

$$\text{Amino acid}_1 + \text{keto acid}_2 \rightarrow \text{keto acid}_1 + \text{amino acid}_2$$

Examples of various transamination reactions are given in Table 13.3.  The production of amino acids as a result of transamination reactions is accomplished with a few enzymes that recognize substrates of similar structure.  For example, *E. coli* has four genes for transamination reactions:  (1) *tryB*, which interacts with aromatic amino acids; (2) *ilvE*, which recognizes branched chain amino acids; (3) *aspC*, which interacts with aspartate and glutamate; and (4) *avtA*, which recognizes alanine and valine.

Table 13.2. Enzyme reactions involving glutamate and glutamine in bacterial assimilation of ammonia.

| Enzyme | Reaction |
|---|---|
| **Major pathway >1 m$M$ NH$_4^+$** | |
| Glutamate dehydrogenase (GDH) | $\alpha$-Ketoglutarate + NH$_4^+$ + NAD(P)H + H$^+$ $\rightarrow$ L-glutamate + NAD(P)$^+$ + H$_2$O |
| **Major pathway <1 m$M$ NH$_4^+$** | |
| Glutamine synthetase (GS) | L-Glutamate + NH$_4^+$ + ATP $\rightarrow$ L-glutamine + ADP + P$_i$ |
| Glutamate synthase (GOGAT) | $\alpha$-Ketoglutarate + L-Glutamine + NAD(P)H + H$^+$ $\rightarrow$ 2 L-glutamate + NAD(P)$^+$ |
| **Alternate pathways** | |
| Asparagine synthetase | L-Aspartate + NH$_4^+$ + ATP $\rightarrow$ L-asparagine + AMP + PP$_i$ |
| Pyruvate dehydrogenase | Pyruvate + NH$_4^+$ + NAD(P)H + H$^+$ $\rightarrow$ L-alanine + NAD(P)$^+$ |

Table 13.3. The use of the amino group in glutamate for building other nitrogen-containing compounds.

| Enzyme required | Reactions |
|---|---|
| **Amino transfer with keto acids:** | |
| Glutamic-pyruvate transaminase (GPT) | L-glutamate + pyruvate → α-ketoglutarate + L-alanine |
| Glutamic-oxalacetate transaminase (GOT) | L-glutamate + oxalacetate → α-ketoglutarate + L-aspartate |
| Glutamic-phenylpyruvate transaminase (GPT) | L-glutamate + L-phenylpyruvate → α-ketoglutarate + L-phenylalanine |
| **Amino transfer with keto kexoses:** | |
| | L-glutamine + keto-glucose → L-glutamate + glucosamine |
| **Amino transfer with nucleotide bases** | |
| | UTP + L-glutamine + ATP → L-glutamate + CTP + ADP + P$_i$ |

In addition, aminotransferase reactions are involved with using the –NH$_2$ group from glutamine as the source of amino groups in CTP, NAD, GTP, and carbamoyl phosphate.

The transamination reactions require pyridoxal phosphate (B$_6$) as the cofactor and these bacteria synthesize their own pyridoxal phosphate. Pyridoxine is a precursor in the formation of pyridoxal phosphate and isoniazid is a structural analog of pyridoxine. The structures of pyridoxine and isoniazid are given in Figure 13.5. When bacteria utilize isoniazid instead of pyridoxine, a nonfunctional analog of folic acid is made and the bacterium is unable to use the aminotransfer reaction and dies. This inability of bacteria to synthesize folic acid in the presence of isoniazid is the basis for the use of isoniazid as a chemotheraputic agent in treatment of mycobacterial infections.

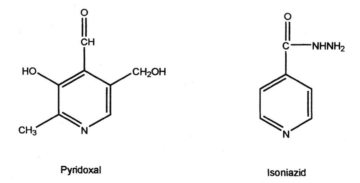

**Pyridoxal**        **Isoniazid**

Figure 13.5. Structures of pyridoxal and isoniazid.

## 13.2.4. Dissimilatory Nitrate Reduction[6,7]

The pathway for dissimilatry reduction, as first proposed in 1968 by Matsubara and Mori, proceeds with nitrogen oxide intermediates:

$$+5 \qquad +3 \qquad +2 \qquad +1 \qquad 0$$
$$NO_3^- \rightarrow NO_2^- \rightarrow NO \rightarrow N_2O \rightarrow 2\,N_2$$

Nitrate  Nitrite  Nitric  Nitrous  Dinitrogen
oxide  oxide

The redox potentials ($E_0'$) for each of the reactions is as follows: $NO_5^-/NO_2^- = +0.430$ V; $NO_2^-/NO = +0.350$ V; $NO/N_2O = +1.175$ V; and $N_2O/N_2 = +1.355$ V. These reactions associated with dissimilatory reduction of nitrate are summarized in Table 13.4.

Although Kluyver recognized in 1926 that denitrification provided bacteria with the opportunity for anaerobic growth, the progress in understanding the fundamental process has been slow and fruitful investigations are assured for some time into the future. It has been relatively recent that nitrate reduction has been recognized as an important respiratory process that charges the plasma membrane and promotes anaerobic oxidative phosphorylation. Thus, it is important to examine the location of the various $NO_x$ reductases in the membrane and understand the potential for proton pumping. Denitrification is found in halophilic and hyperthermophilic archaea as well as in many of the heterotrophic and chemolithotrophic bacteria belonging to the subclasses of *Proteobacteria*. Bacteria have evolved with various levels of oxygen tolerance for denitrification activity. *Paracoccus denitrificans* and *Thiobacillus versutus* will denitrify only in the absence of oxygen, *Hyphomicrobium* X will reduce nitrate in 15% air saturation, *Alcaligenes faecalis* S6 is active in 72% air saturation, and *Thiospaera pantotropha* displays denitrification activity at 95% air saturation.

The molecular biology of denitrification has been examined to varying levels of completeness in *Pseudomonas aeruginosa, Pseudomonas stutzeri, Paracoccus denitrificans, Ralstonia eutropha* (formerly *Alcaligenes eutrophus*), *Sinorhizobium* (formerly *Rhizobium*) *meliloti,* and *Klebsiella oxytoca.* When the genes encoding nitrate respiration (*nar*), nitrite respiration (*nir*), NO respiration (*nor*), and $N_2O$ respiration (*nos*) are identified in bacteria, they are frequently found in clusters. *Pseudomonas*

Table 13.4. Enzymes involved in microbial denitrification.[a]

| Enzymes | Reactions |
|---|---|
| Nitrate reductase (NAR) | $NO_3^- + UQH_2 \rightarrow NO_2^- + UQ + H_2O$ |
| Cu-nitrite reductase (NIR) | $NO_2^- + Cu^{1+} + 2\,H^+ \rightarrow NO + H_2O + Cu^{2+}$ |
| $c_{d1}$-nitrite reductase (NIR) | $NO_2^- + cyto^{2+} + 2\,H^+ \rightarrow NO + H_2O + cyto^{3+}$ |
| NO reductase (NOR) | $2\,NO + 2cyto^{2+} + 2\,H^+ \rightarrow N_2O + H_2O + cyto^{3+}$ |
| $N_2$ reductase (NOS) | $N_2O\ 2cyto^{2+} + 2\,H^+ \rightarrow N_2 + H_2O + cyto^{3+}$ |

[a]From Jetten et al., 1997.[6]

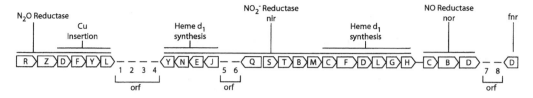

Figure 13.6. Organization of genes for dissimilatory nitrate reduction in *Pseudomonas stutzeri*. Genes with regulatory function include *R*, *y*, and *FnrD*. Orf is an open reading frame. (From Zumft, 1997.[7])

*stutzeri* has the most genes known for denitrification in a single super cluster of 33 genes (see Figure 13.6); however, the *nar* genes for nitrate reduction are not linked to these denitrification genes.

### 13.2.4.1. Nitrate Reductase[8,9]

Three types of nitrate reductases are commonly found in microorganisms: (1) an assimilatory-type enzyme localized in the cytoplasm, (2) a dissimilatory-type enzyme found in the periplasm, and (3) a dissimilatory-type enzyme in the plasma membrane. The dissimilatory nitrate reductases are broadly distributed in facultative and strict anaerobes. The reaction catalyzed by dissimilatory nitrate reductases is as follows:

$$NO_3^- + H_2 \rightarrow NO_2^- + H_2O \qquad (\Delta G^{0'} = -163.2 \text{ kJ/mol})$$

The periplasmic enzyme is produced and functions under aerobic conditions while the membrane-bound enzyme is produced only under anaerobic conditions. It has been suggested that the periplasmic nitrate reductase may function in the cell transition from an aerobic to an anaerobic state and may promote cellular homeostasis by producing nitrite for ammonia formation. A scheme indicating the location of the denitrification enzymes is given in Figure 13.7.

The respiratory nitrate reductases are difficult to isolate from the plasma membrane and the method of solubilization appears to be critical. The prosthetic groups generally reported consist of several Fe-S clusters, a heme group, and a molybdenum-containing cofactor. In some instances, the molybdenum cofactor is molybdopterin guanine dinucleotide (MGD); however, in many of the cases the exact nature of the molybdo-cofactor has not been determined. The structure of MGD is given in Figure 13.8 and it may be the active site for binding nitrate.

The membrane-bound nitrate reductase in *E. coli* consists of four genes: *narG, narH, narJ,* and *narI*. The NarG protein is 112 to 145 kDa and it binds the molybdenum cofactor. The NarH protein is 58 kDa and it contains three [4Fe-4S] clusters and one [3Fe-4S] cluster. The redox potentials of these four Fe-S clusters have been determined to be $+80$, $+60$, $-200$, and $-400$ mv. The NarL protein is 19 to 23 kDa, is transmembrane, and binds the NarH and NarG subunits on the cytoplasmic side. Bound to the NarL complex is heme B.

The periplasmic nitrate reductase from *Paracoccus denitrificans* is encoded on a

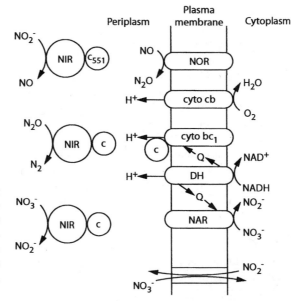

Figure 13.7. A model indicating proposed localization of denitrification enzymes on the cell surface of *Pseudomonas stutz*eri. NIR, Nitrite reductase; NAR, nitrate reductase; NOR, NO reductase; c, cyto c; Q, a quinone. (From Zumft, 1997.[7])

gene cluster containing *napE*, *napD*, *napA*, *napB*, and *napC*. The NapA protein binds the MGD molybdenum cofactor, the NapB is proposed to have two binding sites for heme C, and the NapC protein appears involved in electron transfer from membrane-localized quinone to the periplasmic nitrate reductase. The roles for genes *napE* and *napD* remain unresolved.

As reviewed by Mora in 1997, a diversity of cofactors for the movement of electrons from nitrate reduction are present in bacteria. Nitrate reductases from both eukaryotes and prokaryotes contain a molybdenum cofactor that is assumed to interact with the nitrate anion. Aside from differences in proteins, there is a range in the number and

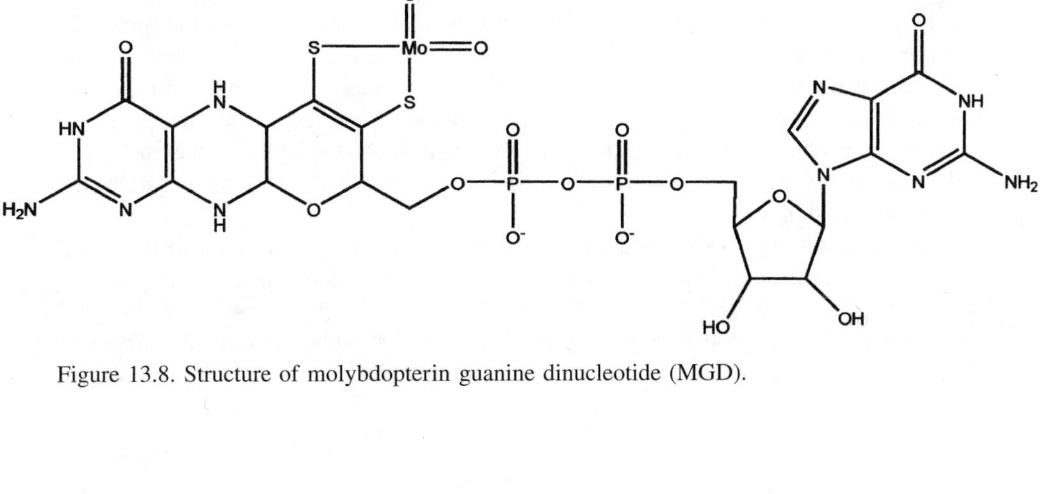

Figure 13.8. Structure of molybdopterin guanine dinucleotide (MGD).

types of iron sulfur clusters present. Many of the assimilatory type nitrate reductases found in the cytoplasm contain two iron sulfur centers: one as 4Fe-4S and another as 2Fe-2S. Respiratory nitrate reductases localized in the plasma membrane contain three iron sulfur centers: two are 4Fe-4S and the other is either a 3Fe-4S or 2Fe-2S. With each molecule of periplasmic nitrate reductases there is found a 4Fe-4S cluster; however, differences are found in the presence and type of heme. The cytoplasmic assimilatory nitrate reductases are without a heme cofactor and may acquire electrons from soluble cytochrome $c$ or other soluble electron carriers. The membrane-bound nitrate reductase found in dissimilatory nitrate reducers contains a heme B. Periplasmic nitrate reductase found in dissimilatory nitrate reducers may either contain a heme C or if the periplasm is rich with cytochrome $c$ no heme groups are found. The general electron flow in nitrate reduction appears to be as follows: heme B or heme C → the Fe-S cluster → molybdenum cofactor, which is commonly MGD, → nitrate.

### 13.2.4.2. Nitrite Reductase[6-9]

Nitrite reductase is present in a limited number of microorganisms and the reaction catalyzed is as follows:

$$2 \, NO_2^- + H_2 + 2 \, H^+ \rightarrow 2 \, NO + 2 \, H_2O \qquad (\Delta G^{0\prime} = -147.0 \, \text{kJ/mol})$$

The nitrite reductase in denitrifying bacteria is of either of two types: a copper-containing or a heme-containing enzyme. With respect to the pseudomonads, in about three fourths of the bacteria isolated heme is the prosthetic group. Because diethyldithiocarbamate (DDC) inhibits the copper-containing enzyme, a presumptive test to distinguish the type of cofactor present is to determine if DDC inhibits nitrite reductase in a cell-free extract.

The nitrite reductases containing heme are also known as cytochrome $cd_1$ or heme $cd_1$-containing nitrite reductase because the enzyme has both a heme C and a heme $D_1$. The synthesis and structures of heme C and heme $D_1$ are given in Chapter 12. The nitrite reductase has been purified from several bacteria including *Pseudomonas aeruginosa*, *Pseudomonas stutzeri*, *Paracoccus denitrificans*, and *Thiobacillus denitrificans*. With these bacteria, the nitrite reductase is a tetraheme protein consisting of a homodimer with 60-kDa subunits in which each subunit binds one heme C and one heme $D_1$. As determined from three-dimensional crystal analysis, the nitrite reductase from *Thiosphaera pantotropha* has heme C covalently bound to one domain with iron of heme C coordinated by two histidine residues. The active site for nitrite reduction is the heme $D_1$, in which the axial ligands for heme $D_1$ are histidine and tyrosine. Depending on the species, electron donors for the cytochrome $cd_1$ nitrite reductase are exogenous cytochrome $c_{550}$ or $_{551}$ or azurin, which is a low molecular weight copper-containing protein.

The Cu-containing nitrite reductase (CuNIR) has been characterized from several bacteria including *Nitrosomonas europaea*, *Bacillus halodenitrificans*, *Rhodobacter sphaeroides*, *Alcaligenes faecalis*, *Achromobacter cycloclastes*, and *Pseudomonas aeruginosa*. The enzyme is composed of three identical subunits around a threefold axis with a central channel of 5 to 6 A. There are two domains in each of the subunits with a copper center in each of the domains. The Cu(II) centers in the enzyme are classified as either type 1 or type 2 based on electron paramagnetic resonance (EPR) spectroscopy and optical properties. The type 1 copper center is bound into domain I of CuNIR and a type 2 copper center is bound into domain II of the enzyme. Nitrite binds to the type 2 copper center. The CuNIR isolated from *Achromobacter cycloclastes* has a midpoint potential of +240 mV for the type 1 copper center and +260 mV for the type 2 copper center.

### 13.2.4.3. Nitric Oxide (NO) Reductase[8,9]

The reaction catalyzed by nitric oxide reductase is as follows:

$$2\ NO + H_2 \rightarrow N_2O + H_2O \qquad (\Delta G^{0'} = -306.1\ kJ/mol)$$

The enzyme has been isolated from several bacteria including *Pseudomonas stutzeri*, *Paracoccus denitrificans*, and *Achromobacter cycloclastes*. The NO reductase is encoded on the *norC* and *norB* genes which produce subunits of 17 kDa and 53 kDa, respectively. The intact NO reductase is 160 kDa to 180 kDa and is synthesized only under anaerobic conditions. Heme C is attached by covalent bond to the NorB protein through the CXXCH motif that is near the N-terminal domain. The NorB protein is highly hydrophobic, consisting of 12 transmembrane helices with considerable similarity to cytochrome *c* oxidase. Heme B is found in the NorC protein and in some of the NO reductases iron is found in the NorB subunit. The mechanism of the NO reduction is not resolved nor is the active site identified because it must accommodate two molecules of NO in the formation of $N_2O$.

The origin of NO in the environment may result from activities other than reduction of nitrite. Nitric oxide can be synthesized by heterotrophic organisms by the oxidation of L-arginine with the formation of L-citrulline as presented in Figure 13.9. There is a two-step reaction that results in an unusual five-electron transfer involving $O_2$ and NADPH. In mammalian systems the initial reaction involves flavins and a heme protein with the formation of hydroxyarginine, which is oxidized by unknown mechanisms to citrulline with the release of NO. The extent to which NO is derived from these heterotrophic processes remains to be established.

### 13.2.4.4. Nitrous Oxide (N₂O) Reductase[8,9]

The nitrous oxide reductase is a multicopper protein encoded on the *nosZ* gene and carries out the following reaction:

$$N_2O + H_2 \rightarrow N_2 + H_2O \qquad (\Delta G^{0'} = -341.4\ kJ/mol)$$

This enzyme has been isolated from several bacteria including *Wolinella succinogenes*, *Rhodobacter sphaeroides*, *Paracoccus denitrificans*, *Alcaligenes xylosoxidans*, *Pseudomonas aeruginosa*, and *Pseudomonas stutzeri*. There is some molecular variation with *Pseudomonas* and *Paracoccus* producing two subunits, each 66 kDa and containing 8 Cu atoms per holoenzyme of 120 kDa. The enzyme from *Wolinella succinogenes* is 162 kDa with subunits of 88 kDa and containing 6 Cu atoms and 2 Fe in each holoenzyme. In *Rhodobacter sphaeroides* a highly unusual enzyme has been

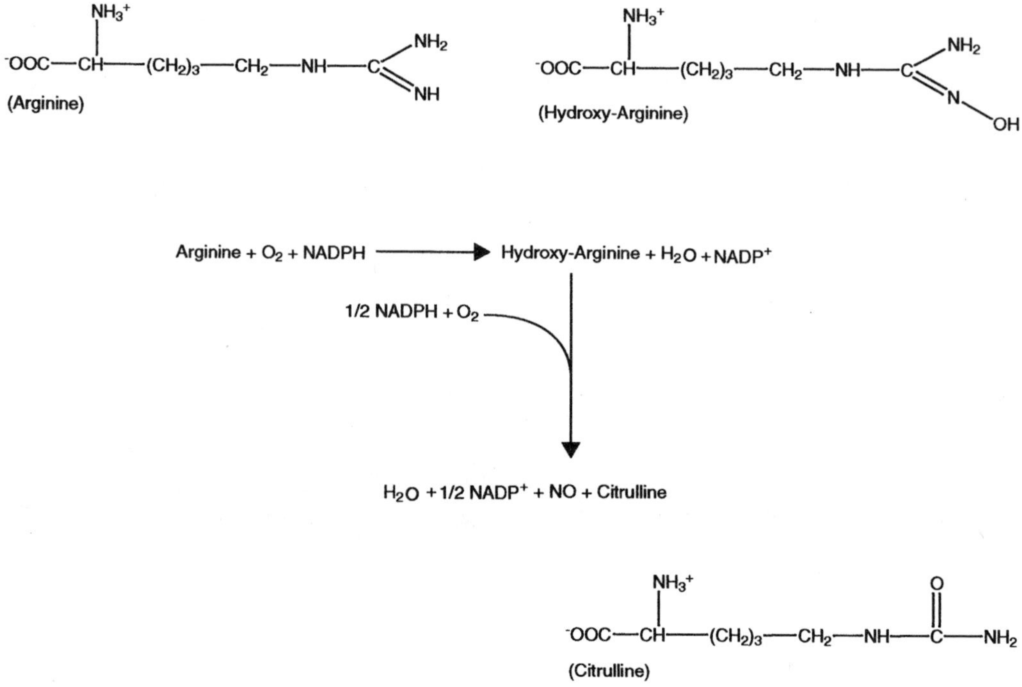

Figure 13.9. Production of NO and citrulline from arginine.

reported that consists of a 95-kDa monomer with 4 Cu, 2 Zn, and 1 Ni per holoenzyme. In the enzyme consisting of two subunits, there are two distinct binuclear Cu centers that have been designated as $Cu_A$ and $Cu_Z$. The midpoint potential or $E_0{'}$ for the $Cu_A$ center is $+260$ mV and electrons are presumed to be transferred to the $Cu_Z$ center which is the catalytic site for reduction of $N_2O$ to $N_2$.

## 13.3. Dissimilatory Nitrite Reduction[8,9]

The reduction of nitrite to ammonia without intermediates is known as dissimilatory nitrite reduction or nitrite ammonification. Several anaerobic species of bacteria are capable of the six-electron reduction process which proceeds according to the following reaction:

$$NO_2^- + 2 H^+ + 4 H_2 \rightarrow NH_4^+ + 2 H_2O \qquad (\triangle G^{0'} = -163.2 \text{ kJ/ reaction})$$

Of the bacteria demonstrated to have a dissimilatory nitrite reductase, the best studied systems are those from *Campylobacter sputorum, Vibrio fischeri, Wolinella succinogenes, Clostridium perfringens, Sulfurospirillum deleyianum, Escherichia coli*, and *Desulfovibrio desulfuricans*. While the enzyme in these organisms is either soluble or membrane bound, the presence of several heme C prosthetic groups per enzyme molecule is a common feature. The enzyme in *Wolinella succinogenes* and *Sulfurospirillum deleyianum* is membrane bound with a molecular mass of 552 kDa with three or four heme C per molecule. The nitrite reductase in *Desulfovibrio desulfuricans* is composed of two subunits of 62 kDa and 19 kDa with a total of six heme C groups for the intact enzyme. The oxidation–reduction potentials for each of the heme C groups has been determined by Mössbauer spectroscopy or EPR analysis to be $+150$ mV, $-80$ mV, $-100$ mV, $-320$ mV, $-400$ mV, and $-480$ mV. The mechanism of the reaction has not yet been established for this multiheme enzyme. It has been shown that $H_2$ oxidation coupled to nitrite reduction generates a proton-motive force that not only couples ATP synthesis but also accounts for solute transport.

## 13.4. Nitrification[10–13]

Ammonia is oxidized to nitrate by a procedure that involves two distinct activities. The first phase is the oxidation of ammonia to nitrite with hydroxylamine as an intermediate. The second phase is the oxidation of nitrite to nitrate. The oxidation of ammonia to nitrate by microorganisms is essential for cultivation of field plants and for nitrogen cycling. The scheme for nitrification is given as follows and the redox potentials for the enzyme reactions are listed under the arrows:

$$
\begin{array}{ccccccc}
-3 & & -1 & & +3 & & +5 \\
NH_3 & \rightarrow & NH_2OH & \rightarrow & NO_2^- & \rightarrow & NO_3^- \\
 & +730 \text{ mV} & & +60 \text{ mV} & & +420 \text{ mV} &
\end{array}
$$

The six-electron transfer is accomplished by the following reactions with the enzymes given in parentheses:

$NH_3 + O_2 + 2 H^+ + 2 e^- \rightarrow NH_2OH + H_2O$ (Ammonia monooxygenase, AMO)

$NH_2OH + H_2O \rightarrow NO_2^- + H^+ + 4 \text{ [H]}$    (Hydroxylamine oxidoreductase, HAO)

$NO_2^- + O_2 \rightarrow NO_3^-$    (Nitrite oxidase, NO)

The spatial arrangement of these enzymes and associated electron transfer in the bacterial membrane and periplasm are given in Figure 13.10. There is considerable specialization by bacteria participating in these reactions, with one group of organisms responsible for the formation of nitrite from ammonia and the second group converts nitrite to nitrate. Thus far, no one single organism has been isolated that can oxidize ammonia to nitrate.

## 13.4.1. Aerobic Ammonia Oxidation[10–13]

The oxidation of ammonia to nitrite occurs by the highly versatile group of nitrifying bacteria which include *Nitrosomonas* and *Nitrosolobus* in the soil, *Nitrosococcus* in marine environments, and *Nitrosospira* in acid soils. The reaction with ammonia oxidation to hydroxylamine is unique in that one oxygen atom is associated with the N atom of ammonia while the other oxygen atom results in the formation of water.

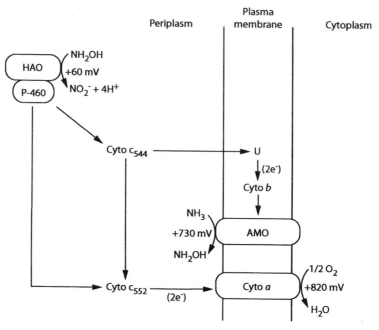

Figure 13.10. A model for location of nitrification enzymes on the cell surface of *Nitrosomonas europaea*. AMO, Ammonia monooxygenase; HAO, hydroxylamine oxidoreductase; U, ubiquinone. (From Hooper et al., 1997.[10])

Ammonia monooxygenase (AMO) is a unique enzyme because it must handle three substrates: ammonia (but not ammonium), $O_2$, and the electron donor. Present in bacteria that oxidize ammonia are internal membranes frequently termed cytomembranes (see Figure 13.11). Cytomembranes are commonly found in nitrifying bacteria, but the location of these memebranes may vary with the species. For example, *Nitrobacter winogradskyi* has cytomembranes that form a polar cap, *Nitrococcus mobilis* has tubular cytomembranes that are distributed through the cytoplasm, and *Nitrospira gracilis* appears to lack cytoplasmic membranes. In several isolates from fresh water or sewage, *Nitrosomonas* has membranes that are stacked along the periphery of the cell next to the plasma membrane. The cytomembranes in *Nitrocystis oceanus* have flattened vesicles that are localized in the central portion of the cell. Additional study is needed to trace the origin of these cytomembranes and the plasma membrane.

An understanding of the mechanism of ammonia oxidation is limited by the absence of a purified enzyme. A molecular biology approach for examination of *Nitrosomonas europaea* has produced the genes for AMO which are in duplicate form with designations of *amoA* and *amoB*. AMO is expected to consist of an A-subunit that has five transmembrane domains and a B-subunit that has two transmembrane domains. Although direct evidence has not yet been obtained, copper is considered to be an important cofactor for AMO.

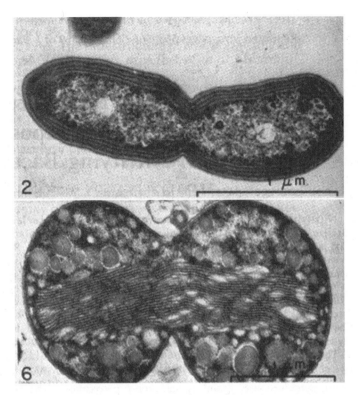

Figure 13.11. Internal membranes in ammonia oxidizing bacteria. The picture identified as "2" is a marine *Nitrosomonas* sp. with peripheral cytomembranes. $\times$ 37,600 (original magnification). The picture identified as "6" is an electron micrograph of *Nitrocystis oceanus* where the cytomembranes are in the central region of the cell. $\times$ 27,1000 (original magnification). (From Watson and Mandel, 1971.[12] Used with permission.)

With *Nitrosomonas* using AMO, growth is not coupled to the oxidation of substrates other than ammonia; however, the low molecular weight carbon products of methane, methanol, and CO oxidation can be used as a carbon source for growth. The methane monooxidase (MMO) enzymes in bacteria exist as one type in the plasma membrane and another type in the cytoplasm and both types are capable of oxidizing ammonia in addition to methane. However, the turnover of methane monooxygenase with methane is about 100 times greater than with ammonia. While both ammonia and methane can be oxidized by nitrifying bacteria and methanotrophs, methanotrophs are incapable of growing autotrophically with ammonia and autotrophic ammonia oxidizers cannot grow with energy from methane oxidation.

There is considerable similarity in substrate utilization by AMO and methane monooxygenase (MMO). It has been well demonstrated that AMO will recognize more than 75 different substrates and many of these substrates are discussed in Chapter 14. MMO is homologous to AMO and MMO also oxidizes many different compounds.

The broad use of substrates by these two enzymes suggests that oxygen is activated and not the substrate in the oxygenase reactions. From numerous evaluations using molecular biology, AMO and MMO have been found to have a common ancestry. In the β-Proteobacteria subdivision, AMO evolved to oxidize ammonia and in the γ-subdivision, AMO evolved to use both methane and ammonia.

With the AMO projected to be in the plasma membrane of *Nitrosomonas europaea*, hydroxylamine is released into the periplasm to be oxidized by a periplasmic enzyme. Organisms displaying high AMO activity have multilayered intracytoplasmic membranes that are considered to contain the ammonia-oxidizing enzyme. It would be expected that the hydroxylamine-oxidizing enzyme would also be associated with the membrane layers because hydroxylamine is a potent mutagen and would be detrimental if accumulated intracellularly.

## 13.4.2. Hydroxylamine Oxidoreductase[10–13]

The conversion of hydroxylamine to nitrite occurs in the periplasm and is catalyzed by hydroxylamine oxidoreductase (HAO). It has been determined that considerable molecular variation occurs in HAO found in different bacteria. Purified HAO from *Nitrosomonas europaea* has a molecular weight of 63 kDa with a P-460 heme and a *c*-type heme functioning in the active site. A gene cluster containing the genes for hydroxylamine oxidoreductase (*hao*) and cytochrome $c_{554}$ (*cycA*) is present in *Nitrosomonas europaea* as a triplicate repeat. These gene clusters are not identical but dispersed within them are genes for membrane-localized tetraheme cytochrome, cytochrome P-460, cytochrome $c_{552}$, and cytochrome *c* peroxidase. The HAO contains eight heme C groups, cytochrome $c_{554}$ contains four *c*-type heme groups per enzyme, and cytochrome $c_{552}$ contains a single *c*-type heme. The redox potential of these *c*-type cytochromes of HAO ranges from $-412$ to $+288$ while cytochrome $c_{544}$, which serves as the electron acceptor, has a redox range of $-276$ to $+47$ mV. A model of the electron transport scheme for HAO in *Nitrobacter* is given in Figure 13.12. The

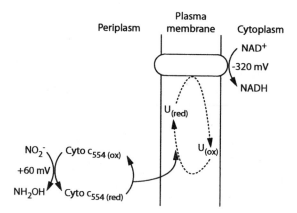

Figure 13.12. Generalized scheme for reverse electron flow in *Nitrobacter* with oxidation of nitrite.

enzyme purified from *Thiosphaera pantoptropha* is 20 kDa and contains no heme groups while the HAO from *Pseudomonas* PB 16 is 68 kDa and is a dimer without heme groups. *Methylococcus capsulatus* has a HAO consisting of two subunits of 39 kDa and 16.4 kDa with one heme (P-460) group per structure.

A most interesting feature of ammonia oxidation to nitrite is the electron flow and energetics of the reactions. The oxidation potential of ammonia/hydroxylamine is $+730$ mV while for hydroxyoxylamine/nitrite the $E_0'$ is $+60$ mV. Because the $E_0'$ for $NAD^+/NADH + H^+$ is $+320$ mV, reverse electron flow is needed to produce reducing equivalents to drive biosynthetic activities. A proposed scheme for this reverse electron flow in *Nitrosomonas* is given in Figure 13.12.

## 13.4.3. Nitrite Oxidation[10–14]

The oxidation of nitrite to nitrate is accomplished by relatively few species of bacteria and these bacteria are generally related to the ammonia oxidizing bacteria. Perhaps the most often discussed organism is *Nitrobacter* which is found in soil and fresh water. All *Nitrobacter* species have intracytoplasmic membranes that are at the polar region of the cell. The best characterized is *Nitrobacter winogradskyi* and the other species of this genera include *hamburgensis* and *vulgaris*. *Nitrococcus mobilis* and *Nitrospira gracilis* are marine organisms with tubular intracytoplasmic membranes. *Nitrospira marina* is a marine isolate that lacks intracytoplasmic membranes.

From an energetics perspective, electron flow from the oxidation of nitrite must be coupled to proton translocation across the plasma membrane to produce a charge on the membrane. The reaction of $NO_2^-/NO_3^-$ ($E_0' = +420$ mV) is sightly more electronegative than the reaction of $O_2/H_2O$ ($E_0' = +820$ mV). For reducing equivalents in the form of NADH or NADPH to be realized from the oxidation of $NO_2^-$ to $NO_3^-$, it would be difficult to explain since the $NAD^+/NADH$ reaction has a markedly negative potential ($E_0' = -320$ mV). Thus, to enable electrons to be deflected from $NO_2^-$ to $NAD^+$, a reverse electron flow is required. As presented in Figure 13.12, The 2 $e^-$ released by nitrite oxidoreductase (NO) in the oxidation of $NO_2^-$ are diverted to the cytochrome $aa_3$ and protons are pumped outward across the plasma membrane and these protons serve as a source for oxidative phosphorylation. Electrons are also directed toward NAD oxidoreductase from NO by an energy-dependent process using a mechanism that is not understood. Only 2% to 10% of the energy released from oxidation of nitrite is converted to biosynthesis and this low energy yield has been proposed to account for the lengthy generation times of 10 hr to 140 hr for *Nitrobacter*. Another way to express the poor yield of biologically usable energy is to note that about 100 mol of nitrite are oxidized for each mole of $CO_2$ assimilated.

While nitrite reductase accounts for a major portion of the total cell protein (i.e., 10% to 30% of protein) in *Nitrobacter*, definitive information about the enzyme has not been presented. A nitrite reductase is also found in *Nitrobacter* and this reductase copurifies with the nitrite oxidoreductase (NO). It appears that the NO contains a heme C and a heme $A_1$ but no flavin moieties. The nitrite reductase is reported to have a molecular weight of about 100 kDa with three subunits.

## 13.4.4. Anaerobic Ammonium Oxidation[15]

In 1990, van de Graal and colleagues presented evidence that ammonia can be oxidized under anaerobic conditions with nitrate as the electron acceptor and this observation is an important contribution to understanding anaerobic microbial growth. Experiments have revealed that a mixed culture of anaerobes in a methanogenic digester will oxidize 5 mol of ammonium with the reduction of 3 mol of nitrate for the production of 4 mol of dinitrogen. Because ammonium and nitrate constitute a serious environmental problem, the conversion of the nitrogen compounds to $N_2$ is attractive and at least one European Patent has been given for the Anammox (*an*aerobic *amm*onium *ox*idation) process. The free energy of the anaerobic reaction is projected to be $-297$ kJ/mol, which is sufficient to support cellular growth and comparable to that observed in aerobic nitrification. Further information about this reaction will require evaluation of pure cultures.

## 13.5. The Sulfur Cycle[1]

As in the case with nitrogen, sulfur is used by bacteria as an essential nutrient for cellular biosynthesis. In each gram dry weight of bacteria, there is about 11 mg of sulfur as cysteine and methionine. In the nonprotein soluble fraction of bacteria, sulfur may constitute up to 3 mg per gram dry weight of cells. Sulfur makes up about 0.1% of the Earth's surface and the stable forms of sulfur available to microorganisms are sulfate, $S^o$, $HS^-$, and sulfhydryl as R-SH. In marine and liminic environments, sulfate is readily available, with an average concentration estimated to be 28.7 m$M$. Through assimilatory processes most microorganisms can acquire sulfur and synthesize the organic sulfur compounds required to sustain cell growth.

In some prokaryotes, sulfur may have an additional role in cellular energetics by serving as either an electron donor or an electron acceptor. The biological transformation through discrete enzymatic action makes up the redox cycle of sulfur as shown in Figure 13.13. Unlike nitrogen, which is found predominately in the atmosphere, sulfur is found only in small quantities in the atmosphere; terrestrial and ocean environments are the reservoirs for sulfur.

## 13.5.1. Sulfate Reduction[16]

The most stable sulfur oxy-ion is sulfate, and most prokaryotes have the gene content to use sulfate to produce sulfur-containing amino acids; a few use sulfate as a final electron acceptor. In both pathways in which sulfate is used, there is an overall eight-electron reduction from sulfate to sulfide. However, sulfate is remarkably stable and before reduction can be initiated, sulfate must be activated by ATP, as shown in the following equation:

$$SO_4^{2-} + 2\ H^+ + ATP \rightarrow APS + PP_i$$

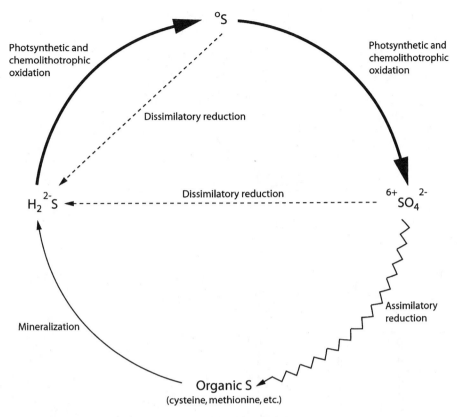

Figure 13.13. Major features of the microbial sulfur cycle.

The initial reaction uses the enzyme ATP sulfurylase to activate sulfate with the consumption of ATP to produce adenosine phosphosulfate (APS) plus inorganic pyrophosphate (PP$_i$). The structure of APS is given in Figure 13.14. The reaction catalyzed by ATP sulfurylase does not favor the formation of APS and the $K_{eq}$ of the reaction is $10^{-8}$. Thus, the formation of APS is not favored and the reaction proceeds only as a result of the hydrolysis of PP$_i$ by a pyrophosphatase. This removal of an end product shifts the equilibrium of the ATP sulfurylase reaction in the direction of APS formation. As indicated in Figure 13.15, APS is the initial molecule for either an assimilatory reduction or dissimilatory reduction process rather than sulfate.

### 13.5.1.1. Assimilatory Sulfate Reduction[17]

Unless a prokaryote has a specific requirement for an organo-sulfur compound, assimilated sulfate is used for synthesis of cysteine, methionine, biotin, lipoic acid, thiamine, coenzyme A, S-adenosylmethionine, and nonheme iron proteins with Fe-S clusters. An additional requirement for sulfate is seen in the acid-fast bacteria which produce

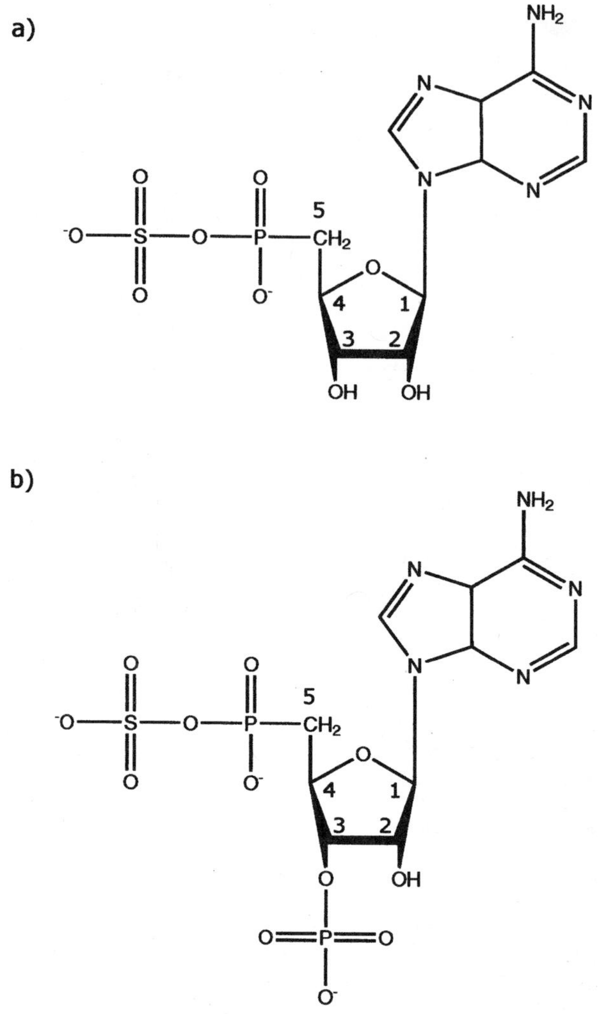

a)

b)

Figure 13.14. High-energy compounds in chemolithotrophic bacteria. Structures of (a) APS and (b) PAPS.

sulfur-containing lipids and the methanogenic archaea have the unique coenzyme M cofactor. In general, 70 $\mu M$ sulfate is required for maximal microbial cell growth and about 225 $\mu mol$ of sulfur are needed for the production of 1 g of $E.\ coli$. The pathway of assimilatory reduction of sulfate culminates with the formation of cysteine, as the central molecule for production of other organo-sulfur compounds is cysteine. This has led many biochemists to designate the assimilatory reduction of sulfate as the pathway for cysteine biosynthesis.

For assimilatory sulfate reduction, APS is not used directly but the molecule is first energized with the addition of another high-energy phosphate. Through the interaction of ATP, APS is converted to phosphoadenosine phosphosulfate (PAPS) by the enzyme

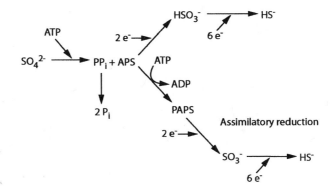

Figure 13.15. Metabolic pathways for dissimilatory and assimilatory sulfate reduction.

APS kinase. The pathway for assimilatory reduction of sulfate to sulfide is given in Figure 13.16 and the structures of APS and PAPS are given in Figure 13.14.

In the assimilatory reduction of sulfate, sulfur is ultimately reduced and incorporated into an organic molecule. To accomplish this, the sulfur atom is reduced to sulfide and without intracellular accumulation of sulfide the sulfide is inserted into the carbon chain to produce cysteine. In the assimilatory system, reduction of APS does not occur but rather PAPS is reduced. The reduction of the sulfur atom in PAPS is accomplished by reduced thioredoxin, a 12-kDa electron transport protein, with the formation of sulfite which is reduced to sulfide. In *E. coli*, cysteine is formed from the addition of sulfide to *O*-acetylserine. The synthesis of cysteine in *Pseudomonas aeruginosa* does not involve *O*-acetylserine but sulfide is added to *O*-succinyl-homoserine with the formation of homocysteine which is converted to methionine and cysteine.

Of the enzymes of the assimilatory sulfate system, sulfite reductase has attracted the most attention because it catalyzes the 6 e⁻ reduction in the formation of sulfide. In *E. coli*, the sulfite reductase is a large complex of 670 kDa with a structure of $\alpha_8\beta_4$. The $\alpha_8$ segment is a 60-kDa flavoprotein that contains 4 FAD, 4 FMN, and a NADPH-cytochrome *c* site. The $\beta_4$ segment is a hemoprotein consisting of four 55-kDa polypeptides, a $Fe_4S_4$ cluster, and an unusual siroheme. It has been proposed that the electron flow is from NADPH → FAD → FMN → the 4Fe-4S cluster → siroheme → sulfite. Owing to the chemical similarity in substrates, purified sulfite reductase will also reduce nitrite; however, the significance of in vivo nitrite reduction is unknown.

If cysteine is present in the environment, the assimilatory sulfate pathway is not used, with the cell conserving ATP and reducing activity. There is a high degree of regulation on this assimilatory-sulfate reduction pathway because both sulfite and sulfide are chemically reactive; therefore neither must be allowed to accumulate in the cytoplasm. Regulation of genes for cysteine biosynthesis is associated with the cys-

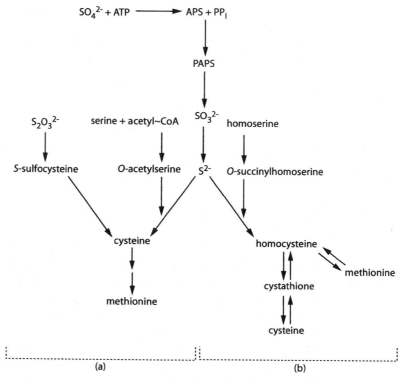

Figure 13.16. Pathways for production of amino acids from sulfate and thiosulfate. Pathway used by (**a**) *Escherichia coli* and by (**b**) *Pseudomonas aeruginosa*.

teine regulon which is under positive control. Genes for the synthesis of cysteine are given in Figure 13.17. In *E. coli* and *Salmonella typhimurium* there is a complex mode of regulation with more than a dozen genes making up the cysteine regulon and these genes are found in five different regions on the chromosome. In some prokaryotes, it appears that the uptake of sulfate is influenced by the cysteine concentrations. The transport of sulfate is discussed in Chapter 10. The system of assimilatory sulfate reduction is controlled by several mechanisms which include: (1) inhibition of enzyme activity by accumulation of cysteine; (2) repression of the cysteine regulon by cysteine; and (3) degradation of ATP sulfurylase and APS kinase, which may be attributed to proteolytic enzymes or proteosomes. Cysteine enters *S. typhimurium* by a periplasmic protein that is part of the cysteine regulon which displays saturable transport system with a $K_m$ of 2 $\mu M$ and a $V_{max}$ of 9 nmol/min/mg of cell protein. In this same bacterium, there is also a nonsaturable transport system for cysteine that can be attributed to passive diffusion.

Other sulfur compounds that can satisfy the sulfur requirements of prokaryotes include thiosulfate, sulfite, or sulfide and this is usually after they have been oxidized to sulfate. Either cysteine or methionine can serve as the sulfur source for many

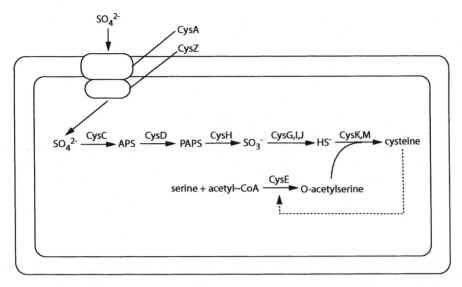

Figure 13.17. Genes for synthesis of cysteine in *Escherichia coli*. The cysB protein is a tetramer of 36-kDa identical subunits that exerts positive control of the transcription of the cysteine regulon (*cysA*, *C*, *D*, *H*, *I*, *J*, *K*, and *M*). The *cysA* product is a 32-kDa periplasmic sulfate-binding protein and a product of cysZ is needed for sulfate transport. Genes *cysC* and *cysD* encode for APS kinase and ATP sulfurylase, respectively. *cysH* encodes for PAPS sulfotransferase. Sulfite reductase is a product of three genes: *cysI*, which produces the β-subunit; *cysJ*, which produces the α-subunit; and *cysG* which produces a protein for the synthesis of the siroheme. *cysK* and *cysM* encode for the *O*-acetylserine sulfhydrase A and B, respectively. *cysE* is the gene for the multisubunit serine transacetylase which has cysteine, ATP, and acetyl-CoA as effectors. In the absence of cysteine, genes of the cysteine regulon are induced but in the presence of cysteine, feedback inhibition influences cysteine synthesis.

prokaryotes; however, this is not universal because *E. coli* and *S. typhimurium* lack the pathway for converting methionine to cysteine.

### 13.5.1.2. Dissimilatory Sulfate Reduction[18-21]

Sulfate is used as the final electron acceptor in growth of the physiological group known as sulfate-reducing bacteria and archaea. Although there are 10 recognized genera of bacteria that display dissimilatory sulfate reduction, the Gram-negative *Desulfovibrio* and the Gram-positive *Desulfotomaculum* are the best characterized. The hyperthermophiles *Archaeoglobus fulgidus* and *Archaeoglobus profundus* are the only members of the archaea that use dissimilatory sulfate reduction.

The dissimilatory sulfate reduction carried out by these organisms is shown in Figure 13.18. As presented in the figure, there are two projected avenues for the reduction of sulfur in APS to produce sulfide. It is generally agreed that the principal physiological pathway is the direct reduction of sulfite to sulfide. To initiate reduction of sulfate, activation of sulfate by ATP sulfurylase is required and it is the same

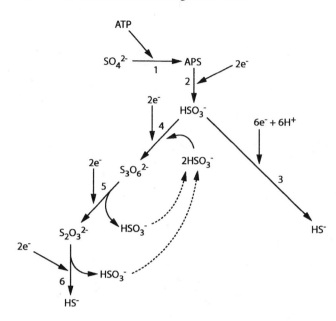

Figure 13.18. Metabolic schemes associated with dissimilatory sulfate reduction. Enzymes in the scheme are as follows: (1) ATP sulfurylase, (2) APS reductase, (3) bisulfite reductase, (4) trithionate-forming enzyme, (5) trithionate reductase, and (6) thiosulfate reductase. (From Fauque et al., 1991.[22])

reaction as used for the assimilatory reduction reaction. The steps in dissimilatory sulfate reduction are given in Table 13.5. The metabolic pathways associated with trithionate and thiosulfate as intermediates in the formation of sulfide are not considered to be used in respiratory-coupled growth on sulfate. However, the use of trithionate and thiosulfate may reflect the ability of sulfate-reducing bacteria to use sulfoxy-anions other than sulfate.

Table 13.5. Enzymes associated with dissimilatory sulfate reduction.

| Enzyme | Reaction | Energetics of the reaction ($\Delta G^{0\prime}$) |
|---|---|---|
| **Sulfate activation** | | |
| ATP sulfurylase | $SO_4^{2-} + 2H^+ + ATP \rightarrow APS + PP_i$ | +46.0 kJ/mol |
| Inorganic pyrophosphatase | $PP_i + H_2O \rightarrow 2\ P_i$ | −21.9 kJ/mol |
| APS kinase | $APS + H_2 \rightarrow HSO_3^- + AMP + H^+$ | −68.6 kJ/mol |
| **Direct reduction of bisulfite** | | |
| Bisulfite reductase | $HSO_3^- + 3\ H_2 \rightarrow SH^- + H_2O$ | −171.7 kJ/mol |
| **Reduction of bisulfite with intermediates** | | |
| Trithionate-forming enzyme | $HSO_3^- + 2\ HSO_3^- + H_2 + H^+ \rightarrow$ $S_3O_6^{2-} + 3\ H_2O$ | −46.4 kJ/mol |
| Trithionate reductase | $S_3O_6^{2-} + H_2 \rightarrow S_2O_3^{2-} + HSO_3^- + H^+$ | −122.9 kJ/mol |
| Thiosulfate reductase | $S_2O_3^{2-} + H_2 \rightarrow HS^- + HSO_3^-$ | −2.1 kJ/mol |

Table 13.6. Characteristics of bisulfite reductases in bacteria and archaea.

| Organism | Mass | Subunit structure | Content of the following/molecule | | |
| | | | Iron | Siroheme | 4Fe:4S cluster |
|---|---|---|---|---|---|
| **Assimilatory-type reductase**[a] | | | | | |
| *Desulfovibrio vulgaris* | 27.2 kDa | — | 4.7 | 0.91 | 0 |
| *Desulfuromonas acetoxidans* | 23.5 kDa | — | 5.3 | 0.94 | 0 |
| *Methanosarcina barkeri* | 23.0 kDa | — | 5.2 | 0.85 | 0 |
| **Dissimilatory-type reductases** | | | | | |
| *Thermodesulfobacterium commune*[b] (desulfofuscidin) | 167 kDa | $\alpha_2\beta_2$ | 20–21 | 4 | 4 |
| *Thermodesulfobacterium mobile*[b] (desulfofuscidin) | 190 kDa | $\alpha_2\beta_2$ | 32 | 4 | 8 |
| *Desulfovibrio gigas*[b] (desulfoviridin) | 200 kDa | $\alpha_2\beta_2$ | 16 | 2[c] | 4 |
| *Desulfuromonas acetoxidans*[b] (desulforubidin) | 225 kDa | $\alpha_2\beta_2$ | 16 | 2 | 4 |

[a]Modified from I. Moura and A.R. Lino, 1994.[20]
[b]Modified from E.C. Hatchikian, 1994.[19]
[c]Siroporphyrin, not siroheme.

Enzymes from the sulfate reduction pathway have attracted considerable interest among biochemists. The ATP sulfurylase purified from the cytoplasm of cells of *Archaeoglobus fulgidus* is 150 kDa with an $\alpha_2\beta$ structure. The $\alpha$-subunit and $\beta$-subunit are 50 kDa and 53 kDa, respectively. The APS reductase, isolated from numerous species of *Desulfovibrio* and from *Archaeoglobus fulgidus*, has similar characteristics in that the molecular mass is 160 kDa to 180 kDa with seven or eight nonheme iron atoms per mole and one FAD per mole.

Two distinct processes can account for sulfite reduction: one is associated with the assimilatory sulfate reduction pathway and the other is associated with dissimilatory sulfate reduction. The assimilatory-type sulfite reductase produces sulfide as the only end product of the reaction and no sulfur compounds are formed as intermediates. The dissimilatory-type sulfite reductases gives sulfide as an end product but, in addition under laboratory assay conditions, will produce considerable quantities of trithionate and thiosulfate. Characteristics of these two sulfite reductases are given in Table 13.6.

### 13.5.1.3. Energetics of Sulfate Activation[22]

In the activation of 1 mol of sulfate, 2 mol of ATP are consumed with the release of 1 mol of AMP. To regenerate APT from AMP, two high-energy compounds are required. With lactate as the electron donor in the culture media, 2 ATP are produced by substrate-level phosphorylation. In 1962 H.D. Peck, Jr. demonstrated that ATP synthesis in *Desulfovibrio gigas* was coupled to electron flow and that the dissimilatory sulfate reducing bacteria displayed a respiratory-coupled oxidative phosphoryla-

Table 13.7. Energetics of sulfate-reducing bacteria considering activation of sulfate.

| | Reactions | |
|---|---|---|
| | 2 lactate + $SO_4^{2-}$ → $H_2S$ + 2 acetate + 2 $HCO_3^-$ | 4 $H_2$ + 2 $H^+$ + $SO_4^{2-}$ → $H_2S$ + 4 $H_2O$ |
| Moles of ATP required to activate 1 mol of sulfate | 2 | 2 |
| Moles of ATP produced by substrate-level phosphorylation in the reaction | 2 | 0 |
| ATP yield by bacteria (substrate-level + oxidative phosphorylation) | >3 | >1 |

tion. From an energy perspective it is feasible because more free energy is produced in the lactate–sulfate reaction (see Table 13.7) than is consumed by activation of sulfate. That is, the lactate–sulfate reaction has a $\Delta G^{0'} = -168.8$ kJ/mol and with $-63.6$ kJ/mol needed to synthesize two ATP molecules there remains $-105.2$ kJ/mol, which is sufficient for synthesis of at least one ATP. With autotrophic growth of *Desulfovibrio* with $H_2$ as the electron donor, it is apparent that the bacteria grow in the absence of substrate level phosphorylation. It is difficult to envision a mechanism to account for oxidative phosphorylation coupled to sulfate respiration in that the membrane must be adequately charged for proton-driven phosphorylation. A system of "hydrogen cycling" was proposed by Odom and Peck and a discussion on this process is covered in Chapter 12.

## 13.6. Dissimilatory Sulfur Reduction[23,24]

Elemental sulfur is used by a diverse group of organisms in the archaeal and bacterial groups as an electron acceptor according to the following reaction:

$$S^o + H_2 \rightarrow HS^- + H^+ \qquad (\Delta G^{0'} = -6.7 \text{ kcal/mol})$$

The substrate for the sulfur reactions is either the sulfur in the polysulfide or the "crystalline" form of sulfur and the structures of these compounds are given in Figure 13.19. Based on physiological and taxonomic considerations, the prokaryotes using elemental sulfur as an electron acceptor can be arranged in three groups (see Table 13.8).

The sulfur-reducing enzyme has been isolated from several sulfate reducers including *Desulfomicrobium baculatum* Norway 4, *Desulfovibrio gigas*, and *Desulfovibrio desulfuricans* Berre-Eau. The sulfur reductase is about 15 kDa and contains four heme C groups. The midpoint potentials of the tetraheme cytochrome from *Desulfomicrobium baculatum* Norway 4 are $-150$ mV, $-300$ mV, $-330$ mV, and $-355$ mV. The sulfur reductase is constitutive and will reduce polythionates but not thiosulfate or trithionate. The tetraheme cytochrome $c_3$ does not function as a sulfur reductase

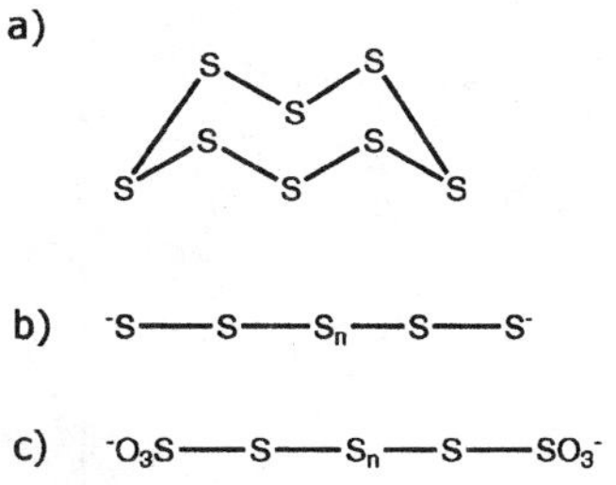

a)

b)  $^-$S——S——S$_n$——S——S$^-$

Figure 13.19. Polymeric sulfur molecules. (**a**) sulfur crown,
(**b**) polysulfide, and (**c**) polythionate.

c)  $^-$O$_3$S——S——S$_n$——S——SO$_3^-$

because it is highly sensitive to hydrogen sulfide, which is the end product of sulfur reduction. Cytochromes from *Desulfovibrio* that are not effective as sulfur reductases include the mono heme, $c_{553}$, and the octaheme $c_3$. Also, the triheme cytochrome from *Desulfomicrobium acetoxidans* is ineffective in reacting with sulfur.

The sulfur reductase, also known as polysulfide reductase, from *Wolinella succinogenes* has been examined by Kröger and colleagues. This enzyme is constitutive and is localized in the plasma membrane with the substrate site exposed on the periplasm side. The enzyme contains nonheme Fe, acid-labile S, and a molybdenum cofactor. The polysulfide reductase operon contains *psrA, psrB,* and *psrC,* which encode for proteins with a mass of 81.2 kDa, 20.9 kDa, and 34 kDa, respectively. The activity and location of the 34-kDa protein remains to be established. The characterization of the sulfur reductase in *Sulfurospirillum deleyianum* has led to the observation that it is an iron-sulfur protein that contains a 4Fe-4S center.

Many of the hyperthermophilic archaea have a requirement for S$^o$. Many of the archaea grow in environments in which H$_2$ accumulates to the extent of being inhibitory for microbial growth. Through the reaction with elemental sulfur, the environment is detoxified from H$_2$. The coupling of H$_2$ oxidation to the reduction of sulfur produces oxidative phosphorylation that supports autotrophic growth. In addition, many of the archaea can grow with S$^o$ using an organic compound as the electron donor. Another group of archaea that metabolize S$^o$ are the mesophilic methanogens. Most of the species of methanogens can replace organic sulfur sources with S$^o$ without any deleterious effect. The sulfur reductase in these archaea has not yet been isolated and the mechanism of reduction is not established.

Table 13.8. Examples of organisms in each of the three groups of prokaryotes that reduce elemental sulfur.

| Sulfate-reducing bacteria | Spirilloid mesophilic, sulfur-reducing bacteria | Hyperthermophilic sulfur-reducing archaea |
|---|---|---|
| *Desulfomicrobium baculatum* | | *Desulfurococcus mobilis* |
| *Desulfovibrio africanus* | *Campylobacter rectus* | *Pyrococcus woesei* |
| *Desulfovibrio desulfuricans* | *Shewanella putrefaciens* | *Staphylothermus marinus* |
| *Desulfovibrio gigas* | *Sulfurospirillum deleyianum* | *Thermodiscus maritimus* |
| *Desulfovibrio salexigens* | *Wolinella succinogenes* | *Thermoproteus tenax* |

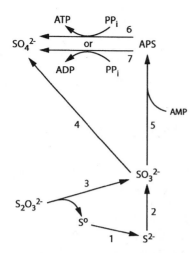

Figure 13.20. Sulfide oxidation in photosynthetic sulfur bacteria. Enzymes/activities are as follows: (1) nonenzymatic reduction, (2) reverse siroheme sulfite reduction, (3) thiosulfate sulfurtransferase, (4) sulfite: acceptor oxidoreductase, (5) adenylyl sulfate reductase, (6) ATP sulfurylase, and (7) ADP-sulfurylase. (From Dahl and Trüper, 1994.[25])

## 13.7. Dissimilatory Sulfide Oxidation by Photosynthetic Sulfur Bacteria[25]

The photosynthetic prokaryotes known as purple-sulfur and green-sulfur bacteria use sulfide, thiosulfate, or elemental sulfur as the electron donor. The key reactions with the substrates oxidized are given in Figure 13.20. The oxidation of sulfide or thiosulfate by *Chromatium* results in accumulation of sulfur globules inside the cell whereas with many other phototrophic sulfur bacteria sulfur colloids accumulate in the culture medium. Electrons are removed from sulfide or elemental sulfur with the formation of sulfite by a sulfite reductase. Once formed, sulfite may be oxidized directly to sulfate by a sulfite oxidoreductase without APS as an intermediate. In some organisms, sulfite is converted to APS by the adenylylsulfate reductase and sulfate is ultimately formed by either ATP-sulfurylase or an ADP-sulfurylase. While oxidation by APS as an intermediate may produce energy in the formation of sulfate, the direct oxidaton of sulfite to sulfate may be a more expedient reaction. Oxidation of thiosulfate by thiosulfate sulfotransferase results in the splitting of the molecule with part being oxidized to sulfite and the sulfane atom of thiosulfate reduced to S°. The thiosulfate molecule has a central sulfur atom onto which two oxygen atoms and one sulfur atom are attached. The outer sulfur atom is referred to as the sulfane atom. Owing to the considerable variability of sulfur oxidizing enzymes in sulfur phototrophs, the distribution and cellular location (i.e., soluble versus membrane) of enzymes cannot be considered diagnostic for a given species.

## 13.8. Autotrophic Oxidation of Sulfur Compounds[26]

Inorganic sulfur compounds are oxidized by a relatively small number of microbial species; however, their activity is important for global cycling of sulfur. The principal

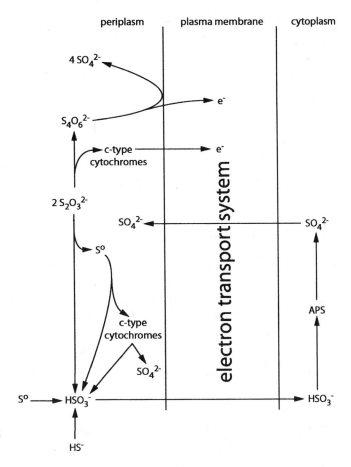

Figure 13.21. Summary of oxidative reactions with various inorganic sulfur compounds as occurs with anaerobic photosynthetic bacteria. (From Kelly, 1988,[26] with permission.)

forms of sulfur that are available in the environment for oxidation are elemental sulfur and thiosulfate. Sulfite, trithionates, or tetrathionates may also be found in special situations and these compounds are also oxidized to sulfate by autotrophic and anoxygenic phototrophic bacteria. A model representing the location of several of these reactions with respect to the plasma membrane is given in Figure 13.21. Electron flow is into the plasma membranes and energy produced from the oxidation of elemental sulfur is used for cell biosynthesis.

## 13.8.1. Oxidation of Elemental Sulfur[26,27]

The prokaryotes that grow on elemental forms of sulfur with the production of sulfate include bacteria (*Acidithiobacillus thiooxidans*, *Thiobacillus thioparus*, and *Acidithiobacillus ferroxidans*) and archaea (*Sulfolobus brierley* and *Desulfurolobus ambivalens*). Elemental sulfur is oxidized by these aerobic microorganisms to sulfuric acid by the following reaction:

$$S^\circ + H_2O + 3\ O_2 \rightarrow SO_4^{2-} + 2\ H^+ \qquad (\Delta G^{0'} = -140.4 \text{ kJmol electron donor})$$

All of the enzymes from the thiobacilli require reduced glutatione (GSH) for oxidation of sulfur; however, the enzymes from the archaea can oxidize sulfur without GSH. The enzyme from *Acidithiobacillus ferrooxidans* is unique in that it can use either $O_2$ or $Fe^{3+}$ as the electron acceptor. Although many of these sulfur-oxidizing bacteria grow in acidic environments, the optimum activity for the enzymes that oxidize sulfur is pH 7 to 8.

## 13.8.2. Oxidation of Thiosulfate and Polythionates[28]

The oxidation of thiosulfate is an energy-yielding reaction that provides sufficient energy for the growth of bacteria. D. Kelly and colleagues proposed that there are two processes employed for the oxidation of thiosulfate. One requires the production of polythionates as an intermediate in the production of sulfate and this pathway is used by chemolithotrophic thiobacilli. The other pathway does not involve polythionates and is used by bacteria such as *Paracoccus* spp. and *Xanthobacter* spp. The basis for establishing these two different mechanisms of utilizing thiosulfate is derived from biochemical evaluations using isolated proteins and from whole cell studies.

The reaction utilized by *Paracoccus* (formerly *Thiobacillus*) *versutus* dealing with the conversion of thiosulfate to sulfate without polythionate intermediates is as follows:

$$S_2O_3^{2-} + 2\,O_2 + H_2O \rightarrow 2\,SO_4^{2-} + 2\,H^+ \quad (\Delta G^{0\prime} = -195.7/\text{mol electron donor})$$

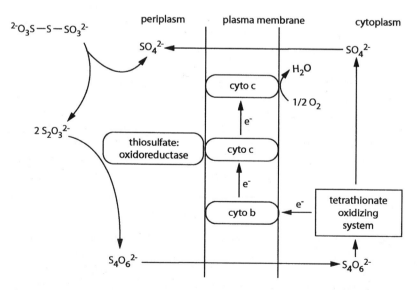

Figure 13.22. A model of thiosulfate oxidation in *Thiobacillus tepidarius*. (From Kelly, 1988,[26] with permission.)

This oxidation is a periplasmic activity in the Gram-negative *P. versutus* and involves a multienzyme complex. The thiosulfate oxidase enzyme has two parts: enzyme A, which is a 16-kDa protein and enzyme B, which is a 60-kDa protein. Thiosulfate initially binds to enzyme A and the product of this action binds to enzyme B. Electrons from thiosulfate are acquired from enzyme A/B by the periplasmic cytochrome $c_{552.5}$ and transferred to cytochrome $c_{552}$ in the plasma membrane. It is interesting to note that the electrons from cytochrome $c_{552}$ are transferred to cytochrome oxidase in the membrane without proceeding through the quinone in the membrane. Thus, the protons expelled from the cytochrome oxidase account for the charge on the membrane and proton-driven activities.

The oxidation of thiosulfate with polythionates as an intermediate is found in *Thiobacillus tepidarius*. Tetrathionate is produced from thiosulfate in the periplasm and tetrathionate is transported into the cell where the oxidation to sulfate occurs. The reactions for this oxidation are as follows:

$$4\ S_2O_3^{2-} + O_2 + 4\ H^+ \rightarrow 2\ S_4O_6^{2-} + 2\ H_2O$$

$$S_4O_6^{2-} + 10\ H_2O + 3\tfrac{1}{2}\ O_2 \rightarrow 4\ SO_4^{2-} + 2\ H^+ + 7\ H_2O$$

The tetrathionate formation from thiosulfate is by a thiosulfate:cytochrome *c* oxidoreductase which is a 138-kDa protein consisting of three subunits. The oxidation of thiosulfate is in the periplasm and this unique reaction is proton consuming. Reactions associated with thiosulfate oxidation are found in Figure 13.22 and further research is required to establish the energetics of these reactions.

## 13.8.3. *Oxidation of Sulfite*[29]

The oxidation of sulfite to sulfate in members of the *Thiobacillus* genus is accomplished by either of two pathways. In one system, sulfite and AMP are combined to produce APS and through the reverse of APS reductase, electrons from APS are withdrawn from sulfur and transferred to a cytochrome or other suitable electron acceptor. A second pathway involves a sulfite oxidase where sulfite is reduced directly without the formation of APS and this enzyme is frequently termed the AMP-independent sulfite oxidase. These two systems for oxidation are illustrated in Figure 13.23. The reaction of sulfite oxidase (sulfite:cytochrome *c* oxidoreductase) is given below:

$$SO_3^{2-} + 2\ \text{cytochrome}\ c_{(ox)} + H_2O \rightarrow SO_4^{2-} + 2\ H^+ + 2\ \text{cytochrome}\ c_{(red)}$$

The sulfite reductase has been purified from *Thiobacillus novellus*, *Thiobacillus thioparus*, and *Acidithiobacillus ferroxidans*. The enzyme from *Thiobacillus novellus* is 40 kDa and contains a *c*-type heme, molybdenum, and molybdopterin.

The production of sulfite is central for the production of sulfate from a variety of reduced sulfur molecules. Summarized in Figure 13.24 are some of the major reactions that result in formation of sulfite. Sulfite is highly reactive and cells could not tolerate the accumulation of it in the cytoplasm or in the periplasm. Thus, the rate of sulfite utilization must be significantly greater than the rate of sulfite formation.

a)

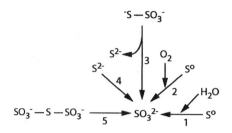

b)

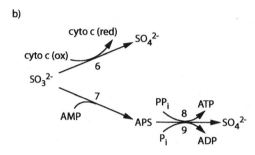

Figure 13.23. Reactions indicating the central role of sulfite in oxidation of sulfur compounds to sulfate. (**a**) Sulfite formation by the following enzymes: (1) sulfur reductase, (2) sulfur oxygenase, (3) thiosulfate reductase, (4) sulfite reductase, (5) trithionate reductase. (**b**) Sulfate production from sulfite by the following enzymes: (6) sulfite: cytochrome $c$ oxidoreductase, (7) adenylylsulfate reductase, (8) ATP sulfurylase, (9) ADP sulfurylase. (From Suzuki, 1994[27]; Kelly, 1988,[26] with permission.)

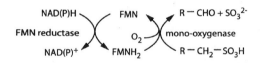

Figure 13.24. A model for oxidation of alkanesulfonate by monooxygenase.

## 13.9. Isotope Fractionation

One of the interesting features of the sulfur cycle is the isotope fraction that occurs as a result of microbial enzyme activity. The most abundant isotope of sulfur in sulfate is $^{32}S$, which is present 95% of the time and the remaining 4% is $^{34}S$. Enzymes will preferentially use the lighter isotope whereas chemical reactions show no preference for either $^{32}S$ or $^{34}S$. This use of one isotopic form over the other is isotope fractionation. In high-velocity reactions, such as with dissimilatory reduction by sulfate-respiring bacteria, a measurable difference in isotopic form of sulfur can be detected in the product. Over long periods of time, sulfide produced by sulfate-reducing bacteria or archaea becomes enriched in $^{32}S$ and sulfur deposits resulting from oxidation of these sulfides would also be enriched with the lighter isotope of sulfur. Using the signature of isotope fractionation, many of the sulfur deposits have been determined to be formed from biological activity.

## 13.10. Fermentation of Thiosulfate and Sulfite[18–21]

Under conditions of limiting electron donor and excess sulfite or thiosulfate, certain chemolithotrophic bacteria are capable of growth by "sulfur fermentation." In this reaction, sulfite is reduced by electrons released as sulfite is oxidized to sulfate. Similarly, thiosulfate can be disproportionated with the sulfane atom (outer S) of thiosulfate released as sulfide and the sulfene (inner sulfur atom attached to oxygen) of thiosulfate oxidized to sulfate. It is unusual to apply the term "fermentation" to inorganic metabolism; however, in this case it is appropriate because the electron acceptor is generated in the metabolic process and the combined end products of the reactions equal that of the substrate. In these fermentation reactions, the single substrate is both the electron donor and the electron acceptor and this change of the substrate is termed "disproportionation." The disproportionation reactions involving sulfite and thiosulfate are given as follows:

$$3\ SO_3^{2-} + SO_3^{2-} \rightarrow 3\ SO_4^{2-} + S^{2-} \qquad (\Delta G^{0\prime} = -58.3\ \text{kJ/mol}\ SO_3^{2-})$$

$$S_2O_3^{2-} + H_2O \rightarrow SO_4^{2-} + S^{2-} + 2\ H^+ \qquad (\Delta G^{0\prime} = -21.9\ \text{kJ/mol})$$

Both of these reactions can be demonstrated in bacteria and the thermodynamics indicates sufficient energy for bacterial growth. *Desulfovibrio sulfodismutans* and *Desulfobacter curvatus* are two sulfate-reducing bacteria that can obtain energy from the fermentation of bisulfite and thiosulfate. These reactions are given in Table 13.9. This is a special type of substrate level phosphorylation in which ATP is formed with APS as the substrate. Because the formation of APS from sulfate is not thermodynamically favored, the reaction catalyzed by ATP sulfurylase favors the formation of ATP. Because APS is reduced in the formation of sulfide, these sulfur disproportionation reactions listed in Table 13.9 are also an example of reversed electron flow in bacteria.

Table 13.9. Reactions associated with fermentation of inorganic sulfur compounds.

| | |
|---|---|
| **Sulfite fermentation:** | $3\ HSO_3^- + 3\ AMP + 3\ H^+ \rightarrow 3\ APS + 6\ e^- + 6\ H^+$  $3\ APS + 2\ PP_i \rightarrow 3\ SO_4^{2-} + 3\ ATP$ |
| | and |
| | $HSO_3^- + 6\ e^- + 6\ H^+ \rightarrow HS^- + 3\ H2O$ |
| **Thiosulfate fermentation:** | $S_2O_3^{2-} + 2\ e^- + 2\ H^+ \rightarrow HSO_3^- + HS^-$  $HSO_3^- + AMP \rightarrow APS + 2\ e^- + 2\ H^+$  $APS + PP_i \rightarrow SO_4^{2-} + ATP$ |

## 13.11. Organic Sulfur Compounds[30]

Mineralization of organic sulfur compounds results in the release of inorganic sulfur, which contributes to the cycling of sulfur. The carbon compounds made available after the removal of sulfur are utilized for synthesis of cellular materials. Organic compounds serve as reservoirs for sulfur, with more than 50% of sulfate in grassland soils covalently bound into humic substances and sizeable quanties of organis sulfur found in marine sediments. Classes of organic compounds containing oxidized sulfur atoms include sulfate esters ($R-C-OSO_3H$), sulfonates ($R-C-SO_3H$), and sulfamates ($R-N-SO_3H$). The enzymes known as sulfatases are responsible for release of sulfur from the organic moiety, and catabolic reactions are responsible for degradation of the remaining carbon skeleton. Based on enzyme activity, the prokaryotic sulfatases have three major classes of substrates: (1) arysulfates, (2) alkylsulfates, and (3) carbohydrate sulfates. The distribution of enzymes that attack these substrates is given in Table 13.10 along with examples of substrates for each enzyme class. These carbohydrate sulfatases, alkylsulfatases, and arylsulfatases are hydrolytic enzymes that add water to the sulfate ester with the release of sulfate.

The bacterial sulfatases, like eukaryotic sulfatases, require posttranslational modification to produce formylglycine (FGly), which is essential for the active site of all sulfatases. These bacterial enzymes have a highly conserved motif of C/S-X-P-X-R-X-X-X-X-T-G and the initial cysteine or serine residue is modified to produce formylglycine (2-oxoalanine). In some cases, as with *Klebsiella*, the sulfatases are only 40% converted to formylglycine and thus only 40% of the protein is an active sulfa-

Table 13.10. Enzymes that hydrolyze organosulfur compounds.

| Enzymes | Bacteria | Substrates |
|---------|----------|------------|
| Arylsulfatases | *Pseudomonas aeruginosa* | Phenolphthalein disulfate |
|  | *Klebsiella pneumoniae* | Tyrosine sulfate |
|  | *Serratia marcescens* | Cerebroside sulfate |
|  | *Mycobacterium* sp. | Estrone sulfate |
|  | *Salmonella typhimurium* |  |
|  | *Alteromonas carrageenovora* |  |
| Alkylsulfatases | *Pseudomonas* sp. | Dodecyl sulfate |
|  | *Bacillus cereus* | Choline sulfate |
|  | *Agrobacterium* sp. | Methyl sulfate |
|  | *Hyphomicrobium* sp. |  |
|  | *Sinorhizobium meliloti* |  |
|  | *Comamonas testosteroni* |  |
| Carbohydrate sulfatases | *Proteus vulgaris* | Mucin sulfate |
|  | *Cytophaga heparina* | Chondroitin sulfate |
|  | *Bacteroides thetaiotaomicron* | Heparin sulfate |
|  | *Burkholderia cepacia* |  |

tase. In other bacteria the conversion approaches 100%. Based on this motif, there are two major groups of sulfatases: serine type and cysteine type.

Bacteria utilize sulfonates by either of two mechanisms and both involve the activation of the carbon atom that is attached onto the sulfur atom. In one case, $O_2$ is added to the carbon atom with the release of sulfite and in a second mechanism the activation of the carbon atoms involves the cofactor of thiamine pyrophosphate with the release of sulfite. For aliphatic sulfonates, mono- and dioxygenases are reported to remove sulfur with the release of sulfite. The desulfurization of taurine in organisms, such as *E. coli*, is by an α-ketoglutarate-dependent taurine dioxygenase that is encoded on the *tauD* gene and catalyzes the following reaction:

Taurine $+ O_2 + $ α-ketoglutarate $\rightarrow$

aminoacetaldehyde $+ SO_3^{2-} + $ succinate $+ CO_2$

In *Alcaligenes* sp. and *Clostsridium pasteurianum*, taurine is metabolized by taurine: pyruvate aminotransferase and thiamine pyrophosphate-dependent sulfoacetaldehyde lyase.

In *E. coli* and *Pseudomonas putida*, the removal of sulfur from alkanesulfonates is attributed to a monooxygenase system. The oxidation of alkanesulfonate is driven by NAD(P)H with FMN as an electron carrier. In *E. coli*, the NAD(P)H-dependent FMN reductase and the FMN-dependent monooxygenase is encoded on the *ssuE* and the *ssuD* genes while in *Pseudomonas putida* these enzymes are encoded on the *msuE* and *msuD* genes. The action of the monooxygenase system on alkanesulfonate is given in Figure 13.24. To achieve the oxidation activity, some of the monooxygenase enzyme systems attacking sulfonates contain a [2Fe-2S] protein for electron transfer. Similarily, toluene monooxygenase systems found in various bacteria also contain ferredoxin.

Anaerobic prokaryotes use additional processes for the metabolism of organosulfur compounds which include those activities that focus on the carbon structure of sulfonates. Taurine is used by a few anaerobic bacteria as a carbon source with sulfoacetaldehyde production following transamination activity attributed to taurine aminotransferase. The sulfoacetaldehyde is hydrolyzed by bacteria to produce acetate and sulfite. Several anaerobes can reduce sulfite and sulfate-reducing bacteria will reduce sulfite to sulfide. Fermentation of cysteate and taurine has been reported for cultures isolated from anaerobic environments. In other anaerobes including *Alcalingenes* and *Paracoccus*, sulfonates are used as electron donors with nitrate as the electron acceptor. Additional research is needed to clarify the activities of anaerobic bacteria in utilizing organosulfur molecules.

# References

1. Brock, T.D., D.W. Smith, and M.T. Madigan. 1984. *Biology of Microorganisms*. Prentice-Hall, Englewood Cliffs, NJ.
2. Howard, J.B. and D.C. Rees. 1994. Nitrogenase: A nucleotide-dependent molecular switch. *Annual Review of Biochemistry* 63:235–264.

3. Lin, J.T. and V. Stewart. 1998. Nitrate assimilation by bacteria. *Advances in Microbial Physiology* 39:2–32.

4. Merrick, M.J. 1988. Regulation of nitrogen assimilation by bacteria. In *The Nitrogen and Sulfur Cycles* (J.A. Cole and S.J. Ferguson, eds.). Forty-Second Symposium of the Society for General Microbiology. Cambridge University Press, Cambridge, pp. 331–361.

5. Schreier, H.J. 1993. Biosynthesis of glutamine and glutamate and the assimilation of ammonia. In Bacillus subtilis *and Other Gram-Positive Bacteria* (A.L. Sonensheim, J.A. Hoch, and R. Losick, eds.). American Society for Microbiology Press, Washington, D.C., pp. 281–298.

6. Jetten, M.S.M., S. Logemann, G. Muyzer, L.A. Robertson, S. deVries, M.C.M. van Loosdrecht, and J.G. Kuenen. 1997. Novel principles in the microbial conversion of nitrogen compounds. *Antonie van Leeuwenhoek* 71:75–93.

7. Zumft, W.G., A. Viebrock, and H. Körner. 1988. Biochemical and physiological aspects of denitrification. In *The Nitrogen and Sulfur Cycles* (J.A. Cole and S.J. Ferguson, eds.). Forty-Second Symposium of the Society for General Microbiology. Cambridge University Press, Cambridge, pp. 245–279.

8. Moura, I., S. Burdakov, C. Costa, and J.J.G. Moura. 1997. Nitrate and nitrite utilization in sulfate-reducing bacteria. *Anaerobe* 3:279–290.

9. Zumft, W.G. 1997. Biology and molecular basis of denitrification. *Microbiology and Molecular Biology Reviews* 61:533–616.

10. Hooper, A.B., T. Vannelli, D.J. Bergmann, and D.M. Arciero. 1997. Enzymology of the oxidation of ammonia to nitrite by bacteria. *Antonie van Leeuwenhoek* 71:59–67.

11. Jetten, M.S.M., P. de Bruijn, and J.G. Kuenen. 1997. Hydroxylamine metabolism in *Pseudomonas* PB16: Involvement of a novel hydroxylamine oxidoreductase. *Antonie van Leeuwenhoek* 71:69–74.

12. Watson, S.W. and M. Mandel. 1971. Comparison of the morphology and deoxyribonucleic acid composition of 27 strains of nitrifying bacteria. *Journal of Bacteriology* 107:563–569.

13. Wood, P.M. 1988. Monooxygenase and free radical mechanism for biological ammonia oxidation. In *The Nitrogen and Sulfur Cycles* (J.A. Cole and S.J. Ferguson, eds.). Forty-Second Symposium of the Society for General Microbiology. Cambridge University Press, Cambridge, pp. 219–243.

14. Bock, E., H.-P. Koops, H. Harms, and B. Ahlers. 1991. The biochemistry of nitrifying organisms. In *Variations in Autotrophic Life* (J.M. Shively and L.L. Barton, eds.). Academic Press, New York, pp. 171–200.

15. Van de Gaaf, A.A., A. Mulder, P. de Bruijn, M.S.M. Jetten, L.A. Robertson, and J.G. Kuenen. 1995. Anaerobic oxidation of ammonium is a biologically mediated process. *Applied and Environmental Microbiology* 61:1246–1251.

16. Barton, L.L. 1992. Sulfur metabolism. In *Encyclopedia of Microbiology*, Vol. 4 (J. Lederberg, ed.). Academic Press, Boca Raton, FL, pp. 135–150.

17. Kredich, N.M. 1987. Biosynthesis of cysteine. In Escherichia coli *and* Salmonella typhimurium: *Cellular and Molecular Biology* (J.L. Ingraham, K.B. Low, B. Magasanik, M. Schaechter, and H.E. Umbarger, eds.). American Society for Microbiology Press, Washington, D.C., pp. 419–428.

18. Akagi, J.M. 1995. Respiratory sulfate reduction. In *Sulfate-Reducing Bacteria* (L.L. Barton, ed.). Plenum Press, New York, pp. 89–112.

19. Hatchikian, E.C. 1994. Desulfofuscidin: Dissimilatory, high-spin sulfite reductase of thermophilic, sulfate-reducing bacteria. In *Inorganic Microbial Sulfur Metabolism. Methods*

*in Enzymology*, Vol. 243 (H.D. Peck, Jr. and J. LeGall, eds.). Academic Press, San Diego, pp. 276–295.

20. Moura, I. and A.R. Lino. 1994. Low-spin sulfite reductases. In *Inorganic Microbial Sulfur Metabolism. Methods in Enzymology*, Vol. 243 (H.D. Peck, Jr. and J. LeGall, eds.). Academic Press, San Diego, pp. 296–302.

21. Peck, H.D. Jr. 1993. Bioenergetic strategies of the sulfate-reducing bacteria. In *The Sulfate-Reducing Bacteria: Contemporary Perspectives* (J.M. Odom and R. Singleton, Jr., eds.). Springer-Verlag, New York, pp. 41–76.

22. Fauque, G., J. LeGall, and L.L. Barton. 1991. Sulfate-reducing and sulfur-reducing bacteria. In *Variations in Autotrophic Life* (J.M. Shively and L.L. Barton, eds.). Academic Press, New York, pp. 271–338.

23. Fauque, G.D. 1994. Sulfur reductase from thiophilic sulfate-reducing bacteria. In *Inorganic Microbial Sulfur Metabolism. Methods in Enzymology*, Vol. 243 (H.D. Peck, Jr. and J. LeGall, eds.). Academic Press, San Diego, pp. 353–367.

24. Fauque, G.D., O. Klimmek, and A. Kröger. 1994. Sulfur reductases from spirilloid mesophilic sulfur-reducing eubacteria. In *Inorganic Microbial Sulfur Metabolism. Methods in Enzymology*, Vol. 243 (H.D. Peck, Jr. and J. LeGall, eds.). Academic Press, San Diego, pp. 367–383.

25. Dahl, C., and H.G. Trüper. 1994. Enzymes of dissimilatory sulfide oxidation in phototrophic sulfur bacteria. In *Inorganic Microbial Sulfur Metabolism. Methods in Enzymology*, Vol. 243 (H.D. Peck, Jr. and J. LeGall, eds.). Academic Press, San Diego, pp. 400–421.

26. Kelly, D.P. 1988. Oxidation of sulphur compounds. In *The Nitrogen and Sulfur Cycles* (J.A. Cole and S.J. Ferguson, eds.). Forty-Second Symposium of the Society for General Microbiology. Cambridge University Press, Cambridge, pp. 65–98.

27. Suzuki, I. 1994. Sulfur-oxidizing enzymes. In *Inorganic Microbial Sulfur Metabolism. Methods in Enzymology*, Vol. 243 (H.D. Peck, Jr. and J. LeGall, eds.). Academic Press, San Diego, pp. 455–462.

28. Kelly, D.P., J.K. Shergill, W.-P. Lu, and A.P. Wood. 1997. Oxidative metabolism of inorganic sulfur compounds by bacteria. *Antonie van Leeuwenhoek* 71:95–108.

29. Suzuki, I. 1994. Sulfite:cytochrome *c* oxidoreductase of thiobacilli. In *Inorganic Microbial Sulfur Metabolism. Methods in Enzymology*, Vol. 243 (H.D. Peck, Jr. and J. LeGall, eds.). Academic Press, San Diego, pp. 447–454.

30. Kertesz, M.A. 1999. Riding the sulfur cycle—metabolism of sulfonates and sulfate esters in Gram-negative bacteria. *FEMS Microbiology Reviews* 24:135–175.

# Additional Reading

Clifton, C.E. 1951. Assimilation by bacteria. In *Bacterial Physiology* (C.H. Werkman and P.W. Wilson, eds.). Academic Press, New York.

## *Nitrogen and Sulfur Metabolism*

Cole, J.A. and S.J. Ferguson (eds.). 1988. *The Nitrogen and Sulfur Cycles.* Forty-Second Symposium of the Society for General Microbiology. Cambridge University Press, Cambridge.

Conrad, R. 1996. Soil microorganisms as controllers of atmospheric trace gases ($H_2$, CO, $CH_4$, OCS, $N_2O$, and NO). *Microbiology Reviews* 60:609–640.

## Nitrogen Metabolism

Butland, G., S. Spiro, N.J. Watmough, and D.J. Richardson. 2000. Two conserved glutamates in bacterial nitric oxide reductase are essential for activity, but not assembly of the enzyme. *Journal of Bacteriology* 183:189–199.

Chen, Y.B., B. Dominic, M.T. Mellon, and J.P. Zehr. 1998. Circadian rhythm of nitrogenase gene expression in the diaotrophic filamentous nonheterocystous *Cyanobacterium trichodesmium* Sp strain IMS101. *Journal of Bacteriology* 180:3598–3605.

Eady, R.R. 1996. Structure–function relationships of alternative nitrogenases. *Chemical Review* 96:3013–3030.

Smith, B.E. 1997. The nitrogenases. In *Transition Metals in Microbial Metabolism* (G. Winkelmann and C.J. Carrano, eds.). Harwood Academic, Amsterdam, pp. 471–496.

Stouthamer, A.H. 1988. Dissimilatory reduction of oxidized nitrogen compounds. In *Biology of Anaerobic Microorganisms* (A.J.B. Zehnder, ed.). John Wiley & Sons, New York, pp. 245–303.

Yates G.M. 1988. The role of oxygen and hydrogen in nitrogen fixation. In *The Nitrogen and Sulfur Cycles* (J.A. Cole and S.J. Ferguson, eds.). Forty-Second Symposium of the Society for General Microbiology. Cambridge University Press, Cambridge, pp. 383–416.

## Sulfur Metabolism

Detmers, J., V. Bruechert, K.S. Habicht, and J. Kuever. 2001. Diversity of sulfur isotope fractionations by sulfate-reducing prokaryotes. *Applied and Environmental Microbiology* 67: 888–894.

Kahnert, A., P. Vermeij, C. Wietek, P. James, T. Leisinger, and M. Kertesz. 2000. The *ssu* locus plays a key role in organosulfur metabolism in *Pseudomonas putida* S-313. *Journal of Bacteriology* 182:2869–2878.

Peck, H.D., Jr. and J. LeGall (eds.). 1994. Inorganic Microbial Sulfur Metabolism. *Methods in Enzymology*, Vol. 243. Academic Press, San Diego, p. 682.

Vermeij, P. and M.A. Kertesz. 1999. Pathways of assimilative sulfur metabolism in *Pseudomonas putida*. *Journal of Bacteriology* 181:5833–5837.

# 14

# Biometals

Our concepts should be flexible, but we need not be skeptical; all that is needed
is an open-minded approach and a willingness to balance, as far as possible,
ideas against one another and in relation to the facts upon which they are based.
C.E. Clifton, *Introduction to Bacterial Physiology*, 1957

## 14.1. Introduction

Inorganic compounds are essential in the support of growth by bacteria and archaea.
Chemolithotrophic organisms use specific oxy-anions and metal cations as electron
donors or electron acceptors. In prokaryotes, inorganic ions have critical roles in
metabolism and these include (1) stabilizing cellular structures, (2) serving as cofactors
for enzymes, (3) participating in electron transfer in oxidation–reduction reactions,
and (4) regulating DNA expression. Chemists group the elements into metals, nonmetals, and metalloids. The nonmetal elements of biological importance include H,
C, N, O, P, and S. Most of the elements are metals and these elements of biological
interest are subdivided into three groups: alkali metals, alkaline earth metals, and
transition metals. Na and K are alkali metals that form compounds with oxygen that
dissolve in water to produce alkaline solutions. Alkali earth metals include Mg and
Ca which are found in soils because the compounds they form with oxygen are insoluble in water. Transition metals of biological importance include Fe, Co, Ni, Mo,
Mn, Cu, and Zn, which are required in trace levels for numerous enzyme reactions.
There is a group of elements that have characteristics of both metals and nonmetals
that are referred to as metalloids. Selenium (Se) is a metalloid that is essential for
some bacteria and arsenic (As) is highly toxic to cells. The concentration of elements
in the environment is crucial because several essential elements are toxic to microbial
cells when present in elevated concentrations. Inhibition of growth by metals is attributed to changes in enzymes and proteins. The action of metals on bacteria that
are associated with toxicity can be attributed to three activities: (1) metals can prevent
an essential group on a molecule from functioning, (2) metal ions can modify the
molecular structure to produce a nonfunctional molecule, and (3) toxic metals can
displace an essential ion with the result of producing a nonfunctional molecule.

Several metals have been discussed in chapters concerning electron transport or solute transport and this discussion builds on that presentation. While this chapter examines some geochemical activities attributed to microorganisms and the role of essential metals for support of cell growth, the principal focus is on the interactions between toxic metals or metalloids and prokaryotes. An emphasis is placed on some of the mechanisms employed to enable prokaryotes to live in these toxic environments. To better understand the biological consequences of cells interacting with metals, it is important to review some relevant aspects of bioinorganic chemistry.

## 14.2. Bacteria–Metal Interaction[1]

Metal ions differ in their affinity for biomolecules and this binding strength is reflected in the hardness scale. According to the Pearson classification, ions that are "hard" acids are relatively small, have a high positive charge, and do not have any unpaired electrons in their outer shell. Their name is derived from the characteristic that the electron configuration is spherical and the electron shells are not distorted when interacting with another charged ion. Common ions that are hard acids include $Na^+$, $Mg^{2+}$, and $Ca^{2+}$. Soft acids are large atoms that have a low positive charge and contain unshared pairs of $p$ or $d$ electrons. The electron shells are easily distorted in soft acids as a result of interaction with another ion. Binding by hard acids is considered to be characterized as ionic while binding attributed to soft acids is more characteristic of a covalent linkage. Soft acids form stable carbon–metal bonds as is the case with methylated tin and organomercury compounds. Hard acids strongly bind with carboxyl ($-COO^-$), amide ($RNH_2$) or alcohol ($ROH$) groups with a preference for ligand atoms as follows: $N >> P$ and $O >> S$. Hard acids bind to hard bases with a preference for groups as follows: $OH^- > RO^- > RCO_2^-$; $CO_3^{2-} >> NO_3^-$; $PO_4^{3-} >> SO_4^{2-} >> ClO_4^-$. Soft acids bind to soft bases with soft bases binding strongly to thiol ($RSH$ and $R$-$S$-$R$) groups. Soft acids have a preference of ligand atoms in soft bases according to the following scheme: $S >> N > O$. Some metals are classified as intermediates on the hardness scale and these include $Ca^{2+}$ and $Zn^{2+}$. Examples of metals and ligands that are distinguished by the hardness scale are given in Table 14.1. A metal-binding site may vary in terms of hardness by the type of organic binding ligands present. For redox active metals, it is important for the environment of the ligand to change to accommodate the different oxidative states of the metal.

### 14.2.1. Biological Requirements for Metals[2–4]

Our understanding of the role of metal ions in nutrition of prokaryotes developed at a relatively slow rate. In 1891, Pasteur used yeast ash to provide minerals for the cultivation of bacteria in a medium known as "Pasteur's fluid." Through the use of media that was partially defined, Winogradsky and Beijerinck in the late 1890s and early 1900s cultivated several different bacteria that are now recognized as important chemolithotrophic organisms. Winogradsky found that soil bacteria could oxidize iron, sulfur, or ammonia and demonstrated autotrophic growth with carbon dioxide as

Table 14.1. Pearson's classification of metals and chemicals relevant to biology into hard or soft acids and bases.

| Acids | | | Bases | | |
|---|---|---|---|---|---|
| Hard | Intermediate | Soft | Hard | Intermediate | Soft |
| $Na^+$ | $Fe^{2+}$ | $Cu^+$ | Hydroxyl $(1,6)$[a] | Amine (1) | Sulfhydryl (thiol,1) |
| $K^+$ | $Co^{2+}$ | $Pb^{2+}$ | Carboxyl (2) | Imine (1) | Thioether (1) |
| $Mn^{2+}$ | $Ni^{2+}$ | $Cd^{2+}$ | Carbonyl | Amide (1) | Imidazole (1) |
| $Fe^{3+}$ | $Cu^{2+}$ | $Tl^+$ | (ketone, 2) | Secondary | Thiosulfate |
| $Co^{3+}$ | $Zn^{2+}$ | $Hg^{2+}$ | Sulfonate (3) | amine (2) | Cyanide |
| $Mg^{2+}$ | $Mn^{2+}$ | $Ag^+$ | Phosphonate (4) | Pyridine | |
| $Ca^{2+}$ | | $Au^+$ | Phosphodiester (5) | | |
| $Al^{3+}$ | | $Sn^{2+}$ | Phosphate | | |
| $Cr^{3+}$ | | $Bi^{3+}$ | Sulfate | | |
| $Fe^{3+}$ | Chloride | | | | |
| $UO_2^{2-}$ | | | | | |

[a]Structures commonly containing these ligands are as follows: (1) amino acids and amino sugars, (2) peptide bonds, (3) sulfonated polysaccharides, (4) phospholipids, (5) teichoic acids and lipopolysaccharides, and (6) polysaccharides.

the carbon source. Beijerinck demonstrated free-living $N_2$-fixing bacteria (*Azotobacter*), symbiotic $N_2$-fixing bacteria (*Rhizobium*), and sulfate-reducing bacteria (*Desulfovibrio*). The attempts to cultivate various types of lithoautotrophs provided a stimulus for the development of media with specified trace elements. The meticulous efforts of Pfennig in cultivating anaerobic photosynthetic bacteria produced trace metal solutions that are still used today. With the advent of bioinorganic chemistry and the development of analytical instruments for measuring metal–protein interactions, considerable focus is now directed at the molecular level to describe the activities of metals as enzyme cofactors.

One of the easiest ways to assess metal ion requirement by bacteria is to demonstrate growth only when a specific metal is present. Based on quantity required, essential metal ions are divided into two groups: macronutrients that are required in considerable quantity and trace elements that are needed only in small quantities. The essential metals to support bacterial growth are listed in Table 14.2. Only four metal ions are included in the macronutrient category: potassium, magnesium, calcium, and sodium. Although there are many different metals that are toxic to bacteria, 10 metal ions are classified as micronutrients and are required for growth.

In generating chemically defined growth media, the metal ions used as macronutrients or trace nutrients are added at different levels. Quantities of metal ions required to support bacterial growth are listed in Table 14.3. Growth dependency works only with those metals required in high concentration because the amount of bacterial biomass is small and the quantity of metal required is also small. Generally, methanogens and bacteria growing in alkaline or high salt environments require $K^+$ and $Na^+$. From the appreciable intracellular concentrations of $K^+$ and $Mg^{2+}$, it is apparent that these two elements are required in considerable concentrations. Of the trace elements, iron is required in greatest level and this reflects the use of $Fe^{2+}$ for redox

Table 14.2. Relationship of elements to microbial growth.

**I. Essential metals: required as macronutrients that may be toxic when present at elevated concentrations**

Sodium[a]
Magnesium
Potassium
Calcium

**II. Essential metals: required as micronutrients that may reach toxic concentrations**

| | |
|---|---|
| Vanadium | Iron |
| Chromium | Cobalt |
| Manganese | Selenium |
| Tungsten | Nickel |
| Zinc | Copper |

**III. Nonessential metals: elements that are not required for bacterial systems and are toxic to cells**

| | | |
|---|---|---|
| Aluminum | Gallium | Strontium |
| Antimony | Lead | Technetium |
| Arsenic | Mercury | Tellurium |
| Barium | Osmium | Thallium |
| Bismuth | Platinum | Tin |
| Cadmium | Ruthenium | Uranium |
| Cesium | Silver | |

**IV. Nonmetals required by bacteria**

Hydrogen
Oxygen
Carbon
Nitrogen
Sulfur
Phosphorus

[a]Sodium is required by methanogens, halophiles, and marine prokaryotes, but not by many bacteria.

proteins and heme synthesis. Other trace elements are required only in small quantities and in many instances these elements are present as contaminants of other chemicals added to the media. Many of the trace elements are difficult to maintain in a dissolved state unless chelators are employed. Ethylenediaminetetraacetic acid (EDTA) and citrate are commonly employed as metal chelators because they are nontoxic. For cells to remove the metal ion from an organic ligand, the biological affinity for the metal ion must be greater than the binding by the chelator. In many instances, elevated concentrations of trace metals are toxic to cells because intracellular concentrations exceed optimal levels. The activities of metal ions essential for bacteria are reviewed in subsequent segments of this chapter and transport features of these ions are found in Chapter 10.

Table 14.3. Concentrations of essential metals required for growth of prokaryotes.

| Metal | Milligrams present[a] | Intracellular concentrations |
|---|---|---|
| **Macronutrient** | | |
| $K^+$ | 1.7 | 250 m$M$ (3–4.5 $M$ in *Halobacterium halobium*) |
| $Mg^{2+}$ | 0.1–0.4 | 10–20 m$M$ |
| $Ca^{2+}$ | 0.1 | 0.1–0.09 μ$M$ |
| $Na^+$ | Varies | (1.4 $M$ in *Halobacterium halobium*) |
| **Trace nutrients** | | |
| $Fe^{n+}$ | 0.015 | 10 n$M$–1 μ$M$ |
| $Mn^{2+}$ | 0.005 | —[b] |
| $Zn^{2+}$ | 0.005 | —[b] |
| $Co^{2+}$ | 0.001 | —[b] |
| $Cu^{2+}$ | 0.001 | —[b] |
| $Mo^{2+}$ | 0.001 | —[b] |
| $SeO_3^{2-}$ | — | —[b] |
| $Ni^{2+}$ | — | —[b] |
| $WO_4^{2-}$ | — | —[b] |

[a]Approximate quantity required to produce 100 g dry weight of bacterial cells.
[b]Although the concentration has not been determined for the amount of metal in solution, several metals are estimated to be <0.01 μ$M$.

## 14.2.1.1. $Mg^{2+}$

$Mg^{2+}$ is one of the major ions required by bacteria to stabilize structures and molecules. While most bacteria have intracellular concentrations of $Mg^{2+}$ that are 10 to 20 m$M$, cells of *Escherichia coli* in the exponential phase may have intracellular concentrations approaching 100 m$M$ $Mg^{2+}$. The magnesium ion is not redox active but always occurs as the divalent positive ion. $Mg^{2+}$ binds to ATP, ADP, and nucleic acids neutralizing charges on the phosphate groups and participating in phosphate transfer. $Mg^{2+}$ is an ideal element to participate in the ATP reactions because as a hard acid it allows for quick release of phosphate from ATP. For optimum stability, the intracellular $Mg^{2+}$ concentration should be 2.2 m$M$ to stabilize ribosomes and when cells are placed in a $Mg^{2+}$-free medium, there is a rapid loss of proteins from the ribosomes.

In Gram-negative bacteria, $Mg^{2+}$ is associated with the outer membrane where it neutralizes charges on the phosphate moieties of the lipids, porin proteins, and lipopolysaccharides. In addition to $Mg^{2+}$, $Ca^{2+}$ is present to participate in the stabilization of the outer membrane but magnesium ion is preferred with a ratio of $Mg^{2+}$:$Ca^{2+}$ of 5: 1. $Mg^{2+}$ is also associated with the plasma membrane; however, there are three times more $Mg^{2+}$ than $Ca^{2+}$ in the plasma membrane. The $F_1$-ATPase on the inner side of the plasma membrane is stabilized on the membrane-bound $F_0$-unit by $Mg^{2+}$ and to isolate the $F_1$-ATPase the plasma membranes are diluted in magnesium free buffer. The $Mg^{2+}$ ion does not interfere with the movement of $H^+$ through the F-type ATPase pore. In addition, several enzymes are stabilized by $Mg^{2+}$ and a few of these are listed in Table 14.4. In most instances, $Mn^{2+}$ can replace $Mg^{2+}$ in the stabilizing of enzymes

Table 14.4. Examples of prokaryotic enzymes requiring metals as macronutrients.[a]

| Element | Enzyme | Bacteria |
|---|---|---|
| $K^+$ | Aspartate kinase | *Bacillus polymyxa* |
| | Carbamoyl-phosphate synthetase | *Escherichia coli* |
| | DNA polymerase | Most bacteria |
| | L-Malate dehydrogenase | *Lactobacillus arabinosus* |
| | Peptidyltransferase | *Bacillus stearothermophilus* |
| | Phosphate acyltransferase | *Clostridium kluyver* |
| | Tryptophan synthase | *Bacillus subtilis, Escherichia coli* |
| $Mg^{2+}$ | Phosphofructokinase | *Escherichia coli* |
| | $F_1$-ATPase | Most bacteria |
| | L-Threonine dehydrogenase | *Escherichia coli* |
| | Glucose isomerase | *Streptomyces griseofuscus* |
| | D-Xylose ketol isomerase | *Lactobacillus brevis, Bacillus coagulans* |
| | Citrate lyase | *Klebsiella aerogenes* |
| $Ca^{2+}$ | Thermolysin (a Zn-endopeptidase) | *Bacillus thermoproteolyticus* |
| | 5' Phosphodiesterase | *Staphylococcus aureus* |
| | Carboxypeptidase | *Streptomyces griseus* |
| | α-Amylases | *Bacillus subtilis* |

[a]From Hughes and Poole, 1989[2]; Wackett et al., 1089.[3]

under in vitro conditions; however, in nature manganese concentrations inside the cell are not high enough to replace $Mg^{2+}$ as the metal for stabilizing enzymes.

### 14.2.1.2.  $Ca^{2+}$

In nonsporulating bacteria, the activities of $Ca^{2+}$ are generally associated with stabilizing molecules or structures outside of the cell. Several extracellular enzymes are stabilized by $Ca^{2+}$ and while removal of calcium atom inactivates the enzyme, $Ca^{2+}$ does not appear to participate in the catalytic process. Several calcium-stabilized enzymes produced by bacteria are listed in Table 14.4. A low intracellular concentration of $Ca^{2+}$ is desirable to avoid precipitates because many of the salts of calcium are insoluble. Several active transport systems that export $Ca^{2+}$ from cells and these include the $Ca^{2+}/H^+$ anteporter, the $Ca^{2+}/Na^+$ anteporter, and an ATP driven exporter of $Ca^{2+}$. In *Bacillus*, there is the active accumulation of calcium for the production of calcium-dipicolinate in the sporulation process and 4% of the cell can be calcium. While calmodulin-like proteins have been proposed to participate in regulation of intracellular processes, $Ca^{2+}$-binding calmodulin molecules have not been isolated from bacteria.

### 14.2.1.3.  $K^+$

A key metal ion in bacterial growth is $K^+$, which is often referred to as an intracellular cation in that it accumulates to 0.25 $M$ in most cells. Intracellular $K^+$ is involved in controlling osmotic pressure in the cell, promoting structural stability, and activating enzymes. In halophilic archaea, the intracellular concentration of $K^+$ may reach 4.5 $M$ and this is to assist the cell in countering the extracellular environment where the salt concentration may exceed 30%. In nonhalophilic organisms, there may be transient changes in osmotic pressure and this can be balanced by the intracellular $K^+$ concentration because active transport systems accumulate $K^+$ up to several thousand times the extracellular level. Negatively charged groups in DNA and other nucleic acids are neutralized by the presence of the $K^+$ ion. Furthermore, $K^+$ contributes to the stability of the ribosome structure and the different conformational forms of the DNA double helix. $K^+$ activates bacterial enzymes and a partial listing is given in Table 14.4. These activities in bacteria are specific for $K^+$ because $Na^+$ or other cations cannot serve as appropriate substitutes.

### 14.2.1.4. $Na^+$

Although the $Na^+$ ion has several functions in the prokaryotic cell, it is difficult to assess an accurate level required for growth because $Na^+$ generally does not accumulate inside cells and growth where $Na^+$ is lacking is not definitive. One of the major activities of $Na^+$ is the establishment of ion gradients across the plasma membrane. For example, a few prokaryotes use a $Na^+$ symport system for the uptake of amino acids, the export of $Ca^{2+}$ is attributed to the $Ca^{2+}/Na^+$ anteport system, and the low intracellular concentration of $Na^+$ is maintained through the action of a $Na^+/H^+$ anteporter. These and other transport systems for $Na^+$ are discussed in Chapter 10. In addition, $Na^+$ is reported to drive flagellar motors in alkaliophilic bacteria. Certain bacteria may use a $Na^+$-driven ATPase that couples uptake of $Na^+$ with ATP synthesis by a system analogous to the $H^+$-coupled ATP synthesis as occurs with the F-type ATPase.

### 14.2.1.5. Transition Metals

Important transition metals in biology include iron, copper, manganese, cobalt, molybdenum, nickel, and zinc. With the exception of zinc, all of the transition metals listed have multiple oxidation states and function in redox catalysis. The role of divalent zinc is to provide structural stability to the enzyme. When present in proteins, the transition metal generally is present as a single species; however, there are examples where multiple species of metals may occur in a protein. This is the case with hydrogenase, which contains both iron and nickel. Other enzymes that have multiple transition metals include cytochrome oxidase with iron and copper while xanthine dehydrogenase contains molybdenum, iron, tungsten, and selenium. Some enzymes produced by prokaryotes require transition metals as cofactors, and a list of some of these enzymes is given in Table 14.5.

Generally a transition metal that functions as a redox cofactor for a specific enzyme cannot be replaced. No metal can replace iron in Fe-S clusters associated with redox centers of dehydrogenases and only $Mn^{2+}$ functions in the catalysis of $O_2$ in photosystem I. In some cases two or more different metal ions may catalyze a reaction;

Table 14.5. Examples of prokaryotic enzymes requiring transition metals.[a]

| Element (biological oxidation state) | Enzyme | Bacteria |
|---|---|---|
| Vanadium (3+, 4+, 5+) | Nitrogenase | *Azotobacter chrococcum* |
| Manganese (3+, 4+, 5+) | Pseudocatalase | *Lactobacillus plantarum* |
|  | 3,4-Dihydroxyphenyl acetate lyase 2,3-dioxygenase | *Bacillus brevis* |
|  | Superoxide dismutase | *Escherichia coli* |
|  | Guanidinoacetate aminohydrolase | *Pseudomonas* sp. |
|  | Allantoate amidinohydrolase | *Pseudomonas aeruginosa* |
|  | α-Isopropylmalate synthetase | *Alcaligenes eutrophus* |
|  | Phosphoglycerate mutase | |
| Iron (2+, 3+) | Catalase/peroxidase | Many bacteria |
|  | Nitrite reductase (24 hemes) | *Nitrosomonas europeae* |
|  | Sulfite reductase | *Escherchia coli* |
|  | ρ-Cresol hydroxylase | *Pseudomonas putida* |
|  | Hydrogenase | *Clostridium pasteurianum* |
|  | Aconitase | Organisms with TCA cycle |
|  | Toluene dioxygenase | *Pseudomonas putida* |
|  | Methane monooxygenase | *Methylococcus capsulatus* |
|  | Catechol 1,2-dioxygenase | *Pseudomonas arvilla* |
|  | Ribonucleotide reductase | *Escherichia coli* |
| Cobalt[a] (1+, 2+, 3+) | Transcarboxylase | *Propionobacterium shermanii* |
|  | Amidino aspartase | *Pseudomonas chlororaphis* |
| Nickel (1+, 2+, 3+) | Hydrogenase | *Methanobacterium thermoautotrophicum* |
|  | CO dehydrogenase | *Clostridium thermoaceticum* |
|  | Urease | *Selenomonas ruminantium* |
| Copper (1+, 2+) | Methylamine oxidase | *Arthrobacter* P1 |
|  | Cytochrome oxidase | Many aerobic bacteria |
|  | Phenylalanine hydroxylase | *Chromobacterium violaceum* |
|  | Superoxide dismutase | *Pseudomonas diminuta* |
|  | Nitric oxide oxidoreductase | *Pseudomonas* sp. |
| Zinc (2+) | Carbonic anhydrase | Most bacteria |
|  | Alkaline phosphatase | *Escherichia coli* |
|  | RNA polymerase | *Escherichia coli* |
|  | Carboxypeptidase | Most bacteria |
|  | Ornithine transcarbamylase | *Escherichia coli* |
| Molybdenum (3+ to 6+) | Nitrogenase | Various prokaryotes |
|  | Formate dehydrogenase | *Escherichia coli* |
|  | Sulfite oxidase | *Thiobacillus novellus* |

[a]From Hughes and Poole, 1989[2]; Wackett et al., 1989.[3]

[b]$B_{12}$-containing enzymes are not included.

however, in these cases different proteins are used. Oxidation of $H_2$ can be accomplished by a Fe-hydrogenase or a Fe, Ni-hydrogenase and the proteins in this case are products of different genes. Many of the enzymes that use $Mg^{2+}$ for structural stability remain active when $Mn^{2+}$ replaces $Mg^{2+}$ and even a few enzymes are active when $Co^{2+}$ replaces $Mg^{2+}$. The substitution of $Mn^{2+}$ or $Co^{2+}$ for $Mg^{2+}$ is readily done with in vitro reactions but it is unlikely that $Mn^{2+}$ or $Co^{2+}$ may be used to substitute for $Mg^{2+}$ in vivo at the time of enzyme synthesis.

## 14.2.2. Geochemical Cycles[5]

Prokaryotic organisms have an important function in mineral transformation, geochemical cycling, and bioaccumulation of metals. Chemolithotrophs or photoautotrophs have a dependancy on inorganic metabolism and require oxidation/reduction reactions for growth. To demonstrate the role of microorganisms in inorganic reactions, several biogeochemical cycles are given in Figure 14.1. At this time, specialized prokaryotes are known to couple the oxidization of Fe, U, Mn, and Se to growth. In several instances, Fe, U, Mn, and Se serve as electron sinks and thereby function as electron acceptors for cell growth. Detoxification process is one of the principal activities of organisms interacting with the As, Se, Hg, and U cycles. Several anaerobic chemolithotrophs display remarkable ability in being capable of interacting with several different cycles and with these redox active elements, the driving force is the flow of electrons (see Figure 14.2).

As a result of oxidation–reduction reactions, these metal cycles are frequently intimately associated with the major nutrient cycles of carbon, nitrogen, and sulfur. It is important to recognize that chemolithotrophs have the capability of interacting with several different biogeochemical cycles. Many of these reactions with inorganic compounds may be highly exerogonic and will support cellular growth. As presented in Figure 14.3, electron donor reactions are generally more electronegative than the electron acceptor reactions. A rough rule is that there is sufficient energy to produce ATP synthesis if the electron donor couple is about 200 mV more negative than the electron acceptor. Coupled reactions that do not provide a release of energy (about $-200$ mV) can support cell growth provided there is an additional input of energy or there is something unique about the equilibrium of the reactions. Most commonly, oxidation reactions use molecular oxygen as the electron acceptor and several of these reactions are listed in Table 14.6. In anaerobic environments, microorganisms frequently use metals and metalloids as terminal electron acceptors (Tables 14.7 and 14.8). A generalization is that anaerobes would have electron donors of about $-400$ mV because of electron donors such as $H_2$, formate, or NADH. Bacteria and archaea use various electron donors and some examples of this activity are listed in Table 14.9. The physiological types of prokaryotic organisms are similar to the biogeochemical cycles.

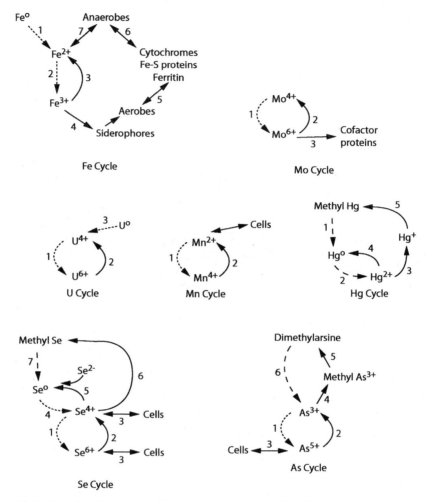

Figure 14.1. The role of microorganisms in some of the geochemical cycles with redox active elements. The oxidative pathway is shown in *dashed lines* while the reduced reactions are shown in *solid lines*. Each cycle is discussed in the text.

### 14.2.2.1. Fe Reactions[6–10]

Iron is the fifth most abundant element in the Earth's crust and is highly involved in biological and geochemical activities. Various interactions between iron and micro-organisms are outlined in Figure 14.1. While the oxidation or reduction of iron is readily demonstrated in microbial systems, it has been difficult to define the enzymes associated with these activities. In nature the oxidation of $Fe^0$ to $Fe^{2+}$ and $Fe^{2+}$ to $Fe^{3+}$ are reactions attributed primarily to prokaryotes. However, it is difficult to obtain sufficient enzyme quantities for characterization of the reactions because the cell yields

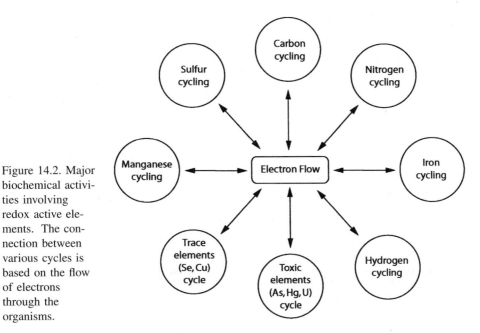

Figure 14.2. Major biochemical activities involving redox active elements. The connection between various cycles is based on the flow of electrons through the organisms.

are poor. In addition, the chemical activity of iron is influenced by environmental conditions and iron minerals may contribute to the complexity of the activities with iron. As indicated in Table 14.9, energetics of the $Fe^{2+}/Fe^{3+}$ couple is approximately +200 mV in circumneutral environments but is +600 mV at pH 2. *Acidithiobacillus ferroxidans* oxidizes $Fe^{2+}$ to $Fe^{3+}$ with $O_2$ as the electron acceptor; however, the energetics of this reaction are not favorable and reverse electron flow is proposed to

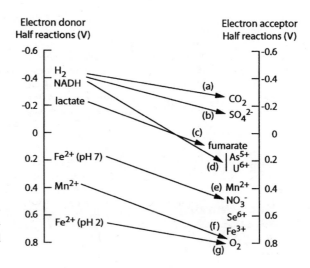

Figure 14.3. Thermodynamic expression of electron donor/electron acceptor couples. (Modified from Nealson, 1997.[12])

Table 14.6. Energetics of oxidation–reduction reactions involving metals.[a]

| Reactions | Free energy/reaction ($\Delta G^{0\prime}$) |
|---|---|
| **Aerobic** | |
| 1. [b]$H_2 + \frac{1}{2} O_2 \rightarrow H_2O$ | −237 kJ |
| 2. $Fe^{2+} + \frac{1}{4} O_2 + H^+ \rightarrow Fe^{3+} + \frac{1}{2} H_2O$ | −29 kJ |
| 3. $Mn^{2+} + \frac{1}{2} O_2 + H_2O \rightarrow MnO_{2(S)} + 2H^+$ | −70.9 kJ |
| **Anaerobic** | |
| 4. $Acetate^- + 8\ Fe^{3+} + 4\ H_2O \rightarrow 2\ HCO_3^- + 8\ Fe^{2+} + 9\ H^+$ | −233 kJ |
| 5. $Acetate^- + SO_4^{2-} + 3\ H^+ \rightarrow 2\ CO_2 + HS^- + 2\ H_2O + H^+$ | −57.5 kJ |
| 6. $Lactate^- + 2\ SeO_4^{2-} \rightarrow Acetate^- + 2\ SeO_3^{2-} + HCO_3^- + H^+$ | −343.1 kJ |
| 7. $Lactate^- + 2\ SeO_3^{2-} + H^+ \rightarrow Acetate^- + 2\ Se^0 + HCO_3^- + H_2O$ | −529.5 kJ |
| 8. $Lactate^- + 2\ AsO_4^{2-} + H^+ \rightarrow Acetate^- + 2\ H_2AsO_3^- + HCO_3^-$ | −140.3 kJ |
| 9. $3\ Lactate^- + 2\ H_2AsO_3^- \rightarrow 3\ Acetate^- + 2\ AsH_3^- + 3\ HCO_3^- + H^+$ | −138.4 kJ |

[a]From Barton et al., 2003[6]; Thauer et al., 1977.[10]
Organisms that are characteristically associated with the reactions: 1. [b]*Ralstonia eutrophus* 2. *Acidithiobacillus ferroxidans*, 3. *Leptothrix discophora*, 4. *Geobacter metallireducens*, 5. various sulfate-reducing bacteria, 6. *Thauera selenatis*, 7. Many bacteria including *Wolinella succonogenes* and *Escherichia coli*, 8. *Escherichia coli*, 9. *Sulfospirillum barnesii* and *Desulfotomaculum auripigmentum*.

Table 14.7. Inorganic anions reduced by selected strains of bacteria.[a]

| Oxidized compound | Reduced compound | Organisms | Reduction supports growth |
|---|---|---|---|
| Arsenate ($As^{5+}$) | Arsenite ($As^{5+}$) | *Bacillus arsenicoselenatis* | Yes |
| | | *Chrysiogenes arsenatis* | Yes |
| | | *Desulfotomaculum auripigmentum* | Yes |
| | | *Sulfospirillum barnesii* | Yes |
| Chromate ($Cr^{6+}$) | Chromite ($Cr^{3+}$) | *Deinococcus radiodurans* | No |
| | | *Desulfotomaculum reducens* | Yes |
| | | *Desulfovibrio vulgaris* | No |
| | | *Geobacter metallireducens* | No |
| Molybdate ($Mo^{6+}$) | Molybdite ($Mo^{4+}$) | *Desulfovibrio desulfuricans* | No |
| Selenate ($Se^{6+}$) | Selenite ($Se^{4+}$) | *Bacillus selenitereducens* | Yes |
| | | *Sulfospirillum arsenophilum* | |
| Selenite ($Se^{4+}$) | Selenium ($Se^0$) | *Desulfovibrio desulfuricans* | No |
| | | *Wolinella succinogenes* | No |
| TcVI[b] | TcII[b] | *Deinococcus radiodurans* | No |
| | | *Geobacter metallireducens* | No |
| | | *Desulfovibrio desulfuricans* | No |
| | | *Paracoccus denitrificans* | No |
| Tellurite | Tellurium ($Te^0$) | *Desulfovibrio desulfuricans* | No |
| | | *Peptococcus aerogenes* | No |
| | | *Veillonella atypica* | No |
| Vanadate | Vanadite | *Desulfovibrio desulfuricans* | No |
| | | *Veillonela atypica* | No |

[a]From Barton et al., 2003.[6]
[b]Technetium.

Table 14.8. Metal cations reduced by selected strains of bacteria.[a]

| Oxidized metal | Reduced metal | Organism | Reduction supports growth |
|---|---|---|---|
| $Co^{3+}$ | $Co^{2+}$ | *Geobacter sulfurreducens* | Yes |
| | | *Shewanella alga* | No |
| $Fe^{3+}$ | $Fe^{2+}$ | *Deinococcus radiodurans* | No |
| | | *Desulfobacter postageti* | No |
| | | *Desulfovibrio desulfuricans* | No |
| | | *Geothrix fermentans* | Yes |
| | | *Geobacter metallireducens* | Yes |
| | | *Shewanella putrefaciens* | Yes |
| $Mn^{4+}$ | $Mn^{2+}$ | *Acinetobacter johnsonii* | No |
| | | *Desulfobacterium autotrophicum* | No |
| | | *Desulfuromonas acetoxidans* | Yes |
| | | *Geobacter metallireducens* | Yes |
| | | *Pseudomonas fluorescens* | No |
| | | *Shewanella oneidensis* | Yes |
| $Np^{5+}$ | ? | *Shewanella oneidensis* | No |
| $Pd^{2+}$ | $Pd^{0}$ | *Desulfovibrio desulfuricans* | No |
| $U^{6+}$ | $U^{4+}$ | *Deinococcus radiodurans* | No |
| | | *Desulfotomaculum reducens* | Yes |
| | | *Desulfovibrio desulfuricans* | No |
| | | *Geobacter metallireducens* | No |
| | | *Shewanella oneidenasis* | No |

[a]From Barton et al., 2003.[6]

provide sufficient energy for growth. As a result, the cell yields of *Acidithiobacillus ferroxidans* may be very low. The electron transport system of *Acidithiobacillus ferroxidans* is discussed in Chapter 9. At pH values near 7, aerobic $Fe^{2+}$ oxidation is commonly associated with *Gallionella ferruginea*, *Sphaerotilus natans*, and *Leptothrix ochracea*. Unfortunately, enzyme systems for iron oxidation have not been isolated from these bacteria growing at neutral pH values. From an energetic perspective, there appears to be sufficient energy released in oxidation of $Fe^{2+}$ with the reduction of $NO_3^{-}$; however, organisms with this activity are poorly characterized. Growing at 85°C, *Ferroglobus placidus*, an archaeon, can oxidize $Fe^{2+}$ with the reduction of nitrate.

The reduction of $Fe^{3+}$ to $Fe^{2+}$ with lactate as the electron donor is accomplished by several bacteria and this reaction provides sufficient energy to support cell growth. Bacteria that can grow with dissimilatory $Fe^{3+}$ reduction include members of *Geobacter*, *Shewanella*, *Desulfuromonas*, and *Pelobacter* while representatives from archaea with this capability include *Pyrobaculum* and *Ferroglobus*. The reduction of $Fe^{3+}$ by *Geobacter metallireducens* is suggested to involve cytochromes and Lovley has proposed a model in which a 41-kDa cytochrome in the outer membrane shuttles electrons to $Fe^{3+}$. Iron would not enter the cell and reduced iron would remain outside

Table 14.9. Half-reactions important for microbial oxidation–reduction of elements.[a]

| Redox couple | $E^{0'}$ (mV) (pH 7 unless specified) |
|---|---|
| **Commonly used as electron donors** | |
| $CO_2$/formate$^-$ | $-432$ |
| $H^+$/$H_2$ | $-410$ |
| $NAD^+$/NADH | $-320$ |
| $CO_2$/acetate | $-290$ |
| $S^0$/HS$^-$ | $-270$ |
| Pyruvate$^-$/lactate$^-$ | $-190$ |
| $Mo^{6+}$/$Mo^{5+}$ ($HMoO_4^-$/$Mo_3O_{8(S)}$) | $-181$[b] |
| **Commonly used as electron acceptors** | |
| Fumarate/succinate | $+33$ |
| $Se^0$/$Se^{4+}$ ($Se_{(S)}$/$SeO_3^{2-}$) | $+175$[b] |
| $Te^0$/$Te^{4+}$ ($Te_{(S)}$/$HTeO_3^-$) | $+194$[b] |
| $Fe^{3+}$/$Fe^{2+}$ (pH 7) | $\sim +200$ |
| $As^{5+}$/$As^{3+}$ ($HAsO_4^-$/$H_4AsO_3$) | $+220$ |
| $U^{6+}$/$U^{4+}$ ($UO_2^{2+}$/$UO_{2(S)}$) | $+227$[b] |
| $Mn^{4+}$/$Mn^{2+}$ ($MnO_{2(S)}$/$Mn^{2+}$) | $+410$ |
| $N^{5+}$/$N^{3+}$ ($NO_3^-$/$NO_2^-$) | $+425$ |
| $Se^{6+}$/$Se^{4+}$ ($SeO_4^{2-}$/$SeO_3^{2-}$) | $+443$[b] |
| $Pd^0$/$Pd^{2+}$ ($Pd_{(S)}$/$PdO_{(S)}$) | $+481$[b] |
| $Cr^{6+}$/$Cr^{3+}$ ($CrO_4^{2-}$/$Cr_2O_{3(S)}$) | $+552$[b] |
| $Fe^{3+}$/$Fe^{2+}$ (pH 2) | $+771$ |
| $O_2$/$H_2O$ (1 atm $O_2$) | $+815$ |
| $Cl^{6+}$/$Cl^-$ ($ClO_4^{2-}$/$Cl^-$) | $+972$[b] |

[a]Data from Barton et al., 2003[6]; and Thauer et al., 1977.[10]
[b]Calculated at 1 $\mu M$.

the cell. Electrons would be shuttled to the cytochrome in the outer membrane by a 9-kDa cytochrome in the periplasm. Ultimately, electrons for the reduction of the cytochrome in the periplasm could be from a 89-kDa cytochrome in the plasma membrane. In *Geobacter metallireducens*, all the cytochromes involved in this electron cascade are of the *c*-type.

For decades scientists have observed that biocorrosion of metallic iron was attributed to biofilm activity. The anaerobic corrosion is often attributed to sulfate-reducing bacteria and the aerobic process has been proposed to be associated with *Thiobacillus* sp. In anaerobic biocorrosion of ferrous materials, the metal would be dissolved as a result of conversion of $Fe^0$ to $Fe^{2+} + 2\ e^-$. Activities that can remove $Fe^{2+}$ or electrons would be considered to promote the corrosion process; however, definitive evidence has been difficult to obtain. Laishley and colleagues have proposed that the electrons from this process can be acquired by cytochromes in the outer membrane of *Desulfovibrio vulgaris* and used for the production of $H_2$ by a periplasmic hydrogenase. Cord-Ruwisch suggests that at pH 7, $H_2$ at a partial pressure of 10 Pa, and $Fe^{2+}$ at 1 m$M$ the following reaction is thermodynamically favorable:

$$Fe + 2 H^+ \rightarrow Fe^{2+} + H_2 \qquad \Delta G^{0\prime} = -39.0 \text{ kJ/mol}$$

In the case where $H_2$ is formed under abiotic conditions, sulfate-reducing bacteria function to remove $H_2$ through the activity of hydrogenase and the end product is $H_2S$.

### 14.2.2.2. Mn Activities[11–13]

Manganese is present in many naturally occurring minerals and the concentration of manganese ranges from 0.002% to 1.0%. Over the years, there have been numerous reports of bacteria having a role in deposition of manganese in freshwater and marine environments. A common feature of bacterial oxidation is the formation of insoluble manganese oxides which readily form manganese deposits. As early as 1901, D.D. Jackson demonstrated that soil bacteria become encapsulated with manganese by oxidation of soluble $Mn^{2+}$. A standard procedure for detection of bacteria that oxidize manganese is to incorporate salts of $Mn^{2+}$ into agar medium, and the deposition of manganese oxide gives the colonies a pronounced dark brown appearance. In some cases the oxidation is attributed to enzyme action of the bacteria and in a few cases the oxidation is associated with products that come from the bacteria. Rocks in deserts throughout the world have a shiny dark brown surface that is referred to as "desert varnish" or "rock varnish," and this covering is considered to be a product of microbial activity. The varnish is produced on rocks that have a neutral pH at the surface and the rocks contain a low nutrient content. As the colonizing photosynthetic or autotrophic organisms on the surface of the rocks die, the released organic compounds are intermittently used by manganese oxidizing bacteria to deposit manganese oxides on the rock surface. It has been proposed that the insoluble manganese protects the underlying bacteria against several environmental hazards including UV damage. Manganese nodules found on the floor of all oceans is another bioproduct of bacterial activity. These nodules are remarkably uniform spherical aggregates that range in size from 1 to 25 cm. Because the nodules of the Pacific Ocean contain about 24% manganese and 14% iron, they are commonly referred to as ferromanganese nodules. There are three types of bacteria present in these nodules: (1) bacteria that oxidize manganese to insoluble manganese oxides, (2) bacteria that have the capability of reducing insoluble $MnO_2$ to soluble $Mn^{2+}$, and (3) bacteria that are unable to either oxidize or reduce manganese but serve to provide an adhesive matrix to secure the manganese metabolizing bacteria into the nodule structure. The bacterial activities of manganese oxidation and manganese reduction are discussed in subsequent sections.

*Manganese Oxidation*

The oxidation of manganese is achieved by bacteria of several different genera and, based on the mechanisms involved, these organisms constitute three groups that are involved in (1) oxidation of soluble manganese, (2) oxidation of bound manganese, and (3) production of hydrogen peroxide to oxidize manganese. The oxidation of soluble $Mn^{2+}$ can be characterized according to the following reaction:

$$Mn^{2+} + \tfrac{1}{2} O_2 + H_2O \rightarrow MnO_2 + 2 H^+$$

## Box 14.1

## Getting to Know Life: From Bioluminescence with Quorum Sensing to "Breathing Metals" by Environmental Bacteria

A Perspective by Kenneth H. Nealson

Growing up on an Iowa farm meant having the luxury of getting to know life firsthand, and it is not surprising to me that this love affair remains strong after all these years. After 17 years in small rural climes, I was fortunate to get a fellowship to the University of Chicago where I studied (after brief flirtations with physics and music) biochemistry and microbiology. During this time I was given a chance to take a summer course in Woods Hole, Massachusetts, where I was "bitten" by the marine sciences bug, and began to study bioluminescence in bacteria, a subject that would keep me busy for more than 25 years. Just how do these diminutive marine bacteria control the light they emit? It seemed a simple enough question at the time, but one that would prove elusive for a while. Doing postdoctoral work at Harvard with Professor Woody Hastings proved key to the success. The mechanism was termed autoinduction, and involved the cell-to-cell communication among the bacteria, allowing light emission only at high cell density. This mechanism, discovered under the guise of basic research, is now called quorum sensing and may have great value for many aspects of microbiology, including medicine and agriculture. This is the joy of basic research, seeing the simple mechanistic facts being elevated to applications undreamed of.

In the late 1970s, still working on bioluminescence at the Scripts Institution of Oceanography, I developed an interest in environmental microbiology, specifically with the interaction between microbes and metals (iron and manganese at the start), an area that is now referred to as geobiology. With the help of a Guggenheim fellowship, I was able to spend a year at the University of Washington, learning the tools of this new trade, and this was my passion for the next two decades. We discovered that microbes "breathe" metals, respiring iron and manganese as other organisms respire oxygen. These bacteria, now known to be abundant around the world, have major impacts on the geology of our planet, and may be of great value for bioremediation, as they are very good at modifying the chemistry of toxic metals like uranium and chromium.

Knowing that life can prosper under such conditions changes the way one feels about looking for life in extreme (even extraterrestrial) environments, and a few years ago I began a third career, this time in astrobiology, leading a group at JPL with the goal of developing new methods for life detection. This new passion is closely connected to that of environmental microbiology, as each new method we develop must be tested on Earth, and extreme environments offer the ideal places.

What one learns from this is that basic research is a true passion, with each day bringing new problems, new insights, and a level of satisfaction matched by few. We also are constantly reminded that we are standing on the shoulders of giants—scientists whose past and present efforts allow us to continue our quest.

This reaction at pH 7.0 has a standard free energy of $-70.9$ kJ/mol which is sufficient energy released to produce at least one ATP molecule. Manganese forms some interesting oxides and in addition to the production of $MnO_2$ there may be the formation of hausmannite ($Mn_3O_4$) and manganite ($MnOOH$). Several organisms display this oxidation of soluble manganese and include *Arthrobacter*, *Bacillus*, *Citrobacter*, *Hyphomicrobium*, *Leptothrix*, and *Pseudomonas*. Essential for the oxidation of $Mn^{2+}$ is the production of a soluble protein that frequently can be found released from the cells. Although some of the manganese-oxidizing bacteria contain sheaths, the enzyme is not bound into the sheath. The enzyme for oxidation of manganese has been shown to divert electrons into the membrane, where electron transport components can couple ATP synthesis to this electron flow. Research progress has been slow because it is difficult to obtain sufficient enzyme for study. At this time, the molecule that has attracted the greatest amount of interest is the 174-kDa protein that has been isolated from *Leptothrix discophora* and has sequence similarity with copper oxidase proteins. Attempts to overexpress the cloned gene in *E. coli* have thus far not been successful because products of the clone are lethal to *E. coli*.

A second group of manganese-oxidizing bacteria prefer to oxidize $Mn^{2+}$ while it is bound as a ferromanganese oxide or onto clay. The oxidation of insoluble manganese proceeds according to the following two reactions and organisms that have this capability are associated with *Arthrobacter*, *Oceanospirillum*, and *Vibrio*.

$$Mn^{2+} + H_2MnO_3 \rightarrow MnMnO_3 + 2\ H^+$$

$$MnMnO_3 + \tfrac{1}{2}\ O_2 + 2\ H_2O \rightarrow 2\ H_2MnO_3$$

The difference between the groups that oxidize soluble manganese and those that oxidize bound manganese may be dependent on the presence of a protein in the outer membrane. Future research is required to better characterize these oxidative processes in bacteria and to determine their presence in archaea.

The third group of bacteria that oxidizes manganese produce $H_2O_2$ and through the action of catalase, $Mn^{2+}$ is oxidized according to the following reaction:

$$Mn^{2+} + H_2O_2 \rightarrow MnO_2 + 2\ H^+$$

Bacteria that oxidize $Mn^{2+}$ by $H_2O_2$ include *Leptothrix pseudoochracea* and *Arthrobacter sidercapsulatus*. Neither of these organisms derive energy from this hydrogen peroxide mediated reaction. Research is required to reveal the stimulus for the production of large quantities of hydrogen peroxide in these manganese-oxidizing bacteria.

An intriguing manganese-oxidizing activity is associated with spores produced by *Bacillus* sp. strain SG-1. There is some evidence that this oxidation of $Mn^{2+}$ proceeds to $Mn^{3+}$ with the production of $Mn_3O_4$ which dismutates slowly to $MnO_2$ and $Mn^{2+}$. Little success has been achieved in isolating a functional enzyme from the bacterial spore. However, through molecular biology a gene has been isolated that encodes for a 138-kDa protein and it is similar to a copper oxidase protein. There are copper oxidases that transfer electrons from an electron donor (e.g., $Mn^{2+}$) to $O_2$ with the formation of $H_2O_2$ and it is possible that the oxidation of $Mn^{2+}$ at the surface of the spore may be attributed to such a copper protein.

*Reduction of Manganese*

The biological reduction of manganese may proceed by either an enzyme reaction coupled to a respiratory system or a chemical reaction with products of bacterial metabolism. The enzyme reduction of $MnO_2$ in an anaerobic environment is considered to occur by the following reaction:

$$MnO_2 + 4 H^+ + 2 e^- \rightarrow Mn^{2+} + 2 H_2O$$

At pH 7.0, the free energy value of the reaction is $-18.5$ kcal/mol ($-77.3$ kJ/mol) and this is sufficient to produce at least one ATP. The reduction of manganese is influenced by the form of manganese oxide present, with greater activity with the amorphous rather than with crystalline form. Numerous bacteria are involved in this reaction and some of the best studied systems concerning manganese respiration include *Shewanella* and *Geobacter*, investigated by the laboratories of K.N. Nealson and D.R. Lovley, respectively. Biochemical evaluations of the isolated proteins in manganese reduction would be useful to understand the molecular events in the electron flow.

The reduction of manganese by end products of metabolism is also an important feature of the manganese cycle. Many fermentative bacteria produce formic acid that reacts with manganese dioxide according to the following reaction:

$$MnO_2 + 2 H^+ + HCOOH \rightarrow Mn^{2+} + CO_2 + 2 H_2O$$

In addition, biologically produced $H_2S$ and $Fe^{2+}$ will react with $MnO_2$ to produce $Mn^{2+}$.

### 14.2.2.3. Mineralization[14–18]

The activities of minerals interacting with microorganisms in the environment are important for crystal growth as well as for catalyzing the life process. As proposed by Waechtershaeuser in 1988, a theory for genesis of biological systems on Earth involves inorganic compounds interacting with charged mineral surfaces. Pyrite is proposed to serve as this nucleation surface for the production of biochemicals. Bacteria display the potential for inducing mineral formation and the resulting materials are referred to as biominerals. Bacteria can induce mineral formation as a result of biologically induced mineralization (BIM) processes and by boundary-organized biomineralization (BOB) activities. With BIM, bacteria produce biominerals as a direct result of microclimate changes of pH and chemical changes in the environment immediately around the bacterial cell. With BIM, the biomineral formed is a result of the cellular surface serving as a support matrix onto which chemicals are deposited. Although both BIM and BOB are important processes in biomineralization, the process employed is perhaps more dependent on the bacterial species involved that on the biomineral formed. Thus, the interfacial dynamics of mineral–cell activity is a fundamental geomicrobiological process.

Although many different bacterial processes have been explored for weathering of minerals, bacterial production of ores and minerals has received relatively little attention from the scientific community. The bacterial surface has been reported to function as a nucleation site for the formation of metal phosphates and metal sulfides. Perhaps one of the best studied processes for BOB mineral formation by bacteria involves the surface layer of the cyanobacterium *Synechococcus* strain GL24. The S-layer of the bacterium has a hexagonally ordered lattice and serves as a template for adsorption of $Ca^{2+}$ ions. In the presence of high concentrations of sulfate, there is the formation of gypsum ($CaSO_4$) on the surface of *Synechococcus*. In experimental conditions when carbonate is added, sulfate is replaced by carbonate and calcite is produced. With $Sr^{2+}$ as the major divalent cation, celestite and strontionite are produced by *Synechococcus*; however, the details for mineralization have not been established.

The production of magnetite ($Fe_3O_4$) is attributed to either geochemical or biogenic activity. Bacteria that produce magnetite, termed magnetogens by Blakemore and Frankel, include the magnetotatic group with magnetite deposited intracellularly as magnetosomes. It is known that when magnetotactic bacteria reduce a quantity of ferrihydrite, magnetite is produced; however, the process whereby magnetosomes are formed from crystals of magnetite remains unknown. Also, the intracellular process accounting for the formation of the inverse spinel crystal structure found in magnetite but not the hexagonal close packed crystal structure found in hematite is unexplained. It is known that the $Eh'$ value must be less than $-100$ mV and it is proposed that membranous vesicles control the magnetite crystal morphology. In *Aquaspirillum magnetotactium*, rectangular horizontal projections are truncated octahedral prisms while rectangular parallelipipeds are truncated hexagonal prisms. In magnetosomes that are tapered or bullet-shaped, a third type of crystal occurs with a hexagonally prismatic structure. It is suggested by Blakemore and Frankel that these crystals could result by selected growth along selected axes from a cubooctahedral form. These crystals of magnetosomes have a distinctive form that has been produced by biogenic processes. Bacterial magnetite may be produced by either the BIB or BOB process and these biomagnetic crystals display a single magnetic domain character ranging from 400 to 1,000 Å along a major axis.

The other group contains the dissimilatory iron-reducing bacteria that use $Fe^{3+}$ as the final electron acceptor of metabolism and produce magnetite external to the cell. In the dissimilatory iron reduction process, $Fe^{3+}$ is reduced at the surface of the bacterial cell, which in the case of Gram-negative bacteria is the outer membrane. *Geobacter* and *Shewanella* are Gram-negative bacteria that are dissimilatory iron reducing bacteria producing extracellular magnetite. Both *Geobacter* and *Shewanella* have cytochromes in the outer membrane that can interact with $Fe^{3+}$. Electrons may be carried from the oxidative reactions in the cytoplasm to the cell surface by a cytochrome cascade process. With iron supplied to *Geobacter* and *Shewanella* as ferric oxyhydroxides, superparamagnetic grains of magnetite are formed with a single magnetic domain that is smaller than that found in magnetosomes. Although these magnetite formations by *Shewanella* appear to be less regular and homogeneous than the magnetosomes, there has not been a careful structural analysis of this magnetite.

Bacteria can promote mineral formation by biologically induced mineralization (BIM) processes and by boundary-organized biomineralization (BOB) activities. With BIM, bacteria produce biominerals as a direct result of microclimate changes of pH and chemical changes in the environment immediately around the bacterial cell. With

BIM, the biomineral formed is a result of the cellular surface that is organic in nature, serving as a support matrix onto which chemicals are deposited. With bacterial growth in the terrestrial environment, there is the release of biomolecules such as capsular material, membrane fragments, and various partially degraded molecules as aged cells disintegrate. These biomolecules will aggregate ions in the soil and complex with minerals that are being crystallized. Some of the molecules, especially the lipids, will have polar groups at the ends of the molecules and when these biomolecules aggregate, the structures formed will reflect the lipophilic/hydrophilic character of the molecules. Thus, the biomolecules as well as the bacterial cells are critical for the development of biominerals. Both BIM and BOB are important in biomineralization and these processes reflect the interfacial dynamics between minerals and biological material.

### 14.2.2.4. Biogenic Formations[1]

As a result of oxidation–reduction activities of chemolithotrophic prokaryotes, there is the potential for geological changes. Evidence for biogenic formation of metal sulfides has been provided in the early work of Baas Becking, Moore, and Miller. Biological sulfide is produced by several different prokaryotes and includes photosynthetic sulfur bacteria, sulfate-reducing prokaryotes, as well as sulfur reducers. These prokaryotes are ubiquitous, and production of hydrogen sulfide occurs in anaerobic environments even in regions where minerals are present. The action of hydrogen sulfide on metal compounds resulting in the formation of metal sulfides is given in Table 14.10. This production of metal sulfides from precursor compounds and biologically produced hydrogen sulfide is a chemical reaction. The low solubility of the metal sulfide as compared to the solubility of the starting material drives these reactions.

The magnitude of prokaryotic reactions with inorganic compounds on Earth are considerable and some of the sulfur deposits are considered to be of biogenic origin. Deposits of elemental sulfur in Texas and Louisiana are the result of natural sulfureta

Table 14.10. Biogenic production of metal sulfides.[a]

| Precursor compound exposed to biogenic $H_2S$ | Product | Solubility product |
|---|---|---|
| 2 $CoCO_3 \cdot 3\ Co(OH)_2$ | Cobalt sulfide | $7 \times 10^{-23}$ |
| 2 $PbCO_3 \cdot Pb(OH)_2$ or $PbSO_4$ | PbS (galena) | $1 \times 10^{-29}$ |
| $Ni(OH)_2$ or $NiCO_3$ | NiS (millerite) | $3 \times 10^{-21}$ |
| $ZnCO_3$ (smithsonite) | ZnS (sphalerite) | $4.5 \times 10^{-24}$ |
| $FePO_4$ (strengite) or $Fe_2O_3$ (hematite) | FeS (pyrrhotite) | $1 \times 10^{-19}$ |
| $CuCO_3 \cdot Cu(OH)_2$ (malachite) | CuS (covellite) | $4 \times 10^{-38}$ |
| $Ag_2Cl_2$ or $Ag_2CO_3$ | $Ag_2S$ (argentite) | $1 \times 10^{-51}$ |
| $MnCO_3$ (rhodochrosite) | MnS (alabandite) | $5.6 \times 10^{-16}$ |
| $Cu_2O$ or malachite plus | $Cu_5FeS_4$(bornite) | — |
| hematitite and lepidochrosite | or $CuFeS_2$ (chalcopyrite) | — |
| $(BiO_2)CO_3 \cdot H_2O$ | Bismuth sulfide | — |

[a]From Ehrlich, 1996.[1]

that occurred about 10 million years ago. Biogenic sulfur deposits are found in various regions of the world including Sicily, Iceland, New Zealand, Russia, Africa, and Egypt. The formation of sulfur in a sulfuretum is attributed to a cycle consisting of anaerobic photosynthetic bacteria providing carbon for sulfate-reducing bacteria to reduce sulfate to sulfide and the photosynthetic bacteria use sulfide as an electron source with the production of elemental sulfur. The biogenic formation was established by analysis of sulfur isotope fractionation in which the lighter isotope of sulfur ($^{32}$S) is preferred by photosynthetic bacteria and sulfate reducers over the heavier isotope of sulfur ($^{34}$S). Thus, the production of biogenic sulfur from sulfate occurs when sulfur is enriched with $^{32}$S and the remaining sulfate is enriched with $^{34}$S.

Sedimentary iron deposits in various regions of the world are considered by some to be produced by bacteria. It is proposed that in early days the Earth did not have an oxygen atmosphere and iron was in the form of $Fe^{2+}$. As plant-like photosynthesis evolved, the presence of $O_2$ in the atmosphere enabled chemolithotrophic bacteria to use $Fe^{2+}$ as the electron donor and $O_2$ as the electron acceptor with the formation of $Fe^{3+}$ which was insoluble and precipitated in stratified form. There is evidence for bacterial involvement in iron deposits in Minnesota and certain islands in the Aegean Sea. Bog iron deposits in New Jersey have also been suggested to be a formation of biogenic iron resulting from acidophilic iron oxidizers producing limonite (FeOOH) that is deposited in the soil.

### 14.2.2.5. Microbial Leaching[1]

Several important activities associated with microorganisms result in the release of metals from ore-bearing materials by a process referred to as microbial mining or microbial leaching. The two most important processes are copper leaching and uranium leaching. The central feature of each of these processes is the release of elements from iron-containing ores through the use of bacteria (e.g., *Thiobacillus thiooxidans* and *Thiobacillus ferroxidans*) or archaea (*Sulfolobus acidocaldarius* and *Acidianus brierleyi*). These aerobic oxidation processes proceed in an acidic environment of about pH 2, which greatly limits the presence of most chemolithotrophic organisms.

*Copper Leaching*

With copper leaching, the principal ores are chalcopyrite, chalococite, and covellite. The biological oxidation of chalcopyrite ($CuFeS_2$) proceeds by the following reaction:

$$2\ CuFeS_2 + 8\tfrac{1}{2}\ O_2 + 2\ H^+ \rightarrow 2\ Cu^{2+} + 4\ SO_4^{2-} + 2\ Fe^{3+} + H_2O$$

For chalococite ($Cu_2S$), the following oxidation of copper occurs by *Thiobacillus* and the reaction is as follows. There is the movement of electrons from $Cu^+$ to $O_2$ in this reaction and the electron flow may contribute to the energetics of the cell as growth proceeds.

$$2\ Cu_2S + O_2 + H^+ \rightarrow 2\ CuS + 2\ Cu^{2+} + 2\ H_2O$$

In the above reaction, half of the copper remains insoluble as CuS. Finally, covellite (CuS) is oxidized according to the following reaction which results in total solubilization of copper from chalococite:

$$CuS + 2\ O_2 \rightarrow Cu^{2+} + 2\ H^+ + SO_4^{2-}$$

*Uranium Leaching*

The release of uranium from low-grade ores can be accomplished by heap, dump, or in situ leaching. In sedimentary deposits, $U^{4+}$ is found as uraninite ($UO_{2+x}$), coffinite ($USiO_4$), and ningyolite [$CaU(PO_4)_2$] while $U^{6+}$ is found as uranophane [$Ca(UO_2)_2(SiO_3OH)_2$]. The reaction described here are with uraninite which is the most common uranium mineral. The principal reactions involve a combination of both biological and chemical reactions. The reaction catalyzed by acidophilic bacteria or archaea is as follows:

$$2\ Fe^{2+} + \tfrac{1}{2}\ O_2 + 2\ H^+ \rightarrow 2\ Fe^{3+} + H_2O$$

The oxidized iron (ferric) reacts with the reduced uraninite according to the following reaction:

$$UO_2 + 2\ Fe^{3+} \rightarrow 2\ Fe^{2+} + UO_2^{2-}$$

*Acidithiobacillus ferroxidans* is known to oxidize insoluble $U^{4+}$ to soluble $U^{6+}$ and to obtain biological energy from this oxidative reaction; however, the extent to which this reaction occurs in a leaching operation is unknown.

*Acid Mine Drainage*

Closely related to uranium and copper leaching is the activity associated with acid mine leaching. The biological reaction involved is aerobic and results in formation of sulfuric acid from the oxidation of metal sulfides. The most common reaction is the oxidation of pyrite ($FeS_2$) according to the following reaction catalyzed by *Acidithiobacillus ferroxidans* or *Acidithiobacillus thiooxidans*:

$$FeS_2 + 3\tfrac{1}{2}\ O_2 + H_2O \rightarrow Fe^{2+} + 2\ H^+ + 2\ SO_4^{2-}$$

*Acidithiobacillus ferroxidans*, but not *Acidithiobacillus thiooxidans*, is capable of oxidizing the free ferrous iron according to the following reaction:

$$2\ Fe^{2+} + \tfrac{1}{2}\ O_2 + 2\ H^+ \rightarrow 2\ Fe^{3+} + H_2O$$

The ferric ion will chemically react with pyrite to generate sulfuric acid and ferrous ion:

$$FeS_2 + 15\ Fe^{3+} + 8\ H_2O \rightarrow 15\ Fe^{2+} + 16\ H^+ + 2\ SO_4^{2-}$$

The oxidation of pyrite in acid mine waste by bacteria results in highly acidic waste streams that frequently will destroy plant and animal life in the immediate vicinity. In addition to pyrite, *Acidithiobacillus*, *Sulfolobus*, and *Acidianus* will oxidize other metal sulfides. It should be noted that not all metal sulfides are metabolized equally. The following is the order in which metal sulfides will be used with the most readily oxidized listed first: pyrrhotite (FeS) > chalcocite ($Cu_2S$) > covellite (CuS) > tetrahedrite ($3Cu_2S \cdot Sb_2S_3$) > bornite ($CuFeS_4$) > galena (PbS) > arsenopyrite (FeAsS) > sphalerite (ZnS) > pyrite ($FeS_2$) > enargite ($3Cu_2S \cdot As_2S_5$) > marcasite ($FeS_2$) > chalcopyrite ($CuFeS_2$) > molybdenite ($MoS_2$). This sequence of metal sufides used as substrates for organisms is attributed to spatial configuration of the atoms in the minerals, toxicity of the metal compound solubilized, and to some extent the solubility of the metal sulfide.

### 14.2.3. Metal Mimicry[19]

Bacteria display great selectivity for metals by inserting the appropriate ion into ligands to produce a functional molecule. The partitioning of metals in the metal–cell interaction is based on the ionic size, oxidation state, and specific electron orbital characteristics of the metal ion. However, some metal ions may be sufficiently similar to replace another cation, especially if the second ion is in elevated concentration. Replacement of metal cations occurs as a result of metal mimicry and the resulting metalloprotein formed is almost always metabolically inactive. Classic examples of inactive enzymes result from $Ni^{2+}$ for $Mg^{2+}$ and $Mg^{2+}$ for $Mn^{2+}$. In addition, unstable molecular structures result when arsenate is substituted for phosphate in ATP or phosphosugars.

Metal mimicry is extremely important in transmembrane uptake of ions from the environment. Because the insertion of a metal ion into an apoprotein is by a self-assembly process, the chemical environment of the cytoplasm is important. In pristine environments, the concentration of trace metals is very low and bacteria acquire trace metals by highly specific transport systems that have a high affinity, low $K_m$. As a result, the concentration of metal ions in the cytoplasm is relatively small. However, if the environment contains a high concentration of a metal ion, the concentration of the metal ion inside the cell may be abnormally high. This entry of metal ions may be attributed to existing transporters of low specificity. Some alternate ions for the transporters are given in Table 14.11. For example, there are two transport systems for $Mg^{2+}$ in *E. coli*, where one system is highly specific and transports only $Mg^{2+}$. A second transport system is less specific for $Mg^{2+}$ and will also transport $Mn^{2+}$, $Ni^{2+}$, and $Co^{2+}$ when these cations are present in concentrations greater than the concentration of $Mg^{2+}$. $Cd^{2+}$ uptake by bacteria is by either the $Zn^{2+}$ or $Mn^{2+}$ transport system. Thallium ($Tl^+$), rubidium ($Rb^+$), and cesium ($Cs^+$) are appropriate replacements for $K^+$ and are transported into the cell on the nonspecific TrkD (Kup) $K^+$ transport system. $Ag^+$ competes with copper and enters the bacterial cell on the copper transporter system. Selenate and chromate mimic the structure of sulfate and enter cells on the sulfate transport system. The chemical properties of arsenate are sufficiently similar to phosphate to enable arsenate to be transported on the phosphate transporter. Generally for metal mimicry to be effective, the toxic metal must be of greater concentration than the nutrient that it mimics. Although there may be some export systems of low specificity in the bacteria to promote metal homeostasis, these

Table 14.11. Selected Substitutions for Metal Ions into Proteins.[a]

| Native cation | Ionic radius $(10^{-10})$ | Replacement Cation | Ionic radius $(10^{-10})$ | Chemical comparisons |
|---|---|---|---|---|
| $K^+$ | 1.33 | $Tl^+$ | 1.40 | Similar characteristics |
| $Mg^{2+}$ | 0.72 | $Ni^{2+}$ | 0.69 | Similar characteristics |
| | | $Mn^{2+}$ | 0.80 | Generally good substitutions |
| $Ca^{2+}$ | 0.99 | $Mn^{2+}$ | 0.80 | Poor size and stereochemistry |
| | | $Ln^{3+}$ | 0.93–1.15 | Binds strongly |
| | | $Cd^{2+}$ | 0.97 | Competes strongly for sites |
| $Fe^{3+}$ | 0.64 | $Ga^{3+}$ | 0.62 | Good but no redox chemistry |
| | | $Tb^{3+}$ | 1.00 | Moderate but size is poor |
| | | $Al^{3+}$ | 0.54 | Needs an oxygen donor, small size causes steric hindrance |
| $Cu^{2+}$ | 0.69 | $Ca^{2+}$ | 0.99 | Poor, cannot participate in redox chemistry |
| $Cu^+$ | 0.96 | $Ag^+$ | 1.26 | Problems with solubility and stereochemistry |
| $Zn^{2+}$ | 0.74 | $Co^{3+}$ | 0.72 | Excellent substitution because of similarity of characteristics |
| | | $Cd^{2+}$ | 0.97 | This results in a toxic substitution |
| $Mo^{n+}$ | — | $W^{n+}$ | — | Moderate |

[a]From Hughes and Poole, 1989.[2]

exporters may not function fast enough to reduce the intracellular concentration of the metal ion and as a result toxicities occur.

## 14.2.4. Detoxification Processes[20,21]

Prokaryotes growing in geological environments may be exposed to inorganic chemicals that are toxic to biological systems. A few bacteria have the capability of interacting with the soluble toxic metal ions to produce compounds of decreased toxicity. There are several different mechanisms whereby the organisms may detoxify the environment and the principal activities are associated with cellular binding or changing the oxidation state of the metal ion. Bacteria have the capability of removing of toxic metals from the area by producing volatile methylated compounds. Environmentalists have explored the possibilty of using some of these bacterial systems for remediation of contaminated sites. Detoxification is also a process used to maintain cell viability. In a few cases, bacteria are capable of growing in toxic metal solutions because the cells have sequestering systems to keep cytoplasmic concentrations low or export toxic ions through the use of energy-dependent systems.

### 14.2.4.1. Binding and Biosorption[22–25]

Surface structures of bacteria have long been considered important in binding metals and this sequestering of metals is often referred to as biosorption because of the complexity of the reactions involved. In the Gram-positive cell walls, anionic sites that are important in binding metals are associated with the following: (1) free carboxyl groups of the peptidoglycan, (2) hydroxyl groups of sugars in both peptidoglycan and teichoic acids, (3) phosphodiesters in teichoic acids, and (4) amide groups in peptides. The relative affinities of the cell wall of *Bacillus subtilis* for metal ions can be divided into three classes. The greatest amount of binding occurs with $Na^+$, $K^+$, $Mg^{2+}$, $Fe^{3+}$, and $Cu^{2+}$ with lesser amounts of binding by $Mn^{2+}$, $Zn^{2+}$, $Ca^{2+}$, and $Ni^{2+}$. Cations that showed either very little or no binding includes $Hg^{2+}$, $Ag^+$, $Sr^{2+}$, $Pb^{2+}$, $Li^+$, $Ba^{2+}$, $Cd^{2+}$, or $Al^{3+}$. In Gram-negative bacteria, the outer membrane binds various metals; however, the structure involved in this metal recognition is the lipopolysaccharide which extends into the extracellular environment. The lipopolysaccharide constitutes about 33% of the outer membrane and has several electronegative sites per molecule. Under normal cultivation procedures, the principal metal used to stabilize lipopolysaccharide in the outer membrane is $Mg^{2+}$ and to a much lesser extent $Ca^{2+}$.

Bacteria in nature have a surface charge that accounts for mobility of cells when placed in an electronic field. At neutral pH, the electrophoretic mobilities for cells of various bacterial species range from $-0.3$ to $-2.3$ (see Table 14.12). With the exposure of bacterial cells to different metal cations, there is relatively little difference in electrophoretic mobility of the bacterial cells, indicating that metals are not binding to the cell wall surface to neutralize the overall charge on the cell. The exposure of cells from *Pseudomonas aeruginosa*, *Bacillus subtilis*, and *Bacillus megaterium* to $Mg^{2+}$, $Hg^{2+}$, $Pb^{2+}$, $Zn^{2+}$, $Ni^{2+}$, or $Cd^{2+}$ has little to no effect on the cell electrophoretic mobility. Only with the exposure of *Bacillus subtilis* to $Cu^{2+}$ or $Cr^{3+}$ was there a noticeable change in the topochemistry of the bacterial cell. These cellular properties of bacteria are critical in the environment because metals bound onto bacterial cells migrate through the terrestrial matrix at a rate greater than if the metal cation was in solution.

Not only is the cell wall of bacteria highly charged but exopolymeric material is also important in binding various types of metal cations. As enumerated by Macaskie and colleagues (see Table 14.13), polysaccharide capsules from various bacteria are highly effective in binding metals because of the acidic units present. The role of

Table 14.12. Electrophoretic mobilities of bacteria at pH 7.0 following exposure to various metal cations.[a]

| Bacterial cells | Electrophoretic mobilities of cells ($\mu$m/sec V/cm) | | | |
|---|---|---|---|---|
| | Native cells (no addition) | Addition of $Mg^{2+}$, $Hg^{2+}$, $Pb^{2+}$, $Zn^{2+}$, $Ni^{2+}$, or $Cd^{2+}$ | Addition of $Cu^{2+}$ | Addition of $Cr^{3+}$ |
| *Bacillus subtilis* | $-2.3$ | $-1.8$ to $-3.2$ | $+1.2$ | $+0.8$ |
| *Bacillus megaterium* | $-1.6$ | $-1.5$ to $-3.2$ | $-1.5$ | $-2.0$ |
| *Pseudomonas aeruginosa* | $-0.9$ | $-0.8$ to $-3.0$ | $-0.8$ | $-3.0$ |
| *Pseudomonas maltophilia* | $-1.8$ | — | — | — |
| *Moraxella bovis* | $-0.3$ | — | — | — |

[a]From Collins and Stotzky, 1989.[23]

Table 14.13. Relative binding of metals by exopolysaccharides and cell wall segments of bacteria.[a]

| Polymer and characteristics | Bacteria producing polymer | Metals bound | Binding data |
|---|---|---|---|
| Emulsan. It has a molecular weight of ~$10^6$ daltons and is polyanionic. The linear polymer is D-galactose and side chains are of fatty acids. | *Acinetobacter calcoaceticus* | $UO_2^{2+}$, $Rb^+$, $Cd^{2+}$, $Ba^{2+}$, and $Pb^{2+}$ | Dissociation constant: U = $1.2 \times 10^{-4} M$ 800 mg U bound/1 g of emulsan |
| Polysaccharide capsule of glucose, galactose, and D-mannuronic acid | *Arthrobacter viscosus* | $Cd^{2+}$ and $Pb^{2+}$ | 3.3 mg $Cd^{2+}$ bound/g of capsule |
| Polysaccharide capsule of glucose, rhamnose, and mannose at a ratio of 2:1:1:1.5. Also present are acetate, pyruvate, and uronic acids. | *Klebsiella* sp. | $Cd^{2+}$, $Ni^{2+}$, $Co^{2+}$, and $Cd^{2+}$ | Stability constants (log $K$) $Cu^{2+} = 7.7$ $Ni^{2+} = 5.5$ $Co^{2+} = 5.5$ $Cd^{2+} = 5.2$ |
| Polysaccharide capsule released into medium. Highly branched with a mass of $10^5$ daltons. Contains glucose, galactose, and pyruvic acid with molar ratios of 11:3.1:1.5. | *Zoogloea ramigera* | $Cu^{2+}$, $Cd^{2+}$, and $UO_2^{2+}$ | 800 mg U/ g of cells 0.17 g Cu/ g of cells |
| Peptidoglycan that contains glutamic acid | *Alcaligenes eutrophus* | $Cd^{2+}$ | 350 μg Cd/ mg of peptidoglycan |
| Lipopolysaccharide containing heptose, galactose, glucose, aminoarabinose, ethanolamine, 2 keto-3-deoxyoctonate (KDO) | *Escherichia coli, Salmonella typhimurium* | $Mn^{2+}$, $Cd^{2+}$ | $Cd^{2+}$ displaces $Ca^{2+}$ with loss of membrane structure |
| Teichoic acid. A linear polymer of glycerol phosphate that accounts for ~50% of the cell wall | *Bacillus subtilis* | $Na^+$, $K^+$, $Ca^{2+}$ and $Mg^{2+}$ preferred but also will bind $Fe^{3+}$, $Cu^{2+}$, $Mn^{2+}$ and other cations | Binding capacity: (μg/ mg of cell wall) $Mg^{2+} = 8.2$ $Fe^{3+} = 3.6$ $Cu^{2+} = 3.0$ $Mn^{2+} = 0.8$ |

[a]From Macaskie and Dean, 1990.[24]

these acidic capsules, peptidoglycan, and lipopolysaccharide in bacteria may be to sequester metals in pristine environments, and binding of metals may be an additional feature that assists organisms in persisting in toxic metal environments.

Other cellular components in addition to bacterial cell walls or capsules are involved in binding metals. Under appropriate environmental conditions, insoluble iron and manganese precipitates become associated with the sheath of *Sphaerotilus natans* and the result is that iron oxide can account for up to 1.8% of the biomass. Various cations will bind to phosphodiesters in dead bacterial cells. For example, $Al^{3+}$ binds to $ATP^{4-}$ about $10^7$ times greater than $Mg^{2+}$ and nmolar amounts of $Al^{3+}$ can out compete millimolar quantities of $Mg^{2+}$. Some have suggested that polyphosphate granules found in bacteria would be suitable for binding different types of metal cations. Finally, it should be considered that sideophores under appropriate conditions will bind various metals in addition to $Fe^{3+}$. For example, mycobactin from *Mycobacterium smegmatis* will complex with $Al^{3+}$, $Cr^{3+}$, $Cu^{2+}$, and $UO_2^{2-}$. Aerobactin produced by *Klebsiella pneumoniae* binds $Cr^{3+}$ while enterochelin produced by *E. coli* will complex with $Ga^{3+}$ and $Sc^{3+}$. There have been claims in patent applications that modified siderophores can bind $Cd^{2+}$, $Cr^{2+}$, $Pb^{2+}$, $Hg^{2+}$, $Ni^{2+}$, $Zn^{2+}$, $Cs^{2+}$, $Th^{2+}$, and $Co^{2+}$; however, generalized information about these modified biochelators is not available.

### 14.2.4.2. Precipitation Reactions[26]

An effective detoxification for bacteria is the precipitation of heavy metals that may be toxic to the cells. One mechanism for producing insoluble metals or metalloids is to reduce the chemical to the elemental state. Bacterial reduction of $Se^{6+}$ (selenate) or $Se^{4+}$ (selenite) results in the formation of $Se^0$ which forms an amorphous red precipitate. Most cultures stabilize amorphous $Se^0$ for indefinite periods of time while a few bacterial strains form a gray trigonal selenium crystal. By a process similar to selenium transformation, some bacteria will enzymatically catalyze the formation of $Te^0$ and $Au^0$ from salts of tellurium and gold. Certain anaerobic bacteria will reduce the soluble uranyl ($UO_2^{2+}$) ion to uraninite ($UO_2$) form, and although uraninite contains both uranium and oxygen, this is an uncharged compound that readily precipitates.

The enzymatically catalyzed production of metal precipitates by bacteria is an example of biomineralization. Biogenic sulfide production by sulfate-, thiosulfate-, or sulfur-reducing bacteria results in production of sulfide which readily precipitates heavy metal cations. Many heterotrophic bacteria release sulfide from cysteine and other organosulfur molecules by putrefaction activities, and this sulfide readily combines with metal cations to produce precipitates. In alkaline environments, the release of $CO_2$ from metabolic activities of cells results in formation of insoluble metal carbonates. Bacterial reduction of $Cr^{6+}$ to $Cr^{3+}$ results in the precipitation of chromium as $Cr(OH)_3$ in neutral or alkaline environments.

Considerable interest has been shown with the biomineralization reactions involving phosphate. Macaskie and colleagues have explored the activities of a *Citrobacter* sp. that has a cell wall–bound phosphatase that releases phosphate from glycerol phosphate and other organophosphate substrates. The oxidized soluble uranyl ion in the environment is readily precipitated in the biomass as hydrogen uranyl phosphate ($HUO_2PO_4 \cdot nH_2O$). A crystal lattice of the hydrogen uranyl phosphate forms in sheets, with water molecules between the sheets forming a network of hydrogen bonds. It is assumed that actinides with a 6+ oxidation state can substitute for $U^{6+}$ in this crystal formation.

### 14.2.4.3. Export Systems

One of the systems for regulating intracellular concentrations of metal ions is to employ energy-linked export of the metal. The first exporter identified was for arsenic and the structure involved has been designated as the A-type ATPase. In addition to the export of arsenite, the A-type ATPase also exports tellurite and antimonite. Information about the arsenic exporter is found in a subsequent section of this chapter. Some bacteria carry a plasmid that encodes for a transmembrane exporter specific for chromate, and thereby reduces the intracellular concentration of chromate. In addition, $Cd^{2+}$, $Zn^{2+}$, and $Co^{2+}$ appear to be exported by specific ATPases in cyanobacteria and several other organisms. *Alcaligenes eutrophus* produces metal resistance pumps that export $Co^{2+}$, $Ni^{2+}$, $Cd^{2+}$, and $Zn^{2+}$. Although the extent to which these metal exporters are found in diverse organisms remains to be established, it is apparent that metal export by energy-coupled processes and ATPases is an important process for bacterial survival. These inducible export systems use specific metal sensing proteins to regulate their expression, and information about metal sensors is discussed in Section 14.2.4.5.

### 14.2.4.4. Methylation Reactions[1]

There are relatively few processes for removal of a toxic metal from the environment except by the formation of a volatile molecular chemical. Monomethyl and dimethyl metals are produced by certain bacteria as they interact with a few metals or metalloids. Many soil bacteria are capable of producing methylated selenium compounds, and these volatile materials have an odor that resembles that of fresh almonds. The volatile selenium compounds produced by bacteria from selenite or selenate reduction include methyl selenide ($CH_3SeH$), dimethyl selenide [$(CH_3)_2 Se$], dimethyl diselenide [$(CH_3)_2 Se_2$], and dimethyl selenone [$(CH_3)_2 SeO_2$]. It appears that dimethyl selenone is a precursor of dimethyl selenide while dimethyl selenide leads to dimethyl diselenide production. Other methylated metals are produced by direct activity of prokaryotes with tellurium, germanium, platinum, gold, thallium, arsenic, mercury, cadmium, lead, and tin. The mechanisms for methylation of mercury by methanogens and sulfate reducers are best characterized and they involve methylcobalamin (methyl-$B_{12}$). Subsequent sections describe methylation of mercury as well as arsenic. In contrast to bacteria, fungi such as *Neurospora crassa* use betaine or choline and not methylcobalamin as the methyl donor. Some marine microorganisms produce methyl halides and methyl iodides that potentially react with palladium, thallium, platinum, and gold to produce unstable methylated compounds. On demethylation of these cations, the metal is released in the elemental state.

### 14.2.4.5. Metallothionein and Metal Sensors[27,28]

Present in many higher eukaryotes, fungi and yeast are metal binding proteins that are called metallothioneins. These small proteins function in metal detoxification in the cytoplasm by binding metals through a number of cysteine moieties. One of the best characterized metallothioneins is from mouse liver and it is a protein of 61 amino

Table 14.14. Characteristics of metallothionein-like proteins produced by bacteria.[a]

| Organism | Protein | Metals bound (with g atoms/mol) | Mass of protein | Cysteine residues (as % of total) |
|----------|---------|-------------------------------|-----------------|-----------------------------------|
| *Escherichia coli* | | Cd (?) | 3,900 | ? |
| *Pseudomonas putida* | Cd-BP1 | Cd (4) Zn (1) Cu (2) | 6,700 | 13 |
| *Pseudomonas putida* | Cd-BP2 | Cd (1) Zn (2) Cu (2) | 6,900 | 10 |
| *Pseudomonas putida* | Cd-BP3 | Cd (3) Zn (2) Cu (2) | 3,600 | 23 |
| *Synechococcus* spp. | | Cd (1.3) Cu (0.6) Zn (0.3) | 8,100 | 15 |

[a]From Robinson et al., 2000.[28]

acids with 20 cysteine moieties that characteristically binds cadmium, zinc, and copper ions. Relatively few prokaryotes have metallothionein-like proteins and the ones reported are listed in Table 14.14. The best characterized bacterial metallothionein-like protein has been isolated from the cyanobacterium *Synechococcus* sp. by Olafson and collaborators. This protein is known as SmtA and contains 56 amino acids with several cysteine-x-cysteine repeats, which is characteristic of metallothioneins from animal cell systems. The primary sequence of metallothioneins from three cyanobacteria species is given in Table 14.15. The amino acid sequence of the SmtA protein from *Synechococcus* is considerably different from the sequence of mammalian metallothionein in that it contains histidine residues in addition to cysteine. The SmtA protein has a single "pocket" for the binding of $Cd^{2+}$ or $Zn^{2+}$ and X-ray characterization reveals that there may be either three or four zinc ions per molecule. Several different strains of *Pseudomonas putida* produce metal-bonding proteins rich in cys-

Table 14.15. Sequence of amino acids in bacterial metallothioneins.[a]

| Protein ( )[b] | Sequence |
|----------------|----------|
| SmtA (56)[c] | MTSTTLVK**C**A**C** EP**C**L**C**NVDPSKAIDRNGLYY**C**SEA**C**ADAG **H**TGGSKG **C**G**H**TG**C**N**C**HG |
| MntA (57)[c] | MTTVTQMK**C**A**C** P**H**C**L**C**IVSLNDAIMVDGKPY**C**SEV**C**ANGT **C**KENSG **C**G**H**AG**C**G**C**GSA |
| *Anabaena* (53)[c] | MTTVTQMK**C**A**C** PS**C**L**C**IISVEDAINKEGKYY **C**SEG**C**AEG **H**KTIKG **C**N**H**NG**C**G**C** |

[a]From Robinson et al., 2001, p. 186.[28]
[b]Number of amino acid residues known or predicted.
[c]SmtA is from *Synechococcus* PCC 7942, MntA is from *Synechococcus vulcanus*, and the sequence for the protein from *Anabaena* is deduced from genome analysis.
Bold and underscore represents amino acid similarities in metallothioneins.

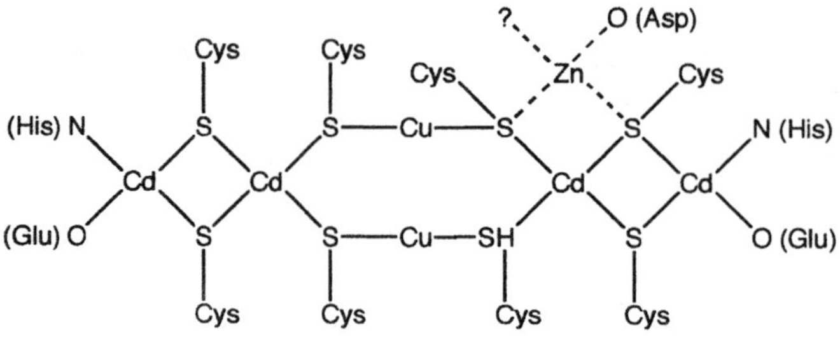

Figure 14.4. Model for Cd-binding metallo-thionein-like protein from *Pseudomonas putida*. (From Rayner and Sadler, 1989.[27])

teine residues and a model is presented to reflect the involvement of cysteine residues in metal binding (see Figure 14.4). Owing to the unstable character of metals associated with the metallothionein protein, it has been a challenge to isolate these bacterial molecules fully loaded with metals.

Located adjacent to the *smtA* gene is *smtB*, which is a zinc-responsive negative regulator of *smtA*. In *Synechococcus*, the SmtB protein contains 122 amino acids and four molecules of SmtB are considered to bind to the operator-promoter region of *smtA*. The SmtB protein contains 5 α-helices and 2-antiparallel β-sheets which has a helix-turn-helix DNA binding motif. The SmtB is a member of the family of metal sensors which influences the gene transcription following metal binding. Several of these metal sensory systems are listed in Table 14.16. Although the sensors may be structurally related, the response that is triggered in the cell is dependent on gene content of the cell. For example, elevated intracellular concentrations of $Zn^{2+}$ in *Synechococcus* PCC result in production of SmtA, the metallothionein while in *Synechcystis* it results in the production of the zinc exporter. At this time, the nature of the metal ion specificity for the sensors has not been established.

Table 14.16. The metal sensor protein family based on sequence similarity.[a]

| Protein | Bacterial strain | Metal sensed | Phenotype |
|---------|------------------|--------------|-----------|
| **Related to SmtB** | | | |
| SmtB | *Synechococcus* PCC | Zinc | Zinc metallothionein (SmtA) |
| ZntR | *Staphylococcus aureus* | Zinc/cobalt | Zinc/cobalt transporter (cation diffusion facilitator family) |
| CmtR | *Mycobacterium tuberculosis* | Cobalt | Cobalt exporting ATPase (CmtA) |
| **Related to ArsR** | | | |
| ArsR | *Escherichia coli* | Arsenic | Arsenic resistance |
| ZiaR/SmtB | *Anabaena* PCC | Zinc | Zinc exporting ATPase |
| **Related to ArsR/SmtB** | | | |
| ZiaR | *Synechocystis* PCC | Zinc | Zinc exporting ATPase (ZiaA) |
| CadC | *Staphylococcus aureus* | Cadmium | Cadmium exporting ATPase (CadA) |

[a]From Robinson et al., 2001, p. 186.[28]

## 14.3. Metabolism of Selected Metalloids and Metals

Various activities of alkali metals and transition metals have been discussed in this chapter. There are a few additional elements that need to be discussed because they present unique activities to cells or because the elements are cytotoxic. Iron and selenium are two elements that are essential for growth of most prokaryotes. Intracellular concentrations of iron are carefully modulated with ferritin to avoid the production of high levels of ferrous which can in certain cases catalyze toxic reactions. Selenium as selenate or selenite is used in trace element levels but at high concentrations toxicities to cells are observed. Some of the heavy metals are among the greatest biological toxins and it is important to address how bacteria have developed systems to grow in the presence of these toxins. Resistance to $Ag^+$, $Cd^{2+}$, $Co^{2+}$, $Cu^{2+}$, $Hg^{2+}$, $Ni^{2+}$, $Pb^{2+}$, $Sb^{3+}$, $Tl^+$, $Zn^{2+}$, $AsO_2^-$, $AsO_4^{3-}$, $CrO_4^{2-}$, and $TeO_3^{2-}$ in bacteria is mediated by unique plasmids and the phenotypes of bacteria with these plasmids detoxify metals through reduction and export processes. From a global perspective, the oxidation–reduction processes discussed here underscore the contribution of prokaryotes to geochemical cycling.

### 14.3.1. Iron

Iron is an essential trace element for prokaryotes except for the lactobacilli, a group of bacteria that grow primarily from energy produced from the glycolytic pathway and lack cytochromes. An important chemical feature of iron is the presence of two stable oxidation states as ferrous ($Fe^{2+}$) and ferric ($Fe^{3+}$). Owing to the oxidation–reduction activity of iron, this element has become a favorite of bacteria to participate in a variety of electron transfer reactions. The diversity of molecules that contain iron include cytochromes in which an iron atom is associated with porphyrins and a diverse group of iron-containing proteins.

The iron-proteins are extremely important because they include the redox proteins of respiration, reductases, and dehydrogenases. The most abundant type of iron-proteins are those that contain Fe-S clusters and these include certain hydrogenases, aconitase, ω-hydroxylase as well as numerous respiratory proteins including the Rieske protein. Several bacteria produce rubredoxin, which has only iron and not sulfide as an essential element. There are some enzymes that have prosthetic groups in addition to the Fe-S cluster and include the following: pyruvate-oxidoreductase contains thiamine pyrophosphate in addition to the Fe-S cluster and dissimilatory sulfite reductase has a heme moiety plus a Fe-S cluster. Several enzymes contain Fe-S plus a flavin and examples of these proteins include succinate dehydrogenase, NADH dehydrogenase, glutamate synthase, and adenylyl sulfate reductase. Some enzymes such as nitrogenase, nitrate reductase, and CO reductase contain molybdenum as well as iron and sulfur. In addition, there are several DNA-binding proteins that contain iron.

A unique DNA base sequence is designated as the "iron box" or the "*fur* box" because constitutive mutants were known as *fur* (*f*erric *u*ptake *r*egulation) mutants. In the 1980s, J. Neilands and colleagues determined that the *fur* gene was an important regulator of genes for biosynthesis and transport of aerobactin. The *fur* gene is found at 15.7' on the *E. coli* chromosome and through cloning with overexpression, the protein has been determined to be 17 kDa. The Fur protein binds to a palindromic consensus sequence of 5'-GATAATGATAATCATTATC provided $Fe^{2+}$ (or perhaps $Zn^{2+}$) is present. The Fur protein interacts with DNA as a dimer, making five contacts with DNA around three turns of the B-DNA helix. The binding of the Fur protein is near the attachment of the RNA polymerase in the promoter and acts as a transcriptional inhibitor. The *fur* system in bacteria is important as a global regulator because there are numerous promoters that bind Fur protein.

### 14.3.1.1. Cytotoxic Reactions

Iron metabolism must be carefully regulated because there can be some detrimental activities associated with intracellular accumulations of $Fe^{2+}$ and $Fe^{3+}$. In aerobic respiration, reactive oxygen species of $O_2^-$ and $H_2O_2$ are transiently produced. Iron cations will react with these oxygen compounds to produce the hydroxyl free radical as outlined below:

$$\text{(Iron reduction)} \quad O_2^- + Fe^{3+} \rightarrow Fe^{2+} + O_2$$

$$\text{(Fenton reaction)} \quad Fe^{2+} + H_2O_2 \rightarrow Fe^{3+} + OH^- + OH^{\cdot}$$

The overall reaction is as follows, where iron is a critical participant in the reaction:

$$\text{(Haber–Weiss reaction)} \quad O_2^- + H_2O_2 \rightarrow OH^- + OH^{\cdot} + O_2$$

The hydroxyl radical ($OH^{\cdot}$) is extremely reactive and produces considerable cell damage. To prevent cell death from these chemical reactions involving iron, cells employ several different iron storage molecules that function to prevent intracellular accumulation of $Fe^{3+}$ and $Fe^{2+}$. All of these iron binding proteins are members of the ferritin–bacterioferritin–rubrerythrin superfamily.

### 14.3.1.2. Ferritin[29,30]

The prokaryotic ferritin has many characteristics similar to the ferritin of eukaryotic cells, and therefore it is useful to review the characteristics of ferritin from animals. Eukaryotic ferritins consist of 24 individual protein subunits that are approximately 19 kDa and associate to produce a spherical protein shell with a diameter of 120 Å. The central core of the ferritin molecule is about 70 Å and contains approximately 4,500 iron atoms as ferri-oxy-hydroxyphosphate which accounts for more than 30% of the weight of the ferritin structure. As presented in Figure 14.5, $Fe^{2+}$ enters the core of the molecule from the cytoplasm and is oxidized to $Fe^{3+}$ which is ultimately incorporated into the iron crystal. The first approximately 50 atoms of $Fe^{2+}$ are oxidized to $Fe^{3+}$ by the protein coat and thereafter the $Fe^{3+}$ ions are oxidized at the surface of the ferrihydrite crystal. In addition, the protein coat of the ferritin serves to keep the insoluble iron aggregate in a relatively soluble form in the cytoplasm. In eukaryotes, the ferritin protein consists of two subunits or chains: a larger one referred to as the heavy (H) chain and a smaller one called the lighter (L) chain. The ferritin subunit composition can range from $H_0L_{24}$ to $H_{24}L_0$ depending on the specific animal

Figure 14.5. Cross section of ferritin showing the protein core and iron center. The model of P. Harrison and colleagues indicates through the use of *arrows* the points where $Fe^{2+}$ enters the ferritin structure. $Fe^{2+}$ enters at specific points (**a**) and is oxidized at ferroxidase centers (**b**). Nucleation of the ferrihydride microcrystal occurs (**c, d**) with $Fe^{2+}$ added at the surface of the growing crystal. (Reprinted from Andrews, S.C. 1998. Iron storage in bacteria. *Advances in Microbial Physiology* 40:281–351. Copyright 1998, with permission from Elsevier.)

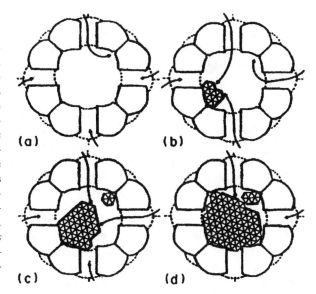

tissue. The L-rich ferritin is considered to be important for iron storage while the H-rich ferritin is involved in both short-term iron storage and intracellular iron flux. Because of physical-chemical stability of ferritin, it can be isolated from a bacterial cell-free extract by denaturing the other proteins with either 8 $M$ urea or by heating at 80°C. Some characteristics of ferritin are given in Table 14.17.

The ferritin from prokaryotes appears to be similar to the H-chain of eukaryotic ferritin. *E. coli* has a ferritin gene (*ftnA*) with a sequence that is about 25% identical to the human H-chain of ferritin. In 1992, Rocha and colleagues reported the isolation of ferritin from *Bacteroides fragilis* and this molecule has an N-terminal amino acid sequence that is 37% identical to human ferritin. The amino acid sequence of ferritin

Table 14.17. Characteristics of iron storage proteins in bacteria.

| Protein | Characteristics |
|---|---|
| Ferritin | Nonheme-containing protein with a small subunit and a large subunit<br>From *Bacteroides fragilis*: subunits = 16.7 kDa and 400 kDa<br>From *Helicobacter pylori*: subunits = 19.3 kDa and 350 kDa with 1,700 Fe atoms/molecule and a 91 Å diameter core |
| Bacterioferritin | Heme-containing protein with a single small protein subunit.<br>From *Azotobacter vinelandii*: 24 protein subunits and 12 hemes/molecule<br>All heme molecules are B-type, protein subunits are 17 kDa, and there are 1,100 to 1,800 Fe atoms per molecule of bacterioferritin |

from *Helicobacter pylori* has 42% identity to that of *E. coli* and contains 1,700 Fe atoms per molecule with a molar ratio Fe atoms to phosphate ions of 1.4:1. Generally, bacterial ferritins have a ratio of Fe:phosphate of 1:1 to 1:2 while animal ferritins have a Fe:phosphate ratio of 1:8. Ferritin is widespread in prokaryotes in that it is found in both bacterial strains and archaeal strains but is absent in several of the prokaryotes examined. At this time it appears that ferritin is likely to be found in anaerobic or microaerophilic prokaryotes but not in aerobic microorganisms. This may reflect the potential for a high intracellular concentration of $Fe^{2+}$ resulting from the high solubility of $Fe^{2+}$ under anaerobic conditions but the relative insolubility of $Fe^{3+}$ in an aerobic environment. At pH 7, the solubility of $Fe^{2+}$ is approximately 0.1 *M* while the solubility for $Fe^{3+}$ is only approximately $1 \times 10^{-18}$ *M*.

### 14.3.1.3. Bacterioferritin[20,21]

The first iron-storage molecule isolated from bacteria was bacterioferritin. In 1979, Stiefel and Watt isolated a protein from *Azotobacter vinelandii* that had subunits of 17 kDa, a central core of 1,100 to 1,800 Fe atoms, and a heme moiety. Assuming the bacterioferritin is a molecule with 24 subunits, there would be 12 B-heme groups per ferritin molecule with a heme located between each protein subunit. The heme groups may function in reducing the $Fe^{3+}$ cation as it is released from the iron core of the bacterioferritin with the soluble $Fe^{2+}$ used for metabolic events. When the iron core is present as is the native molecule, the redox potential of the heme in the *Azotobacter vinelandii* is −475 mV. Many different species of bacteria produce bacterioferritin and these have a high quantity of phosphate in the core with a molar ratio for Fe:P of approximately 2:1. Some bacteria have a bacterioferritin with heteropolymer subunits with one subunit involved in binding the heme while the other subunit has the ferroxidase activity. The distribution of ferritin and bacterioferritin in some of the prokaryotes is given in Table 14.18. Not all prokaryotes have an iron binding

Table 14.18. Distribution of ferritin and bacterioferritin in prokaryotes.

| Organism | Ferritin | Bacterioferritin |
| --- | --- | --- |
| *Archaeoglobus fulgidus* | 1 | 0 |
| *Clostridium acetobutylicum* | 1 | 1 |
| *Escherichia coli* | 2 | 1 |
| *Helicobacter pylori* | 1 | 0 |
| *Haemophilus influenzae* | 2 | 0 |
| *Magnetobacterium magnetotactium* | ? | 2 |
| *Methanobacterium autotrophicum* | 1 | 0 |
| *Methanococcus jannaschii* | 0 | 0 |
| *Mycobacterium tuberculosis* | 1 | 1 |
| *Mycoplasma pneumoniae* | 0 | 0 |
| *Pseudomonas aeruginosa* | 0 | 2 |
| *Thermotoga maritima* | 1 | 0 |
| *Treponema pallidum* | 0 | 0 |
| *Vibrio cholerae* | 1 | 1 |

protein, and there is some suggestion that bacterioferritin may also have a role in oxygen stress response.

## 14.3.2. Selenium[22-24]

As an element, selenium is similar to sulfur in that both have oxidation states of +6, +4, 0, and −2. The oxy-anions of selenium, selenate and selenite, are used by a few bacteria as electron acceptors in the dissimilatory reduction of selenium. In the seleno-organic compounds, selenium usually replaces sulfur and selenium is at the −2 oxidation state. The seleno-amino acids of seleno-cysteine and seleno-methionine have an essential role in some prokaryotes and they cannot be replaced by the corresponding sulfur-containing analog. In comparison to a thiol (R-SH) group, seleno (R-SeH) groups are better nucleophiles and this unique chemistry of selenium is the basis for its presence in oxidoreductases and seleno-tRNA nucleotides. Generally selenium occurs at trace levels in the environment; however, in a few areas, selenate or selenite concentrations are appreciable and selenium becomes toxic to bacteria.

### 14.3.2.1. Selenoproteins

Selenium is present in enzymes of several prokaryotes as either selenocysteine or as selenide in oxidoreductases. In both cases, selenium has a −2 oxidation state. The enzymes that have senocysteine in the polypeptide backbone are known as selenoproteins and there are three groups of selenoenzymes: glycine reductase, formate dehydrogenase, and hydrogenases. Enzymes that have selenium as a cofactor but not as a selenium amino acid include xanthine dehydrogenase, nicotinic acid dehydrogenase, carbon monoxide dehydrogenase, and hydrogenase specifically from *Bradyrhizobium japonicum*.

One of the best studied selenoproteins is glycine reductase (see Chapter 12), which is found in a series of clostridia and participates in purine fermentation and in various other anaerobes in the "Stickland reaction" with the fermentation of amino acids. The glycine reductase complex consists of three polypeptide subunits (protein A, protein B, and protein C) with the selenoamino acid in protein A. The exposed selenide moiety of HS–protein–Se⁻ serves as a carrier for the acetyl group which is acquired from either glycine, sarcosine, or betaine. The use of the selenide carrier instead of the sulfhydral group reflects the unique chemical properties of selenium. The reaction showing activity of glycine reductase is given in Figure 14.6.

Formate dehydrogenase is a selenoprotein found in several enterobacteria, clostridia, and in the archaeon, *Methanococcus vannielii*. In formate dehydrogenase, selenocysteine is important at the active site in that it carries the one-carbon unit (see Figure 14.7). It is proposed that selenocysteine interacts with a molybdenum cofactor in which formate is bound to molybdenum and the result is the oxidation of formate with molybdenum being reduced from $Mo^{6+}$ to $Mo^{4+}$. The reduced cofactor would release electrons to a Fe-S cluster, resulting in regeneration of oxidized Mo.

Several prokaryotes with a selenium-containing hydrogenases include *Methanococcus vannielii*, *Methanococcus voltae*, *Desulfovibrio vulgaris*, *Desulfovibrio salexigens*, and *Desulfovibrio baculatum* (now *Desulfomicrobium*). These hydrogenases have an

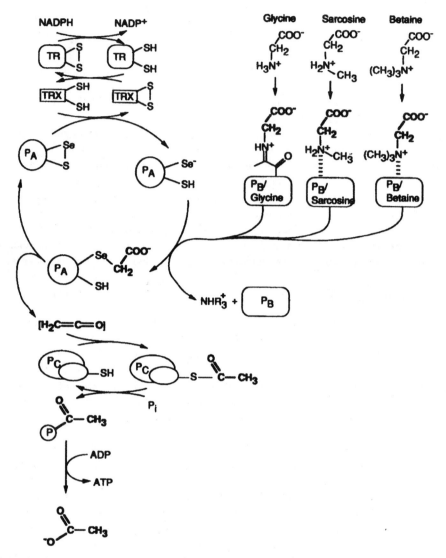

Figure 14.6. Glycine reductase reactions of *Eubacterium acidaminophilum*. Three polypeptides make up the enzyme complex: $P_A$, $P_B$, and $P_C$. Thioredoxin reductase (TR) and thioredoxin (TRX) participate in electron transfer from NADH to oxidized $P_A$. Glycine, sarcosine, or betaine is carried by $P_B$ to reduced $P_A$. Carboxyl methyl groups are attached to $-Se^-$. Phosphorylation of acetyl group occurs on $P_C$, resulting in production of ATP and acetic acid. (Reprinted from Heider, J. and A. Böck. 1993. Selenium metabolism in micro-organisms. *Advances in Microbial Physiology* 35:71–109. Copyright 1993, with permission from Elsevier.)

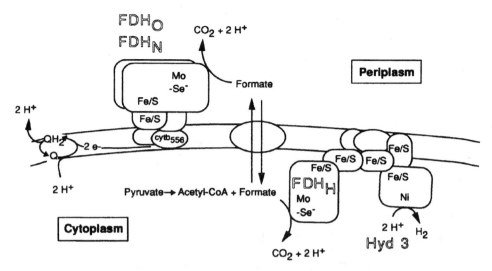

Figure 14.7. Formate dehydrogenase (FDH) isoenzymes of *Escherichia coli*. Fe/S = iron-sulfur centers of redox proteins that transfer electrons to $FDH_O$, $FDH_N$, and $FDH_H$. The presence of Mo and Se in the enzymes is as indicated in the large polypeptide of FDH. The presence of formate transport remains to be established. (Reprinted from Heider, J. and A. Böck. 1993. Selenium metabolism in micro-organisms. *Advances in Microbial Physiology* 35:71–109. Copyright 1993, with permission from Elsevier.)

$\alpha_1\beta_1$ structure with a 1:1 molar ratio of selenocysteine and nickel atom. With the selenium atom being less electronegative than sulfur, electrons are readily lost from selenium. It has been observed that selenium containing hydrogenases functions primarily in $H_2$ evolution and to a lesser extent in $H_2$ uptake. Seleno-hydrogenases are present in cells as a supplement to other hydrogenases and if bacteria contain a single hydrogenase it is not a seleno-hydrogenase.

The biosynthesis of selenoproteins is dependent on the insertion of selenocysteine at the appropriate position. In the mRNA is UGA, which is the unique codon that functions in recognition of selenocysteyl-tRNA[Sec]. The synthesis of this specific tRNA for selenocystine requires four genes for production of selenocysteyl-tRNA[Sec] from serine tRNA[Sec].

### 14.3.2.2. Selenium-Containing tRNA

Several bacterial and archaeal species produce a unique tRNA nucleotide. This molecule is 2-seleno-5-methylaminomethyl-uridine ($Se^2$-mam[5]-U) and it resides at position 34 in the tRNA. The structure is given in Figure 14.8. The selenium-containing tRNA may serve to stabilize the codon–anticodon recognition with a preference for G. Because the codon for glutamate is GAG/A and for lysine it is AAG/A, the "wobble" of the tRNA at the third position of the codon would be reduced with the codons of GAG and AAG.

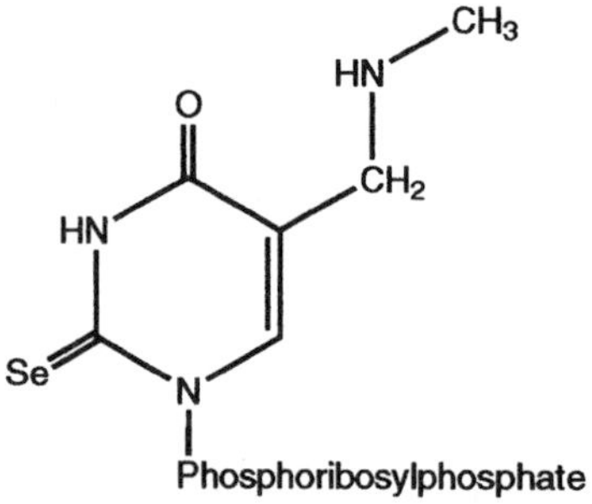

Figure 14.8. Structure of selenium-modified tRNA nucleotide, $Se^2$-mam$^5$-U.

### 14.3.2.3. Selenocysteine Synthesis

Selenocysteine has been called the 21st amino acid and it has a relatively unique metabolic sequence for synthesis. The metabolic pathway for the synthesis of selenocysteine is given in Figure 14.9. Key for the formation of the seleno-amino acid is the formation of selenophosphate from selenide and ATP by selenophosphate synthetase (see reaction 15 in Figure 14.9). Less energy is required to stabilize the selenium–phosphate interaction than for sulfur–phosphate interaction, thereby allowing the selenium atom to be readily added to a carbon unit in the production of selenocysteine. This synthesis of selenocysteine occurs on the carrier (stem) region of the tRNA$^{Sec}$ and because this tRNA has the anticodon for UGA, the seleno-amino acid will be appropriately added to the polypeptide chain. It is possible that selenocysteine is also produced by reaction 19 in Figure 14.9 from selenide and *O*-acetylserine but this would not lead to appropriate addition of selenocysteine to the polypeptide chain. High concentrations of selenocysteine in the cytoplasm could lead to the addition of selenocysteine to the tRNA$^{Cys}$ with selenocysteine substituted for cysteine in the protein.

### 14.3.2.4. Selenomethionine Synthesis

Selenomethionine is produced from selenocysteine by the same reactions used to produce methionine from cysteine. The methyl donor for the formation of methionine and selenomethionine is *S*-adenosylmethionine. Selenomethionine can be added to the tRNA$^{Met}$ with selenomethionine replacing methionine in the protein. In *Escherichia coli*, the presence of selenomethionine in proteins is not toxic to cells.

### 14.3.2.5. Dissimilatory Reduction

A few specialized bacteria have the capability of growing with selenate as the terminal electron acceptor. Bacteria with respiratory coupled selenate reductase include *Thauera selenatis*, *Bacillus arsenicoselenatis*, *Bacillus selenitireducens*, *Sulfurospirillum barnesii*, and *Sulfurospirillum arsenophilum*. Dissimilatory selenate reduction in bacteria has been described with the selenate reductase interfacing with cytochromes in the plasma membrane. In *Thauera selenatis*, the arsenate reductase complex con-

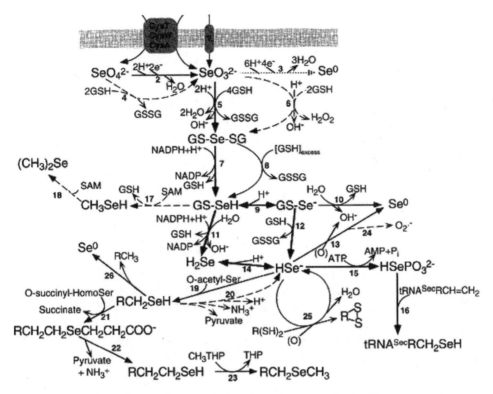

Figure 14.9. Pathway for selenium metabolism in prokaryotes. *Dashed lines* are reactions yet to be established. *Dotted line* is detoxidication reaction not in biosynthesis reactions. Transport of selenate and selenite into the cell is by CysT, CysW, and CysA polypeptides. SAM, *S*-adenosyl-L-methionine; THP, tetrahydropteryl (Glu)$_n$; R in reactions 25 and 26 is $(NH_3)CHOO^-$; GSH, reduced glutathionine; GSSG, oxidized glutathionine; tRNA$^{Sec}$RCH$_2$SeH, selenocstyl-tRNA$^{Sec}$. (Reprinted from Turner, R.J., J.H. Weiner, and D.E. Taylor. 1998. Selenium metabolism in *Escherichia coli. BioMetals* 11:223–227. Copyright © 1998, Kluwer Academic Publishers, Dordrecht, The Netherlands. With kind permission of Kluwer Academic Publishers.)

tains an $\alpha_1\beta_1\gamma_1$ heterotrimer with subunits of 96 kDa, 40 kDa, and 23 kDa, respectively. The metal content of this reductase complex is interesting in that it contains molybdenum and two Fe-S clusters. Because the selenate reductase of *Thauera selenatis* is localized in the periplasm and selenite production occurs outside of the plasma membrane, the cell does not have to contend with high levels of selenite in the cytoplasm. The selenate reductase complex in *Sulfurospirillum barnesii* is a heterotetramer with subunits of 82 kDa, 53 kDa, 34 kDa, and 21 kDa. The selenate reductase of *Thauera selenatis* recognizes only selenate while the enzyme complex from *Sulfurospirillum barnesii* also reduces nitrate, fumarate and thiosulfate. Because the selenate reductase activity is in the plasma membrane of *Sulfurospirillum barnesii*, there is the possibility that the other reductases are associated with the selenate complex isolated from this organism.

### 14.3.2.6. Toxicity

Selenate may enter bacteria by way of the sulfate transporter or by a specific transport system. In bacteria without a specific selenate reductase, selenate may be reduced to selenite by nitrate reductase. The toxicity of selenate and selenite in aerobic bacteria is attributed to reactions associated with glutathionine. In *Escherichia coli*, glutathionine (L-γ-glutamyl-L-cysteinyl-L-glycine, GSH) reacts with selenite to produce selenide and elemental selenium, $Se^0$. Glutathionine will react with selenate but the rate is considerably less than with selenite. In bacteria such as *Escherichia coli*, glutathione concentration in cells is approximately 27 μmol/g cell dry weight and accounts for about 90% of the total reduced thiol concentration. If the concentration of GSH is low, $H_2O_2$ is formed as indicated in reaction 6 of Figure 14.9. The oxidation of selenide by $O_2$ results in the generation of free radicals as indicated in the following reaction:

$$2\ HSe^- + 2\tfrac{1}{2}\ O_2 \rightarrow 2\ Se^0 + H_2O + 2\ O_2^{\cdot -}$$

The production of $H_2O_2$ and $O_3^{2-}$ is highly toxic to the cell. With the addition of selenate or selenite to cells, red-amorphous elemental selenium is produced and gray trigonal crystals of selenium are formed. Perhaps the production of $Se^0$ is intracellular and with the generation of superoxide radical, the cell disintegrates, releasing the elemental selenium into the environment. The reduction of tellurite ($TeO_3^{2-}$) by GSH-based reactions leads to the production of elemental tellurium, $Te^0$.

## 14.3.3. Mercury[26]

Many strains of bacteria are resistant to mercury and have the capability of detoxifying both organic and inorganic compounds. Since the early work of S. Silver and A. Summers in the 1970s, it has been known that bacteria have the enzymatic capability of reducing water-soluble $Hg^{2+}$ to lipid soluble $Hg^0$. Some bacteria add methyl groups to $Hg^{2+}$ with the formation of mono- and dimethyl mercury. $Hg^0$ and dimethyl mercury leave the terrestrial environment and become dispersed in the atmosphere. Owing to photochemical activities in the upper atmosphere, the mercury compounds are converted to $Hg^{2+}$, which gradually settles over the Earth's surface. Organomercury compounds (i.e., merbromin, thiomersal, ethylmercuric chloride, *p*-hydroxymercuribenzoate, phenylmercuric acetate, and fluorescein mercuric acetate) are converted to $Hg^{2+}$ in the terrestrial environments by an organomercurial lyase that is produced by certain bacteria. Most of the research has been done with plasmid-containing strains of *Pseudomonas* sp., *Staphylococcus aureus*, and *Escherichia coli*. *Acidithiobacillus ferroxidans* reduces $Hg^{2+}$ to $Hg^0$ and the genes for this activity are carried in the genomic DNA and not on a plasmid.

### 14.3.3.1. Reduction and the *mer* Operon

There has been considerable interest in the molecular biology of mercury reduction in aerobic bacteria and especially those systems in which plasmid-carried genes are present. A model for the *mer* operon is given in Figure 14.10. The genes for $Hg^{2+}$

metabolism are attributed to the *mer* operon, which contains genes for mercury volatilization. The operon is regulated by *merR*, which in the presence of $Hg^{2+}$ acts as a transcriptional activator and in the absence of mercury ions is a repressor. The following genes are present in Gram-negative bacteria: *merD*, a repressor for the *mer* operon; *merB*, which encodes for the organomercurial lyase; *merF*, which encodes for a transport protein; *merP*, which encodes for a $Hg^{2+}$ periplasmic binding protein; *merT*, which encodes for a transport of $Hg^{2+}$; *merA*, which encodes for mercuric reductase; and *mer C*, a transport protein in the plasma membrane.

MerR is a small regulatory protein of 144 amino acids and it has several functional roles that must be performed by the dimeric protein structure. In the absence of mercury, MerR binds to the O/P region by the helix-turn-helix motif. The presence of the MerR at the O/P region blocks the attachment of a RNA polymerase to the O/P region and no structural genes are transcribed. In the presence of mercury, $Hg^{2+}$ is bound to the MerR protein through cysteine residues because there are two cysteine

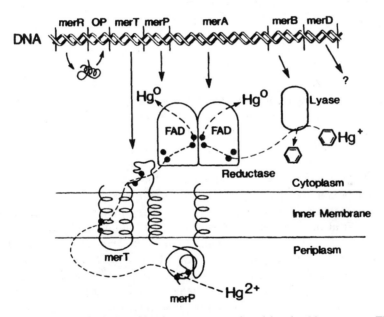

Figure 14.10. Model of $Hg^{2+}$ detoxification system produced by the *Mer* operon. The *merR* gene produces a 144 amino acid polypeptide that interacts at the operator/promoter (OP) site. In the presence of $Hg^{2+}$, transcription occurs with the production of a 72 amino acid polypeptide (MerP) that binds $Hg^{2+}$ in the periplasm and transfers the mercury ion to a 1,000 amino acid polypeptide (MerT) for transport into the cell. Reduction is accomplished by MerA which in *Staphylococcus aureus* is a 547 amino acid polypeptide that has binding sites for FAD and NADH in addition to $Hg^{2+}$. MerB is an organomercurial lyase of about 215 amino acid residues and it releases $Hg^{2+}$ from the organic compound. The reduction of $Hg^{2+}$ produces $Hg^0$ which diffuses out of the cell. (Reprinted from Silver, S., R.A. Laddaga, and T.K. Misra. 1989. Plasmid-determined resistance to metal ions. In *Metal–Microbe Interactions* (R.K. Poole and G.M. Gadd, eds.). Oxford University Press, Oxford, pp. 49–63. Copyright © 1989, John Wiley & Sons, Inc. This material is used by permission of John Wiley & Sons, Inc.)

**Box 14.2**
**Measuring Mercuric Reductase**

The mercuric reductase is a cytoplasmic flavoprotein that is similar in many respects to a FAD-containing disulfide reductase. The reaction catalyzed is as follows:

$$NADPH + H^+ + RS\text{-}Hg\text{-}SR \rightarrow NADP^+ + 2\ RSH + Hg^0$$

Volatilization of mercury can be determined by following either the mercury atoms produced or the amount of NADPH oxidized. The quantity of $Hg^0$ volatilized can be determined by sparging air from the reaction through benzene or toluene because $Hg^0$ is soluble in organic solvents. Analytical measurements can be made to determine the amount of $Hg^0$ in the benzene or toluene. Another procedure for establishing the activity of this enzyme is to follow the oxidation of NADPH by spectrophotometric analysis.

residues per monomer. If an organomercurial is present, it binds at the carboxyl end of the MerR monomer at a site that is distinct from the binding of $Hg^{2+}$. The binding of either $Hg^{2+}$ or an organomercurial to the MerR produces conformational restraints, which results in leaving the O/P region of the operon available to interact with RNA polymerase.

Several proteins are used for the detoxifying of mercury and in each reaction, $Hg^{2+}$ is held in the protein by a pair of cysteine residues. Encoded on *merP* is a periplasmic binding protein that is specific for $Hg^{2+}$. The binding protein is only 72 amino acids in length but it also has a leader sequence that is 19 amino acids in length. Using a pair of cysteine residues, $Hg^{2+}$ is initially sequestered as it enters the periplasm and it is transferred through the plasma membrane by a small transport protein which is a product of the *merT* gene. The MerT has three transmembrane loops and it consists of 119 amino acids with two pair of cysteine residues. As shown in Figure 14.10, there is one pair of cysteine residues in the first transmembrane loop and one pair of the cytoplasmic side of the membrane between the second and third transmembrane loop. The function of MerT is to quickly move $Hg^{2+}$ through the membrane and into the mercuric reductase which is MetT. In Gram-negative bacteria, the mercuric reductase system is a dimer of 67 kDa and each monomer has two pair of cysteine for binding $Hg^{2+}$. There is an 80 amino acid segment at the N-terminal region of the mercuric reductase that has considerable homology with the binding protein (MerP). Other highly conserved regions MerA include the NADPH and FAD binding sites. Also, considerable similarity exists between the bacterial mercuric reductase and glutathione reductase isolated from eukaryotes.

### 14.3.3.2. Methylation Reactions

The reactions of methylation and demethylation of mercury are distinct from mercury reduction. Methyl mercury is produced from inorganic $Hg^{2+}$ by *Clostridium cochlearium,* methanogenic bacteria, *Desulfovibrio,* and *Desulfotomaculum.* The most definitive information concerning detoxification of mercury by anaerobic bacteria has been conducted with *Desulfovibrio desulfuricans.* In this case, the methyl group is transferred from methyl tetrahydrofolate to cobalamin (vitamin $B_{12}$) and ultimately to $Hg^{2+}$. It has been proposed that methane is released from methyl mercury by a methyl mercury lyase system.

There are several interesting ramifications involving hydrogen sulfide from sulfate-respiring bacteria interacting with $Hg^{2+}$ to produce red, hexagonal crystalline $\alpha$-HgS (cinnabar) or black, cubic crystalline $\beta$-HgS (metacimmabarite). In a hydrogen sulfide environment, $Hg^{2+}$ is precipitated as mercury sulfide, and enzymatic reactions are 100 to 1,000 times slower than when mercury is in solution. Thus, when sulfate is not available or when sulfate-reducing bacteria are inhibited from producing sulfide, detoxification of $Hg^{2+}$ would appear to proceed with the formation of methyl mercury. The chemical reaction between $H_2S$ and methyl mercuric chloride ($CH_3HgCl$), results in the formation of insoluble dimethyl mercury sulfide ($(CH_3Hg)_2S$), which spontaneously decomposes to $\beta$-HgS and $(CH_3)_2Hg$. Under mild acidic conditions, dimethyl mercury spontaneously degrades to methane and $Hg^{2+}$.

## *14.3.4. Arsenic*[29,30]

Arsenic is a metalloid that has important biological oxidation states of $+5$ as in arsenate, $AsO(OH)_3$; $+3$ as in arsenite, $As(OH)_3$; $+1$ as in cacodylic acid, $(CH_3)_2 AsO(OH)$; and $-3$ as with dimethylarsine, $(CH_3)_2 AsH$. Arsenate is an analog of phosphate and entrance into the bacterial cell is by either of two transport systems. With an affinity of 25 $\mu M$ for anions, the Pit transport system in bacteria has equal specificity for arsenate and phosphate. Arsenate can also enter the cell by the phosphate specific system (Pst) but the entry is slow because the Pst has an affinity of 0.25 $\mu M$ for phosphate but an affinity of only 25 $\mu M$ for arsenate. With arsenate being an analog for phosphate, arsenate can compete with phosphate in synthesis of ATP or in the formation of sugar phosphates. The organoarsenate compounds are unstable and undergo spontaneous hydrolysis of arsenylated sugars or arsenylated ADP. Thus, arsenate inhibits growth because the cells are unable to generate energy in the form of ATP or other high-energy phosphate compounds.

### 14.3.4.1. Arsenate Detoxification

One mechanism of arsenate detoxification is the reduction of arsenate to arsenite with arsenite pumped out of the cell by a special transporter. The arsenite export system also transports antimonite from the cell. This type of arsenic detoxification has also been observed in *Bacillus subtilis* and *Staphylococcus aureus*; however, the best characterized system is the *Escherichia coli* system, which is plasmid encoded. The arsenic resistance plasmid in *E. coli* carries the arsenic (*ars*) operon which contains five genes: *arsR*, *D*, *A*, *B*, and *C*. A brief description of the activities of these are listed in Table 14.19 and a model of the arsenite system is given in Figure 14.11. The arsenite exporter is similar to an ABC transporter for numerous other solutes in bacteria.

ArsR interacts with DNA by a helix–turn–helix domain and through negative control regulates expression of the *ars* operon. In the presence of arsenite or antimonite, the ArsR protein complexes with the metalloid through three cysteine thiolates. This causes a conformational change of the ArsR and also changes the interaction with DNA. The ArsR has the conserved region of $ELC_{32}VC_{34}DL$ near the N-terminus and the cys 32 and cys 34 are two of the cysteine ligands binding arsenite or antimonite. ArsD controls the amount of structural genes formed from the *ars* operon because overexpression of the transmembrane ArsB may be toxic to the cell. ArsA is a large transport protein that is divided into two halves: ArsA1 and ArsA2. Each half of ArsA contains a Walker A and a Walker B motif which are located near the N-terminus of the polypeptide. The Walker A motif is the nucleotide binding domain which has a sequence of GKGGVGKTS/T for binding ATP while the Walker B motif binds $Mg^{2+}$ and has a sequence of DTAPTGHTIRLLI. The Walker B site recognizes the $Mg^{2+}$ cation that bridges the β-phosphate and γ-phosphate of ATP. In this portion of the ArsA there is a link between the γ-phosphate of ATP and the aspartate residue of the Walker B motif. The binding site for the arsenite/antimonite ion is between the two halves of the polypeptide (e.g., ArsA1 and ArsA2) and consists of three cysteine residues that along with a 40 amino acid linker segment pull the two halves of the ArsA molecule together.

ArsB is a lipophilic protein with 12 transmembrane helices that are crucial for the exit of arsenite or antimonite across the plasma membrane. The ArsA protein is closely associated with the membrane-bound ArsB. ArsC is a cytoplasmic protein that functions as a reductase for arsenate and it appears to be loosely affiliated with ArsA/B to enable rapid export of arsenite. The arsenate reductase does not use NAD(P)H as the immediate reducing agent because it requires glutathione and glutaredoxin for reducing activity.

Table 14.19. Characteristics of the arsenic (*ars*) operon in *Escherichia coli*.

| Gene | Protein mass (kDa) | Characteristic of protein |
|---|---|---|
| R | 13.2 | Regulatory protein that contains a helix–turn–helix domain. |
| D | 13.2 | Controls the amount of Ars A, B, and C produced. |
| A | 63.2 | Contains two Walker A, two Walker B motifs, and one arsenite binding site. |
| B | 45.6 | Proteins Ars A and Ars B constitute the ATP arsenite pump. Ars B has 12 transmembrane helices. |
| C | 15.8 | Arsenate reductase which is in the cytoplasm but is loosely associated with the arsenite pump. |

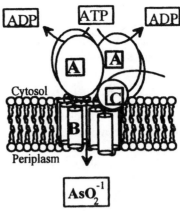

Figure 14.11. Model of the arsenite transport system encoded on the plasmid of *Escherichia coli*. ArsA and ArsB are polypeptides that constitute the arsenite pump while ArsC reduces arsenate to arsenite. (Reprinted from Kaur, P. 2001. Microbial arsenite and antimonite transporters. In *Microbial Transport Systems* (G. Winkelmann, ed.). Wiley-VCH, New York, pp. 377–395. Used with permission.)

### 14.3.4.2. Arsenite Metabolism

Arsenite is about 100 times more toxic than arsenate because of the high affinity of arsenite for sulfhydral groups. Arsenite is a metabolic toxin in that it will inhibit several enzymes including pyruvate dehydrogenase, α-ketoglutarate dehydrogenase, and dihydrolipoate dehydrogenase. Several heterotrophic bacteria (e.g., *Achromobacter, Xanthomonas*, and *Pseudomonas*) are capable of oxidizing arsenite to arsenate. The arsenite oxidase in *Alcaligenes faecalis* is a 85-kDa protein containing a molybdopterin cofactor and several Fe-S clusters. This arsenite oxidase uses azurin and cytochrome *c* as electron carriers with molecular oxygen as the final electron acceptors. *Acidithiobacillus ferrooxidans*, a chemolithotrophic bacterium, oxidizes arsenic in enargite ($Cu_3AsS_4$), orpiment ($As_2S_3$), and arsenopyrite (FeAsS).

### 14.3.4.3. Reduction

One process used by bacteria to remove arsenate from the environment is to volatilize arsenate through methylated reactions. McBride and Wolfe in 1971 demonstrated that *Methanobacterium* converts arsenate to dimethylarsine through the use of coenzyme M as the methyl donor. Other bacteria that methylate arsenate include *Serratia marinorubra, Pseudomonas* spp., *Desulfovibrio desulfuricans*, and mixed cultures of sewage bacteria.

Only a few bacteria are capable of using arsenate as a terminal electron acceptor in the support of bacterial growth. In *Chrysiogenes arsenatis*, the reduction of arsenate to arsenite is by an inducible arsenate reductase (Arr) which is an $\alpha_1\beta_1$ heterodimer of 123 kDa with subunits of 87 kDa and 29 kDa, respectively. The Arr is specific for arsenate and will not reduce nitrite, sulfate, or selenate. In *Sulfurospirillum barnesii*, the Arr is an $\alpha_1\beta_1\gamma_1$ heterotrimer of 100 kDa with subunits of 65 kDa, 31 kDa, and 22 kDa, respectively. These arsenate reductases appear to have molybdenum, acid-labile sulfur, and zinc as cofactors. Both of these enzymes that participate in the reduction of arsenate (AsV) to arsenite (AsIII) appear to be coupled to the respiratory chain; however, the specific electron transport system has not been established for either *Chrysiogenes arsenatis* or *Sulfurospirillum barnesii*. Other bacteria that reduce arsenate to arsenite are given in Table 14.7.

## 14.3.5. Chromium[31,32]

Chromium is a transition metal that in undisturbed environments occurs in low concentrations but industrial processes have significantly increased chromium levels in specific areas. The natural abundance of chromium varies with most soils having $<0.1$ mg/g of soil, $<0.5$ ppm in fresh water and $<0.005$ ppm on marine environments. The stable oxidation states are $+3$ and $+6$. $Cr^{3+}$ is highly insoluble in that it readily precipitates as a hydroxide where the environment is more alkaline than pH 5.5. The common oxyanions of $Cr^{6+}$ include chromate ($CrO_4^{2-}$) which is highly mobile in the environment and dichromate ($Cr_2O_7^{2-}$) which is a very strong oxidizing agent that readily reacts with organic material. Chromate is of concern as a potential contaminant because it is highly mobile with cytotoxic characteristics.

### 14.3.5.1. Transport Systems

Intracellular accumulation of chromate occurs as a result of chemical mimicry with chromate acquired by sulfate transport systems and not to transport of $Cr^{3+}$. When characterizing sulfate transport in bacteria, it was determined that chromate is a competitive inhibitor. Energy-dependent uptake of chromate is by the sulfate transport systems in *Salmonella typhimurium*, *Escherichia coli*, and *Pseudomonas aeruginosa*. At neutral pH, $Cr^{3+}$ derivatives readily bind to bacterial cell walls and γ-glutamyl capsular polysaccharides. This binding by $Cr^{3+}$ greatly diminishes the establishment of a soluble chromium gradient across the cell periphery. It may appear that in some environments that $Cr^{3+}$ may be acquired by $Fe^{3+}$ transport systems; however, mechanisms for such transmembrane transport of $Cr^{3+}$ remains to be established.

Some bacteria are capable of reducing the quantity of intracellular chromate through the action of an exporter specific for chromate. A plasmid carrying chromate resistance was isolated from *Pseudomonas aeruginosa* and *Ralstonia eutrophus*, and this plasmid produces a protein designated as ChrA that participates in chromate export. Based on the hydropathic profile of the amino acid sequence, the ChrA has 10 or 12 transmembrane helices, which is similar to other plasma membrane transporters. ChrA has slightly more than 400 amino acids and is similar in size and hydrophobic amino acids (62%) with the arsenic resistance membrane protein encoded on the *E. coli* plasmid. Using everted membrane vesicles, the efflux of $Cr^{6+}$ by ChrA has been defined by Michaelis–Menten kinetics and the $K_m$ of 0.12 m*M* for chromate efflux is similar to the $K_m$ of 0.14 m*M* for arsenite efflux by the ArsB protein.

### 14.3.5.2. Chromate Reduction

Several bacteria have the capability of reducing $Cr^{6+}$ (chromate) to $Cr^{3+}$ but none to date have the capability of coupling electron flow to cellular growth. Chromate is the substrate and not the highly reactive dichromate for enzymatic reduction of $Cr^{6+}$. Prokaryotes of both Gram-negative and Gram-positive bacteria reduce outer surface of the plasma membrane to $Cr^{3+}$ which readily precipitates as $Cr(OH)_3$. Chromate reduction occurs at either the outer surface of the plasma membrane or in soluble regions of the cell. A membrane-bound chromate reductase is present in *Pseudomonas fluorescens, Enterobacter cloacae*, and *Shewanella putrefaciens* while a soluble enzyme is demonstrated in *E. coli, Desulfovibrio vulgaris, Pseudomonas putida*, and *Pseudomonas ambgua*. The soluble enzyme purified from *Pseudomonas putida* and *Pseudomonas ambgua* uses either NADH or NADPH as the electron donor. In *Desulfovibrio vulgaris*, cytochrome $c_3$ reduces chromate with electrons supplied by $H_2$ by way of a hydrogenase.

The enzymatic reduction of $Cr^{6+}$ to $Cr^{3+}$ is unique because this three-electron reduction process is difficult to reconcile with NAD(P)H being a two-electron donor. Chemical analysis suggests that chromate reduction is a two-step process with the first step being a one-electron reduction with the formation of $Cr^{5+}$ as an intermediate. The reduction of $Cr^{5+}$ to $Cr^{3+}$ by NAD(P)H is by the traditional two-electron reduction step.

### 14.3.5.3. Toxicity

Chromate is considered to be highly toxic to cells because of high oxidative reactivity, high solubility, and potential for uptake by cells. In contrast, $Cr^{3+}$ in the environment forms insoluble complexes, is not readily transported into the cell, and the toxicity is markedly less than with $Cr^{6+}$. It has long been known that chromate is highly inhibitory for protein and nucleic acid biosynthesis. Chemical analysis has demonstrated that with the reduction of chromate to $Cr^{5+}$ there is the concomitant formation of hydroxyl radical ($\cdot$ OH). It has been speculated that hydrogen peroxide is also a product of the reaction in the reduction of $Cr^{6+}$ and that hydrogen peroxide is an intermediate in the hydroxyl radical produced by a mechanism similar to the Fenton reaction. The hydroxyl radical is extremely toxic in that it indiscriminately oxidizes organic material.

## *14.3.6. Cadmium*[14–18,33–35]

Several different species of bacteria are resistant to cadmium, and this activity is necessitated by the accumulation of cadmium in the environment to toxic levels. Cadmium is found in agricultural soil at 1 to 14 ppm but can accumulate to 75 ppm in domestic sewage sludge. $Cd^{2+}$ enters the bacterial cells by way of the $Mn^{2+}$ transporter and the exit from cells is by a $Cd^{2+}$ transporter (see the model in Figure 14.12). Considerable progress has been made in understanding the resistance of bacteria to $Cd^{2+}$. Simon Silver's laboratory has examined the plasmid carrying this resistance and from his research efforts, a model of a $Cd^{2+}$ exporter has been described (see Figure 14.13).

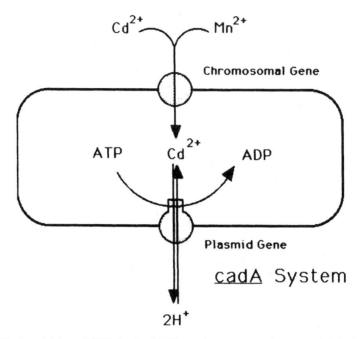

Figure 14.12. Acquisition of $Cd^{2+}$ by the $Mn^{2+}$ uptake system and export of $Cd^{2+}$ by the *cadA* system. The resistance to $Cd^{2+}$ in *Staphylococcus aureus* is encoded on a plasmid and consists of a 727 amino acid polypeptide known as CadA and it functions as a $Cd^{2+}/H^+$ exchange. The coupling of $Cd^{2+}$ export to ATP hydrolysis is unresolved. (Reprinted from Silver, S., R.A. Laddaga, and T.K. Misra. 1989. Plasmid-determined resistance to metal ions. In *Metal–Microbe Interactions* (R.K. Poole and G.M. Gadd, eds.). Oxford University Press, Oxford, pp. 49–63. Copyright © 1989, John Wiley & Sons, Inc. This material is used by permission of John Wiley & Sons, Inc.)

The sites for $Cd^{2+}$ binding and for ATPase activities have been identified as well as the essential amino acids in the transmembrane protein export system. The plasmid of *Staphylococcus aureus* carries a single gene, *cadA*, for production of ATPase-driven $Cd^{2+}$ efflux. The CadA protein has 727 amino acids but there are only four cysteine residues for the movement of $Cd^{2+}$ across the membrane. The $Cd^{2+}$ exporter is proposed to have six transmembrane helices with major cytoplasmic extensions. Two cysteine residues are on one of the transmembrane helix segments, and these may participate in forming a channel for export of cadmium. One of the cytoplasmic segments of the CadA proteins contains two cysteine residues that is for binding the $Cd^{2+}$ ion. Another intracytoplasmic segment of CadA has ATP binding regions and with the hydrolysis of ATP, there is conformational changes in the protein to account for driving $Cd^{2+}$ export process. The polypeptide loops in the periplasm are relatively small, and this may be important to prevent proteolytic attack. Verification of this model requires direct laboratory experimentation.

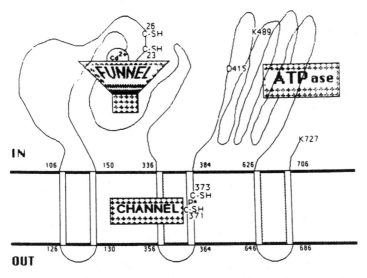

Figure 14.13. Model of the $Cd^{2+}$ export system by a specific ATPase. Through the two cysteine residues at position 23 and 26 at the N-terminus, the metal export system recognizes the $Cd^{2+}$ ion. The central channel of the ATPase has two cysteines at positions 371 and 373 with a proline residue, and the presence of the proline is typical of this class of transporter. A lysine residue is found at position 489 that presumably binds ATP and donates phosphate to aspartate at position 415. (Reprinted from Silver, S., R.A. Laddaga, and T.K. Misra. 1989. Plasmid-determined resistance to metal ions. In *Metal–Microbe Interactions* (R.K. Poole and G.M. Gadd, eds.). Oxford University Press, Oxford, pp. 49–63. Copyright © 1989, John Wiley & Sons, Inc. This material is used by permission of John Wiley & Sons, Inc.)

## 14.3.7. Uranium

In recent years, the interaction of bacteria with uranium has become important owing to the abundance of uranium in mill tailings sites and related environments. Uranium exists in several oxidation states, with the uranyl ion ($U^{6+}$, $UO_2^{2+}$) being highly soluble and mobile in groundwater while urannite ($U^{4+}$, $UO_2$) is insoluble and therefore immobile in the environment. Owing to unique chemical features of uranium, two atoms of oxygen per uranium atom are present in both $U^{4+}$ and $U^{6+}$. There are two major types of bacterial activities to remove toxic uranium from the environment: cellular binding and reduction of uranium. Lovley and colleagues have energized the study of bacteria in the reduction of $U^{6+}$. As indicated in Table 14.7, various strains of iron-reducing bacteria and sulfate-reducing bacteria reduce uranium by mechanisms that are poorly defined. This reduction appears to occur at the cell surface because sections of cells do not reveal intracellular accumulation of uranium. In contrast, dead cells that have lost membrane integrity will bind $U^{6+}$ to the cytoplasmic materials. Lovley and colleagues have demonstrated that cytochrome $c_3$ from *Desulfovibrio vulgaris* will mediate electron transfer from $H_2$ to $U^{6+}$. In most cases, the reduction of $U^{6+}$ is not an example of energy conservation but *Desulfotomaculum desulfuricans*

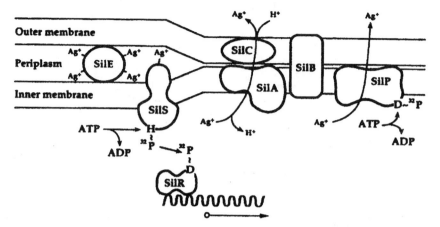

Figure 14.14. Model for plasmid-encoded silver resistance system of *Salmonella*. The silver operon consists of eight genes: SilS and SilR are sensory kinase and responder protein, respectively. SilE is a periplasmic binding protein. SilA, SilB, and SilC are components of the transmembrane system for $Ag^+/H^+$ exchange. SilP is a P-type ATPase for export of $Ag^+$. (Reprinted from Silver, S. 1997. The bacterial view of the periodic table: Specific functions for all elements. *Reviews in Mineralogy* 35:345–360. Used with permission.)

can grow with energy produced by dissimilatory reduction of $U^{6+}$ to $U^{4+}$. As indicated in Table 14.8, several metal ions similar to $U^{6+}$ are also reduced by bacteria.

A second process for the detoxification of $U^{6+}$ is attributed to binding of this uranyl ion to bacterially produced inorganic phosphate. Macaskie and colleagues found that a specific strain of *Citrobacter* has a cell-bound phosphatase that catalyzes the liberation of phosphate from organic substrates such as glycerol phosphate. In the presence of uranyl ion ($UO_2^{2+}$), insoluble uranyl phosphate ($HUO_2PO_4$) is produced. Through the use of electron microscopy, it was determined that uranyl phosphate ($HUO_2PO_4$) is produced as a polycrystal at the surface of the cell. Considerable loading of the cell with uranium can occur and it has been reported that the amount of uranium accumulating can approach 900% of the cellular dry weight.

## 14.3.8. Silver Resistance

Resistance to $Ag^+$ is carried on a plasmid that has been isolated from *Salmonella*. There are eight genes that account for export of $Ag^+$ from the cytoplasm and the proteins associated with this resistance are presented in the model of Figure 14.14. This system of $Ag^+$ export is reminiscent of plasmid-borne resistance to divalent cations such as $Cd^+$. As the silver ion penetrates the outer membrane, it is immobilized by a periplasmic binding proteins designated as SilE. The $Ag^+$ ion is transferred to SilS which functions as a sensory kinase that accepts $Ag^+$ from SilE and activates a responder molecule, SilR. Finally, transcription of the silver operon results in production of $Ag^+$ export. There are two different systems involved in removal of $Ag^+$

from the cytoplasm. One system is a P-type ATPase specific for $Ag^+$ and the second is a $H^+/Ag^+$ anteporter. In the anteport system, three proteins are involved, with SilA located in the plasma membrane, SilC positioned in the outer membrane, and SilB traversing both the outer and plasma membranes. It should be noted that this resistance mechanism does not involve a reduction step with the formation of $Ag^0$ as occurs with mercury resistance. $Ag^+$ resistance in bacteria is important in the clinical setting as well as in environmental sites involving biogeochemical activities.

# References

1. Ehrlich, H.L. 1996. *Geomicrobiology*. Marcel Dekker, New York.
2. Hughes, M.N. and R.K. Poole. 1989. *Metals and Micro-organisms*. Chapman & Hall, New York.
3. Wackett, L.P., W.H. Orme-Johnson, and C.T. Walsh. 1989. Transition metal enzymes in bacterial metabolism. In *Metal Ions and Bacteria* (T.J. Beveridge and R.J. Doyle, eds.). John Wiley & Sons, New York, pp. 165–206.
4. Brock, T. 1975. *Milestones in Microbiology*. American Society for Microbiology Press, Washington, D.C.
5. Nealson, K.H. and D.A. Stahl. 1997. Microorganisms and biogeochemical cycles. *Reviews in Mineralogy* 35:5–34.
6. Barton, L.L., R.M. Plunkett, and B.M. Thomson. 2003. Reduction of metals and nonessential elements by anaerobes. In *Biochemistry and Physiology of Anaerobic Bacteria* (L.G. Ljungdahl, M.W. Adams, L.L. Barton, J.G. Ferry, and M.K. Johnson, eds.). Springer-Verlag, New York, pp. 220–234.
7. Lovley, D.R. 2000. Fe(III) and Mn(IV) reduction. In *Environmental Microbe–Metal Interactions* (D.R. Lovley, ed.). American Society for Microbiology Press, Washington, D.C., pp. 3–30.
8. Ehrlich, H.L., W.J. Ingledew, and J.C. Salerno. 1991. Iron- and manganese-oxidizing bacteria. In *Variations in Autotrophic Life* (J.M. Shively and L.L. Barton, eds.). Academic Press, New York, pp. 147–170.
9. Emerson, D. 2000. Microbial oxidation of Fe(II) and Mn(II) at circumneutral pH. In *Environmental Microbe–Metal Interactions* (D.R. Lovley, ed.). American Society for Microbiology Press, Washington, D.C., pp. 31–52.
10. Thauer, R.K., K. Jungermann, and K. Decker. 1977. Energy conservation in chemotrophic anaerobic bacteria. *Bacteriological Reviews* 41:100–180.
11. Tebo, B.M., W.C. Ghiorse, L.G. van Waasbergen, L.P. Siering, and R. Caspi. 1997. Bacterially-mediated mineral formation: Insights into manganese(II) oxidation from molecular genetic and biochemical studies. *Reviews in Mineralogy* 35:225–266.
12. Nealson, K.H. 1997. Sediment bacteria: Who's there, what are they doing, and what's new? *Annual Review of Earth and Planetary Science* 25:403–434.
13. Nealson, K.H., R.A. Rosson, and C.R. Myers. 1989. Mechanisms of oxidation and reduction of manganese. In *Metal Ions and Bacteria* (T.J. Beveridge and R.J. Doyle, eds.). John Wiley & Sons, New York, pp. 383–412.
14. Blakemore, R.P. and R.B. Frankel. 1989. Biomineralization by magnetogenic bacteria. In *Metal–Microbe Interactions* (R.K. Poole and G.M. Gadd, eds.). Oxford University Press, New York, pp. 85–98.
15. Beveridge, T.J., J.D. Meloche, W.S. Fyfe, and R.G.E. Murry. 1983. Diagenesis of metals

chemically complexed to bacteria: Laboratory formulation of metal phosphates, sulfides, and organic condensates in artificial sediments. *Applied and Environmental Microbiology* 45:1094–1108.

16. Schultze-Lam, S. and T.J. Beveridge. 1994. Nucleation of celestite and strontionite on a cyanobacterial S-layer. *Applied and Environmental Microbiology* 60:447–453.

17. Schultze-Lam, S., G. Harauz, and T.L. Beveridge. 1992. Participation of a cyanobacterial S-layer in fine-grain mineral formation. *Journal of Bacteriology* 174:7971–7981.

18. Waechtershaeuser, G. 1988. Before enzymes and templates: Theory of surface metabolism. *Microbiology Reviews* 52:452–484.

19. Hughes, M.N. and R.K. Poole. 1989. Metal mimicry and metal limitations in studies of metal–microbe interactions. In *Metals and Micro-organisms* (R.K. Poole and G.M. Gadd, eds.). Oxford University Press, New York, pp. 1–18.

20. Barton, L.L., H.E. Nuttall, W.C. Lindemann, and R.C. Blake III. 1994. Biocolloid formation: an approach to bioremediation of toxic metal wastes. In *Remediation of Hazardous Waste Contaminated Soils* (D.L. Wise and D.J. Trantolo, eds.). Marcel Dekker, New York, pp. 481–486.

21. Volesky, B. (ed.). 1990. *Biosorption of Heavy Metals*. CRC Press, Boca Raton, FL.

22. Brierley, C.L., J.A. Brierley, and M.S. Davidson. 1989. Applied microbial processes for metals recovery and removal from wastewater. In *Metal Ions and Bacteria* (T.J. Beveridge and R.J. Doyle, eds.). John Wiley & Sons, New York, pp. 359–382.

23. Collins, Y.E. and G. Stotzky, 1989. Factors affecting the toxicity of heavy metals to microbes. In *Metal Ions and Bacteria* (T.J. Beveridge and R.J. Doyle, eds.). John Wiley & Sons, New York, pp. 31–90.

24. Macaskie, L.E. and A.C.R. Dean. 1990. Metal-sequestering biochemicals. In *Biosorption of Heavy Metals* (B. Volesky, ed.). CRC Press, Boca Raton, FL, pp. 199–247.

25. Turner, R.J., J.H. Weiner, and D.E. Taylor. 1998. Selenium metabolism in *Escherichia coli*. *BioMetals* 11:223–227.

26. Hobman, J.L., J.R. Wilson, and N.L. Brown. 2000. Microbial mercury reduction. In *Environmental Microbe—Metal Interactions* (D.R. Lovley, ed.). American Society for Microbiology Press, Washington, D.C., pp. 177–197.

27. Rayner, M.H. and P.J. Sadler. 1989. Cadmium accumulation and resistance mechanisms in bacteria. In *Metals and Micro-organisms* (R.K. Poole and G.M. Gadd, eds.). Oxford University Press, New York, pp. 39–47.

28. Robinson, N.J., S.K. Whitehall, and J.S. Cavet. 2001. Microbial metallothioneins. *Advances in Microbial Physiology* 44:183–214.

29. Andrews, S.C. 1998. Iron storage in bacteria. *Advances in Microbial Physiology* 40:281–351.

30. Matzanke, B.F. 1997. Iron storage in microorganisms. In *Transition Metals in Microbial Metabolism* (G. Winkelmann and C.J. Carrano, eds.). Harwood Academic, Amsterdam, pp. 117–158.

31. Park, C.H., M. Keyhan, B. Wielinga, S. Fendorf, and A. Matin. 2000. Purification to homogeneity and characterization of a novel *Pseudomonas putida* chromate reductase. *Applied and Environmental Microbiology* 66:1788–1795.

32. Wang, Y.-T. 2000. Microbial reduction of chromate. In *Environmental Microbe–Metal Interactions* (D.R. Lovley, ed.). American Society for Microbiology Press, Washington, D.C., pp. 225–235.

33. Heider, J. and A. Böck. 1993. Selenium metabolism in micro-organisms. *Advances in Microbial Physiology* 35:71–109.

34. Schröder, I., S. Rech, T. Kraft, and J.M. Macy. 1997. Purification and characterization of

the selenate reductase from *Thauera selenatis*. *Journal of Biological Chemistry* 272:23765–23768.

35. Stolz, J.F. and R.S. Oremland. 1999. Bacterial respiration of arsenic and selenium. *FEMS Microbiology Reviews* 23:615–627.

36. Kraft, T. and J.M. Macy, 1988. Purification and characterization of the respiratory arsenate reductase of *Chryiogenes arsenatis*. *European Journal of Biochemistry* 255:647–653.

37. Newman, D.K., D. Armann, and F.M.M. Morel. 1998. A brief review of microbial arsenate reduction. *Geomicrobiology* 15:255–268.

38. Silver, S., R.A. Laddaga, and T.K. Misra. 1989. Plasmid-determined resistance to metal ions. In *Metal–Microbe Interactions* (R.K. Poole and G.M. Gadd, eds.). Oxford University Press, Oxford, pp. 49–63.

39. Kaur, P. 2001. Microbial arsenite and antimonite transporters. In *Microbial Transport Systems* (G. Winkelmann, ed.). Wiley-VCH, New York, pp. 377–395.

40. Cervantes, C., J. Campos-Garcia, S. Devars, F. Gutiérrez-Corona, H. Loza-Tavera, J.C. Torres-Guzmán, and R. Moreno-Sánchez. 2001. Interactions of chromium with microorganisms and plants. *FEMS Microbiology Reviews* 25:335–247.

41. Chen, J.M. and O.J. Hao. 1998. Microbial chromium (VI) reduction. *Critical Reviews in Environmental Science and Technology* 28:219–251.

42. Lovley, D.R., E.E. Roden, E.J.P. Philips, and J.C. Woodward, 1993. Enzymatic iron and uranium reduction by sulfate-reducing bacteria. *Marine Geology* 113:41–53.

43. Silver, S. 1997. The bacterial view of the periodic table: Specific functions for all elements. *Reviews in Mineralogy* 35:345–360.

# Additional Reading

Clifton, C.E. 1957. *Introduction to Bacterial Physiology*. McGraw-Hill, New York.

## *Cells and Metal Interactions*

Beveridge, T.J., M.N. Hughes, H. Lee, K.T. Leung, R.K. Poole, I. Savvaidis, S. Silver, and J.T. Trevors. 1997. Metal–microbe interactions: Contempary approaches. *Advances in Microbial Physiology* 38:178–243.

Fenchel, T.M., G.M. King, and T.H. Blackburn. 1998. *Bacterial Biogeochemistry: The Ecophysiology of Mineral Cycling*. Academic Press, New York.

Hamilton, W.A. 2003. Microbially influenced corrosion as a model system for the study of metal microbe interactions: A unifying electron transfer hypothesis. *Biofouling* 19:65–76.

Stumm, W. and Morgan, J.J. 1996. *Aquatic Chemistry*. John Wiley & Sons, New York.

## *Specific Elements*

Lloyd, J.R. and L.E. Macaskie. 2000. Bioremediation of radionuclide-containing wastewaters. In *Environmental Microbe-Metal Interactions* (D.R. Lovley, ed.). American Society for Microbiology Press, Washington, D.C., pp. 277–328.

Morel, F.M.M. and N.M. Price. 2003. The biogeochemical cycles of trace metals in the oceans. *Science* 300:944–947.

Mielczarek, E.V. and S.B. McGrayne. 2000. *Iron, Nature's Universal Element*. Rutgers University Press, New Brunswick, NJ.

Neilands, J.B. 1993. Overview of bacterial iron transport and siderophore systems in *Rhizobia*. In Iron Chelation in Plants and Soil Microorganisms (L.L. Barton and B.C. Hemming, eds.). Academic Press, San Diego, pp. 180–196.

Oremland, R.S. and J.F. Stolz. 2003. The ecology of arsenic. *Science* 300:939–944.

Self, W.T. 2003. Selenium-dependent enzymes from clostridia. In *Biochemistry and Physiology of Anaerobic Bacteria* (L.G. Ljungdahl, M.W. Adams, L.L. Barton, J.G. Ferry, and M.K. Johnson, eds.). Springer-Verlag, New York, pp. 157–170.

Wiegel, J., J. Hanel, and K. Aygen. 2003. Chemolithotrophic thermophilic iron(III)-reducer. In *Biochemistry and Physiology of Anaerobic Bacteria* (L.G. Ljungdahl, M.W. Adams, L.L. Barton, J.G. Ferry, and M.K. Johnson, eds.). Springer-Verlag, New York, pp. 235–251.

# Appendix

## Abbreviations for the Twenty Amino Acids Found in Proteins

| Amino acids | Three-letter abbreviation | One-letter abbreviation |
|---|---|---|
| Alanine | Ala | A |
| Arginine | Arg | R |
| Asparagine | Asn | N |
| Aspartic acid | Asp | D |
| Cysteine | Cys | C |
| Glutamic acid | Glu | E |
| Glutamine | Gln | Q |
| Glycine | Gly | G |
| Histidine | His | H |
| Isoleucine | Ile | I |
| Lysine | Lys | K |
| Methionine | Met | M |
| Phenylalanine | Phe | F |
| Proline | Pro | P |
| Serine | Ser | S |
| Threonine | Thr | T |
| Tryptophan | Trp | W |
| Tyrosine | Tyr | Y |
| Valine | Val | V |

## Study Questions

The following is a series of questions relevant for each chapter. The purpose of these questions is to promote an understanding of the material presented in the text. These questions are suitable for self-review or for class discussion. Each chapter has at least one question that directs the reader to the Internet to search for a publication addressing a specific topic in prokaryotic physiology. The written paper resulting from the review of a scientific paper may be used either for self-improvement or could be a requirement for the class. To provide a guide for students to make a written report of a paper published in a scientific journal, a guide, "Reviewing Papers," is presented at the end of this section.

## *Chapter 1. The Cellular System*

1. With extraterrestrial exploration becoming a reality, what type of life would scientists be seeking to demonstrate? Would this life form be expected to be a prokaryotic cell? What types of requirements would be needed to establish a new life form that does not fit into either the *Bacteria* or *Archaea* domains?

2. Briefly describe the following processes for disruption of bacterial cells: (a) osmotic disruption, (b) French Pressure cell, and (c) ultrasonic treatment.

3. Construct flow diagrams for the isolation of each of the following prokaryotic cell structures: peptidoglycan, plasma membrane, filaments of the flagellum, ribosomes, enzymes in the periplasm, capsule, pili, and outer membrane. If you are provided with only one quantity of bacterial cells, how many of these structures could be isolated with a single processing of the cell mass?

4. Following physical isolation of cell wall structures (such as ribosomes, peptidoglycan, plasma membrane, outer membrane, and capsule), why is it necessary to establish purity of isolated structure through the use of chemical analysis? Could you use electron microscopy to establish purity? Explain.

5. Is it possible to follow enzyme activity through the use of electron microscopy? Describe a procedure to assess immunological differences in the flagella produced by two closely related motile prokaryotes. If you have difficulty with this question, look for a published procedure using colloidal gold immobilized on antibodies.

6. Through the use of the Internet, find a recent report describing the genome of a prokaryote. What physiological activities constitute the greatest number of genes? What is the closest relative of this organism, and how was this relationship determined? What percentage of the genomic DNA of this organism has no known function?

7. All of the following are characteristics used to correctly describe members of the archaea except

    a. The cell wall contains glycoproteins

    b. Subunits of flagellar filament consist of glycoproteins

    c. *N*-formylmethionine is the first amino acid in protein synthesis

    d. Ribosomal-mediated protein synthesis is not inhibited by streptomycin

8. In centrifugation of disrupted bacterial cells, what force is required to pellet membrane–bound enzymes?

    a. 100,000*g* for 1 hr

    b. 200,000*g* for 2 hr

    c. 100,000 rpm for 2 hr

    d. 10,000*g* for 10 min

9. Phosphotungstic acid is

    a. a unique uronic acid in the peptidoglycan of certain bacteria

    b. used as a stain for transmission electron microscopy (TEM)

    c. part of the cell wall structure of archaea

    d. a unique phospholipid found in the plasma membrane of archaea

## *Chapter 2. The Plasma Membrane*

1. Even though prokaryotes and eukaryotes have considerable structural differences, the thickness of the plasma membrane (or cell membrane) in both of the cell types is remarkable similar. What could be some features that account for this similarity in membrane thickness between bacteria and eukaryotic cells? Would this also be the reason why the plasma membranes in bacteria and archaea have a similar thickness?

2. How do the lipids differ in the plasma membrane of archaea and bacteria?

3. In assessing the turnover of molecules in organisms, it is common to use radioisotopes at tracer levels. How would one measure the turnover of phospholipids in the plasma membrane? Propose a process for using radioisotopes to measure turnover rates of proteins in the plasma membrane.

4. What would be the difficulties in synthesis of the membrane structure if the plasma membrane contained covalent cross-bridge bonds, as in the case of the peptidoglycan?

5. Since compounds such as phospholipids are not soluble in water, where would the synthesis of phospholipids occur?

6. Through the use of the Internet, find papers that study each of the following in prokaryotes: (a) synthesis of plasma membrane and (b) an unusual fatty acid or phospholipids in the plasma membrane. Prepare a brief synopsis of each of the papers.

7. Would you expect the outer side of the plasma membranes of different bacteria to vary considerably, or would you expect the outer side of the membranes to be highly similar? Describe an experiment that would be used to test your theory.

8. It is proposed that the bacterial plasma membrane has an asymmetric distribution of lipids. Would you expect to find the plasma membrane of archaea to also have

an asymmetric distribution of lipids? Propose a series of experiments to test the asymmetric distribution of lipids in plasma membranes of archaea.

9. What are some of the proposed roles of phospholipids in the plasma membrane? What is the action of Polymyxin on the bacterial plasma membrane?

## *Chapter 3. Cell Wall Structure*

1. The peptidoglycan structure in Gram-positive bacteria is exposed to various enzymes produced by eukaryotic cells and proteolytic bacteria. Why is the four amino acid peptide extending from NAM not degraded by these proteolytic enzymes in the environment?

2. Why is the synthesis of the bacterial peptidoglycan not inhibited by streptomycin or tetracycline?

3. Cell wall synthesis has been described as an activity attributed to "controlled" autolysis. Would you agree to this? Defend your answer.

4. Based on cell wall composition in prokaryotes, what selective pressures may have contributed to the bacteria characteristically having peptidoglycan while there is a diversity of cell wall structures in the archaea?

5. Since S-layers are found on a large number of microbial species, why do you think there has there been little to no mention in the literature about S-layers and their role in establishing diseases in humans?

6. Why is it necessary for the peptidoglycan in the spore to be different from the peptidoglycan in the mother cell?

7. Draw the structures of NAM and NAG as they are in the peptidoglycan polymer and indicate the attachment of teichoic acids and of amino acids making up the peptide segment. What are some segments of the peptidoglycan that are highly conserved in the bacteria and what parts of the structure are variations permitted? Indicate the bonds hydrolyzed by lysozyme, muramyl endopeptidase, carboxypeptidase, and *N*-acetylglucosaminidase.

8. Which antibiotic/chemotherapeutic agent is associated with inhibiting the addition of the NAG–NAM dimer into the growing region of the peptidoglycan?

   a. Bacitracin

   b. D-cycloserine

   c. Vancomycin

   d. Cephalosporin

9. Of the following, which is found in both bacteria and fungi?

   a. Diaminopimelic acid

   b. *N*-acetylmuramic acid

   c. D-glutamic acid

   d. *N*-acetylglucosamine

10. In bacteria, the third amino acid attached to the *N*-acetyl muramic acid of the peptidoglycan may be

    a. L-glutamic acid or D-glutamic acid

    b. Diaminopimelic acid or Lysine

    c. D-alanine

    d. Glycine

11. Which of the following statements about inhibitors of peptidoglycan synthesis is not correct?

    a. Bacitracin inhibits the release of $P_i$ from the P-P-$C_{55}$ lipid in the plasma membrane and prevents the regeneration of the P-$C_{55}$ lipid carrier

    b. Penicillin inhibits the D,D-carboxypeptidase

    c. Polymyxin inhibits the release of NAG–NAM dimer from the $C_{55}$ lipid

    d. Inhibition of D-alanine racemace would prevent cross-bridge formation in cell walls of Gram-positive bacteria

## *Chapter 4. The Cell Wall Matrix*

1. The cell wall of the mycobacteria is relatively unique. What characteristics would lead some to consider that the mycobacteria have a "Gram-positive type" cell wall, and what characteristics would support the idea that mycobacteria have a "Gram-negative type" cell wall?

2. Describe the ribitol and glycerol types of teichoic acids in bacterial cell walls. How are these attached to the cell? What are teichuronic acids, and when are they produced?

3. Describe the Braun lipoprotein in the cell wall of Gram-negative bacteria and indicate how it is attached to the peptidoglycan and to the outer membrane. Why is the Braun molecule referred to as a "lipoprotein"?

4. At the molecular level, what accounts for antigenic differences of lipopolysaccharide in different strains of *E. coli*?

5. List several activities of teichoic acids in the cell walls of Gram-positive bacteria.

6. Through the use of the Internet, find a paper that examines the activity of a specific porin. Briefly summarize the results of this scientific contribution.

7. Describe the role of protein M and surface amphiphiles as virulence-enhancing factors for bacteria.

8. Heptose is a special type of sugar that is found in

    a. the peptidoglycan of archaea

    b. the lipopolysaccharide of Gram-negative bacteria

    c. the capsule of oral streptococci

    d. the peptidoglycan of bacteria

9. 2-keto-3-deoxyoctonic acid (KDO) is found in

   a. bacterial flagella

   b. teichoic acid of Gram-negative bacteria

   c. the core polysaccharide of lipopolysaccharide

   d. the pseudopeptidoglycan of archaea

10. Which of the following is found in the O-specific chain of the lipopolysaccharide of *Salmonella typhimurium*?

   a. D-glutamic acid

   b. Levan

   c. Abequose

   d. Tallosamine

## Chapter 5. Capsules, Pili, and Internal Structures

1. Distinguish between homopolymer and heteropolymer polysaccharide in capsules of bacteria. What chemical features of the capsule accounts for antigenic differences of bacterial capsule from organisms of the same species? What are uronic acids, and what is their role in the capsule? What unique chemical features are displayed by amino acids in the "peptide capsules"?

2. What criteria would be used to determine that the internal granules of magnetosomes or polyphosphate are not surrounded by a typical membrane? Is there an advantage to the cell in not having a membrane surrounding these internal granules?

3. Search the Internet for a recent paper concerning the role of pili in promotion of a specific disease. Prepare a review of this report.

4. Gas vacuoles have considerable structural detail but lack an impermeable coat that encloses the hollow chamber. In view of the absence of an "airtight" covering of the chamber, how are gases contained inside the vacuole? In terms of evolution, is there another structure in bacteria that may share an origin with gas vacuoles?

5. In capsule synthesis it is fairly easy to envision the random addition of precursor molecules onto the capsule; however, with the synthesis of crystalline cellulose, the addition of precursors is an ordered event. Propose a model for synthesis of extracellular crystalline cellulose by bacteria.

6. Devise an experiment that would determine if the amino acids in the peptide capsule is attached to the peptidoglycan.

7. Compare and contrast the synthesis of bacterial pilus with the synthesis of bacterial flagellum.

8. Some have proposed that type 4 pillus contributes to bacterial mobility. Devise an experiment that would test if these pili are responsible for cell movement. If you have difficulty in devising an experiment, check the Internet to see how others have attempted to demonstrate this phenomenon of movement.

9. The purified peptide capsule of *Bacillus anthrasis*

   a. would have an absorption peak at 280 nm

   b. would have an adsorption peak at 260 nm

   c. would contain five different amino acids

   d. would consist of a polymer of D-glutamic acid

10. Bacteria that have a peptide capsule include which one of the following?

    a. *Streptococcus pyogenes*

    b. *Bacillus anthrasis*

    c. *E. coli*

    d. *Pseudomonas aeruginosa*

11. Which of the following has a composition of polyfructose?

    a. dextran

    b. crystalline cellulose

    c. levan

    d. lipid granules

12. Which of the following types of pili are involved in the transfer of DNA by the process of conjugation?

    a. type 1

    b. type 4

    c. type F

    d. type P

## *Chapter 6.  Cell Motion, Sensing, and Communication*

1. Describe the mechanism for spirochete movement.  In comparison to the *E. coli*–type motility, is the spirochete type of movement more suited for the highly viscous environment where the spirochetes are found?

2. Bacterial sporulation occurs at greatest levels at the onset of the stationary phase and in colonies on agar plates.  Is this consistent with quorum sensing?  Defend your answer.

3. Is quorum sensing a desired feature for bacteria?  Why is this considered to be an example of cell–cell communication?

4. Distinguish between the following resistive structures of bacteria:  Spores, Cycts, and Conidia.  The basis of comparison is heat resistance, presence of a cortex, and presence of dipicolinic acid plus calcium ions.

5. Describe the two-component signal response in bacteria by drawing a model and indicate the regions of the cell where each activity is displayed.

6. In the differentiation of *Caulobacter* there are several places where two-component sensing may be important.  Give one of these possible differentiations and describe the phenotypic changes that would occur.

7. What is the role of Tsr, Tar, Trg, Tap, CheW, CheB, CheY, and CheA in bacterial chemotaxis? How is chemotaxis regulated in *E. coli*?

8. Compare the composition of the bacterial flagella with the archaeal flagella.

9. The bacterial flagellar system is a nanomotor that displays circular rotation. Are there other examples of structures in bacteria with similar motor characteristerics? If so, could they have had a common evolutionary origin?

10. Spore formation in bacteria is an important process. Examine Figure 6.20 and propose a reason why peptides serve as the signal instead of another biomolecule. What would be the rationale for bacteria to use such an elaborate type of signal for initiation of sporulation?

11. Through the use of the Internet, find a recent review of bacterial chemotaxis and examine the methylation/demethylation activities associated with flagellar rotation.

## Chapter 7. Cellular Growth and Reproduction

1. Discuss global regulation in bacteria and identify the role of cAMP and ppGpp in global regulation.

2. List and discuss the changes in cellular activities that occur as a culture makes the transition from lag to log phase and from log to stationary phase.

3. What structures or molecules require renewed synthesis and become important in cell maintenance energy? When is the level of maintenance energy highest and when is it lowest in a bacterial culture?

4. Describe the benefits to the bacterial culture in the stationary phase that produces antibiotics, toxins, and bacteriocins. Why is it not of benefit to the culture to produce antibiotics and toxins while the culture is in the log phase?

5. Use the Internet to find recent papers on the role of proteins in the cell division process. Prepare a short synopsis of a paper that discusses the formation of the Z ring at the time of cell division.

6. The bacterial cells are highly dynamic, and numerous phenotypic changes occur as cells progress through the growth phase. List 6–10 examples that would illustrate the point that oscillations and periodic changes are characteristic of individual bacterial cells.

7. Discuss the following activities with respect to germination in bacteria:

   a. cytoplasmic changes leading to formation of the forespore

   b. formation of wall around the developing endospore

   c. formation of the cortex and dehydration of the developing endospore

   d. use of different sigma factors in sporulation

   e. formation of cortex wall and exosporium

   f. development of SASP

8. Discuss the regulatory signals in germination. Discuss the activities in the three stages of germination.

9. Distinguish between regulon, modulon, and alarmones.

10. Using the *pho* regulon as a model, propose components that may be present in the SSI response.

11. What are two factors that would influence the rate of bacterial growth? (This is the log phase measurement of the growth curve.) How can you extend the length of the lag phase or the length of the stationary phase?

12. List the various factors that enable bacteria to have a faster cell growth rate than fungi or other eukaryotes.

## *Chapter 8. Physiological Basis for Growth in Extreme Environments*

1. Discuss oxygen response or oxygen stress observed in bacteria. Clearly distinguish between the following enzymes: superoxide dismutase, rubrerythrin oxidoreductase, catalase, and alkyl hydroperoxide reductase.

2. What are the "Arc system" and the "FNR system," and what phenotypic changes are associated with these systems?

3. List some of the major physiological problems that bacteria must solve to be able to grow in each of the following conditions: (a) extreme high temperatures, (b) alkaline conditions, (c) extremely cold environments, (d) high osmotic conditions.

4. What are the conditions under which certain bacteria can tolerate desiccation?

5. List the molecular and phenotypic changes resulting with temperature stress response to heat shock and to cold shock.

6. Mixed cultures involved in decomposition produce a variety of end products, and some of these contribute to stress on other bacteria. List the stress response expected as a compost pile shifts from an aerobic to anaerobic environment with the production of organic acids followed by the production of ethanol and butanol.

7. With respect to thermophiles growing at high temperatures, provide information on the following:

   a. characteristics of proteins that lead to thermostability

   b. changes in lipids to promote membrane fluidity

   c. small molecules in the cytosol to protect cofactors and nucleic acid

8. Enumerate the factors that appear important for *Deinococcus* to resist exposure to high levels of ionizing radiation.

9. Describe how some halophilic bacteria have adjusted to growth in saturating levels of salt.

## *Chapter 9. Electron Transport and Coupled Phosphorylation*

1. Using the following information, construct an electron transport chain for this bacterium. There were five different experiments conducted with anaerobic cuvettes in the spectrophotometer and these are the results:

   a. Lactate → Oxygen    This electron flow is not inhibited by atebrin but is inhibited by HQNO with the production of a spectral peak of 556 nm. Electron transport is inhibited by azide and cyanide with the production of reduced cytochrome peaks at 556 and 598 nm.

   b. Lactate → Nitrite    This electron flow is not inhibited by atebrin or cyanide but is inhibited by light at 390 nm. Electron flow is inhibited by HQNO and Antimycin A with the production of a reduced cytochrome peak at 556 nm. From other experiments, there is evidence that a cytochrome that adsorbs at 602 nm is involved.

   c. Lactate → Fumarate    This electron transport is inhibited by light at 390 nm. When electron transport is inhibited by HQNO there is the emergence of a spectral peak at 390 nm.

   d. NADH → Fumarate    Electron flow is inhibited only by atebrin and light at 390 nm.

   e. NADH → Oxygen    Electron flow is inhibited by light at 390 nm. With HQNO, there is inhibition of electron flow with the emergence of a spectral peak at 560 nm. With the addition of cyanide there is the production of spectral peaks at 560 nm and 598 nm.

   There is molar growth yield values obtained for the bacterium giving the above reactions, and this information is as follows:

   | Electron donor | Electron acceptor | $Y_{ATP}$ |
   | --- | --- | --- |
   | Formate | Nitrate | 10.5 |
   | Formate | Nitrite | 21.5 |
   | Formate | Fumarate | 11.0 |
   | Succinate | Fumarate | 0 |

   What additional information is needed to construct a complete electron transport scheme?

2. Distinguish between ionophores and uncoupling agents. Are all ionophores uncouplers of oxidative phosphorylation?

3. What is reverse electron flow in prokaryotes? Why is it required? Give a specific example of a reaction and a bacterial species that may use it.

4. Distinguish between the following sets of words or phrases:

    a. plasma membrane vesicles and liposomes

    b. ferredoxin and flavodoxin

    c. F-type ATPase and V-type ATPase

    d. Cytochrome $o$, cytochrome $d$, cytochrome $a$, and cytochrome $c_{co}$

    e. PQQ and $NAD^+$ as electron carriers

    f. TMANO reductase and DMSO reductase

5. What is the benefit of using both $\Delta pH$ and $\Delta \psi$ in calculation of proton-motive force? Describe an experiment for the measurement of $\Delta pH$ across the plasma membrane under a prescribed condition. Describe another experiment for measuring the $\Delta \psi$ across a plasma membrane under a prescribed condition.

6. Describe the proton transduction sites that may be used by *E. coli* with NADH as the electron donor and molecular oxygen as the terminal electron acceptor. What about the "Q-cycle"? Could it be important in transferring of protons to the external surface of the plasma membrane in bacteria?

7. Describe cyclic and noncyclic (linear) electron transport in cyanobacteria. Would you expect the green and purple phototrophs to also display cyclic and noncyclic electron flow?

8. What is the composition and the function of the phycobilisome in cyanobacteria? What is the benefit to the various phototrophic bacteria in that they do not absorb light at the same wavelength?

9. Draw a model of a bacterial respiratory system with formate as the electron donor and nitrite as the terminal electron acceptor. Calculate the free energy of the flow of electrons from formate to nitrite. Is there sufficient energy in this reaction to enable ATP to be produced? In your model, include how protons would be pumped outward by this electron flow.

10. Use the Internet to find a recent article that discusses electron flow in a chemolithotrophic bacterium. Prepare a short review of the paper.

11. Some bacteria are suggested to have cytochromes in the outer membrane. Search the Internet for a paper concerning cytochrome of the c-type in the outer membrane of bacteria and review (a) the criteria used to establish that the cytochrome was indeed in the outer membrane and (b) what type of function would be expected to be involved with this outer-membrane cytochrome.

12. What are the differences between bacterial and mitochondrial electron transport systems?

13. What are the various electron sources for photosynthesis in bacteria, and how can these electron donors be used to group the bacteria?

## Chapter 10. Transmembrane Movement: Mechanisms and Examples

1. All of the following are associated with movement of water in bacteria. Briefly describe the activity of each: aquaporins, gated channels, and compatible solutes.

2. What are siderophores, and what role do they play in bacterial metabolism? Why does there seem to be a variety of different types of siderophores produced by bacteria? Do all bacteria produce siderophores? If not, which ones would not be expected to produce them Is ferrous taken up by a siderophore system in bacteria? Make a list of iron compounds that can serve as an iron source for bacteria.

3. Clearly distinguish between the following sets of word groups:

   a. P-Type ATPase and A-Type ATPase

   b. ABC importers and ABC exporters

   c. Passive diffusion and Facilitated diffusion

4. Why are only a few bacteria capable of using proteins as a carbon source for growth, even though many bacteria have an extensive system of peptide transport?

5. Give a few specific examples where cations can be taken up by bacteria using multiple uptake systems. Give a few examples where anions can be taken up by multiple systems.

6. Many of the metabolic important metals are needed at only trace levels. Since many of these metals can be taken up by active systems, what prevents an over-accumulation of these trace metals and thereby prevents toxic reactions?

7. Search the Internet and find a specific uptake system for iron, nickel, cobalt, or zinc. Prepare a report on one of these active transport systems.

8. While many transporters are found in bacteria, there are a few exchange systems that account for movement of essential nutrients. Characterize a few of these exchange reactions in bacteria.

9. Enumerate the components of the phosphotransferase system for the uptake of glucose into cells. Which components are constitutive and which are specific? What is the energy source for this active transport system?

10. Summarize the system for DNA uptake in bacteria. Include in your discussion the differences in DNA uptake by Gram-negative bacteria and by Gram-positive bacteria.


## Chapter 11. Pathways of Carbon Flow

1. Consider that you have glucose with each carbon radio labeled (for example, $^1C-^2C-^3C-^4C-^5C-^6C$). (a) Which carbon would be as $CO_2$ if the pathway of alcoholic glycolysis is used? (b) Which carbon would be as $CO_2$ if the pathway

of Entner–Douderoff is used? (c) Which carbon would be as $CO_2$ if the pathway of phosphoketolase is used and endproducts are lactic acid and ethanol plus $CO_2$? (d) Which carbon would be as $CO_2$ if the pathway of glycolysis is followed by the phosphoroclastic reaction? (e) Which carbon(s) would be as $CO_2$ if the pathway of glycolysis is followed by the TCA cycle?

2. Prepare a table and supply the appropriate information:

| Metabolic pathway | Key enzymes for the pathway | Key intermediate molecules characteristic of the pathway | Important function associated with each pathway |
|---|---|---|---|
| Glycolysis | | | |
| TCA cycle | | | |
| Glyoxylate cycle | | | |
| Entner–Douderoff pathway | | | |
| Stickland Reaction | | | |
| Phosphoroclastic reaction | | | |
| Phospentose pathway | | | |

3. Outline the steps that bacteria would take in the oxidation of methane to carbon dioxide, and list the enzymes used at each step. Which of these steps provides for the greatest amount of energy when molecular oxygen is the electron acceptor?

4. Pyruvic acid has often been called the "hub" in bacterial metabolism. Please explain why this designation is appropriate.

5. Distinguish between isofunctional enzymes and regulation of metabolic pathways by concerted feedback.

6. Carbon monoxide is toxic to most animals, but several different bacterial groups are capable of oxidizing CO to $CO_2$. How have the bacteria been able to use the CO dehydrogenase reaction to support growth?

7. There are a variety of unique coenzymes in methanogens. Are these coenzymes alternates to those found in bacteria, or do these cofactors participate in reactions that cannot be supported by the bacterial cofactors? Speculate on the evolutionary origin of these cofactors.

8. Describe the enzymatic pathway that enables acetogens to produce acetate from $CO_2$. Would you expect that the genes from the acetogens could be transferred to *E. coli* with expression so that *E. coli* could make acetate from $CO_2$? If so, how many genes would be required?

9. In the oxidation of methanol by bacteria, formaldehyde is produced. Why is PQQ involved in methanol dehydrogenase and not $NAD^+$? Formaldehyde is used as a carbon source for these bacteria by either the RuMP pathway or the serine pathway. Briefly characterize the RuMP and serine pathways by giving some critical reactions of each.

10. List the enzyme for the initial oxidation of the following substrates:

    a. trimethylamine

    b. dimethylamine

    c. methylamine

11. $CO_2$ fixation can be accomplished in bacteria by several different pathways. Give the substrate, enzyme, and product for each of the following $CO_2$ fixing reactions:

    a. Calvin–Benson–Bassham cycle

    b. The reductive TCA cycle

    c. The acetyl~CoA pathway

    List examples of bacteria that use the above three $CO_2$-fixing reactions.

12. Search the Internet for catabolic pathways in archaea. Prepare a report from a paper that you have found and mention if the catabolic pathway is unique for that organism or is also found in bacteria.

## Chapter 12. Organization and Cellular Processing

1. Discuss how the different sigma factors can contribute to regulation of unique cellular processes.

2. Briefly outline the steps for protein export by the type II secretion system.

3. What are lantibiotics, and what is their biological role? Which bacteria produce them, and what scheme is used for their production?

4. Describe the "self-assembly" process used to establish the structure of the bacterial 30S ribosomal subunit. How many other systems of self-assembly are present in bacteria?

5. Draw a model to give the sequence of activities resulting in assembly of the 70S ribosome on the mRNA for the initiation of protein synthesis.

6. Search the Internet to find papers describing the activity or presence of type III and type IV secretion systems in bacteria. Prepare papers to review the published reports.

7. Search the Internet for publications concerning global regulation in prokaryotes. Prepare a report on one of the papers that you have read on the Internet.

8. Glycoproteins are commonly present in eukaryotes, and for many years it was considered that prokaryotes did not produce glycoproteins. It is now known that prokaryotes do produce a few glycoproteins. Why do you think prokaryotes have not developed the processes for glycoprotein synthesis?

9. List the various sites where antibiotics can inhibit protein synthesis.

10. The pathway for heme synthesis is rather extensive in bacteria. Would you expect to find heme synthesis in archaea? Defend your answer. Are more steps required for *E. coli* growing aerobically on glucose than for *E. coli* to grow anaerobically with dissimilatory nitrate reduction? Does heme synthesis require a major expenditure of energy for the bacteria?

## *Chapter 13.  Metabolism of Nitrogen and Sulfur*

1. Compare the ATP requirement, reducing equivalents, and oxygen lability of the $N_2$ fixation and $CO_2$ fixation by the Calvin–Benson–Bassham cycle.

2. Compare the dissimilatory reduction of sulfate with the dissimilatory reduction of nitrate. As a point of comparison, provide information on ATP requirements, nature of the electron carriers and the number of electrons consumed, the number of enzymatic steps, and where the activity is located in the bacterial cell.

3. Propose a rationale as to why internal membrane structures are present in aerobic nitrifying bacteria but not in bacteria metabolizing sulfur compounds. Compare ammonia oxidation with methane oxidation by bacteria. Would you expect to find internal membranes in bacteria that oxidize ammonium by an anaerobic process? Are internal membranes required for nitrogen fixation?

4. What is isotope fractionation by microorganisms? Which elements are commonly involved, and how can isotope fractionation be used to predict geological formations?

5. Through the use of a diagram, construct a sulfuretum (microbial ecosystem with interdependent sulfur oxidizing and sulfate reducing bacteria) providing the reactions and names of organisms that may occur there. Based on energetics, which reactions are the least favored?

6. Compare and contrast the prokaryotic enzyme systems involved in oxidation of elemental sulfur with the enzyme systems for reduction of elemental sulfur.

7. Mineralization occurs in both the nitrogen and sulfur global cycles. Describe the reactions and the organisms that would be associated with these mineralization events.

8. Distinguish between GOGAT, GS, and GDH as they function in the assimilation of ammonia. Are these reactions distributed broadly in bacteria and, if so, what are the conditions that control their use? Are these ammonia assimilation reactions dependent on membranes in the cytoplasm? Describe the role of glutamate in the synthesis of amino sugars, nucleotide bases, and other amino acids.

9. Compare assimilatory sulfate reduction with assimilatory nitrate reduction as occurs in prokaryotes. How do they compare in terms of energetics and the nature of the electron donor?

10. An important aspect of agriculture is the supply of nitrogen to plants. Bacterial fixation of nitrogen by association with plant roots is an important aspect of agriculture. Use the Internet to obtain a recent review describing the reactions in the root nodule of a legume associated with nitrogen fixation. Prepare a report on this article.

11. An organism of unknown genus is growing slowly in a defined medium. The medium contains sulfate and cysteine. Hydrogen sulfide is being produced by the culture. How would you determine if the hydrogen sulfide is a result of anaerobic respiration or from cleavage of –SH from cysteine?

## Chapter 14.  Biometals

1.  What is the role of magnesium, potassium, and selenium in microbial metabolism, and how do bacteria maintain homeostasis of each ion?

2.  Describe the bacterial activity associated with Mn oxidation and with Mn reduction.  What are some of the difficulties in conducting Mn-associated research?

3.  Give several examples of metal mimicry experienced by prokaryotes.  Is metal mimicry an advantage or disadvantage to bacteria?  Are there special conditions required for metal mimicry to occur?

4.  Metallothionein is important for metal detoxification in animal systems.  Do prokaryotes use metallothionein and, if so, how broad is metallothionein distribution in prokaryotes?  What other systems are used by bacteria for metal detoxification?

5.  Distinguish between bacterioferritin and ferritin.  Are both of these systems found in prokaryotes and, if so, what is their role?

6.  Describe the systems that bacteria use for mercury detoxification.  Is this highly specific for mercury or can other metals be detoxified by a similar system?

7.  Describe the systems that bacteria use for the detoxification of chromium, cadmium, arsenic, and silver.  What similarities and what differences are characteristic of these four systems?

8.  Microbial leaching of metals has been attributed to both bacteria and archaea.  Describe the reactions resulting in the solubilization of copper and uranium from various ores.  What are some of the environmental consequences associated with these microbial leaching operations?

9.  Describe some of the passive processes (binding and precipitation) that result in detoxification of metals.  Are these processes widely distributed throughout the prokaytotes, or are they limited to a few organisms?

10.  Microorganisms are intimately associated with the geochemical cycles.  Use the Internet to locate a paper that addresses the role of a specific prokaryotic species in one of the geochemical cycles.  Prepare a report that appropriately summarizes the publication.

11.  The public press frequently issues short reports on the use of bacteria for environmental restoration of metal-contaminated sites.  Use the Internet to find some of these reports in newspapers or public magazines.  Is there a focus in the public press on a specific metal or a specific process?

## Reviewing Papers

The Internet provides access to almost all of the recent scientific papers published in the various journals.  By placing a key word in one of the popular search engines, numerous electronic reports can be viewed.  Some of the responses to a key word will include information from the popular presses as well as from different home pages. However, the best scientific information (because it is well reviewed before publication) is found in the journal articles. Some of these publications in the scientific journals are initial reports on an observation; these are referred to as the primary literature. Some papers are reviews of a topic, and these would not contain the methods of research. It is useful to follow scientific reports that are both primary literature and of the review type. Public news articles are of use to persons not in a science area, and the wording is often to explain information to a general audience and not to those with a strong biology or biochemical background.

In the preparation of reports from the primary literature or from scientific review articles, there are several points that must be considered. Style is paramount in preparation of a review that is to be handed to the professor or completed as part of the response to questions at the end of chapters. The reference is to be complete so another person can take the information and readily find the article. The following items must be present in the report:

*Title*: A complete title without abbreviations is to be used.

*Authors*: A listing of all authors must be included. If there are over 10 authors, it is then permissible to include a few and then list "et al." for the unmentioned authors. It is not acceptable to list only the first author on any report.

*Source*: It is best to spell out the journal title; however, if abbreviations are given in the article, then it is appropriate to list them. In addition, the volume number, the year of publication, and the pages of the article are to be included. Frequently an issue number is listed, but this is not required. The following example provides the appropriate information:

Person, A. 2004. Structure and function in bacteria. *Journal of Scientific Information* 200: 2004–2016.

The review should follow, and it should be separated into the following categories. The first section is the *Introduction*. Information for this part is found in the introduction of the paper that you have read. You should indicate the relevance of this research and the specific goals of this paper. Generally the goals of the paper are found in the last paragraph of the introduction. The next section is *Methods*, and unless this paper establishes a new procedure, the methods section is covered in general terms. The next section constitutes the largest portion of the review because it includes the *Results and Discussion*. The significance of the results is discussed using the information provided in the discussion section of the original paper. Frequently, the original paper will compare results obtained with those obtained by others in the field. It is always useful in your critique to state if the authors were able to prove their initial objective.

# Index